ROUTLEDGE HANDBOOK OF TOURISM CITIES

The *Routledge Handbook of Tourism Cities* presents an up-to-date, critical and comprehensive overview of established and emerging themes in urban tourism and tourist cities. Offering socio-cultural perspectives and multidisciplinary insights from leading scholars, the book explores contemporary issues, challenges and trends.

Organised into four parts, the handbook begins with an introductory section that explores contemporary issues, challenges and trends that tourism cities face today. A range of topics are explored, including sustainable urban tourism, overtourism and urbanisation, the impact of terrorism, visitor–host interactions, as well as reflections on present and future challenges for tourism cities. In Part II the marketing, branding and markets for tourism cities are considered, exploring topics such as destination marketing and branding, business travellers and exhibition hosting. This section combines academic scholarship with real-life practice and case studies from cities. Part III discusses product and technology developments for tourism cities, examining their supply and impact on different travellers, from open-air markets to creative waterfronts, from social media to smart cities. The final Part offers examples of how urban tourism is developing in different parts of the world and how worldwide tourism cities are adapting to the challenges ahead. It also explores emerging forms of specialist tourism, including geology and ecology-based tourism, socialist heritage and post-communist destination tourism.

This handbook fills a notable gap by offering a critical and detailed understanding of the diverse elements of the tourist experience today. It contains useful suggestions for practitioners, as well as examples for theoretical frameworks to students in the fields of urban tourism and tourism cities. The handbook will be of interest to scholars and students working in urban tourism, heritage studies, human geography, urban studies and urban planning, sociology, psychology and business studies.

Alastair M. Morrison, PhD is Research Professor in Marketing, Events and Tourism at the Business School, University of Greenwich, UK. Formerly, he was Distinguished Professor Emeritus at Purdue University, USA, specialising in the area of tourism and hospitality marketing, and President of the International Tourism Studies Association. Alastair is the Co-Editor-in-Chief of the *International Journal of Tourism Cities*. He is a member of the Experts Committee of the World Tourism Cities Federation.

J. Andres Coca-Stefaniak is Associate Professor of Tourism and Events at the University of Greenwich, UK, deputy leader of the university's Tourism Research Centre, and Co-Editor-in-Chief of the *International Journal of Tourism Cities*. Andres serves on the editorial boards of various journals, including the *Journal of Hospitality and Tourism Research*; *Sustainability*; *Journal of Tourism Futures*; and the *Journal of Place Management and Development*, among others. He is a member of the Experts Committee of the World Tourism Cities Federation.

ROUTLEDGE HANDBOOK OF TOURISM CITIES

*Edited by Alastair M. Morrison and
J. Andres Coca-Stefaniak*

LONDON AND NEW YORK

First published 2021
by Routledge
2 Park Square, Milton Park, Abingdon, Oxon OX14 4RN

and by Routledge
52 Vanderbilt Avenue, New York, NY 10017

Routledge is an imprint of the Taylor & Francis Group, an informa business

British Library Cataloguing-in-Publication Data
A catalogue record for this book is available from the British Library

Library of Congress Cataloging-in-Publication Data
A catalog record has been requested for this book

ISBN: 978-0-367-19999-9 (hbk)
ISBN: 978-0-429-24460-5 (ebk)

Typeset in Bembo
by Wearset Ltd, Boldon, Tyne and Wear

To Anna Maria, Alick and Andy

CONTENTS

FIGURES

TABLES

CONTRIBUTORS

Prof. Alastair M. Morrison, PhD is Research Professor in Marketing, Events and Tourism at the Business School, University of Greenwich. Formerly, he was Distinguished Professor Emeritus at Purdue University, USA specialising in the area of tourism and hospitality marketing. He has published approximately 300 academic articles and conference proceedings and is the author of five books on tourism marketing and development, *Marketing and Managing Tourism Destinations*, 2nd edition (Routledge, 2019); *The Tourism System*, 8th edition, (Kendall/Hunt Publishing Company, 2018); *Global Marketing of China Tourism* (China Architectural & Building Press, 2012); *Hospitality and Travel Marketing*, 4th edition (Delmar Publishers, 2010); and *Tourism: Bridges across Continents* (McGraw-Hill Australia, 1998). Alastair is the Co-Editor-in-Chief of the *International Journal of Tourism Cities*.

Dr J. Andres Coca-Stefaniak is Associate Professor of Tourism and Events at the University of Greenwich (UK), deputy leader of the university's Tourism Research Centre, advisor to the World Tourism Cities Federation and Co-Editor-in-Chief of the *International Journal of Tourism Cities*. His research interests include urban tourism, place marketing and branding, smart tourism destinations, smart cities and the management of sustainability in events and tourism destinations. Andres serves on the editorial boards of various journals, including the *Journal of Hospitality and Tourism Research*; *Sustainability*; *Journal of Tourism Futures*; and the *Journal of Place Management and Development*, among others.

Dr Rafael Anaya-Sánchez has been an assistant professor of marketing in the Faculty of Economics and Business of the University of Malaga (Spain) since 2009. His research interests include consumer behaviour, online communities, social-commerce and e-tourism. His articles are published in leading academic journals such as *Tourism Management, Electronic Commerce Research and Applications, Journal of Marketing Management, Computers in Human Behavior, Journal of Retailing and Consumer Services,* among others.

Dr Vladimir Antchak is Senior Lecturer in Applied Management at the University of Derby, UK. His research interests focus on event portfolio design and management, place experience, destination branding and strategic storytelling. He has over ten years of experience in events management, including organisation of business forums and conferences, cultural exhibitions, international business visits and presentations.

Dr Elisa Backer (Zentveld) is an associate professor of tourism and management and chair of the academic board at Federation University, Australia. She was previously at Southern Cross University; and prior to working in academe, Elisa acquired industry experience through working in management positions at destination marketing organisations. Elisa has published extensively in leading international journals and has received 14 awards to recognise her contributions to research and education and is on the editorial board for 15 journals. Her major research interests are VFR travel, family tourism, quality of life and family violence.

Hugo Robarts Bandeira is on the faculty at the Macau Institute for Tourism Studies, in the Macau SAR, China, where he teaches about wine, food and beverages in general. Born in Macau from a Portuguese father and Macanese mother, he moved to Portugal when only two years old. He would return to Macau when he was 16 and remain until the present date. He holds a Master's degree in international wine management, a Bachelor's degree in tourism business management and various wine-related certificates. He is a member of several wine associations and the founding president of the Macanese Gastronomy Association.

Dr Amanda Belarmino is an assistant professor at the William F. Harrah College of Hospitality at the University of Nevada, Las Vegas. She holds a PhD in hotel and restaurant management from the University of Houston. Her research interest is the impact of social and cultural influences on consumer behaviour and strategic management decisions in hospitality and tourism. Her research specifically examines peer-to-peer accommodations, fan-driven tourism and online reviews. She has 15 years of hospitality management experience in restaurants, hotels and casinos.

Professor Stephen W. Boyd, PhD is Professor of Tourism in the Department of Hospitality and Tourism Management at Ulster University, Northern Ireland. He received his PhD in geography from the University of Western Ontario, Canada in 1995 and has since held academic posts in Canada, England and New Zealand before joining Ulster University in 2004. He has published extensively in many areas of tourism, including heritage tourism, tourism and national parks, tourism and trails, and tourism and political change. His current research examines resident attitudes to tourism development for small islands, rural food tourism networks and post-conflict tourism development.

Dr Blanca A. Camargo is Associate Professor of Tourism in the Business School at Universidad de Monterrey, Mexico. Originally from Colombia, she received her doctoral degree from Texas A&M University and her research is oriented towards sustainable tourism with a focus on ethics and justice in tourism development, marketing and management. She is also interested in examining pedagogical aspects of sustainable tourism and ethics. She also holds a Master's degree in hotel and tourism management from Purdue University and in international tourism management from AILUN (Italy).

Dr María L. Chávez is an associate professor in the Department of Administration at Universidad de Monterrey (UDEM), Mexico. She holds a degree in business administration, with an MBA from ITESM, and is currently pursuing a doctorate in administration at the Universidad Autónoma de Nuevo León. Her dissertation focuses on the impact of local people's hospitality in the satisfaction with the destination. At UDEM she teaches marketing, communication and sales promotion, service marketing and marketing for tourism services, and also advises on undergraduate theses for the international marketing and the international tourism programmes. In addition, she has developed and led consulting projects in relation to the development of business plans, satisfaction models, marketing strategies, market segmentation analysis, and strategic planning for private and government institutions.

Jun Chen is currently a second-year Master's student of hospitality and tourism management at Purdue University, USA. She received a Bachelor's degree in tourism management from Sichuan University, Chengdu, China. Her research interests include urban tourism, sustainable tourism, cultural tourism, and tourist experience. At present, Jun is working on the walkability design in tourism as her Master's thesis. In the future, she hopes to pursue a PhD and finally be a professor. Jun spends her spare time visiting museums and appreciating art. She is also an avid lover of ballet and classical Chinese dance.

Dr Rob Davidson is the Managing Director of MICE Knowledge, a consultancy specialising in research, education and training services for the meetings industry. His principal area of expertise is business events, and he has written seven books on that topic. In addition to his position as visiting fellow at the University of Greenwich, he also teaches as a visiting professor in three mainland European universities, in France, Austria and Switzerland. In 2015, he was awarded a doctorate from the University of Greenwich for his thesis, "Technological and Demographic Factors as Agents of Change in the Development of Business Events".

Dr Jonathon Day is an associate professor in Purdue's School of Hospitality and Tourism Management, and is committed to ensuring tourism is a force for good in the world. In addition to over 30 academic articles and chapters, he is the author of *An Introduction to Sustainable Tourism and Responsible Travel* and co-author of *The Tourism System*, 8th edition. Dr Day's research interests focus on sustainable tourism, responsible travel and strategic destination governance within the tourism system. He is interested in the role of business in solving grand challenges through corporate social responsibility programmes and social entrepreneurship.

Dr Deborah Edwards is an associate professor in the Business School, University of Technology Sydney. Her research interests include sustainability, tourist's spatial behaviour, urban tourism, business events and visitor experiences. She has undertaken extensive research collaboration with industry and government, and built strong linkages in research and innovation nationally and internationally.

Dr Mahender Reddy Gavinolla is Assistant Professor in Tourism and Coordinator-International Relations at National Institute of Tourism and Hospitality Management, Hyderabad, India and PhD candidate at Indira Gandhi National Tribal University, India. Previously, he earned an MBA in tourism and hospitality management. He is an avid reader, a travel enthusiast and exhibits keen research interest in the areas of sustainable and responsible tourism, heritage tourism, tourism planning and development.

María del Carmen Ginocchio is an associate professor of marketing in the Business School at Universidad de Monterrey (UDEM), Mexico. She holds an MBA from the Universidad de Monterrey. She has been teaching marketing for over 30 years and also coordinates undergraduate thesis projects in management.

Dr Valentina Gorchakova is Senior Lecturer and Programme Leader of an online business and management programme at the University of Derby, UK. She returned to academia after working for more than ten years in business, not-for-profit organisations and the UN Development Programme. Her current research interests lie within destination marketing, cultural and event tourism, visitor economy and service and arts marketing. She is also researching the areas of online learning and learning in adulthood.

Dr Gaitree Gowreesunkar holds a PhD which cuts across three disciplines namely tourism management, communication and marketing. She serves as Head of Department for Hospitality and Tourism at the University of Africa (Bayelsa State, Nigeria). She is a senior lecturer with over a decade of teaching experience in a number of international universities/institutions. Vanessa is an editorial board member of several scientific journals, and she has a number of publications in international peer-refereed journals. Her research interests include island tourism, tourism management and marketing, informal tourism economy, women entrepreneurship and sustainable tourism.

Professor Maria Gravari-Barbas, PhD is Professor of Cultural and Social Geography at Paris 1 Panthéon-Sorbonne. She holds a degree in architecture and urban design and a PhD in geography and planning from Sorbonne Université. She was Fellow at the Urban Program of Johns Hopkins University, Baltimore, USA. She is the director of EIREST, a multidisciplinary research team dedicated to tourism studies and the director of the UNESCO Chair "Tourism, Culture, Development". She is the author of several books and papers related to tourism, culture and heritage with special focus on gentrification, tourismification and heritagisation of urban spaces.

Dr Ulrike Gretzel is a senior fellow at the Center of Public Relations, University of Southern California and serves as director of research at Netnografica. She received her PhD in communications from the University of Illinois at Urbana-Champaign. Her research focuses on the impact of technology on human experiences and the structure of technology-mediated communication. She studies social media marketing, influencer marketing and the emerging reputation economy, as well as smart tourism development, technology adoption in tourism organisations, tourism in technological dead zones and the quest for digital detox experiences.

Dr Antonio Guevara-Plaza is Dean of the Faculty of Tourism at the University of Malaga, and professor in the Computer Sciences Department and member of the Institute of Tourism Intelligence and Innovation at the University of Malaga. He is currently president of the Inter-University Network for Tourism (REDINTUR) involving 28 Spanish universities with studies in tourism. His research interests focus on the field of information technology and communication (ICT) applied to tourism. He has been the principal investigator of several projects for R&D related to the implementation of ICT in tourism. His research is published in top ranked journals.

Dr Sandra Guinand is an urban planner and geographer. She is a post-doc researcher at the Institute of Geography and Regional Studies, Vienna University, Austria and an associate researcher at EIREST Paris 1 – Sorbonne, France. Her research interests focus on urban regeneration projects, socio-economical transformations of urban landscape, with a specific focus on heritage processes, public–private partnerships and tourism.

Dr Joan C. Henderson obtained a PhD from Edinburgh University and recently retired from full-time teaching. Her last post was as an associate professor at Nanyang Business School, Nanyang Technological University, in Singapore where she taught on the tourism and hospitality management programme for over 20 years. She has published in a wide range of international tourism and hospitality journals, as well as edited books, with a focus on South East Asia. Research interests include urban tourism, tourism and heritage and the politics of tourism.

Professor Sotiris Hji-Avgoustis, PhD is chairperson of the Ball State University Department of Management in the Miller College of Business overseeing programmes in business administration, human resources management, entrepreneurial management, hospitality and food management, and residential property management. He joined Ball State in 2013 after a successful academic career at Indiana University – Indianapolis. His research interests are in the area of city tourism with emphasis on cultural and sports tourism. He has published in national and international journals on these topics and he has been an invited presenter at national and international tourism conferences. He received an MS from Purdue University and a PhD from Indiana State University.

Dr Katia Iankova is an urban studies PhD holder from the University of Quebec in Montreal. She is currently Tourism Operations Management Programme Coordinator at Higher Colleges of Technologies – Abu Dhabi Women's Campus, UAE. Her research interests are in the domain of the processes of human development in relation with the concept of time and in the psychology of human interaction with places.

Dr Rami K. Isaac is currently a senior lecturer in tourism teaching at the Academy for Tourism at Breda University of Applied Sciences in the Netherlands. In addition, he is an assistant professor at the Faculty of Tourism and Hotel Management at Bethlehem University, Palestine. Born in Palestine, Rami Isaac did his undergraduate studies in the Netherlands, graduate studies in the UK and earned his PhD in spatial sciences from the University of Groningen, Netherlands. He has published numerous articles and edited volumes on tourism and political (in)stability, tourism and occupation, tourism and war, dark tourism and transformational tourism.

Dr Ece Kaya joined the University of Technology Sydney Management Department in 2018. Ece is a creative and passionate academic with over eight years of experience teaching students from various social and cultural backgrounds. Previously, she worked at Western Sydney University as subject coordinator and lecturer where she completed her PhD in heritage and tourism studies at the Institute for Culture and Society. She holds a MSc in urban planning from Istanbul Technical University. Her research interests include placemaking and creative space, art, culture and creativity engagement, digital applications in teaching and learning practices and revitalisation of industrial spaces.

David Kerr (BA general studies, Dip. project management) has managed a business consultancy for over 30 years with a diverse client base nationally across a range of industry sectors and is currently the CEO of the Dolphin Discovery Centre in Bunbury, Western Australia. For the last ten years he has overseen the redesign and redevelopment of the Dolphin Discovery Centre, a not-for-profit social enterprise focused on conservation, education and research. He is particularly interested in the social value and provision of nature-based experiences in urban environments and the catalytic impact tourism can have on regional economies.

Dr Dae-Young Kim is an associate professor in the Department of Hospitality Management at the University of Missouri, USA. He has a doctoral degree from Purdue University, USA. His research interests include information technology, consumer behaviour, agritourism and cognitive psychology in hospitality and tourism management.

Dr Tamara Klicek has a PhD from the University of Novi Sad, Serbia in tourism and geography. She is a Slovenian (EU) and Serbian citizen with residency in Taiwan where she has been working as an assistant professor for a few years. Dr Klicek's research interests are urban issues like quality of life in cities for all, city diplomacy, city management, urban tourism and city branding.

Dr Weng-Hang Kong is an assistant professor at the Macao Institute for Tourism Studies (IFTM). She received her PhD from Nottingham Trent University, UK. She has delivered papers at international conferences and published journal articles. Her research interests include destination management, tourism planning and development.

Dr Chi Fung Lam has worked in several hospitality and tourism industries in a few different countries including Hong Kong and Canada. His research area focuses on tourism demand analysis and economic impacts. His recent publications have appeared in highly prestigious international academic journals.

Uyen Le is from Hanoi, Vietnam and is currently a junior student pursuing a Bachelor's degree in hospitality and tourism management with a minor in management at Purdue University, USA. Her major research interests are tourism planning and development, hospitality and customer satisfaction and service quality management.

Ye-Jin Lee is a PhD student and a research and teaching assistant in hospitality management at the University of Missouri. She holds a Master's degree in tourism from Kyung Hee University in South Korea. Her research interests cover applications of big data analytics, information and emerging technologies, sustainable tourism and destination marketing.

Professor Xinran Y. Lehto, PhD is a professor at the School of Hospitality and Tourism Management, Purdue University. She is an associate editor of the *Journal of Hospitality and Tourism Research*. She is currently the president of the International Tourism Studies Association. Dr Lehto worked in the travel and tourism industry as a marketing officer for China National Tourism Administration and a planning executive for Chan Brothers Travel, Singapore. her research addresses how travellers can better leverage vacation's restorative value for health and wellness and how destinations can effectively market vacation products to unique segments such as family travellers.

Professor John J. Lennon, PhD is the Dean of Glasgow School for Business and Society, Glasgow Caledonian University and is responsible for the management and leadership of 170 academic staff and the education of over 5,000 undergraduate and postgraduate students. He is also the director of the Moffat Centre for Travel and Tourism Business Development which is responsible for the production of international consumer and market research in tourism. In the commercial sector of travel and tourism, he has undertaken over 700 tourism and travel projects in over 50 nations on behalf of private sector and public sector clients.

Qing Li is an MS candidate who has been studying at Peking University since 2017. He obtained a Bachelor of Technology degree from Sun Yat-sen University in 2017 in urban planning. His major research interests include urban tourism and tourism cities.

Bingyu Lin is a MS candidate in the Graduate School of Architecture, Planning and Preservation at Columbia University. Lin completed her undergraduate studies at Peking University where she majored in human geography and urban–rural planning. Her research interests lie in urban history and architectural history of the late imperial and modern China, with a focus on the architectural encounter between the East and West from the late nineteenth and twentieth centuries

Qingyin Liu is an editor at Peking University Press and earned a Master's degree in landscape architecture from Peking University. Her research interest is in urban spaces.

Dr Kim-Ieng Loi studied tourism business management at IFTM (with a specialisation in hotel management). She subsequently received a PhD in tourism from James Cook University in Australia after pursuing a Master's degree in financial management from the University of London. She is the coordinator for hotel management and tourism event management programmes at the Macao Institute for Tourism Studies. With research interests in areas such as tourist behaviour, destination marketing, service quality, tourism product management, results of her research appear frequently in international academic journals and in conference proceedings in the field of tourism and hospitality.

Dr Jian Ming Luo, Doctor of Hotel and Tourism Management, is an associate professor in the Faculty of International Tourism and Management at City University of Macau. He has extensive international and multinational tourism working experience. His teaching and research interests focus on urbanisation, tourism development, entertainment and consumer behaviour.

Feiya Ma is a Master's student in city and regional planning at the International Center for Recreation and Tourism Research (iCRTR) of Peking University, Beijing. She has a Bachelor's degree from the University of California Berkeley. Her major research interest is city tourism.

Dr Yan (Mary) Mao is Vice Dean of the School of Tourism and Hotel Management, Hubei University of Economics in Wuhan. She is a "Tourism Young Expert" of the Ministry of Culture and Tourism of China and a "Hubei Province Decision-making Expert of Tourism Development" as appointed by Hubei Provincial Government. Her main research fields are tourism enterprise management, urban tourism management and regional tourism spatial development. She has presided over and participated in a number of national, provincial and ministerial research projects.

Dr Cristina Maxim is Senior Lecturer in Tourism at University of West London. Cristina has over eight years of teaching experience in various higher education institutions in the UK. Cristina completed her PhD at Cities Institute, London Metropolitan University, as the recipient of a Vice Chancellor's Scholarship, with a thesis that examined the planning and management of tourism in London. She also holds an MA in public administration and a BA in economics. Cristina is a member of the Editorial Review Board for the *International Journal of Tourism Cities* and a peer reviewer for a number of international academic journals including *Tourism Management* and the *International Journal of Tourism Cities*. Her research interests include destination management, sustainable development, world tourism cities, tourism planning and management, tourism impacts, regional development and local government.

Professor Sonia Mileva, PhD is a full professor at Sofia University "St. Kliment Ohridski" – Faculty of Economics and Business Administration. She holds a doctor of science in economics from Sofia University "St. Kliment Ohridski" and a PhD in marketing from the University of National and World Economy – Sofia. Her major research interests include tourism marketing, strategic development, specialised (niche) tourism, professional skills and competencies in tourism, etc. She is Editor-in-Chief of the journal *Science & Research* and a member of other editorial boards. She is fluent in English, Russian and Portuguese and is annually a visiting lecturer in Portugal, Finland, UK, Cyprus among others.

Dr Sebastian Molinillo earned his PhD in business management from the University of Malaga (Spain). He has been an associate professor of marketing in the Faculty of Economics and Business at that university since 2003. His key research interests lie in the domain of consumer behaviour, social media, technology adoption and branding in the tourism and retail sectors. In these topics, he has published articles in leading academic journals such as *Tourism Management*, *International Journal of Contemporary Hospitality Management*, *Journal of Destination Marketing & Management*, *Journal of Business Research*, *Computers in Human Behavior* and *Technological Forecasting and Social Change*.

Dr Blake Morris is an artist and independent scholar based in the United Kingdom. His first book, *Walking Networks: The Development of an Artistic Medium*, was published as part of Rowman and Littlefield International's Radical Cultural Studies series in 2019, and offers an overview of the current field of walking art, and its potential to create meaningful exchanges at a distance. Prior to relocating to the UK, Blake founded the New York City-based Walk Exchange, a cross-disciplinary walking group exploring walking as a critical and artistic practice.

Dr David Newsome is an associate professor at Murdoch University College of Science, Health, Engineering and Education, Environmental and Conservation Sciences Group, Perth, Australia. His academic interests focus on nature-based tourism with a particular emphasis on the environmental impacts of recreation and tourism, the sustainability of tourism in national parks and nature reserves, evaluation of the quality of ecotourism operations, geological tourism, wildlife tourism and enhancement of the social value of biodiversity and pro-tected areas globally and especially in Asia.

Sello Samuel Nthebe holds a BTech in hospitality manage-ment, which he obtained from the Central University of Technology, South Africa. He also holds a Master's cum laude degree in tourism management which he obtained from the University of South Africa and is currently studying towards a PhD in tourism management. He spent over nine years in the hotel industry and has held various positions in front office operations. He will be joining the University of South Africa as a lecturer in tourism management from February 2020. His research focuses on the hotel services relating to tourist attractions. He has presented his research at both local and international tourism research conferences. He has published a research article in an accredited African journal and is currently in the process of publishing another research article.

Dr Hera Oktadiana earned her PhD from the School of Hotel and Tourism Management, the Hong Kong Polytechnic University. She also received CHE (Certified Hospitality Edu-cator) from the American Hotel and Lodging Educational Institute. She joined James Cook University, Australia as a vis-iting scholar (tourism research) in 2017, then later as an adjunct senior lecturer. She is also affiliated with the Master's in tourism at Trisakti School of Tourism, Indonesia. Her research interests include tourism and hospitality education, particularly in cur-riculum design and philosophy, and tourist behaviour, espe-cially in the area of Muslim tourists. She is presently the Regional Vice President Southeast Asia of the International Tourism Studies Association (ITSA).

Dr Irem Önder is an associate professor at the Department of Hospitality and Tourism Management at University of Massachusetts Amherst. She obtained her PhD from Clemson University, South Carolina, where she worked as a research and teaching assistant from 2004 until 2008. She obtained her Master's degree in information systems management from Ferris State University, Michigan. Her main research interests include information technology and tourism economics, specifically big data analysis, smart destinations, decision support systems, blockchain and tourism demand forecasting. She serves on the editorial boards of *Journal of Travel Research*, *Tourism Economics* and *Journal of Information Technology and Tourism*.

Dr Odete Paiva has a PhD in tourism, leisure and culture and a Master's in museology and cultural heritage from Coimbra University. She has been an invited professor at the Polytechnic Institute of Viseu, Higher School of Technology and Management, since 2000, in the graduation of tourism. She is affiliated with CEGOT – Geography and Spatial Planning Research at Coimbra University and the Center of Education, Technology and Health at the Polytechnic Institute of Viseu. Currently she is the director of the Grão Vasco National Museum. Odete's current research is in cultural tourism and heritage.

Dr Claire Papaix is Senior Lecturer of Transport and Business Logistics at the University of Greenwich (UK), programme leader of the MA Logistics and Supply Chain Management, founding member of the Connected Cities Research Group and associate of King's College's Urban Science and Progress (CUSP). She has experience in business creation and events management and is a mentor of the University's Enterprise Challenge. Her research interests include sustainable urban mobility, policy implementation, transport wellbeing appraisal, community engagement, citizen science, public health and urban wellness.

Jessica Patroni graduated from Murdoch University with a BSc (wildlife and conservation biology) and First Class Honours in environmental science. Having gained knowledge of the biological, ecological and human dimensions of wildlife conservation through her tertiary studies, Jessica is focused on contributing to the long-term sustainability of wildlife populations. Now an avid and experienced snorkeler and diver, Jessica is pursuing her SCUBA diving and boat handling qualifications and is working towards building a career in conserving and researching wildlife in marine and coastal environments

Professor Philip L. Pearce, DPhil (University of Oxford) is Foundation Professor of Tourism, James Cook University, Townsville, Australia. He is interested in all aspects of tourist behaviour and has developed key approaches to tourist motivation and tourist experience. His work is built on core ideas in social and cognitive psychology and he places a key emphasis on the differences among tourists to offer solutions for enhancing positive experiences, reducing undesirable behaviours and limiting negative outcomes on host societies. In this work he has undertaken studies mainly in Australia, Asia and to a lesser extent the United Kingdom. He has published 17 books in tourism and 300 refereed publications.

Samantha Richards, BSc (Environmental) is a Murdoch University Honours student, environmental management and sustainability. Samantha's current research focus is urban geotourism in Southwest Western Australia, particularly the Noongar people's cultural connections to the geology of the region.

Dr Cláudia Seabra is an assistant professor in the University of Coimbra. She has a PhD in tourism and she is doing her post-doc on "Terrorism and the EU28: Impact on Citizens and Organizations" in NOVASBE. She has published in *Journal of Business Research, Tourism Management, Annals of Tourism Research, International Journal of Tourism Cities, European Journal of Marketing, Journal of Marketing Management, ANATOLIA, Journal of Hospitality and Tourism Technology* among others. She is affiliated with CEGOT – Geography and Spatial Planning Research Centre; NOVASBE – Nova School of Business and Economics; CISeD – Research Center in Digital Services. Her research interests are safety and terrorism, and tourism.

Professor Gildo Seisdedos, PhD is a marketing professor at IE Business School, where he combines teaching, research and consulting activities in the fields of urban planning, local policies and city marketing. Cities are his passion. He has participated in diverse international research projects centred on competitive urban strategy, urban management indicators and consulting projects. He holds a PhD in urban economy from Madrid's Autonoma University, a Bachelor's in business administration (E3) from ICADE, Madrid and an MBA in sales and marketing management from IE Business.

Dr Hugues Séraphin is a senior lecturer in event and tourism management studies. He is also an associate researcher at La Rochelle Business School (France). He holds a PhD from the Université de Perpignan Via Domitia (France) and joined the University of Winchester Business School in 2012. He was the programme leader of the event management programme between 2015 and 2018. He has expertise and interests in tourism development and management in post-colonial, post-conflict and post-disaster destinations. He has recently published in *International Journal of Culture, Tourism, and Hospitality Research*; *Current Issues in Tourism*; *Journal of Policy Research in Tourism, Leisure and Events*; *Journal of Business Research*; *Worldwide Hospitality and Tourism Themes*; *Tourism Analysis*; and *Journal of Destination Marketing & Management*.

Greg Simpson is an emerging graduate researcher in environmental and conservation sciences at Murdoch University. Reflecting the duality of his undergraduate studies (BA sustainable development and BSc Honours 1 environmental science), Greg's research interests combine his commitment to the conservation, protection and innovative reintroduction of urban nature with his passion for connecting people to nature through nature-based tourism and recreation to enhance the liveability of urban environments. As exemplified by the "Dolphins in the City" chapter, Greg strives to deliver his research through equitable collaborations with community members, students, industry partners and his growing academic network.

Dr Melanie Kay Smith, PhD is an associate professor, researcher and consultant whose research includes urban planning, cultural tourism, health tourism and the relationship between tourism and wellbeing. She has lectured in the UK, Hungary, Estonia, Germany, Austria and Switzerland as well as being an invited keynote speaker in many countries. She was chair of ATLAS (Association for Tourism and Leisure Education) for seven years and has undertaken consultancy work for UNWTO and ETC.

Professor Costas Spirou, PhD is Professor of Sociology and Public Administration and currently serves as Provost and Vice President for Academic Affairs at Georgia College and State University, Georgia's public liberal arts university. In 2017, Spirou was named a fellow by the American Council of Education (ACE). His research interests and scholarship centre on mayoral leadership, public policy, technology and political sociology, and urban tourism and sustainability. He has also written about downtown revival, the politics of stadiums and convention centres, and urban affairs and governance. He holds a PhD from Loyola University Chicago.

Dr Magdalena Petronella (Nellie) Swart holds a DCom. in leadership performance and change, is an associate professor in tourism at the University of South Africa, and a certified meeting professional. Nellie has authored and co-authored accredited journal articles, book chapters and conference proceedings related to business tourism and tourism education. The training of tourism teachers is among her community engagement projects. She is an executive committee member of a number of tourism committees and serves on the editorial board of the *International Journal of Tourism Cities*. Nellie has organised national and international conferences and is often an invited speaker at business events.

Professor Cina van Zyl is a professor in tourism management at the University of South Africa. She holds a Doctorate Commercial, and previously gained management experience by chairing the department. Her primary research interest is in tourism marketing, although she also dabbles in the fields of transport economics and higher education practice. She is a member of the editorial board for the *Journal of Transport and Supply Chain Management* as well as theme editor of the *International Journal of Tourism Cities*. Cinà is currently heading a registered community engagement project empowering small and medium enterprises (SMEs) in the tourism field.

Dr Jennifer Verduin is a senior lecturer in oceanography and marine pollution and the Head of Environmental and Conservation Sciences in the College of Science, Health, Engineering and Education at Murdoch University. She has many years of experience in research on the role of high latitude oceans in decadal climate variability and thermo-haline circulation. Jennifer's current interests focus on the impact of physical processes on biological processes and ecosystem dynamics including the effects of coastal hydrodynamics on benthic vegetation (seagrass and seagrass restoration), particles and structures.

Professor Fang Wang, PhD is a professor at the College of Architecture and Landscape, Peking University, the Chinese director of the NSFC-DFG Sino-German Cooperation Group on Urbanization and Locality (UAL), and a registered urban planner. She has conducted research on "characterisation, evolution and prediction of the locality of built environment in complex adaptive systems". She pays special attention to the relationship between city and water, and focuses on areas such as the Yellow River basin and the Grand Canal region to carry out research. Her representative accomplishments include works on geoarchitecture and landscape, urban and rural memory, and cultural landscape security patterns.

Ting Wang is a graduate student at the College of Urban and Environmental Sciences, Peking University. Her research focuses on tourism geography and cultural geography.

Dr Craig Webster is an associate professor in the Department of Management at Ball State University, USA. He received an MA and PhD in political science from Binghamton University in New York State and an MBA at Intercollege, Cyprus. He has taught at Ithaca College, the College of Tourism and Hotel Management and the University of Nicosia. His research interests include the political economy of tourism, event management and robots in tourism and hospitality. Craig has published in many peer-reviewed journals internationally and is co-editor of two books.

Jacqueline Wichnewski is an alumna of the Breda University, Netherlands, holding a Bachelor's degree in international tourism management and consultancy. Her major research interests include the psychology of mobility in tourism, destination management and marketing, cross-cultural research and brand management.

Professor Bihu Wu, PhD is a professor and director in the International Center for Recreation and Tourism Research at the College of Urban and Environmental Sciences, Peking University. He specialises in the areas of tourism geography, city and regional tourism planning, destination management and marketing, cultural heritage rejuvenation, national park recreation and history of world experoutination (exploration and experience en route and at destination).

Dr Bozana Zekan is an assistant professor in the Department of Tourism and Service Management at Modul University Vienna, Austria. She holds a Doctor of Social and Economic Sciences (with Honours) degree from the Vienna University of Economics and Business. Her research interests are mainly within the field of destination management (e.g. key performance indicators, competitiveness, benchmarking, efficiency studies with the application of data envelopment analysis (DEA), etc.).

INTRODUCTION

City tourism and tourism cities

Alastair M. Morrison and J. Andres Coca-Stefaniak

Background and aims

As the Co-Editors-in-Chief of the *International Journal of Tourism Cities* (IJTC), the editors of the *Routledge Handbook of Tourism Cities* have a deep and long-term interest in urban tourism research and tourism cities. The main aims of the *Routledge Handbook of Tourism Cities* are to:

- Review contemporary issues, trends and challenges in urban tourism and tourism cities.
- Present practical approaches and solutions for marketing and branding tourism in urban environments.
- Describe key markets for urban tourism.
- Elaborate on tourism product development innovations and trends for cities.
- Examine the impacts of technology on tourism cities including smart city destination dimensions.
- Explore the unique characteristics of marketing and development of urban tourism in different regions of the world.

The *Routledge Handbook of Tourism Cities* covers key issues, trends and challenges for urban tourism destinations worldwide as well as contemporary debates related to research and practice in this field. Topics discussed include the marketing and branding of tourism cities, the growth of smart city tourism destinations, sustainability in urban tourism management, overtourism, sharing economy influences on urban destinations, cultural-heritage tourism in cities, business tourism, urbanisation, terrorism, among others. The *Routledge Handbook of Tourism Cities* merges the latest academic research with insights drawn from practice in urban destinations internationally to provide recommendations for tourism management professionals as well as researchers. Unlike other texts, the *Routledge Handbook of Tourism Cities* adopts a multidisciplinary approach to tourism drawing from fields such as sociology, psychology, urban management, business and critical management perspectives. In addition to this, a balance is provided between urban destinations in emerging economies and more established tourism cities in G20 countries.

Urbanisation

It can be argued quite convincingly that cities as tourism destinations have been neglected by scholars in comparison to rural and resort areas. This is perhaps not that surprising since academic researchers have focused mainly on holiday, vacation and leisure travellers. They have regarded cities more as a source of tourists, rather than themselves being tourism destinations. However, business tourism is significant worldwide and is concentrated mainly in urban areas. While university investigators have tended to short-change urban tourism until recently, it has not been neglected by city governments and destination management organisations (DMOs). Cities are now placing greater emphasis on the planning, development, marketing and branding of tourism than they did previously. Another reason that urban areas have not been in the research limelight is because greater attention has been given to national-level tourism and to international travellers rather than domestic. Many cities have far greater volumes of domestic visitors than people coming from abroad. However, there is likely in the future to be much greater focus on urban tourism as the trend toward greater urbanisation gains momentum.

> More than one half of the world population lives now in urban areas, and virtually all countries of the world are becoming increasingly urbanized.
>
> *(United Nations, 2020)*

As more people live and work in the world's cities, these urban areas will surely become more powerful magnets for visitors, as well as representing a greater portion of the tourism sector. This will also put greater pressure on cities' resource capacities and may exacerbate current issues and challenges that growing tourist numbers are causing. Hence, the need for this publication to bring together some of the best thinking on urban tourism and tourism cities.

The collection of outstanding chapters in the *Routledge Handbook of Tourism Cities* is a testament to the rapidly growing academic and public interest in urban tourism. Rather than casting the net broadly, the authors were hand-picked by the editors due to their significant recognition and interest in city tourism and tourism cities.

Need for a balanced perspective

Not all that is associated with tourism is good and beneficial to cities, so there is a need in any discourse such as this for a balanced assessment of the phenomenon. Having more than five years of publishing the *International Journal of Tourism Cities*, and themselves being active tourism researchers in this field, the editors have become familiar with the positive and negative sides of the urban tourism phenomenon. First, in taking an objective view of urban tourism, it is important to identify and review the major contemporary issues and challenges. In combing the recent literature, four of the most discussed issues are sustainability, overtourism, terrorism and the sharing economy. Also, a positivist approach is adopted in chapters on city destination marketing, market segments, branding, product development and technology applications.

What is urban?

Let's start with the basics then. As this is a text about tourism cities, it follows that we are talking about urban areas and this raises the simple question of "what is urban"? According to Ritchie and Roser (2019), "there is currently no universal definition of what 'urban' means". The most quoted figures on urban areas are from the United Nations (UN), which generally uses populations of 150,000 and 300,000 and over for its urban statistics. For example, in its 2019 reports, the UN quoted a figure of 4,220 urban settlements in the world, of which 529 were megacities of 10 million or more, 325 were large cities of five to 10 million, 926 were medium-sized cities of one to five million, 415 were cities with 500,000 to one million, 275 were cities with 300,000 to 500,000, and 1,750 were urban settlements with less than 300,000 inhabitants. The combined populations of these urban settlements represented 55% of the world's total with the remaining 45% living in rural areas (United Nations, 2019a; 2019b, p. 55). Not everybody agrees with the UN definitions and statistical estimates, and that includes the European Commission (European Union, 2016), which projected the world's urban population at a much higher proportion (84%) than did the UN (Cox, 2018). The gap between 55% and 84% is huge and the controversy revolves around what the European Commission classifies as being urban under the 300,000-population level. Another source, the *Atlas of Urban Expansion*, uses a cut-off of 100,000 population for urban areas and estimates 4,231 such areas in the world in 2016 (Atlas of Urban Expansion, 2016). One of the main problems in measurement is that individual countries use different population figures for what they define as being urban.

Without delving deeper into the intricacies of urban area measurement, it suffices to say that even at 4,000 there are many more urban areas in the world than countries. Hence, there is also great scope for urban tourism and tourism cities. Additionally, there is a significant diversity of cities and that is the topic to next be discussed.

City rating systems

Not all cities and urban areas are alike. However, they all attract residents, visitors, investors, companies, students and other groups and individuals. Due to the magnetism of cities, several ranking schemes have developed to rate them on various criteria. Most of these include some measures of tourism and the cultural offerings within specific cities, as will now be highlighted.

The *A.T. Kearney 2019 Global Cities* report uses several metrics to rank many of the world's most recognised urban areas. In the following quote about the firm's 2019 report, use of the words "vibrancy" and "competitive" is noticeable, as is the bottom-line statement that global cities attract people and businesses:

> The vibrancy of the world's most competitive cities – places such as London, New York, Singapore, and San Francisco – is no happy coincidence. With a focus on human capital, thoughtful municipal policies, smart corporate investment, and a commitment to building a technology pathway into the future, these cities have become bustling, global hubs that attract people and businesses alike.
>
> *(A.T. Kearney, 2020)*

The 2019 report ranked 130 cities with 27 metrics divided into five categories – business activity (30%), human capital (30%), information exchange (15%), cultural experience (15%) and political engagement (10%). London ranked first for cultural experience which had the six metrics of museums, visual and performing arts, sporting events, international travellers, culinary offerings and sister cities. London received the highest scores for four of these six criteria (museums, sporting events, international travellers and culinary offerings).

The *Global Power City Index* (GPCI) is operated by the Institute for Urban Strategies of the Mori Memorial Foundation in Japan. It measures city magnetism defined as their "power to attract people, capital, and enterprises from around the world". The six criteria that GPCI covers are economy, research and development, cultural interaction, liveability, environment and accessibility. Within cultural interaction, the measurements include several tourism-related items including tourism resources (tourist attractions; proximity to World Heritage sites; night-life options), visitor amenities (number of hotel rooms; number of luxury hotel rooms; attractiveness of shopping options; attractiveness of dining options), cultural facilities (number of museums; number of theatres; number of stadiums), international interaction (number of foreign visitors; number of foreign residents), and trendsetting potential (number of international conferences; number of cultural events; cultural content export value; art market environment). These are great indicators for assessing tourism in cities; however, the GPCI only reports on 48 cities worldwide. GPCI had London, New York, Tokyo, Paris and Singapore in the top five slots for 2019.

> A great place to live is also a great place to visit.
>
> *(Dion, 2014)*

Liveability is a quality often ascribed to cities and there are also several polls that measure this construct. Before describing these rankings, the quote shown above needs some thought. It was said at the 2014 ITSA (International Tourism Studies Association) 5th Biennial Conference in Perth, Australia by Jim Dion of *National Geographic*. He argued that places that followed sustainable development were great places to live and they were also very attractive to tourists.

The *Global Liveability Index* is published annually by the Economist Intelligence Unit (EIU). Thirty criteria (quantitative and qualitative) are used in five categories (stability, healthcare, culture and environment, education and infrastructure) to rank 140 cities around the world. Within the culture and entertainment category, there are nine indicators, three of which are cultural availability, food and drink, and sporting availability. There are transportation-related indicators within the infrastructure category. Ranked highest in 2018 and in order were Vienna (Austria), Melbourne (Australia), Osaka (Japan), Calgary (Canada), Sydney (Australia), Vancouver (Canada), Toronto (Canada), Tokyo (Japan), Copenhagen (Denmark) and Adelaide (Australia).

The *Mercer's Quality of Living Ranking* was produced in its 21st annual edition for the year 2018 (Mercer LLC, 2020). Some 231 cities were ranked and once again Vienna came out on top and Baghdad was 231st. The other top-ranked cities, in order, were Zurich (Switzerland), Vancouver, Munich (Germany), Auckland (New Zealand), Düsseldorf (Germany), Frankfurt (Germany), Copenhagen, Geneva (Switzerland) and Basel (Switzerland). The ranking criteria used by Mercer are not easily determined; however, in their rankings for 2018, the company measured personal safety and provided separate statistics for it. Helsinki (Finland), Luxembourg, Bern, Basel and Zurich were ranked as the best for personal safety (Mercer LLC, 2019).

These ranking systems, most if not all owned by commercial interests, demonstrate a high level of interest in cities as well as the desire for city governments to have competitive benchmarks. Their limitation is that they only cover a small portion of the world's urban areas and do not offer much guidance on how to manage and improve tourism in cities. Also, they do not classify urban areas by their economic emphasis such as say on tourism. Now, therefore, we need to look at urban tourism and that special type of place that is termed as a tourism city.

What is urban or city tourism?

It is easy to answer this by saying that it is tourism that occurs in a city or urban area. However, it is much more than just that and we can say that it is a tourism system that exists in a city or urban area (Morrison et al., 2018). These authors describe a tourism system as consisting of four interconnecting parts – destination, marketing, demand and travel. They also stress that tourism systems are open and are affected by external environment factors. Destinations, including cities, have products that include attractions and events, built facilities, transportation, infrastructure, and service quality and friendliness. Apart from these destination products, cities may have tourism policies and plans, purpose-specific tourism organisations (DMOs), legislation and regulations, and sustainable tourism guidelines. Marketing involves communications with targeted markets to convince people to visit the city, and this effort is often coordinated by city DMOs. Positioning, image and branding (PIB) are critical to successful marketing (Morrison, 2019). Demand includes the factors that influence people in making travel decisions, as well as how they make bookings, travel and recall their memories of trips. The fourth system part, travel, encompasses online and offline travel distribution channels and modes of transportation.

Thus, city tourism is a system that involves a great deal more than just what visitors do in urban areas. This system needs careful management as there are multiple stakeholders to be considered, including visitors, residents and tourism sector owners and staff. Also, there needs to be a priority in making tourism sustainable for the longer term.

Is tourism important to cities? From all the information available, tourism is of great significance to many cities around the world. This point is confirmed by statements made by the World Conference of Mayors, which includes tourism in its "7 T" goals – "*to stimulate increased tourist travel between cities of the world*". The other six Ts are trust, trade, technology transfer, twin cities, treasury and training (World Conference of Mayors, 2019).

What is a tourism city?

This is a very tough question to answer as all cities welcome some level of visitors. It could be argued that tourism cities are places where tourism is important and in which city governments put a high priority on this economic sector (McClain, 2015). Additionally, it could be said that these urban areas are the ones that get the most tourists. Using these parameters, it is easy to say that London, New York and Paris are tourism cities; however, what about less famous "provincial" cities like Anqing in China, Portland in the USA and Dundee in the UK?

There are undoubtedly both quantitative and qualitative criteria that determine tourism cities. Considering the volumes of tourists and expenditures is an example of one of the potential quantitative metrics. As an example of this, a ranking of the world's top tourism cities is produced each year in Mastercard's *Global Destination Cities Index* (GDCI). Some 167 cities were in the results for 2019 and Bangkok, London and Paris were the top three. The GDCI ranks the cities by their total volumes of international overnight visitors. GDCI also reports on the spending of international visitors in cities. Bangkok in 2019 welcomed 22.78 million international visitors.

Euromonitor International publishes an annual ranking of the *Top 100 City Destinations*, which also uses the total number of international tourist arrivals. The company's 2019 rankings had a top ten that included five cities in Asia (Hong Kong as #1 in the world; Bangkok, #2; Macau, #4; Singapore, #5; Kuala Lumpur, #9). The other top-ranking cities were London (#3), Paris (#6), Dubai (#7), New York City (#8) and Istanbul (#10) (Euromonitor International, 2020).

On the business tourism side, GainingEdge, a consulting company based in Melbourne, Australia, has developed a city-based *Competitive Index* with respect to international conventions. For 2019, the Index ranked 103 cities worldwide, with Paris, Barcelona, Singapore, Tokyo, New York and Beijing receiving the highest scores (GainingEdge, 2019). Additionally, the International Congress and Convention Association (ICCA), based in Amsterdam, and the Union of International Associations (UIA) (Brussels) produce annual statistics on international conferences and conventions by city.

Another potential measurement, and one that is more subjective, is to examine the polls of popular and "bucket list" city destinations prepared by travel magazines and guidebooks, and other sources. Entities that prepare these include *Travel + Leisure, Condé Nast, National Geographic Traveler, Frommer's, TripAdvisor* and *Lonely Planet*. Getting on these lists is surely an indicator that the cities are popular and desirable for visitors and that these authoritative sources are recommending them as destinations. However, as with most of the aforementioned city ranking systems, some of the results are quite predictable as the most famous and highly visited cities are usually listed such as London, Paris, Rome, Tokyo and Venice. Lesser well-known cities seldom feature in these rankings.

Greater legitimacy to the term tourism city arrived in 2012. Since then, tourism cities have had their own representational body through the World Tourism Cities Federation (WTCF) headquartered in Beijing. Through self-selection, the cities that belong to WTCF consider themselves to be tourism cities.

> World Tourism Cities Federation (WTCF), voluntarily formed by famous tourism cities and tourism-related institutions in the world under the initiative of Beijing, is the world's first international tourism organization focusing on cities. Established on 15 September 2012 in Beijing, the headquarters and Secretariat of WTCF are based in Beijing, and Chinese and English are its official languages.
>
> *(World Tourism Cities Federation, 2020)*

Information on WTCF's website at the time of writing indicated that it had 145 members. While this number is growing, it still represents only a minor proportion of the world's urban areas.

Another group that definitely has tourism city members is the BestCities Global Alliance. Eleven cities around the world are cooperating in this initiative through their DMOs and they include Berlin, Bogotá, Cape Town, Copenhagen, Dubai, Houston, Madrid, Melbourne, Singapore, Tokyo and Vancouver (BestCities Global Alliance, 2020). This alliance focuses on serving international business events.

Surely, it might also be said that tourism cities must have many hotels and rooms, convention facilities, attractions for tourists and significant transport access and capacity. It will be recalled from earlier that the GPCI uses some of these metrics in its ranking systems. Nowadays, cities like Las Vegas, Orlando, Beijing and Shanghai are thought to have the largest hotel capacities.

The academic literature is not particularly helpful in answering the question although many articles can be found when searching by "tourism city" in the Web of Science, Scopus and Google Scholar. For example, there are now a significant number of research articles on "smart tourism cities" as well as many case examples about individual places identified as "tourism cities".

So, after all of this has been said, what is a tourism city? Once more, let's be honest and say that all urban areas are potentially tourism cities. However, some cities demonstrate qualities and characteristics that make them stronger tourism cities and the editors consider these to be:

- Have a formal or official priority on tourism (such as tourism policy statement or tourism plan).
- Have a dedicated and official organisation with responsibility for city destination and management.

- Have significant numbers of visitors (business, holiday/vacation, visiting friends and relatives – VFR, and other) to the cities.
- Tourism supports significant levels of employment.
- Have significant levels of tourism capacity in respect to, for example, hotel rooms, convention/conference facilities, attractions and transport.
- Have a concerted effort in marketing, branding and promotion to attract visitors.
- "Bucketability" or the popularity of cities as expressed by the media, influencers and others.

The *Routledge Handbook of Tourism Cities* has many case studies and examples on cities that meet these criteria. Included are Bangkok, Beijing, Budapest, Dubai, Hangzhou, Hong Kong, Hyderabad, Jakarta, Kuala Lumpur, Las Vegas, Lima, London, Macau, Medellin, Melbourne, New Orleans, New York City, Paris, Prague, Seoul, Shanghai, Singapore, Tijuana, Vancouver and many others.

Issues, trends and challenges for tourism cities

One of the main aims of this *Routledge Handbook of Tourism Cities* is to raise awareness of the contemporary issues, trends and challenges with respect to urban tourism and tourism cities. An issue is defined as an important topic or problem that people are discussing, or talking or writing about related to markets, supply or the external environment (politics, economy, society, technology, environment and legislation). The topics or problems are unresolved, and people have differing opinions on them. Issues often have opponents and proponents, and can be short-, medium- or long-term. Trends are new developments or changes in tourism related to markets, supply and external factors. Trend lines can be increasing or decreasing. The changes brought about by trends may be considered as positive, negative or neutral. Trends can be measurable (quantitative) or non-measurable (qualitative). Trends are short-, medium- or long-term. The intensity of trends in tourism may vary for different parts of the world.

What is a challenge then? Challenges in tourism often relate to various forms of resource capacity including funding, human talent, management and working expertise and experiences, available time, natural and cultural-heritage resources, for example. Challenges frequently involve shortages in the quantity and/or quality of such resources. It is often difficult to separate out issues, trends and challenges, and indeed it is possible for something to belong to all three.

Overtourism is a term that first appeared in the media in 2016 (Ali, 2018) and was associated with certain cities in Europe. One can say it definitely is a supply-side issue as it is unresolved. It could also be argued that it is an increasing phenomenon and therefore could be called a trend. As it involves crowding and over-capacity situations that are resource related, overtourism might also be classified as a challenge.

The "smartening" of cities would seem to be primarily a trend, although it may pose challenges for certain urban areas in terms of resources and expertise. The long-term growth in worldwide international tourism arrivals is also mainly trend-oriented; however, some may argue that this trend is causing issues and posing resource challenges. The expansion of cruise travel is definitely a trend, although cruise ships causing ocean pollution is an issue.

The *International Journal of Tourism Cities* at the end of 2016 asked selected members of its editorial board to write about urban tourism issues, trends and challenges (Coca-Stefaniak & Morrison, 2016). In the editorial, Deborah Edwards of the University of Technology Sydney identified the issue of homogenisation among cities; in other words, cities are becoming more alike one another. She alluded to the issue of some cities becoming over-reliant on tourism, which is undesirable economically and can also strain resources. Safety within cities and the need for more urban greening were two other issues raised by her (Edwards, 2016, p. 273).

Prof. Nelson Graburn of the University of California Berkeley also highlighted safety concerns as an issue for cities. He discussed climate change and its negative impacts on coastal cities. On the positive side, he mentioned "contents tourism" in northeast Asia and particularly in South Korea and Japan. This tourism is based on aspects of religion, mythology, folklore and popular literature (Graburn, 2016, pp. 273–274). Claire Liu of Auckland University of Technology pointed to the sustainable development of cities as the key issue for the coming decade. She also discussed the impact of social media on city destination marketing. Urban tourism being under researched was another issue identified (Liu, 2016, p. 274). Prof. Philip Pearce of James Cook University also raised the increasing influence of technology on city tourism. He stressed the growth in tourism demand from Asian countries, particularly China. The trend to more experienced travellers in the market was an interesting phenomenon that he highlighted (P. Pearce, 2016, p. 275). Prof. Can Seng Ooi, now with the University of Tasmania, pinpointed social media and the sharing economy as two important and connected trends for urban tourism. As did Claire Liu, he highlighted sustainability as a major issue for cities (Ooi, 2016, p. 275). Prof. Douglas Pearce of Victoria University of Wellington said a major challenge was bringing urban tourism practitioners and academics closer together. In addition, he noted for New Zealand a realignment of DMOs into broader regional economic initiatives (D. Pearce, 2016, pp. 275–276). Svetlana Stepchenkova of the University of Florida stated that cities were increasingly diversifying their tourism offers to grow tourism. She mentioned the Pokémon Go craze as an example of cities having to be ready for trends that emerge very quickly in today's society (Stepchenkova, 2016, p. 276). Prof. Greg Richards of Tilburg University remarked on the increasing involvement of local residents in promoting tourism to their cities. He stated as well that more small- and medium-sized cities were trying to increase tourism (Richards, 2016, p. 276). Amy So of the University of Macau discussed the increasing impacts of Millennials and Generation Z on tourism. Like others, she talked about the influence of new technologies, social media and big data (So, 2016, pp. 276–277). Prof. Costas Spirou of Georgia College & State University identified two supply-side trends in city tourism. One of these was the expansion of infrastructure and other amenities in downtowns and specialised districts; the other was the greater emphasis by cities on image-building and branding. He also argued that public–private partnerships (PPPs) will increase in importance (Spirou, 2016, p. 277). Keith Dinnie, now attached to the University of Dundee, described the growing emphasis on smart destinations. He suggested that city mayors may play a more prominent role in the future as some of them become better known by the public (Dinnie, 2016, pp. 277–278). The late John Heeley reviewed the trend of the Internet and social media linking tourism demand and supply. Additionally, he stated that consumer expectations were increasing and becoming more deal-oriented. The influence of TripAdvisor was mentioned as was that of sharing economy platform, Airbnb (Heeley, 2016, p. 278). Prof. László Puczkó of Budapest University of Economics concentrated on the growth of budget or low-cost airline carriers (LCCs) and their use of secondary or tertiary cities for their airports. He cited the impacts of large crowds associated with music festivals and other events on city centres and the night-time economy, as well as the emergence of creative industries (Puczkó, 2016, p. 278). Prof. Han Shen of Fudan University focused on new information technologies and especially social media. She suggested that access to this information through apps was reshaping patterns of travel and other aspects of city tourism (Shen, 2016, p. 279). Martin Selby of Coventry University cited a number of issues and trends in urban tourism, including overtourism, the sharing economy, increasing use of mobile apps and growth in independent travel. He pinpointed the technology-enabled trend of city tourists seeking more authentic and creative experiences, as well as comradeship, and being more concerned with their well-being and having a sense of place (Selby, 2016, p. 279). Prof. Hong-Bumm Kim of Sejong University emphasised the greater priority by cities on their destination positioning and branding. He illustrated this with the example of the new city branding for Seoul (Kim, 2016, pp. 279–280). Prof. Guoqing Du of Rikkyo University mentioned the growth in international travel and posited that it was making cities more global. She also noted the increasing use of technology including smartphones (Du, 2016, p. 280).

These 17 experts put forward a substantial set of trends, issues and challenges to urban tourism and tourism cities. These demonstrate the complexity of urban tourism systems which are being buffeted by the winds of change from many different directions. The changes are occurring in tourism demand, supply and external environmental factors. The effect of technologies was a trend mentioned by several of the authors and particularly relative to social media, smartphone and app usage.

While these statements on city tourism issues, trends and challenges are authoritative, they are not exhaustive and there are certainly others that must be contemplated. Table I.1 provides a list of 50 trends prepared by the editors that covers most of the ones identified by the 17 experts, plus several others.

Table I.2 is a list of selected issues developed by the editors. Again, this is not a complete listing and more items could be added. However, it highlights the diversity of problems facing urban destinations.

Table I.1 Trends in tourism demand and supply

Adventure tourism	LBGT travel
Artificial intelligence (AI)	LCCs (low-cost air carriers)
Augmented and virtual reality (AR-VR)	Medical and wellness tourism
Bleisure	Millennials and Gen Z
Breakationing	Multi-destination travel
Bucket-list travel	Multi-generational travel
Casinos and gaming	New accommodation types
China and India outbound market growth	Online travel booking
Co-creation	Quest for authentic experiences
Creative industries and tourism	Sharing economy
Crowdsourcing and crowdfunding	Slum or ghetto tourism
Cruising growth	Smart tourism
Domestic tourism	Smartphone and app use
Driverless vehicles	Social media use
Ecotourism	Solo tourism
Event festivalisation	Space tourism
Flashpacking	Special-interest tourism
Food/gastronomic/culinary tourism	Staycationing
Genealogy and roots tourism	Super long-haul flights
Geotourism	User-generated content (UGC)
Glamping	Vegetarianism
High-speed rail travel	Voluntourism
Independent travel	Wildlife tourism
Influencers	Wine tourism
Last-minute travel	Worldwide tourism growth

Table I.2 Issues with tourism demand and supply

Accessible tourism	Labour supply
All-inclusive resorts	Loss of authenticity/homogenisation
Animal rights	Marijuana/cannabis/pot tourism
Climate change	Orphanage tourism
Commodification	Overtourism
Competition	Pollution
Crimes and scams	Problem gambling
Crises	Safety and security
Crowding	Sex tourism
Drinking/pub crawl/stag tourism	Sharing economy
Funding for tourism	Slum or ghetto tourism
Gender inequality	Sustainable tourism
Gentrification	Terrorism
Global warming	Touristification
Globalisation	Urbanisation
Governance	Water shortages/inequalities

Future of urban tourism research

Tina Šegota of the University of Greenwich prepared an editorial for the *International Journal of Tourism Cities* under the title of "Future Agendas in Urban Tourism Research" (Šegota, 2019a). Thirteen experts contributed their thoughts and opinions on the topic. Marianna Sigala of the University of South Australia picked out liveability, sustainability, quality of life and well-being of tourists but also residents, security and safety, avoidance of commercialisation/Disneyfication of cultural resources and spaces as being among the top issues facing tourism cities. She emphasised new technology applications, smart tourism, the sharing economy and big data. There are many research questions still to be addressed into the tourism applications of new technology, for example with respect to drones and driverless vehicles (Sigala, 2019, pp. 109–110).

Ulrike Gretzel of the University of Southern California also chose to highlight the technological influences on tourism cities. Her focus was on social media and how these platforms and apps were affecting travellers. Interesting here was her assertion that social media was creating "vacation envy" and that cities were great places for images to share (Gretzel, 2019, pp. 110–111).

Jonathon Day of Purdue University took a different tack and asserted that the big story for sustainable tourism in the future will be about cities and not as much on rural areas as before. He stated that the sustainable development of tourism for cities was particularly complex and advised that more research was needed using multidisciplinary approaches (Day, 2019, pp. 111–112).

Tina Šegota herself wrote about the challenges to residents' quality of life in tourism cities. In mentioning the resident protests against tourism in Barcelona, Mallorca and Venice, she called for more research not only related to resident's well-being but also tourists' well-being. Also, she stressed that quality of life must be given a higher priority in urban tourism research (Šegota, 2019b, pp. 112–113).

Jithendran Kokkranikal of the University of Greenwich also cited the anti-tourism movements in several cities, along with noting that urban areas were becoming more important destinations. As with Jonathon Day, he recommended that more research on sustainability in urban tourism was needed, especially given the growing visitor volume pressures on cities (Kokkranikal, 2019, p. 113).

Melanie Smith of Budapest Metropolitan University reviewed the transitions of post-socialist cities particularly from the tourism standpoint. The role of budget airlines and the development of the night-time economies were discussed, sometimes resulting in undesirable activities such as drunk tourism. She urged for more research on image construction of these cities (Smith, 2019, p. 114).

Joan Henderson, formerly of Nanyang Technological University, reviewed the relationship of urban tourism and crisis management. She cited capital cities, places of political and financial power, as being especially susceptible to riots and other forms of dissent, and acts of terrorism. A call was made for more research in the city context on the causes and catalysts of crises (Henderson, 2019, pp. 114–115).

Cláudia Seabra of the University of Coimbra also addressed the topic of crises and specifically those resulting from terrorism. She alluded to the high value of tourists as terrorist targets and tourists were becoming increasingly aware and uncomfortable as a result of potential threats. More research is needed on how terrorism affects travellers (Seabra, 2019, pp. 115–116).

Philip Pearce of James Cook University, who discussed tourist behaviour, identified four areas for more future research – wayfinding, dealing with others, appraising satisfaction and storytelling. He cited new smart technologies, overcrowding situations, changes in the outcomes people want from their travels, and posting travel experiences on social media as influences that required the rethinking of previous explanations of tourist behaviour (P. Pearce, 2019, pp. 116–117).

Rob Davidson of MICE Knowledge had a focus on business events and noted an increase in the research on business tourism. However, he pointed out that this research was under-represented in the tourism academic literature and needed to be given more attention. Investigating business tourists' experiences in cities was a research direction much required (Davidson, 2019, pp. 117–118).

Cina van Zyl of the University of South Africa looked at special-interest tourism within cities, and identified heritage, culture, events, gastronomy and cuisine as particularly important in this context. She called for more multidisciplinary research on these topics, both conceptual and empirical. The management and research requirements for advancing special-interest tourism were recommended from different stakeholder perspectives (van Zyl, 2019, pp. 118–119).

David Newsome of Murdoch University and James Hardcastle of the International Union for Conservation of Nature (IUCN) took up the topic of "nature in the city" and its increasing importance for residents and visitors. However, they noted, citing TripAdvisor, that natural areas seldom feature among the top ten attractions for cities. They envisaged a need for new international standards and criteria for the conservation of urban natural and cultural resources (Newsome & Hardcastle, 2019, pp. 119–120).

Tijana Rakić of the University of Brighton contemplated the use of visual methodologies in tourism research. She particularly focused on the analysis of photos and videos recorded by city tourists. With the rapidly increasing influence of social media and user-generated content, tourism researchers are advised to make greater use of these research approaches (Rakić, 2019, p. 120).

The views of these 30 experts across the two IJTC editorials are a rich source of information on city issues, trends, challenges and future research directions. There appears to be a consensus on several highest priority topics including the influence of technology, increasing importance of sustainability, safety and security, overtourism and quality of life, sharing economy and new types of tourist behaviour. These data informed the editors, along with their own research and practical experiences, in developing the outline for the *Routledge Handbook of Tourism Cities* and in inviting the most relevant contributions based on the foregoing materials.

Organisation of the handbook

The *Routledge Handbook of Tourism Cities* is structured into four parts. Implementing the balanced approach mentioned earlier, Part I addresses "Contemporary Issues, Challenges and Trends in Urban Tourism". Issues are discussed including terrorism, sustainability, the sharing economy, urbanisation, gentrification, overtourism, micro-shocks and public outrage. Part II adopts a positivist approach with "Marketing, Branding and Markets for Tourism Cities". It covers city destination management and marketing, and urban branding with case studies from Melbourne (Australia) and Vancouver. Key markets of business tourism, Millennials, families, visiting friends and relatives (VFR), dark tourism and birdwatching are reviewed. The positivist theme continues with Part III on "Product and Technology Developments for Tourism Cities". Topics discussed are culture and heritage, outdoor and indoor city markets, touristic urban spaces, attractions, old and new sections of cities, coastal city development, smart urban tourism destinations, eTourism challenges, social media, urban transport and the artistic medium of walking. Part IV presents "Worldwide Tourism Cities and Urban Tourism" and demonstrates there are differences around the globe related to these concepts. The geographic areas featured are Australia, Europe, United States, Latin America, China, ASEAN and former socialist countries.

Summary

There is a clear message from several observers than urban tourism and tourism cities require greater attention. In introducing the *Routledge Handbook of Tourism Cities*, the editors set out to define and highlight the importance of these two connected concepts. They also wanted to clearly isolate many of the major trends, issues and challenges confronting tourism cities. In so doing, another message becomes clear and that is that city destination management and marketing are extremely complex, although they are professionalising at a healthy pace.

While everything has been done to make this text as current as possible, the volatility of tourism worldwide inevitably changes the circumstances of certain cities. Hong Kong's tourism, for example, was decimated after several months of civil unrest. The coronavirus outbreak in China caused the curtailing of travel within China and the closure of several cities including Wuhan, and as the pandemic spread to other parts of the world, the tourism sector was devastated. The wildfires in Eastern Australia had a major detrimental effect on tourism. These three cases highlight the fragility of tourism and its susceptibility to external influences.

This handbook fills a significant void in the tourism literature on urban tourism and tourism cities by bringing together some of its top experts. Over 70 of these thought leaders have offered many valuable insights and case examples on these concepts.

References

Ali, R. (2018). The genesis of overtourism: Why we came up with the term and what's happened since. *Skift*, 14 August. https://skift.com/2018/08/14/the-genesis-of-overtourism-why-we-came-up-with-the-term-and-whats-happened-since/.

A. T. Kearney. (2020). A question of talent: How human capital will determine the next global leaders. *2019 Global Cities Report*. www.kearney.com/global-cities/, accessed 19 January 2020.

Atlas of Urban Expansion. (2016). *Atlas of Urban Expansion 2016*. New York University, Lincoln Institute of Land Policy. www.atlasofurbanexpansion.org/, accessed 14 January 2020.

BestCities Global Alliance. (2020). Overview. www.bestcities.net/about-us/, accessed 20 January 2020.

Coca-Stefaniak, J. A., & Morrison, A. M. (2016). Views from the editorial advisory board. *International Journal of Tourism Cities*, 2(4), 273–280.

Cox, W. (2018). European Commission exaggerates urbanization. *New Geography*. www.newgeography. com/content/006057-eu-exaggerates-urbanization, accessed 13 January 2020.

Davidson, R. (2019). Research into business tourism: Past, present and future. *International Journal of Tourism Cities*, 5(2), 117–118.

Day, J. (2019). The diversity of approaches to sustainability in tourism cities: Initiating new research on dynamic and adaptive responses. *International Journal of Tourism Cities*, 5(2), 111–112.

Dinnie, K. (2016). Views from the editorial advisory board of the *International Journal of Tourism Cities*. *International Journal of Tourism Cities*, 2(4), 277–278.

Dion, J. (2014). 5th ITSA Biennial Conference, Perth, Australia.

Du, G. (2016). Views from the editorial advisory board of the *International Journal of Tourism Cities*. *International Journal of Tourism Cities*, 2(4), 280.

Economist Intelligence Unit. (2020). *The Global Liveability Index 2019*. www.eiu.com/public/topical_report.aspx?campaignid=liveability2019, accessed 14 January 2020.

Edwards, D. (2016). Views from the editorial advisory board of the *International Journal of Tourism Cities*. *International Journal of Tourism Cities*, 2(4), 273.

Euromonitor International. (2020). *Top 100 City Destinations: 2019 Edition*. https://go.euromonitor.com/white-paper-travel-2019-100-cities.html, accessed 20 January 2020.

European Union. (2016). *Atlas of the Human Planet 2016*. Luxembourg: European Union.

GainingEdge. (2019). *Competitive Index 2019: International Convention Destinations*. https://gainingedge.com/2019-competitive-index-offers-free-tool-for-destination-benchmarking/, accessed 20 January 2020.

Graburn, N. (2016). Views from the editorial advisory board of the *International Journal of Tourism Cities*. *International Journal of Tourism Cities*, 2(4), 274.

Gretzel, U. (2019). Social media and the city: Mediated gazes and digital traces. *International Journal of Tourism Cities*, 5(2), 110–111.

Heeley, J. (2016). Views from the editorial advisory board of the *International Journal of Tourism Cities*. *International Journal of Tourism Cities*, 2(4), 278.

Henderson, J. C. (2019). Looking beyond the negative: Crises in tourism cities as a tool for learning and positive change. *International Journal of Tourism Cities*, 5(2), 114–115.

International Congress & Convention Association (ICCA). (2019). *2018 ICCA Statistics Report: Country & City Rankings*. www.iccaworld.org/knowledge/benefit.cfm?benefitid=4036, accessed 20 January 2020.

Kim, H.-B. (2016). Views from the editorial advisory board of the *International Journal of Tourism Cities*. *International Journal of Tourism Cities*, 2(4), 279–280.

Kokkranikal, J. (2019). Informing tourism policy and development with responsible tourism and ethnoscapes. *International Journal of Tourism Cities*, 5(2), 113.

Liu, C. (2016). Views from the editorial advisory board of the *International Journal of Tourism Cities*. *International Journal of Tourism Cities*, 2(4), 274.

McClain, J. (2015). City tourism brings great profits to the communities. *Tourism Review News*, 30 June. www.tourism-review.com/travel-tourism-magazine-why-is-tourism-so-important-to-cities-article2601, accessed 20 January 2020.

Mastercard. (2019). *Global Destination Cities Index Report 2019*. https://newsroom.mastercard.com/documents/global-destination-cities-index-report-2019/.

Mercer LLC. (2019). Vienna tops Mercer's 21st Quality of Living ranking. www.mercer.com/newsroom/2019-quality-of-living-survey.html, accessed 19 January 2020.

Mercer LLC. (2020). *Quality of Living City Ranking*. https://mobilityexchange.mercer.com/Insights/quality-of-living-rankings, accessed 14 January 2020.

Mori Memorial Foundation. (2018). What is the GPCI? http://mori-m-foundation.or.jp/english/ius2/gpci2/index.shtml, accessed 13 January 2020.

Morrison, A. M. (2019). *Marketing and managing tourism destinations*, 2nd ed. London: Routledge.

Morrison, A. M., Lehto, X. Y., & Day, J. (2018). *The tourism system*, 8th ed. Dubuque, Iowa: Kendall Hunt Publishing.

Newsome, D., & Hardcastle, J. (2019). Conservation of "nature in the city" and its importance for city tourism. International Journal of Tourism Cities, 5(2), 119–120.

Ooi, C. S. (2016). Views from the editorial advisory board of the *International Journal of Tourism Cities*. *International Journal of Tourism Cities*, 2(4), 275.

Pearce, D. (2016). Views from the editorial advisory board of the *International Journal of Tourism Cities*. *International Journal of Tourism Cities*, 2(4), 275–276.

Pearce, P. L. (2016). Views from the editorial advisory board of the *International Journal of Tourism Cities*. *International Journal of Tourism Cities*, 2(4), 275.

Pearce, P. L. (2019). Resetting and re-exploring the foundations of tourists' behaviour in tourism cities. *International Journal of Tourism Cities*, 5(2), 116–117.

Puczkó, L. (2016). Views from the editorial advisory board of the *International Journal of Tourism Cities*. *International Journal of Tourism Cities*, 2(4), 278.

Rakić, T. (2019). Opening new research avenues with visual research methodologies. *International Journal of Tourism Cities*, 5(2), 120.

Richards, G. (2016). Views from the editorial advisory board of the *International Journal of Tourism Cities*. *International Journal of Tourism Cities*, 2(4), 276.

Ritchie, H., & Roser, M. (2019). Urbanization. OurWorldInData.org. https://ourworldindata.org/urbanization, accessed 13 January 2020.

Seabra, C. (2019). On terrorism and its challenges to freedom, mobility and way of life. *International Journal of Tourism Cities*, 5(2), 115–116.

Šegota, T. (2019a). Future agendas in urban tourism research: Special editorial. *International Journal of Tourism Cities*, 5(2), 109–124.

Šegota, T. (2019b). For residents or for tourists? The quality of life nexus in tourism cities. *International Journal of Tourism Cities*, 5(2), 112–113.

Selby, M. (2016). Views from the editorial advisory board of the *International Journal of Tourism Cities*. *International Journal of Tourism Cities*, 2(4), 279.

Shen, H. (2016). Views from the editorial advisory board of the *International Journal of Tourism Cities*. *International Journal of Tourism Cities*, 2(4), 279.

Sigala, M. (2019). eTourism and on how to address the smart city complexity. *International Journal of Tourism Cities*, 5(2), 109–110.

Smith, M. K. (2019). Lessons learnt and future challenges from tourism planning in post–communist countries. *International Journal of Tourism Cities*, 5(2), 114.

So, A. (2016). Views from the editorial advisory board of the *International Journal of Tourism Cities*. *International Journal of Tourism Cities*, 2(4), 276–277.

Spirou, C. (2016). Views from the editorial advisory board of the *International Journal of Tourism Cities*. *International Journal of Tourism Cities*, 2(4), 277.

Stepchenkova, S. (2016). Views from the editorial advisory board of the *International Journal of Tourism Cities*. *International Journal of Tourism Cities*, 2(4), 276.

Union of International Associations. (2020). *International Meetings Statistics Report*. https://uia.org/publications/meetings-stats, accessed 20 January 2020.

United Nations. (2019a). *The world's cities in 2018*. New York: United Nations.

United Nations. (2019b). *World urbanization prospects: The 2018 revision*. New York: United Nations.

United Nations. (2020). Urbanization. www.un.org/en/development/desa/population/theme/urbanization/index.asp, accessed 23 April 2020.

van Zyl, C. (2019). The need for different approaches to understanding special interest tourism in the urban context. *International Journal of Tourism Cities*, 5(2), 118–119.

World Conference of Mayors. (2019). Services. https://theworldconferenceofmayors.org/services, accessed 20 January 2020.

World Tourism Cities Federation. (2020). Introduction to WTCF. https://en.wtcf.org.cn/About/WhoWeAre/, accessed 19 January 2020.

PART I

Contemporary issues, challenges and trends in urban tourism

This part provides a framework for the book by outlining some of the main challenges that tourism cities around the world face today. Although the list of topics is by no means exhaustive, it does cover challenges such as the development and management of sustainable tourism in urban environments, terrorism and its impact on the planning and image of tourism cities, the growing phenomena of overtourism and urbanisation, the proactive role of residents in visitor–host interactions that often add a key element of authenticity against a backdrop of progressive gentrification, and reflections on present and future challenges for tourism cities and urban tourism based on current trends.

Summary of chapters

Cristina Maxim begins this part by outlining the challenges faced by world tourism cities in their capacity as centres for business, cultural excellence, connectivity hubs, key players in the visitor economy of destinations, homes to world-class tourist attractions and key players in economic development at regional, national and global levels. She examines how, despite these advantages, or perhaps because of them, world tourism cities face a myriad of interconnected issues and trends, which are often intimately linked to the complex economic, social and political functions they exhibit, as well as the diversity of people they attract (e.g. long-term residents, immigrants, visitors). Based on an analysis of examples drawn from London, Paris, Hong Kong, New York, Singapore and Dubai, recommendations are made for policy makers in cities, including the need for a more integrated understanding of the factors shaping their development (e.g. sustainability, traffic congestion, pollution) combined with global trends (e.g. global tourism market competitiveness, conflict between visitors and residents) when planning and managing tourism in these destinations.

Cláudia Seabra and **Odete Paiva** consider the tragically growing threat of terrorism to tourism cities, and especially to those that host UNESCO World Heritage Sites. The authors do this by considering the impact of this growing phenomenon on residents and tourists as well as the global image of these destinations. Given that a key objective of terrorism today is to maximise the level of media coverage linked to major disruption to Western values, way of life and culture, World Heritage Sites have become prime targets in this context as cultural identity icons to their host cities and sources of authentic and unique experiences for tourists. This

15

chapter provides evidence of this growing threat and discusses its implications for key decision makers in the planning and development of urban tourism destinations, as well as further academic research needed in this field.

Jonathon Day discusses the important intersection of urban tourism and sustainable development. A systems-thinking approach is applied to addressing the issues facing sustainable tourism in cities. The author begins by defining sustainability and sustainable tourism. Then, the author describes achieving sustainability in both cities and tourism systems as a "wicked problem". System hierarchies, independent actors, feedback loops, adaptation, resilience and tipping points as characteristics of systems are reviewed in the context of urban tourism. The chapter concludes by stating that it is critical for researchers and practitioners to apply systems-thinking approaches to understanding how sustainability can become standard practice in urban tourism.

Amanda Belarmino examines the effects of the sharing economy on cities, particularly with respect to peer-to-peer accommodation and ridesharing. A timeline of the contemporary sharing economy is provided beginning with the introduction of eBay in 1995. The author provides three mini-case studies of the impact of the sharing economy on Seoul (South Korea), Las Vegas (USA), and Dubai (UAE). Emerging segments of the sharing economy are identified and discussed including crowdfunding, third-party food apps, craft beer exchanges, office space sharing and pay-by-the-month living accommodations. The chapter concludes by stating that the sharing economy is providing increased access to tourism and increasing income for city residents. The author proposes areas for future research on the sharing economy and urban tourism.

Jian Ming Luo and **Chi Fung Lam** consider the effects of urbanisation on the development and management of urban tourism destinations, especially in China. The authors focus specifically on the relationship between urbanisation and tourism development as well as the impacts of urbanisation on tourism with its implications for the sustainable development of city tourism in China. It is suggested that policy makers should adopt an integrated long-term outlook on urban planning, which ensures that urban development is carried out in line with the idiosyncrasy of each city.

Maria Gravari-Barbas and **Sandra Guinand** ponder the effects of gentrification processes on tourism cities, including their impacts on the socio-economic fabric of communities and local economies. Although a direct link between gentrification per se and tourism remains a contested debate among scholars, the authors argue that the first documented evidence of tourism-led gentrification dates to 2005 in the context of New Orleans' French Quarter. This chapter provides a critical retrospective analysis of the historical and intellectual evidence of this debate in order to establish the boundaries of current knowledge on this topic – including links to overtourism in some urban tourism destinations – and outline an agenda for further research in a topic, which may be a manifestation of the growing vulnerabilities, frustrations and resistances developing in global cities, which tourism demand may tend to exacerbate further.

Gaitree (Vanessa) Gowreesunkar and **Mahender Reddy Gavinolla**, with Hyderabad, India as a case example, investigate the relationship between urbanism and overtourism. The chapter begins with a literature review on city tourism, urbanism and overtourism. Then, Hyderabad's function as a tourism destination is described. The impacts and implications of urbanism and overtourism for the city are then identified. The authors conclude that urbanism and overtourism are inevitable in an era characterised by globalisation, sophistication and emancipation.

Craig Webster and **Sotiris Hji-Avgoustis** review the literature on political shocks and discuss how they impede the flow of tourists to cities. With a focus on the United States, typologies of shocks and how they differ in terms of impacting upon the decision to visit a particular city destination are defined. The authors discuss how political authorities and tourism managers can work to decrease the negative impacts of shocks on city tourism inflows. It is concluded that there is reason to believe that some shocks, but not all, may have negative impacts on cities' tourism images in the minds of potential tourists.

1

CHALLENGES OF WORLD TOURISM CITIES

London, Singapore and Dubai

Cristina Maxim

Introduction

The number of people living in cities increases from year to year, with the latest figures produced by the UN Department of Economic and Social Affairs (2018) showing that 55% of the world's total population can now be found in urban areas. This percentage is expected to continue to rise and is projected to reach 68% by 2050, meaning that an additional 2.5 billion people will live in towns and cities. The most urbanised regions are North America, Latin America and the Caribbean, and Europe, where at least three-quarters of the total population lives in cities, while Asia and Africa are the two regions with a relatively lower level of urbanisation. This move from rural to urban environments has contributed, together with other factors, to the increasing importance of the phenomenon of urban tourism.

Urban tourism is considered to be one of the earliest forms of tourism that re-emerged in the 1980s as a result of an increasing interest from tourists in heritage and cultural activities found in cities (European Communities 2000, Maxim 2016). Yet, it has started to be recognised as a separate area of study only recently, with a number of authors pointing out the limited research available on the topic (Edwards *et al.* 2008, Maxim 2013). One of the most influential studies worth noting is the work produced by Ashworth (1989), titled "Urban Tourism: An Imbalance in Attention", which ignited the interest in urban tourism. In this paper the author highlighted a double neglect of the topic – tourism studies neglected large cities, while scholars who studied large cities overlooked the important role played by the tourism industry in their economy. Since then, however, more progress has been made in discussing urban tourism and the different aspects related to tourism development in cities (Pearce 2001, Sharpley and Roberts 2005, Edwards *et al.* 2008, Maitland and Newman 2009, Ashworth and Page 2011, Maitland 2012, 2013, Miller *et al.* 2015, Maxim 2016, 2019). Moreover, in 2015 a new multidisciplinary journal has emerged that focuses on tourism within urban areas – the *International Journal of Tourism Cities*.

A selection of studies, together with their theoretical and conceptual contribution in the field of urban tourism, and their implications for cities, is included in Table 1.1. This shows how the focus of researchers has changed over time, from understanding the phenomenon of urban tourism and its importance, towards the current debates such as sustainability, smart destinations, augmented reality and overtourism.

Table 1.1 Theoretical and conceptual contributions to the study of urban tourism (selection)

Author(s)	Year	Contribution
Jansen-Verbeke	1986	Elements of tourism
Ashworth	1989, 2003	Urban tourism: imbalance in attention
Ashworth and Tunbridge	1990	The tourist-historic city
Burtenshaw *et al.*	1991	Users of the city
Garreau	1991	Edge city as centres for services consumption
Mullins	1991, 1994	Tourism urbanization
Law	1992, 2002	Urban tourism and economic regeneration; Urban tourism synthesis
Getz	1993	The tourism business district
Page	1995	Urban tourism as a system
Castells	1996	The rise of the network city
Zukin	1996	The culture of cities and post-modern environment
Thrift	1997	Cities without modernity, cities with magic
Mazanec and Wöber	1997, 2009	Management of cities for tourism
Hannigan	1998	Fantasy city
Dear and Flusty	1999	Engaging post-modern urbanism
Page and Hall	2002	Modelling tourism in the post-modern city
Pearce	2002	Integrated framework for urban tourism research
Mommaas	2004	Cultural clusters and the post-industrial city
Beedie	2005	The adventure of urban tourism
Pearce	2007	Capital city tourism
Mordue	2007	Tourism, urban governance and public space
Edwards *et al.*	2008	Research agenda for Australian urban tourism
McNeill	2008	The hotel and the city
Maitland and Ritchie	2009	National capital tourism (expanding the knowledge)
Maitland and Newman	2009	World tourism cities
Ashworth and Page	2011	Urban tourism research progress and paradoxes
Richards	2014	Creativity and tourism in the city
Maxim	2015, 2016	Sustainable tourism implementation in urban areas
Gretzel *et al.*	2015	Smart tourism ecosystems, smart cities
Gutiérrez *et al.*	2017	Airbnb in tourist cities
Su *et al.*	2018	Urban heritage tourism (expanding the knowledge)
tom Dieck and Jung	2018	Mobile augmented reality in urban tourism
Koens *et al.*	2018	Overtourism and impact of tourism in cities
Maxim	2019	World tourism cities (expanding the knowledge)
Cohen and Hopkins	2019	Autonomous vehicles and urban tourism

Source: Based on the work of Ashworth and Page (2011, p. 12).

Still, not much has so far been written on world tourism cities, environments that attract a large number of visitors. Some of the latest works on the subject are Maxim (2019), who identifies a number of challenges faced by policy makers in London; Maitland (2016), who looks at how tourists experience world tourism cities; and Simpson (2016), who discusses "tourist utopia" in three post-world cities – Las Vegas, Dubai and Macau. Worth noting is that half of the top 20 most visited cities in the world are now located in Asia (Mastercard 2019), which has led to an increase in studies that focus on different aspects related to tourism development in Asian and Middle

Eastern cities. These include Gong, Detchkhajornjaroensri and Knight (2019), who discuss responsible tourism in Bangkok; Kotsi, Pike and Gottlieb (2018), who look at travellers' perceptions of Dubai as an international stopover destination; Bhati and Pearce (2017), who link vandalism at tourist attractions in Bangkok and Singapore with the site characteristics; or Tolkach, Pratt and Zeng (2017), who focus on the ethics of the Chinese and Western tourists in Hong Kong.

The concept of world tourism cities can be understood to refer either to those cities that depend on tourism for their global profile such as Venice (Ashworth 2010), or to world cities as environments were tourism occurs (Maitland and Newman 2009). The present work adopts the latter meaning and views world tourism cities as "large polycentric cities offering a range of experiences and, as visitors move between and around established centres, they offer apparently seamless opportunities for adding new desirable places to explore to already crowded and diverse tourism possibilities" (Maitland and Newman 2009, p. 2). These cities therefore perform multiple functions, such as centres for business, as well as cultural excellence, and are home to many world-class tourist attractions (Law 2002). They are also important players in the world economy and offer easy access through better connectivity (Maxim 2019). In addition, these cities play an important role in the visitor economy of a destination, with the success of the tourism industry in a country often reliant on their success.

Many governments and policy makers encourage tourism growth in large cities as it contributes to their economic development (Simpson 2016). Yet, there are a number of studies that highlight the negative impacts associated with tourism development in cities, such as overcrowding, conflicts between visitors and locals, property conflicts created by peer-to-peer platforms such as Airbnb, or the worsening of existing congestion in busy areas, to name but a few (Law 2002, Gutiérrez *et al.* 2017). All these factors add to the challenges faced by policy makers in world tourism cities when trying to balance the benefits of tourism for the local economy with negative consequences such as these. On top of this, the mission of policy makers is made even more difficult by the complex economic, social and political functions of such cities, and the diversity of people they attract (Maitland 2012, Maxim 2019).

According to Mastercard's (2019) *Global Destination Cities Index*, the top ten cities that attract the most international overnight visitors are Bangkok, Paris, London, Dubai, Singapore, Kuala Lumpur, New York, Istanbul, Tokyo and Antalya. Of these, the current chapter focuses on London, Singapore and Dubai, briefly discussing their particularities and the challenges they face. These three cities also rank among the top five most visited in the world, with each belonging to a different region, i.e. Europe, Asia and the Middle East (see Figure 1.1 for the change in visitor numbers in the three cities, over the past decade).

London

London, the capital of the United Kingdom, has been one of the world tourism cities for many years, and is ranked third after Bangkok and Paris in terms of the number of international overnight visitors, which for 2018 stands at 19.09 million (Mastercard 2019). The total number of visitors is, however, significantly higher when including domestic visitors – estimated at 12 million per year, and day visitors – estimated at 274 million per year (London & Partners 2015). The city is also one of the largest in Europe, with a total population of 8.53 million (Office for National Statistics 2016), and an important gateway for the UK, as three-quarters of the country's overseas visitors arrive through one of its six airports (DCMS 2016). These figures illustrate the vital importance of the tourism industry for the economy of the city, as the second most important sector after financial services (Maitland and Newman 2009), with a contribution of 11.6% to its GDP (London & Partners 2017).

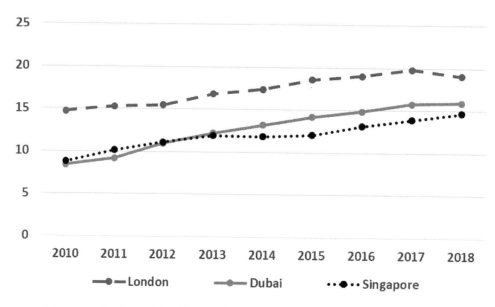

Figure 1.1 International overnight visitor numbers for London, Dubai and Singapore.

Source: Author's own work, based on data from the Mastercard index (2014, 2017, 2019).

The popularity of the city among international visitors is due to its diverse offer and the variety of its attractions, such as historic buildings (e.g. the Tower of London, Westminster Palace, Westminster Abbey), museums (e.g. the British Museum, Tate Modern, the National Gallery), theatres and other cultural establishments (e.g. the Royal Opera House, Shakespeare's Globe, the Royal Albert Hall), beautiful parks and promenade areas (e.g. Hyde Park, Regent's Park, the Royal Botanic Gardens Kew), as well as numerous restaurants, pubs and clubs. In addition, London is one of the most diverse cities in the world, being home to more than 50 ethnic groups, with over 200 different languages spoken on its streets.

The organisation responsible with the governance of tourism in the capital is the Greater London Authority (GLA), the strategic administrative body for Greater London. This is an elected body, consisting of the Mayor of London and the London Assembly, and represents all 32 London boroughs and the City of London. The GLA produces the London Plan, the spatial development strategy for the city that provides the policy context for the local planning policies of the London boroughs. Its latest version (GLA 2016, p. 155), presents the Mayor's vision for tourism development in the capital, which focuses on three key aspects:

- to develop the quality of accommodation facilities;
- to enhance visitor perception of value for money;
- to improve the inclusivity and accessibility of the visitor experience.

Another important organisation involved in the development of tourism is London & Partners, which is the official agency that focuses on promoting London internationally (London & Partners 2019).

As with many other large city destinations, policy makers in London face a number of challenges when planning and managing tourism in the UK capital. These were identified and discussed in a recent study by Maxim (2019), and are summarised below.

To start with, there are limited planning provisions to guide the 33 local authorities in London in their efforts to promote policies for tourism development in the capital. In addition, even though most London boroughs consider tourism as a strategic industry, there is a lack of tourism-specific policies and strategies to help manage this activity at the local (borough) level. This may be linked to the lack of resources allocated by the national and local government for tourism development in the capital, especially after the 2007–2008 financial crisis. Considering that local authorities are seen as key players in managing tourism development in a destination (Ruhanen 2013, Maxim 2016), the current situation may impact on the city's competitiveness and its ability to sustainably accommodate increasing numbers of visitors.

Another important challenge faced by policy makers in London is to ensure the sustainable development of tourism. Aspects such as the need to improve public transport and alleviate existing traffic congestion have been highlighted by planners and researchers (Maxim 2016), while conflicts between hosts and visitors have been noted in the busiest parts of the city. At the same time, the protection and conservation of the natural environment and the built heritage remains an important concern expressed by policy makers, and is also recognised in the latest planning document for the city (GLA 2016).

In their efforts to attract additional funds to help manage tourism in the capital, many local authorities also face a challenge in working in partnership with other public and private organisations (Maxim 2019). Yet, this can be an important driver that contributes to the implementation of sustainable tourism practices in destinations, helping to maximise the availability of human and financial resources (Devine and Devine 2011, Maxim 2015, 2016).

Researchers have also highlighted the growing popularity of the sharing economy in large cities, in particular of platforms such as Airbnb, which threatens the traditional accommodation sector (Guttentag 2015, Zervas *et al.* 2016). To address this challenge, in 2015 a 90-day rule was introduced in London requiring specific planning permission when renting a property for more than three months in a year (Hickey and Cookney 2016). Even so, the Residential Landlords Association points out that over half (61%) of all Airbnb listings in the capital do not observe this rule, and are in fact available for more than 90 days per year (Simcock and Smith 2016).

Nevertheless, it should be noted that in large cities such as London, not every borough (local authority district) faces the same challenges. While inner boroughs, such as Westminster and Camden, suffer from overcrowding and high crime levels, outer boroughs struggle to attract more visitors and to develop their accommodation facilities. To address this aspect, a number of policies were promoted by the Greater London Authority (2016), aimed at easing the pressure on central London and spreading visitor numbers across the city.

London therefore endeavours to keep its status as one of "the best cities in the world to visit" (GLA and CTC 2015, p. 4) by encouraging new visitor attractions and developing new accommodation facilities, while protecting the built and natural heritage it offers. Yet, the local authorities in London (i.e. at borough level) have to deal with limited resources, as well as limited planning provisions to help them manage tourism in a sustainable manner. Moreover, the long-term implications of Brexit for the travel and tourism industry in the capital are still unclear and should not be overlooked, with some organisations fearing that the industry will suffer if the strong connections that currently exist between the economy of the UK and that of the EU are affected (ABTA and Deloitte 2016). The EU is currently the main market for overseas travel in the UK, and potential changes in the free movement of people, goods and services between the two could have significant implications for the travel and tourism industry in London.

Singapore

Singapore is one of the top five world tourism cities, attracting 14.67 million international overnight visitors (Mastercard 2019). This makes tourism one of the most important service sectors in the city (Meng *et al.* 2013), with a total contribution of 10.2% to its GDP and accounting for 8.8% of the total employment (WTTC 2018). Even though over the years the city experienced periods of negative growth following events such as the 9/11 terrorist attacks in the US, the 2003 outbreak of the Severe Acute Respiratory Syndrome disease in Asia, and the 2007–2008 global financial crisis (Meng 2014), tourism ranks among the fastest growing industries in Singapore, recording an average growth rate of 15% over the last few decades (Al-Shboul and Anwar 2017).

Due to its geographical location and well-developed infrastructure, Singapore is the main gateway to Southeast Asia, playing an important role as a regional aviation and sea hub (Saunders 2004). The city has also positioned itself as a regional, as well as global financial hub (Henderson 2007a), a leading destination for business and leisure tourism, as well as an important player in the health tourism economy (Lohmann *et al.* 2009). Most of the visitors attracted by the city are from other Asian countries (e.g. China, Indonesia, Japan, Malaysia and South Korea) who are enticed by its shopping facilities, food diversity and high-quality medical services, while for the other international visitors Singapore represents an attractive transit stop (Al-Shboul and Anwar 2017). Among the most popular attractions are the Singapore Flyer, Universal Studios Singapore, Singapore Zoo, Art Science Museum, Gardens by the Bay and the Esplanade.

The organisation responsible with tourism development in the city is the Singapore Tourism Board (STB), which was established in 1964. Even though Singapore has limited natural tourism attractions and historic sites (Al-Shboul and Anwar 2017), through the active policies promoted by the STB, the city managed to increase the number of visitors from 91,000 in 1964 (Meng *et al.* 2013), to nearly 14.7 million international visitors at present (Mastercard 2019). Some examples of the tourism strategies implemented by the STB over the years are the "garden attractions" and modern hotels in the 1970s; appreciating the local heritage and revitalising former ethnic enclaves such as Chinatown, Little India and Kampong Glam in the 1980s; master planning and focusing on tourism growth in the 1990s; and investing in infrastructure, product development (e.g. integrated resorts) and attracting major events such as the Singapore Grand Prix after 2000. Other factors that contributed to this extraordinary performance are the city's strategic geographical position, its stability, good connectivity, stable tropical weather, a diverse tourism product, the public–private sector partnership, innovative management in developing natural and cultural resources, cultural diversity, well-trained workforce, as well as world-class venues and facilities (Meng *et al.* 2013, Ganguli and Ebrahim 2017, Lim and Zhu 2018).

Even though Singapore is one of the most visited cities in the world and the number of international visitors is forecasted to continue to rise over the coming years, the city is facing a number of challenges related to the development of tourism. First, its small size means there is limited space for expansion, with policy makers struggling to reconcile different demands such as more houses, better infrastructure or water supply systems (Henderson 2005). This makes long-term planning a necessity, an aspect also recognised by the Centre for Liveable Cities (2015) in the latest tourism plan for the city.

Competition is another important challenge that policy makers in Singapore need to consider (Lim and Zhu 2018). There are a number of other important world tourism cities in Asia, such as Hong Kong, Kuala Lumpur or Macau, that are looking to attract business and leisure travellers (Henderson 2007a) and are stepping up their marketing campaigns. To stay competitive in this challenging environment, Singapore will require renewed efforts to "better capitalise on fast-changing trends and realise opportunities", as pointed out by the Singapore Tourism Board (2016, p. 4) in their latest tourism marketing strategy for the city.

Some authors argue that in their efforts to turn the city into one of the most modern and successful in Asia, Singapore's authentic heritage, and its indigenous cultures have suffered (Saunders 2004, Henderson 2007a). Saunders (2004) for example, notes that many of the traditional arts and local crafts which were common in the late 1970s and 1980s had disappeared by the 1990s. According to the latest Singapore tourism plan, however, efforts were made by the government to find a balance between maximising the economic potential of the land (by allowing new developments) and protecting the local heritage (Centre for Liveable Cities 2015). Yet, the document notes that starting with the 1980s the government promoted an adaptive use of conserved buildings, allowing many of them to be given new uses instead of being preserved.

Another two challenges worth noting are that Singapore is usually perceived as an expensive destination when compared to some other cities in the region (Henderson 2007b), and many international visitors will see it only as a stopover location to spend a few days instead of their final destination. For Singapore to remain a popular destination among international visitors and to encourage longer stays, Al-Shboul and Anwar (2017) argue that the city needs to continue to improve its existing infrastructure and attractions. This, however, needs to be done very carefully to avoid destroying what is left of its local heritage and community identity, and thus gather the support of the locals. Ultimately, this would contribute to a sustainable development of the tourism industry in the city.

Dubai

With 4,114 square kilometres, including land reclaimed from the sea, and a population of 3.19 million, which has doubled over the past decade (Dubai Statistics Center 2018), Dubai is the second largest of the seven sheikhdoms that make up the United Arab Emirates (Henderson 2007a). The city proved to be "one of the fastest growing tourism destinations of the early twenty-first century" (Ryan and Stewart 2009, p. 288), evolving rapidly from a small fishing village and open desert, as it was in the 1900s, to become the first global city in the region in the early 2000s (Akhavan 2017). This is in spite of its hot and dry climate that lasts nearly all year round, with temperatures in the high 30 degrees Celsius, and high level of humidity in late summer (Ryan and Stewart 2009).

In the 1990s, when its oil production started to decline, Dubai decided to diversify its economy and turned its attention to the service sector (Sutton 2016). Tourism was therefore recognised as an important industry and the city started to invest heavily in its infrastructure, including air and maritime transport, to help attract and accommodate more visitors. This strategy led to the number of international tourists arriving in Dubai to increase considerably in the 1990s, and by 2002 the city became one of the world's fastest growing destinations (Sharpley 2008). That year the tourism industry also overtook the oil and gas industry (in terms of its contribution to the GDP), which used to be the main source of revenue for the emirate (Henderson 2007a). Over the past two decades, the expansion of the tourism industry in Dubai has continued, with the city currently ranking as one of the top five global destination cities and attracting a total of 15.93 million international overnight visitors (Mastercard 2019).

The Department of Tourism and Commerce Marketing (DTCM) established in 1989, and which is under the supervision of the Crown Prince, is the main organisation responsible with managing tourism in the city. They recognise the important role that tourism plays for the city's economy and aim to make Dubai the first choice as a leading global destination for travel, tourism and events – their goal is to attract 20 million visitors a year by the end of the decade (Visit Dubai 2019). This ambitious figure represents almost twice the number of visitors accommodated in 2012, when the city attracted a total of 10.95 million international visitors (Mastercard 2017). To achieve this considerable growth, the organisation recognises that the city needs to continue to invest in its transport infrastructure and accommodation facilities, as well as to broaden its offering when it comes to events and attractions.

Having a short history and few historic sites, Dubai has tried to differentiate itself from other global cities "by a sustained focus on grandeur" (Nadkarni and Heyes 2016, p. 214) and by offering a variety of iconic attractions to its visitors. Among these are the Burj al Arab Jumeirah, considered the world's most luxurious hotel; the Burj Khalifa, which is the world's tallest building; the Palm Island, which is the largest offshore artificial island; high-end shopping malls such as the Dubai Mall, which is the second largest mall in the world, and the Mall of the Emirates, which accommodates a 400-metre-long indoor ski slope; as well as high-tech trade and convention centres. Currently, Dubai is also on track to develop the world's largest airport – Al Maktoum International Airport at Dubai World Centre, which is expected to accommodate over 200 million passengers a year. Furthermore, to expand its offer and attract visitors from emerging markets (e.g. China, India and the Asia-Pacific region), in particular during the low season, a number of festivals and events were organised, such as the Dubai Shopping Festival in February/March and the Dubai Summer Surprises in July/August. Tourism perception has therefore changed, from the initial sun and sand destination, to a leading destination for luxury, shopping, leisure, global meetings, conventions and events. In addition, Dubai will soon host the 2020 World Expo Trade Convention that is expected to have a huge economic and social impact, and to attract a record 25 million visitors (Sutton 2016).

The city, however, faces a number of challenges that were highlighted by researchers, among them being the loss of its heritage and authenticity as a result of the excessive modernisation (Henderson 2007a, Nadkarni and Heyes 2016). Steiner (2009) describes Dubai as a "hyper-real destination" characterised by "cultural fluidity" (Stephenson 2014, p. 728), where the past history and culture tend to be forgotten. As Stephenson notes, only 10% of Dubai's historic buildings have survived its rapid expansion and modernisation. However, Nadkarni and Heyes (2016) point out that visitors are now looking for authentic experiences even when choosing luxury vacations. As such, a focus should be placed on the indigenous-based activities, in particular those which are historically linked to the desert life of the Bedouins (Stephenson 2014), and thus on promoting what is left of the local cultural heritage.

The rapid expansion of the city and its tourist infrastructure means that Dubai also faces a challenge in sourcing the large numbers of qualified workforce that are needed to cater for the much anticipated growth of its tourism industry. The city is already dependent on foreign workers to fill the increasing number of job vacancies available in the travel and tourism sector (Stephenson 2014). With the city soon to host the 2020 World Expo Trade Convention, and with the number of hotels expected to increase considerably in the near future, there is a huge demand for qualified employees. This puts pressure on the hotel industry to attract, train and retain talented individuals able to deliver excellent services and maintain Dubai's reputation as a luxury destination (Brien *et al.* 2019).

According to Stephenson (2014), another challenge noted by the Department of Tourism and Commerce Marketing is to convince visitors to stay longer in the city. With an average length of stay of 3.7 days (Pike and Kotsi 2018) and occupancy rates slightly decreasing, building more hotels may result in an imbalance in supply and demand that could impact the profitability of the hotel industry. Moreover, Nadkarni and Heyes (2016, p. 216) argue that increasing the luxury accommodation capacity risks "diluting the rarity and exclusivity elements from the consumer experience". This may lead to visitors becoming unwilling to pay premium prices during their Dubai stay due to over-exposure and ease of access to such luxury experiences. Considering these aspects, and in the face of increasing competition from its neighbouring emirates such as Abu Dhabi and Sharjah (Zaidan 2016), Dubai needs to think its strategy carefully if it is to remain the top destination in the region.

A number of authors also point out the environmental concerns and socio-cultural consequences linked to the development of tourism infrastructure in the city (Sharpley 2008, Stephenson and Ali-Knight 2010, Stephenson 2014). Even though Dubai, and indeed the entire United Arab Emirates, face a significant water shortage, they have one of the highest per capita water consumption rates worldwide. On top of that, the continuous focus on infrastructure development, and particularly large-scale projects, is likely to further impact the city's natural environment (or what is left of it). At the same time, a continuous increase in the number of international tourist arrivals will likely lead to increased tensions between visitors and the local community, and may for example give rise to cultural conflicts, or cause low-income families to be displaced from certain areas due to the increasing price of land and accommodation (e.g. the Garden City Project).

All these challenges represent in fact different aspects that should be considered for the sustainable development of tourism in a destination, and would need to be addressed by the policy makers in Dubai sooner rather than later. This is reinforced by the findings of a recent study conducted by Martens and Reiser (2019), which notes that prospective visitors perceive the city as not environmentally conscious and unsustainable, although the government tries to promote the image of a sustainable destination.

Conclusion

World tourism cities are among the most visited destinations in the world due to their distinctive characteristics, the world-class attractions they accommodate and the accessibility offered by their better connectivity (Maxim 2019). Yet, research published on this topic to date is relatively limited. This chapter therefore advances the current body of knowledge by discussing the particularities and challenges faced by three top world tourism cities from three different regions – London, Singapore and Dubai.

In a very competitive world, where many new as well as traditional destinations try to attract ever more visitors, policy makers in large cities need to better understand the challenges they face so they can implement sustainable measures when planning and managing tourism in such destinations. This chapter showed that while some of the challenges faced by world cities may be specific to each destination due to their particular political situation or geographical location (e.g. the impact of Brexit on the future of tourism development in London, or the limited space available for expansion in Singapore), others are likely to be relevant to many such destinations (e.g. protecting and conserving the built and natural environment, or the increasing competition from other world cities). Without understanding and addressing the challenges they face, world tourism cities could experience numerous negative consequences that may impact on both the residents' quality of life, and the quality of the visitor experience. It may also lead to overtourism (Milano *et al.* 2019), nowadays a term often used in connection with popular city destinations such as Barcelona, Amsterdam or Venice, which are struggling to manage the growth of tourism sustainably.

Further research is needed to understand how well world tourism cities are prepared to respond to the challenges they face, and whether governments are promoting measures to address such challenges. In addition, creating a common framework to facilitate the exchange of existing knowledge from academics and practitioners, and sharing best practices found in large cities around the world may help policy makers and destination managers in their efforts to manage tourism sustainably in these complex environments.

References

ABTA and Deloitte, 2016. *What Brexit might mean for UK travel*. London.

Akhavan, M., 2017. Development dynamics of port-cities interface in the Arab Middle Eastern world: The case of Dubai global hub port-city. *Cities*, 60, 343–352.

Al-Shboul, M. and Anwar, S., 2017. Long memory behavior in Singapore's tourism market. *International Journal of Tourism Research*, 19 (5), 524–534.

Ashworth, G.J., 1989. Urban tourism: An imbalance in attention. *In*: C.P. Cooper, ed. *Progress in Tourism, Recreation and Hospitality Research*. London: Belhaven, 33–54.

Ashworth, G.J., 2010. Book review: World Tourism Cities: developing tourism off the beaten track. *Tourism Management*, 31 (5), 696–697.

Ashworth, G.J. and Page, S.J., 2011. Urban tourism research: Recent progress and current paradoxes. *Tourism Management*, 32 (1), 1–15.

Bhati, A. and Pearce, P., 2017. Tourist attractions in Bangkok and Singapore: Linking vandalism and setting characteristics. *Tourism Management*, 63, 15–30.

Brien, A., Anthonisz, A. and Suhartanto, D., 2019. Human capital in the Dubai hotel industry: A study of four- and five-star hotels and the HR challenges they face. *Journal of Human Resources in Hospitality & Tourism*, 1–19.

Centre for Liveable Cities, 2015. Planning for Tourism: Creating a Vibrant Singapore.

DCMS, 2016. *Tourism Action Plan*. London.

Devine, A. and Devine, F., 2011. Planning and developing tourism within a public sector quagmire: Lessons from and for small countries. *Tourism Management*, 32 (6), 1253–1261.

Dubai Statistics Center, 2018. Population by gender – Emirate of Dubai.

Edwards, D., Griffin, T. and Hayllar, B., 2008. Urban tourism research: Developing an agenda. *Annals of Tourism Research*, 35 (4), 1032–1052.

European Communities, 2000. Towards Quality Urban Tourism: Integrated Quality Management (IQM) of Urban Tourist Destinations.

Ganguli, S. and Ebrahim, A.H., 2017. A qualitative analysis of Singapore's medical tourism competitiveness. *Tourism Management Perspectives*, 21, 74–84.

GLA, 2016. *The London Plan: The Spatial Development Strategy for London Consolidated with Alterations since 2011*. London.

GLA and CTC, 2015. *A Cultural Tourism Vision for London 2015–2017*. London.

Gong, J., Detchkhajornjaroensri, P. and Knight, D.W., 2019. Responsible tourism in Bangkok, Thailand: Resident perceptions of Chinese tourist behaviour. *International Journal of Tourism Research*, 21 (2), 221–233.

Gutiérrez, J., García-Palomares, J.C., Romanillos, G. and Salas-Olmedo, M.H., 2017. The eruption of Airbnb in tourist cities: Comparing spatial patterns of hotels and peer-to-peer accommodation in Barcelona. *Tourism Management*, 62, 278–291.

Guttentag, D., 2015. Airbnb: Disruptive innovation and the rise of an informal tourism accommodation sector. *Current Issues in Tourism*, 18 (12), 1192–1217.

Henderson, J.C., 2005. Planning, changing landscapes and tourism in Singapore. *Journal of Sustainable Tourism*, 13 (2), 123–135.

Henderson, J.C., 2007a. Destination development: Singapore and Dubai compared. *Journal of Travel & Tourism Marketing*, 20 (3–4), 33–45.

Henderson, J.C., 2007b. Hosting major meetings and accompanying protestors: Singapore 2006. *Current Issues in Tourism*, 10 (6), 543–557.

Hickey, S. and Cookney, F., 2016. Airbnb faces worldwide opposition: It plans a movement to rise up in its defence. *The Guardian*, 29 October.

Kotsi, F., Pike, S. and Gottlieb, U., 2018. Consumer-based brand equity (CBBE) in the context of an international stopover destination: Perceptions of Dubai in France and Australia. *Tourism Management*, 69, 297–306.

Law, C.M., 2002. *Urban Tourism: The Visitor Economy and the Growth of Large Cities*. 2nd Edition. London: Continuum.

Lim, C. and Zhu, L., 2018. Examining the link between meetings, incentive, exhibitions, and conventions (MICE) and tourism demand using generalized methods of moments (GMM): The case of Singapore. *Journal of Travel & Tourism Marketing*, 35 (7), 846–855.

Lohmann, G., Albers, S., Koch, B. and Pavlovich, K., 2009. From hub to tourist destination: An explorative study of Singapore and Dubai's aviation-based transformation. *Journal of Air Transport Management*, 15 (5), 205–211.

London & Partners, 2015. *London Tourism Report 2014–2015*. London.

London & Partners, 2017. A Tourism Vision for London.

London & Partners, 2019. London & Partners' 2019/20 Business Plan [online]. Available from: https://files.londonandpartners.com/l-and-p/assets/business-plans-and-strategy/london-and-partners-business-plan-201920.pdf [Accessed 23 Oct. 2019].

Maitland, R., 2012. Global change and tourism in national capitals. *Current Issues in Tourism*, 15 (1–2), 1–2.

Maitland, R., 2013. Backstage behaviour in the global city: Tourists and the search for the "Real London". *Procedia – Social and Behavioral Sciences*, 105, 12–19.

Maitland, R., 2016. Everyday tourism in a world tourism city: Getting backstage in London. *Asian Journal of Behavioral Studies*, 1 (1), 13–20.

Maitland, R. and Newman, P., eds., 2009. *World Tourism Cities: Developing Tourism off the Beaten Track*. New York: Routledge.

Martens, H.M. and Reiser, D., 2019. Analysing the image of Abu Dhabi and Dubai as tourism destinations: The perception of first-time visitors from Germany. *Tourism and Hospitality Research*, 19 (1), 54–64.

Mastercard, 2014. *Global Destination Cities Index 2014*.

Mastercard, 2017. *Global Destination Cities Index 2017*.

Mastercard, 2019. *Global Destination Cities Index 2019*.

Maxim, C., 2013. Sustainable tourism planning by local authorities: An investigation of the London Boroughs. Ph.D. London Metropolitan University.

Maxim, C., 2015. Drivers of success in implementing sustainable tourism policies in urban areas. *Tourism Planning and Development*, 12 (1), 37–47.

Maxim, C., 2016. Sustainable tourism implementation in urban areas: A case study of London. *Journal of Sustainable Tourism*, 24 (7), 971–989.

Maxim, C., 2019. Challenges faced by world tourism cities: London's perspective. *Current Issues in Tourism*, 22 (9), 1006–1024.

Meng, S., 2014. The role of inbound tourism in the Singaporean economy: A computable general equilibrium (CGE) assessment. *Journal of Travel & Tourism Marketing*, 31 (8), 1071–1089.

Meng, X., Siriwardana, M. and Pham, T., 2013. A CGE assessment of Singapore's tourism policies. *Tourism Management*, 34, 25–36.

Milano, C., Cheer, J.M. and Novelli, M., eds., 2019. *Overtourism: Excesses, Discontents and Measures in Travel and Tourism*. Wallingford, Oxfordshire; Boston, MA: CABI.

Miller, D., Merrilees, B., and Coghlan, A., 2015. Sustainable urban tourism: Understanding and developing visitor pro-environmental behaviours. *Journal of Sustainable Tourism*, 23 (1), 26–46.

Nadkarni, S. and Heyes, A., 2016. Luxury consumption in tourism: The case of Dubai. *Research in Hospitality Management*, 6 (2), 213–218.

Office for National Statistics, 2016. Subnational population projections for England: 2014-based projections.

Pearce, D., 2001. An integrative framework for urban tourism research. *Annals of Tourism Research*, 28 (4), 926–946.

Pike, S. and Kotsi, F., 2018. Stopover destination image: Perceptions of Dubai, United Arab Emirates, among French and Australian travellers. *Journal of Travel & Tourism Marketing*, 35 (9), 1160–1174.

Ruhanen, L., 2013. Local government: Facilitator or inhibitor of sustainable tourism development? *Journal of Sustainable Tourism*, 21 (1), 80–98.

Ryan, C. and Stewart, M., 2009. Eco-tourism and luxury: The case of Al Maha, Dubai. *Journal of Sustainable Tourism*, 17 (3), 287–301.

Saunders, K.J., 2004. Creating and recreating heritage in Singapore. *Current Issues in Tourism*, 7 (4–5), 440–448.

Sharpley, R., 2008. Planning for tourism: The case of Dubai. *Tourism and Hospitality Planning & Development*, 5 (1), 13–30.

Sharpley, R. and Roberts, L., 2005. Managing urban tourism. *In*: L. Pender and R. Sharpley, eds. *The Management of Tourism*. London: Sage, 161–174.

Simcock, T. and Smith, D., 2016. *The Bedroom Boom: Airbnb and London*. Residential Landlords Association.

Simpson, T., 2016. Tourist utopias: Biopolitics and the genealogy of the post-world tourist city. *Current Issues in Tourism*, 19 (1), 27–59.

Singapore Tourism Board, 2016. STB Marketing Strategy: Of Stories, Fans and Channels.

Steiner, C., 2009. From heritage to hyperreality? Prospects for tourism development in the Middle East between Petra and the Palm. Presented at the conference Traditions and Transformations: Tourism, Heritage and Cultural Change in the Middle East and North Africa Region, Amman, Jordan.

Stephenson, M.L., 2014. Tourism, development and "destination Dubai": Cultural dilemmas and future challenges. *Current Issues in Tourism*, 17 (8), 723–738.

Stephenson, M.L. and Ali-Knight, J., 2010. Dubai's tourism industry and its societal impact: Social implications and sustainable challenges. *Journal of Tourism and Cultural Change*, 8 (4), 278–292.

Sutton, J., 2016. From desert to destination: Conceptual insights into the growth of events tourism in the United Arab Emirates. *Anatolia*, 27 (3), 352–366.

Tolkach, D., Pratt, S. and Zeng, C.Y.H., 2017. Ethics of Chinese and Western tourists in Hong Kong. *Annals of Tourism Research*, 63, 83–96.

UN Department of Economic and Social Affairs, 2018. The 2018 Revision of World Urbanization Prospects.

Visit Dubai, 2019. Tourism Vision [online]. Available from: www.visitdubai.com/en/department-of-tourism/about-dtcm/tourism-vision [Accessed 23 Oct. 2019].

WTTC, 2018. Travel & Tourism Economic Impact 2018 Singapore.

Zaidan, E.A., 2016. Tourism shopping and new urban entertainment: A case study of Dubai. *Journal of Vacation Marketing*, 22 (1), 29–41.

Zervas, G., Proserpio, D., and Byers, J.W., 2016. The rise of the sharing economy: Estimating the impact of Airbnb on the hotel industry. Boston University School of Management Research Paper no. 2013–16.

2

GLOBAL TERRORISM IN TOURISM CITIES

The case of World Heritage Sites

Cláudia Seabra and Odete Paiva

Introduction

"Heritage draws on the past and is intimately related to our identity requirements in the present" (MacDowel 2016, p. 49). Heritage is itself a symbol of a nation's own identity, it is the selective use of the past as a resource for the present and future (Ashworth 2017) while preserving the memories of communities.

Since the first classification, that took place in 1978, the list of UNESCO's World Heritage Sites (WHS) is generally considered to be an excellent contribution to preserving history through the protection of any monument and cultural landscape that deserves to be preserved. However, this classification is often part of the local development process aiming at increasing the site's attractiveness, especially those aspects related with its tourist assets. "World Heritage" status plays a crucial role in attracting visitors to sites and presents itself as a magnet for tourism (Shackley 1998). Tourism attractiveness is, in fact, strongly connected with heritage. Heritage tourism is an important segment, promoting places of national inheritance and of heritage interest (Saha and Yap 2014). According to previous studies, nearly 60% of visitors consider history and culture to be major influences when they have to choose their tourist destination (Southall and Robinson 2011). The UNESCO list has thus become very popular and attracts the interest of different stakeholders: tourists, the general public, policy makers and organisations (Steiner and Frey 2012).

WHS classification draws attention to places that are true witnesses of a nation's history and identity. Unfortunately, these sites are also attracting the attention of terrorist groups, not only because of their heritage value, but also because of their tourist importance. In fact, when terrorists choose high-profile targets like tourist destinations, they achieve the publicity they crave and attract global attention to their cause (Seabra 2019). When those destinations are also important cultural centres, due to their WHS classification, the terrorists' message has an even stronger impact. The recent attacks on important Heritage Sites, like the Buddhas of the Bamiyan Valley in Afghanistan or the demolition of the city of Nimrud in Iraq, show how terrorists are committed to destroying the memories of the past and mankind's most revered places just to send a message of power and impunity (Smith *et al.* 2015).

Most studies agree that terrorism has a negative impact on the tourism industry. However the extent of its effects on UNESCO's WHS has been marginally studied (Yap and Saha 2013). On the other hand, past research has addressed the destruction of cultural heritage as part of modern conflict (Meskell 2002, Bevan 2006, Stone 2012, Van der Auwera 2012, Smith *et al.* 2015); nevertheless, there is an urgent need to analyse the role played by the destruction of high-profile and worldwide recognised cultural heritage in terrorist strategies (Smith *et al.* 2015).

This chapter wishes to contribute to this discussion by drawing the public's attention to the threats that those precious sites may attract, just because they are cultural and identity symbols of nations and humankind, places which are highly attractive to heritage tourists but that have also become relevant terrorist targets.

The rise of a new terrorism era

Terrorism has been a known fighting strategy since the beginning of history. However, over these last few years, this political weapon has become increasingly common and its impact on the places affected has reached a level never seen before (Seabra 2019). Terrorism attacks have increased from 650 events in 1970 to more than 15,000 in 2015. Overall, there were more than 185 thousands terrorist incidents over the last 45 years. Terrorist attacks rose from an average of 983 incidents each year in the 1970s to about 2,500 in the first decade of the twenty-first century. In only five years (2010–2015) there were nearly as many terrorist attacks as those perpetrated in the 1970s, 1980s and 1990s put together (START 2019). Currently, the public fears terrorism more than ever. In 2014 the death toll from terrorism increased 84%. The Global Terrorism Index considered 2015 as the worst of the last 16 years in OECD countries (IEP 2016).

There is also evidence of the so-called geographic dispersion of terrorism incidents. Clearly, one can notice a special shift in the geography of world terrorism. With large variations among the different regions of the world and fluctuations over time, terrorism has been spreading from the underdeveloped regions, such as Africa and South America in the 1970s and 1980s, to more developed regions like Europe, Northern America and Asia over the last decades (IEP 2016, START 2019).

During the 1990s, terrorism entered a new and distinct phase. Many researchers revealed the new face of terrorism: a new type of terrorists (Stern 1999) who emerged during the post-Cold War (Hudson 1999), a new generation of terrorists (Hoffman 2002) with different characteristics bringing a new wave of terrorism (Jenkins 2001, Rapoport 2001). In fact, those new terrorist organisations were substantially different from their previous incarnations as they began to break away from a traditional structured and rigid shape to adopt a more flexible and anonymous nature. Until the 1980s, terrorist groups clearly had a nationality. Currently terrorist organisations are transnational groups with members and cells from different nationalities. The responsibility for the attacks is not always claimed and when terrorists do claim responsibility for an incident, their demands are vague or even inexistent. But the main differences are the terrorists' motivations and the power of the message of current terrorist organisations. In the past, terrorists' motivations were clearly political and had a local and national impact; currently motivations are clearly religious and cultural and seek a global impact (Bergesen and Han 2005, Seabra 2019). This is the so-called "media-oriented terrorism" (Weimann 2005, p. 382).

Nowadays, violence seems to be happening randomly (Seabra 2019). Collateral damage is higher than it has ever been before. Terrorist attacks have increased and hit all types of targets but private citizens and property, as well as religious and tourism institutions, are those which seem to have gained terrorists' preference since the 1970s (START 2019).

Tourism as a terrorism target: crises and disasters in the tourism industry

Terrorism involves violent actions that aim to generate broad emotional and psychological impact that extends beyond the immediate audience (Smith *et al.* 2015). According to this idea, when terrorist organisations choose tourism and heritage targets they will be more likely to achieve their goals, since they will generate a strong emotional impact that is capable of reaching global audiences (Seabra *et al.* 2012).

Over the last decades, terrorist organisations have chosen tourist targets due to their high value (Combs 2017). In fact, attacks on tourism targets, such as tourist destinations, hotels, resorts or airports can help terrorists achieve several objectives: advertising, economic threat or ideological opposition, among others (Sönmez 1998). Attacks on tourist targets give terrorists international media attention (Baker 2014). Whenever tourists are injured, kidnapped or killed, especially if they come from foreign countries, media coverage helps terrorists gain global exposure which will, in turn, empower their message (Seabra *et al.* 2012). The tourism industry is extremely vulnerable to the effects of terrorism. According to the Institute for Economics and Peace, the negative economic effects of terrorism are especially strong in countries which have suffered terrorist attacks (IEP 2016).

The frequency and duration of the terrorist attacks, their intensity and severity will define the extent of the impact of terrorism on the tourism industry (Bassil *et al.* 2019). Frequent terrorist incidents have a negative impact on destination image and discourage potential tourists from visiting (Alvarez and Campo 2014, Walters *et al.* 2019). The immediate consequence is a decrease in tourist arrivals and receipts. This decrease has a direct impact not only on the tourism sector, but also on its related industries. The effects are very negative to countries and lead to a drastic drop in employment and to a decrease in the country's GDP (Seabra 2019). In addition to the loss of national revenue, terrorism has other severe consequences for tourism: additional investment in advertising will be needed to help recover the destination's damaged image; security-related expenses will have to be provided to prevent future attacks and a considerable sum of money will have to be made available to rebuild and repair the damages inflicted by the terrorist attacks (Bassil *et al.* 2019).

The tourism industry is highly vulnerable to negative events and the constant crises triggered by internal or external factors anywhere in the world put the sector under a permanent threat of yet another imminent crisis (Pforr and Hosie 2008). Terrorism and war are the main factors responsible for the most serious crises that have affected tourism activity over the last few decades (Hall 2010, Alvarez and Campo 2014, Baker 2014, Saha and Yap 2014, Walters *et al.* 2019). While many crises in tourism activity take place over a short, limited time, others are caused by ongoing armed conflicts and terrorist events that create a prolonged period of uncertainty that will strongly impact on the destination image and make it hard for the countries affected to recover from the damage inflicted (Mansour *et al.* 2019). This is the reason why terrorism and war are commonly singled out as the external factors with the highest negative impact on the destination image of a given country or of an entire region due to the strong spillover effect they trigger (Seabra *et al.* 2020).

Due to the vulnerability of the tourism sector to external shocks and considering its massive economic value, several efforts and research work have emerged over the last few decades focusing on how the tourism sector should deal with threats like terrorism, war or political instability (Ioannides and Apostolopoulos 1999, Mansfeld 1999, Sönmez *et al.* 1999, Stafford *et al.* 2002, Taylor and Enz 2002, Beirman 2003, Blake and Sinclair 2003, Henderson 2003, Cushnahan 2004, Hitchcock and Darma Putra 2005, Roach and Kemish 2006, Paraskevas and Beverley 2007, Morakabati 2013, Avraham 2015, Mansour *et al.* 2019).

Researchers agree that terrorism incidents can provoke both crises and disasters depending on the severity and on the level of damage inflicted "with the latter implying a situation where there is severe loss of life and long-term damage to the society" (Prayag 2018, p. 133) and require high-quality management crisis planning and response strategies in the destinations affected by those events (Pforr and Hosie 2008).

The increasing frequency of terrorism incidents exerts more and more pressure on managers and forces them "to consider the impact of crises and disasters on the industry and develop strategies to deal with the impacts to protect tourism business and society in general" (Ritchie 2004, p. 670) developing actions that will allow them to better analyse and understand the incidents and to lay down strategies to deter or limit their impacts on tourism (Laws and Prideaux 2006). More importantly, terrorism impacts on the tourism industry stress the need to "integrate crisis management with strategic planning processes, prepare detailed contingency plans, define decisional roles and responsibilities, and to retain a degree of flexibility" (Pizam and Smith 2000, p. 135).

Existing research conducted on tourism crisis management focus on how tourism organisations and businesses should respond to crises, but especially on the destination's ability to respond (Mansour *et al.* 2019). Some authors state that tourism should be considered an open system as tourism destinations are constantly changing and influenced by external and internal factors, especially negative incidents that can damage their image like terrorism, crime, war, conflicts or political instability, among others (Coca-Stefaniak and Morrison 2018).

To deal with crises and survive instability, Mansour and his colleagues (2019) argue that successful tourism firms and destinations should develop dynamic crisis management capabilities that will have to include two dimensions: (i) crisis assessment (cognition) and (ii) crisis response (behaviour). According to the authors' results, crisis management assessment (cognition) involves the development of strong relationships: in a business context, supportive relationships between management and employees are needed, while in a wider context those relationships will involve organisations, government, residents and all the other players that take part in the tourism process. At this stage, the collection and analysis of information covering the initial steps of the crisis are extremely important in order to determine the risks. The crisis response (behavioural) dimension focuses on developing the adequate operational responses to counter the risks identified during the crisis assessment phase. In both these stages, cooperation is essential (Mansour *et al.* 2019).

This perspective is consistent with research and with the main strategic frameworks for tourism crisis and disaster management (Mansour *et al.* 2019). Crisis assessment and response dimensions are in accordance with the 4Rs model (Becken and Hughey 2013): reduction, readiness, response and recovery; with the three-stage crisis event model (Paraskevas and Arendell 2007): pre-crisis, crisis event and post-crisis; and with the framework for tourism disaster management (Ritchie 2004).

Destroying heritage: a strong message from terrorism and war

Heritage and cultural sites are also becoming increasingly popular targets for different terrorist organisations. In a recent study Smith and her colleagues (Smith *et al.* 2015), identified three strategies used by terrorists to gain high publicity through the destruction of cultural heritage: shock, awe and censure. By destroying important cultural heritage, terrorists are using a "new grammar of violence" (McDonald 2014). Terrorist organisations achieve a strategic goal by proving that Western countries and institutions like UNESCO are powerless to protect humankind heritage they consider as their own, either by destroying them in real-time shocking videos broadcasted on social media or by selling small, movable antiquities to improve internal sources of funding (Meskell 2002, Keatinge 2014).

Heritage destruction is considered a political act and serves several goals: (i) to express an actor or group's view on these tensions (Shahab and Isakhan 2018); (ii) to erase the ideology and memories of a particular community represented by the targeted site (Layton *et al.* 2001, Mitchell 2016); (iii) to destroy the symbolic value of a given community, their history, customs and identity (Herscher 2010); (iv) to achieve a broader genocidal campaign directed at the eradication of a specific people and their past (Bevan 2006); (v) to perform a ritual aiming at creating a break with previous social customs, creating new ties and mobilising new members to engender group loyalty (Shahab and Isakhan 2018); (vi) as a form of "cultural cleansing" that will cause the loss of identity and belonging and that will culminate in social disintegration (Bokova 2015); (vii) to send a message of power to a global audience (Klein 2018, Shahab and Isakhan 2018).

Nevertheless, and on an even more important note, by destroying cultural heritage sites terrorists are humiliating everyone since they are erasing the identity of entire civilisations, preventing us from acknowledging these symbols of ancient communities while making it impossible for humankind to perpetuate those memories (Bokova 2015).

Assessing the threat of terrorism

The World Heritage List, created by UNESCO, reflects the richness and diversity of the world's cultural and natural heritage and draws attention to its protection and to the importance that must be attached to its transmission (UNESCO 2020). This list is the result of an international agreement aimed at identifying, recognising and protecting places of global value. The importance given to the transmission of this treasure is evidenced in Article 4 of the World Heritage Convention which states that one of the main obligations is to ensure the conservation of WHS to make sure that their meaning is made available to the public and to future generations (Li *et al.* 2008). However, according to Article 11 of the World Heritage Convention, UNESCO has identified 54 properties threatened by serious and specific dangers. Among all these dangers, the threat of disappearance caused by large-scale development projects, the outbreak of armed conflicts, poor management systems or changes in the legal protective status or gradual changes due to geology, climate or environmental factors have to be highlighted. The sites included in the List of World Heritage in Danger (LWHD) require major conservation operations for which "assistance has already been requested" (UNESCO 2019).

Table 2.1 shows important WHS that are being damaged and destroyed due to the actions and attacks perpetrated by terrorist groups that choose those places in an attempt to destroy humanity's "culture by affecting directly the identity, dignity and future of people, and moreover their ability to believe in the future" (Bokova 2015, p. 289). For methodological reasons, the WHS presented in Table 2.1 are those included in the UNESCO List of World Heritage in Danger that have been targeted by acts of terrorism or war.

Unfortunately, there are too many examples of the trail of destruction left by terrorism and war on culture and heritage. In this chapter the authors focused only on the UNESCO List of World Heritage in Danger destroyed or endangered by terrorism and war.

The sites are located in a very sensitive geographic area marked by great political turmoil and conflicts, especially since the Arab Spring that was followed by a strong decline in the number of international tourist arrivals (Mansour *et al.* 2019). Afghanistan, Iraq, Libya, Mali, Syria and Yemen are territories where ongoing political instability and conflicts are aggravated by the presence of terrorist groups that are occupying, destroying and looting important heritage sites and pieces. Some of those treasures are irretrievably lost.

Table 2.1 Heritage destroyed or endangered by terrorism

Country	Cultural Site	Description	Destruction	Danger and Preservation
Afghanistan	Archaeological remains of the Bamiyan Valley (nominated and inscribed on the LWHD in 2003)	The Buddhas of Bamiyan statues were massive structures in the northwest of Kabul at an elevation of 2,500 metres, carved out of the cliffs of Afghanistan's Bamiyan Valley by Buddhist monks who lived and meditated there 1,500 years ago. This was a location of cultural, historical, religious and archaeological significance. The giant guards were visited by historians, religious groups, visiting tourists, among many other people.	The giant Buddhas were dynamited and destroyed in March 2001 by Taliban soldiers during their genocidal campaign against the West (Klein 2018).	Several attributes considered of Outstanding Universal Value by UNESCO, such as Buddhist and Islamic architectural forms and their setting in the Bamiyan landscape, remain intact, including the vast Buddhist monastery in the Bamiyan Cliffs. Since 2003, UNESCO has been leading a safeguarding plan to protect the site and the artefacts that survived the destruction of the Buddha statues.
Iraq	Ancient City of Nimrud (submitted to UNESCO Tentative List since 2000)	That city had a very important religious and cultural meaning for Assyrian Christians. The most important and well-known artefacts and archaeological remains were the winged-bull deity statues that guarded the palace gates and the royal tombs filled with gold antiquities. It was a matter of pride for the Iraqis especially when tourists came to the country.	In mid-2014, the Islamic State of Iraq and the Levant (ISIL) occupied the territory surrounding Nimrud. In March 2015, all the history and culture were erased by the Islamic State militants who bulldozed the ancient city to the ground (Klein 2018).	It is estimated that 80% of the city has been destroyed. However, the territory is not safe enough to evaluate the real damages.

Ashur (nominated and inscribed on the LWHD in 2003)	The ancient city of Ashur dates back to the third millennium BCE, being the first social and religious capital of the Assyrian Empire between the fourteenth and the ninth centuries BCE. The city was destroyed by the Babylonians but was rebuilt during the Parthian period in the first and second centuries CE. Artefacts, statues, buildings and temples from the Assyrian Empire and the Parthian Empire periods could be admired at this site.	The territory of the ancient site was occupied by the Islamic State of Iraq and the Levant (ISIL) in 2015. According to some sources, the citadel of Ashur was badly damaged in May 2015 by explosive devices fired by members of ISIL (Associated Press 2017).	Since February 2017, the terrorist group has no longer controlled the site, however the territory is not safe enough to allow visits and an assessment of the damage suffered.
Hatra (1985 in the LWHD since 2015)	Hatra was the capital of the first Arab Kingdom and lies 290 km northwest of Baghdad and 110 km southwest of Mosul. Possibly built by the Assyrians, the city flourished under the Parthians during the first and second centuries CE as a religious and trading centre. It was also an important city for the Greeks, the Romans and the Persians. With a circular plan encircled by big inner and outer walls supported by more than 160 towers, it was known for its Hellenistic and Roman temples blended with Eastern decorative features. Ashur and Hatra are two sites included in the International Committee for Tourism Attractiveness and Sustainable Tourism Development (2019a, 2019b) and in the World List of Tourist Attraction Sites since 2017, due to their heritage, interest and historical importance.	The territory was occupied by ISIL in mid-2014. In early 2015 they announced their intention to destroy many artefacts and the ruins of Hatra. The pro-Iraqi government Popular Mobilization Forces (PMF) captured the city in 2017. ISIL destroyed the sculptures and engraved images but its walls and towers are still standing. PMF units also stated that ISIL had mined the site's eastern gates (Yacoub 2015).	Preliminary reports confirm that the city was partially destroyed since it was taken in 2015. As soon as security conditions are ensured, UNESCO will send an emergency assessment mission to evaluate the damages so that a safeguarding plan can be laid down.

(Continued)

Table 2.1 Continued

Country	Cultural Site	Description	Destruction	Danger and Preservation
	Samarra Archaeological City (1985 in the LWHD since 2007)	Located 130km north of Baghdad, it stands on both sides of the River Tigris. It was a powerful Islamic capital ruling the Abbasid Caliphate in medieval times. With great religious importance granted by its several Shi'i holy sites, that included the tombs of several Shi'i Imams, it is the only remaining Islamic capital that retains its original plan, architecture and artistic relics, mosaics and carvings.	Since the Iraqi war, the site has been occupied by military forces that used it as a military operations base. Sunni units linked to Al-Qaeda attacked the city in 2007 destroying the mosque compound and its minarets. In June 2014, the city was once again attacked by ISIL, but it seems that its integrity is ensured (Hassan 2014).	The state of conflict does not allow the responsible authorities to ensure the protection and management of the property. Also, because of the increase in violence in the country, UNESCO helped build an Emergency Response Action Plan to protect Iraqi's cultural heritage.
Israel	Old City of Jerusalem and its Walls (1981 in the LWHD since 1982)	Jerusalem is considered the holy city for Judaism, Christianity and Islam. The old city accounts for 220 historic monuments, such as the Dome of the Rock, the Wailing Wall, the Church of the Holy Sepulchre, the Christ's Tomb and several gates, among many others. The city is visited for religious and tourist purposes by millions of people every year.	Being the theatre of several territorial disputes between Jordan, Israelis and Palestinians, the Old City has been threatened by continuous fighting.	Today, the Israeli government controls the entire area, however East Jerusalem is now regarded by the international community as part of the occupied Palestinian territory. The City of Jerusalem and its Walls are in a challenging context and that is why the World Heritage Centre is building an action plan to ensure the protection of the Old City (Benvenisti 2004).
Libya	Archaeological Site of Cyrene (1982 in the LWHD since 2016)	It was one of the main cities in the Hellenic world. It was the oldest and most important of the five Greek cities located in the region of Libya. Cyrene is located in a lush valley in the Jebel Akhdar uplands. It was consecrated to Apollo and has numerous temples and statues from the Greek period. It was later Romanised and remained a great capital until the earthquake of 365. Thousands of years of history are written into its ruins, famous since the eighteenth century.	In May 2011, many objects, excavated from Cyrene in 1917 and that were kept in a vault of the National Commercial Bank in Benghazi were stolen and are currently missing (Yates 2012).	Since Colonel Qaddafi's downfall, locals have been trying to organise themselves to protect the site. Cyrene Friends Society is trying to protect, preserve and promote Cyrene's heritage and is striving to preserve its education, its tourism, archaeology and culture (Elkin 2012).

Archaeological Site of Leptis Magna (1982 in the LWHD since 2016)	This was a prominent city of the Carthaginian and Roman Empires. Its ruins, 130 km east of Tripoli, are among the best-preserved Roman sites in the Mediterranean. Before 2011, the site was visited by many tourists attracted by its reputation for being one the "best-preserved" Roman cities in the world.	There were unfounded reports that the sites were used as a cover for tanks and military vehicles by pro-Qaddafi forces during the 2011 Libyan civil war (CNN Wired Staff 2011).
Archaeological Site of Sabratha (1982 in the LWHD since 2016)	It lies on the Mediterranean coast about 70 km west of Tripoli. It was a Phoenician trading centre for African products and part of the Numidian Kingdom of Massinissa. It was Romanised and rebuilt in the second/third centuries CE. Its main archaeological remains are the Roman Theatre, temples, a Christian basilica or its public baths, among others. Some treasures are kept in the national museum in Tripoli which is visited by thousands of tourists (mainly before the outbreak of the Libyan Civil War).	The World Heritage Committee inscribed these sites on the List of World Heritage in Danger due to the high political instability of the country and to the fact it still harbours armed groups, some of them terrorist groups.
Old Town of Ghadames (1986 in the LWHD since 2016)	Known as "the pearl of the desert", Ghadames is an oasis Berber town in northwestern Libya, considered one of the oldest pre-Saharan cities and an outstanding example of a traditional settlement. The small town of around 11,000 people was a key destination for tourists who came to Libya before the uprisings known as the Arab Spring.	

(Continued)

Table 2.1 Continued

Country	Cultural Site	Description	Destruction	Danger and Preservation
	Tadrart Acacus Rock Art Sites (1985 in the LWHD since 2016)	It is a mountain range in the desert of the Ghat District in western Libya, about 100 km from the border with Algeria. The area has a particularly rich array of prehistoric rock art of thousands of cave paintings and carvings in very different styles, dating from 12,000 BCE to 100 CE. The paintings reflect the fauna and flora changes and the different ways of life of the Sahara populations.		Besides vandalism, some reports draw people's attention to the looting of ancient artefacts and of portions of the cave paintings.
Mali	Timbuktu (1988 in the LWHD since 2012)	Located 20 km north of the Niger River, Timbuktu is the home of the prestigious Koranic Sankore University and other madrasas. It was an intellectual and spiritual capital and a propagation centre for Islam in the fifteenth and sixteenth centuries and also a regional trade centre for salt, gold, ivory, and slaves. Besides thousands of manuscripts, three great mosques, Djingareyber, Sankore and SidiYahia, remind us of Timbuktu's golden age. The city's reputation for mystery and richness has attracted thousands of tourists over the years.	The Al-Qaeda Organization in the Islamic Maghreb started kidnapping groups of tourists in the Sahel region in 2008. In November 2011, terrorists attacked tourists who were staying at a hotel in Timbuktu. Over the following years, tourism activity suffered a serious drop. In 2012, Timbuktu was captured by the Tuareg rebels (CNN 2011). In January 2013, French and Malian government troops retook the city. The Islamist groups had already destroyed the Ahmed Baba Institute and many of its important manuscripts. In 2012, Ansar Dine destroyed several shrines, including the mausoleum of Sidi Mahmoud (Shamil 2013).	Because of the situation of armed conflict in the northern region of Mali and of the takeover of Timbuktu by MNLA and by Ansar Dine, UNESCO considered the sites at risk.

Tomb of Askia (2004 in the LWHD since 2012)	The 17-metre mud-pyramidal structure is the tomb of Askia Mohamed, the most prolific emperor of the Songhai Empire and was built in the fifteenth century. The complex includes the pyramidal tomb, two flat-roofed mosque buildings, the mosque cemetery and the open-air assembly ground and is the largest pre-colonial architectural monument in the region. UNESCO's plans for conservation and maintenance include the site's promotion for educative and tourism purposes.	The highest risk that this site faces is the looting of artefacts and arts pieces that are being sold on the black market to finance criminal and terrorist groups.
Palestine Hebron/Al-Khalil Old Town (nominated and inscribed on the LWHD since 2017)	Located 30 km south of Jerusalem in the Judaean Mountains, Hebron claims to be one of the oldest cities in the world, dating from the Chalcolithic period or from more than 3,000 years BCE. It was conquered by Romans, Jews, Crusaders and Mamluks. This place became a site of pilgrimage for the three monotheistic religions: Judaism, Christianity and Islam because of its association with Abraham. Hebron's old town shelters, since the first century CE, the tombs of the patriarch Abraham/Ibrahim and his family in which Isaac and Jacob are included. Judaism ranks Hebron the second holiest city after Jerusalem, while some Muslims regard it as one of the four holy cities.	"Hebron provides a stark illustration of Israeli–Palestinian conflict. A few hundred Israelis live closed off in several small settlements, protected by hundreds of Israeli soldiers. The Palestinians are largely banned from entering and using nearby streets and say the settlements make their lives impossible. Israel seized the West Bank in the 1967 war in a move considered illegal by the United Nations and most of the world regards the settlements in Hebron as illegal" (Agence France-Presse 2017). Another site that is a victim of a continuous conflict between Israel and Palestine (Reuters 2017).

(*Continued*)

Table 2.1 Continued

Country	Cultural Site	Description	Destruction	Danger and Preservation
Syrian Arab Republic	Ancient City of Aleppo (1986 in the LWHD since 2013)	The historic city centre remained essentially unchanged since its construction that took place between the twelfth and the sixteenth centuries. Because of the constant invasions, its inhabitants were forced to build cell-like quarters. Each district was characterised by the religious and ethnic features of its inhabitants. The monumental Citadel of Aleppo is mainly marked by souqs, mosques, madrasas, residences, khans and public baths which give the city centre a unique aspect that used to attract many visitors before the outbreak of the civil war.	Many sections in the ancient city were destroyed in the fighting between the Syrian Arab Army and the rebel forces of Jabhat al-Nusra, in the Battle of Aleppo that took place in September 2012. Two years later, the Islamic Front's opposition groups claimed responsibility for destroying many important historic buildings that were used as military bases by the Syrian Army (United Nations Institute for Training and Research 2017).	UNESCO experts estimate that 30% of the Ancient City of Aleppo has been destroyed (United Nations Institute for Training and Research 2017).
	Ancient City of Bosra (1980 in the LWHD since 2013)	Bosra is located in southern Syria. This important city on the caravan route to Mecca was in times the capital of the Roman province of Arabia. It is a major archaeological site with Roman, Byzantine and Muslim ruins and possesses a magnificent Roman theatre, early Christian ruins and several mosques. Before the civil war, the Roman Theatre hosted every year an important national music festival.	In 2013 it was reported that the citadel was used by the army for military purposes. In January 2015 the Syrian Army engaged in a battle against a rebel contingent near the famous Roman Theatre. In March 2015, Syrian rebels reoccupied the territory, banishing Syrian soldiers (Middle East Online 2015). Bosra was recaptured by the Syrian Arab Army in July 2018.	The six Syrian World Heritage Sites were added to UNESCO's List of World Heritage in Danger in 2013; the organisation continues to lead international efforts to protect and restore cultural heritage across the country.

(Continued)

Ancient City of Damascus (1979 in the LWHD since 2013)	The city centre contains numerous archaeological sites, including city walls, gates, 125 protected monuments: churches, mosques, the Umayyad Mosque, madrasas, khans, the Citadel and private houses. It is one of the most ancient cities in the Middle East and was founded in the third millennium BCE. Many cultures (Hellenistic, Roman, Byzantine and Islamic, among others) have left their mark on the city. It was one of the country's most important tourist attractions before the civil war.	The conflicts caused by the Syrian Civil War, along with other threats qualifies this city as one of the world's most endangered sites.	As soon as security conditions are ensured, UNESCO will send an emergency assessment mission in order to evaluate the damages in order to prepare a safeguarding plan.
Ancient Villages of Northern Syria (2011 in the LWHD 2013)	Forty villages dating from the first to the seventh century are situated between Aleppo and Idlib. These settlements offer an important understanding of rural life in Late Antiquity of the Byzantine period. There are several remains of dwellings, pagan temples, churches, cisterns, bathhouses among others.	The conflicts of the civil war in Syria left destruction marks all over these sites.	The Syrian Arab Army recaptured the territories in the southern and western regions. "Major tourist sites were damaged and are inaccessible due to the conflict. Teams at damaged UNESCO World Heritage sites such as The Old City of Aleppo, the ruins of Palmyra and Krak des Chevaliers have begun the arduous task of restoration and reconstruction following years of conflict and devastation" (Cascone 2018).
Crac des Chevaliers and Qal'at Salah El-Din (2006 in the LWHD 2013)	These two castles are situated 40 km west of the city of Homs, close to the Lebanese border. They are one of the most important and well-preserved medieval castles in the world, representing significant examples of the Crusades period (eleventh to thirteenth century).		

Table 2.1 Continued

Country	Cultural Site	Description	Destruction	Danger and Preservation
	Site of Palmyra (1980 in the LWHD since 2013)	Situated in an oasis in the Syrian desert, it lies 215 km northeast of Damascus. It is a monumental city and one of the most important cultural centres of the Ancient World. With remains from the Neolithic period, it was occupied by a number of civilisations and empires and became part of the Roman Empire in the first century CE. The mixture of these many civilisations left many architectural and artistic traits all over the city influenced by Graeco-Roman and Persian cultures. The discovery of the ruined city by travellers in the seventeenth and eighteenth centuries brought many other visitors and explorers to the site.	In 2015 many remains were destroyed by ISIL militants. Among them were the Lion of Al-lāt and other statues; the Temple of Baalshamin, the Temple of Bel, three of the best preserved tower tombs including the Tower of Elahbel, the Tetrapylon and part of the theatre (Shaheen and Levett 2015). In March 2017, the Syrian Army captured the city and the Director of Antiquities and Museums stated that the damage to ancient monuments might have been less serious than what had initially been foreseen (Makieh and Francis 2017).	
Yemen	Historic Town of Zabid (1993 in the LWHD since 2000)	Zabid was the capital of Yemen from the thirteenth to the fifteenth century. The urban plan, the domestic and military architecture, the Great Mosque that occupies a prominent place in the town and the vestiges of its university make this city an important archaeological and historical site.	Saudi coalition bombed Zabid in May 2015. The bombs hit a restaurant and the explosions' impact damaged several historic houses (Khalidi 2017).	UNESCO built an Action Plan to coordinate an international response through the Global Coalition and the "Unite4Heritage" campaign to protect and preserve cultural heritage in Yemen.

Old Walled City of Shibam (1982 in the LWHD since 2015)	This city situated in the central–western area of the Hadhramaut Governorate lies in the desert of Ramlat al-Sab'atayn. It is a sixteenth-century picturesque city surrounded by a fortified wall and is known as the Manhattan of the desert for its 14-storey mudbrick-made towers. It is one of the oldest and best examples of urban planning based on the principle of vertical construction. This architectural style was used to protect residents from Bedouin attacks.	Since March 2015 many archaeological sites, museums and other important cultural and religious sites were damaged or intentionally destroyed by Saudi coalition bombs and raids or explosive devices detonated by Al-Qaeda and ISIL. The city was damaged by Al-Qaeda in November 2016 (Khalidi 2017).
Old City of Sana'a (nominated in 1986 and inscribed on the LWHD since 2015)	Located in a mountain valley at an altitude of 2,200 m, this city is one of the largest urban centres in Yemen and one of the highest capital cities in the world. The city has been inhabited for more than 2,500 years. It has a very important religious, political and social heritage with its 103 mosques, 14 hammams and over 6,000 houses built prior to the eleventh century. Traditionally Yemen was a tourist centre and tourism had been a fundamental activity for centuries. However, since the beginning of the twenty–first century several terrorist attacks against tourist targets provoked a drop in tourism.	In June 2015 the old City of Sana'a was hit in a raid conducted by the Saudi coalition provoking direct and collateral damage across the whole city. An ancient mosque, Qubbat al-Mahdi, was also damaged by ISIL (BBC News 2015). UNESCO built an Action Plan to coordinate an international response through the Global Coalition and the "Unite4Heritage" campaign to protect and preserve cultural heritage in Yemen.

Source: Adapted from UNESCO (2019).

Recent events in Syria, Iraq, Libya and Mali have highlighted the multiple threats to cultural heritage during crisis, including deliberate attacks, destruction as collateral damage in fighting, the greed of unscrupulous traders and collectors, vandalism of factions that seek to erase the achievements of past cultures.

(UNESCO Media Services 2014)

Most of those territories are not safe enough to be visited, so the assessment of the real existing damage is impossible.

UNESCO is helping local authorities to build Emergency Response Action Plans to safeguard the heritage. In some cases, World Heritage Fund assistance was triggered to rebuild and preserve the endangered sites. Also, some local organisations and the international community are trying to rebuild and protect the heritage. This endeavour is unquestionably a massive and formidable task. The Middle East and North Africa (MENA) region has been experiencing ongoing turmoil and instability for decades and that instability had a deterrent effect on the amount of international tourist arrivals to the region (Morakabati 2013). These territories offer special insight over the history, the culture and over humankind. They share this role with UNESCO World Heritage Sites, sea resorts, desert tourism, with the culture and the uniqueness of the tourist products and destinations. In fact, tourism was one of the most important economic sectors for some of these countries (Mansour *et al.* 2019). However, these conflicts have weakened an already fragile tourism sector and have diverted governments' attention from supporting and helping tourism enterprises. As a result, tourism activity across the region experienced an inexorable decline (Morakabati 2013).

Conclusions

"The destruction of heritage undermines wellsprings of identity and belonging, paving the way to social disintegration. Eliminating the layers of history, cities, and homes affects people's perceptions of the past and present" (Bokova 2015, p. 291). In fact, "WHS belong to all the peoples of the world, irrespective of the territory on which they are located" (UNESCO 2019).

The destruction of cultural heritage was in the past a strategy used to defeat and subdue communities and territories. Nowadays terrorist organisations use the same strategy to humiliate governments and universal organisations and achieve their goals through actions carried out to shock, manipulate and persuade global audiences (Smith *et al.* 2015). These destructive acts have a global impact "the visual, symbolic nature of cultural heritage makes this heritage particularly vulnerable. In any war, symbolic heritage is a target for destruction, but within a modern, post 9/11 terrorist framework, iconic cultural heritage is more at risk than ever" (Smith *et al.* 2015, p. 18).

The Parliamentary Assembly Council of Europe (2004) recognised the link between terrorism and the destruction of cultural heritage. In 2004 this Assembly claimed:

Culture … is becoming increasingly a target of terrorism. Beyond the physical damage or destruction of monuments, temples or symbols of a given culture and way of life, such terrorist acts target the very cultural identity of a people or of a population. They also harm a cultural heritage that is common to all peoples of the world.

(Parliamentary Assembly Council of Europe 2004)

The present work is an attempt to draw people's attention to this dangerous connection: terrorism against heritage culture and tourism. The UNESCO classification of WHS makes these sites even more important to the local and national communities to which they belong but also to humankind, as they portray and preserve cultural identity and heritage.

But UNESCO also plays an important role in promoting tourist destinations while sustaining national heritage. WHS have become important tourist destinations, especially for cultural and heritage tourists that aim to "experience places of historical importance and significance" (Southall and Robinson 2011, p. 177) that give visitors the opportunity to discover the identity of a given country and of its people and the symbols of their pride (UNESCO 2019).

Over the last few decades, the most serious disasters and crises affecting tourism were due to terrorist acts and turmoil (Prayag 2018). In fact, insecurity and unsafety have become critical issues and have raised serious global concern within the tourism sector. Terrorism events create a generalised perception of unsafety that impacts negatively on the image of a destination and affects tourist flows (Seabra *et al.* 2020). Moreover, when terrorism and instability are severe and last too long, the negative impacts are stronger and the destination image and the tourism activity are seriously damaged. In a worst-case scenario, destinations suffering from ongoing turmoil, acts of terrorism and wars may experience the total demise of their tourism industry (Walters *et al.* 2019) making it hard for these regions to recover from successive crises. Thus, managers and governments of these affected destinations should focus on the preparation and implementation of crisis management plans and strategies.

Bearing in mind the heavy importance that users' positive perceptions have on tourist destinations' image, the implementation of an effective crisis communication plan whose role will be to provide relevant stakeholder groups with precise and up-to-date information is imperative (Park *et al.* 2019). In the immediate post-crisis phase the focus is crisis assessment by managing a good network of stakeholders to understand the crisis' dimensions and features, which should be followed by the crisis response strategy where the different stakeholders build a plan with the operational responses to recover from the crisis (Mansour *et al.* 2019). Also of major importance is the implementation of a strategic crisis communication plan to manage as much as possible the destination image in the global media (Park *et al.* 2019). In this context, destination management organisations play a very important role by helping local and national authorities to deal with the bad publicity (Coca-Stefaniak and Morrison 2018), also helping them to identify strategies to help destinations to re-establish tourists' and international community confidence.

In the cases presented in this chapter, several local, national and international organisations are gathering efforts to safeguard the heritage and to revive tourism activity. UNESCO is playing a very important role by calling international attention to those sites destroyed and in danger, also working with the local organisations to restore and protect the sites in order to once again attract tourism flows. Tourism activity driven by world heritage can be of major importance to recover the damaged economy of those countries (Mansour *et al.* 2019). As stated by Frey and colleagues (Frey *et al.* 2007, cited in Bassil *et al.* 2019) "The exceptionality of a destination may entice tourists to tolerate the risk when making travel choices". Tourism activity itself can be a strategic tool to create more resilient destinations, giving power and financial support to the local governments to deal with the terrorist threat.

In the present study only sites in the List of World Heritage in Danger were included. In future studies, a wider analysis should be considered. Also, future research should focus on crisis management strategies for local and international organisations and communities.

This work was funded by national funds through FCT – the Portuguese Foundation for Science and Technology (UID/ECO/00124/2013 and Social Sciences DataLab, Project 22209), POR Lisboa (LISBOA-01–0145-FEDER-007722 and Social Sciences DataLab, Project 22209), POR Norte (Social Sciences DataLab, Project 22209) and under the projects UID/Multi/04016/2019 and UID/GEO/04084/2019. Also, the first author was supported by FCT with a scholarship SFRH/BPD/109245/2015. Furthermore, we would like to thank the Nova School of Business and Economics, CEGOT – Geography and Spatial Planning Research Centre, and CI&DETS – Centre for the Study of Education, Technologies and Health for their support.

References

Agence France-Presse, 2017. UNESCO puts Hebron on endangered heritage list. *Agence France-Presse*, 7 July.

Alvarez, M. D. and Campo, S., 2014. The influence of political conflicts on country image and intention to visit: A study of Israel's image. *Tourism Management*, 40, 70–78.

Ashworth, G., 2017. *Senses of Place: Senses of Time*. 2nd ed. Aldershot: Routledge.

Associated Press, 2017. Iraqis seek funds to restore cultural artifacts recovered from ISIS. *CBS News*, 24 February.

Avraham, E., 2015. Destination image repair during crisis: Attracting tourism during the Arab Spring uprisings. *Tourism Management*, 47, 224–232.

Baker, D., 2014. The effects of terrorism on the travel and tourism industry. *International Journal of Religious Tourism and Pilgrimage*, 2 (1), 58–67.

Bassil, C., Saleh, A. S., and Anwar, S., 2019. Terrorism and tourism demand: A case study of Lebanon, Turkey and Israel. *Current Issues in Tourism*, 22 (1), 50–70.

BBC News, 2015. UNESCO condemns Yemen heritage site "air strike". *BBC News*, 12 June.

Becken, S. and Hughey, K., 2013. Linking tourism into emergency management structures to enhance disaster risk reduction. *Tourism Management*, 36, 77–85.

Beirman, D., 2003. Marketing of tourism destinations during a prolonged crisis: Israel and the Middle East. *Journal of Vacation Marketing*, 8, 167–176.

Benvenistî, E., 2004. *The international law of occupation*. Princeton, NJ: Princeton University Press.

Bergesen, A. and Han, Y., 2005. New directions for terrorism research. *International Journal of Comparative Sociology*, 46, 133–151.

Bevan, R., 2006. *The Destruction of Memory: Architecture at War*. London: Reaktion Books.

Blake, A. and Sinclair, M., 2003. Tourism crisis management: US response to September 11. *Annals of Tourism Research*, 30 (4), 813–832.

Bokova, I., 2015. Culture on the front line of new wars. *Brown Journal of World Affairs*, 22 (1), 289–296.

Cascone, S., 2018. Nearly destroyed by ISIS, the Ancient City of Palmyra will reopen in 2019 after extensive renovations. *Artnet*, 27 August.

CNN, 2011. Mali kidnapping: One dead and three seized in Timbuktu. *BBC News*, 25 November.

CNN Wired Staff, 2011. South African president blasts NATO actions in Libya. *CNN*, 14 June.

Coca-Stefaniak, A. and Morrison, A., 2018. City tourism destinations and terrorism: A worrying trend for now, but could it get worse?. *International Journal of Tourism Cities*, 4 (4), 409–412.

Combs, C., 2017. *Terrorism in the Twenty-first Century*. London: Routledge.

Cushnahan, G., 2004. Crisis management in small-scale tourism. *Journal of Travel and Tourism Marketing*, 15 (4), 323–338.

Elkin, M., 2012. Hope for future of Libyan tourism in sprawling Greek ruins. *New York Times*, 15 February.

Hall, C., 2010. Crisis events in tourism: Subjects of crisis in tourism. *Current Issues in Tourism*, 13 (5), 401–417.

Hassan, G., 2014. Iraq dislodges insurgents from city of Samarra with airstrikes. *Reuters*, 05 June.

Henderson, J., 2003. Managing the aftermath of terrorism: The Bali bombings, travel advisories and Singapore. *International Journal of Hospitality and Tourism Administration*, 4, 17–31.

Herscher, A., 2010. *Violence Taking Place: The Architecture of the Kosovo Conflict*. Stanford, CA: Stanford University Press.

Hitchcock, M. and Darma Putra, I., 2005. The Bali bombings: Tourism crisis management and conflict avoidance. *Current Issues in Tourism*, 8 (1), 62–76.

Hoffman, S., 2002. Clash of globalizations. *Foreign Affairs*, 81, 104–115.

Hudson, R., 1999. *Who Becomes a Terrorist and Why: The 1999 Government Report on Profiling Terrorists*. Guilford, CT: The Lyons Press.

IEP – Institute for Economics and Peace, 2016. *Global Terrorism Index: Measuring and Understanding the Impact of Terrorism*. Sydney: Institute for Economics and Peace.

International Committee for Tourism Attractiveness and Sustainable Tourism Development, 2019a. *World Tourist Attraction*. [Online] Available at: http://worldlist.travel/eurasia/iraq/index.phtml [Accessed March 2019].

International Committee for Tourism Attractiveness and Sustainable Tourism Development, 2019b. *World List Travel*. [Online] Available at: http://worldlist.travel/index.phtml [Accessed February 2019].

Ioannides, D. and Apostolopoulos, Y., 1999. Political instability, war, and tourism in Cyprus: Effects, management, and prospects for recovery. *Journal of Travel Research*, 38 (1), 51–56.

Jenkins, B., 2001. Terrorism and beyond: A 21st century perspective. *Studies in Conflict and Terrorism*, 24, 321–327.

Keatinge, T., 2014. Defeating ISIS: How financial liabilities will undo the Jihadists. Commentary. [Online] Available at: www.rusi.org/analysis/commentary/ref:C544E693A1171D/#.VfljfXan7No [Accessed January 2019].

Khalidi, L., 2017. The destruction of Yemen and its cultural heritage. *International Journal of Middle East Studies – Special Issue 4 (Forced Displacement and Refugees)*, 49 (3), 735–738.

Klein, A., 2018. Negative spaces: Terrorist attempts to erase cultural history and the critical media coverage. *Media, War and Conflict*, 11 (2), 265–281.

Laws, E. and Prideaux, B., 2006. Crisis management: A suggested typology. *Journal of Travel and Tourism Marketing*, 19 (2–3), 1–8.

Layton, R., Stone, P., and Thomas, J. eds., 2001. *Destruction and Conservation of Cultural Property*. London: Routledge.

Li, M., Wu, B., and Cai, L., 2008. Tourism development of World Heritage Sites in China: A geographic perspective. *Tourism Management*, 29 (2), 308–319.

McDonald, K., 2014. ISIS jihadis' use of social media and "the mask" reveals a new grammar. *The Conversation*, 24 June.

MacDowel, S., 2016. Heritage, memory and identity. In: B. Graham and P. Howard, eds. *The Ashgate Research Companion to Heritage and Identity*. Farnham, Hampshire: Ashgate Publishing, 37–53.

Makieh, K. and Francis, E., 2017. Less damage to ancient Palmyra than feared, Syrian antiquities chief says. *Reuters*, 03 March.

Mansfeld, Y., 1999. Cycles of war, terror and peace: Determinants and management of crisis and recovery of the Israeli tourism industry. *Journal of Travel Research*, 38 (1), 30–36.

Mansour, H. E., Holmes, K., Butler, B., and Ananthram, S., 2019. Developing dynamic capabilities to survive a crisis: Tourism organizations' responses to continued turbulence in Libya. *International Journal of Tourism Research*, 21, 493–503.

Meskell, L., 2002. Negative heritage and past mastering in archaeology. *Anthropological Quarterly*, 75 (3), 557–574.

Middle East Online, 2015. Syria rebels seize ancient town of Busra Sham. *Middle East Online*, 25 March.

Mitchell, W., 2016. *Image Science: Iconology, Visual Culture, and Media Aesthetics*. Chicago, IL: University of Chicago Press.

Morakabati, Y., 2013. Tourism in the Middle East: Conflicts, crises and economic diversification, some critical issues. *International Journal of Tourism Research*, 15 (4), 375–387.

Paraskevas, A. and Arendell, B., 2007. A strategic framework for terrorism prevention and mitigation in tourism destinations. *Tourism Management*, 28 (6), 1560–1573.

Paraskevas, A. and Beverley, A., 2007. A strategic framework for terrorism prevention and mitigation in tourism destinations. *Tourism Management*, 28 (6), 1560–1573.

Park, D., Kim, W. G., and Choi, S., 2019. Application of social media analytics in tourism crisis communication. *Current Issues in Tourism*, 22 (15), 1810–1824.

Parliamentary Assembly Council of Europe, 2004. Combating terrorism through culture. [Online] Available at: http://assembly.coe.int/nw/xml/XRef/Xref-XML2HTML-en.asp?fileid=17278andlang=en [Accessed March 2019].

Pforr, C. and Hosie, P., 2008. Crisis management in tourism: Preparing for recovery. *Journal of Travel and Tourism Marketing*, 23 (2–4), 249–264.

Pizam, A. and Smith, G., 2000. Tourism and terrorism: A quantitative analysis of major terrorist acts and their impact on tourism destinations. *Tourism Economics*, 6 (2), 123–138.

Prayag, G., 2018. Symbiotic relationship or not? Understanding resilience and crisis management in tourism. *Tourism Management Perspectives*, 25, 133–135.

Rapoport, D., 2001. The fourth wave: September 11 in the history of terrorism. *Current History*, December, 419–424.

Reuters, 2017. UNESCO puts Hebron on its World Heritage in Danger List. *Reuters*, 7 July.

Ritchie, B., 2004. Chaos, crises and disasters: A strategic approach to crisis management in the tourism industry. *Tourism Management*, 25 (6), 669–683.

Roach, J. and Kemish, I., 2006. Bali bombings: A whole of government response. In: J. Wilks, D. Pendergast and P. Leggat, eds. *Tourism in Turbulent Times: Towards a Safe Experience for Visitors*. London: Elsevier, 277–289.

Saha, S. and Yap, G., 2014. The moderation effects of political instability and terrorism on tourism development: A cross-country panel analysis. *Journal of Travel Research*, 53 (4), 509–521.

Seabra, C., 2019. Terrorism and tourism consumption revisited. In: A. Correia, A. Fyall and M. Kozak, eds. *Experiential Consumption and Marketing in Tourism: A Cross-Cultural Context*. Oxford: Goodfellow Publishers, 58–75.

Seabra, C., Abrantes, J., and Kastenholz, E., 2012. TerrorScale: A scale to measure the contact of international tourists with terrorism. *Journal of Tourism Research and Hospitality*, 1 (4), 1–8.

Seabra, C., Reis, P., and Abrantes, J., 2020. The influence of terrorism in tourism arrivals: A longitudinal approach in a Mediterranean country. *Annals of Tourism Research*, 80, 102811.

Shackley, M., ed., 1998. *Visitor Management: A Strategic Focus*. London: Focal Press.

Shahab, S. and Isakhan, B., 2018. The ritualization of heritage destruction under the Islamic State. *Journal of Social Archaeology*, 18 (2), 212–233.

Shaheen, K. S. G. and Levett, C., 2015. Palmyra – what the world has lost. *The Guardian*, 05 October.

Shamil, J., 2013. Timbuktu's treasure trove of African history. *BBC News*, 29 January.

Smith, C., Burke, H., Leiuen, C., and Jackson, G., 2015. The Islamic State's symbolic war: Da'esh's socially mediated terrorism as a threat to cultural heritage. *Journal of Social Archaeology*, 0 (0), 1–26.

Sönmez, S., 1998. Tourism, terrorism and political instability. *Annals of Tourism Research*, 25 (2), 416–456.

Sönmez, S., Apostolopoulos, Y., and Tarlow, P., 1999. Tourism and crisis: Managing the effects of terrorism. *Journal of Travel Research*, 38 (1), 13–18.

Southall, C. and Robinson, P., 2011. Heritage tourism. In: P. Robinson, S. Heitmann and P. Dieke, eds. *Research Themes for Tourism*. London: CABI, 176–187.

Stafford, G., Yu, L., and Armoo, A., 2002. Crisis management and recovery: How Washington, DC, hotels responded to terrorism. *Cornell Hotel and Restaurant Administration Quarterly*, 43 (5), 27–40.

START – National Consortium for the Study of Terrorism and Responses to Terrorism, 2019. *Global Terrorism Database*. [Online] Available at: www.start.umd.edu/gtd/ [Accessed March 2019].

Steiner, L. and Frey, B. S., 2012. Correcting the imbalance of the world heritage list: Did the UNESCO strategy work. *Journal of International Organizations Studies*, 3, 25–40.

Stern, J., 1999. *The Ultimate Terrorists*. Cambridge, MA: Harvard University Press.

Stone, P., 2012. Human rights and cultural property protection in times of conflict. *International Journal of Heritage Studies*, 18 (3), 271–284.

Taylor, M. and Enz, C., 2002. Voices from the field: GMs' responses to the events of September 11, 2001. *Cornell Hospitality Quarterly*, 43 (1), 7–20.

UNESCO, 2019. *List of World Heritage in Danger*. [Online] Available at: https://whc.unesco.org/en/danger/ [Accessed March 2019].

UNESCO, 2020. *Global Strategy*. [Online] Available at: https://whc.unesco.org/en/globalstrategy/ [accessed 25 April 2020].

UNESCO Media Services, 2014. UNESCO strengthens action to safeguard cultural heritage under attack. [Online] Available at: www.unesco.org/new/en/media-services/single-view/news/unesco_strengthens_action_to_safeguard_cultural_heritage_und/ [Accessed March 2019].

United Nations Institute for Training and Research, 2017. *Five Years of Conflict: The State of Cultural Heritage in the Ancient City of Aleppo*. Geneva: United Nations Institute for Training and Research.

Van Der Auwera, S., 2012. Contemporary conflict, nationalism, and the destruction of cultural property during armed conflict: A theoretical framework. *Journal of Conflict Archaeology*, 7 (1), 49–65.

Walters, G., Wallin, A., and Hartley, N., 2019. The threat of terrorism and tourist choice behavior. *Journal of Travel Research*, 58 (3), 370–382.

Weimann, G., 2005. The theater of terror: The psychology of terrorism and the mass media. *Journal of Aggression, Maltreatment and Trauma*, 9 (3–4), 379–390.

Yacoub, S., 2015. IS destroying another ancient archaeological site in Iraq. *ArmyTimes. Associated Press*, 07 March.

Yap, G. and Saha, S., 2013. Do political instability, terrorism, and corruption have deterring effects on tourism development even in the presence of UNESCO heritage? A crosscountry panel estimate. *Tourism Analysis*, 18 (5), 587–599.

Yates, D., 2012. Benghazi treasure. *Trafficking Culture Encyclopedia*. [Online] Available at: https://trafficking culture.org/encyclopedia/case-studies/benghazi-treasure [Accessed March 2019].

3

SUSTAINABLE TOURISM IN CITIES

Jonathon Day

Introduction

Urbanisation has been a defining feature of the past 30 years. Since 1990 the portion of people living in cities has grown from 43% to an expected 56% in 2020 (UN_DESA, 2018) with over two billion more people living in cities. During that same period, international tourism has grown from just over 400 million to over 1.3 billion travellers, with domestic travel considered to account for at least four times that number (UNWTO, 2018). These two important social phenomena have interacted in complex, often complementary ways throughout the years. As the world faces increasing challenges, cities and the tourism industry are working to respond and support sustainable development goals. Achieving sustainability in both cities and tourism systems is a wicked problem, and this chapter applies a systems-thinking approach to addressing the issues facing sustainable tourism in cities.

Growth of cities and urban tourism

We have experienced extraordinary growth in cities during the past 50 years. Today more people live in cities than ever before. Over half the world's population lives in cities. And cities are growing larger and more complex. There are now 33 megacities, cities with more than ten million inhabitants, with another ten cities expected to meet the threshold of megacity by the end of the decade (United Nations, 2018).

Cities face challenges as they grow and adapt to new circumstances. Much of the growth in urbanisation has taken place in developing countries. Cities all around the world face challenges, but many of the problems are most acute in developing-world cities. Cities are at the frontline of many of the twenty-first century's major challenges. Environmental issues are perhaps most acute in cities. Water shortages in Cape Town, South Africa, air pollution in Beijing, and the fight against plastic waste in cities like Seattle are merely some of the more prominent stories of a wide-reaching set of environmental issues facing cities. Cities are confronting the impacts of climate change. Beyond environmental issues, cities are facing challenges with a wide range of economic, social and cultural issues. These range from social challenges like the impacts of poverty and inequality to the loss of heritage and cultural identity. The World Cities (UN-Habitat, 2016) report identifies a number of problems, including the growth of slums and informal settlements, the challenge of providing municipal services, issues of inequality and exclusion, rising insecurity and crime, and climate change. The importance of ensuring that cities are sustainable is a global priority and has been highlighted in Sustainable Development Goal (SDG) 11: "Make cities and human settlements inclusive, safe, resilient and sustainable", one of the 17 Sustainable Development Goals identified by the United Nations.

Cities have always been attractions for visitors. From the earliest times, cities have drawn a wide range of visitors, from business travellers to leisure travellers to pilgrims. Today, cities are growing in importance to tourism. WTTC (2018) notes that the top 300 cities account for approximately 45% of all arrivals and the market share of urban travel is expected to continue to rise. As cities attract new visitors, their tourism growth creates new issues. Although often touted as an economic benefit to cities, tourism comes with a range of benefits and costs for the host destination. Some issues, like "overtourism", have risen to the top of public consciousness in recent years, but tourism presents a range of challenges – and opportunities – for cities. Despite the importance of sustainability and urban tourism, research on the topic is scarce. As early as 1996, it was noted that "despite the emerging literature on sustainable cities, the role of tourism within them is largely ignored" (Hinch, 1996). While there has been some improvement since 1996, much still needs to be done.

Sustainable urban tourism systems

Given this context, it is important that tourism be sustainable. Yet, achieving such an outcome – indeed, even determining what sustainable tourism means for cities – is challenging. This chapter will examine sustainability in urban destinations through several lenses, each of which requires discussion and exploration. Sustainability in cities and sustainable urban tourism are complicated, complex problems. While there has been considerable progress in elements of these problems, there is value in examining these issues from a systems perspective.

Sustainability and sustainable tourism

It is worthwhile defining both sustainability and sustainable tourism. Sustainability, a term used frequently in society, has been acknowledged as a term that is poorly defined and not well understood (Day, 2016). Much of the discussion over the last three decades about sustainability has its beginnings in new approaches to sustainable development initiated in the late 1980s. Sustainable development is commonly defined as "to ensure that (development) meets the needs of the present without compromising the ability of future generations to meet their own needs" (WCED, 1987). A related concept, frequently associated with sustainability, is the "triple-bottom-line" (TBL). The approach, first proposed by Elkington (1997), borrows from accounting and recognises the costs and benefits of action against three broad categories of impact – environmental, social and economic. A third key theme of sustainable development is stakeholder engagement. Freeman (1984) described stakeholders as "any individual or organization who can affect or is affected by the achievement of an organization's objectives" (Freeman, 1984). While Freeman's work focused on corporations, the concept of stakeholder engagement and management has extended into broader society. While these three core themes – focused on present and future use, optimising the TBL and engaging stakeholders in the sustainable development process – are adopted in a wide range of academic and institutional literature, each theme is open to significant variations in interpretation. For example, many researchers and practitioners focus primarily on environmental issues when discussing sustainability. Perhaps as a result, agreement on what sustainability means in practice varies widely.

Definitions of sustainable tourism incorporate the three definitional themes of sustainability. For example, the United Nations World Tourism Organization (UNWTO) defines sustainable tourism as "Tourism that takes full account of its current and future economic, social and environmental impacts, addressing the needs of visitors, the industry, the environment and host communities" (UNWTO, 2005). It is important to note the principles of sustainable tourism are relevant to cities and are not limited to small-scale tourism operations in environmentally sensitive locations. Since the Mohonk Agreement (2000) there has been an explicit understanding that sustainable tourism is a set of principles that can be applied to any tourism organisation or system. Nevertheless, discussions of sustainability in urban tourism have been lacking (Maxim, 2016). Too often discussions of sustainable tourism focus on rural and natural areas, overlooking the sustainability issues associated with urban centres. As a result, little research has focused on sustainable tourism in cities (Maxim, 2016). This knowledge gap must be addressed if tourism is to contribute to more sustainable cities.

Sustainable tourism management: it's complicated

While these definitions may seem straightforward enough, it is important to recognise the range of activities required to implement these principles. Although conceptually the TBL is simple, determining what must be done to achieve optimal outcomes in the social, environmental and economic pillars has proven a challenge. Some are frustrated that there is a gap between "sustainability doctrine and actual achievements" (Ruhanen, 2008). While the idea of sustainable tourism has been widely adopted, many consider the implementation of sustainable tourism programmes to be disappointing (Hall, 2011). While there are many factors contributing to this frustration, one is the need to determine what specifically must be done to achieve sustainable tourism.

There are efforts to move sustainable tourism from a mere philosophical base (Ruhanen, 2008) to a set of activities that are applicable to tourism-related organisations. Operationalising sustainable tourism is complicated; a large number of activities must be managed over an extended period of time. Each TBL pillar requires specific sets of actions and even determining what activities must be undertaken has proven challenging. As noted previously, while there is general consensus on the definition of sustainability, there is variation on how its key themes are interpreted and, as Bricker and Schultz (2011) note, definitions of sustainability have implications for the component activities and related metrics of many programmes. A number of researchers have presented sets of indicators to measure the performance of specific elements of sustainable tourism (Schianetz and Kavanagh, 2008, Torres-Delgado and Palomeque, 2014, Franzoni, 2015). The growth of certifications for sustainable tourism reflects the need for a comprehensive set of sustainable tourism measures to understand how destinations, including cities, perform. The UNWTO is working with governments around the world to develop a set of indicators (2017). Although significant progress has been made at a national level, regional and subnational metrics are still being developed. Of course, the identification of criteria for achieving sustainable tourism and indicators for measuring progress are only the first steps in managing for performance improvement.

The rhetoric of sustainable tourism envisages systematic policies and programmes, with multiple organisations working in unity toward comprehensive goals. Significant focus on the implementation of sustainable tourism has focused on the destination. For example, the Global Sustainable Tourism Council has identified 42 criteria and 108 performance indicators for destinations (GSTC, 2013). These criteria are organised around four pillars, including the three TBL components and management. A handful of destinations have been certified for their comprehensive approach to sustainable tourism management. Regions with towns and cities like Kaikoura, New Zealand and Huatulco, Mexico as well as cities, Vail and Sedona have implemented comprehensive sustainable tourism plans (GSTC, 2019) but these destinations remain the exception rather than the rule. Rather than comprehensive programmes, many sustainable tourism initiatives focus on specific activities – such as reducing waste or buying locally – and sustainable tourism programmes tend to be somewhat piecemeal in their implementation in at least two ways. Sustainable tourism programmes tend to be "isolated and limited" (Maxim, 2016).

As we will address later, the sustainability of a tourism system implies that the various subsystems and individual actors each contribute to the system's sustainability. The focus on destination-level sustainable tourism programmes sometimes overlooks activities that must be undertaken by businesses and other actors. Sustainability cannot be achieved in cities without businesses and other organisations adopting sustainable tourism activities. These actions, often described as "corporate social responsibility", are also comprehensive sets of actions a business may undertake to demonstrate its corporate citizenship. Some of these recommendations for corporate behaviour come from outside tourism, while others are specific to tourism. Influential programmes from outside tourism include the Global Reporting Initiative (GRI), a comprehensive reporting system for corporations that has been adopted by hotel companies (GRI, 2013), and the LEED programme offered by the US Green Building Council. With tourism, GSTC has a set of criteria for tourism enterprises. Finally, individuals in the urban destination must adopt a set of pro-TBL behaviours to contribute to sustainability. Programmes like the Travel Care Code (TCCI, 2019) identify sets of behaviours appropriate for responsible travellers seeking to support sustainability in the host destination. Individuals, businesses and organisations, destinations – each are elements of the tourism system and each must contribute to the sustainability of the city. The systemic nature of sustainable tourism is the topic of the following section.

Tourism systems | urban systems

Recognising the systemic nature of both tourism and cities and applying a systems-thinking approach to understanding them is useful. Systems thinking has been described as "the discipline for seeing wholes. It is a framework for seeing interrelationships rather than things, seeing patterns of change rather than static 'snapshots'" (Senge, 1990). Perhaps more specifically, Arnold and Wade (2015) define systems thinking as "a set of synergistic analytic skills used to improve the capability of identifying and understanding systems, predicting their behaviors, and devising modifications to them to produce desired effects. These skills work together as a system" (Arnold and Wade, 2015). Rather than taking a reductionist approach to a significant problem and breaking it down to solve specific elements of the issue, systems thinking looks at the whole (Baggio, 2013). The application of systems thinking is growing in a variety of fields. For instance, in business – in the years since Senge (1990) popularised systems thinking within a business – it has become common to talk about "eco-systems" for entrepreneurs and emerging industries.

The value of considering tourism as a system is growing in recognition. Although researchers have identified tourism's systemic nature for over 30 years (Leiper, 1990, Mill and Morrison, 2002), it is rare that the implications of tourism system characteristics are explored. Understanding this term and its implications provides important insights into the nature of the challenges faced in cities seeking to achieve sustainable tourism. Meadows describes systems as "a set of things – people, cells, molecules, or whatever – interconnected in such a way that they produce their own pattern of behaviors" (2008). She goes on to say "a system is an interconnected set of elements that is coherently organized in a way that achieves something …

[a] system must consist of three kinds of things: elements, interconnections and a function or purpose" (Meadows, 2008). The tourism system meets these definitional requirements; it is composed of a wide range of actors – organisations and individuals – working through a complex network of relationships to deliver hospitality-related experiences to visitors.

There are several aspects of the nature of systems that are worth noting. First "a system is more than the sum of its parts" (Meadows, 2008). The tourism experience is often perceived as more than the sum of the products and services consumed. For example, destination image researchers refer to the "gestalt" of a destination image. Second, a system "may exhibit adaptive, dynamic, goal seeking, self-preserving, and sometimes evolutionary behavior" (Meadows, 2008). Third, both the tourism system and cities can be described as Complex Adaptive Systems (CAS). A CAS is "a large collection of diverse parts interconnected in a hierarchical manner such as the organization persists and grows over time without centralized control". We will explore the term "complexity" in more detail in the next section. Fourth, CAS tend to be open systems; that is, a CAS is "continually interacting with the external environment, and able to dynamically maintain its integrity and function" (Baggio et al., 2010). Fifth, "systems are dynamic, adapting and evolving over time. The brain, the immune system, an ant colony, and human society are often presented as examples" (Eidelson, 1997). Farrell and Twining-Ward (2005) highlight the importance of understanding tourism as a complex adaptive system as a critical step toward sustainable tourism. Discussing sustainable tourism as a system can provide a new perspective. The saying "can't see the forest for the trees" describes a challenge that actors in the system face. Actors within the system are not necessarily aware of their role in the system or the way their actions impact the performance of the system. Stepping back and considering the functions of the whole system can provide useful perspective for actors in the system.

Complexity in the city and the wicked problem of sustainable tourism

If sustainable tourism programmes are complicated, then urban and tourism systems are best described as complex. Complexity is difficult to define (Johnson, 2007, Heylighen, 2008); nevertheless, Heylighen (2008) provides a summary of commonly identified characteristics:

> Complexity is situated between order and disorder ... complex systems are neither random and chaotic ... nor regular and predictable.... Complex systems consist of many parts that are connected via their interactions ... they are distinct and connected, both autonomous and mutually dependent.
>
> *(Heylighen, 2008, p. 4)*

Heylighen also notes that the activity of one actor may impact other players, and the impacts may be both local and global. Another important characteristic of complexity is that a change in one part of the system may lead to "non-linear" changes in other parts of the system, but their effects may not be proportional to their causes, which make systems difficult to control. The "Butterfly Effect", the idea that a butterfly flapping its wings may cause a hurricane on the other side of the world, describes non-linearity well. Non-linear changes in systems are difficult to anticipate, and there are many examples of unintended consequences resulting from non-linear effects of policy changes.

Cities are complex systems. "Complexity is a function of the number and diversity of the players who are involved" (Conklin, 2005), and cities have a large number of diverse players. So, too, does the tourism system. Several authors (Farrell and Twining-Ward, 2004, Day, 2016) have recognised the significance of complexity in addressing sustainable tourism. Recognising the complexity of the system provides a foundation for analysis (Farsari et al., 2011). The challenge of implementing sustainable tourism practices in a city can best be described as a complex or "wicked" problem. Implementing sustainable tourism shares characteristics with other complex, wicked problems, such as reducing poverty, addressing climate change or reducing opioid usage. Wicked problems share a number of characteristics: "There is no simple solution to a wicked problem. Proposed solutions are neither wholly right nor completely wrong, and every wicked problem is unique" (Morrison et al., 2019). Rittel and Webber (1973) further expand on the characteristics of wicked problems. They note that there is no "stopping rule" for wicked problems; addressing wicked problems is an ongoing endeavour. There is no test to determine the end to a wicked problem, and consequences – both intended and unintended – can extend into the future. There are no black and white answers to wicked problems; solutions are either better or worse or good enough. Even so, every attempt at solving a wicked problem counts; there are no inconsequential trials in working to solve these problems. Every wicked problem is unique, and each solution has various unique characteristics. Every wicked problem can be considered a symptom of another problem. And finally, how the problem is framed will determine its solution (Rittel and Webber, 1973).

Recognising that implementing sustainable tourism in a tourism system is a wicked problem suggests that, while there are lessons to be learned from other's experiences, each city's approaches will be unique. There is no single answer to the challenge. The complexity of sustainable tourism implies that, while it is useful to develop standardised sets of sustainable tourism indicators, the implementation of sustainable tourism is going to be different in each tourism system.

Applying systems thinking to tourism and cities

Although the systems nature of tourism is recognised, the implications are rarely unpacked. It is worthwhile to examine some of the fundamental characteristics of systems and system thinking. In considering urban tourism from a systems perspective, we will examine a number of distinct characteristics including system hierarchies, system management and governance, feedback loops, resilience, adaptation and tipping points.

System hierarchies: A powerful insight from systems thinking is that all systems are composed of other systems, or subsystems. In the body, a cell is a system, within an organ, within an animal, within an ecosystem. Meadows (2008) notes that the world is created by subsystems aggregating into larger systems and that "systems evolve from the bottom up". Several tourism researchers have adopted a hierarchy of systems to address tourism issues. Kline et al. (2013) recognise these hierarchies and note that ecological systems theory defines the hierarchies in terms of microsystems, mesosystems and macrosystems. Similarly, Baggio (2017) highlights three scales – micro, meso and macro – as common units of network analysis. Farrell and Twining-Ward (2004) describe complex adaptive tourism system hierarchy as a tourism "panarchy", composed of a core tourism system, comprehensive tourism system and regional tourism system, which is ultimately embedded in the larger global/earth system. Arnold and Wade (2015), in defining systems thinking, highlight the importance of understanding systems at different scales. It is important to note that much of the conversation about sustainable tourism refers to tourism destinations. Destinations, like the tourism system and cities, are complex adaptive systems (Baggio et al., 2010, Day, 2016).

Applying this approach to sustainability within the tourism system, it becomes clear that effecting change throughout the hierarchy and in a variety of subsystems is necessary. In each city destination, change must occur at a (micro) individual behaviour, at the (meso) business or enterprise level, and in the destination (macro) system as a whole.

It is noteworthy that much of the research described as "sustainable tourism" focuses on destinations. In the same way that we lack a full picture of sustainability because of our focus on specific actions, our understanding of the research undertaken addressing sustainable tourism can be limited by the narrow range of work defined by the term. Research examining sustainable tourism in tourism systems is being done under a variety of themes, and much is not identified as sustainable tourism per se.

This is not to say that significant work has not been undertaken that contributes to sustainability in the tourism system. There is a significant body of research in the environmental performance of hotels and other types of businesses within tourism. There is also a significant research theme addressing the ways in which enterprises and other organisations address social justice issues. Collectively, these studies addressing elements of corporate social responsibility are addressing elements of sustainable tourism at the meso-level of the system.

Studies of individual behaviours as they relate to sustainability must also be considered in terms of the sustainable tourism system as a whole. Indeed, there are at least two streams of behaviour to consider – individual traveller behaviour and organisational behaviour. Individual travellers' pro-sustainable tourism behaviours extend from pro-environmental behaviours to other behaviours that support positive TBL outcomes, such as cultural awareness and conscious consumption. With few exceptions – Miller et al. (2014) examination of pro-environmental behaviours in Melbourne, for example – these studies are rarely considered in the context of their relationship to the urban tourism system as a whole. The adoption of sustainable tourism in organisations and other social structures requires organisational change, and while this is an important field of study in management, sociology and political science, there is still much to learn. Although much of the discussion on sustainable tourism focuses on macro-level issues, it is important not to overlook these micro-level issues. As Meadows notes, systems "evolve from the bottom up" (2008), so we must realise that the hierarchy of systems supporting sustainable tourism is driven by individual (micro-level) concerns for better global outcomes.

Perhaps one of the most important insights from a systems-thinking approach is the recognition that tourism is an embedded system within a larger social system. The tourism destination system is embedded in the larger system of the city itself. As Hunter (1995) notes,

> it is extraordinary that to find in literature the earnest construction of "plans", or "strategies" or "frameworks" which purport to chart a sustainable tourism development in a destination area or wider region, with little or no explicit or explanatory recourse to the actual or potential interactions with other sectors.
>
> *(Hunter, 1995)*

This is a critical issue as many of the services that facilitate sustainable tourism systems are provided by organisations that do not consider themselves part of tourism. For example, tourism systems cannot use renewable energy unless it is available through energy utilities, and recycling at hotels or convention centres requires sanitation services that provide recycling.

A result of the traditional reductionist approach to science is the lack of examination of issues across the hierarchy levels. One exception is a study by Aydin and Emeksiz (2018) that provides success factors for urban sustainable tourism based on the economic performance of 330 small tourism enterprises. Such examination, applying a systems-thinking approach, can expose important paradoxes. For example, at an enterprise level, linen reuse programmes reduce energy and water consumption as well as labour costs, but at the destination level these programmes may equate to fewer jobs and reduced benefits in the community.

Independent actors, "management" and governance in the urban tourism system: Tourism systems are composed of many actors, each with a unique role in the system and each acting independently. An important insight in system governance is that management within the system may require a wide range of techniques. Individual businesses, each a meso-level system within the larger tourism system, may operate on a hierarchal, "command and control" basis. In the simplest sense, in such systems, senior managers may direct actions with staff enacting them based on a variety of direct rewards and punishments to enforce compliance. Of course, even in traditional businesses, implementing change is never as straightforward as this and there are libraries of management books on managing organisational change.

In higher-level systems, other techniques are necessary to steer independent actors toward specific objectives. For example, Destinations require somewhat different forms of organising and steering than hierarchically structured firms. It is for this reason that destination governance, rather than destination management, is the preferred term for many researchers (Baggio et al., 2010, Beaumont and Dredge, 2010). Governance has several noteworthy characteristics that differentiate it from traditional management. Network governance tends to be stakeholder-oriented and participatory rather than top-down management (Volgger and Pechlaner, 2015). Collaboration and cooperation skills are critical in achieving sustainability in tourism systems (Aydin and Emeksiz, 2018). Volgger and Pechlaner (2015) note that there are numerous means of governance in destination networks including budgets, knowledge, trust, themes and brands. Several researchers have identified means of improving sustainable tourism performance, including knowledge management (Ruhanen, 2008), policy development and implementation (Hall, 2011).

As noted previously, much discussion of sustainable tourism focuses on destination systems. There is an assumption in much of the literature that destination marketing organisations will play a major role in implementing sustainable tourism. Although there is evidence of destination management organisation (DMO) involvement in some destinations, there are a wide variety of organisations that may lead in the process. Beaumont and Dredge (2010) explore the effectiveness of three possible structures for destination governance, each with their own strengths and weaknesses. One of the important characteristics of CAS is that they are self-organising. Systems' members form structures – both formal and informal – to address system needs. The growth of tourism industry associations, collaborations and partnerships are all evidence that structures and subsystems are evolving to meet system needs.

The adoption of sustainable tourism practices through the destination system requires systemic change, and change agents must recognise the need to engage with a wide range of actors. By definition, sustainable tourism acknowledges the importance of stakeholder engagement. Appreciating the role of stakeholders, each an actor/agent in the system, is critical to understanding how the system functions. How change agents define the system will significantly impact their effectiveness. For example, we see increasing awareness of the importance of community perceptions of tourism's function, as Zamfir and Corbos (2015) note in their evaluation of sustainable tourism in Bucharest.

Understanding the relationships within complex systems and identifying ways to effectively support the system change to meet sustainability goals is still in early stages. One promising theme in tourism is network analytics, "the analysis of the patterns of connections between the elements of a system" (Baggio, 2017), but it is clear that new tools and techniques will be required to assist tourism researchers improve their understanding of system functioning.

Feedback loops and their impact on urban tourism systems: Two important concepts in systems thinking are stocks and feedback loops. Stocks are the "accumulation of material or information that has built up in a system over time" (Meadows, 2008). Stocks change with flows – both in and out – of the system. It is important to note that stocks may be physical or intangible. For instance, goodwill may be considered a stock. Meadows illustrates this when she states, "a stock is the memory of the history of changing flows in the system" (Meadows, 2008). Stocks often act as buffers or "shock absorbers" in systems. For instance, it may take some time for a decline in cocoa bean production to impact chocolate prices because of the stock – or inventory – of chocolates in factories and stores around the world. A feedback loop is a "closed chain of causal connections from a stock, through a set of decisions or rules or physical laws or actions that are dependent on the stock level, and back again through a flow to change the stock" (Meadows, 2008). Some feedback loops stabilise the level of stocks and stabilise the system. Other feedback loops reinforce change and are self-enhancing. When the change is positive, they can be considered "virtuous cycles", but when the change is negative, they are "doom loops".

The speed and efficiency of feedback within a system is critical to understand many of the challenges with which we are presented in sustainable tourism. Delays may have a significant impact on system performance. With imperfect information about the functioning of the system as a whole, or delayed information from other parts of the system, individual actors overreact to stimuli, creating oscillations in the system (Senge, 1990, Meadows, 2008). Significant challenges in the tourism system can be understood through the lens of these system oscillations.

Adaptation, resilience and tipping points in urban tourism systems: The tourism system demonstrates its adaptive capacity in a variety of ways. In recent years, technology has enabled several disruptive industries and transformed urban destinations. Airbnb has burst the "tourism bubble" and enabled visitors to extend beyond the tourism precinct and into communities. Systems are resilient – until they aren't. Systems, like cities or tourism, are able to respond and adapt to changes, evolving in response to changing circumstances. Holling defines resilience as a "measure of the persistence of systems and of their ability to absorb change and disturbance and still maintain the same relationships between populations or state variables" (Holling, 1973). It is amazing to consider the changes to tourism in recent years, from the impacts of new technologies and products – such as Airbnb and Uber – on existing business models to changing consumer preferences as new generations enter the marketplace. With each change, the tourism system changes and adapts, proving its resilience. Yet, there are always limits to resilience. Understanding the thresholds at which the system can no longer "bounce back" has been a rich area of study in tourism literature. Limits of acceptable change research may be applied to tourism cities (Nasha and Xilai, 2010).

It would be wrong to assume that change in systems is slow. There are times when a relatively small change may cause a rapid change in a system. In recent years, the factors that contribute to moving beyond thresholds or "tipping points" (Gladwell, 2000) have become an important consideration in system change. Gladwell (2000) identifies specific types of relationships within the system, the nature of the message, and the context in which the message is conveyed, which contribute to the likelihood a message will move through a system – like an epidemic moves through a population.

Recognising system characteristics in tourism challenges

Applying systems thinking provides useful insights to tourism researchers and practitioners alike. At a relatively simple level, the boom or bust of hotel development is a clear example of a delayed feedback loop. Even more complex tourism phenomena can be analysed through the prism of systems thinking. Overtourism, an issue that has become part of public discourse in recent years, is only one aspect of the pressures from rising urbanisation and increased tourism. While there are many dimensions to the issue of overtourism, some of them can be understood through a systems lens. Consider the following system-related issues: delayed feedback loops allowed resentment against the growth of tourism to fester. At the same time, reinforcing feedback loops, in which "success breeds success", at the affected tourism destinations or attractions led to extreme demand for these experiences. During this time, the system changed as new players, notably Airbnb, entered and changed the relationships between actors in the system. Throughout this period, cities exhibited a resilience to changes taking place until a threshold was met and a tipping point reached, at which time the cries of overtourism erupted into the international headlines.

Changing the urban tourism system

The challenge of sustainable tourism in cities is not only in the identification of actions required for sustainability, but also in the implementation of sustainable tourism programmes across the system. The value of applying systems thinking to sustainable tourism in urban systems is the ability to use knowledge from other systems to address the current issue. Systems thinking provides some insights into where intervention in the system is likely to make the greatest impact. Once again, Meadows (2008) provides useful recommendations. A partial list, in increasing order of importance, includes establishing standards, reinforcing feedback loops, improving information flows, establishing rules and policies, self-organisation, establishing goals and redefining the purpose of the system, changing paradigms from the which the system arises, and ultimately transcending paradigms (Meadows, 2008). Although complex, systems can be "managed" and influenced to support specific goals.

To date there has been a focus on establishing policy frameworks to require systemic change. The slow implementation and limited effectiveness of approaches, relying on the traditional hierarchical view of management, have frustrated researchers and practitioners who recognise a number of barriers to be overcome, including "political, cultural, economic, social and psychological change" (Dodds and Butler, 2010), for greatest impact. Systems are resilient, and new policies can be resisted for a variety of reasons. Overcoming policy resistance demands a commitment to both collaboration between the stakeholders in the system and commitment to learning and adaptation by policy makers.

Each element of the system brings its own assets: skills and abilities, knowledge and resources, to name a few. Change agents, like those looking to improve sustainability across the system, must consider how these assets may be used, or enhanced, to achieve goals. Examining emerging fields of research, such as application of community-based assets (Dolezal and Burns, 2015) or community capitals (Kline, 2017) to sustainable tourism development, are promising. So too is taking a systems approach to workforce development and education, including training and other capacity building. As noted previously, knowledge management (Ruhanen, 2008) and enhancing knowledge assets within a system are important tasks for sustainable tourism advocates.

It is clear that important skills for working in systems are collaboration and cooperation. Given the independent nature of these systems, finding common tasks in which stakeholders can work together to improve the system will be an important foundation for future success. Understanding how to initiate and support such collaborations will be critical skills for implementing sustainable tourism in urban systems. While there are many examples of such collaborations, the recent UNWTO Mayors Forum for Sustainable Tourism (2019) provides an interesting roadmap for collaboration between cities that are competing for the same consumers. Their agreement includes commitment to contributing to SGD 11 – Make cities and human settlements inclusive, safe, resilient and sustainable – through adopting a common code of ethics, including tourism in wider urban planning, increasing communications between stakeholders, encouraging adoption of sustainable practices, using data to manage and improve sustainable tourism performance, and engaging stakeholders – from residents to mayors – in the sustainable tourism process. These types of commitments utilise many of the techniques known to be effective in implementing change in systems.

Beyond sustainability in the city

While sustainability has been a focus of much research since the 1980s, concern is growing that sustainability is not enough (Pollock, 2015, Day, 2016). Regenerative approaches to development are becoming more widely adopted (Mang and Haggard, 2016), and interest in regenerative approaches to tourism and adoption of circular economy principles in cities are growing. These concepts build on concepts of sustainability but add a proactive, restorative dimension. There is no doubt that many cities face challenges of decline and need regeneration, and tourism may play an important role in addressing these issues (Wise, 2016). Once again, if these new approaches are to be effective, they must be adopted throughout the system, at all levels of the system hierarchy.

The future of sustainable urban tourism

Urbanisation and the growth of tourism have been defining phenomena of the past 30 years. Their development is intertwined. Tourism and cities are complex adaptive systems, adapting and evolving to meet new challenges and a changing environment. As we proceed into the twenty-first century, there is an imperative for both tourism and cities to become more sustainable. While conceptually sustainability and sustainable tourism appear simple, the implementation of comprehensive sustainability plans is complicated, with a range of actions required. Tourism systems are complex, and implementing sustainable tourism is a wicked problem.

To achieve the goals of urban sustainability we must deeply understand each aspect of the process. Researchers must continue to delve into each component of these plans. At the same time, it is critical that researchers and practitioners apply systems-thinking approaches to understanding how sustainability can become standard practice in urban tourism. It is clear that such an approach requires new tools for analysis. Fortunately, research examining new ways to manage systems is emerging (Zheng-Xin and Pei, 2014, Baggio et al., 2010), but great opportunity to learn more remains. In addition, many of our current research themes may benefit from a systems perspective. Sustainable tourism, as a concept, has been a great success, yet there remains hard work to ensure we achieve its promise in our urban tourism systems.

References

Arnold, R. D. and Wade, J. P. 2015. A definition of systems thinking: A systems approach. *Procedia Computer Science*, 44, 669–678.

Aydin, B. and Emeksiz, M. 2018. Sustainable urban tourism success factors and the economic performance of small tourism enterprises. *Asia Pacific Journal of Tourism Research*, 23, 975–988.

Baggio, R. 2013. Oriental and occidental approaches to complex tourism systems. *Tourism Planning & Development*, 10, 1–11.

Baggio, R. 2017. Network science and tourism: The state of the art. *Tourism Review*, 72, 120–131.

Baggio, R., Scott, N. and Cooper, C. 2010. Improving tourism destination governance: A complexity science approach. *Tourism Review of AIEST – International Association of Scientific Experts in Tourism*, 65, 51–60.

Beaumont, N. and Dredge, D. 2010. Local tourism governance: A comparison of three network approaches. *Journal of Sustainable Tourism*, 18, 7–28.

Bricker, K. and Schultz, J. 2011. Sustainable tourism in the USA: A comparative look at global sustainable tourism criteria. *Tourism Recreation Research*, 36, 215–229.

Conklin, J. 2005. Wicked problems and social complexity *In:* J. Conklin (ed.) *Dialogue Mapping: Building Shared Understanding of Wicked Problems*. Chichester, England; Hoboken, NJ: Wiley.

Day, J. 2016. *An Introduction to Sustainable Tourism and Responsible Travel*. West Lafayette, IN: Placemark Solutions.

Dodds, R. and Butler, R. 2010. Barriers to implementing sustainable tourism policy in mass tourism destinations. *Tourismos: An International Multidisciplinary Journal of Tourism*, 5, 35–53.

Dolezal, C. and Burns, P. M. 2015. ABCD to CBT: Asset-based community development's potential for community-based tourism. *Development in Practice*, 25, 133–142.

Eidelson, R. 1997. Complex adaptive systems in the behaviorial and social sciences. *Review of General Psychology*, 1, 42–71.

Elkington, J. 1997. *Cannibals with Forks: The Triple Bottom Line of 21st Century Business*. Oxford: Capstone Publishing.

Farrell, B. and Twining-Ward, L. 2004. Reconceptualizing tourism. *Annals of Tourism Research*, 31, 274–295.

Farrell, B. and Twining-Ward, L. 2005. Seven steps towards sustainability: Tourism in the context of new knowledge. *Journal of Sustainable Tourism*, 13, 109–122.

Farsari, I., Butler, R. W. and Szivas, E. 2011. Complexity in tourism policies: A cognitive mapping approach. *Annals of Tourism Research*, 38, 1110–1134.

Franzoni, S. 2015. Measuring the sustainability performance of the tourism sector. *Tourism Management Perspectives*, 16, 22–27.

Freeman, R. E. (ed.) 1984. *Strategic Management: A Stakeholder Perspective*. Boston, MA; Englewood Cliffs, NJ: Prentice-Hall.

Gladwell, M. 2000. *The Tipping Point: How Little Things Can Make a Big Difference*. Boston, MA: Back Bay Books.

GRI. 2013. *G4 Sustainability Reporting Guidelines: Reporting Principles and Standard Disclosures*. Amsterdam: Global Reporting Initiative.

GSTC. 2013. *Global Sustainable Tourism Council Criteria for Destinations* [Online]. Global Sustainable Tourism Council. Available: www.gstcouncil.org/en/gstc-criteria/criteria-for-destinations.html [Accessed 28 February 2016].

GSTC. 2019. *Certified Sustainable Destinations* [Online]. Global Sustainable Tourism Council. Available: www.gstcouncil.org/certified-sustainable-destinations/ [Accessed].

Hall, C. 2011. Policy learning and policy failure in sustainable tourism governance: From first- and second-order to third-order change? *Journal of Sustainable Tourism*, 19, 649–671.

Heylighen, F. 2008. Complexity and self organization *In:* M. Bates and M. Mack (eds.) *Encyclopedia of Library and Information Sciences*. Boca Raton, FL: Taylor & Francis.

Hinch, T. D. 1996. Urban tourism: Perspectives on sustainability. *Journal of Sustainable Tourism*, 4, 95–110.

Holling, C. S. 1973. Resilience and stability of ecological systems. *Annual Review of Ecology and Systematics*, 4, 1–23.

Hunter, C. J. 1995. On the need to re-conceptualise sustainable tourism development. *Journal of Sustainable Tourism*, 3, 155–165.

Johnson, N. F. 2007. *Two's Company, Three Is Complexity: A Simple Guide to the Science of All Sciences*. Oxford: Oneworld.

Kline, C. 2017. Applying the community capitals framework to the craft heritage trails of western North Carolina. *Journal of Heritage Tourism*, 12, 489–508.

Kline, C., Gard McGehee, N., Paterson, S. and Tsao, J. 2013. Using ecological systems theory and density of acquaintance to explore resident perception of entrepreneurial climate. *Journal of Travel Research*, 52, 294–309.

Leiper, N. 1990. Tourist attractions systems. *Annals of Tourism Research*, 17, 367–384.

Mang, P. and Haggard, B. 2016. *Regenerative Development and Design: A Framework for Evolving Sustainability*. Hoboken, NJ: Wiley.

Maxim, C. 2016. Sustainable tourism implementation in urban areas: A case study of London. *Journal of Sustainable Tourism*, 24, 971–989.

Meadows, D. 2008. *Thinking in Systems: A Primer*, ed. D. Wright. White River Junction, VT: Chelsea Green Publishing.

Mill, R. and Morrison, A. 2002. *The Tourism System.* Dubuque, IA: Kendall/Hunt Publishing Company.

Miller, D., Merrilees, B. and Coghlan, A. 2014. Sustainable urban tourism: Understanding and developing visitor pro-environmental behaviours. *Journal of Sustainable Tourism,* 23, 1–21.

Mohonk Agreement. 2000. *Mohonk Agreement: Proposal for an International Certification Program for Sustainable Tourism and Ecotourism* [Online]. The International Ecotourism Society. Available: www.rainforest-alliance.org/tourism/documents/mohonk.pdf [Accessed].

Morrison, E., Hutcheson, S., Nilsen, E., Fadden, J. and Franklin, N. 2019. *Strategic Doing: Ten Skills for Agile Leadership.* Hoboken, NJ: John Wiley and Sons.

Nasha, Z. and Xilai, Z. 2010. Conceptual framework of tourism carrying capacity for a tourism city: Experiences from national parks in the United States. *Chinese Journal of Population Resources and Environment,* 8, 88–92.

Pollock, A. 2015. *Social Entrepreneurship in Tourism: The Conscious Travel Approach.* TIPSE – Tourism Innovation Partnership for Social Entrepreneurship.

Rittel, H. and Webber, M. 1973. Dilemmas in a general theory of planning. *Integrating Knowledge and Practice to Advance Human Dignity,* 4, 155–169.

Ruhanen, L. 2008. Progressing the sustainability debate: A knowledge management approach to sustainable tourism planning. *Current Issues in Tourism,* 11, 429–455.

Schianetz, K. and Kavanagh, L. 2008. Sustainability indicators for tourism destinations: A complex adaptive systems approach using systematic indicator systems. *Journal of Sustainable Tourism,* 16, 601–628.

Senge, P. 1990. *The Fifth Discipline: The Art and Science of the Learning Organization.* New York: Currency – Doubleday.

TCCI. 2019. *Travel Care Code* [Online]. Travel Care Code Initiative. Available: http://travelcarecode.org/ [Accessed].

Torres-Delgado, A. and Palomeque, F. L. 2014. Measuring sustainable tourism at the municipal level. *Annals of Tourism Research,* 49, 122–137.

UN_Desa. 2018. *World Urbanization Prospects: The 2018 Revision,* custom data acquired via website. [Online]. United Nations, Department of Economic and Social Affairs, Population Division. Available: https://population.un.org/wup [Accessed].

UN-Habitat. 2016. *World Cities Report 2016: Urbanization and Development: Emerging Futures,* abridged edition. UN-Habitat.

United Nations. 2018. *The World's Cities in 2018: Data Booklet.* United Nations.

UNWTO. 2005. *Making Tourism More Sustainable: A Guide for Policy Makers.* Madrid: UNEP UNWTO.

UNWTO. 2017. Measuring sustainable tourism: A call to action. *In:* UNWTO, ed. *6th International Conference on Tourism Statistics 2017 Manilla, Philippines.* UNWTO.

UNWTO. 2018. *UNWTO Tourism Highlights,* 2018 edition. Madrid: UNWTO.

UNWTO Mayors Forum for Sustainable Urban Tourism. 2019. *Lisbon Declaration on Cities for All: Building Cities for Citizens and Visitors.* Lisbon: UNWTO Mayors Forum for Sustainable Urban Tourism.

Volgger, M. and Pechlaner, H. 2015. Governing networks in tourism: What have we achieved, what is still to be done and learned? *Tourism Review,* 70, 298–312.

WCED. 1987. *Our Common Future.* Oxford: Oxford University Press.

Wise, N. 2016. Outlining triple bottom line contexts in urban tourism regeneration. *Cities,* 53, 30–34.

WTTC. 2018. *Travel and Tourism: City Travel and Tourism Impact 2018.* London: World Travel and Tourism Council.

Zamfir, A. and Corbos, R. 2015. Towards sustainable tourism development in urban areas: Case study on Bucharest as tourist destination. *Sustainability,* 7, 12709–12722.

Zheng-Xin, W. and Pei, L. 2014. A systems thinking-based grey model for sustainability evaluation of urban tourism. *Kybernetes,* 43, 462–479.

4

THE SHARING ECONOMY IN TOURISM CITIES

Amanda Belarmino

Introduction

The sharing economy is currently the single largest disrupter traditional tourist firms have ever experienced (Sundararajan, 2016). Two segments of the sharing economy, in particular, are impacting tourism cities: peer-to-peer (P2P) accommodation (e.g. Airbnb) (Belarmino & Koh, 2020) and ridesharing (e.g. Uber) (Schneider, 2017). Through an examination of current academic literature as well as examination of recent popular press accounts of changes and challenges to the sharing economy, this chapter will provide a summary of the history of the sharing economy in tourism cities including a timeline and a series of mini-case studies reflecting on the impact of the sharing economy in different parts of the world. Through an examination of the legal challenges and changes in regulations, this chapter will provide a series of recommendations for policy makers.

In addition to describing the impact of peer-to-peer accommodations and ridesharing, this chapter also includes a brief examination of new segments of the sharing economy that may impact tourism. This chapter will examine the nascent food-sharing industry dominated by companies like Eatwith (Seright, 2018) and examine the potential impact to restaurants from sharing economy food delivery apps like Uber Eats and Postmates (Cho et al., 2019). In addition to food, this chapter will also examine the influence of crowdfunding on hospitality entrepreneurs as well as the impact of emerging niche markets on hospitality and tourism. This chapter proposes ideas for future research related to the sharing economy in tourism cities.

The sharing economy has profoundly changed the way people around the world are consuming goods and services (Sundararajan, 2016). In the hospitality industry, P2P accommodations (Belarmino & Koh, 2020) and ridesharing have become multi-million dollar business (Schneider, 2017). The concept of the sharing economy has been related back to game theory and the tragedy of the commons. The tragedy of the commons is the concept that when individuals act in their own self-interest, they will deplete a commonly used item and it will no longer be available for anyone to use (Binmore, 2007). Botsman and Rogers' (2010) book popularised the term "collaborative consumption" to describe the concept of sharing access to goods rather than trying to purchase them.

The rise of the sharing economy coincided with the Great Recession, but the concept of the sharing economy as it exists today, namely a web platform allowing users to sell, exchange or

Figure 4.1 Timeline of sharing economy development.

share items, dates back to the founding of eBay in 1995 (eBay, 2019) (Figure 4.1). The convenience and efficiency of the website has been shown to have a significant impact on the intention to continue to sell on this platform (Murphy & Liao, 2013) and can be attributed to the success of this first web-based sharing economy resource. In 1999, Napster was the first music sharing website; it allowed users to share files for free (Scharf, 2011) and in turn revolutionised not only music (e.g. the rise of digital music) but the ability of users to share files directly (Green, 2002). P2P accommodations began in 2004 with HomeAway and Airbnb was founded in 2008. The founding of Uber in 2009 led the beginning of ridesharing. These two phenomena have had the most impact on hospitality and tourism of the different segments of the sharing economy.

Peer-to-peer accommodations

The impact of P2P accommodations on cities has been vast and varied. Researchers have examined a variety of topics related to P2P accommodations including the impact of P2P accommodations on affordable housing (Barron et al., 2018), legal issues (McNamara, 2015) and taxation (Guttentag, 2015) as well as the economic impact of these accommodations on individual markets (Airbnb, 2014). While the impact on traditional lodging options, namely hotels, has been investigated by multiple researchers (Zervas et al., 2017; Blal et al., 2018), this impact has been difficult to measure in large part due to the lack of publicly available data on P2P accommodations and the difficulty in applying traditional hotel metrics (i.e. occupancy, ADR and RevPAR) to a market where many owners only have one accommodation and where lower occupancy may be desirable for work/life balance (Lane & Woodworth, 2016). Researchers agree that P2P accommodations are a lodging disrupter (Guttentag & Smith, 2017). In certain markets, like Barcelona and Budapest, P2P accommodations are clustered around the city centre and tourist attractions (Gutiérrez et al., 2017; Boros et al., 2018), while in other markets, like Paris, they are often located in the residential parts of the city as well (Heo et al., 2019). Indeed, in some locations the P2P accommodations are actually closer to the tourist attractions than the hotels (Eugenio-Martin et al., 2019).

The economic impact of P2P accommodations has been hard to measure due to the micro entrepreneurial nature of the business; however, both industry and academic researchers have endeavoured to measure the impact of P2P accommodations on tourism cities. Airbnb's studies have found that P2P accommodations tend to have longer lengths of stay (2.1 times as long); additionally, 42% of guest spending is done in the neighbourhood (Airbnb, 2014). Economic impact studies undertaken by Airbnb have estimated the direct spending to be in millions with an estimated economic impact on Berlin for 2012–2013 at €100 million and for San Francisco at $56 million (Airbnb, 2014). However, the United States Economic Policy Institute's study found that only 2%–4% of the travellers indicated that their stay at a P2P accommodation was a new trip; the rest replaced a stay at a hotel with a stay at a P2P accommodation (Bivens, 2019).

Researchers have investigated the relationship between affordable housing and P2P accommodation rentals; in fact, ten articles on the subject were published between 2010 and 2017

(Belarmino & Koh, 2020). The majority of the studies support the hypothesis that P2P accommodations are correlated with increased rental rates (Horn & Merante, 2017) and housing prices (Barron et al., 2018). For example, an increase in P2P accommodations in Boston correlated with an increase in rent (Horn & Merante, 2017). A three-year study of P2P accommodations in New York City found that P2P accommodations in that city are a form of gentrification without the normal benefits of property re-development inherent in traditional forms of gentrification (Wachsmuth & Weisler, 2018). In Los Angeles, apartment buildings in several highly sought neighbourhoods have been converted into P2P accommodations and while P2P accommodations alone are not responsible for the housing crisis in that city, they have added to the problem (Lee, 2016). When it comes to the price of homes, a 1% increase in Airbnb listings has been found to be correlated with a 0.026% increase in home prices in the United States (Barron et al., 2018).

However, the impact on affordable housing does not seem to be consistent across all markets. A study in the United States of the impact of P2P accommodations on rental rates in New Orleans did not find a significant relationship between P2P accommodation presence and rental rates (Levendis & Dicle, 2016). The cost of housing, both rental costs and mortgage, have been found to have a significant impact on the number of P2P accommodations available in cities in the United States (Yang & Mao, 2019). A study conducted by the United States government suggested that P2P accommodations may be a solution to middle-class stagnation by assisting homeowners with additional income to pay their mortgage (Sperling, 2015). Jefferson-Jones (2015) argues that when homeowners are allowed to rent their space as P2P accommodations, it actually provides them with an income that allows them to maintain their homes and, in the long term, prevent foreclosure, both of which will increase property values for the P2P accommodation owner and their neighbours.

In most jurisdictions, when P2P accommodations began, there were no laws which directly regulated P2P accommodations. As such, the spread of P2P accommodations has led to increased laws related to a variety of issues including health and safety regulations, licenses and zoning. The lodging industry has been particularly vocal in its criticism of the lack of health safety standards for P2P accommodations especially those concerning fire and life safety (McNamara, 2015). Many municipalities have argued that P2P accommodation owners should adhere to the same standards as owners of bed and breakfasts (AAP, 2017). The most common response has been to create licenses for P2P accommodations and fine properties that do not have a license, and many cities have reduced the number of listings through restricted licenses (Leshinsky & Schatz, 2018). When licensing has been implemented, Airbnb has responded by delisting properties not in compliance and refunding travellers, most notably in Tokyo (Deahl, 2018). Finally, in response to both issues with affordable housing (Weiser & Goodman, 2019) and issues arising from guest behaviour (Sokolowsky, 2019), many municipalities have responded by implementing new zoning restrictions for P2P accommodations (Leshinsky & Schatz, 2018). For example, in New Orleans the city has restricted the number of P2P accommodations allowed in non-residential areas and requires the owners to live on site (Stein, 2019).

A final challenge related to P2P accommodations for tourist cities arises from taxation. When P2P accommodations first emerged as a new lodging sector, occupancy taxes were not imposed on them like they were on hotels (Guttentag, 2015). In the United States, individual owners are required to claim the income on their taxes (Airbnb, 2019); however, each city and state has had to pass new laws regarding charging and collecting occupancy tax (Airbnb, 2019). As each municipality has passed taxation laws, Airbnb has facilitated collection of occupancy taxes; for example, in 2018 Airbnb collected $15.3 million in occupancy tax revenue just for the state of Texas (Cargo, 2019). HomeAway, while not as proactive as Airbnb, also collects occupancy on their websites (Sokolowsky, 2018). The European Union does not currently have laws that govern the sharing economy and each country has its own laws regarding occupancy tax (Pantazatou, 2018). Australian researchers have proposed the necessity of taxing P2P accommodations in that country (Gogarty & Griggs, 2018). While most Asian countries have taken the approach of regulating or outlawing P2P accommodations, Japan has partnered with Airbnb for the Rugby World Cup for accommodations and has agreed on taxation as well (Sugiura, 2019).

Despite these issues, consumer interest in P2P accommodations continues to increase. In large part, the interaction with the host and successful monetisation of trust have been credited with the rise of P2P accommodations (Belarmino et al., 2019). As the P2P accommodation segment has grown, the role of user-generated content (UGC) has evolved. The interactions on several P2P sites utilise social media sites; Airbnb requests that the users log-in with their Facebook account (Airbnb, 2020) and HomeAway allows travellers to sign-up using their Facebook log-in (HomeAway, 2020). Both these sites utilise Facebook for promotions and advertising as well as interaction with guests and hosts. This reflects the trend noted by Xiang and Gretzel (2010) towards traveller reliance on social media for travel information. In fact, Yannopoulou, Moufahim, and Bian (2013) noted that the emergence of Airbnb and the free P2P accommodation site Couchsurfing are examples of user-generated brands, created by both the company and the UGC. The UGC for these brands reflects the access to the private sphere and a level of authenticity unavailable in other types of accommodations (Yannopoulou et al., 2013). One of the unique aspects of P2P sites is the reliance on UGC. Unlike hotel websites which include accolades from professional organisations, i.e. JD Powers, P2P sites do not include this type of information. This reflects the consumer trend towards higher trust in UGC than in expert reviews (Zhang et al., 2010).

An added level of complexity in the P2P environment is the ability of hosts to review their guests. Initiated by Airbnb (Porges, 2014), HomeAway adopted the same policy in 2016 (HomeAway, 2020). The intent of the reviews is to ensure that hosts are fulfilling guest expectations and guests are behaving properly (Porges, 2014). Airbnb changed their system to simultaneous reviews, meaning that neither review was posted until both guest and host had written a review (Porges, 2014). This system has led to few negative reviews; however, neutral or lacklustre reviews can signal a negative experience (Porges, 2014). In the research done by Fradkin and colleagues (2016), using Airbnb review data, over 70% of the sample size gave a five-star rating but the content was not consistently positive.

One of the controversies regarding UGC, however, has been its credibility (Ayeh et al., 2013). For P2P accommodations, UGC can only be left on its website if the user is a registered user and has a confirmed stay (Airbnb, 2020; HomeAway, 2020). This gives the UGC on P2P sites a credibility lacking on third-party sites, although third-party sites do include the ability to review P2P accommodations. Clearly, understanding the language of online reviews for P2P sites has a certain level of complexity. Unlike hotel reviews, the star rating cannot be used to measure the experience (Fradkin et al., 2016), and yet the restricted access to writing the reviews adds a level of authenticity to P2P sites lacking on third-party sites (Porges, 2014).

Ridesharing

While the P2P accommodation segment is the part of the sharing economy which most directly competes with a traditional hospitality segment, ridesharing is no less impactful on the tourist experience. Ridesharing is defined as individual car owners using their private vehicles to provide a taxi service which is ordered through a mobile app (Cramer & Krueger, 2016).

Passengers have the ability to ride in the car alone or to carpool with other riders (Crammer & Krueger, 2016). Consumers choose ridesharing over traditional taxi services because it is easier to use and the price is displayed before the passenger accepts the trip (Young & Farber, 2019).

Another key difference between these two aspects of the sharing economy is the role online reviews play in the relationships (Tham, 2016). While ridesharing apps allow for a simultaneous review system similar to that on P2P accommodation websites (Wan et al., 2016), the effect of these reviews has not been thoroughly examined in academic research. However, the companies use the reviews. Uber drivers with an average rating below 4.6 are in danger of losing their job, and customers often feel a sense of guilt about giving riders less than five stars (Miller, 2019), echoing the same rate inflation seen in P2P accommodations but with very different consequences.

Documented benefits of ridesharing range from having a higher capacity utilisation than taxi cabs (Cramer & Krueger, 2016) to convenience and low price (Young & Farber, 2019). Ridesharing has been shown to be more efficient than taxi cabs; riders are more willing to share an Uber or Lyft ride with another passenger than they are to share a taxi (Cramer & Krueger, 2016). Uber has been cited as changing the way people travel, making them more adventurous in foreign countries where they do not speak the language (Dickinson, 2018). While hailing a cab is easy in large cities like New York City, it is much more difficult in most other cities in the United States; ridesharing allows travellers to be more adventurous while on vacation because of how easy it is to get a ride to and from their destination (Clampet, 2014). Ridesharing allows passengers to know the price before they accept the ride, taking out the ambiguous pricing that had previously surrounded trips in taxi cabs (Dickinson, 2018). Additionally, the availability of ridesharing apps has decreased drunk driving (Wang et al., 2017).

Ridesharing has not been without its controversies. Taxation issues have been explored, in particular, issues arising from whether or not drivers are employees and therefore subject to payroll tax (Heinemann & Shume, 2015). Uber drivers have been accused of a variety of assaults along with racist and homophobic behaviour (Nuzzi, 2017). However, the drivers are also often victims of attacks, as are taxi drivers (Keister, 2016). Safety and the impact of risk perceptions are an under-explored topic regarding ridesharing in academic literature. Finally, the impact on tourist cities is under-researched although highlighted in the profiles of three cities that demonstrate the array of impacts of the sharing economy on tourist cities: Seoul, Paris and Las Vegas.

Sharing economy profiles

Seoul, South Korea

Seoul, South Korea is an example of a city whose government has fully embraced the sharing economy (Johnson, 2014). Starting 2012, Seoul has sought to implement legislation which promotes sharing (Bernardi & Diamantini, 2018). These include 77 distinct projects involved in sharing which range from car sharing services, parking lots, tourism and food (Bernardi & Diamantini, 2018). In part as a result of Seoul's openness to sharing, P2P accommodations have been thriving in this market with 2.9 million visitors using Airbnb throughout South Korea in 2018 (Shin & Minu, 2019). However, proponents of P2P accommodations have argued that growth in South Korea will slow unless laws are changed to allow Koreans to rent P2P accommodations in the large cities; as of 2019, locals were restricted to renting them in small cities and rural areas (Shin & Minu, 2019). Uber, however, has faced more difficulty breaking into this market. After their initial entry in 2013, they were barred because of the Passenger Transport Service Act which makes it illegal to charge someone for transportation in a private car (Park,

2019). However, in 2018 Uber partnered with International Taxi, a company in Seoul which provides multi-lingual taxi drivers for foreign visitors; travellers can now book this service through their Uber app (Park, 2019).

Paris, France

Paris has been host to sharing economy-style services for decades. Companies like ParisAddress have been renting out entire apartments as short-term rentals to travellers for over 50 years (Schechner & Verbergt, 2015); however, the onset of the digital sharing economy has disrupted the established sharing economy. An examination of the Paris P2P accommodation found that growth and seasonality for P2P accommodations and hotels are dramatically different and do not attempt to be in direct competition (Heo et al., 2019). Ridesharing companies, however, met with violent protests. In 2016, taxi drivers held a protest across all of France which included blocking traffic and hurling tires into moving vehicles (Tasch & Slater-Robins, 2016).

Las Vegas, Nevada, United States

Las Vegas has emerged as a battleground between traditional tourism industries and the sharing economy. Part of the uniqueness of this location is not only that its primary industry is tourism but that the main attraction is the hotels themselves. The sharing economy has become something of a battleground in this city. The emergence of ridesharing in the city has been embraced by the casino-hotels on the Las Vegas Strip and the local airport, both of which have designated areas for ridesharing (Contreras & Paz, 2018). Ridesharing has been found to have a significant, negative impact on taxi ridership in Las Vegas (Contreras & Paz, 2018). However, the well-established taxi industry has responded by strategically boycotting hotels in an effort to get the casino-hotels to force the ridesharing companies off the Strip (Ventura, 2018). On the lodging side, P2P accommodations went essentially unregulated until 2018. While condominium developments connected to casino-hotels like the Palms Place and MGM Signature (Elwood, 2017) originally promoted their owners utilising short-term rentals during the recession, the lack of growth in major hotel metrics like occupancy and ADR led the major companies to support regulating P2P accommodations. Laws passed in 2018 require that only primary residences can be posted on P2P accommodation sites (Wilson, 2019); however, a recent shop of listings on Airbnb indicate that there are still listings on the Las Vegas Strip which are most likely not the primary residence of the owners.

Emerging industries

Within the sharing economy, there are several emergent industries which can potentially impact the hospitality industry and the way consumers interact with tourist cities. First, innovations in crowdfunding have had an impact on new hospitality businesses (Honisch et al., 2019). Restaurants have been able to successfully finance their openings through crowdfunding sources, and researchers have found that the quality of the video pitch, brief textual description, team composition considerations, appropriate funding targets, shorter funding duration, specific rewards and campaigns with broad reaching advertising are the most successful (Honisch et al., 2019).

Additionally, there have been emergent sharing economy food and beverage services which can impact hospitality and tourism. Third-party food apps such as DoorDash, GrubHub, Postmates and UberEats have changed the way consumers utilise restaurants (Roh & Park, 2019). Preliminary studies examining consumers in China (Cho et al., 2019), Jordan (Alalwan, 2020) and South Korea (Roh & Park, 2019) have found that price, trustworthiness, food choices, app design and convenience as well as moral obligation to make dinner all impacted behavioural intention in this emergent technology.

Finally, emergent sharing economy segments in other industries may impact the way consumers experience hospitality and tourist products. Craft beer exchanges have recently emerged in the United States as a way for craft beer enthusiasts to trade beers they are unable to purchase due to limited distribution by small producers (Whalen et al., 2019). These exchanges allow consumers to try different beers and may thusly influence both their travel planning behaviour and consumption behaviour when they do travel (Whalen et al., 2019). Sharing office space companies, like WeWork, also have the potential to impact hospitality. WeWork's spin-off, WeLive, offers pay-by-the-month living accommodations as well as hotel space (Kessler, 2016), yet this segment has yet to be explored by tourism and hospitality researchers.

Conclusion

The sharing economy has had a profound impact on the way we live our lives. For tourism cities, this impact has been felt by providing increased access to tourism and increasing income for residents. However, this has not come without a price. Dredge and Gyimóthy (2015) called for a balanced perspective of the impact of the sharing economy; the sharing economy should not simply strip away all regulations put in place to protect both businesses and consumers (Dredge & Gyimóthy, 2015) but help services that are currently unmet in the traditional economy.

Future researchers are encouraged to examine the multi-faceted impacts of the sharing economy. Consumer behaviour and antecedents of choice related to P2P accommodations have been well researched in academic literature (Belarmino & Koh, 2020). However, the reasons behind these choices and the impact of hotel design decisions have not been explored. When Marriott choose to eliminate bathtubs from most of their properties because of injuries to employees and guests (Watkins, 2016), were they in fact eliminating an amenity that was desirable to travellers with small children? Home benefits are one of the most often mentioned antecedents of behavioural intention for choosing a P2P accommodation; are these consumers actually families who feel disenfranchised by traditional hotel accommodations? While most of the focus has been on consumers, only a few studies have measured the motivations of hosts. What are the impacts of hosting on the homeowners, both positive and negative? Additionally, the literature has yet to examine the impact of P2P accommodations on cities hosting mega-events. Airbnb was the official alternative accommodation provider for the Summer Olympics in Rio de Janeiro and helped the city prevent overbuilding (Ting, 2016). Researchers are encouraged to investigate if these accommodations can help cities who plan on hosting events prevent overbuilding hotel rooms.

In terms of ridesharing, the impact on tourism and tourist cities is under-researched. Most of the academic research regarding ridesharing has focused on the antecedents of the drivers to participate in ridesharing or the technology used in these services (Standing et al., 2019). Researchers are encouraged to examine the impact of ridesharing on consumer behaviour in tourist cities. In particular, how does use of ridesharing vary by gender and age? Are consumers more adventurous than they would have been otherwise? Researchers are encouraged to investigate the impact of ridesharing on destination choice and choice of tourist activities.

Finally, while issues of regulation and taxation have been investigated (Belarmino & Koh, 2020), their impacts on consumers and consumers' image of sharing economy brands has not been researched. Issues of brand loyalty have yet to be investigated, namely what happens when a consumer loyal to a sharing economy brand can no longer access that brand in certain destinations? How does regulation impact consumer behaviour? The assumption would be that the consumer would switch to a traditional economy company but that may not be the case. Researchers are also encouraged to understand the role regulations play in minimising risk perception on consumers in the sharing economy. Finally, researchers are encouraged to examine the effect regulations have on the suppliers of sharing economy services and the impact changing regulations have on their behavioural intentions to continue to be a supplier in the sharing economy.

References

AAP (2017). NSW to regulate Airbnb and homeshare sites. *SBS News*. Retrieved from: www.sbs.com.au/news/nsw-to-regulate-airbnb-and-homeshare-sites. Accessed 15 October 2019.

Airbnb (2014). The economic impacts of home sharing around the world. *Airbnb*. Retrieved from: www.airbnb.com/economic-impact. Accessed 15 October 2019.

Airbnb (2019). In what areas is occupancy tax collection and remittance by Airbnb available? *Airbnb*. Retrieved from: www.airbnb.com/help/article/2509/in-what-areas-is-occupancy-tax-collection-and-remittance-by-airbnb-available. Accessed 14 October 2019.

Airbnb (2020). Login. www.airbnb.com/login?redirect_url=%2Finbox www.airbnb.com/login?redirect_url=%2Finbox. Accessed 25 April 2020.

Alalwan, A. A. (2020). Mobile food ordering apps: An empirical study of the factors affecting customer e-satisfaction and continued intention to reuse. *International Journal of Information Management, 50*, 28–44.

Ayeh, J., Au, N., & Law, R. (2013). "Do we believe in TripAdvisor?" Examining credibility perceptions and online travelers' attitude toward using user-generated content. *Journal of Travel Research, 52*(4), 437–452.

Barron, K., Kung, E., & Proserpio, D. (2018). The sharing economy and housing affordability: Evidence from Airbnb. Retrieved from: https://papers.ssrn.com/sol3/papers.cfm?abstract_id=3006832.

Belarmino, A., & Koh, Y. (2020). A critical review of research regarding peer-to-peer accommodations. *International Journal of Hospitality Management, 84*, 102315.

Belarmino, A., Whalen, E., Koh, Y., & Bowen, J. T. (2019). Comparing guests' key attributes of peer-to-peer accommodations and hotels: Mixed-methods approach. *Current Issues in Tourism, 22*(1), 1–7.

Bernardi, M., & Diamantini, D. (2018). Shaping the sharing city: An exploratory study on Seoul and Milan. *Journal of Cleaner Production, 203*, 30–42.

Binmore, K. (2007). *Game theory: A very short introduction*. Oxford: Oxford University Press.

Bivens, J. (2019). The economic costs and benefits of Airbnb. *Economic Policy Institute*. Retrieved from: www.epi.org/publication/the-economic-costs-and-benefits-of-airbnb-no-reason-for-local-policymakers-to-let-airbnb-bypass-tax-or-regulatory-obligations/. Accessed 14 October 2019.

Blal, I., Singal, M., & Templin, J. (2018). Airbnb's effect on hotel sales growth. *International Journal of Hospitality Management, 73*, 85–92.

Boros, L., Dudás, G., Kovalcsik, T., Papp, S., & Vida, G. (2018). Airbnb in Budapest: Analysing spatial patterns and room rates of hotels and peer-to-peer accommodations. *GeoJournal of Tourism and Geosites, 21*(1), 26–38.

Botsman, R., & Rogers, R. (2010). *What's mine is yours: The rise of collaborative consumption*. New York: Harper Business.

Cargo, K. (2019). City of Corpus Christi to start collecting hotel occupancy taxes from Airbnb, Homeaway.com. *Caller Times*. Retrieved from: www.caller.com/story/news/local/2019/10/15/city-corpus-christi-now-collect-taxes-airbnb-rentals/3985806002/. Accessed 16 October 2019.

Cho, M., Bonn, M. A., & Li, J. J. (2019). Differences in perceptions about food delivery apps between single-person and multi-person households. *International Journal of Hospitality Management, 77*, 108–116.

Clampet, J. (2014). 6 ways Uber has changed how we get from Point A to Point B. *Skift*. Retrieved from: https://skift.com/2014/08/04/6-ways-uber-has-changed-how-we-get-from-point-a-to-point-b/. Accessed 19 October 2019.

Contreras, S. D., & Paz, A. (2018). The effects of ride-hailing companies on the taxicab industry in Las Vegas, Nevada. *Transportation Research Part A: Policy and Practice, 115*, 63–70.

Cramer, J., & Krueger, A. B. (2016). Disruptive change in the taxi business: The case of Uber. *American Economic Review*, *106*(5), 177–182.

Deahl, D. (2018). Airbnb cancels bookings under new Japan law. *The Verge*. Retrieved from: www.theverge.com/2018/6/8/17442230/airbnb-cancels-bookings-under-new-japan-law. Accessed 18 October 2019.

Dickinson, G. (2018). From Ubercopters to Uberboats: How the ride-sharing app changed the way we travel. *The Telegraph*. Retrieved from: www.telegraph.co.uk/travel/comment/where-can-you-get-uber-overseas-map/. Accessed 16 October 2019.

Dredge, D., & Gyimóthy, S. (2015). The collaborative economy and tourism: Critical perspectives, questionable claims and silenced voices. *Tourism Recreation Research*, *40*(3), 286–302.

eBay (2019). Who we are. *eBay*. Retrieved from: www.ebayinc.com/our-company/who-we-are/. Accessed 14 October 2019.

Elwood, M. (2017). Why Airbnb can't crack Las Vegas. *Conde Nast*. Retrieved from: www.cntraveler.com/story/why-airbnb-cant-crack-las-vegas. Accessed 16 October 2019.

Eugenio-Martin, J. L., Cazorla-Artiles, J. M., & González-Martel, C. (2019). On the determinants of Airbnb location and its spatial distribution. *Tourism Economics*, 1354816618825415.

Fradkin, A., Grewal, E., Holtz, D., & Pearson, M. (2016). Bias and reciprocity in online reviews: Evidence from field experiments on Airbnb. MIT Sloan School of Management.

Gogarty, B., & Griggs, L. (2018). Collaborative consumption, AirBNB and land tax-time for a national (Australian) approach. *Curtin Law and Taxation Review – Consumer Law Special Issue*, *4*.

Green, M. (2002). Napster opens pandora's box: Examining how file-sharing services threaten the enforcement of copyright on the Internet. *Ohio State Law Journal*, *63*, 799–831.

Gutiérrez, J., García-Palomares, J. C., Romanillos, G., & Salas-Olmedo, M. H. (2017). The eruption of Airbnb in tourist cities: Comparing spatial patterns of hotels and peer-to-peer accommodation in Barcelona. *Tourism Management*, *62*, 278–291.

Guttentag, D. (2015). Airbnb: Disruptive innovation and the rise of an informal tourism accommodation sector. *Current Issues in Tourism*, *18*(12), 1192–1217.

Guttentag, D. A., & Smith, S. L. (2017). Assessing Airbnb as a disruptive innovation relative to hotels: Substitution and comparative performance expectations. *International Journal of Hospitality Management*, *64*, 1–10.

Heinemann, F., & Shume, M. (2015). Uber, Airbnb, Netflix … Australia's steps to tax the sharing and digital economies. *TPIIT*, *13*, 13–15.

Heo, C. Y., Blal, I., & Choi, M. (2019). What is happening in Paris? Airbnb, hotels, and the Parisian market: A case study. *Tourism Management*, *70*, 78–88.

HomeAway (2020). Traveler login. www.homeaway.com/. Accessed 25 April 2020.

Honisch, E., Harrington, R. J., & Ottenbacher, M. C. (2019). Crowdfunding: Preparation considerations and success factors for the German restaurant sector. *International Journal of Hospitality & Tourism Administration*, *20*(2), 182–205.

Horn, K., & Merante, M. (2017). Is home sharing driving up rents? Evidence from Airbnb in Boston. *Journal of Housing Economics*, *38*, 14–24.

Jefferson-Jones, J. (2015). Can short-term rental arrangements increase home values? A case for Airbnb and other home sharing arrangements. *The Cornell Real Estate Review*, *13*(1), 12–19.

Johnson, C. (2014). Sharing city Seoul: A model for the world. *Shareable*. Retrieved from: www.shareable.net/sharing-city-seoul-a-model-for-the-world/. Accessed 21 October 2019.

Keister, A. (2016). The safety of taxi and rideshare drivers – Part 1: Heightened concerns. *On Labor*. Retrieved from: https://onlabor.org/the-safety-of-taxi-and-rideshare-drivers-part-1-heightened-concerns/. Accessed 15 October 2019.

Kessler, S. (2016). WeWork quietly opens hotel rooms in New York City. *Fast Company*. Retrieved from: www.fastcompany.com/3060537/wework-quietly-opens-hotel-rooms-in-new-york-city. Accessed 20 October 2019.

Lane, J., & Woodworth, R. M. (2016). The sharing economy checks in: An analysis of Airbnb in the United States. *CBRE Hotel's Americas Research*.

Lee, D. (2016). How Airbnb short-term rentals exacerbate Los Angeles' affordable housing crisis: Analysis and policy recommendations. *Harvard Law & Policy Review*, *10*(1), 229–253.

Leshinsky, R., & Schatz, L. (2018). "I don't think my landlord will find out": Airbnb and the challenges of enforcement. *Urban Policy and Research*, *36*(4), 417–428.

Levendis, J., & Dicle, M. F. (2016). The neighborhood impact of Airbnb on New Orleans. Retrieved from: https://ssrn.com/abstract=2856771.

McNamara, B. (2015). Airbnb: A not-so-safe resting place. *Colorado Technology Law Journal, 13*, 149–170.

Miller, M. (2019). How bad Uber ratings affect drivers' careers – and why you shouldn't be scared to report bad behavior. *Mic.* Retrieved from: www.mic.com/p/how-bad-uber-ratings-affect-drivers-careers-why-you-shouldnt-be-scared-to-report-bad-behavior-17865617. Accessed 19 October 2019.

Murphy, S., & Liao, S. (2013). Consumers as resellers: Exploring the entrepreneurial mind of North American consumers reselling online. *International Journal of Business and Information, 8*(2), 183–228.

Nuzzi, O. (2017). The definitive list of Uber horror stories. *The Daily Beast.* Retrieved from: www.thedailybeast.com/the-definitive-list-of-uber-horror-stories. Accessed 10 October 2019.

Pantazatou, K. (2018). Taxation of the sharing economy in the European Union. In: N. M. Davidson, M. Finck, & J. J. Infranca (eds.), *Cambridge handbook of the law of the sharing economy.* Cambridge: Cambridge University Press, 368–380.

Park, T. (2019). Uber to provide taxi service for foreigners in Seoul. *Hani.* Retrieved from: http://english.hani.co.kr/arti/english_edition/e_business/879739.html. Accessed 21 October 2019.

Porges, S. (2014). The strange game theory of Airbnb reviews. *Forbes.* Retrieved from: www.forbes.com/sites/sethporges/2014/10/17/the-strange-game-theory-of-airbnb-reviews/#1b0dfe935e56. Accessed 6 June 2016.

Roh, M., & Park, K. (2019). Adoption of O2O food delivery services in South Korea: The moderating role of moral obligation in meal preparation. *International Journal of Information Management, 47*, 262–273.

Scharf, N. (2011). Napster's long shadow: Copyright and peer-to-peer technology. *Journal of Intellectual Property Law & Practice, 6*(11), 806–812.

Schechner, S., & Verbergt, M. (2015). Paris confronts Airbnb's rapid growth. *Wall Street Journal, 25.*

Schneider, H. (2017). *Creative destruction and the sharing economy: Uber as disruptive innovation.* Cheltenham: Edward Elgar Publishing.

Seright, S. (2018). Sharing economy pioneers: 15 companies disrupting industries left and right. *Neighbor Storage.* Retrieved from: www.neighbor.com/storage-blog/sharing-economy-pioneers/. Accessed 25 April 2020.

Shin, H., & Minu, K. (2019). Airbnb users in South Korea up 56% on year in 2018. *Pulse.* Retrieved from: https://pulsenews.co.kr/view.php?year=2019&no=91290. Accessed 21 October 2019.

Sokolowsky, J. (2018). Homeaway steps up lodging collection efforts. *Avalara.* Retrieved from: www.avalara.com/mylodgetax/en/blog/2018/08/homeaway-steps-up-lodging-tax-collection-efforts.html. Accessed 14 October 2019.

Sokolowsky, J. (2019). New Orleans City Council passes stricter Airbnb rules. *Avalara.* Retrieved from: www.avalara.com/mylodgetax/en/blog/2019/05/new-orleans-city-council-passes-stricter-airbnb-rules.html. Accessed 20 October 2019.

Sperling, G. (2015). How Airbnb combats middle class stagnation. 1–17.

Standing, C., Standing, S., & Biermann, S. (2019). The implications of the sharing economy for transport. *Transport Reviews, 39*(2), 226–242.

Stein, M. (2019). City council passes tighter restrictions on short-term rentals. *The Lens.* Retrieved from: https://thelensnola.org/2019/08/08/short-term-rental-restrictions-fly-through-final-council-vote-advocates-partially-pleased/. Accessed 16 October 2019.

Sugiura, W. (2019). World Cup gives embattled Airbnb a second chance in Japan. *Asian Review.* Retrieved from: https://asia.nikkei.com/Spotlight/Cover-Story/World-Cup-gives-embattled-Airbnb-a-second-chance-in-Japan. Accessed 15 October 2019.

Sundararajan, A. (2016). *The sharing economy: The end of employment and the rise of crowd-based capitalism.* Cambridge, MA: The MIT Press.

Tasch, B., & Slater-Robins, M. (2016). French taxi drivers shut down Paris as protests over Uber turn violent. *Business Insider.* Retrieved from: www.businessinsider.com/uber-protests-in-paris-2016-1. Accessed 19 October 2019.

Tham, A. (2016). When Harry met Sally: Different approaches towards Uber and AirBnB – an Australian and Singapore perspective. *Information Technology & Tourism, 16*(4), 393–412.

Ting, D. (2016). Airbnb has a golden moment at the Rio Olympics. *Skift.* Retrieved from: https://skift.com/2016/08/11/airbnb-has-a-golden-moment-at-the-rio-olympics/. Accessed 22 October 2019.

Ventura, L. (2018). Taxi drivers continue to boycott the Strip … on some days. *Las Vegas Sun.* Retrieved from: https://lasvegassun.com/news/2018/jul/05/taxi-drivers-continue-to-boycott-the-strip-on-some/. Accessed 17 October 2019.

Wachsmuth, D., & Weisler, A. (2018). Airbnb and the rent gap: Gentrification through the sharing economy. *Environment and Planning A: Economy and Space, 50*(6), 1147–1170.

Wan, W. N. A. A. B., Mohamad, A. F. M. F., Shahib, N. S., Azmi, A., Kamal, S. B. M., & Abdullah, D. (2016). A framework of customer's intention to use Uber service in tourism destination. *International Academic Research Journal of Business and Technology, 2*(2), 102–106.

Wang, X., Ardakani, H. M., & Schneider, H. (2017). Does ride sharing have social benefits? *Association for Information Systems.* Retrieved from: https://aisel.aisnet.org/amcis2017/DataScience/Presentations/13/. Accessed 22 October 2019.

Watkins, E. (2016). Guest preference, brands push shift to walk-in showers. *Hotel News Now.* Retrieved from: http://hotelnewsnow.com/Articles/74922/Guest-preference-brands-push-shift-to-walk-in-showers. Accessed 20 October 2019.

Weiser, B., & Goodman, J. D. (2019). Judge blocks New York City law aimed at curbing Airbnb rentals. *New York Times.* Retrieved from: www.nytimes.com/2019/01/03/nyregion/nyc-airbnb-rentals.html. Accessed 19 October 2019.

Whalen, E. A., Belarmino, A., & Taylor Jr, S. (2019). Share and share alike? *Journal of Hospitality and Tourism Insights.*

Wilson, M. (2019). Las Vegas makes progress in enforcing short-term rental laws. *Las Vegas Sun.* Retrieved from: https://lasvegassun.com/news/2019/mar/18/las-vegas-makes-progress-in-enforcing-short-term-r/. Accessed 17 October 2019.

Xiang, Z., & Gretzel, U. (2010). Role of social media in online travel information search. *Tourism Management, 31*(2), 179–188.

Yang, Y., & Mao, Z. (2019). Welcome to my home! An empirical analysis of Airbnb supply in US cities. *Journal of Travel Research, 58*(8), 1274–1287.

Yannopoulou, N., Moufahim, M., & Bian, X. (2013). User-generated brands and social media: Couch-surfing and Airbnb. *Contemporary Management Research, 9*(1), 85–90.

Young, M., & Farber, S. (2019). The who, why, and when of Uber and other ride-hailing trips: An examination of a large sample household travel survey. *Transportation Research Part A: Policy and Practice, 119*, 383–392.

Zervas, G., Proserpio, D., & Byers, J. W. (2017). The rise of the sharing economy: Estimating the impact of Airbnb on the hotel industry. *Journal of Marketing Research, 54*(5), 687–705.

Zhang, Z., Ye, Q., Law, R., & Yi, L. (2010). The impact of e-word-of-mouth on the online popularity of restaurants: A comparison of consumer reviews and editor reviews. *International Journal of Hospitality Management, 29*(4), 694–700.

5

URBANISATION AND ITS EFFECTS ON CITY TOURISM IN CHINA

Jian Ming Luo and Chi Fung Lam

Global urbanisation

City tourism, a part of tourism, has shown tremendous growth internationally with influential impact in many destinations (Bock 2015). Shao et al. (2017) argued that to plan urban tourism successfully, city planners should carefully design the spatial structures of cities. Meanwhile, people have been migrating from rural to urban areas for a long time. The rural population decreased from 70% in the 1950s to 45% in 2018 and is expected to decrease further to 32% by 2050. At the same time, urban population reached 3.2 billion in 2018 and is expected to reach over six billion in 2050 (UN 2018). These growth patterns are mainly concentrated in the under-developed countries, whereas urban areas in the developed countries increased simultaneously. Over 50% of the world's population are now living in urban areas, generating over 80% of the global GDP. However, the above statistics do not necessarily show the whole picture. According to the McKinsey Global Institute (2011), 60% of the global GDP is actually generated by 600 urban cities and driven by one-fifth of the population. Figure 5.1 shows the urban population of the world from 1950 to 2050.

Urbanisation and industrialisation were two distinguishing characteristics of economic development in the world during the nineteenth century. Arguably, these characteristics are considered as the drivers of the cities' economic, demographic, functional and extensive growth (Bähr and Jürgens 2005). During the twentieth century, many cities in the developing countries became more urbanised as several countries in Africa and Asia declared independence. According to Henderson et al. (2009), urbanisation and industrialisation come hand-in-hand and are the drivers of income growth. Many manufacturing factories and services have become more efficient once they learned and applied the best practices in terms of accessing technology and management in the industry. They have also been able to hire skilled labour easily and reduce transportation costs due to convenient locations. As innovations are incubated and sophisticated skills are developed, large cities become the drivers of economic growth as emerging financial and business centres. Meanwhile, factories tend to move to smaller cities specialising in special streams of production, such as steel, autos, electronics, textiles, apparel and wood products, as well as specialised services, such as entertainment, insurance, certain forms of health care and tourism (Henderson et al. 2009).

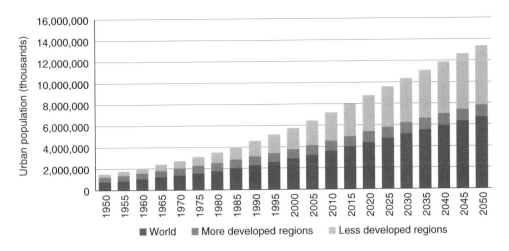

Figure 5.1 Urban population from 1950 to 2050.

Source: United Nations (2018).

Urbanisation in China

Urbanisation is a sophisticated socioeconomic phenomenon that converts existing rural areas into urban areas by changing the spatial distribution of populations. During this process, not only location, but also dominant occupations, lifestyle, culture and behaviours are changed drastically. The demographic and social structures are also altered simultaneously. Prominent effects of urbanisation include the increase in land size and population in urban areas and the decrease of population in rural areas (McKinsey Global Institute, 2018). "Urbanisation", according to the UN (2010), is defined as the level, proportion or the increase of this proportion of total population in an area or place. Urbanisation affects spatial and urban planning as well as public and private investments in infrastructures. When more economic activities become concentrated in cities, corresponding activities, such as transportation, information and services, will be developed, thus improving the quality and accessibility of both private and public services (McKinsey Global Institute 2018).

China, the largest developing country in the world, has experienced tremendous urbanisation development since the 1990s (Li and Yao 2009). China adopts the same urbanisation definition as the UN (National Bureau of Statistics of China (NBSC) 2018). The urbanisation growth rate in China is remarkable at an annual growth rate of around 4%. From 1990 to 2017, the urbanisation level in China doubled and is expected to continue in the coming years. In fact, the UN (2018) has predicted that the urban population in China will reach 1 billion in 2050. Together with India and Nigeria, these three countries will represent more than 30% of the growth of urban population in 2050.

More than 100 cities in China developed from 1978 to 2017 (NBSC 2018), and the urbanisation level increased over three times during the same period. The annual urban population growth is expected to reach 20 million in the next 20 years. While only 23 cities in Europe have a population of over five million, more than 200 cities in China have populations exceeding one million (McKinsey Global Institute 2009, 2018). The McKinsey Global Institute (2011) also forecasted that 136 new cities from the developing countries and 100 cities from China shall comprise the top 600 cities in the world by 2025. Moreover, the share of the urban population

Table 5.1 China's urbanisation level from 1950 to 2025

Year	Urban population (thousands)	Percentage urban (%)	Urban annual growth rate (%)
1950	65,437	11.8	5.20
1955	84,640	13.9	4.71
1960	106,562	16.2	3.72
1965	130,685	18.1	1.97
1970	143,513	17.4	2.32
1975	160,244	17.4	3.57
1980	192,392	19.4	4.78
1985	244,946	22.9	4.51
1990	310,022	26.4	4.32
1995	383,902	31.0	3.84
2000	460,377	35.9	4.00
2005	561,983	42.5	3.44
2010	669,354	49.2	2.85
2015	775,353	55.6	2.11
2020	875,076	61.0	1.49
2025	956,554	65.4	1.00
2050	1,091,948		

Source: United Nations (2018).

will increase from 36% (360 million) in 2018 to almost 50% (950 million) in 2038. This increase in population is larger than the entire US population today (McKinsey Global Institute 2016). Meanwhile, 90% of China's GDP would be generated from the urban areas (McKinsey Global Institute 2009). The abovementioned figures are consistent with previously mentioned urbanisation growth rates. The urbanisation growth rate in China in the last century is nothing short of phenomenal, with the urban population increasing from 65 million in 1950 to over one billion in 2050, catapulting the country to the top spot among all countries in the world. Table 5.1 shows the urbanisation levels in China from 1950 to 2050.

Urbanisation in China has received enormous attention from Chinese researchers due to its tremendous growth. Although many researchers have studied this phenomenon for a long time, many specific questions remain unresolved (Qiu 2007). The success of urbanisation in China is mainly due to the demise of the centrally planned political economy, which enables local entrepreneurs to utilise different resources, such as national economic policies, local initiatives and production and capital restructuring, to match specific development targets (Qian et al. 2012). Using secondary data collected from 31 provinces, Zhang (2002) found significant differences in urbanisation levels among different regions and identified the relationships between urbanisation and other important economic variables, such as GDP and foreign investment. Liu (2004) found similar results indicating that the growth rates in the eastern and southern regions are higher than those in the western and northern regions. Such a difference in growth rate (also known as "regional disparity"), as the author argued, is due to variations in natural resources, policies, population growth rates and economic development in the corresponding regions. Thus, the regional disparity changed the policy focus from northern–southern to eastern–western.

Meanwhile, Demurger (2000) argued that the Open-Door Policy and other reform policies were not implemented in all cities/provinces/regions simultaneously. The Chinese government tended to select several cities/provinces/regions as "testing points". For instance, the Open-Door Policy was initially implemented in Guangdong and Fujian; as a result, these provinces eventually became the drivers of economic development in China. As these provinces became more developed, they attracted foreign investments. In comparison, those cities/provinces/ regions without policy support were left behind, thereby broadening the regional disparity. Nevertheless, the Chinese government did not leave the disparity unresolved. Since 2000, inter-regional disparity in urbanisation has been reduced, as reported by Zhang and Han (2009), who argued that one of the reasons for such reduction is the change of state development policies.

Tourism development and urbanisation in China

Urbanisation in China, despite its remarkable success, also experienced setbacks. For instance, the urbanisation growth rate only ranged from 3% to 4% from 1990 to 2004. This rate is significantly lower than the growth rates of other developing countries during the same period (Renaud 1981). At that time, China was experiencing the most difficult part of its urban transition. However, this was also the period when many tourism facilities and services, such as international and domestic airlines, theme parks, tourist attractions, hotels and so on, were being developed. In the mid-1980s, international visitors were allowed to visit over 250 cities and counties. At present, there are more than 600 cities that can be visited by tourists.

Ever since China implemented the Open-Door Policy in 1978, its tourism industry has experienced extraordinary growth. Tourism is one of the pillar industries of the country's economy (Tsang and Hsu 2011), not only bringing tourists but also many economic benefits (Xiao 2006). Zhang et al. (2005) discussed the three stages of tourism and hospitality development in China. The first stage, from 1949 to 1978, began with the founding of the People's Republic of China. During this stage, tourism was used as a political tool aimed to increase China's political influence and international recognition. Therefore, tourists mainly consisted of foreign government officials and politicians (Zhang et al. 1999; Han 1994). The second stage, from 1978 to 1985, began with the Open-Door Policy in 1978. During this period, the purpose of tourism was changed from political influence to economic gain. Hence, tourism in China began to expand at a remarkable pace. Many hotels and accommodation facilities were built during this period, and these infrastructures were built with the aim of serving tourists from around the world. However, the review provided by Tisdell and Wen (1991) shows that the investments made during this period focused mainly on hotels. The authors argued that this unusual concentration may be due to the preconception of government officials instead of the actual aim of satisfying real tourism demand. Another important event during this period was China's accession into the UNWTO in 1983 (Zhang et al. 2005). The last stage, from 1986 to the present, began with the 7th Five-Year National Plan, which was the first time tourism was included in the national development plan. In the 1990s, international tourists began to recognise China as a potential tourism destination, thus increasing actual visits or intentions to visit the country. China was the beneficiary of such massive expansion of tourism, and the economic benefits were tremendous. As the Chinese government began to realise the economic benefits, it identified tourism as the new economic driver and pillar industry at the provincial and national levels. During this stage, the original purpose of tourism, political influence, was diminished and gradually changed to economic benefits, mainly about earning foreign exchange (Han 1994).

In 2009, the State Council officially included tourism in the list of pillar industries in the 12th Five-Year Plan (2011–2015). The State Council further confirmed the contribution of tourism to the country's economy, propelling the tourism industry onto the next level (Yang 2011). Gradually, China gained more influence in the international tourism market, not only as a destination but also as a source market of tourists. The country soon became one of the world's most important inbound tourist markets. In 2017, Chinese tourists spent over USD 8 billion, making them the top spenders in the world (UNWTO 2018).

Table 5.2 shows the Total Population, Urban Share in Total Population, Number of Overseas Visitor Arrivals, Foreign Exchange Earnings from International Tourism and Number of Domestic Visitor Arrivals and Earnings from Domestic Tourism from 1978 to 2017. In 2017 alone, over 700 million international tourists visited China. Together with 5 billion domestic tourists in the same year, they generated over US$4 trillion in tourism revenues. According to Table 5.2, all six variables increased over the selected period. The relationship between the Number of Overseas Visitor Arrivals and Foreign Exchange Earnings is obvious: as the number of international tourists increased, this also led to an increase in foreign exchange. A similar explanation can also be applied to the relationship between the Number of Domestic Visitor Arrivals and Earnings from Domestic Tourism. Unfortunately, data on the Number of Domestic Visitors are not available. However, it is apparent that Urban Share in Total Population and the Number of Overseas Visitor Arrivals increased simultaneously during the same period. Luo et al. (2016a) have suggested that urbanisation in China is likely to increase international and domestic tourists despite the limited data on domestic tourists.

Tourist spending can be considered as a different form of export and can contribute to the balance of payments via foreign exchange earnings. These earnings can provide an important income source for the economy (Balaguer and Cantavella-Jorda 2002) as well as purchase capital to provide more goods and services (McKinnon 1964), thus driving more economic growth and urbanisation. This provides a positive cycle of economic development. Spence et al. (2009) believed that the relationship between urbanisation and national income is strong and stable. Almost all countries that became middle-income countries reached an urbanisation rate of above 50% and those that achieved high-income status reached an urbanisation rate of above 70%. A rapidly growing tourism expansion follows a similar path.

The relationship between urbanisation and tourism development

Urbanisation is considered one of the drivers of modernisation and the transformation of rural areas to urban areas with high living standards. Urbanisation not only increased economic growth, it also created drastic structural changes. For example, during the process of urbanisation in China, the service industry has become increasingly important. The proportion of service industry in China increased from 13% in 1970 to over 70% in 2017 (NBSC 2018). Huang and Bouis (1996) agreed that urbanisation transformed the economic composition and employment structure of China. Spence et al. (2009) explained the economic composition in further detail, arguing that urbanisation narrowed the productivity difference between rural and urban areas and increased the productivity in cities. Many researchers also agreed that urbanisation has a significant impact on economic growth (Kasarda and Crenshaw 1991).

Meanwhile, Lewis (1997) argued that until urbanisation exceeds 60%, then it would have a minimal impact on individual incomes. In other words, only when the above threshold is achieved can individual incomes increase rapidly. Davis and Henderson (2003) attempted to examine the relationships among urbanisation, economic growth and agriculture. Using total population percentage and GDP per capita as proxy variables for urbanisation and economic growth, respectively, they found that the relationship between urbanisation and economic growth is positive. A similar relationship also exists between urbanisation and agricultural products (Davis and Henderson 2003). Henderson (2003) found similar relationships between urbanisation and economic development; specifically, the correlation coefficient between the two variables is 0.85, indicating an extremely strong relationship. Bradshaw and Fraser (1989)

Table 5.2 Urbanisation and tourism growth in China (1978–2017)

Year	Total population (millions)	Urban share in total population (percentage)	Number of overseas visitor arrivals (million)	Foreign exchange earnings from international tourism (billion RMB)	Number of domestic visitors (million)	Earnings from domestic tourism (billion RMB)
1978	962.59	17.92	1.81	0.45	–	–
1979	975.42	18.96	4.20	0.70	–	–
1980	987.05	19.39	5.70	0.92	–	–
1981	1,000.72	20.16	7.77	1.38	–	–
1982	1,016.54	21.13	7.92	1.57	–	–
1983	1,030.08	21.62	9.48	1.86	–	–
1984	1,043.57	23.01	12.85	2.64	–	–
1985	1,058.51	23.71	17.84	3.67	–	–
1986	1,075.07	24.52	22.82	5.29	–	–
1987	1,093.00	25.32	26.90	6.86	–	–
1988	1,110.26	25.81	31.69	8.36	–	–
1989	1,127.04	26.21	24.50	7.00	–	–
1990	1,143.33	26.41	27.46	10.61	–	–
1991	1,158.23	26.94	33.35	15.14	–	–
1992	1,171.71	27.46	38.11	21.77	–	–
1993	1,185.17	27.99	41.53	26.98	410.00	86.40
1994	1,198.50	28.51	43.68	63.11	524.00	102.35
1995	1,211.21	29.04	46.39	72.93	629.00	137.57
1996	1,223.89	30.48	51.13	84.80	639.00	163.84
1997	1,236.26	31.91	57.59	100.09	644.00	211.27
1998	1,247.61	33.35	63.48	104.33	694.00	239.12
1999	1,257.86	34.78	72.80	116.72	719.00	283.19
2000	1,267.43	36.22	83.44	134.31	744.00	317.55
2001	1,276.27	37.66	89.01	147.29	784.00	352.24
2002	1,284.53	39.09	97.91	168.73	878.00	387.84
2003	1,292.27	40.53	91.66	145.72	870.00	344.23
2004	1,299.88	41.76	109.04	213.04	1102.00	471.07
2005	1,307.56	42.99	120.29	239.98	1212.00	528.59
2006	1,314.48	44.34	124.94	270.63	1394.00	622.97
2007	1,321.29	45.89	131.87	318.75	1610.00	777.06
2008	1,328.02	46.99	130.03	283.66	1712.00	874.93
2009	1,334.50	48.34	126.48	271.02	1902.00	1018.37
2010	1,340.91	49.95	133.76	310.14	2103.00	1257.98
2011	1,347.35	51.27	135.42	313.02	2641.00	1930.54
2012	1,354.04	52.57	132.41	323.13	2957.00	2270.62
2013	1,369.72	53.73	129.08	333.69	3262.00	2627.61
2014	1,367.82	54.77	128.50	367.59	3611.00	3031.19
2015	1,374.62	56.10	133.82	734.04	4000.00	3419.51
2016	1,382.71	57.35	138.44	775.05	4440.00	3939.00
2017	1,390.08	58.52	139.48	797.12	5001.00	4566.08

Source: National Bureau of Statistics of China (2018).

Note
The data for "Number of Domestic Visitors" and "Earnings from Domestic Tourism" were not available until 1993.

also found a similar relationship and a consistent result between urbanisation and income. In addition, they found that urbanisation does not only increase income and economic growth, but also increases the quality of life, which they measured using infant mortality, death rate and illiteracy (Bradshaw and Fraser 1989). However, this result is different from that reported by Bertinelli and Black (2004) who found that certain attitudes and values are required for urbanisation to affect economic development.

Burgess and Venables (2004) reported that the service industry showed distinguishing features of urbanisation. However, many researchers did not incorporate urbanisation into their research on economic growth and development. Specifically, Kasarda and Crenshaw (1991) argued that, surplus from the society is a strict requirement that would allow society to utilise a sophisticated method of production and develop a spatial concentration of customer and labour markets. Tourism is an important section of the service industry; therefore, urbanisation has a significant impact on tourism development. Several researchers examined tourism development and argued that it is influenced by urbanisation as the former is part of the service industry. Lu (2008) studied the relationship between urbanisation and sport tourism, reporting that, although sport tourism is fundamentally driven by demand and supply forces, government policy is an important explanatory variable of sport business. Hence, future studies should consider urbanisation as part of China's development strategies. Likewise, tourism and many other service industries are important contributors of economic growth in China.

Gao et al. (2013) examined the relationship between urbanisation and the tourism industry in Xi'an. They found that despite the comprehensive functional value of urbanisation increasing along with the tourism industry in Xi'an, the increasing trend generally slowed down from 2003 to 2009. Their results confirmed that urbanisation level is important to the development of a tourism centre as the development of attractions and tourism facilities, such as hotels, theme parks and so on, is related to the population (Kastarlak 1971). In other words, when the demand for food and beverages, hospitality facilities and transportation increases, producers will exert more resources and construct more facilities to satisfy these demands. This is exactly how the prototype of a city begins.

Meanwhile, McCroskey (1990) found a positive relationship between urbanisation and hotel development in the United States. The author argued that hotel development is further enhanced by changes in people's habits. For example, as automobiles become more accessible and the highways connecting cities become more developed, people would prefer to travel individually more frequently. In turn, this change of habit has led to the increased demand for accommodation services. One would expect China would follow such pattern. This is confirmed by the research conducted by Shen (2000) who found that many people who permanently stay in the city usually choose hotels as their accommodation. Luo et al. (2016b) found that urbanisation can be used to explain tourism development via data from Guangdong. The central government identified that tourism is another pillar industry and a focal point of the economic development, not only on the national level but also on the local level. As mentioned above, after the Open-Door Policy, the purpose of tourism focused on obtaining foreign exchange. As tourism is a comprehensive economic activity (Han 1994), policymakers should consider policy goals and the existing urbanisation level when deciding tourism development policies for each city.

The positive effects of urbanisation on the tourism industries of cities

Urbanisation is an economic development or social phenomenon wherein people in the rural areas move to the urban areas. Urbanisation, living standard and income usually increase simultaneously (Luo 2016). When people become wealthier, the demands for other services or products, such as transportation, food and beverages, hospitality and so on, are expected to increase at the same time. These industries will benefit from more consumers, higher revenues, easier access to labour and lower labour costs. Moreover, these industries will not only attract workers, but also attract residents. When more people are concentrated in the city, demand for other services will increase and eventually lead to a sustainable economy. Therefore, urbanisation does not only increase economic activities, but also enhance environment protection. For example, as the population and the number of visitors increase, the demand for transportation increases. Hence, to accommodate these needs, the Chinese government recently enhanced the transportation system. This enhancement does not only make the transportation more efficient and more accessible, but also reduces people's demand to drive their own vehicles. The reduction of car usage, in turn, reduces pollution. Furthermore, demand and supply of tourism increases when people move to the cities. When people live in denser locations, they travel more frequently to escape. In addition, many tourists – both domestic and international – use metropolitan cities as a medium of transportation. Many cities in China, such as Beijing and Tianjin , possess rich and abundant cultural and historical backgrounds. These backgrounds are very attractive for many international and domestic tourists, hence creating greater tourism demand and supply.

Population migration is important for the development of a sustainable economy and urbanisation. This is because the corresponding demands for food, shelter, transportation and other services increase simultaneously when people move to the city. When there are greater demands for goods and services, the supply of the corresponding goods and services will develop at the same time to stabilise the economy. Similarly, as the demand of tourism increases, transportation in the cities will also improve to satisfy the corresponding needs of the growing population. In addition, as the demand of tourism increases, more high-quality tourist attractions and resources will be developed. As more people are now located in the urban areas, businesses will be able to hire skilled labour more easily with lower costs. Hence, the standard of living will improve. Zhang et al. (2013) found that urbanisation positively affects hotel growth. Luo and Lam (2017) found a similar relationship between urbanisation and occupancy rate. Both results are consistent with the existing marketing literature. Charles and Anderson (2016) and Steenkamp (2001) also found a significant relationship between urbanisation and marketing strategy. These results imply that hotel performance is also affected by the urbanisation level of the corresponding area.

The negative effects of urbanisation on the tourism industries of cities

When people migrate to the city, the population in the city becomes denser. The cost of living, including housing, daily and transportation expenses, all increase at the same time. Apart from the increased cost of living, other issues also emerge, such as pollution, crime, regional disparity and so on (Zhang 2011). This means that, although urbanisation attracts more tourists, the influx of tourists can also create certain "side effects" that can endanger the sustainable development of the city. This is exactly what is happening in China's major cities. China has developed rapidly in the last few decades. However, many cities and tourist attractions are operating beyond their limits. The pressure of over-capacity is worsened by the uncivilised behaviour of many tourists and the mis-management of the site operators. These problems deteriorate the attractions and the surrounding environment as well as the host–visitor relationship. Russo (2002) described the above situation as a vicious cycle. People are less likely to stay overnight when the city is over-crowded. This means they have less time to explore the city. However, because the attractions in the city are famous ("must-see" sites), more people will become excursionists. Hence, by definition, they will not be able to provide revenues, such as those from hotel accommodations, to the city; instead, they would try to fit all sight-seeing attractions into a day. This means many people will come to the same attraction at the same time because there is a logical sequence of visiting based on the location of the attractions. This behaviour will worsen the existing over-crowding situations in those attractions or sites. As a result, people will have less intention to stay in the city overnight, hence creating the vicious cycle.

Gentrification is another negative effect of tourism. Gentrification is defined as the phenom-enon wherein wealthy people settle in a city and create extra demand for property or other assets. Hence, the property or asset values increase, which also alters the characteristic and culture of the urban district. Although one would understand gentrification in a positive manner, generally, it translates into a negative impact. The literature on gentrification has shown that this concept is really complicated. Hence, the results are contradictory and the real impact of the phenomenon changes from time to time and location to location. On the one hand, Atkinson (2004) concluded that gentrification is generally harmful as it creates household displacement and conflicts in community. On the other hand, one could easily argue that gentrification is desirable. Higher property or asset values mean higher income for many individuals in the com-munity. However, these benefits are not equally distributed to the neighbourhood; hence, these benefits are often neglected and many residents will soon find themselves economically and socially marginalised. For example, Gotham (2005) studied the socio-spatial transformation of New Orleans' Vieux Carre (French Quarter) from the 1980s to 2000s and found that, along with the income and property value increase during this period, the rent of property also increased at a much higher rate. Hence, the accelerating increasing rate of rent has driven relat-ively poor people, particularly African Americans, away from the community. Moreover, many existing resources are replaced by tourist attractions and giant entertainment complexes. The author argued that the changing flows of capital into the real estate market, combined with the growth of tourism, enhance the significance of consumption-oriented activities in residential space and encourage gentrification (Gotham 2005). Similarly, in China, when Shenzhen was still in the development stage in the 1990s, the government provided affordable housing and jobs to the low-income group with cautious consideration of the hasty and large-scale redevel-opment of these villages (Wang et al. 2009).

As for the impact of urbanisation on the authenticity of visitor experiences (Stephen and Bosak 2019), the basis of the criterion of "authenticity" would require more than the concept of collective memory to justify its use. Of course, it remains true that trust in the authenticat-ing process is a precondition for the authentication being accepted by others as true (Benson 2018). Heynen (2006) argued that practical problems may emerge when one maintains the authenticity of rebuilt heritage. For example, many rebuilt heritages fail to restore the original authentic theme and this can be attributed to the questionable rebuilding process being imple-mented. Many stakeholders have different opinions on which section of the heritage is signi-ficant. This conflict of opinions deteriorates the information received by the management teams, which renders them incapable of making appropriate decisions (Sjöholm 2017). Meng et al. (2019) investigated the subjective well-being of Chinese rural–urban migrants by exam-ining the effects of nostalgia and perceived authenticity in the context of rural tourism. The results showed that the preservation of rural authenticity could increase the social and cultural welfare of hosting communities and the subjective well-being of tourists. Hence, when designing policies for heritage rebuilding, urban re-infrastructure and other similar projects, city planners should be aware of the tourist resources' authenticity and the preservation of such authenticity.

Conclusion and future research directions

The purpose of this chapter was to provide a better understanding of the link between urbanisation and tourism development. In particular, this chapter focused on the urbanisation factors in China. Different tourism sectors are affected by urbanisation in varying ways. Policymakers should consider the current level of urbanisation in the city before they develop any tourism products. Urbanisation is a broad topic and it is beyond the scope of this chapter to cover every aspect of this huge social phenomenon. Many urban characteristics, such as intensity, labour structure and population structure, were not discussed in this study, and these could be excellent topics to be discussed by future studies on city tourism. Practitioners (e.g. destination management organisation (DMO) managers) and policymakers, with a view to tackling the challenges posed by the rapid urban development of Chinese cities on tourism development, can also benefit from a thorough understanding of the development of rural enterprises, management of the household registration system and the management of urban land-use. Urbanisation can be beneficial or harmful to the city. Hence, a suitable and sustainable management of urbanisation is necessary for the development of tourism in a city. Similarly, poorly managed urbanisation can only create disaster. Therefore, city planners should assess the distinct characteristics of their cities and design suitable policies to further promote the level of urbanisation.

References

Atkinson, R., 2004. The evidence on the impact of gentrification: New lessons for the urban renaissance? *European Journal of Housing Policy*, 4(1), 107–131.

Bähr, J. and Jürgens, U., 2005. *Stadtgeographie II – Regionale Stadtgeographie*. Braunschweig: Westermann.

Balaguer, L. and Cantavella-Jorda, M., 2002. Tourism as a long-run economic growth factor: The Spanish case. *Applied Economics*, 34, 877–884.

Benson, C., 2018. "Authenticity" for the visited or for the visitors? "Collective memory", "collective imagination" and a view from the future. In *ICOMOS University Forum* (pp. 1–11). ICOMOS International.

Bertinelli, L. and Black, D., 2004. Urbanization and growth. *Journal of Urban Economics*, 56, 80–96.

Bock, K., 2015. The changing nature of city tourism and its possible implications for the future of cities. *European Journal of Futures Research*, 3(1), article 20.

Bradshaw, Y.W. and Fraser, E. 1989. City size, economic development, and quality of life in China: New empirical evidence. *American Sociological Review*, 54(6), 986–1003.

Burgess, B. and Venables, A.J., 2004. Towards a microeconomics of growth. World Bank Working Paper 3257, April, Washington, DC.

Charles, G. and Anderson, W., 2016. *International Marketing: Theory and Practice from Developing Countries*. Newcastle upon Tyne: Cambridge Scholars Publishing.

Davis, J.C. and Henderson, J.V., 2003. Evidence on the political economy of the urbanization process. *Journal of Urban Economics*, 53(1), 98–125.

Demurger, S., 2000. *Economic Opening and Growth in China*. OECD development centre studies. Paris: OECD.

Gao, N., Ma., F.-Y., Li, T. and Bai, K., 2013. Study on the coordinative development between tourism industry and urbanization based on coupling model: A case study of Xi'an. *Tourism Tribune/Lvyou Xuekan*, 28(1), 62–68.

Gotham, K.F., 2005. Tourism gentrification: The case of new Orleans' Vieux Carre (French Quarter). *Urban studies*, 42(7), 1099–1121.

Han, K.H., 1994. *China: Tourism Industry*. Beijing: Modern China Press.

Henderson, J.V., Quigley, J. and Lim, E., 2009. Urbanization in China: Policy issues and options. China Economic Research and Advisory Programme, Center for International Development at Harvard University, Cambridge. MA.

Henderson, V., 2003. The urbanization process and economic growth: The so-what question. *Journal of Economic Growth*, 8, 47–71.

Heynen, H. 2006. Questioning authenticity. *National Identities*, 8(3), 287–300.

Huang, J. and Bouis, H., 1996. Structural changes in demand for food in Asia. Food, Agriculture and the Environment Discussion Paper. Washington, DC, International Food Policy Research Institute.

Kasarda, J.D. and Crenshaw, E.M., 1991. Third World urbanization: Dimensions, theories, and determinants. *Annual Review of Sociology*, 17, 467–501.

Kastarlak, B., 1971. Planning tourism growth. *Cornell Hotel and Restaurant Administration Quarterly*, 11, 26–33.

Lewis, W.A., 1997. The evolution of the international economic order. Discussion paper 74, Research Program in Development Studies, Woodrow Wilson School, Princeton University, Princeton, NJ.

Li, B. and Yao, R., 2009. Urbanisation and its impact on building energy consumption and efficiency in China. *Renew Energy*, 34, 1994–1998.

Liu, S., 2004. Causal analysis on the regional disparities of urbanization in china. *Resources and Environment in the Yangtze Basin*, 13(6), 530–535.

Lu, J., 2008. The factor analysis in the process of urbanization in the growth of sports business. *Urban Studies*, 15(5), 21–23.

Luo, J.M., 2016. *Urbanization and Tourism Development in China*. New York: Nova Science Publishers.

Luo, J.M. and Lam, C.F., 2017. Urbanization effects on hotel performance: A case study in China. *Cogent Business & Management*, 4(1), 1412873.

Luo, J.M., Qiu, H. and Lam, C.F., 2016a. Urbanization impacts on regional tourism development: A case study in China. *Current Issues in Tourism*, 19(3), 282–295.

Luo, J.M., Qiu, H., Goh, C. and Wang, D., 2016b. An analysis of tourism development in China from urbanization perspective. *Journal of Quality Assurance in Hospitality and Tourism*, 17(1), 24–44.

McCroskey, M.L. 1990. Arizona's community-built hotels. *Cornell Hotel and Restaurant Administration Quarterly*, 31(1), 26–33.

McKinnon, R., 1964. Foreign exchange constraints in economic development and efficient aid allocation. *Economic Journal*, 74, 388–409.

McKinsey Global Institute, 2009. *Preparing for China's Urban Billion*. [Online]. Available from www.mckinsey.com/insights/urbanization/preparing_for_urban_billion_in_china [Accessed 1 February 2019].

McKinsey Global Institute, 2011. *Urban World: Mapping the Economic Power of Cities*. [Online]. Available from www.mckinsey.com/featured-insights/urbanization/urban-world-mapping-the-economic-power-of-cities [Accessed 1 February 2019].

McKinsey Global Institute, 2016. *Urban World: Meeting the Demographic Challenge*. [Online]. Available from www.mckinsey.com/featured-insights/urbanization/urban-world-meeting-the-demographic-challenge-in-cities [Accessed 1 May 2019].

McKinsey Global Institute, 2018. *World Urbanization Prospects 2018*. [Online]. Available from https://population.un.org/wup/Publications/ [Accessed 1 May 2019].

Meng, Z., Cai, L.A., Day, J., Tang, C.H., Lu, Y. and Zhang, H., 2019. Authenticity and nostalgia-subjective well-being of Chinese rural–urban migrants. *Journal of Heritage Tourism*, 1–19.

National Bureau of Statistics of China (NBSC), 2018. *China Statistical Yearbook*. Beijing: China Statistics Press.

Qian, J., Feng, D. and Zhu, H. 2012. Tourism-driven urbanization in China's small town development: A case study of Zhapo Town, 1986–2003. *Habitat International*, 36, 152–160.

Qiu, Z., 2007. Study on the globalization of the Chinese movie industry. *Journal of Beijing Administrative College*, 1.

Renaud, B., 1981. *National Urbanization Policy in Developing Countries*. London: Oxford University Press.

Russo, A.P., 2002. The "vicious circle" of tourism development in heritage cities. *Annals Tourism Research*, 29(1), 165–182.

Shao, H., Zhang, Y. and Li, W., 2017. Extraction and analysis of city's tourism districts based on social media data. *Computers, Environment and Urban Systems*, 65, 66–78.

Shen, J. 2000. Chinese urbanization and urban policy. In C.M. Lau and J. Shen (eds) *China Review 2000* (pp. 455–480). Hong Kong: Chinese University Press.

Sjöholm, J. 2017. Authenticity and relocation of built heritage: The urban transformation of Kiruna, Sweden. *Journal of Cultural Heritage Management and Sustainable Development*, 7(2), 110–128.

Spence, M., Annez, P.C. and Buckley, R.M., 2009. *Urbanization and Growth*. Washington, DC: The World Bank.

Steenkamp, J.B.E., 2001. The role of national culture in international marketing research. *International Marketing Review*, 18(1), 30–44.

Stephen, F.M. and Bosak, K., 2019. *A Research Agenda for Sustainable Tourism*. Cheltenham: Edward Elgar.

Tisdell, C. and Wen, J., 1991. Investment in China's tourism industry: Its scale, nature, and policy issues. *China Economic Review*, 2(2), 175–193.

Tsang, N.K.F. and Hsu, C.H.C., 2011. Thirty years of research on tourism and hospitality management in China: A review and analysis of journal publications. *International of Hospitality Management*, 30, 886–896.

United Nations, 2010. *World Urbanization Prospects: The 2009 Revision.* New York: United Nations, Department of Economic and Social Affairs, Population Division.

United Nations, 2018. *2018 Revision of World Urbanization Prospects.* New York: United Nations, Department of Economic and Social Affairs, Population Division.

United Nations World Tourism Organization, 2018. *Tourism Highlights.* [Online]. Available from http://cf.cdn.unwto.org/sites/all/files/pdf/unwto_barom18_02_mar_apr_excerpt__0.pdf [Accessed 1 February 2019].

Wang, Y.P., Wang, Y. and Wu, J., 2009. Urbanization and informal development in China: Urban villages in Shenzhen. *International Journal of Urban and Regional Research*, 33(4), 957–973.

Xiao, H., 2006. The discourse of power: Deng Xiaoping and tourism development in China. *Tourism Management*, 27(5), 803–814.

Yang, Y., 2011. Number of hotel rooms set to soar. *China Daily* [Online]. Available from www.chinadaily.com.cn/cndy/2011-11/08/content_14053498.htm [Accessed 1 February 2019].

Zhang, H., Chong, K. and Ap, J., 1999. An analysis of tourism policy development in modern China. *Tourism Management*, 20(4), 471–485.

Zhang, H., Luo, J.M., Xiao, Q. and Guillet, B., 2013. The impact of urbanization on hotel development: Evidence from Guangdong Province in China. *International Journal of Hospitality Management*, 34, 92–98.

Zhang, H.Q., Pine, R. and Lam, T., 2005. *Tourism and Hotel Development in China: From Political to Economic Success.* New York: Psychology Press.

Zhang, J., 2011. Several problems in the course of urbanization in China and planning responses. In X. Qu and Y. Yang (eds) *International Conference on Information and Business Intelligence* (pp. 494–499). Berlin; Heidelberg: Springer.

Zhang, L. and Han, S.S., 2009. Regional disparities in China's urbanisation: An examination of trends 1982–2007. *International Development Planning Review*, 31(4), 355–376.

Zhang, S., 2002. Analysis on the difference of the developing level of the regional urbanization in China. *Population Journal*, 5, 37–42.

6

TOURISM AND GENTRIFICATION

Maria Gravari-Barbas and Sandra Guinand

Introduction

For the last ten years, international tourism has steadily been growing at an average rate of 4% per year. Years 2017 and 2018 have recorded the highest growth rate (7% and 6% respectively) since 2010 (UNWTO, 2019). Increasing tourism mobilities and development affect places in different ways. A large body of literature has been analysing the changes induced by tourism flows in different places around the world (Colomb and Novy, 2016; Bures and Cain, 2008; Fainstein and Judd, 1999a). Tourism impacts different types of spaces: urban centres (Fainstein and Gladstone, 1999), islands (Herrera et al., 2007) rural (Guimond and Simard, 2010; Donaldson, 2009; Gonzalez, 2017) or coastal areas (Mullins, 1994) and in different ways. Those more traditional residential, educational or commercial functions are challenged by new tourism (tourist rentals, shops catering to exclusive and international customers, or revamped and heritage-led urban areas). As the tourism economy can bring-in revenues for the country, city or local community, it can also harm the socio-economic composition of these places by introducing displacement or transforming the local economy. Interestingly, it is only recently that scholars in social sciences, in the fields of both gentrification and tourism studies, have paid attention to tourism and tourist mobility as main triggers for gentrification (Gotham, 2005a). In the scientific debate, this phenomenon has been coined as "tourism gentrification" (Gotham, 2005a, 2018; Hiernaux and Gonzalez, 2014; Gravari-Barbas and Guinand, 2017; Cócola-Gant, 2018), Gotham being the first to introduce the concept in the field of urban studies with his seminal work on the French Quarter Vieux Carré in New Orleans (2005a, 2005b). Since then, changes have occurred in different places and within different contexts. With the growth in tourism, this phenomenon can today be considered as both a widespread and a global one (Gotham, 2018). However, the concept still raises much debate among scholars in terms of conceptualisation and epistemological implication. This chapter aims to examine the historical and intellectual debate on the concept of tourism gentrification, in order to highlight the key contributions that have been identified so far. This critical retrospective should help to identify the limits and the tensions this concept holds and thus set the agenda for further research.

Looking at tourism gentrification

Tourism gentrification: the historical and intellectual development of a concept

From its early ages, tourism has been accompanied with structural, social, cultural and even political changes. For instance, concurrent with the implementation of new infrastructures or services, are the confrontation of otherness for tourists and locals and a new and fresh tourist "gaze" (Urry, 2002) on people, places and landscapes. The work of Mullins (1994) on class structure of an area mainly devoted to tourism on Australia's Golden Coast resulted in interesting findings. He evidenced the rise of a small bourgeoisie clearly linked to the tourism economy. This change in the social structure of an area affected by tourism was a ground-breaking result and raised attention to social/class transformation influenced by tourism. However, it is with the work of Gotham (2005a, 2005b) that the link between tourism and gentrification was expressly stated. When the author first coined the concept, it defined "the transformation of a middle-class neighbourhood into a relatively affluent and exclusive enclave marked by a proliferation of corporate entertainment and tourism venues" (Gotham, 2005a, p. 1102). The major interest with this definition lies in the dual processes of globalisation and localisation embedded in the urban redevelopment processes. In Gotham's case, tourism is expressed by international global actors (hotel chains, car rentals, tour operators, etc.), linked to the service industry (communications, finance, etc.) while at the same time investing on the local level by developing local culture, products and places for consumption that will appeal to visitors. For Gotham, the nexus between the global and the local in tourism (Milne and Ateljevic, 2001) cannot be separated. Gotham provided the missing conceptual link between the production-side and the demand-side explanations of gentrification, while avoiding one-sided conclusions (Gotham, 2005a, p. 1103). He offered a new way to theorise and analyse tourism as a set of practices that has causal impacts on urban forms, socio-spatial patterns and processes of urban development (Gotham, 2018).

In the recent years, scientific literature has shown that the relationship between tourism and gentrification is of a more complex and diverse nature. Works by researchers look at the process and investigate which comes first, gentrification or tourism. The analysis, for instance, has traditionally been focusing on the impact of tourism over population and territories as in the case for Mullins and Gotham's work. Indeed, in some cases, projects are planned and designed from their inception to cater to the visitors' economy (Fainstein and Judd, 1999b). In other cases, it is during a later stage that new owners, renters and consumers, as well as other institutional and collective social actors (real estate agents, developers, mortgage lenders, etc.) invest in urban areas (Hamnett, 1991). These actors are attracted by the services and the culture created and promoted, in order to attract local, regional or international visitors. This can be the case of derelict but centrally located places (former industrial, port or warehouse areas) that have experienced capital and human disinvestment since the second part of the twentieth century. These areas, often transformed into "tourism playgrounds" (Judd, 2003), can attract (often with the help of public intervention) residential gentrification. This was for instance the case in Baltimore, where the private/public-led Inner Harbor redevelopment generated gentrification phenomena that manifested with new constructions adjacent to the Harbor East area. However, another body of works has demonstrated that tourism actually follows urban gentrifiers (Bridge, 2007; Schlichtman and Patch, 2014; Schlichtman et al. 2017). Tourists are attracted to gentrified and gentrifying neighbourhoods. In Le Marais, Paris, for instance, tourism invested in the neighbourhood after its heritagisation and the general upgrading of its image and accessibility (Gravari-Barbas, 2017). This can be also understood by both the physical and the symbolic changes that these areas experienced after their gentrification: heritagisation, improvement of the urban infrastructure and public spaces, with simultaneous creation of shops that cater to new residents and are also very attractive to tourists (farmers' markets and gourmet shops, designer retail shops, specialised bookstores, art galleries, etc.).

Key issues on tourism gentrification

Short-term accommodations as "gentrifying machines"

Tourists are searching for contact and interaction with the local population (Stor and Kagermeier, 2017). They are particularly attracted by the possibilities offered by social and business operators to "share" the living quarters of the "locals". "Living like a local" becomes the ultimate value for international elites, since it offers a more distinctive and insider's approach to the places they visit. Far from being part of the so-called sharing economy (Hamari et al., 2016) tourism rentals exemplify the desire of tourists to go beyond traditional "commercial" accommodations in order to reach the status of a *cognoscente*, feeling and living "at home" at any longitude and latitude.

Tourism rentals represent in the recent years, a real urban phenomenon – a "new gentrification battlefront" (Cócola Gant, 2016), the consequences of which produced a considerable body of works in an increasingly large number and types of places, ranging from established tourist destinations in Europe (Amsterdam (Pinkster and Boterman, 2017), Athens (Balampanidis et al., 2019); Barcelona (Lopez-Gay and Cócola Gant, 2016; Qualgieri and Scarnato, 2017), Berlin (Novy, 2017), Reykjavik (Mermet, 2017), Vienna (Gunter, 2018)) and North America (Dudás et al., 2017)) to other, less common places in the world (González Pérez, 2017; Piñeros, 2017). Literature also tends to show that the phenomenon increasingly applies to emerging destinations such as Sofia, Bulgaria (Roelofsen, 2018).

In most cases, tourism rentals have been viewed as a powerful yet fearsome "gentrifying machine" converting housing into accommodation for visitors. In the case of Athens, Balampanidis et al. (2019) demonstrated that tourism is more pervasive than "classic" residential gentrification, since short-term tourism rentals tend also to apply to low-quality and low-return properties, considered until recently disadvantaged for conventional rental (and therefore resulting in residential gentrification). The authors showed (2019, p. 14) that, adequately refurbished, such properties can compete with lower-class hotel rooms as they have the advantage of their central location. This trend impacts the social composition of central and ethnically diverse neighbourhoods since it removes a long-standing barrier to gentrification in all the high-rise neighbourhoods of the city centre.

Researchers have indicated that apartment conversion into short-term tourism rentals causes an out-migration of residents, a shortage in housing and price increase which also excludes other residents from the possibility of moving into the area (Cócola Gant, 2016). According to Gravari-Barbas (2017) tourism rentals gentrification can be compared to "super-gentrification" phenomena (Lees, 2003). As Cócola Gant (2016, p. 7) showed, the tourism rental phenomenon corresponds to "a snowball process.... It leads to a form of collective displacement never seen in classical gentrification, that is to say, to a substitution of residential life by tourism".

This is also the conclusion of Gladstone and Préau (2008, p. 145):

> the growth of tourism is a major factor increasing potential land rents in inner-city neighbourhoods, with the resultant stock of gentrified spaces and neighbourhoods serving in turn to attract even more visitors, further increasing land values and leading

to even more gentrification and sweeping changes in the demographic composition of neighbourhoods.

Tourism-induced commercial gentrification

Tourism visitation tends to modify commercial and business landscape. This goes beyond the stereotypical image of the souvenir, postcard and T-shirt shops, which usually flourish in tourist areas. Tourism development brings in bigger and more international markets with specific needs in terms of catering and other types of consumption. Tourism demand impacts the existing restaurants, bars or shops, catering for locals, which either adapt their products to their new customers' demands and expectations or are being replaced by tourism-targeting businesses.

Tourists desire to be part of the local, share everyday experiences, visit "ordinary" places (Condevaux et al., 2019), go off the beaten track (Gravari-Barbas and Delaplace, 2015; Delaplace and Gravari-Barbas, 2016) contribute to investing in traditional local food markets (Gonzalez and Waley, 2013) as showed for Barcelona or Lisbon (Guimarães, 2018). Flea markets, such as Saint-Ouen in the north of Paris (Gravari-Barbas and Jacquot, 2019) are also illustrative of the changes brought by tourism: small objects, easier to carry, or design artefacts tend to substitute the usual market objects (second-hand or antiques). This evolution and adaptation of the offered products and services are symptomatic of the markets' transformation into tourist attractions. In the barrio Boqueria of Barcelona, for example, visitors can take part in cooking experiences or participate in guided tours (Crespi Vallbona and Domínguez Pérez, 2016).

If many existing restaurants or businesses adapt to serving tourism and are packaged and commodified for tourism, many others who do not support the tourism business have to eventually close down. This is especially the case of everyday shops, which are progressively replaced by others, mainly catering to tourism. Gazillo (1981) showed the impact of mass tourism on the bars and restaurants of Old Québec. However, the departure of everyday shops is not only attributed to the arrival of tourists, but also to the departure of significant numbers of local residents due to the touristification of the visited areas, as is the case of Old Québec (Berthold, 2012).

The role of tourism in commercial gentrification in places that have already experienced extensive residential gentrification (such as Le Marais in Paris or the Lower East Side in New York) is more difficult to analyse. The exclusive cultural spots, contemporary art galleries (Mathews, 2010), state-of-the-art temporary exhibitions, gourmet restaurants, trendy bars or "starchitectural" museums (Gravari-Barbas and Renard-Delautre, 2015), cater primarily to the neighbourhood and to the metropolitan customers and are analysed as artist-induced gentrification phenomena (Mathews, 2010). However, located in areas that are current premium urban destinations, and made part of the global tourist, leisure, culture and art offer of the area, they are also shaped by (and cater to) the international tourism demand. Commercial gentrification is expressed by new relationships between previously unrelated domains: gastronomy, fashion, art, architecture, design, etc. closely interacting with each other and offering a total and exclusive experience (Lipovetsky and Serroy, 2013).

Although commercial gentrification follows and reflects residential gentrification, often commercial and residential gentrification are parallel and reciprocally supported processes. Tourism-induced commercial gentrification may even precede residential gentrification phenomena. In Saint-Ouen's flea market for example, commercial gentrification can be viewed as the pioneering front of future changes of the residential patterns (Cousin et al., 2015).

Tourism gentrification and displacement

Quoting García and Claver (2003), Häussermann and Colomb (2003), Hoffman (2003), Terhorst et al. (2003) or Gotham (2005a), Cócola Gant (2015) underlines the conflicting nature of commerce gentrification and shows the increasing conflicts between how affluent visitors and residents use the city and the needs of lower income residents. Tourism commercial gentrification leads to displacement not only of the commercial venues but also of the local residents to whom those venues used to cater.

Using Marcuse's conceptualisation, Cócola Gant (2015) distinguishes between "direct displacement" and "indirect displacement".

> While "direct displacement" refers to the out-migration from the neighbourhood or the moment of eviction, "indirect displacement" is a long-term process that results in a set of pressures that makes it progressively difficult for low-income residents to remain over time.
>
> *(Cócola Gant, 2015, p. 8)*

Tourism-induced commercial gentrification "signifies that residents have lost their battle to remain". The author draws on Davidson (2008, 2009) and Davidson and Lees (2010) who suggest that "the pressure of indirect displacement leads residents to experience what they call 'loss of place': a forced dispossession and dislocation from their places that leads them to a form of 'displacement' into a new colonised social context" (Cócola Gant, 2015, p. 9).

Self- and bottom-up tourism gentrification

Tourism gentrification goes beyond the forces of the corporate tourism industry that influences space development and consumption. Authors have particularly paid attention to local actors, and residents, as well as the tourists themselves, as main contributors to the gentrification phenomena. For instance, the work of Herzfeld (2017) on the Old Town of Rethymnon on the island of Crete and the community of Pom Mahakan in the old historic centre of the Thai capital Bangkok, has shown that local actors do act, cope and structure their environment in order to take advantage of the tourism economy leading to transformation in the urban environment. This analysis has also been drawn by Chan et al. (2016) in their investigation of the Honghe Hani Rice Terraces in Hunan, China. The authors noticed that some Indigenous people were taking advantage of the UNESCO title and the tourism flow as a mean to improve their socio-economic standing and reach middle-class standards, particularly through adopting entrepreneurial strategies gleaned from their encounters with outside-gentrifiers and tourists. The authors' position on the concept of "self-gentrification" is interesting as they consider it a reaction (to help conserve both heritage landscapes and Indigenous ways of life) to the process of "external" gentrification. At another level of analysis, Freytag and Bauder (2018) show that in Paris tourists can, through mundane practices such as soft mobility, induce bottom-up touristification and contribute to the transformation of urban areas. For example, cycling (using rental bikes) and walking as tourist practices (Lorimer, 2011), the authors argue, induce bottom-up touristification by interconnecting tourist hotspots and the accommodation locations. The cyclists and walkers, both tourists and local residents, contribute, with their presence, to the urban transformation (tangible and intangible) and to the touristification of the places that are located along the way (Jensen, 2009). Bottom-up approaches to tourism gentrification are very much connected to new patterns of behaviours, consumption and commodification that are, consequently connected to financial incentives and opportunities (Condevaux et al., 2019). Different authors and bodies of work show that the tourist alone cannot be solely blamed for the gentrification process; instead tourism and gentrification are much more complex phenomena (Füller and Michel, 2014).

Vieux Québec – mini case study

On the UNESCO World Heritage list since 1985 for its "living and inhabited" character, the Vieux Québec, has since then been loosing inhabitants, public services and daily amenities. On the other hand, this historical core has witnessed significant and regular tourism growth with an increasing internationalisation of visitors and a growing flow of cruise passengers. This important increase has generated a set of diversified offers to cater to the visitors' demands (new events, shops, restaurants, cultural facilities and amenities). But it has also destructured the housing market notably in terms of rentals with a strong demand for short-term rentals and real estate speculation. This has given rise to a boom in tourist residences and Airbnb-type accommodations in Old Québec with growing discontent from the residents but also from the hotel industry.

In this difficult social-economical environment and increasing tensions, the city of Québec implemented in 2012 a consultation process bringing together the main stakeholders of the historic centre. The objective was to work to reverse the current trend of residents' decline and to reflect on the coexistence of different uses in Old Québec(residential, commercial, institutional, administrative, touristic, etc.). This was also a consequence of the mobilisation of engaged citizens of Old Québec (Comité des citoyens du Vieux-Québec) who, together with the Québec City area hotel association, have also pushed for a population consultation on short-term rentals in 2018. The city has since then reformed the tax rating system for short-term rentals, required them to be officially declared as touristic accommodation and limited the permit to transform or construct condominiums for tourism purposes.

Current scope of interests and future perspectives on tourism gentrification

An abundant, recent and burgeoning literature examines the field of tourism gentrification today. Some of the numerous emerging issues pose questions and deserve more attention. They also call for more comparative studies – the risk being to multiply the individual cases and the monographs.

Post-colonial approach to tourism gentrification

According to Cócola-Gant (2018), in recent years, a new set of literature coming from the non-Western countries has emerged among the scientific debate on tourism gentrification (Hiernaux and Gonzalez, 2014; Janoschka et al., 2013). It comes from a posture of "epistemological resistance" (de Sousa Santos, 2012) to Western scientific paradigms and attempts to reach a new definition of "cosmopolitanism" (Hiernaux and Gonzalez, 2014, p. 59). This posture sets a new light and brings in new approaches. For instance, a broad set of literature from Chinese scholars has focused on the radical transformation of the socio-economic urban landscape – from lower-class (or, even rural) to upper-class spaces (due to the fast pace of the phenomena) (Su and Teo, 2009; Zhao et al., 2009) through the implementation of residential tourism leisure areas (Liang, 2017). In these cases, tourism becomes the main driving force in economic, social, cultural and lifestyle transformation (Liang and Bao, 2015). In Latin America scholars such as Hiernaux and Gonzalez (2014) have challenged the idea of the "rent gap" (Smith, 1987) in the reconfiguration of certain historic centres or certain places, to replace it with the idea of a reconfiguration of the land market. The main claim being that the differential in income from the land in these Latin American urban settings is being sustained by symbolic values that can cater to potential uses and imaginaries of visitors. Emphasis is put on the intangible dimensions of these specific areas being transformed by and for tourism. Following the same trend, other researchers, as in the case of Singapore's little India (Chang, 2016), have questioned Western expressions of commodification of culture such as "gentrification aesthetics", as they draw on different perspectives depending on the local policy and state ideology which influence urban redevelopment. This broad range of literature marks the beginning of a much wider and promising set of new perspectives on the tourism gentrification phenomenon.

Tourism gentrification in a post/hyper/trans-tourism era

The increasingly blurred lines between the ideal/typical figures of tourists and residents, their utilisation of space, and the city at large, in a closer interaction, the exchange and sharing of places and leisure activities, impacts gentrification phenomena. Tourists and visitors not only reproduce the implicit rules in the guided journey but also experience the city and its spaces following their own initiatives (Judd, 2003). The figures of the "post-tourist" (Urry, 2002; Feifer, 1985), the "hyper-tourist" (Viard, 2000) or the "trans-tourist" (Ateljevic, 2009; Corneloup, 2009), discussed by Bourdeau (2018), provide an overview of the changing status and practices of tourism. Respectively referring to "play, oblique posturing and ludic transgression interact(ing) with artifice, assumed inauthenticity and pastiche, the clash of opposites, provocation and even cynicism" (post-tourism), the "technological, geographical and cultural intensification of the meaning and forms of tourism" (hyper-tourism) and the "transcendence of the usual borders and categories with the aim of exploring neglected or repressed othernesses" (trans-tourism) (Bourdeau, 2018), these renewed tourism forms show the capacity and will of tourists to escape from the "tourist bubble" while creating the conditions for tourism gentrification. As Gravari-Barbas and Guinand (2017) stressed, these post/hyper/trans situations go hand-in-hand with gentrification: as visitors are increasingly searching for new and "off-the-beaten-track" places, cities try to capture visitor (local or foreign) imagination, offering an array of activities and places to explore. Thus, gentrification affects increasingly new destinations, which are reputed to be "more authentic" as well as "off-the-beaten-track" areas in established destinations.

The image of the post/hyper/trans-tourist finds its mirror in the cosmopolitan "polytopical" resident (Stock, 2007) who has multiple residential attachments. Literature connects both expressions to gentrification phenomena (Guinand, 2017). But as the tourist's profile (increasingly behaving "as a local") and the local inhabitant's profile (increasingly cosmopolitan and multi-resident) tend to be blurred, the literature referring to the phenomena particularly relating to tourism or to residential gentrification, also tends to be blurred, thus calling for necessary clarifications and theoretical frames by researchers (Sequera and Nofre, 2018).

Tourism as a "super-gentrifier"

In several central metropolitan areas that have experienced residential gentrification phenomena since the 1960s, tourism may act as a super-gentrifier. Lees (2003) used the term of "super-gentrification" to describe the real estate market evolution in Brooklyn Heights, NY: "the transformation of already gentrified, prosperous and solidly upper-middle-class neighbourhoods into much more exclusive and expensive enclaves". She localised this intensified re-gentrification in a few select areas of global cities such as London (Butler and Lees, 2006) and New York "that have become the focus of intense investment and conspicuous consumption by a new generation of super-rich 'financifiers' fed by fortunes from the global finance and corporate service industries" (Lees, 2003, p. 2487). According to the author, this new stage of gentrification, which started in New York in the 1990s, does not follow a disinvestment stage. It rather corresponds to a new wave of gentrification and housing renewal to even higher standards that have the potential to partially evict the early gentrifiers contradicting the Marxist model of capital investment and disinvestment cycles in the urban space. Unlike gentrification that occurs in different phases and, therefore, suggests that the process will finally reach a state of stabilisation, super-gentrification emphasises the ability of the process to constantly renew itself. This means that its end cannot be imagined. Much more than a residential real estate market, short-term tourist real estate tends to act as a super-gentrifier. In the Marais, Paris, for example, short-time tourism rentals tend to substitute first and second-wave gentrification of this central, historic and highly attractive Parisian neighbourhood, since tourism rents generate more value than classic residential rents that are already fairly expensive (Gravari-Barbas, 2017). The capacity of tourism to constantly renew the building stock without experiencing intermediary disinvestment stages represents a fertile ground for further research.

Hyper-tourismification, tourism gentrification and the emerging tourism governance

According to Gravari-Barbas (2017) "hyper-tourismification" designates tourism phenomena observed in areas used for leisure purposes by both tourists and local residents, characterised by the embedment of tourism into everyday life; from globalisation of the real estate tourism markets and the intermingling of residential and tourism rentals; by a global *culturescape* and *brandscape*; by creativity, design and contemporary art; by architectural iconicity; by hyper-aestheticisation; and by the existence of a tourism governance. These areas are *infused* by tourism, which plays an important role on neighbourhood planning issues, on the nature of commercial functions and on the tourists' and locals' place imaginary.

In these hyper-tourismified areas, residential and commercial super-gentrification tend to produce specific urban environments designed to cater to a large range of "gentrifiers", from "permanent" (often with "polytopic" and "multiresidential" attachments (Stock, 2007)) and to more ephemeral (tourist) populations. This situation of different strata of gentrifiers is two-fold: on the one hand, it results in conflicts, which have been abundantly studied in recent works (Colomb and Novy, 2016); on the other hand, it leads to the development of specific tools and regulations which tend to produce, in these super-gentrified areas, a specific "tourism governance". The local decision-makers are aware that tourism is not only a "cash windfall" but also an increasingly constitutive component of the hyper-tourismified areas. Consequently, in these areas, and more specifically in central neighbourhoods of large metropolises, a new governance model is taking place that involves permanent and secondary residents (a growing proportion of whom come to these areas for similar reasons as the tourists), tourists (for whom residents form part of the attractiveness of the place) and local decision-makers who integrate, or even use tourism, as a part of cultural differentiation in their neighbourhood.

A promising research field is therefore the evolution of those super-gentrified areas, in which tourism experienced though tourism rentals, boutique hotels or other hybrid residential formats, tends to transform the nature of neighbourhoods.

Conclusion and further agenda for research

As we have seen in this chapter, tourism gentrification has, since it was first coined in 2005 (Gotham, 2005a, 2005b), been widely used to define different and complex phenomena inducing spatial, social and economic change linked to the tourism economy. This wide spread of the concept calls for further research as well as recommendations for practitioners and policy makers in the field of tourism. First, one should be careful not to fall into the trap of calling "tourism gentrification" any transformation set in a tourism context. As we have attempted to underline in this chapter, tourism gentrification is of a complex and sometimes counterintuitive nature. It does not follow logic or linear patterns and goes beyond traditional dichotomies (Sequera and Nofre, 2018). For instance, speculation does not always cause displacement, as the first ones to speculate could be the residents themselves. For some (Sequera and Nofre, 2018), the overuse of the term tourism gentrification implies an epistemic gap between touristification and gentrification. According to the authors, both theoretical and conceptual tools need to be used especially in the field of urban studies. Sequera and Nofre mobilise the example of protests against mass tourism in central urban cores that have been held by popular classes, but also by the middle class and the elite, calling for a more mundane and ordinary landscape setting, in order to show that when touristification occurs, it is not always linked to gentrification (2018, p. 851). This means that policy makers need to be attentive to the mechanisms of gentrification and not be too quick to link them to tourism. For instance, as we have seen, the post-tourism phenomenon is one outcome of the call for a more "authentic" experience that concerns the whole set of actors who participate in the tourism economy and sometimes lead to a gentrification outcome. This is also one of the reasons why, in the scientific field, tourism, touristification and gentrification still need further research and theoretical debate, since an understanding of the process has not been fully achieved yet (Cócola-Gant, 2018).

If further investigation is needed on the epistemic ground, other fields and perspectives also need to be addressed to further the research agenda. For instance, little has been said on urban landscape (Duarte Paes, 2017) and tourism gentrification. Scientific research needs to go beyond the commodification of culture and postmodern landscape debate (Zukin, 1992) and investigate the changes developing most notably within its intangible dimension, and how people (residents, tourists, locals, etc.) relate to these changes. Does it, as Cócola-Gant (2018) puts it, provoke a loss of place? This question could be better investigated by practitioners and public authorities with the help of academia. Another limit that has been highlighted in the field of tourism gentrification is the dichotomy between mobility and immobility: the ones who move and the ones who settle and stay. This categorisation then tends to essentialise categories of "residents" or "locals" as being static and rooted (Franquesa, 2011). However, as we all know, residents are also visitors and could be so in their own city. Further academic research should be undertaken to unravel these categories, their attributes and what they entail in a time of increasing multiresidential individuals and public authorities should be more attentive to multiple "residents" they host in their territories in order to better understand the socio-economic and spatial dynamics but also better respond to the needs of these individuals.

Finally, in a context of growing tourism economy, growing vulnerabilities, frustrations and resistances, there is a crucial need to push the research agenda on tourism gentrification further in order to better grasp and unravel the complexity of this phenomenon. In fact, rather than a symptom, tourism gentrification should be used as an instrument and a lens through which to read the complex balance of social, economical and political power.

References

Ateljevic, I. (2009) "Transmodernity: Remaking our (tourism) world? Theories of modern and postmodern tourism". In J. Tribe (ed.) *Philosophical Issues in Tourism*. Channel View Publications, pp. 278–300.

Balampanidis, D., Maloutas, T., Papatzani, E. and Pettas, D. (2019) "Informal urban regeneration as a way out of the crisis? Airbnb in Athens and its effects on space and society", *Urban Research & Practice*. DOI: 10.1080/17535069.2019.1600009.

Berthold, E. (ed.) (2012) *Les quartiers historiques. Pressions, enjeux, actions*. Québec: Presses de l'Université Laval.

Bourdeau, P. (2018) "After-tourism revisited", *Via Tourism Review*, 13, online: https://journals.open edition.org/viatourism/1971.

Bridge, G. (2007) "A global gentrifier class?, *Environment and Planning A*, 39(1), pp. 32–46.

Bures, R. and Cain, C. (2008) *Dimensions of Gentrification in a Tourist City*, online: http://paa2008. princeton.edu/papers/81623.

Butler, T. and Lees, L. (2006) "Super-gentrification in Barnsbury, London: globalization and gentrifying global elites at the neighbourhood level", *Transactions of the Institute of British Geographers*, NS 31, pp. 467–487.

Chan, J.H., Iankova, K., Zhang, Y., McDonald, T. and Qi, X. (2016) "The role of self-gentrification in sustainable tourism: Indigenous entrepreneurship at Honghe Hani Rice Terraces World Heritage Site, China", *Journal of Sustainable Tourism*, 24(8–9), pp. 1262–1279. DOI: 10.1080/09669582.2016.1189923.

Chang, T.C. (2016) "'New uses need old buildings': Gentrification aesthetics and the arts in Singapore", *Urban Studies*, 53(3), pp. 524–539. DOI: 10.1177/0042098014527482.

Cócola Gant, A. (2015) "Tourism and commercial gentrification". Paper presented at the RC21 International Conference on *The Ideal City: Between Myth and Reality. Representations, Policies, Contradictions and Challenges for Tomorrow's Urban Life*, Urbino (Italy), 27–29 August 2015, online: www.rc21.org/en/conferences/urbino2015/.

Cócola Gant, A. (2016) "Holiday rentals: The new gentrification battlefront", *Sociological Research Online*, 21(3), online: www.socresonline.org.uk/21/3/10.html.

Cócola-Gant, A. (2018) "Tourism gentrification". In L. Lees and M. Phillips (eds.) *Handbook of Gentrification Studies*. London: Edward Elgar Publishing, pp. 281–296.

Colomb, C. and Novy, J. (eds.) (2016) *Protest and Resistance in the Tourist City*. New York: Routledge.

Condevaux, A., Gravari-Barbas, M. and Guinand, S., (eds.) (2019) *Lieux ordinaires, avant et après le tourisme*. Paris: PUCA.

Corneloup, J. (2009) "Comment aborder la question de l'innovation?", *Revue de Géographie Alpine*, Dossier 97–1, online: http://rga.revues.org/index828.html.

Cousin, S., Djament, G., Gravari-Barbas, M. and Jacquot, S. (2015) "Contre la métropole créative … tout contre/Les politiques patrimoniales et touristiques de Plaine Commune, Seine-Saint-Denis", *Metropoles*, 17, online: https://metropoles.revues.org/5171.

Crespi Vallbona, M. and Domínguez Pérez, M. (2016) "Los mercados de abastos y las ciudades turísticas", *Pasos*, 14(2), pp. 401–416.

Davidson, M. (2008) "Spoiled mixture: Where does state-led positive gentrification end?", *Urban Studies*, 45(12), pp. 2385–2405.

Davidson, M. (2009) "Displacement, space and dwelling: Placing gentrification debate", *Ethics Place and Environment*, 12(2), pp. 219–234.

Davidson, M. and Lees, L. (2010) "New-build gentrification: Its histories, trajectories, and critical geographies", *Population, Space and Place*, 16(5), pp. 395–411.

de Sousa Santos, B. (2012) "Public sphere and epistemologies of the South", *Africa Development*, 37(1), pp. 43–67.

Delaplace, M. and Gravari-Barbas, M. (2016) "On the margins of tourism: Utopias and realities", *Via Tourism Review*, 9, online: https://journals.openedition.org/viatourism/417.

Donaldson, R. (2009) "The making of a tourism-gentrified town: Greyton, South Africa", *Geography*, 94, pp. 88–121.

Duarte Paes, M.T. (2017) "Gentrificação, preservação patrimonial e turismo: os novos sentidos da paisagem urbana na renovação das cidades", *Geousp – Espaço e Tempo*, 21(3), pp. 167–184. DOI: 10.11606/issn.2179-0892.geousp.2017.128345.

Dudás, G., Vida, G., Kovalcsik, T. and Boros, L. (2017) "A socio-economic analysis of Airbnb in New York City", *Regional Statistics*, 7(1), pp. 135–151; DOI: 10.15196/RS07108.

Fainstein, S. and Gladstone, D. (1999) "Evaluating urban tourism". In D. Judd and S. Fainstein (eds.) *The Tourist City*. London: Yale University Press, pp. 21–34.

Fainstein, S. and Judd, D. (1999a) "Global forces, local strategies and urban tourism". In D. Judd and S. Fainstein (eds.) *The Tourist City*. London: Yale University Press, pp. 1–17.

Fainstein, S. and Judd, D. (1999b) Cities as place to play. In D. Judd and S. Fainstein (eds.) *The Tourist City*. London: Yale University Press, pp. 261–272.

Feifer, M. (1985) *Going Places*. London: Macmillan.

Franquesa, J. (2011) "'We've lost our bearings': Place, tourism, and the limits of the 'mobility turn'", *Antipode*, 43(4), pp. 1012–1033. http://doi.org/10.1111/j.1467-8330.2010.00789.x.

Freytag, T. and Bauder, M. (2018) "Bottom-up touristification and urban transformations in Paris", *Tourism Geographies*, 20(3), pp. 443–460. DOI: 10.1080/14616688.2018.1454504.

Füller, H. and Michel, B. (2014) ""Stop being a tourist!" New dynamics of urban tourism in Berlin-Kreuzberg", *International Journal of Urban and Regional Research*, 38(4), pp. 1304–1318. http://doi.org/10.1111/1468-2427.12124.

García, M. and Claver, N. (2003) "Barcelona: Governing coalitions, visitors and the changing city center". In L. Hoffman, S. Fainstein and D.R. Judd (eds.) *Cities and Visitors: Regulating People, Markets, and City Space*. Oxford: Blackwell, pp. 113–125.

Gazillo, S. (1981) "The evolution of restaurants and bars in Vieux-Québec since 1900", *Cahiers de géographie du Québec*, 25(64), pp. 101–118.

Gladstone, D. and Préau, J. (2008) "Gentrification in tourist cities: Evidence from New Orleans before and after Hurricane Katrina", *Housing Policy Debate*, 19(1), pp. 137–175.

Gonzalez, P.A. (2017) "Heritage and rural gentrification in Spain: The case of Santiago Millas", *International Journal of Heritage Studies*, 23(2), pp. 125–140.

Gonzalez, S. and Waley, P. (2013) "Traditional retail markets: The new gentrification frontier?", *Antipode: A Radical Journal of Geography*, 45(4), pp. 965–983.

González Pérez, J. (2017) "A new colonisation of a Caribbean city: Urban regeneration policies as a strategy for tourism development and gentrification in Santo Domingo's Colonial City". In M. Gravari-Barbas and S. Guinand (eds.) *Tourism and Gentrification in Contemporary Metropolises: International Perspectives*. Oxford: Routledge, pp. 25–41.

Gotham, K.F. (2005a) "Tourism gentrification: The case of New Orleans' Vieux Carré (French Quarter)", *Urban Studies*, 42(7), pp. 1099–1121.

Gotham, K.F. (2005b) "Tourism from above and below: Globalization, localization, and New Orleans's Mardi Gras", *International Journal of Urban and Regional Research*, 29(2), pp. 309–326. DOI: 10.1111/j.1468-2427.2005.00586.x.

Gotham, K.F. (2018) "Assessing and advancing research on tourism gentrification", *Via Tourism Review*, (13), online: http://doi.org/10.4000/viatourism.2169.

Gravari-Barbas, M. (2017) "Super-gentrification and hyper-tourismification in Le Marais, Paris". In M. Gravari-Barbas and S. Guinand (eds.) *Tourism and Gentrification in Contemporary Metropolises: International Perspectives*. Oxford: Routledge, pp. 299–329.

Gravari-Barbas, M. and Delaplace, M. (2015) "Le tourisme urbain 'hors des sentiers battus'. Coulisses, interstices et nouveaux territoires urbains", *TEOROS*, (34), pp. 1–2, online: https://journals.openedition.org/teoros/2729.

Gravari-Barbas, M. and Guinand, S. (eds.) (2017) *Tourism and Gentrification in Contemporary Metropolises: International Perspectives*. Oxford: Routledge.

Gravari-Barbas, M. and Jacquot, S. (2019) "Mechanisms, actors and impacts of the touristification of a tourism periphery: The Saint-Ouen flea market, Paris", *Journal of Tourism and Cultural Change* (forthcoming).

Gravari-Barbas, M. and Renard-Delautre, C. (2015) *Figures d'architectes et espace urbain/Celebrity Architects and Urban Space*. Paris: L'Harmattan.

Guimarães, P.P.C. (2018) "The transformation of retail markets in Lisbon: An analysis through the lens of retail gentrification", *European Planning Studies*, 26(7), pp. 1450–1470. http://doi.org/10.1080/09654313.2018.1474177.

Guimond, L. and Simard, M. (2010) "Gentrification and neo-rural populations in the Québec countryside: Representations of various actors", *Journal of Rural Studies*, 26(4), pp. 449–464.

Guinand, S. (2017) "Post-tourism on the waterfront: Bringing back locals and residents at the Seaport". In M. Gravari-Barbas and S. Guinand (eds.) *Tourism and Gentrification in Contemporary Metropolises: International Perspectives*. Oxford: Routledge, pp. 207–232.

Gunter, U. (2018) "Austria determinants of Airbnb demand in Vienna and their implications for the traditional accommodation industry", *Tourism Economics*, 24(3), pp. 270–293.

Hamari, J., Sjöklint, M. and Ukkonen, A. (2016) "The sharing economy: Why people participate in collaborative consumption", *Journal of the Association for Information Science and Technology*, 67(9), C1–C1, pp. 2045–2306.

Hamnett, C. (1991) "The blind men and the elephant: The explanation for gentrification", *Transactions of the Institute of British Geographers*, 16, pp. 173–189.

Häussermann, H. and Colomb, C. (2003) "The new Berlin: Marketing the city of dreams". In L. Hoffman, S. Fainstein and D.R. Judd (eds.) *Cities and Visitors: Regulating People, Markets, and City Space*. Oxford: Blackwell, pp. 200–218.

Herrera, L.M.G., Smith, N. and Vera, M.Á.M. (2007) "Gentrification, displacement, and tourism in Santa Cruz de Tenerife", *Urban Geography*, 28(3), pp. 276–298. DOI: 10.2747/0272–3638.28.3.276.

Herzfeld, M. (2017) "Playing for/with time: Tourism and heritage in Greece and Thailand". In M. Gravari-Barbas and S. Guinand (eds.) *Tourism and Gentrification in Contemporary Metropolises: International Perspectives*. Oxford: Routledge, pp. 233–252.

Hiernaux, D. and Gonzalez, C.I. (2014) "Turismo y gentrificación: pistas teóricas sobre una articulación", *Revista de Geografía Norte Grande*, 58, pp. 55–70.

Hoffman, L.M. (2003) "Revalorizing the inner city: Tourism and regulation in Harlem". In L.M. Hoffman, S. Fainstein and D.R. Judd (eds.) *Cities and Visitors: Regulating People, Markets, and City Space*. Oxford: Blackwell, pp. 91–112.

Janoschka, M., Sequera, J. and Salinas, L. (2013) "Gentrification in Spain and Latin America: A critical dialogue", *International Journal of Urban and Regional Research*, 38(4), pp. 1234–1265.

Jensen, O.B. (2009) "Flows of meaning, cultures of movements: Urban mobility as a meaningful everyday life practice", *Mobilities*, 4(1), pp. 139–158.

Judd, D.R. (2003) *The Infrastructure of Play: Building the Tourist City*. Cleveland, OH: Cleveland State University.

Lees, L. (2003) "Super-gentrification: The case of Brooklyn Heights, New York City", *Urban Studies*, 40(12), pp. 2487–2509.

Liang, Z.X. (2017) "The rent gap re-examined: Tourism gentrification in the context of rapid urbanisation in China". In M. Gravari-Barbas and S. Guinand (eds.) *Tourism and Gentrification in Contemporary Metropolises: International Perspectives*. Oxford: Routledge, pp. 276–298.

Liang, Z.-X. and Bao, J.-G. (2015) "Tourism gentrification in Shenzhen, China: Causes and socio-spatial consequences", *Tourism Geographies*, 17(3), pp. 461–481. DOI: 10.1080/14616688.2014.1000954.

Lipovetsky, G. and Serroy, J. (2013) *L'esthétisation du monde. Vivre à l'âge du capitalisme artiste*. Paris: Gallimard.

Lopez-Gay, A. and Cócola Gant, A. (2016) "Cambios demográficos en entornos urbanos bajo presión turística: el caso del barri Gòtic de Barcelona". In J. Domínguez-Mújica, R. Díaz-Hernández (eds.) *XV Congreso Nacional de la Población Española*. Fuerteventura: Asociación de Geógrafos Españoles, pp. 399–413.

Lorimer, H. (2011) "Walking: New forms and spaces of pedestrianism". In T. Cresswell and P. Merriman (eds.) *Geographies of Mobilities: Practices, Spaces, Subjects*. London: Ashgate, pp. 19–34.

Mathews, V. (2010) "Aestheticizing space: Art, gentrification and the city geography", *Compass*, 4/6, pp. 660–675.

Mermet, A.-C. (2017) "Airbnb and tourism gentrification: Critical insights from the exploratory analysis of the 'Airbnb syndrome' in Reykjavík". In M. Gravari-Barbas and S. Guinand (eds.) *Tourism and Gentrification in Contemporary Metropolises: International Perspectives*. Oxford: Routledge, pp. 52–74.

Milne, S. and Ateljevic, I. (2001) "Tourism, economic development and the global local nexus: Theory embracing complexity", *Tourism Geographies*, 3(4), pp. 369–393.

Mullins, P. (1994) "Class relations and tourism urbanization: The regeneration of the petite bourgeoisie and the emergence of a new urban form". *International Journal of Urban and Regional Research*, 4(18), pp. 591–608.

Novy, J. (2017) "The selling (out) of Berlin and the *de-* and *re-politicization* of urban tourism in Europe's 'Capital of Cool' ". In C. Colomb and J. Novy (eds.) *Protest and Resistance in the Tourist City*. London: Routledge, pp. 51–72.

Piñeros, S. (2017) "Tourism gentrification in the cities of Latin America: The socio-economic trajectory of Cartagena de Indias, Colombia". In M. Gravari-Barbas and S. Guinand (eds.) *Tourism and Gentrification in Contemporary Metropolises: International Perspectives*. Oxford: Routledge, pp. 75–103.

Pinkster, F. and Boterman, W. (2017) "When the spell is broken: Gentrification, urban tourism and privileged discontent in the Amsterdam canal district", *Cultural Geographies*, 24(3), pp. 457–472.

Qualgieri, A. and Scarnato, A. (2017) "The Barrion Chino as last frontier: The penetration of everyday tourism in the dodgy heart of the Raval". In M. Gravari-Barbas and S. Guinand (eds.) *Tourism and Gentrification in Contemporary Metropolises: International Perspectives*. Oxford: Routledge, pp. 107–133.

Roelofsen, M. (2018) "Performing 'home' in the sharing economies of tourism: The Airbnb experience in Sofia, Bulgaria", *Fennia*, 196(1), pp. 24–42. https://doi.org/10.11143/fennia.66259.

Schlichtman, J. and Patch, J. (2014) "Gentrifier? Who, me? Interrogating the gentrifier in the mirror", *International Journal of Urban and Regional Research*, 38(4), pp. 1491–1508.

Schlichtman, J.J., Patch, J. and Hill, M.L. (2017) *Gentrifier*. Toronto: University of Toronto Press.

Sequera, J. and Nofre, J. (2018) "Shaken, not stirred", *City*, 22(5–6), pp. 843–855. DOI: 10.1080/13604813.2018.1548819.

Smith, N. (1987) "Gentrification and the rent gap", *Annals of the Association of American Geographers*, 77(3), pp. 462–465.

Stock, M. (2007) "Théorie de l'habiter. Questionnements". In T. Paquot (ed.) *Habiter, le propre de l'humain. Villes, territoire et philosophie*. Paris: La Découverte, « Armillaire », pp. 103–125, online: www.cairn.info/ habiter-le-propre-de-l-humain-9782707153203-page-103.htm.

Stor, N. and Kagermeier, A. (2017) "The sharing economy and its role in metropolitan tourism". In M. Gravari-Barbas and S. Guinand (eds.) *Tourism and Gentrification in Contemporary Metropolises: International Perspectives*. Oxford: Routledge, pp. 181–206.

Su, X. and Teo, P. (2009) *The Politics of Heritage Tourism in China: A View from Lijiang*. New York: Routledge.

Terhorst, P., Ven, J. van de, and Deben, L. (2003) "Amsterdam: It's all in the mix". In L. Hoffman, S. Fainstein and D.R. Judd (eds.) *Cities and Visitors: Regulating People, Markets, and City Space*. Oxford: Blackwell, pp. 75–90.

UNWTO – United Nations World Tourism Organization. (2019) *International Tourism Results 2018 and Outlook 2019*, online: http://cf.cdn.unwto.org/sites/all/files/pdf/unwto_barometer_jan19_presentation_ en.pdf accessed 31 March 2019.

Urry, J. (2002) *The Tourist Gaze*. London: Sage Publications.

Viard, J. (2000) *Court traité sur les vacances: les voyages et l'hospitalité des lieux*. Paris: L'Aube.

Zhao, Y.Z., Kou, M., Lu S. and Li, D.H. (2009) "The characteristics and causes of urban tourism gentrification: A case of study in Nanjing", *Economic Geography*, 29(8), pp. 1391–1396.

Zukin, S. (1992) "Postmodern urban landscapes: Mapping culture and power". In S. Lash and J. Friedman (eds.) *Modernity and Identity*. London: Basil Blackwell, pp. 221–247.

7

URBANISM AND OVERTOURISM

Impacts and implications for the city of Hyderabad

Gaitree (Vanessa) Gowreesunkar and Mahender Reddy Gavinolla

Introduction

Urbanism and overtourism are new realities in many popular tourism cities (see Sommer, 2018; Kiralova and Hamarneh, 2018; Séraphin et al., 2018; Ashworth and Page, 2011). Studies show that because tourism cities perform multiple functions that are conducive for urbanised tourists (Maxim, 2017), a new breed of travellers driven by globalisation and sophistication are increasingly opting for this form of tourism (see Maxim, 2017; Hall, 2006; Gowreesunkar, 2019). As such, in modern tourism settings, research on urban places and their impacts on economy, society and environment have become conspicuous (Kuščer and Mihalič, 2019; Tribe, 1997; Farrell and Twining-Ward, 2004). Urban areas are usually places with a dense population, a major transport hub and a gateway for further travel in the region, as well as commercial, financial and industrial centres, and they offer a variety of recreational and cultural experiences (Ashworth and Tunbridge, 1990; Kiralova and Hamarneh, 2018). As a result, when from one side, urbanism is an evolving phenomenon that is rapidly spreading over the tourism industry, in parallel, overtourism, as another emerging phenomenon, has also started manifesting in many urbanised cities (see Séraphin et al., 2019). Therefore, urbanism and overtourism cannot be treated using the same tool; rather, these are two distinct phenomena which deserve separate attention, as the former is a consequence derived from environmental development (Bock, 2015) while the other is the outcome of too many visitors at a tourism place (Gowreesunkar, 2019). However, in the context of tourism, it would seem that urbanism is linked to tourism and both may be managed jointly. Sommer's (2018) study on "what begins at the end of urban tourism" is clear evidence that overtourism follows urbanism. Raising this point to another level, it would seem that consumerism has created a new breed of tourists characterised by emancipation, sophistication and urbanisation (Gowreesunkar, 2019), whereas technology (among other factors) has influenced the phenomenon of overtourism (see Kuščer and Mihalič, 2019; Séraphin, 2018, 2019). If the root of urbanisation stems from technology, then it would be plausible to suggest that overtourism and urbanism are both connected. Urbanism is obviously a consequence of sophistication (Bock, 2015) whereas overtourism is a consequence of poor destination management (see Séraphin et al., 2018). This point is also supported by Smith et al. (2010) who assert that urban tourism is one of the most complex forms of tourism to manage, as urban areas are not exclusively used by the visitors only but also by residents and working people who are connected (Kiralova and Hamarneh, 2018). Therefore, it becomes necessary to study both phenomena in parallel to better understand their impacts and implications for the tourism industry.

Additionally, in modern settings, research on urban cities is now gaining increased attention due to the impacts it carries along in its process (Tribe, 1997; Farrell and Twining-Ward, 2004). Studies in tourism from the view point of urbanisation have indeed been limited (see Hall, 2006; Sommer, 2018; Ashworth and Page, 2011). With this as background, this chapter aims at exploring the concept of urbanism and overtourism in an Indian city. Hyderabad (India) is chosen as a case study, as it is a well-known "high tech" tourism city having embraced urbanisation, but facing tourism challenges (see Das, 2015; Leonard, 2013; Brito, 2013). India is now a melting pot for cultural tourism and its long-term brand "Incredible India" used at the embryonic stage of its tourism development might no longer be effective with emerging phenomenon like overtourism and urbanism spreading rapidly across tourism cities. The methodological approach is based on the integrative research method and a case study. These are respectively inspired from Torraco (2016) and Yin et al. (2009). The study was solely based on written records, semi-structured interviews and a case study. The outcome of this exploratory study proposes some interesting information on the implications of urbanism and overtourism in Hyderabad. The findings show contrasting outcomes. It is found that urbanisation attracts a large number of visitors which in turn impacts on the social carrying capacity and hence, leads to overtourism. Signs of overtourism are found to be subtle in Hyderabad, but the possibility of the development of anti-tourism movements is high. In contrast, the constantly increasing level of urbanisation has encouraged urban tourism, which, in turn, contributes to the repositioning and diversification of the tourism industry in the state of Telangana.

The outcome of this exploratory study provides an improved understanding of the implications of overtourism and urbanism in Indian cities. Practitioners may draw from its results to make informed decisions on the management and development of urban tourism product based on emerging realities like overtourism. In light of prevailing methodological limitations, the outcome of the study might be indicative, but not necessarily reflective of trends and realities in other Indian cities.

The structure of the chapter is built around the following: The first part of the chapter will throw some theoretical insights on city tourism, urbanism and overtourism. The next part will draw a portrait of Hyderabad as a tourism destination. The third part of the chapter will provide some generic discussions derived from the findings. The focus will be on impacts and implications of urbanism and overtourism in the city of Hyderabad. The final section will close with a concluding note.

Literature review

City tourism

City tourism is one of the fastest growing travel segments worldwide and the changing nature of city tourism becomes increasingly apparent in many cities (Bock, 2015). It is regarded as one of the oldest and fastest growing forms of tourism where tourists travel for pleasure in cities other than their own (Sommer, 2019; Maxim, 2015). A wide variety of existing tourism resources in the urban area is a precondition of a city to attract tourists with various goals and motivations. If the city offers a greater variety of attractions and complementary capabilities, it can become more competitive in the tourism market. Examples of cities' attractions are dining out and shopping facilities, performances, events and nightlife together with a wide range of accommodation capacities (often in a different design or thematic ones). Inner-city leisure spaces, waterfront developments, festival marketplaces, casinos, museums, conference centres and sports stadiums are the physical manifestations of a wave of new local economic development initiatives for urban tourism and economic regeneration (Rogerson, 2002).

According to the United Nations World Tourism Organization (UNWTO), city tourism (also referred to as urban tourism) is considered as trips taken by travellers to cities or places of high population density (UNWTO, 2012, cited in Bock, 2015). Tourism cities perform multiple functions and exhibit various characteristics that influence tourism development within their boundaries. For instance, Maxim (2015) describes London as the capital offering a large variety of attractions, including historic buildings, cityscapes, parks and promenade areas, cultural establishments, numerous restaurants, pubs and clubs, and hosts various cultural and sporting events. Martens and Reiser (2019) portray the city of Dubai as the most urbanised tourism hub in the Middle East. Their study was based on a cognitive image attribute of Dubai as a city destination. Likewise, Jutla (2000) paints an impressive picture of Simla as a tourism city by highlighting its natural and cultural landscape. Tourism cities in fact perform multiple functions and exhibit various characteristics that influence tourism development in these destinations (Simpson, 2016). They accommodate world-class attractions (Law, 2002) and are centres of business and cultural excellence; they offer visitors a number of benefits, such as easier accessibility through better connected airports, better scheduled tourism services, diverse accommodation facilities and a variety of entertainment options (Edwards et al., 2008).

City tourism is mainly driven by emancipation, sophistication and globalisation (Gowreesunkar, 2019). The proliferation of information and communication technologies (ICTs) has had a significant impact on the travel industry, as well as on tourist behaviour, and is also transforming the nature of travel and the actual tourist experience (Neuhofer et al., 2015). The new tourists make decisions based on "word of mouse" rather than the traditional "word of mouth". This implies that choice of destinations is usually made by looking for online information (Séraphin and Gowreesunkar, 2019). According to Bock (2015), what makes city tourism distinct from other types of tourism is that cities have a high density of diverse cultural offerings in a relatively small area, attracting different types of tourists. For instance, the Calabar carnival in Nigeria yearly attracts worldwide tourists to its city called Calabar. African cultures have had a massive impact on carnival celebrations around the world. Despite being perceived as an unsafe and PCCD (post-colonial, conflict and disaster) destination (Séraphin and Gowreesunkar, 2019), this Nigerian city is a main gateway for tourists visiting the country and its success has a direct impact on the visitor economy of the destination. Eventful and happening cities have usually remained successful in offering its people escape routes from their hectic lives (Richards and Wilson, 2006; Quinn, 2005). In his study on city tourism, Bock (2015, p. 3) attributes a number of factors to the rising popularity of city tourism. These are detailed below:

- First, urbanisation is believed to reinforce the trend towards city tourism as people living in cities are more likely to associate with cities and the more they are inclined to visit other cities (UNWTO, 2019).
- Second, without doubt the proliferation of low-cost carriers has had a major impact on the popularity of city trips, mainly due to the fact that they made flights more affordable to the masses, but also because they expanded and improved flight networks, thus offering more city destination options and making them more accessible from a growing number of departure points than in the past, considerably decreasing travel times. As a consequence, this development has made a wide range of cities available to tourists at lower costs (Dunne et al., 2010).

- Third, as the proportion of the population taking several trips per year continues to grow, there is a tendency towards an increasing number of shorter holidays rather than just one main holiday per year (TripAdvisor, 2015).
- Fourth, due to the increasing availability and penetration of internet-based services during the travel cycle, information can easily be accessed, and the ease of making bookings online and retrieving a wide range of information while in a destination has greatly facilitated city tourism. As the biggest proportion in terms of the booking value of a city trip usually consists of the two elements transport and accommodation, it is relatively easy and not as risky to book online as, for example, a multi-country trip, a round trip or even a beach holiday where the hotel is not easily accessible from the airport by public transport. Moreover, as will be discussed in more detail later, the proliferation of ICTs, and in particular of mobile technologies, increasingly empowers consumers to create and plan further components of their city trip while in the destination. Due to the density of cultural offerings, the quantity of options to choose from in a city surpasses those of other destination types.
- Finally, peoples' perceptions of cities as tourist destinations have been changing. Nowadays, travellers no longer regard a city merely as an entry, exit or transit point, but as a destination in its own right (Dunne et al., 2010).

Urbanism in tourism

Urbanism is a "major force" that contributes to the development of towns and cities (Page and Connell, 2009, p. 471), and this is a significant factor for tourist destinations (Edwards et al., 2008). Urbanisation has therefore influenced the phenomenon of city tourism, and has contributed to the repositioning of the tourism industry within national economies (Ashworth and Page, 2011). Urban areas are usually places with a dense population, a major transport hub and a gateway for further travel in the region, as well as commercial, financial and industrial centres. They offer a variety of recreational and cultural experiences (Ashworth and Tunbridge, 1990; Page, 1995). Urban tourism is an important and one of the most dynamic forms of tourism; it is one of the leading factors of economic increase of European cities (Delitheou et al., 2010).

Urban tourism is sometimes complex and difficult to define, as it depends on many factors such as the size of the town, its history and heritage, its morphology and its environment, its location, its image. Law (2002) characterises urban tourism merely as tourism in urban areas. Ashworth (1992) noted that tourists seeking to visit urban areas consider the experience of urban tourism as closely related to visitor satisfaction and the standard of services based on visitors' demands, a point also shared by Page (1995). Based on his studies on the relationship between tourism and urban areas, Law (2002) defined three elements of city resources. Primary elements provide the main reasons why tourists visit cities and consist of facilities for activities such as cultural sport and leisure facilities, and factors of recreation and relaxation such as physical characteristics and socio-cultural characteristics. Secondary elements are related to accommodation services, restaurants, shops and other services. The additional items are designed to facilitate access to primary and secondary elements through accessibility e.g. transportation, parking places, tourist information offices, leaflets and maps, and are not the primary attractor of visitors. Urban tourism can generate income and employment in the urban area. Schofield (2011) states that urban tourism can create jobs and revenues for a government that are often higher than the income from other types of destinations. The global options for travel destinations are extensive, so cities compete alongside all others, for visits longer than short breaks, or single visits (Dwyer et al., 2009).

Consequences of urban tourism

While the growth of tourism in cities is generally encouraged by policy makers as it brings economic and social benefits to an area (Simpson, 2016), there are also a number of negative consequences which should not be overlooked. For example, existing congestion could get worse due to increased numbers of tourists; certain areas may become overcrowded, and conflicts may arise between the needs of visitors and locals (Gutiérrez et al., 2017; Law, 2002). Other challenges include protection of the environment, conservation of heritage and preservation of the local culture, while improving the quality of life of residents. Therefore, urban tourist destinations face significant challenges. Furthermore, in cities "leisure tourism is now just one of many different mobilities that bring people" to these areas (Maitland, 2016, p. 14), with other less visible forms of tourism also present, such as the VFR market, educational and health tourists, or even internal tourists (visitors from the city itself). The presence of these various forms of mobilities in cities make it difficult to distinguish between touristic and non-touristic behaviour, and thus to understand tourists' consumption demands. In addition, the sharing economy and peer-to-peer platforms such as Airbnb put pressure on the traditional tourist accommodation model and can create property conflicts (Gutiérrez et al., 2017), adding to the challenges of managing tourism in cities.

In a globalised world that affects tourism development in most cities, these environments face a number of challenges, including pressures from standardisation as they "need to negotiate the challenges of updating their appeal to visitors" while trying to maintain their distinctiveness (Maitland, 2012, p. 1). They are centres of corporate headquarters, business services, transnational institutions and they control "the flows of information, cultural products and finance that, collectively, sustain the economic and cultural globalisation of the world" (Knox, 2005). Despite their advantages, these cities are as vulnerable as other urban destinations to ecological, social and developmental problems (Ng and Hills, 2003). More recently, Maitland (2016) published a paper focusing on how tourists are experiencing world tourism cities, using evidence from London. Worth noting is also the work of Simpson (2016, p. 27) who discusses "tourist utopia" in three "post-world cities" – Las Vegas, Dubai and Macau – destinations with the common characteristic of being enclaves within larger states. Besides the complexities in terms of economic, social or political functions, these destinations have to deal with the diversity of the people experiencing such places either as residents, visitors or migrants (Stevenson and Inskip, 2009). Hence, it can be observed that world tourism cities display a number of characteristics which add to the challenges of planning and managing tourism in urban environments.

Overtourism

Overtourism, as a phenomenon, has been described from various perspectives. In plain terms, it refers to a destination suffering the strain of tourism (Richardson, 2017). In broader terms, it relates to a situation when a popular tourism destination no longer wishes to entertain tourists due to the negative consequences caused by tourism activities. For instance, across Europe, many destinations (Barcelona, Cambridge, Dubrovnik, Florence, Oxford, Rome, Venice, and York) are voicing their concern regarding the development of the tourism industry via protests, graffiti and physical intimidation (see Tapper, 2017; Séraphin et al., 2018). The claim is that overtourism is harming the landscape, damaging beaches, putting infrastructure under enormous strain and pricing residents out of the property market. A common thread in this early work was that excessive tourism concentrations led to harm to the local environment and negative attitudes among residents living at popular tourism destinations. Kuščer and Mihalič (2019), on overtourism in Europe, explain that the phenomenon is related to economic, social and environmental unsustainability. According to their study, overtourism implies too many tourists at a place and in an unsustainable way so that sustainable tourism can never be achieved. Indeed, overtourism represents an excessive growth of visitors leading to overcrowding in areas where residents suffer the consequences through enforced permanent changes in their lifestyles, access to amenities, changing physical environment, increase in property price and other economic discomfort which negatively affect their quality of life (Milano et al., 2018). These observations coincide, to a great extent, with the characteristics of urbanism and city tourism discussed above. For instance, countries like Nigeria, Germany and the United Arab Emirates use the world's online platforms to market their city tourism product like trade fairs and exhibitions (Dubai), Calabar Carnival (Nigeria) and beer festival (Munich) (see www.expo2020dubai.com, www.carnivaland. net/calabar-carnival/, www.oktoberfest-besuch.de/). As a result, over-exposure and visibility of the destination along with successful marketing campaigns conducted by destination management organisations (DMOs) and image building attract more tourists. Because tourism is an information-intensive industry, the search for information is guided by various models (Gowreesunkar and Dixit, 2017), and the internet is considered to be the most powerful because it can quickly alter a tourist's image of a destination. The image of a destination plays a significant role in its ability to attract visitors; the more positive the image is, the more it attracts tourists, and hence overtourism; the less exposed is a destination, the less it attracts tourists, and hence is less visited (Séraphin and Gowreesunkar, 2019). In today's technology-mediated environment, word of mouse travels quicker than word of mouth. This implies that tourists share their experience within a click of their computer mouse and their impression counts more than marketing experts. Based on the fact that online reviews play a major role in the way a destination is perceived, it becomes quite obvious that tourists will tend to choose specific, and often, the same destinations.

Over-mobility triggered by new tourism trends (rental websites, low-cost tourism, technology and online information sources, packaged holidays) and the substantial and fortuitous increase in demand for some destinations are also at the origin of the phenomenon of overtourism (cited in Séraphin et al., 2019). With sophistication, globalisation and emancipation, people are indeed increasingly indulging in tourism and this is well supported by indicators from the UNWTO, which reports that the number of international trips taken yearly has increased from 25 million in the 1950s to 1.4 billion in 2018 (UNWTO, 2019). Likewise, the work of Séraphin et al. (2018) on the "Fall of Venice", clearly explains that current and recent issues of overtourism in Europe are mainly due to a laissez-faire economy which allows policy makers and entrepreneurs to grow their businesses and expand exponentially due to the growth of digital bookings, which promote and offer ceaseless experiences over sustainable travel.

Contextual framework

Hyderabad

The contextual framework sets out the geographic limits of the study. For the current study, the city of Hyderabad is chosen as a case study. Hyderabad is the capital of the Indian state of Telangana and it comprises two main cities, namely Hyderabad and Secunderabad (Madhusudhan, 2016). With a population of about 8 million and a metropolitan population estimated above 9 million, it is the fourth most populous city and sixth most populous urban agglomeration in India (Das, 2015). Hyderabad is an urbanised city often referred to as the "high tech" city and it is a major centre for public sector enterprises, education, defence establishments and the phar-maceutical industry and has been on the global map in the last couple of decades due to its emergence as the global hub for information technology, allied services and for highly skilled, trained manpower (Ramachandraiah and Bawa, 2000). Figure 7.1 locates the city of Hyderabad in India.

Tourism in Hyderabad

Hyderabad has emerged as one of India's world-class tourist cities and according to the Telangana State Tourism Development Corporation (TSTDC), domestic visitors in the Hyderabad circle accounted for approximately 1.7 million visits in the fiscal year 2018 (www.statista.com, 2019). Foreign tourist arrivals rose from 2.33 million in 2016 to 2.71 million in 2017. As a result, tourism is a booming business, due to its rich culture and heritage sites. Hyderabad is a cosmopolitan city gifted with varied cultures, communities, art and architecture. Today this cosmopolitan city has preserved the wealth of heritage as traditional art and architecture in the form of monuments, buildings and lakes, granite rock formations that exhibit a mixture of different cultures and natural landscape. There are a multitude of religious buildings, galleries and shopping streets in the city. Hyderabad comprises both natural tourism and man-made attractions such as the Ramoji Film City, Golconda Fort, Hussain Sagar Lake, Chaar Minaar, Qutubshahi Tombs, Salarjung Museum and State Archaeological Museums, and urban tourists are increasingly visiting Hyderabad for the urbanised aspects of the city (Guntuka et al., 2017). These resources provide a wide variety of attractions for tourists, and over a period of time the city has emerged as a diverse and important tourism destination in India (Karan, 1979). The city is famous for its theme parks and monuments, which include the masterpiece Chaar Minaar, Golconda Fort and the UNESCO Asia Pacific Heritage Site of the Chowmahalla Palace

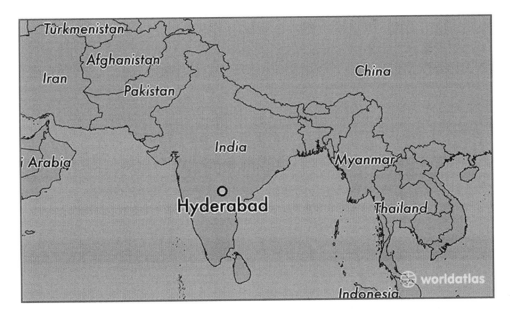

Figure 7.1   Location of Hyderabad in India.

Source: maps.google.com (2019).

(UNESCO, 2010). Moreover, Hyderabad has been included in the list of the world's creative cities for gastronomy for the year 2019; the city is famous for local dishes like *biryani* (flavoured spicy rice cooked with meat and potatoes), *haleem* (a special spicy soup made with local ingredients and meat) and *khoobani* (sweet made from a unique local fruit). The area of study, Chaar Minaar, a mosque monument with four pillars (Chaar means four and Minaar means pillar in the local Hindi language), is a global icon of Hyderabad. In this respect, many tourism businesses have been established around this area as it has the strong potential of attracting tourists and investors. Chaar Minaar is also known as the "city of pearls" dues to its numerous pearl and diamond trading centres and it has attracted lots of formal and informal tourism businesses. Slocum et al. (2011) define the informal economy as all those individuals and businesses that engage with tourists and the tourism industry, but are not registered with any formal association or trade organisation. Informal business in the Chaar Minaar area comprises street food vendors, local tour guides, transport providers, music performers, artisans, prostitutes, providers of homestays, tourist helpers and beggars. These form part of the Chaar Minaar landscape and urban tourists are increasing around this area (see Malik and Roy, 2012; Pathirana and Gnanapala, 2015; Hailu, 2017).

According to the TSTDC, the increase in the number of urban tourists in Hyderabad Minaar is favoured by factors such as increased infrastructure, accommodation facilities, transport services and connectivity, particularly the development of international airports, the MICE (meetings, incentives, conferences and exhibitions) industry and the promotion of tourist attractions. Moreover, rapid urbanisation, increased connectivity, change in the socio-economic condition of the middle-classes, globalisation, the technology business, healthcare and the MICE industry have all increased mass and domestic tourism. With the propagation of globalisation and the development of technology in Hyderabad, the cities of the Global South and North are increasingly connected and more easily able to communicate (Sudhira et al., 2007). With increased demand for growth of IT and IT-enabled services and the changing socio-economic and political situation in Hyderabad, the city took advantage of the developing services sector (Das, 2015) such as the international airport, highways, hotels, convention centres, ultra-modern shopping malls and multiplexes. State government gave immense importance to the development of the tourism infrastructure in the city, particularly at the potential heritage sites, and it was mentioned in the Vision 2020 tourism development and management plan. Over a period of time, and after implementation of the projects proposed in the Vision 2020 plan, the city of Hyderabad has come to generate a global space and flow for tourists. In 2014 the National Geographic's *Traveller* magazine listed Hyderabad as the "second best place in the world that one should see in 2015" (Telangana Tourism, 2015).

Methodology

Literature reviewed shows that urbanism and overtourism have links as they both have people as a common factor to manage. Therefore, both phenomena need to be researched jointly in order to understand their impacts and implications on the tourism industry. As overtourism is an emerging topic and urbanism, an evolving one, it was deemed appropriate to use the integrative review method and a case study. These are drawn from the work of Whittemore and Knafl (2005) and Yin et al. (2009) respectively. According to Torraco (2016), integrative literature reviews address two common types of topics, namely mature topics and new and emerging topics. The integrative review method chosen seeks to undertake an interdisciplinary discussion about urbanism and overtourism in the context of the city of Hyderabad. As such, content analysis (websites and online reviews), desk-based research (research publications and newspaper articles) and a case study were used in order to produce a meaningful synthesis of the overall situation. According to Yin et al. (2009), case studies are employed in many situations when little is known about a topic and when the scope of research is to contribute to the current knowledge, in this case on urbanism and overtourism in Hyderabad. Veal (2011) also underlines the merits of case study methodology in tourism research as it helps in understanding complex phenomena by analysing individual examples.

This exploratory study therefore took place in various phases. The first step comprised a literature review conducted by the two academic researchers whereas the local collaborator was not involved. The aim was to highlight the literature gap. The second stage was undertaken by the second researcher who is a resident of Hyderabad. The study took place from January 2019 to March 2019. This period was chosen, as the main researcher was attending a training session in the city of Hyderabad, and hence, she used the opportunity to collect data; this saved travelling and accommodation costs. Data were collected during weekdays (18 days scattered over ten weeks) and the main investigator was accompanied by the local collaborator. The duration of the survey was approximately eight hours and on a few occasions went beyond nine hours. Since the focus was mainly on urbanism and overtourism, the popular tourist site of Chaar Minaar was chosen for observation.

By focusing on Chaar Minaar (Figure 7.2), the chapter provides rich information on the characteristics of urbanism and overtourism in this tourism site. The study being exploratory in nature, it was deemed appropriate to adopt a qualitative approach which made possible the investigation of the research topic through the analysis of multiple sources of evidence (Yin et al., 2009).

A total of six semi-structured interviews (Altinay and Paraskevas, 2008) were conducted with top management and mid-management representatives of the department of Telangana Tourism in the city of Begampet. This technique is one of the most important sources of data collection in case study research (Yin et al., 2009) and has been widely used in tourism studies (Pizam, 1994). The officers were chosen based on their position in the planning and marketing department of Telangana Tourism. They willingly accepted to participate in the interview which was less than one hour in duration. Appointments were sought and the exercise was spread over two weeks (three interviews per week). Questions were formulated around: impression of Hyderabad as a tourism city; strength and weakness of the city; urbanisation and tourism development in the city; the impacts on economy, society and environment; opinion on overtourism; the link between overtourism and urbanism; the future of Hyderabad as an evolving "hi-tech" urbanised tourism destination. The interviews proved particularly useful in gaining rich data on the implications of urbanism and overtourism in Hyderabad. To thank the interviewees, a key ring from Mauritius (the main researcher's country) was presented as a token of appreciation for their participation in the survey.

To co-ordinate data from observations and interviews, the main researcher kept field notes. Content analysis was done through data selection, a data reduction process, structuring, coding and interpretation of meaning. The use of secondary sources was helpful to identify some similarities and differences with other tourism destinations and cities of India. The primary data were then analysed in conjunction with secondary data. During the final stage, analysis of the written descriptions of both phenomena was undertaken. In so doing, data included in the notebook were read and re-read, bracketed and compared to draw the final implications and conclusion. The final stage also involved the assembling, cross-validating and content analysis of data collected from those methods.

One of the limitations of this study is inherent to case study research, this method being often criticised because it provides little basis for generalisation – "scientific generalisation" (Maxim,

Figure 7.2   Chaar Minaar.

Source: Authors.

2015, p. 14). Although the findings from case studies cannot be generalised when compared with those obtained from random sample surveys for example, a number of inferences can be made and these may apply to other contexts. Indeed, Yin et al. (2009) argue that in case study research another type of generalisation applies "analytical generalisation", which is oriented towards theoretical propositions rather than enumerating frequencies. Another limitation relates to the representativeness of the officers that took part in the interviews. Even though not every stakeholder was covered in this study, efforts were made to include more representatives from diverse sectors to gain a better understanding of the implications of urbanisation in tourism development in Hyderabad.

Overall findings and discussion

The link between urbanism and overtourism

The overall findings show that city tourism is a complex phenomenon that overlaps with urbanism and overtourism in Hyderabad. Overtourism and urbanism have consequences on people and people are the causes of overtourism and urbanism. Driving factors for city tourism in Hyderabad are mainly an increase in purchasing power of the middle class, improved connectivity, affordable accommodation, development and promotion of the tourist sites, increased concern over safety and security from the government, technology and business tourism. The mentioned factors have boosted tourism demand ultimately leading to overtourism, and overtourism in the city is causing serious threats to the environment and monuments. The same outcome was observed in Lisbon by Richard and Marques (2019). The authors explained the term "overtourism" as an overdose of travellers and the solution was wise cultural management. Indeed, Chaar Minaar might also go through the same phase if timely action is not taken. For instance, statistics from Telangana Tourism show that popular tourist site Golconda Fort attracted around 775,810 tourists in 2004, which increased to 17.2 million in 2017 (Telangana Tourism, 2019). This shows that there is an increase which can lead to an "overdose" as mentioned by Richards and Marques (2019). Due to the rapid urbanisation, the city is receiving an excessive number of visitors, and this has led to Chaar Minaar and Golconda Fort having issues with toilet facilities, water, and sound pollution (Pulla et al., 1985). The impacts of urbanism are also reflected in the consequences of overtourism. For instance, increased traffic congestion due to the increased number of tourists creates noise and air pollution at Chaar Minaar, causing damage to the monument, following which conservation becomes an issue. Similarly, an increase in the number of tourists at the monuments causes drainage and garbage problems, as the cities are not well planned and the streets are not pedestrian friendly. Furthermore, during the season and festival time, it is difficult to manage the number of tourists which blend with the locals and cause friction and anti-tourism behaviour. Interviews from officers revealed that traditionally, tourism was regarded as a profitable activity in Hyderabad, but nowadays, tourists are infiltrating spaces and places so much that the industry is seen as a pain rather than gain, a point also noted by Séraphin et al. (2018) with regard to Venice. To exemplify the context, heritage tourism is ambitiously promoted in Hyderabad, and this has caused those places to be overcrowded not only with tourists, but also with locals (Singh and Gowreesunkar, 2019), thus leading to overtourism. The growing number of tourists to the city is imposing increased pressure on the destination – for example traffic, congestion, pollution, crime and threats to the natural environment. Moreover, it was also noted that during festivals like Ramzaan, the city is overcrowded to the point that locals show resentment. For instance, they do not allow visitors to park their cars near their premises.

Overtourism is relatively new to India and recently started in places like Goa, Agra and Kerala (Routledge, 2001; Sreekumar and Parayil, 2002). Sustaining urban tourism while controlling overtourism with its various social and environmental implications (public transport improvements, traffic congestion, conflicts between hosts and visitors, limited space, street vendors), was recognised as a challenge by interviewees. Sustainability, together with planning and managing tourism, and working in partnership are, however, among the key drivers of success identified by researchers if an urban destination is to succeed in the long term (Paskaleva-Shapira, 2007). Therefore, in order to stay competitive on the tourism market, the Telangana Tourism Department needs to better understand the complex environments in which they operate and to take a leading role in bringing together the key stakeholders involved in tourism development in order to address the challenges. Hyderabad as a sustainable city should have concern over its cultural and natural environmental assets, and minimise impacts on locals. One of the major objectives of the Vision 2020 tourism development is not only to generate socio-economic and employment benefits but also to provide quality life to the local community and environmental sustainability (Dwyer, 2000). Although the vision is fulfilled to some extent in the context of economic benefits, the increase in the number of tourists at the major heritage sites has led to the phenomenon of overtourism. However, ultimately the issue is to provide a better quality of life to the local community and a better quality of experience to the tourists, but this can also lead to issues of environmental sustainability. This view is also shared by Pearce (2011), who argues that in most cases, tourism is part of a broader urban context and cannot be separated from other phenomena arising in the tourism environment.

Impact of technology on urbanism

Overall findings also reveal that changes in consumerism and technology have mainly fuelled the phenomenon of urbanism. Looking at Chaar Minaar area through the lens of urbanism, it would seem that the type of tourism product available suits corporate time-conscious visitors who for a short period of time have all possible realisable options to choose from (Bock, 2015). Tourists of the twenty-first century are busy and they look for facilities to capitalise on time and money (Gowreesunkar et al., 2018). Obviously, urbanism has not only boosted tourism, but also created a new breed of tourists whose characteristics can be explained along a spectrum of emancipation and sophistication. The "new urban tourists" visiting Hyderabad are savvy and knowledgeable and their decision-making process is influenced mostly by online reviews. Tourists' perception of Hyderabad is shaped by many factors, among which, online reviews dominate, a point highlighted in the work of Gowreesunkar and Dixit (2017). Undoubtedly, technology has a major role to play in converting destinations to overvisited places and hence overtourism. Taking the example of Haiti (a post-colonial, post-conflict and post-disaster destination), Séraphin (2016) has explained how technology can completely alter the image of a destination. Technology-savvy tourists draw extensively on the internet, and these interactive media not only yield information but also provide feedback, and hence influence the image formation and decision making of tourists. In most cases, common people are seen as experts, as they post about their experience and tourists mostly trust their comments as compared to dedicated and qualified marketers. Exposure of destinations on social media and TV shows completes the traditional campaigns of promotion and communication from official tourism boards and maximises their visibility (Gowreesunkar and Dixit, 2017), eventually causing more people to visit and hence, over-visitation.

Cultural tourism and overtourism

Festivals play a major role in the promotion of the city tourism and attract large numbers of tourists. Interviews revealed that in Hyderabad, festivals are tourism resources attracting both locals and tourists. For instance, in September and October, the state celebrates a floral festival (in local Telugu language it is called "Bathukamma", which means "Live Forever"). The festival is celebrated for nine days and on the ninth day Bathukamma is immersed in a pond or lake or other body of water. Although the festival is seasonal and is celebrated for nine days, with the floral immersion taking place on the final day, it has created several environmental issues in the city due to the lack of control, mismanagement and the unsustainable way the celebrations proceed. However, floral immersion is eco-friendly, having several medicinal values. The festival has started being celebrated on a large scale in the city. Bonalu is a major Hindu festival of the city celebrated annually either in July or August, when Goddess Mahakali is worshiped and given thanks. Mahakali exists in various forms. Celebrations begin at Golconda Fort followed by other parts of the city. Thousands of devotees throng the temples to pay obeisance to Mahakali with folk dance and music. The festival attracts a large number of visitors to the Golconda Fort (Telangana Tourism, 2019). Likewise, Ganesh Chaturti, a Hindu festival, is celebrated annually in August or September with the installation of Ganesha idols. Generally, the festival in Hyderabad ends on the tenth day and the idols are carried in a public procession to Hussain Sagar Lake to be immersed in the water. On the one hand, these festivals promote cultural tourism, but on the other hand, they attract mass tourists (domestic and international). As a result, locals suffer the consequences of overtourism and overcrowding. The festivals cited above are witnessed by thousands of visitors and on immersion day at Hussain Lake, roads are closed and many non-Hindu residents face challenges to use the public infrastructure. Traffic restrictions and congestion are witnessed and access to basic amenities and facilities, water, toilets disturb the locals' routine lives. Moreover, there are no public toilet facilities in the vicinity of the Hussain Sagar Lake. Therefore, the lake and other water bodies in the city are polluted due to mass tourism activities. Use of plastic bags and carrier bags at the lakes are also leading to an increase in solid waste, creating problems for the aquatic animals and affecting the quality of water. Additionally, loss of originality, authenticity and commodification has also been observed (Madhusudhan, 2016). These consequences have obviously caused an antagonistic sentiment among the locals.

In recent years, terms like overtourism became unfortunately common in a number of tourism destinations and these situations gave rise to further new phenomena like tourismphobia and anti-tourism movements. Cities such as Barcelona, Venice and London are voicing their concern regarding the development of the tourism industry via protests, graffiti and physical intimidation (see Smith et al., 2010; Tapper, 2017). These destinations under anti-tourist anger are already planning to monitor tourists and tourism more closely possibly by limiting the number of visitors. In the context of Hyderabad, signs of overtourism are found to be subtle, but the possibility of an anti-tourism movement remains high. Locals tend to manifest their discontent by troubling tourists. For instance, in the interview gathered from Telangana Tourism officers, it was noted that locals misguide beggars in order to discourage them to wander in their local environment. For instance, a few observations indicated that locals were expressing discontent in the preferential treatment given to foreigners in shopping centres. The tourism industry is based on the people and places and the interaction between them and the industry is extremely sensitive to the social and physical conditions of the destination's micro and macro environment The reasons for anti-tourism are due to the fact that the large number of tourists visiting Hyderabad are affecting the quality of life of locals and they do not really buy from locals. Rather, they use several natural and man-made resources and cause undue pressure on some infrastructures, on the quality of residents' daily lives, their mobility and in some areas on the price and rent of resident accommodation, goods and services. This has been witnessed in the work of Mathew and Sreejesh (2017) who investigated the impact of responsible tourism on destination sustainability and the quality of life of the community in the state of Kerala (India).

Since urbanism attracts urban tourism, the state is interested to further invest for profitability while the locals' concerns are not prioritised. Hence, this kind of development slowly leads to residential areas losing their unique identity and character due to businesses increasingly adapting their offerings to the lucrative tourist demand. The "high tech" city is an example. But, from another perspective, it can also present development opportunities for deprived areas like Borabanda, Kamala Yadgar and Macca Maszid.

Urbanism and re-invented beggarism

Interviewees reveal that urbanism has also attracted a new form of tourism as recently researched by Gowreesunkar et al. (2019). Historically, cities have been major tourist attractions and potential sources for both formal and informal businesses (Ashworth and Page, 2011). For instance, in Chaar Minaar, among the informal businesses are counted beggars, as studies show that begging is an informal profession in Chaar Minaar and beggars are often involved in some forms of trade with locals (see Malik and Roy, 2012; Pathirana and Gnanapala, 2015; Hailu, 2017). Khan (2017) has reported that there are about 15,000 beggars (of which 1,500 are children) in the city of Hyderabad. The survey was undertaken by a voluntary organisation in the context of a governmental mission to make Hyderabad a beggar-free city. The study further reveals that approximately 14,000 beggars roam around tourism cities, of which 98% of them are professionals earning as much as 24 crore Indian rupees (US$3,374,400) per annum. This gives an indication of the amount of money yielded from the begging business and therefore provides legitimacy to the work of Riaz and Baloch (2019) who argue that a growing number of educated people are now increasingly joining the begging business rather than respectable professions. As a result, the new smart and savvy beggars are now increasingly tapping into city tourism (see Delap, 2009; Brito, 2013; Gösling et al., 2004; Andriotis, 2016). They play on the sympathy of tourists and try to capitalise on this encounter by establishing contact with the tourists in order to maximise benefit, even after the end of their visit. Beggarism is found to be a networking activity, where beggars capture attention by using physically disabled children, pathetic-looking old people and pity-evoking mothers carrying children in order to establish a contact to be used in the long term. For instance, connected beggars operate in a cartel and from key locations to maximise their prospects. As such, they are found near important public places like bus stops, metro stations, public gardens, religious sites, tourist markets, taxi stands among others. With progress in technology, begging has turned out to be a lucrative form of business in many tourism countries (see Bukoye, 2015; Andriotis, 2016; Qiao et al., 2017) and educated people are found to be more interested in joining the begging business rather than respectable professions (see Riaz and Baloch, 2019). From an urban tourism perspective, begging is seen as part of the tourism process, as engaging with beggars is a way of experiencing the local lifestyle and hence it contributes to the tourists' experience (Brito, 2013; Kotler et al., 1993). For instance, Brito (2013) identified a type of tourist in India who regularly visited and had privileged relations with individuals belonging to the begging community. A similar observation was also noted in the city of Heraklion (Greece); beggars were successful in evoking pity and they established contact with a small number of sympathetic tourists (Andriotis, 2016). Moreover, performing beggars also attract tourists (Figure 7.3). From a tourism point of view, begging can be a very powerful pull

Figure 7.3 Performing beggars in Chaar Minaar.

Source: Authors.

factor because of its authenticity, a point also shared by Qiao et al. (2017, p. 282): "begging can be transformed into an activity that is seen as part of local culture or traditions, 'begging' could be seen as adding to the flavor of destinations".

Destination management

According to Minguzzi (2006), destination management organisations (DMOs) are responsible for the planning, marketing, management, product development, industry advocacy and coordination of the tourism destination. The DMO is the nerve centre at the destination and as such, its role is not only focused on cost minimisation or customer satisfaction, but rather on relationship optimisation (Gowreesunkar, 2019). City tourism is an opportunity for communities and their people to share the benefits of tourism, which is why the relations between the sector and the communities need to be strengthened. In contrast, overtourism creates frustrations among stakeholders and degradation of tourism resources while urbanisation does not always favour locals as per data collected in Hyderabad. To tackle both issues and related perverse impacts, the role of the DMO is key, as it is the authority which regulates the tourism system and works for the benefit of the destination and its people. Prioritising the welfare of residents above the needs of the global tourism supply chain is as important as satisfying the increasing demand of tourism and sustaining tourism stakeholder businesses. Furthermore, it is observed that one of the leading strategies to be adopted in pursuit of overtourism reduction is pushing tourism from over-demanded to under-demanded spaces. While such a policy may make some sense, it also encapsulates several risks and, if not well managed and controlled, may end up shifting overtourism to other locations instead of reducing the consequent stress. The DMO's role is therefore to devise pre-emptive strategies to ensure a balance is maintained and destinations have their share in tourism.

Conclusion

This chapter sought to provide some insights on the implications of urbanism and overtourism in the city of Hyderabad. Observations and opinions drawn together, it would seem that urbanism and overtourism are inevitable in an era characterised by globalisation, sophistication and emancipation. Because of the overlapping nature of overtourism and urbanism, it is important to re-think city tourism from a refreshed perspective. Overtourism and urbanism have consequences on people who nurture contrasting and conflicting interests and people are the cause of overtourism and urbanism. One of the critical success factors in this vicious circle lies in the role played by DMOs. DMOs need to ensure that product development achieves a balance between the optimal tourist experience and a commensurate local benefit. Tourists must also play their part by making travel choices that are sensitive to the places they visit and those who live in and around them. Tailor-made management strategies like creative tourism may be developed to cope with the current situation. Research, planning and ongoing dialogue between stakeholders like tourism operators, regulators, civil society groups and local residents are essential. Prioritising the welfare of local residents above the needs of the global tourism supply chain is obviously vital, as residents form an integral part of the tourism experience. Moreover, prime consideration needs to be given to ensure that the level of visitation fits within a destination's capacity. Last, but not least, it is important to learn from lessons derived from destinations facing challenges of urbanism, overtourism and anti-tourism.

References

Altinay, L., and Paraskevas, A. (2008). *Planning Research in Hospitality and Tourism*. Oxford: Elsevier.

Andriotis, K. (2016). "Beggars–tourists' interactions: An unobtrusive typological approach", *Tourism Management*, 52, 64–73.

Ashworth, G. J. (1992). "Is there an urban tourism?", *Tourism Recreation Research*, 17(2), 3–8. DOI: 10.1080/02508281.1992.11014645.

Ashworth, G., and Page, S. J. (2011). "Urban tourism research: Recent progress and current paradoxes", *Tourism Management*, 32(1), 1–15.

Ashworth, G. J., and Tunbridge, J. E. (1990). *The Tourist-Historic City*. London: Belhaven Press.

Bock, K. (2015). "The changing nature of city tourism and its possible implications for the future of cities", *European Journal of Futures Research*, 3(1), article 20.

Brito, O. (2013). "International tourism and reinvention of begging: Portrayal of child-centred begging in Bangkok and Bombay", unpublished doctoral dissertation, University Paris Ouest Nanterre La Defense, France.

Bukoye, R. (2015). "Prevalence and consequences of streets begging among adults and children in Nigeria, Suleja Metropolis", *Procedia – Social and Behavioral Science*, 17, 323–333.

Das, D. (2015). "Hyderabad: Visioning, restructuring and making of a high-tech city", *Cities*, 43, 48–58.

Delitheou, V., Vinieratou, M., and Touri, M. (2010). "The contribution of public and private investments to the growth of conference tourism in Greece", *Management Research and Practice*, 2(2), 165–178.

Delap, E. (2009), "Begging for change: Research findings and recommendations on forced child begging in Albania/Greece, India and Senegal", London: Anti-Slavery International.

Dunne, G., Flanagan, S., and Buckley, J. (2010). "Towards an understanding of international city break travel", *International Journal of Tourism Research* 12, 409–417.

Dwyer, L. (2000). "Economic contribution of tourism to Andhra Pradesh", *Tourism Recreation Research*, 25(3), 1–11.

Dwyer, L., Edwards, D., Mistilis, N., Roman, C., and Scott, N. (2009). "Destination and enterprise management for a tourism future", *Tourism Management*, 30(1), 63–74.

Edwards, D., Griffin, T., and Hayllar, B. (2008). "Urban tourism research: Developing an agenda". *Annals of Tourism Research*, 35(4), 1032–1052.

Farrell, B. H., and Twining-Ward, L. (2004). "Reconceptualizing tourism", *Annals of Tourism Research*, 31(2), 274–295.

Gössling, S., Schumacher, K., Morelle, M., Berger, R., and Heck, N. (2004). "Tourism and street children in Antananarivo, Madagascar", *Hospitality and Tourism Research*, 5(2), 131–149.

Gowreesunkar, V. (2019). "African Union (AU) Agenda 2063 and tourism development in Africa: Contribution, contradiction and implications", *International Journal of Tourism Cities*. https://doi.org/10.1108/IJTC-02-2019-0029.

Gowreesunkar, V. G. B., and Dixit, S. (2017). "Consumer information-seeking behaviour". In S. B. Dixit (ed.) *The Routledge Handbook of Consumer Behaviour in Hospitality and Tourism*. London: Routledge, pp. 55–68.

Gowreesunkar, V. G., Séraphin, H., and Morrison, A. (2018), "Destination marketing organisations: Roles and challenges". In D. Gursoy and C. G. Chi (eds) *Routledge Handbook of Hospitality Marketing*. London and New-York: Routledge. Also available at www.routledge.com/The-Routledge-Handbook-of-Destination-Marketing/Gursoy-Chi/p/book/9781138118836.

Gowreesunkar, V., Séraphin, H., and Teare, R. (2019). "Reflections on the theme issue outcomes", *Worldwide Hospitality and Tourism Themes*, 11(5), 634–640. https://doi.org/10.1108/WHATT-07-2019-0046.

Guntuka, D., Sravani, K., and Vijaya Bhole (2017). "Urban tourist attractiveness index: A case study of greater Hyderabad", *International Journal of Recent Scientific Research*, 8(10), 20877–20879. DOI: http://dx.doi.org/10.24327/ijrsr.2017.0810.0977.

Gutiérrez, J., García-Palomares, J. C., Romanillos, G., and Salas-Olmedo, M. H. (2017). "The eruption of Airbnb in tourist cities: Comparing spatial patterns of hotels and peer-to-peer accommodation in Barcelona", *Tourism Management*, 62, 278–291.

Hailu, D. (2017). "Begging and urbanization: A sociological analysis of the impacts of begging for urban security, sanitation and tourism", *Research on Humanities and Social Sciences*, 7(3).

Hall, C. M. (2006). "Tourism urbanization and global environmental change". In S. Gössling and C. M. Hall (eds) *Tourism and Global Environmental Change: Ecological, Economic, Social and Political Interrelationships*. London: Routledge, pp. 142–156.

Jutla, R. S. (2000). "Visual image of the city: Tourists' versus residents' perception of Simla, a hill station in northern India", *Tourism Geographies*, 2(4), 404–420.

Karan, M. (1979). "Hyderabad", *Tourism Recreation Research*, 4(2), 31–33.

Khan, Y. A (2017). "Begging is banned in Hyderabad", *Deccan Chronicle*. [Online] Available at www. deccanchronicle.com/nation/current-affairs/081117/begging-banned-in hyderabad.html. Accessed on 08 November 2019.

Kiralova, A., and Hamarneh, I. (2018). "Urban tourism and regional development". In *21st International Colloquium on Regional Sciences: Conference Proceedings*. Brno: Masarykova univerzita, pp. 1–5. [Online] Available at www.researchgate.net/publication/325852578.

Knox, P. L. (2005). "Creating ordinary places: Slow cities in a fast world". *Journal of urban design*, 10(1), 1–11.

Kotler, P., Haider, D. H., and Rein, I. (1993). *Marketing Places: Attracting Investment, Industry, and Tourism to Cities, States, and Nations*. New York: The Free Press.

Kuščer, K., and Mihalič, T. (2019). "Residents' attitudes towards overtourism from the perspective of tourism impacts and cooperation: The case of Ljubljana", *Sustainability*, 11(6), 1823.

Law C. (2002). *Urban Tourism: The Visitor Economy and the Growth of Large Cities*. London: Continuum.

Leonard, K. (2013). "Political players: Courtesans of Hyderabad", *The Indian Economic and Social History Review*, 50(4), 423–448.

Madhusudhan, L. (2016). "Floral festival: A culture of Telangana", *Arts and Social Sciences Journal*, 7, 227–231.

Maitland, R. (2012). "Global change and tourism in national capitals", *Current Issues in Tourism*, 15(1–2), 1–2.

Maitland, R. (2016). "Everyday tourism in a world tourism city: Getting backstage in London", *Asian Journal of Behavioural Studies*, 1(1), 13–20.

Malik, S., and Roy, S. (2012). "A study on begging: A social stigma", *Journal of Human Values*, 18, 187–199.

Martens, H. M., and Reiser, D. (2019). "Analysing the image of Abu Dhabi and Dubai as tourism destinations: The perception of first-time visitors from Germany", *Tourism and Hospitality Research*, 19(1), 54–64.

Mathew, P. V., and Sreejesh, S. (2017). "Impact of responsible tourism on destination sustainability and quality of life of community in tourism destinations", *Journal of Hospitality and Tourism Management*, 31, 83–89.

Maxim, C. (2015). "Drivers of success in implementing sustainable tourism policies in urban areas", *Tourism Planning & Development*, 12(1), 37–47.

Maxim, C. (2017). "Challenges faced by world tourism cities: London's perspective", *Current Issues in Tourism*. DOI: 10.1080/13683500.2017.1347609.

Milano, C., Cheer, J. M., and Novelli, M. (2018). "Overtourism: A growing global problem", *The Conversation*. [Online] Available at https://theconversation.com/overtourism-a-growingglobal-problem-100029. Accessed on 25 April 2020.

Minguzzi, A. (2006). "Destination competitiveness and the role of destination management organization (DMO): An Italian experience". In L. Lazzeretti and C. S. Petrillo (eds) *Tourism Local Systems and Networking*. London: Routledge, pp. 197–208.

Neuhofer, B., Buhalis, D., and Ladkin, A. (2015). Technology as a catalyst of change: Enablers and barriers of the tourist experience and their consequences. In *Information and Communication Technologies in Tourism 2015*. Cham: Springer, pp. 789–802.

Ng, M. K., and Hills, P. (2003). "World cities or great cities? A comparative study of five Asian metropolises", *Cities*, 20(3), 151–165.

Page, S. (1995). *Urban Tourism*. London: Routledge.

Page, S. J., and Connell, J. (2009). *Tourism: A Modern Synthesis*, 3rd edition. Andover: Cengage Learning EMEA.

Paskaleva-Shapira, K. A. (2007). "New paradigms in city tourism management: Redefining destination promotion", *Journal of Travel Research*, 46(1), 108–114.

Pathirana, D., and Gnanapala, A. (2015). "Tourist harassment at cultural sites in Sri Lanka", *Tourism, Leisure and Global Change*, 2, 42–56.

Pearce, D. (2011). "Tourism, trams and local government policy making in Christchurch: A longitudinal perspective". In D. Dredge and J. Jenkins (eds) *Stories of Practice: Tourism Policy and Planning*. Farnham: Ashgate Publishing, pp. 57–78.

Pizam, A. (1994). "Planning a tourism research investigation". In J. R. Ritchie and C. R. Goeldner (eds) *Travel, Tourism and Hospitality Research: A Handbook for Managers and Researchers*, 2nd edition. John Wiley & Sons, pp. 1–104.

Pulla, V. R., Jafri, S. A., and Rao, N. (1985). *City Report, Hyderabad: The State of Art of Physical Environment, a Citizens' Report*. Secunderabad, India: Centre for Environment Concerns.

Qiao, G., Chen, N., and Prideaux, B. (2017). "Understanding interactions between beggars and international tourists: The case of China", *Asia Pacific Journal of Tourism Research*, 22(3), 272–283. DOI: 10.1080/10941665.2016.1233891.

Quinn, B. (2005). "Arts festivals and the city", *Urban Studies*, 42(5–6), 927–943.

Ramachandraiah, C., and Bawa, V. K. (2000). "Hyderabad in the changing political economy", *Journal of Contemporary Asia*, 30(4), 562–574.

Riaz, S., and Baloch, M. A. (2019). "The socio-cultural determinants of begging: A case study of Karachi city", *Journal of Economics and Sustainable Development*, 10(11), 75–88.

Richard, G. and Marques, L. (2019). "Overtourism in Lisbon: Is culture the salvation?" [Online] Available at www.researchgate.net/publication/332111673. Accessed on 25 April 2020.

Richards, G., and Wilson, J. (2006). "Developing creativity in tourist experiences: A solution to the serial reproduction of culture?", *Tourism Management*, 27(6), 1209–1223.

Richardson, D. (2017). "Suffering the strain of tourism". [Online] Available at TTG@wtm. Accessed on 25 April 2020

Rogerson, C. M. (2002). "Urban tourism in the developing world: The case of Johannesburg", *Development Southern Africa*, 19(1), 169–190.

Routledge, P. (2001). "Selling the rain, resisting the sale: Resistant identities and the conflict over tourism in Goa", *Social & Cultural Geography*, 2(2), 221–240.

Schofield, P. (2011). "City resident attitudes to proposed tourism development and its impacts on the community", *International Journal of Tourism Research*, 13(3), 218–233.

Séraphin, H. (2016). "Impacts of travel writing on post-conflict and post-disaster destinations: The case of Haiti". In Z. Roberts (ed.) *River Tourism: The Pedagogy and Practice of Place Writing*. Plymouth: TKT, 17–35.

Séraphin, H., and Gowreesunkar, V. (2019). "What marketing strategy for destinations with a negative image?", *Worldwide Hospitality and Tourism Themes*, 11(2).

Séraphin, H., Gowreesunkar, V., Roselé-Chim, P., Duplan, Y. J., and Korstanje, M. (2018). Tourism planning and innovation: The Caribbean under the spotlight. *Journal of Destination Marketing & Management*, 9, 384–388.

Séraphin, H., Gowreesunkar, V., Zaman, M., and Lorey, T. (2019). "Limitations of trexit (tourism exit) as a solution to overtourism", *Worldwide Hospitality and Tourism Themes*, 11(5), 566–581. https://doi.org/10.1108/WHATT-06-2019-0037.

Séraphin, H., Sheeran, P., and Pilato, M. (2018). "Over-tourism and the fall of Venice as a destination", *Journal of Destination Marketing & Management*, 9, 374–376.

Simpson, T. (2016). "Tourist utopias: Biopolitics and the genealogy of the post-world tourist city", *Current Issues in Tourism*, 19(1), 27–59.

Singh, T. V., and Gowreesunkar, V. (2019). "Transformation of Himalayan pilgrimage: A sustainable travel on the wane". *Journal on Tourism and Sustainability*, 2(2).

Slocum S., Backman, K., and Robinson K. (2011). "Tourism pathway to prosperity: Perspective of informal economy of Tanzania", *Tourism Analysis*, 16(1), 45–55.

Smith, M., MacLeod, N., and Robertson, M. H. (2010). *Key concepts in tourist studies*. Los Angeles, CA: Sage.

Sommer, C. (2018). "What begins at the end of urban tourism, as we know it?" [Online] Available at www.researchgate.net/publication/324894914. Accessed on 15 December 2019.

Sommer, C. (2019). "What begins at the end of urban tourism, as we know it? *Europe now*. [Online] Available at www.europenowjournal.org/2018/04/30/what-begins-at-the-end-of-urban-tourism-as-we-know-it/. Accessed on 25 April 2020.

Sreekumar, T. T., and Parayil, G. (2002). "Contentions and contradictions of tourism as development option: The case of Kerala, India", *Third World Quarterly*, 23(3), 529–548.

Stevenson, N., and Inskip, C. (2009). "Seeing the sites: Perceptions of London". In R. Maitland and B. W. Ritchie (eds) *City Tourism: National Capital Perspectives*. Wallingford: CABI, pp. 94–109.

Sudhira, H. S., Ramachandra, T. V., and Subrahmanya, M. B. (2007). "City profile", *Cities*, 24(5), 379–390.

Tapper, J. (2017). "As touting for punt trips becomes a crime, is tourism overwhelming Britain's cities", *The Guardian*, 30 July. [Online] Available at www.theguardian.com/uk-news/2017/jul/29/cambridge-tourist-boom-ruins-city. Accessed on 10 December 2019.

Telangana Tourism (2015). "Global recognition for Hyderabad, 2nd best place to see in 2015". [Online] Available at www.telanganatourism.gov.in/partials/more/news.html. Accessed on 10 December 2019.

Telangana Tourism (2019) "Bonalu, the State Festival of Telangana!" Telangana Tourism. [Online] available at https://telanganatourism.gov.in/blog/06-07-2019.html. Accessed on 8 November 2019.

Torraco, R. J. (2016). "Writing integrative literature reviews: Using the past and present to explore the future". *Human Resource Development Review*, 15(4), 404–428.

Tribe, J. (1997). "The indiscipline of tourism", *Annals of Tourism Research*, 24(3), 638–657.

TripAdvisor (2015). "TripBarometer 2015: Five key traveler trends". [Online] Available at www.tripadvisor.com/TripAdvisorInsights/n2582/tripbarometer-2015-five-key-travelertrends. Accessed 15 December 2019.

UNESCO (2010). "Accolade for Singapore landmark: Pressure to pass prompts revision of evaluation systems", *VOICES UNESCO in the Asia-Pacific*, No. 24, October – December 2010. [Online] Available at https://unesdoc.unesco.org/ark:/48223/pf0000189826. Accessed November 2019.

United Nations World Tourism Organization (2019). "Overtourism? New UNWTO report offers case studies to tackle challenges", Press Release 19016, 6 March.

Veal, A. J. (2011). *Research Methods for Leisure and Tourism: A Practical Guide*, 4th edition. Harlow: Prentice Hall.

Whittemore, R., and Knafl, K. (2005). "The integrative review: Updated methodology", *Journal of Advanced Nursing*, 52, 546–553.

Yin, Y., Zhang, X., Peng, D., and Li, X. (2009). "Model validation and case study on internally cooled/heated dehumidifier/regenerator of liquid desiccant systems", *International Journal of Thermal Sciences*, 48(8), 1664–1671.

8

MICRO SHOCKS AND PUBLIC OUTRAGE

City tourism in a turbulent world

Craig Webster and Sotiris Hji-Avgoustis

Introduction

The size and importance of tourism and its related industries is noted widely and contributes a great deal (over 10%) to the global GDP (World Travel and Tourism Council 2019, p. 1). As such, it is an important aspect of the global economy to investigate, especially since the most recent data shows that the travel and tourism component of the global economy is growing more rapidly than the global GDP as a whole (World Travel and Tourism Council 2019). What is also of note from the most recent data is that more than three-quarters of global travel and tourism spending is for leisure (World Travel and Tourism Council 2019). The composite picture to those concerned with the study of the world economy is that travel and tourism is a large chunk of the global economy; it is growing; and that leisure tourism is the largest component of travel and tourism spending. The important aspect to focus on in a free market economy, then, is the role of consumer choice, as a large part of the travel and tourism spending is voluntarily spent; consumers have a choice of tourism destinations and a great deal of choice in terms of tourism products. Thus, it seems that the large academic literature that is devoted to the study of tourism is warranted, as it is a business that is vital to the global economy and is based largely upon consumer choice.

Traditionally, there is a stress on the push and pull factors that influence tourism, a framework that has a long history and has been used as a framework for understanding tourism flows for decades (see, for example, Dann 1977, 1981, Crompton 1979, Giddy 2018, Jang and Cai 2002, Hsu and Lam 2003). How it is generally conceptualised is that the push factors are the internal characteristics of a traveller that would make the person want to travel, while pull factors are those characteristics of a destination or impressions of a destination that would encourage a person to visit the destination (Gnoth 1997). For this reason, a great deal of the tourism literature has either focused upon the internal drives of individuals, aspects of destinations, or individual impressions of destinations that would influence destination choice.

One of the key characteristics of destinations is the image of the destination. This has been a lively field of research for decades, since a destination's image will have a direct impact upon the pull factor that will influence an individual to choose to travel to a specific destination, whether it is a foreign or domestic destination. Britton (1978) pioneered much of this work to delve into how individuals have versions of a destination that may be somewhat different from the reality experienced on the ground in a destination. Others have continued to research how individuals form their impressions of destinations (Chon 1990, Pike and Page 2014). A review of the literature (Pike 2002) from the early 1970s showed a growing field that has developed over time to be what it now is, a fairly robust and interesting strand of the scholarship in the field of tourism.

There are many political and social forces that impact upon the flows of tourists to various destinations to create a vibrant environment for visitors. While there are elements of the natural and built environments that would influence the flow of tourism to a destination, there are social elements that are economic and political in nature that would be expected to influence the flow of visitors to a destination, among them political shocks of one sort or another. In essence, the political shock may be considered the "push" among the pull factors, as a political shock is a political, social or economic characteristic of a destination and this aspect typically will act in ways to undermine the attractiveness of a destination to many visitors, ensuring that many will not want to visit a specific destination. However, there may be a certain niche market that may find the political or other shock as something that is attractive, pulling in the dark tourist.

In this work, we will delve into political shocks and how they impact the pull of tourists to destinations, especially cities. To do this, we will start by reviewing the literature on shocks. Then, we discuss the typologies of shocks and how these shocks may pertain to tourism in cities, one typology of destinations. We then discuss some cases in which political shocks of various types have threatened tourism in destinations and how this would be expected to impact tourism in cities, if at all. Finally, we discuss how cities can manage political shocks in ways to allow them to mitigate the negative impact that many political shocks may have on a city's tourism economy.

Literature review

The literature on hospitality and tourism is replete with literature that deals with the impact of political and economic shocks on hospitality and tourism. The most extreme political shock that a destination can ever face is war and there are edited volumes in recent years that have focused upon how wars impact destinations (Butler and Suntilkul 2013, Moufakkir and Kelly 2010, Ryan 2007,). While much of the literature on the role of war in tourism deals with reconstruction of the economy following warfare, there are some who look upon war as a pull factor, attracting the dark tourist (Bigley et al. 2010, Henderson 2000, Ioannides and Apostolopoulos 1999). As a matter of fact, there has been substantial thought to how war can actually work in ways to attract tourists to a destination, such as Yuill's (2003) discussion of the push (heritage, history, guilt, curiosity, death and dying, and nostalgia) and pull factors (education, remembrance, artefacts and site sacralisation) in the relationship between war and tourism. Much of the literature on war and tourism is generally irrelevant with regard to an investigation of lesser political shocks to tourism, as most of it pertains to using the heritage from war as a platform for tourism development, presuming that the war is over.

There is also a subset of the literature on war and tourism that views travel and tourism as platforms that can be utilised to create peace. So, rather than viewing war/violent conflict as an aspect of the built environment that would create a draw for tourists, it envisions the tourism industry as something that can build an economic and social foundation that will encourage social and economic interactions that will create a sustained peace among populations or countries that have had violent political conflicts. These authors (see, for example, Askjellerud 2003, Anson 1999, D'Amore 1988, 2009; Khamouna and Zeiger 1995, Levy and Hawkins 2009) base their assumptions upon the liberal idea that contacts that are economically beneficial for two parties will create an environment in which individuals (and by extension firms, ethnicities and countries) will see the benefits of cooperating in a liberal economy. The grounding idea is that these economic interactions will lead to a more peaceful world, based upon the simple understanding that cooperation pays off and that sustained and beneficial economic contacts will lead to social and political cooperation. Webster and Ivanov (2014) analysed the political and economic situations in several destinations that have had sustained problems with ethnic/religious conflict and their findings do support the notion that in some instances, there is a great deal of validity to the liberal approach, showing that cooperation in tourism can lead to peacebuilding between religious groups/ethnicities.

However, in the hierarchy of political shocks war is the most bombastic, meaning that there are types of political shocks short of war that could also have a possible impact upon flows to destinations for tourists. There is significant literature that does not really deal with war but deals with other types of shocks and characteristics that would make a destination less attractive to many visitors. Probably the most influential and significant of these investigations is Neumayer's (2004) study to determine those political factors that seem to dissuade tourism flows to some destinations. In his research, he used a large sample of countries to determine if there is evidence that political factors would negatively impact tourism flows. His findings illustrated substantial evidence that there are many political factors that seem to dissuade tourism from destinations, such as conflict history, political violence and human rights issues. The findings illustrate that there seem to be political characteristics, even if not shock (such as political violence) that seem to work in ways to undermine tourism to destinations. In a much more substantial piece of research, Llorca-Vivero (2008) analysed over 130 destinations to determine which political events or characteristics seem to dissuade tourism flows. The findings were generally consistent with the findings of Neumayer (2004), showing that there is substantial quantitative evidence that tourists are dissuaded from travelling to destinations because of political characteristics that many would consider either unpleasant (systematic problems with human rights abuse) or dangerous for a visitor (terror attacks or ethnic conflicts).

While Neumayer (2004) and Llorca-Vivero (2008) both demonstrate in separate analyses and using different databases that there are a multitude of political and social elements of a destination that appear to influence tourism flows, there is also a vast literature that investigates the impacts of specific types of shocks to a destination and its attractiveness to tourists. In the days since September 11, 2001, there has been a substantial literature that has investigated the impact of political violence and terror attacks to look into how these types of political events seem to impact tourism destinations (see, for example, Aimable and Rosselló 2009, Araña and León 2008, Björk and Kauppinen-Räisänen 2011, Causevic and Lynch 2013, Drakos and Kutan 2003, Feridun 2011, Ingram et al. 2013, Larsen et al. 2011, O'Connor et al. 2008, Saha and Yap 2014, de Sausmarez 2013, Solarin 2015, Wolff and Larsen 2014, Zenker et al. 2019). The bulk of these studies deal with specific cases or multiple cases but have findings generally consistent with the findings of Neumayer (2004) and Llorca-Vivero (2008), showing that many political or social characteristics of a country or unpleasant political happenings make many destinations seem less attractive than they otherwise would be. So, "bad" attributes or political events are generally associated with a more negative perception of a destination. One finding (Wolff and Larsen 2014) that seemed to go counter to the prevailing findings was that in some cases, a destination could have such a good reputation that a one-off terror attack would not have a deleterious impact on the reputation/image of the destination. But this finding (that a terror attack in Norway did not undermine the positive attitude most had towards Norway as a destination) seems to be the exception to the rule. Clements and Georgiou (1998), however, conclude that even if the public has the perception of a potential ethnic clash, it is enough to seriously undermine a tourism destination's reputation, suggesting that even the threat of a potential ethnic clash is enough to harm the flow to tourists to a destination.

While the destination literature has shown that there is a great deal of evidence to support the notion that negative characteristics of destinations largely dissuade tourist flows from those destinations, there is also literature that is devoted to how travel and tourism industries can respond in strategic ways to various political shocks that they face. These approaches (see, for example, Ivanov et al. 2016a, 2016b, 2016c, Jallat and Shultz 2010, Purwomarwanto and Ramachandran 2015, Webster et al. 2017). Ivanov et al. (2016a, 2016b, 2016c) and Webster et al. (2017) are concerned with the management of the outcomes of the Crimean Crisis of 2014. They investigate how Russian hoteliers responded to the political and economic shocks of the Crimean Crisis (Ivanov et al., 2016c), how the hoteliers of Ukraine responded to the crisis (Ivanov et al. 2016a) and how the hoteliers of Crimea responded to the crisis (Ivanov et al. 2016b). In addition, Webster et al. (2017) gave an overview of how the crisis was dealt with in a strategic way for the hoteliers in Crimea, Russia and Ukraine. Jallat and Shultz (2010) focus on how multinational corporations (MNCs) in the tourism sector in Lebanon can cope with the historical and current challenges of the political characteristics of that destination. Purwomarwanto and Ramachandran (2015) analysed how the tourism sector in Indonesia, Malaysia and Singapore adjusted to the shocks of the 2008 financial crisis. Paraskevas (2013) developed actionable plans for how the industry should respond to terrorist threats.

The most critical articles in the subfield of the study of shocks and the response to shocks (Neumayer 2004, Llorca-Vivero 2008) illustrate what seems to be intuitive; that tourists do not want to go to tourism destinations associated with poor human rights, war, conflict, violence or poverty. The other literature that is less universal also shows much the same; that the tourism image of a country or tourism flows can be damaged by empirical realities that the public is aware of. There is some indication (Wolff and Larsen 2014) that, if a destination's image is sufficiently good, a one-time event will not have a major impact on the image of a destination nor the willingness of people to travel to that destination. However, not all countries have such a positive image as Norway, so this may simply be the exception that proves the general rule.

In terms of cities and shocks to impact tourism in cities, there is little literature that would suggest that political shocks would impact city destinations, probably since the majority of the literature on tourism and shocks deals with destinations as countries, rather than cities. However, Webster et al. (2016) write about the political shock of the passing of a piece of legislation (Religious Freedom Restoration Act – RFRA) in Indiana that threatened the entire tourism and hospitality industry in the state and the biggest city in the state (Indianapolis) disproportionately. The authors delved into the ways that the bloodless crisis impacted specifically the tourism economy in Indianapolis, finding that there was no indication that the crisis was something that had an impact on the hospitality and tourism industry in Indianapolis. The authors posit that the crisis did not have a meaningful impact upon the city's tourism, possibly because of the aggressive campaign of the city's destination management organisation (DMO) and partly because the political authorities partially backtracked on the legislation that had the potential to be used to discriminate against members of the LGBTQ+ community.

Otherwise, there is little in the literature that relates specifically to tourism in cities and crisis management. In the following section, we now deal with typologies of shocks that can have an impact upon the destination image or tourism flows to a destination and discuss recent shocks and how they could threaten tourism.

Types of shocks and recent political shocks

By definition, a shock is an unexpected event that would impact the economy. As such, shocks can be physical, environmental or political in nature. For the purposes of our discussion, we focus upon those that are political in nature. While there have been shocks in recent history that are not necessarily caused by political forces, there are a number of recent environmental or ecological shocks that may have had an impact upon the flow of tourists. For example, the outbreak of SARS (a respiratory infection) in 2003 had a major impact upon travel in tourism in that year (Wilder-Smith 2006). But not all things that seem like they should be shocks that would lead to a negative outcome for tourism are. Surprisingly, the 2010 eruption of Iceland's volcano Eyjafjallajökull seems to have not led to a reduction in tourism flows to the island and instead is used as a platform to encourage tourism (Troxler and van Holle 2018).

In terms of recent political shocks that could have had an impact on tourism in cities, there are several examples from the USA that should be discussed. These political shocks were passed by states in the USA and caused an outcry from some who opposed the political actions of the state. Much like the RFRA case (mentioned above) they were laws that could impact populations. Responses from various parts of the country led to calls to boycott travel and tourism to the states that had passed the laws that had been interpreted as possibly being offensive to some. Table 8.1 summarises the political shocks.

During 2018 and 2019, a number of states in the United States passed laws limiting the ability of women to get abortions, something that could have implications for tourism to those states. The laws range from outright bans of abortion to a redefinition of when it would be legal for a woman to have an abortion, usually limiting it after a period in which a heartbeat can be detected. Louisiana in May 2019 passed a bill to ban abortions after the detection of a foetal heartbeat. This makes Louisiana the ninth state in 2019 to pass abortion restrictions that could challenge the Supreme Court's *Roe* v. *Wade* decision in 1973, a decision that legalised abortion during the first trimester of a pregnancy. Also during May 2019, Alabama legislators voted to ban abortions in nearly all cases. Georgia, Kentucky, Louisiana, Missouri, Mississippi and Ohio stopped short of outright bans, instead passing so-called "heartbeat bills" that effectively prohibit abortions after six to eight weeks of pregnancy. All these bans will face lengthy court battles that may even reach the Supreme Court.

Such decisions often get the attention of companies who fear retaliation from their own customers if they choose to remain on the sidelines. Three of the world's biggest entertainment companies (Netflix, Disney and Warner Media) came out against Georgia's bill and are threatening to stop producing movies and TV shows in Georgia if the state's new abortion law takes effect. Comcast's NBC Universal said the spread of these anti-abortion bills would affect their decision-making on where they produce future content. Georgia became a hub

Table 8.1 Recent political shocks in states in the USA

Year	State(s)	Intention
2019	Alabama	Ban abortions in nearly all cases
2018–2019	Georgia, Kentucky, Louisiana, Missouri, Mississippi and Ohio	Heartbeat bills that effectively prohibit abortions after six to eight weeks of pregnancy
2016	North Carolina	A law that banned transgender people from using a bathroom aligning with their gender identity
2015	Indiana	Religious Freedom Restoration Act (RFRA)

for entertainment industry production because of generous tax incentives it offers to film-makers and producers. The industry employs some 92,000 people and generates billions of dollars in economic impact, according to statistics cited by the city. Alabama's decision also touched off calls to boycott the state and its products among many prominent women and Democrats, including officials in Colorado and Maryland who said that they would work to divest state spending from Alabama. Some of the more vocal proponents of the ban have come from the world of Hollywood. Actors Sean Penn, Alec Baldwin, Mia Farrow, Don Cheadle, Jason Bateman and Ben Stiller have all signed a petition calling for boycotting the state.

Other states faced economic boycotts because of their legislation on certain social issues. In 2016, North Carolina faced pressure from the National Collegiate Athletic Association (NCAA) after it passed a law that banned transgender people from using a bathroom aligning with their gender identity. The intention of the law was to ensure that the gender of a person from their birth certificate would be the way in which usage of public restrooms would be segregated, rather than the perception of the subjective gender identity of the individual. The NCAA threatened not to host its championship in the state and the state eventually repealed the law, something that would have substantial impact economically on a state with a little over ten million people. The Trump administration's plans to redefine gender by removing any mention of transgender drew the attention of many businesses. Companies like Google, Coca-Cola, Apple, Amazon, Pepsi, JPMorgan Chase, Dow Chemical and Uber were among many com-panies that signed a letter condemning the proposed changes while expressing support for the transgender community. The companies who signed the letter, identified as Business for Trans Equality, represent more than $2.4 trillion in annual revenue, suggesting a real threat to the economy of North Carolina.

In 2016, Disney and its Marvel affiliate rallied behind a planned boycott after Georgia signed a religious freedom bill, perceived as antigay by many. The Religious Freedom Restoration Act (RFRA) enacted in the state of Indiana faced similar criticism and it negatively affected the city's tourism and convention industry. According to Visit Indy, the marketing arm of Indianapolis, the capital of Indiana, since April 2015, Indianapolis has lost more than $60 million in future convention business because of the RFRA controversy. Passed at a time when same-sex mar-riage was about to be legalised in every state, the RFRA was portrayed by supporters as a way to protect individuals and businesses with a religious objection to same-sex weddings. Oppon-ents saw it as a license to discriminate against LGBT people. Immediately after the law passed, public reaction was fast and mainly negative, with cancellations of planned events and business expansions, travel bans and denunciations from across the spectrum. Companies including Sales-force, Apple, Eli Lilly and Angie's List; sports leagues including the NCAA, NBA (National Basketball Association) and WNBA (Women's National Basketball Association); states and municipalities across the country; rock concerts; and comedy shows all threatened to boycott the city. Under this mounting pressure, the RFRA was amended about a week later with lan-guage adding sexual orientation and gender identity to the list of factors that could not be cause for denial of service. Soon after many organisations rescinded their travel bans to the state (Webster et al. 2016).

These recent examples from the USA show that there are issues that seem to attract dispro-portionate public reactions relative the numbers of those who the legal changes would seem to impact. Although the examples are all at the state level, the changes at the state level would negatively impact the tourism industry in the entire state and all its cities. The lesson is that laws that may seem to be unjust can be passed and that national and international condemnation can threaten the tourism product of the political unit's cities. However, there is indication (Webster et al. 2016) that in terms of impact, the threats to the tourism economy may remain just that, threats. But they could also be more than threats.

Discussion and conclusion

While there are many shocks that can have an impact upon tourism flows to cities, there is an interesting class of political shocks that seem to be minor, in some sense, but nevertheless attract significant opposition from companies and public figures. This would seem to be a new type of shock: a melange of a public outrage by influential individuals and support from companies that are eager to be seen as socially and politically progressive. What is interesting is that the political shocks are far short of war, armed conflict, systematic human rights issues with governments and other factors that are usually seen as major political shocks that should impact upon tourism flows. However, they are issues of discussion and bring up the issue of how outrage culture has entered into the public discussion and may have an impact upon tourism and tourism flows.

While the reaction of outrage to political actions may impact tourism flows, there may be some reason to believe that outrage following some political decisions may remain simply rhetoric. For example, Le (2017) speculated that there would be a serious drop in tourism from the UK to the USA, presumably in protest to the election of Donald Trump to the presidency in the USA. The data cited by Le (2017) suggested that one in five Britons would not even consider the USA as a destination for their travels following the unexpected election of the former reality TV star to the US presidency. The data, though, does not suggest a substantial drop in tourism from the UK to the USA since the election of Donald Trump to his first elected position and the US tourism industry seems to be doing fine. It may be that the outrage and talk about how the tourism industry of the USA would suffer because of the US public's election of a president seems to have remained largely just talk and has not really impacted the US and its tourism negatively, as Webster (2017) predicted.

To some extent, a zeitgeist is at work. We are living in an era of microaggressions and outrage culture. The idea of microaggressions is that some spoken words are "brief, everyday exchanges that send denigrating messages to certain individuals because of their group membership" (Sue 2010). The key to that is the "micro", that is that people are conditioned in educational institutions into taking small political statements as being indicative of deeply offensive thinking. As such, political socialisation suggests massive reactions to policy changes that are perceived as being indicative of deeply offensive or dangerous attitudes. Thus, outrage is expected in these political changes, otherwise people and organisations may be seen as being complicit in the biases/racism that could be inferred by the policy changes. Thus, it seems that outrage and outrage culture contributes to the inflation of a political shock that in another era would have been seen as peripheral to something central to the economy of a state. To some extent, micro shocks can become or may threaten to become macro-shocks. The outrage culture then is the catalyst that will permit the elevation of tensions and elevation of risks to the tourism industry.

In closing, political change and political shocks are inevitable in any political economy and there is reason to believe that some shocks may have a negative impact upon a city's tourism image in the mind of a potential tourist. However, there is also some reason to believe that such threats can be mitigated against to minimise losses. Webster et al. (2016) illustrate some of the intelligent mitigation of Visit Indy, an organisation that promoted Indianapolis as a welcoming city, despite the state government's passing of a law that could be understood as being anti-LGBT+. Talk from those who oppose the passage of laws that may have a miniscule direct impact upon major companies or commercial interests in tourism may lead to threats and actions that can harm tourism in cities. While threats may simply remain talk and talk can be cheap, what is to be noted is the way that commercial interests and individuals feel compelled to react to the passage of laws in ways which make them potential shocks to tourism in cities. Thus, talk can be cheap, but it can also be the harbinger of political decisions that are serious and have substantial negative impacts upon the tourism products of many cities.

References

Aimable, E., and Rosselló, J., 2009. The short-term impact of 9/11 on European airlines demand. *European Journal of Tourism Research*, 2 (2), 145–161.

Anson, C., 1999. Planning for peace: The role of tourism in the aftermath of violence. *Journal of Travel Research*, 38 (1), 57–61.

Araña, J. E., and León, C. J., 2008. The impact of terrorism on tourism demand. *Annals of Tourism Research*, 35(2), 299–315.

Askjellerud, S., 2003. The tourist as messenger of peace? *Annals of Tourism Research*, 30 (3), 741–744.

Bigley, J., Lee, C.K., Chon, J., and Yoon, Y., 2010. Motivations for war-related tourism: A case of DMZ visitors in Korea. *Tourism Geographies*, 12 (3), 371–394.

Björk, P., and Kauppinen-Räisänen, H., 2011. The impact of perceived risk on information search: A study of Finnish tourists. *Scandinavian Journal of Hospitality and Tourism*, 11 (3), 306–323.

Britton, R. A., 1978. The image of the Third World in tourism marketing. *Annals of Tourism Research*, 6 (3), 318–329.

Butler, R., and Suntilkul, W., 2013. *Tourism and War*. Abingdon: Routledge.

Causevic, S., and Lynch, P., 2013. Political (in)stability and its influence on tourism development. *Tourism Management*, (34), 145–157.

Chon, K., 1990. The role of destination image in tourism: A review and discussion. *The Tourist Review*, 45 (2), 2–9.

Clements, M. A., and Georgiou, A., 1998. The impact of political instability on a fragile tourism product. *Tourism Management*, 19 (3), 283–288.

Crompton, J., 1979. Motivations for pleasure vacation. *Annals of Tourism Research*, 6 (4), 408–424.

D'Amore, L., 1988. Tourism: The world's peace industry. *Journal of Travel Research*, 27 (1), 35–40.

D'Amore, L., 2009. Peace through tourism: The birthing of a new socio-economic order. *Journal of Business Ethics*, 89 (4), 559–568.

Dann, G., 1977. Anomie, ego-enhancement and tourism. *Annals of Tourism Research*, 4 (4), 184–194.

Dann, G., 1981. Tourist motivation: An appraisal. *Annals of Tourism Research*, 8 (2), 187–219.

de Sausmarez, N., 2013. Challenges to Kenyan tourism since 2008: Crisis management from the Kenyan tour operator perspective. *Current Issues in Tourism*, 16 (7–8), 792–809.

Drakos, K., and Kutan, A. M., 2003. Regional effects of terrorism on tourism in three Mediterranean countries. *The Journal of Conflict Resolution*, 47 (5), 621–641.

Feridun, M., 2011. Impact of terrorism on tourism in Turkey: Empirical evidence from Turkey. *Applied Economics*, 43 (24), 3349–3354.

Giddy, J., 2018. Adventure tourism motivations: A push and pull factor approach. *Bulletin of Geography. Socio-economic Series*, 42 (42), 47–58.

Gnoth, J., 1997. Tourism motivation and expectation formation. *Annals of Tourism Research*, 24 (2), 283–304.

Henderson, J., 2000. War as a tourist attraction: The case of Vietnam. *International Journal of Tourism Research*, 2 (4), 269–280.

Hsu, C. H. C., and Lam, T., 2003. Mainland Chinese travellers' motivations and barriers of visiting Hong Kong. *Journal of Academy of Business and Economics* [online], 2 (1), 60–67.

Ingram, H., Tabari, S., and Watthanakhomprathip, W., 2013. The impact of political instability on tourism: Case of Thailand. *Worldwide Hospitality and Tourism Themes*, 5 (1), 92–103.

Ioannides, D., and Apostolopoulos, Y., 1999. Political instability, war, and tourism in Cyprus: Effects, management, and prospects for recovery. *Journal of Travel Research*, 38 (1), 51–56.

Ivanov, S., Gavrilina, M., Webster, C., and Ralko, V., 2016a. Impacts of political instability on the hotel industry in Ukraine. *Journal of Policy Research in Tourism, Leisure and Events* [online], 9 (1), 100–127.

Ivanov, S., Idzhilova, K., and Webster, C., 2016b. Impacts of the entry of the Autonomous Republic of Crimea into the Russian Federation on its tourism industry: An exploratory study. *Tourism Management*, 54, 162–169.

Ivanov, S., Sypchenko, L., and Webster, C., 2016c. International sanctions and Russia's hotel industry: The impact on business and coping mechanisms of hoteliers. *Tourism Planning & Development* [online], 14 (3), 430–441.

Jallat, F., and Shultz, C. J., 2010. Lebanon: From cataclysm to opportunity – Crisis management lessons for MNCs in the tourism sector of the Middle East. *Journal of World Business*, 46 (4), 476–486.

Jang, S. C., and Cai, L. A., 2002. Travel motivations and destination choice: A study of British outbound market. *Journal of Travel & Tourism Marketing*, 13 (3), 111–133.

Khamouna, M., and Zeiger, Z. B., 1995. Peace through tourism. *Parks & Recreation*, 30(9), 80–86.

Larsen, S., Brun, W., Øgaard, T., and Selstad, L., 2011. Effects of sudden and dramatic events on travel desire and risk judgments. *Scandinavian Journal of Hospitality and Tourism*, 11 (3), 268–285.

Le, A., 2017. Trump's presidency: The future of American tourism industry. *Journal of Tourism Futures*, 3 (1), 8–12.

Levy, S. E., and Hawkins, D., 2009. Peace through tourism: Commerce based principles and practices. *Journal of Business Ethics*, 89 (4), 569–585.

Llorca-Vivero, R., 2008. Terrorism and international tourism: New evidence. *Defence and Peace Economics*, 19 (2), 169–188.

Moufakkir, O., and Kelly, I. (Eds.), 2010. *Tourism, Progress and Peace*. Wallingford: CABI.

Neumayer, E., 2004. The impact of political violence on tourism: Dynamic cross-national estimation. *The Journal of Conflict Resolution*, 48 (2), 259–281.

O'Connor, N., Stafford, M. R., and Gallagher, G., 2008. The impact of global terrorism on Ireland's tourism industry: An industry perspective. *Tourism and Hospitality Research*, 8 (4), 351–363.

Paraskevas, A., 2013. Aligning strategy to threat: A baseline anti-terrorism strategy for hotels. *International Journal of Contemporary Hospitality Management*, 25 (1), 140–162.

Pike, S., 2002. Destination image analysis: A review of 142 papers from 1973–2000. *Tourism Management*, 23 (5), 541–549.

Pike, S., and Page, S., 2014. Destination marketing organizations and destination marketing: A narrative analysis of the literature. *Tourism Management*, 41, 202–227.

Purwomarwanto, Y. L., and Ramachandran, J., 2015. Performance of tourism sector with regard to the global crisis: A comparative study between Indonesia, Malaysia, and Singapore. *The Journal of Developing Areas*, 49 (4), 325–339.

Ryan, C. (Ed.), 2007. *Battlefield Tourism: History, Place, and Interpretation*. Oxford: Elsevier.

Saha, S., and Yap, G., 2014. The moderation effects of political instability and terrorism on tourism development: A cross-country panel analysis. *Journal of Travel Research*, 53 (4), 509–521.

Solarin, S. A., 2015. September 11 attacks, H1N1 influenza, global financial crisis and tourist arrivals in Sarawak. *Anatolia*, 26 (2), 298–300.

Sue, D. W. (Ed.). 2010. *Microaggressions and Marginality: Manifestation, Dynamics, and Impact*. Hoboken, NJ: John Wiley & Sons.

Troxler, S., and van Holle, S., 2018. Eyjafjallajökull: The volcano that caused an eruption in Icelandic tourism. *Hospitality Insights* [online]. Available from: https://hospitalityinsights.ehl.edu/iceland-tourism-boom [Accessed 2 October 2019].

Webster, C., 2017. Political turbulence and business as usual: Tourism's future. *Journal of Tourism Futures*, 3 (1), 4–7.

Webster, C., and Ivanov, S., 2014. Tourism as a force for political stability. In C. Wohlmuther and W. Wintersteiner (Eds.), *The International Handbook on Tourism and Peace*. Klagenfurt/Celovec: Drava Verlag, 167–180.

Webster, C., Ivanov, S., Gavrilina, M., Idzhylova, K., and Sypchenko, L., 2017. Hotel industry's reactions to the Crimea Crisis. *e-Review of Tourism Research* (eRTR) [online], 14 (1–2), 57–71. Available from: https://ertr.tamu.edu/files/2017/12/4.-eRTR_ARN_Vol.14-No1.2_Webster-et-al.FINAL_.pdf [Accessed 2 October 2019].

Webster, C., Yen, C.-L., and Hji-Avgoustis, S., 2016. RFRA and the hospitality industry in Indiana: Political shocks and empirical impacts on Indianapolis' hospitality and tourism industry. *International Journal of Tourism Cities*, 3 (2), 221–231.

Wilder-Smith, A., 2006. The severe acute respiratory syndrome: Impact on travel and tourism. *Travel Medicine and Infectious Disease*, 4 (2), 53–60.

Wolff, K., and Larsen, S., 2014. Can terrorism make us feel safer? Risk perceptions and worries before and after the July 22nd attacks. *Annals of Tourism Research*, 44, 200–209.

World Travel and Tourism Council, 2019. *Travel and Tourism Economic Impact 2019* [online]. Available from: www.wttc.org/-/media/files/reports/economic-impact-research/regions-2019/world2019.pdf [Accessed 20 September 2019].

Yuill, S. M., 2003. Dark tourism: Understanding visitor motivations at sites of death and disaster. Master's dissertation, Department of Parks, Recreation and Tourism Sciences, Texas A&M University. Available from: https://oaktrust.library.tamu.edu/bitstream/handle/1969.1/89/etd-tamu-2003C-RPTS-Yuill-1.pdf?sequence=1&isAllowed=y [Accessed 25 September 2019].

Zenker, S., von Wallpach, S., Braun, E., and Vallaster, C., 2019. How the refugee crisis impacts the decision structure of tourists: A crosscountry scenario study. *Tourism Management*, 71, 197–212.

PART II

Marketing, branding and markets for tourism cities

Marketing is critical to the success of tourism in cities. Part II considers destination marketing, branding and management as they are applied within urban areas. Significant markets including business travellers, exhibitions, families, Millennials and those visiting friends and relatives (VFR) are discussed in detail. One of the goals here was to provide a good blending of academic scholarship with real-life practice and case studies from cities.

Marketing, or at least sales and promotion, has long been done by cities to build their images and attract visitors. A major paradigm change came in the 1990s with Web 1.0 and later with Web 2.0 and social media. Now those that carry out city tourism marketing and branding, mainly consisting of destination management organisations (DMOs) are increasingly spending more on online or digital marketing. These new technological applications are discussed later in Part III. It is not just the communication platforms and tools that are changing, so too are the visitors. For example, in Part II the impacts of Millennials and Generation Z are considered, and these are two generational cohorts that drive and are driven by new technologies. Also covered are the family and business tourism markets where the influences of technology are significant.

Mainstay, misunderstood and overlooked – that's what we might call three of the markets considered in Part II. The mainstay market is business tourism, which has long been a solid core of the demand in cities. The misunderstood is the VFR market, which has been neglected by most practitioners and only until recently by academic scholars. The overlooked are the avian species in urban areas and the lure they represent to birdwatchers. Another concept being that of dark tourism is probably better understood by academics than the general public. However, it represents a product for many cities and there is demand for the experiences involved.

This handbook is loaded with good case studies and examples, and Part II has two outstanding cases on marketing and branding from Melbourne, Australia and Vancouver.

The following is a description of the nine chapters:

Alastair M. Morrison opens Part II with a discussion on marketing and managing city tourism destinations. The main purpose of the chapter is to situate the tourism marketing, branding and product development of urban areas within the context of destination marketing and management. A descriptive research approach is followed using literature reviewing and expert opinion on the themes. The chapter begins with a mini literature review on the topic, comprising an analysis of related articles in tourism and non-tourism journals. The author also discusses review articles on urban tourism. The second part of the chapter is a brief review of the historical evolution of destination marketing and management. Professional processes and practices in city destination marketing and management are then presented. The chapter's main conclusion is that city destination marketing has now transformed into destination management that covers multiple roles.

Rob Davidson sets out to examine the emergence of the global market for business tourism in urban areas. The chapter uses literature reviewing, expert opinion and selected case examples. It is suggested that business tourism has received comparatively scant attention in academic circles. The author examines the terminology issues in this segment of the market and especially surrounding business events. He looks at the use of the business tourism market as a tool of urban development, identifying the traditional tourism and also the "beyond tourism" benefits. The chapter then looks at the demand- and supply-side characteristics of developing business tourism in cities and discusses the present and future roles of city DMOs in these processes.

Valentina Gorchakova and **Vladimir Antchak** provide an excellent case study on Melbourne, Australia and how a portfolio of major international cultural exhibitions has contributed to the development of city eventfulness and positively affected the brand of the destination and the overall attractiveness of the place. The chapter draws together the idea of an "eventful" city with the concepts of storytelling and place identity and examines events as their key agents. In addition to literature reviewing, data were collected from semi-structured interviews with destination managers, event planners, representatives of museums and other cultural institutions. The authors emphasise the importance of storytelling, creativity and eventfulness to success in city branding, citing Copenhagen, Edinburgh, Rotterdam, Auckland, Macau, Canberra and Manchester as places that have succeeded.

Rami K. Isaac and **Jacqueline Wichnewski** evaluate the new destination brand of Vancouver, British Columbia, Canada. This study aims to identify whether the communicated brand promise of Vancouver's newly established destination brand is in line with the brand experience perceived by tourists visiting the city. It begins by describing the most important elements of this branding including the brand identity, brand promise and brand experience. The authors used qualitative field research and semi-structured interviews with industry experts and tourists to determine if visitor experiences matched up with the brand promise. Generally, the authors found that the brand rang true for tourists to the city; however, they also found some issues with the new brand and suggested improvements.

Xinran Lehto, **Jun Chen** and **Uyen Le** consider the family travel market for cities. The chapter aims to provide a better understanding of the research topics, theoretical thrusts, methodological approaches and the evolving patterns. This outstanding and comprehensive chapter uses data from a literature review of 159 studies related to family tourism that were gathered and analysed through content analysis. The results reveal that family tourism is a well-defined field; however, this domain of research is in need of improved depth and scope, creating numerous research opportunities.

Dae-Young Kim and **Ye-Jin Lee** have the main goal of providing urban tourism scholars and practitioners with a better understanding of how marketing and programming related to travel can be tailored to Millennial audiences. A combination of literature reviewing and selected case studies are used for the data. The authors discuss the overall patterns of Millennial travellers in how they search for travel information, behavioural characteristics on trips, and trip activities to the specific context of urban tourism. Three city case examples of Hong Kong, Seoul and Barcelona are reviewed in the context of Millennial travel.

Elisa Backer (Zentveld) has made a huge contribution to VFR (visiting friends and relatives) research in recent years. In this chapter, she reviews VFR trends for tourism cities and especially for older people. The chapter aims not only to explore quality of life for older people, it also discusses housing and transport trends related to the aging in tourism cities. Literature reviewing and selected case examples are employed to explore the themes of elderly quality of life in the context of city VFR. This chapter makes a significant contribution to future research and development by raising a new tourism segment and a new opportunity for tourism city planning – that of planning commercial accommodation to support and encourage visiting older friends and relatives (VOFR).

John J. Lennon, one of the creators of the dark tourism concept, analyses dark tourism in urban settings. The author looks in detail at two tragic historical events, one in Phnom Penh, Cambodia and the other in the Lety, Czech Republic. Literature reviewing and personal visits to the sites were the sources of research data. A short discussion of contrasting approaches in these two locations is used to illustrate the fundamental challenge for urban locations of a dark or dissonant heritage. The two dark tourism sites are controversial; one associated with the genocide perpetrated by the Khmer Rouge and the other with mistreatment of Roma people. The author finds a disregard for such heritage, its selective interpretation and the limited conservation and attributes this to a historical repression of these events and period, as government and citizens focus on the future

David Newsome and **Greg Simpson** present a fascinating view of cities as bird-watching destinations, surely a new potential market for many urban areas. They explore the topic with real examples from New York City, Singapore, Dubai and Perth, Australia. The authors focus on the green spaces offered by tourism cities, such as urban parks and gardens, including more innovative sustainability projects such as living green-skins planted on the walls and roofs of buildings. In this sense, the presence of birds is associated not only with images of sustainable and healthy urban spaces but also with resident and visitor well-being benefits such as contemplation and peaceful relaxation. They conclude that cities with green spaces suitable for bird watching are more likely to be perceived as highly liveable and more attractive by residents and visitors alike, making them more competitive as global urban tourism destinations.

9

MARKETING AND MANAGING CITY TOURISM DESTINATIONS

Alastair M. Morrison

Introduction

Cities are critical to tourism in all countries of the world. They are often important transportation hubs and contain extensive arrays of daytime and night-time attractions, activities and experiences. The main purpose of this chapter is to situate the tourism marketing, branding and product development of urban areas within the context of destination marketing and management. A descriptive research approach is followed using literature reviewing and expert opinion on the themes. In so doing, the author acknowledges that two streams of urban tourism research have developed, one stream within tourism journals, books and association professional development activities; the other stream, one sub-stream of which can be called place marketing and branding, appears in similar venues related to urban studies and planning, city management, sustainable development, transportation and other. Furthermore, it is recognised that there has been a considerable gap between city marketing practice and related academic scholarship, and both have developed rather separately.

To say that city tourism marketing is something recent is far from the truth. In fact, 1896 saw the establishment of the first city convention promotion bureau in Detroit, Michigan (Gartrell, 1988, p. 4; Travel Michigan, 2016). However, the marketing of cities has changed quite dramatically in the ensuing approximately 125 years, becoming more professional and broad-reaching, and has transformed with Web 1.0 and 2.0, the greater emphasis on destination/place branding, more concern with sustainability and the advent of smart cities.

This chapter begins with a mini academic literature review on city tourism marketing, branding and product development. The literature review highlights among other things that city tourism needs to be managed and not just marketed. Second, the chapter provides a short history of destination marketing and management. It demonstrates the transformation from destination marketing to destination management. The third part of the chapter is devoted to best processes and practices in city destination marketing and management. This discourse explains the multiple roles of destination management beyond just marketing and branding. The chapter ends with a short summary drawing together the strands from the previous four parts.

A brief review of the academic literature

Chronology of the literature

The academic literature on urban tourism and city tourism marketing is characterised by a broad range of contributions from multiple disciplines. This review begins by examining two major streams of this literature, from tourism and from place marketing and branding. The review is indicative rather than comprehensive and is designed to give an overview of the relevant academic literature.

Tourism journal research stream: The literature in English on urban tourism stretches back to the early 1980s, with Vandermey (1984) being one of the first to use the expression of "urban tourism" in an article title. Based on the example of Calgary, Alberta, Canada, he proposed a model of the urban tourism system with a destination management organisation (DMO) at its core. This article measured the scope and economic impact of tourism in Calgary, mirroring the focus of many of the earlier contributions in the 1960s and 1970s covering tourism planning, development and impact measurement (e.g. An Foras Forbartha, 1966; Archer and Owen, 1971, Butler, 1974; Gunn, 1977). Other important contributions to urban tourism research during the 1980s were from Buckley and Witt (1987) (tourism in difficult areas), Law (1985) (selected British urban tourism case studies), Riley (1984) (hotels and city identities), and Smith (1985) (locations of urban restaurants). The 1980s also saw an increase in consumer-based tourism and community resident research, which at times had a focus on urban areas. For example, Haywood and Muller (1988) evaluated the city tourism experiences of visitors to Toronto, Ontario, Canada. Um and Crompton (1987) measured resident attitudes toward tourism in New Braunfels, Texas.

The 1990s was the decade when significant publishing began on city and urban tourism marketing. Most of the articles had a focus on the supply side of urban tourism and were based on specific city cases. Bramwell and Rawding (1994) examined the tourism marketing organisations in Birmingham, Bradford, Manchester, Sheffield and Stoke-on-Trent, all traditional industrial cities in England. They noted the trend toward forming public–private partnerships (PPPs) to govern city tourism marketing. The authors also highlighted the role of city destination branding in competitive differentiation. Jansen-Verbeke and van Rekom (1996) determined the motivation of visitors to museums in Rotterdam. Bramwell and Rawding (1996) contrasted the marketing images being used by Birmingham, Bradford, Manchester, Sheffield and Stoke-on-Trent. Lawton and Page (1997), who asserted that urban tourism was not adequately recognised as a concept, found that the image being promoted of Auckland, New Zealand did not match what the city really offered to visitors. Qu and Zhang (1997) analysed the marketing approaches and markets of Hong Kong as an urban tourism destination. Spotts (1997) found that urban destinations had the greatest impacts on variations in visitor spending within Michigan. Bramwell (1998) contributed an article, based on Sheffield, where he examined methods for measuring user satisfaction with city tourism products. Also noteworthy was his recommendation that cities should adopt a "place marketing framework". Van Limburg (1998) investigated the attributes that attracted visitors to Den Bosch in the south of the Netherlands.

The new millennium signalled a major increase in article and book publishing on urban tourism and its marketing and branding. The first decade was when the concepts of place branding and destination branding became popular topics for academic authors. In particular, there was greater focus on city destination branding (Paskaleva-Shapira, 2007; Pearce, 2007; Chacko and Marcell, 2008), specific city tourism market segments (Hughes, 2003; McKercher, Okumus and Okumus, 2008), stakeholder relationships and cooperation (Sheehan, Ritchie and Hudson,

2007), city products and attractions (Leslie and Craig, 2000; Chang and Lai, 2009; Griffin and Hayllar, 2009), and competitiveness (Asprogerakas, 2007; Paskaleva-Shapira, 2007). Hughes (2003) examined Manchester's promotion aimed at the gay market and found it could have an undesirable effect on gay residents of the city. Paskaleva-Shapira (2007) outlined new paradigms of city tourism management as competitiveness, cultural heritage tourism, city branding, visitor perception and urban quality of life. Pearce (2007) concluded that there was further scope for Wellington, New Zealand to profit from its unique selling point as a capital city. Sheehan, Ritchie and Hudson (2007) examined the relationship among city governments, hotels and the DMO in marketing and promoting urban tourism destinations. As will be seen in the next sub-section, there was also an upsurge in publishing on urban tourism research in the non-tourism journals during 2000–2009.

For 2010–2019, the volume and diversity of publishing on urban tourism continued to increase. As in the previous decade, the popular topics included branding (Wu, Funck, and Hayashi, 2014; Apostolopoulou and Papadimitriou, 2015; Roult, Adjizian, and Auger, 2016; Chigora and Hoque, 2019), competitiveness (Valls, Sureda, and Valls-Tuñon, 2014) and market segments (Irimiás, 2012; Dai, Hein, and Zhang, 2019). Other topics covered were city attractiveness (Rogerson and Rogerson, 2017; Boivin and Tanguay, 2019), consumer behaviour (Su, Hsu, and Marshall, 2014; Caldeira and Kastenholz, 2018), economic diversification (Erkuş-Öztürk and Terhorst, 2018), future of city tourism (Postma, Buda, and Gugerell, 2017), Olympic Games legacies (Sant, Mason, and Hinch, 2013; Roult and Auger, 2016), online tourism information (Lee, Yoon, and Park, 2017), policy (Amore and Hall, 2017; Maxim, 2019), public transport (Le-Klähn and Hall, 2015), rural versus urban media coverage (Lahav, Mansfeld, and Avraham, 2013) and urbanisation (Luo et al., 2016).

Publishing on urban tourism research was an increasingly popular theme for authors in tourism journals since the 1980s. However, until the arrival of the *International Journal of Tourism Cities* in 2015, there was no single dedicated venue in tourism for contributions on urban tourism research and tourism cities. Having said that, several tourism journals published such content, notably *Tourism Management* and *Annals of Tourism Research*, and more recently the *Journal of Destination Marketing & Management*. Journals outside of tourism also accepted these works and they are now discussed with particular reference to place marketing and branding.

Non-tourism journal research stream: The foregoing literature was placed in tourism journals; however, a parallel stream developed in non-tourism journals. These include journals with a focus on urban and regional studies and planning such as *Cities* (since 1983), *International Journal of Urban and Regional Research* (since 1977), *Journal of Place Management and Development* (since 2008), *Place Branding and Public Diplomacy* (since 2007), *Journal of Urban Affairs* (since 1981) and *Urban Studies* (since 1964). The relevant articles in these journals either partly or primarily dealt with city tourism. Generally, but not exclusively, the contributions were from researchers not considered as "mainstream" tourism scholars.

Cities is a journal that has published many articles related to tourism. For example, Barker and Page (2002) reviewed visitor safety in Auckland, New Zealand. They argued that there was generally a poor understanding of perceived visitor safety during special events held in cities. Rabbiosi (2015) based upon Paris, examined how leisure shopping could be used as a city destination branding approach. She found there to be ambivalence in the city about using such a commercialised theme to brand Paris. Molinillo et al. (2019) investigated the extent of visitor engagement through popular social media platforms in Spanish smart cities. They concluded that these smart cities needed to improve their communications and branding through Facebook, Twitter and Instagram.

The *International Journal of Urban and Regional Research* (IJURR) has published several articles in urban tourism research and city destination marketing. Husbands and Thompson (1990) analysed resident attitudes in Livingstone, Zambia. They found that residents' perceptions were affected by their social status and by educational levels attained. Hoffman (2003) considered the marketing of ethnic diversity within Harlem in New York City. She characterised tourism as being an economic development strategy for inner city ghettos combined with cultural and political objectives (p. 297). Dürr (2012) later raised the issue of slum tourism and analysed the case of Mazatlán in Mexico. She found that tours to slums had ambiguous implications, positive and negative, and did not particularly benefit those living in poverty there.

Associated with the Institute of Place Management, the *Journal of Place Management and Development* has published a significant number of city-tourism-related articles. Jackson (2008) studied residents' perceptions of the social, economic and environmental impacts of special events in the destinations in which they lived. He determined that residents were generally supportive of special events if they contributed socially and economically to their communities. However, they were not unaware of the negative effects of certain special events. Martínez-Ruiz, Martínez-Caraballo and Amatulli (2010) conducted an analysis of luxury goods stores in Venice, Italy with a view to determining the characteristics of business success in a tourism destination. The most successful fashion retailers, by length of store operation, were Italian. They intimated that all luxury brand retailers were not taking full advantage of their location in a famous tourism destination. Duignan (2019) examined the legacies of the London Summer Olympics on small retailers in Greenwich. He found that locally based small retailers were increasingly failing due to higher commercial rents and indifference.

The *Journal of Urban Affairs* has featured several topics related to urban tourism. Boyd (2000) covered racial heritage tourism being promoted by African American neighbourhoods in the United States. She acknowledged the role of this form of tourism in redeveloping these parts of cities. Gladstone and Fanstein (2001) compared the development of tourism in Los Angeles and New York City based upon labour market effects. Turner and Rosentraub (2002) investigated the roles of centre cities in tourism, culture, sport and entertainment. Russo and Scarnato (2018) examined the development of tourism in Barcelona and the impacts of tourism growth on the city and its political regime.

Place Branding and Public Diplomacy has published a significant volume of city destination branding articles. Bouchon (2014) reviewed the positioning and branding of Kuala Lumpur, Malaysia. He identified the challenges in Kuala Lumpur having a global city image. Lai and Ooi (2015) considered the political difficulties ensuing the designations of World Heritage listings for George Town and Melaka, Malaysia. Alonso and Bea (2012) attempted to quantify the brand images of Spanish cities on the Internet. They found that cities with high Internet visibility enjoyed considerable strength in cultural tourism.

In *Urban Studies* Gotham (2002) described the gentrification that had occurred in New Orleans, Louisiana as a result of the city's tourism marketing and development. Using Patras, Greece as their case study, Daskalopoulou and Petrou (2009), also in *Urban Studies*, used data envelopment analysis (DEA), to determine the tourism competitiveness of the city. Later in *Urban Studies*, Hartal (2019) explored the background to Tel Aviv's marketing as a place for LGBT vacations.

Journals with a focus on sustainability and sustainable development also publish articles related to tourism and some of these concern urban tourism. These include *Sustainable Development*, *Sustainable Cities and Society* and *Sustainability*. Timur and Getz (2009) in *Sustainable Development* investigated the goals and challenges for sustainable tourism in Calgary and Victoria, Canada. Kapera (2018) in *Sustainable Cities and Society* examined the sustainable tourism development efforts of local authorities in Poland. Aall and Koens (2019) in *Sustainability* called for broader discussion in urban sustainable tourism than just about overtourism.

Other non-tourism journals have published articles related to city tourism marketing and branding. For example, Ashworth and Voogd (1988) had an article on city marketing in *The Town Planning Review*. Biagi and Detotto (2014) wrote on crime as a tourism externality in *Regional Studies*. This chapter's reference list further demonstrates that scholarly article publishing on urban tourism and city destination marketing is scattered over many journals, within and outside of tourism.

This is just a small sampling of tourism-related articles in these non-tourism journals. However, it is obvious from the foregoing that there is a rich history of publishing on city tourism in non-tourism journals. Although not all contributed by tourism scholars, these articles offer many valuable perspectives for the marketing and branding of urban areas. Some are critical of tourism's impacts on cities and, therefore, contribute to a more balanced view of the phenomenon. Moreover, these articles and their contents indicate that it is insufficient to only consult the tourism journals when analysing these topics in the literature.

Review articles on urban tourism: In addition to the specific literature from these two streams on city destination marketing and management, there have been several important review articles on city tourism that include related content. The most cited work among these is Ashworth and Page (2011), which identified 12 sub-themes of urban tourism research. Marketing and place imagery represented one of the sub-themes. The authors noted an increased focus on city tourism marketing including greater attention being given to branding. Pearce (2001) proposed an integrative framework for urban tourism research with one of the eight themes being marketing. Selby (2004) noted that the recent research on urban tourism had improved the understanding of city marketing and management. However, he recommended a tentative research agenda for furthering urban tourism knowledge. Edwards, Griffin, and Hayllar (2008), based on research conducted in Australia, developed a somewhat broader agenda for urban tourism research. Once again, city destination marketing emerged as a significant topic and issue. Dupré (2019) conducted a 25-year review (1991–2016) of the literature to identify trends and gaps in the research on urban development and tourism. Importantly, she found a lack of adequate synergy between urban development and tourism in relation to place-making.

Together, these review articles delivered several important messages. First, they suggested that urban tourism had not yet received the recognition it deserved. Second, the review articles recommended that a multidisciplinary perspective for research was required for urban tourism. Third, they outlined schemes for comprehensively researching the urban tourism phenomenon.

Having examined the two main streams of the academic literature and then useful review articles, several predominant research themes in urban tourism marketing and markets are now reviewed. Due to space limitations, the following does not represent a complete coverage of all the literature themes associated with city destination marketing and management.

Destination and place marketing and branding

Particularly in the new millennium, many articles were published on destination and place branding with respect to cities. Previously, the cases of Paris (Rabbiosi, 2015), Kuala Lumpur (Bouchon, 2014), George Town and Melaka (Lai and Ooi, 2015) and Spanish cities (Alonso and Bea, 2012) were mentioned. Other examples are Balakrishnan (2008) who reviewed Dubai's success as a case study in destination branding; Uysal (2013) analysed the role of religion in the branding of Istanbul; Bellini and Pasquinelli (2016) investigated how Florence's fashion companies influenced its destination branding. Noteworthy here is that all these contributions were in non-tourism journals and fuelled by the greater emphasis on place marketing and branding in the past 20 years.

Although the tourism literature has tended to have a greater focus on the branding of countries, states and provinces, there have been many articles and books with a focus on city destination branding. For example, Heath and Kruger (2010) contributed a chapter on the branding of Tshwane (Pretoria), South Africa in the book, *City Tourism: National Capital Perspectives*. Sahin and Baloglu (2014) used a city brand advocacy model in determining the effects of word-of-mouth generation in various trip-purpose segments. Kavaratzis (2017) advocated a participative place branding process including the views of residents. Roult, Adjizian and Auger (2016) examined the recognition of Montréal's Olympic Park and its stadium among international visitors to the city. Apostolopoulou and Papadimitriou (2015), using Patras, Greece as a case example, investigated the influence of destination personality on the behaviour of tourists. Once again, these represent only a small sample of the city destination branding articles in the tourism literature; however, they indicate the recent growing interest in the topic and the types of research studies that are being conducted.

Hankinson (2010) discussed the growth of published research on place branding. However, he argued that it had made no reference to the development of mainstream branding research, nor did it adequately reflect the views of practitioners. Gertner (2011) reviewed 212 articles on place marketing and place branding published from 1990 to 2009. He is highly critical of the rigour of the research in this body of literature.

Markets and types of tourism

Scholars have produced significant amounts of journal articles and books related to specific markets for cities and types of tourism within urban areas. Within this *Routledge Handbook of Tourism Cities*, there are several related articles including ones on business tourism (Davidson), events (Gorchakova and Antchak), cultural and heritage tourism (Boyd), visiting friends and relatives (VFR) (Backer), dark tourism (Lennon), geotourism (Richards, Simpson, and Newsome), bird watching (Simpson and Newsome), walking tourism (Morris), families (Lehto, Chen, and Le) and Millennials (Kim). As these markets for and types of city tourism are so numerous, only a small slice of the literature is presented here on: (1) history, heritage, culture and creativity; (2) events and festivals; (3) business tourism and events; (4) sport; (5) shopping, entertainment and dining; (6) built contemporary leisure attractions; (7) architecture; and (8) ethnicity, diasporas and VFR.

History, heritage, culture and creativity: With respect to types of tourism, history, heritage and culture have attracted great attention in the literature. Jansen-Verbeke and van Rekom (1996) studied the motivations of museum visitors in cities in the Netherlands, while recognising that museums were an effective way of attracting visitors for urban tourism (p. 373). Ashworth and Tunbridge (2000) published the book, *The Tourist-Historic City*, and highlighted the expanding role of heritage and cultural products in city tourism. Later, Richards and Palmer (2010) authored the book, *Eventful Cities: Cultural Management and Urban Revitalization*, which reviews the development of culture, events and creativity in enhancing the attractiveness of city destinations. The fusion of history, heritage and culture with creativity is becoming a major drawing card for many cities and their tourism sectors. There are numerous examples of cultural and arts districts dotted around the world, including the Distillery District in Toronto, Bangkok River, 798 Art District in Beijing, Pier 2 in Kaohsiung (Taiwan), Dublin's Creative Quarter in Ireland and Westergasfabriek in Amsterdam. Roberts (2001, p. 12) noted that the creative industries were increasingly being used as the foundation for urban cultural development. Other authors considered the roles of the creative industries in specific destinations (Rogerson, 2006) and the interactions between tourism and the creative industries (Long and Morpeth, 2016; Long, 2017). Richards (2011) argued that traditional cultural tourism was transforming and shifting from tangible to intangible heritage along with greater involvement with the everyday life of cities.

Events and festivals: There is a substantial literature on events in tourism and academic journals devoted to this topic. Getz (2008, p. 403) said that events are "important motivators of tourism" and proposed a typology of planned events that includes cultural celebrations, political and state, arts and entertainment, business and trade, educational and scientific, sport competition, recreational and private events. Cities are hubs for all these categories of events and celebrations and a main reason for urban vibrancy. Business and trade events (often referred to as the MICE markets) are a significant source of tourism for major cities, as just discussed, as well as a key focus for city destination marketing and branding. Sport competition events are also of great significance to urban tourism and sport "mega events" in cities have attracted attention from many scholars (e.g. Gursoy and Kendall, 2006; Preuss, 2007).

Festivals also play a significant role in city tourism worldwide. In Getz's taxonomy (2008) these include cultural celebrations and arts and entertainment. O'Sullivan and Jackson (2002) investigated the contribution of festivals to sustainable local economic development using three case studies from Wales. They found that the opportunities to enhance sustainability were not being fully capitalised upon. Gotham (2005a) examined festivals in New Orleans and found that their effects were not all positive. Quinn (2006) analysed two arts festivals in Wexford and Galway, Ireland and found that tourism had a strong influence on their growth.

Business tourism and events: Business tourism and events are exceptionally important to most cities in terms of volume and economic impacts. However, as Davidson remarks in this handbook, this type of city tourism has received scant attention compared with leisure tourism. The author concurs with this viewpoint and from personal experience with city DMOs in North America observes that they tend to place greater emphasis on business tourism in their sales and marketing. This contradiction is another indication of the wide chasm between academia and practice, and it is also not surprising that leisure tourism-oriented journals achieve high metrics on SSCI and SCImago, while meeting and event journals struggle to achieve even modest ratings.

Draper, Thomas and Fenich (2018), in a review article of event management research, highlight an upsurge in publishing on the topic beginning in the 1990s. They classified events into two main categories, business and leisure. Business events are organised by associations, companies and governments for educating, motivating, selling and/or networking with their employees, members and/or customers to attain business goals (Fenich, 2012).

Incentive travel also generates events for urban areas. However, it is one aspect of business tourism that has not attracted as much scholarly attention as from other authors, despite a flurry of academic articles in the early 1990s (e.g. Ricci and Holland, 1992; Sheldon, 1995; Shinew and Backman, 1995). Mair (2005) argued that there was not enough research about why travel was such a powerful motivator. However, Fenich, Vitiello, Lancaster and Hashimoto (2015) concluded that companies have been using incentive travel effectively to motivate employees toward higher sales.

Sport: Sport competition events have a significant influence on urban tourism. Cities are important venues for professional and amateur sport, and for hosting sport events. Gibson (1998) in a review of the sport tourism literature to date noted that cities were increasingly using sport events to attract tourists. Kurtzman (2005) concluded that sport tourism has a very high economic value throughout the world. Most of the largest sport installations are in urban areas. Gratton, Shibli and Coleman (2005) investigated UK city investment in sporting infrastructure as a strategy for economic regeneration in former industrial urban areas. They noted that sport events were increasingly being used to raise city profiles and enhance their images. Another research focus with sport tourism is on city resident support for its development. For example, Hritz and Ross (2010) surveyed Indianapolis residents and found that social and economic benefits were predictors of support for future sport tourism development.

Shopping, entertainment and dining: Shopping, entertainment and dining in all formats are hallmarks of city tourism. Their clustering within urban areas is a magnet for local residents and out-of-town visitors. Timothy (2005) and Tosun et al. (2007) highlight the importance of shopping to travel and tourism, and cities as the main "shopping hubs" of countries are the principal beneficiaries.

Cities are at the heart of the experience economy (Richards, 2001). Lorentzen (2009, p. 833) identified the creative branches generating experiences in cities as tourism, fashion, visual arts, radio/television, publishing firms, toys/entertainment, sports, architecture, design, film/video, advertising, edutainment, events, computer games and cultural institutions.

Cities are great places for dining. Franck (2005) characterises urban areas as dining rooms, markets and farms for residents and visitors. Du Rand and Heath (2006), rather surprisingly, found that food was not being featured significantly in destination marketing in South Africa and globally. McKercher, Okumus and Okumus (2008) concluded that food consumption was an "ubiquitous activity" for most visitors and not a special interest phenomenon. In contrast to du Rand and Heath, Henderson (2009), who published a literature review on food tourism, found that it was a common theme in destination and business marketing.

Culinary and gastronomic tourism are other terms found in the literature related to food tourism. Ignatov and Smith (2006, p. 238) defined culinary tourism as "tourism trips during which the purchase or consumption of regional foods (including beverages), or the observation and study of food production (from agriculture to cooking schools) represent a significant motivation or activity". Gastronomic experiences, according to Quan and Wang (2004, p. 302) can be one of the major motivations for travel. Kivela and Crotts (2006) found that gastronomy plays a major role in the way that tourists experience destinations.

The performing arts are a sub-set of culture and cultural tourism, as well as often being associated with festivals. However, they are a key part of city entertainment and the night-time economy (Lovatt and O'Connor, 1995) in many urban areas. Some cities like New York and London are famous for their theatre districts, others for their music including Nashville, New Orleans and Memphis. There is some research linking tourism and the performing arts (e.g. Barbieri and Mahoney, 2010; Lim and Bendle, 2012); however, it seems to be a fertile theme for more analysis. Bars and nightclubs are found in most cities as well and they have attracted some attention because of drunkenness and associated social problems (e.g. Roberts, 2006).

Casinos are significant tourism magnets for several cities including Macau, Las Vegas, Singapore, Incheon and Manila. There has been a significant amount of research on casino tourism and gaming, with much attention being given to resident reactions to the phenomenon. Other scholars have focused their attention of the development and impacts of tourism on specific cities, including Atlantic City (e.g. Braunlich, 1996), Detroit (Wiley and Walker, 2011), Incheon, Korea (e.g. Choi et al., 2019), Las Vegas (e.g. Ritzer and Stillman, 2001), Macau (e.g. Loi and Kim, 2010) and Singapore (e.g. Wu and Chen, 2015).

Built contemporary leisure attractions: Large urban and metropolitan areas are popular locations for a variety of leisure and entertainment attractions, including theme parks (Bigné, Andreu, and Gnoth, 2005; Milman, 2009), family recreation centres and water parks (Jin, Lee, and Lee,

2015), museums (Plaza, 2006; Silberberg, 1995), science centres (Lipardi, 2013), aquaria (Ballantyne et al., 2007; Cater, 2010), zoos (Mason, 2000; Frost, 2011), performing arts centres (Hale and MacDonald, 2005; Quinn, 1967), botanic gardens (Ballantyne, Packer, and Hughes, 2008; Henderson, 2014) and others. The significant local and regional populations are the main economic justification for selecting these sites, while out-of-town tourism may also contribute substantially.

Architecture: Cities are known for having stunning architecture that adds to their attractiveness and memorability. There are of course many styles of architecture ranging from the ancient to ultra-modern. Some of the cities known for older architecture are Athens (Greece), Budapest (Hungary), Florence (Italy), Istanbul (Turkey), Oxford (England), Paris (France), Rome (Italy) and St. Petersburg (Russia). Modern or contemporary architecture is well represented in cities such as Barcelona (Spain), Brasilia (Brazil), Columbus (Indiana, USA), Dubai (UAE), Shanghai (China) (from *Condé Nast Traveler* (2017) and Travel Channel (2019) ratings of cities).

Specht (2014) published the book, *Architectural Tourism: Building for Urban Travel Destinations*, signalling this to be a special form of tourism. He argued that architecture plays a critical role in every area of tourism (p. 2), including offering numerous venues for leisure activities. Lasansky and McLaren (2004) produced an edited book, *Architecture and Tourism: Perspective, Performance and Place*, in which they noted the "reciprocal relationship between the modern practice of tourism and the built environment" (p. 1). Both sets of authors cite the Guggenheim Museum Bilbao (Spain) as a success story in how postmodern architecture positively influenced tourism to a city, as is the Sydney Opera House. In fact, there are so many iconic historic and modern structures that are associated with cities and their tourism including the Eiffel Tower (Paris), Statue of Liberty (New York), London Eye, Atomium (Brussels), Taj Mahal (Agra, India), Great Wall (China), Voortrekker Monument (Pretoria, South Africa) and Burj Al Arab Hotel (Dubai) to name just a few.

Ethnicity, diasporas and VFR: People from specific ethnic groups and national origins have tended to cluster in certain parts of cities. These clusters, many as a product of immigration, have in many instances become attractions for visitors as well as leisure places for residents. Jan Rath (2007) edited the book, *Tourism Ethnic Diversity and the City* (Routledge Contemporary Geographies on Leisure, Tourism and Mobilities) that included chapters on New Orleans, New York, Miami, Lisbon and Boston discussing these neighbourhoods' roles in tourism. The Greek-towns in Detroit, Michigan and Toronto, Ontario and Chicago's Polish Downtown are great examples, as are the Chinatowns in many major cities.

Developing and marketing these urban areas is not without particular challenges and problems. For example, Conforti (1996) using Little Italy as a case example identified issues with using ethnic ghettos as tourist attractions and as an element of city tourism marketing and branding. He pointed out that historically these were places of oppression and restrictions, and that their residents were victims of derogatory stereotypes.

Another aspect of immigration that influences city tourism might be called the diaspora effect, where substantial communities of people originating from other countries live and work in particular urban areas (Bruner, 1996; Scheyvens, 2007). These immigrant communities often become magnets for inbound VFR travel from the origin countries. For example, several cities in the GCC (Gulf Cooperation Council) have significant numbers of immigrants and immigrant workers from South Asia and benefit from many family and friend visits from Bangladesh, India, Pakistan and Sri Lanka.

Smart tourism in cities

According to Lopez de Avila (2015, n.p.), a smart destination is

> an innovative tourist destination, built on an infrastructure of state-of-the-art techno-
> logy guaranteeing the sustainable development of tourist areas, accessible to everyone,
> which facilitates the visitor's interaction with and integration into his or her surround-
> ings, increases the quality of the experience at the destination, and improves residents'
> quality of life.

Information communication technologies and big data are often quoted as being at the heart of smart destinations and cities (Buhalis and Amaranggana, 2013; Gretzel et al., 2015; Boes, Buhalis, and Inversini, 2016; Gretzel, Zhong, and Koo, 2016).

Cities as transportation hubs

Many cities are important transportation hubs which tends to enhance their significance in tourism. Their strategic locations are the reason for airlines, high-speed railway systems, cruise ships and ferry companies, and other transport providers to choose them as hubs. Some of the cities known for being excellent and multi-modal hubs are Amsterdam, Dubai, Miami, Shang-hai and Singapore.

Surprisingly, there is not a substantial body of literature linking transportation and tourism. However, the extant studies show strong synergies between the two. For example, as early as in 1985, Mescon and Vozikis explored the significant economic impact of cruise tourism on Miami and Dade County. Later, Lohmann, Albers, Koch and Pavlovich (2009) analysed how Dubai and Singapore, with coordinated efforts from the airlines, airports and DMOs, used their air hub status to become more viable tourism destinations. Pagliara, La Pietra, Gomez and Vassallo (2015) examined the impact of the high-speed rail system on tourism to Madrid, Spain. Tang, Weaver and Lawton (2017), with Singapore Changi Airport as their focus, investigated whether airport users could be convinced to return as stayover visitors in the city. All of these studies highlighted the significant impact of transport hubbing on city tourism; however, this is an area of research that requires more attention in the future.

Negative aspects of city tourism development and marketing

There have been many articles and books that have criticised tourism due to its negative impacts on urban areas. As noted earlier, the non-tourism journal stream tends to have signi-ficant negative assessments of city tourism. Some of the issues of concern include overtourism (Ali, 2018; Séraphin, Sheeran, and Pilato, 2018; Pinke-Sziva, Smith, Olt, and Berezvai, 2019), sustainability (Hinch, 1996; Timur and Getz, 2009), globalisation (Gotham, 2005b; Dupré, 2019), urbanisation (Mullins, 1991); gentrification (Gotham, 2002; Gravari-Barbas and Guinand, 2017) and touristification (Freytag and Bauder, 2018), terrorism (Coca-Stefaniak and Morrison, 2018) and slum tourism (Meschkank, 2011; Dürr, 2012; Frenzel and Koens, 2012), most of which are discussed in other chapters in the *Routledge Handbook of Tourism Cities*.

Criticisms of public investment in sport facilities and convention and exhibition centres reflect another negative view on certain aspects of city tourism (Whitson and MacIntosh, 1996; Laslo and Judd, 2005; Long, 2005; Sanders, 2014; Tomlinson, 2014). Whitson and MacIntosh

(1996) argued that large events create significant benefits for tourism and the construction sector; however, the public sector costs are often understated.

Another aspect of criticism on tourism in cities has a focus mainly on the night-time activities of certain tourists. These include the negative impacts of noise (Zaeimdar and Bahmanpour, 2014; Rouleau, 2017), and crime, drunkenness and stag tourism (Vesey and Dimanche, 2003; Thurnell-Read, 2011, 2012; Biagi and Detotto, 2014).

Management issues

The research literature suggests that cities face other issues and challenges in marketing and managing tourism. Governance is one of these issues and fundamentally involves the choice of which body should assume the responsibility for tourism and its marketing and management. Schmallegger and Carson (2010), for example, highlight the limitations of strong government "patronage" of tourism for Darwin in the Northern Territory of Australia.

Cities are the beneficiaries of historical urban planning (Shoval, 2018) as well as being the inheritors of newer urban planning strategies (Maxim, 2019). While city tourists may be unaware of the fruits and follies of urban planning, they undoubtedly interact with many of its facilities and services, whether these be modern rapid urban transport systems, cultural performance venues or historical monuments.

The advances of the sharing economy have had a profound influence on cities and urban tourism as well. "Living like a local" has become an important aspect of staying in peer-to-peer accommodations such as Airbnb (Paulauskaite et al., 2017; Stone, 2018).

This brief review of the literature demonstrates a growing academic interest in urban tourism research and city tourism marketing and branding. Not all of these commentaries have portrayed tourism in a positive light and underline the need for city destination management, and that it is not just sufficient to focus on urban economic development through city destination marketing and branding. Now, the third part of the chapter provides a historical background on destination marketing and management. The progression in actual practice from destination marketing to a fuller scope of destination management is noteworthy in this discourse.

A short history of destination marketing and management

The brief review of the academic literature reflected that scholars began to show greater interest in urban tourism research and city tourism marketing in the 1980s and 1990s. Destination marketing has been discussed as a concept for more than 30 years. The pioneering book in English on destination marketing was published in 1988 by Richard Gartrell, *Destination Marketing for Convention and Visitor Bureaus*, and it was released under the banner of the International Association of Convention & Visitor Bureaus (IACVB) (now Destinations International). It is noteworthy that Ashworth and Voogd (1990) published their book, *Selling the City: Marketing Approaches in Public Sector Urban Planning*, just two years later. After that, Eric Laws published a text with the title of *Tourist Destination Management: Issues, Analysis and Policies* (Routledge) in 1995. The introduction of the Certified Destination Management Executive (CDME) programme in 1992 by the then IACVB was a watershed for the field of destination marketing and management. It recognised that destination marketing and management were not just topics; they more importantly represented a profession. Core and elective classes were offered, and the participants were senior DMO executives and managers, mainly from city destinations.

Later in 2004, Destinations International established a Performance Measurement Team to begin the process of identifying DMO performance measurement benchmarks. A *Handbook* of measures was released in 2005. Also, in 2005, Destinations International released a book about destination branding, called *Destination BrandScience*, co-authored by Duane Knapp and Gary Sherwin. During 2010–2011, Destinations International appointed a Performance Reporting Task Force, and the Task Force's efforts resulted in an updated edition of the *Standard DMO Performance Reporting: A Handbook for DMOs*. Additionally, in 2012 Destinations International produced a revised edition of the *DMO Uniform System of Accounts (Standard Financial Reporting Practices for DMOs)*.

The Destination Marketing Accreditation Program (DMAP) was another breakthrough on the professional side of destination marketing. Unlike the CDME programme, DMAP focuses on DMOs as organisations rather than on individual DMO professionals. DMAP was especially important in identifying 16 "domains" for measuring the performance of a DMO (governance; finance; human resources; technology; marketing; visitor services; group services; sales; communications; membership; management and facilities; brand management; destination development; research/marketing intelligence; innovation; and stakeholder relationships).

During the first two decades of the new millennium, there was a surge in published books, academic articles and practice-oriented manuscripts on destination marketing and management. One of the most influential of the new books was *Destination Branding* (2004) by Nigel Morgan, Annette Pritchard and Roger Pride. This new publication seemed to spur many academic researchers into doing research, writing articles and arranging conferences around the topic of destination branding. It should be recognised that many tourism scholars beginning in the early 1970s were producing valuable research contributions on destination image and its measurement. This research undoubtedly provided a valuable platform for what was to come later about destination branding and positioning.

Another benchmark was UNWTO's *A Practical Guide to Destination Management* published in 2008. Prepared for UNWTO by TEAM Tourism Consulting of the UK, this was the first practical guide on all aspects of destination management. UNWTO and the European Travel Commission (ETC) later co-sponsored two additional practical guides: *Handbook on Tourism Destination Branding* (2009, prepared by Tom Buncle) and *Handbook on Tourism Product Development* (2011, prepared by Tourism Development International).

Several related books from academic authors have been added in recent years. These have included two books by Stephen Pike: *Destination Marketing Organisations: Bridging Theory and Practice* (Elsevier Science, 2005) and *Destination Marketing: An Integrated Communication Approach* (Butterworth-Heinemann, 2008). Two other books were published in 2011: *Destination Marketing and Management: Theories and Applications* by Youcheng Wang and Abraham Pizam (CABI) and *Managing and Marketing Tourist Destinations: Strategies to Gain a Competitive Edge* by Metin Kozak and Seyhmus Baloglu (Taylor & Francis). All four of these books were welcome additions to the scholarship on destination marketing and management and offered a variety of different perspectives on the topics.

There has been an evolution within cities from sales and selling convention capacity to employing a broader range of promotional techniques, and from there to destination marketing that in part includes urban tourism product development. That journey has continued to the present day, in which city destination management is the focus and where there is a concern for sustainable tourism development. Having discussed this transformation, the chapter now reviews the best approaches to city destination marketing and management.

Best processes and practices in city destination marketing and management

It is obvious from the foregoing that city destination marketing has matured during recent decades and that it remains an important role within the broader concept of destination management. Useful guidelines for destination marketing and management have been produced by trade associations, inter-governmental organisations and NGOs, consultants, authors of books and others.

Roles of city destination management

Destination marketing is just one of several components of destination management. Morrison (2019) identifies the other roles of destination management as community relationships and involvement, leadership and coordination, partnerships and team-building, planning and research, product development and visitor management (Figure 9.1).

Although historically, the marketing and promotion role was the primary and most important one, it now represents just one of several in city destination management. A short description of all these roles follows:

- *Leadership and coordination*: Setting the agenda for tourism and coordinating all stakeholders' efforts toward achieving that agenda.
- *Partnership and team-building*: Fostering cooperation among government agencies and within the private sector and building partnership teams to attain the destination vision and specific goals and objectives.
- *Community relationships and involvement*: Involving local community leaders and residents in tourism and monitoring resident attitudes towards tourism.
- *Visitor management*: Managing the flows, impacts and behaviours of visitors to protect resources and to enhance visitor safety, experiences and satisfaction.
- *Planning and research*: Conducting the essential planning and research needed to attain the destination vision and goals.
- *Product development*: Planning and ensuring the appropriate development of physical products and services for the destination.

Figure 9.1   The destination management roles in city tourism.

- *Marketing and promotion*: Creating the destination positioning and branding, selecting the most appropriate markets and promoting the destination.

Within city destination marketing, there are specific processes that require significant attention: (1) marketing planning process; (2) marketing strategy; (3) brand development; (4) marketing plan development; and (5) performance evaluation. Before discussing these, the success factors for city tourism destinations must be discussed.

City destination success factors

The 10 As Framework: The 10 As Framework was first introduced by Professor Alastair M. Morrison of Purdue University when working on the team to develop the Tourism Master Plan for the city of Hangzhou in Zhejiang Province, PR China in 2003–2004. It has since been in continuous use within international tourism policy and planning exercises in several countries and was first published by Morrison in *Marketing and Managing Tourism Destinations* by Routledge in 2013 and subsequently updated in the 2nd edition of this book in 2019. Prior to its application in tourism policy and planning, the 10 As Framework was validated by DMO leaders in the Certified Destination Management Executive (CDME) programme, offered by Destinations International from the early 1990s.

The main purpose of the 10 As Framework is to provide criteria for policy, planning, management and marketing leading to successful destinations. It is a parsimonious typological process model that focuses a destination's attention on critical aspects ranging from awareness and accessibility to action and accountability.

Determining the success of city tourism destinations: How can it be determined if a city tourism destination is successful or not? If the destination is judged to be successful, can the DMO take the sole credit for this great achievement? These are hugely difficult questions to answer but nevertheless they should be tackled.

- *Quantity or quality?* One answer to the first question is that the successful destinations are the ones with the most tourists. So, you will often see the "world's top destinations" identified as the ones with the most tourist arrivals according to UNWTO. These would include countries such as France, USA, China, Spain, Italy and the UK. However, many will argue that this is a choice of "quantity" over "quality" and that smaller destinations are not necessarily inferior because they have fewer visitors. Additionally, these are countries and there are many more destinations and DMOs below the country level.
- *Can we believe the rating schemes?* Some travel magazines and guidebooks publish "top destination" lists each year. There are many of these "top destination" lists but it is interesting to note that not many destinations appear twice on these. But more importantly no specific and detailed criteria are given for the selections. Some of the ratings are done by editors and journalists and others by consumers.
- *What about other destination rating systems?* The World Centre of Excellence for Destinations (CED), located in Montréal, Canada, developed the *System of Measures for Excellence in Destinations* (SMED). Established in 2007, CED has evaluated several destinations around the world with SMED. A panel of SMED experts visits and assesses each destination that applies, and the destination pays a fee for this service. The destinations that have been evaluated successfully include Abitibi-Témiscamingue (Canada), Andorra, Cantons de L'Est (Canada), Chengdu (China), Crete (Greece), Douro Valley (Portugal), Jeddah (Saudi Arabia), Madeira (Portugal), Mexico City, Montréal (Canada), Riviera Maya (Mexico),

Samos (Greece) and Tela (Honduras). This system was a breakthrough for destination management and was created with the support of UNWTO. However, the criteria for approval of destinations under SMED have not been made public.

- *Description of the 10 As Framework*: The 10 As Framework outlines a set of attributes for judging the success of tourism destinations. Each of these ten attributes begins with the letter "A" (Figure 9.2).

The following is a short explanation of each of the ten A attributes:

- *Awareness*: This attribute is related to tourists' level of knowledge about the city destination and is influenced by the amount and nature of the information they receive. The main question to be asked here is: *Is there a high level of awareness of the city among potential tourists?*
- *Attractiveness*: The number and geographic scope of appeal of the city destination's attractions comprise this attribute. The main question is: *Does the city offer a diversity of attractions that are appealing to tourists?*
- *Availability*: This attribute is determined by the ease with which bookings and reservations can be made for the city destination, and the number of booking and reservation channels available. The main question is: *Can bookings and reservations for the city be made through a variety of distribution channels?*
- *Access*: The convenience of getting to and from the city destination, as well as moving around within the destination, constitutes this attribute. The main questions are: *Is there convenient access to and from the destination by all modes of transportation? Is there convenient transportation within the destination?*
- *Appearance*: This attribute measures the impressions that the city destination makes on tourists, both when they first arrive and then throughout their stays in the city. The main questions are: *Does the city make a good first impression? Does the city make a positive and lasting impression?*

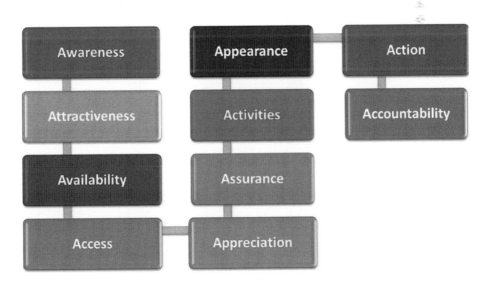

Figure 9.2 The ten As of successful city tourism destinations.

- *Activities*: The extent of the array of activities and experiences available to tourists within the city destination is the determinant of this attribute. The main questions are: *Does the city offer a wide range of activities and experiences in which tourists want to engage? Does the city offer authentic experiences that are of interest to tourists?*
- *Assurance*: This attribute relates to the safety and security of the city destination for tourists. The main question is: *Is the city clean, safe and secure?*
- *Appreciation*: The feeling of the levels of welcome and hospitality contribute to this attribute. The main question is: *Do tourists feel welcome and receive good service in the city?*
- *Action*: The availability of a long-term tourism plan and a marketing plan for tourism are some of the required actions. The main questions are: *Have appropriate tourism policies been developed? Is the tourism development and marketing in the city well planned?*
- *Accountability*: This attribute is about the evaluation of performance by the city DMO. The main question is: *Is the DMO measuring the effectiveness of its performance?*

These ten attributes can be useful for all city destinations, but they need to be expressed in greater detail than that shown above. Additionally, there are other criteria that could be added to this list of ten. For example, the economic contributions of tourism to the city might also be included, as well as the degree to which the city is following a sustainable tourism agenda.

Marketing planning process

While academics dissect urban tourism in seemingly unending ways, it is left to city practitioners to accomplish marketing in the most professional fashion possible. Undoubtedly, most of them have never read many of the previously cited articles or books, however they do their level best to use the right marketing processes and practices. While academics are trying harder now to have meaningful impact on the practice of tourism, the chasm between university research writing and industry implementation remains very great. Industry associations for city tourism marketing practitioners likewise have achieved much on professionalising destination marketing and management for cities.

All cities must plan their tourism marketing. This planning consists of long-term (strategic) and short-term (tactical) time periods. For city tourism marketing, the strategic time period is three to five up to ten years into the future, while the tactical time period is one to two years ahead. A core part of the marketing planning process is a time-ordered hierarchy of vision, marketing goals and objectives. Figure 9.3 shows this hierarchy including the destination vision, DMO vision, marketing goals and objectives, and a description of these four follows.

- *Destination vision*: This is a super-long-term goal for tourism in the city. It is expressed as a future "picture in words" or a verbal, aspirational image of the city.
- *DMO vision*: The future desired status and characteristics of the DMO that support the attainment of the destination vision.
- *Marketing goals*: Long-term (three to five years) measurable results for city tourism marketing to be achieved.
- *Marketing objectives*: Short-term (one to three years) measurable results for city tourism marketing to be achieved.

Reaching these goals and objectives requires answering the five questions shown in Figure 9.3 and using the processes that go along with each of them, beginning with an environmental scan and situation analysis for "Where are we now?" The main outcome of this stage is the definition of the city's unique selling propositions (USPs) for tourism.

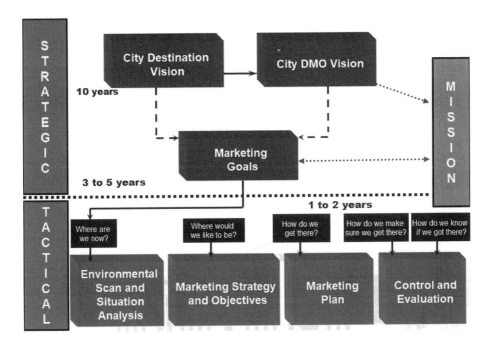

Figure 9.3 City destination marketing planning process.

Marketing strategy process

Addressing the "Where would we like to be?" question constitutes the marketing strategy process. Basically, a marketing strategy is a combination of the city's targeted tourism markets and its chosen approach to positioning, image and branding (Figure 9.4).

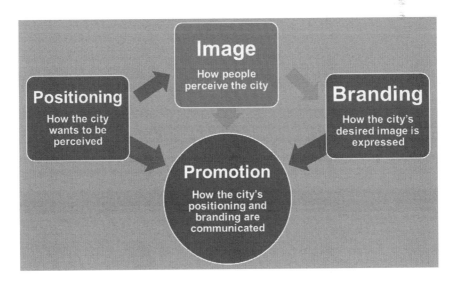

Figure 9.4 City positioning, image and branding.

Brand development process

Earlier in the chapter, the academic literature on destination and place branding was reviewed and it was pointed out that the brand development for city tourism was getting greater attention. There are four parts in the brand development process for city tourism. Figure 9.5 shows the first part of destination branding as the situation analysis (destination, competitive, market, destination image, resident and past marketing). The destination image analysis determines the existing perceptions of the destination among past and potential tourists.

The second part is where input is gathered from tourism sector stakeholders and residents about the image and positioning of the city. In particular, they should express what they see as being the most unique features of the city. This input feeds into the third part of the brand development process, where the city's tourism unique selling propositions (USPs) are identified. These USPs are crucial to city destination branding, as they spell out what is different in the city in comparison with competitors.

The fourth part is where the city tourism branding is designed, implemented and evaluated, which is accomplished in six sequential steps as shown in Figure 9.5.

- *Brand strategy development*: Defining the branding objectives, brand positioning and target markets.
- *Brand identity development*: Designing a creative approach that normally will include a new logo, colour scheme and other visual image guidelines, slogan, musical score and other elements.
- *Brand launch and introduction*: Revealing the new city tourism brand to the public for the first time, usually at some sort of special event. For the brand introduction, a variety of materials (e.g. a brand manual) are prepared for the use of the brand by the city DMO and by tourism sector stakeholders
- *Brand implementation*: Embedding the brand within the city such that it appears in every communication and interaction with tourists. This also includes delivering on the brand promise.

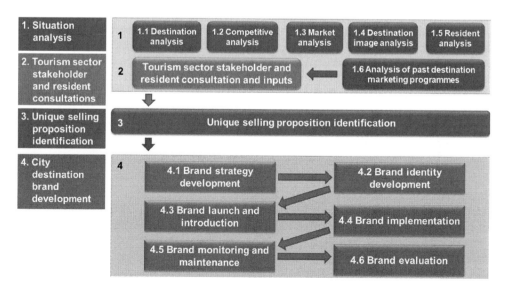

Figure 9.5 City destination brand development process.

- *Brand monitoring and maintenance*: Tracking implementation of the city tourism brand and assessing the progress towards achieving the branding objectives. Tweaking and refreshing the brand are steps that can occur in brand maintenance.
- *Brand evaluation*: Determining the effectiveness of the brand and its implementation in reaching its objectives.

Marketing plan and implementation

Every city should regularly develop a marketing plan for tourism. This plan is a written document that details what will be done to accomplish the marketing objectives. An executive summary, marketing plan rationale and implementation plan are the three parts of the marketing plan.

- *Executive summary*: A brief summary of the key highlights and major initiatives outlined in the plan.
- *Rationale*: The reasons and assumptions behind the choices of city marketing activities.
- *Implementation plan*: The marketing objectives, activities and programmes, marketing budget, timetable, assignment of responsibilities and evaluation procedures and measures.

Performance evaluation process

All good city marketing will be for naught if it is not evaluated against the goals and objectives. This evaluation is accomplished in two stages, monitoring progress and measuring performance. The monitoring of progress takes place when implementing the marketing plan; the performance measurement when the plan is completed. City DMOs must demonstrate their marketing effectiveness and performance, as this indicates accountability for the funds invested in city marketing.

Summary

Cities play a vital role in tourism worldwide. Deservedly, urban destination marketing and management are receiving higher priorities in practice as well as in the academic literature. Although city promotion has historically been a core focus for tourism in urban areas, the remit has significantly broadened into city destination management. It is now recognised that there is more to city tourism than just marketing and promotion and that all three concerns of the triple-bottom-line of sustainability (economic, social-cultural and environmental) must be considered. The coining of overtourism in 2016 (Ali, 2018) has put city tourism under a media and public microscope, questioning the wisdom of too much economic reliance and marketing of urban tourism.

Figure 9.6 is a summary diagram of the three main parts of this chapter. This discourse has highlighted the research on city tourism marketing, branding and product development. These are aspects of tourism that are attracting more attention; however, there is much more scope for future research.

City destination marketing and management would gain greater attention and recognition if its various stakeholders and research strands coalesced. There is a chasm between academics and practitioners and even within academia various parties are not coordinating and integrating their efforts. There are multiple agendas for future academic research as well as self-criticism of the quality of research being done on urban tourism.

Figure 9.6 Summary diagram of chapter.

References

Aall, C., and Koens, K. (2019). The discourse on sustainable urban tourism: The need for discussing more than overtourism. *Sustainability, 11*, 4228.

Ali, R. (2018). The genesis of overtourism: Why we came up with the term and what's happened since. *Skift*, 14 August. https://skift.com/2018/08/14/the-genesis-of-overtourism-why-we-came-up-with-the-term-and-whats-happened-since/.

Alonso, I., and Bea, E. (2012). A tentative model to measure city brands on the Internet. *Place Branding and Public Diplomacy, 8*, 311–328.

Amore, A., and Hall, C. M. (2017). National and urban public policy in tourism. Towards the emergence of a hyperneoliberal script? *International Journal of Tourism Policy, 7*(1), 4–22.

An Foras Forbartha. (1966). *Planning for amenity and tourism*. Dublin: An Foras Forbartha.

Apostolopoulou, A., and Papadimitriou, D. (2015). The role of destination personality in predicting tourist behaviour: Implications for branding mid-sized urban destinations. *Current Issues in Tourism, 18*(12), 1132–1151.

Archer, B. H., and Owen, C. B. (1971). Towards a tourist regional multiplier. *Regional Studies, 5*(4), 289–294.

Ashworth, G. J., and Page, S. J. (2011). Urban tourism research: Recent progress and current paradoxes. *Tourism Management, 32*(1), 1–15.

Ashworth, G. J., and Tunbridge, J. E. (2000). *The tourist-historic city*. London: Routledge.

Ashworth, G. J., and Voogd, H. (1988). Marketing the city: Concepts, processes and Dutch applications. *The Town Planning Review, 59*(1), 65–79.

Ashworth, G. J., and Voogd, H. (1990). *Selling the city: Marketing approaches in public sector urban planning*. London: Bellhaven Press.

Asprogerakas, E. (2007). City competition and urban marketing: The case of tourism industry in Athens. *TOURISMOS: An International Multidisciplinary Refereed Journal of Tourism, 2*(1), 89–114.

Balakrishnan, M. S. (2008). Dubai – a star in the east: A case study in strategic destination branding. *Journal of Place Management and Development, 1*(1), 62–91.

Ballantyne, R., Packer, J., Hughes, K., and Dierking, L. (2007). Conservation learning in wildlife tourism settings: Lessons from research in zoos and aquariums. *Environmental Education Research, 13*(3), 367–383.

Ballantyne, R., Packer, J., and Hughes, K. (2008). Environmental awareness, interests and motives of botanic gardens visitors: Implications for interpretive practice. *Tourism Management, 29*(3), 439–444.

Barbieri, C., and Mahoney, E. (2010). Cultural tourism behaviour and preferences among the live-performing arts audience: An application of the univorous-omnivorous framework. *International Journal of Tourism Research, 12*, 481–496.

Barker, M., and Page, S. J. (2002). Visitor safety in urban tourism environments: The case of Auckland, New Zealand. *Cities, 19*(4), 273–282.

Bellini, N., and Pasquinelli, C. (2016). Urban brandscape as value ecosystem: The cultural destination strategy of fashion brands. *Place Branding and Public Diplomacy, 12*(1), 5–16.

Biagi, B., and Detotto, C. (2014). Crime as tourism externality. *Regional Studies, 48*(4), 693–709.

Bigné, J. E., Andreu, L., and Gnoth, J. (2005). The theme park experience: An analysis of pleasure, arousal and satisfaction. *Tourism Management, 26*, 833–844.

Boes, K., Buhalis, D., and Inversini, A. (2016). Smart tourism destinations: Ecosystems for tourism destination competitiveness. *International Journal of Tourism Cities, 2*(2), 108–124.

Boivin, M., and Tanguay, G. A. (2019). Analysis of the determinants of urban tourism attractiveness: The case of Quebec City and Bordeaux. *Journal of Destination Marketing & Management, 11*, 67–79.

Bouchon, F. A. L. (2014). Truly Asia and global city? Branding strategies and contested identities in Kuala Lumpur. *Place Branding and Public Diplomacy, 10*, 6–18.

Boyd, M. (2000). Reconstructing Bronzeville: Racial nostalgia and neighborhood redevelopment. *Journal of Urban Affairs, 22*(2), 107–122.

Bramwell, B. (1998). User satisfaction and product development in urban tourism. *Tourism Management, 19*(1), 35–47.

Bramwell, B., and Rawding, L. (1994). Tourism marketing organizations in industrial cities: Organizations, objectives and urban governance. *Tourism Management, 15*(6) 425–434.

Bramwell, B., and Rawding, L. (1996). Tourism marketing images of industrial cities. *Annals of Tourism Research, 23*(1), 201–221.

Braunlich, C. G. (1996). Lessons from the Atlantic City casino experience. *Journal of Travel Research, 34*(3), 46–56.

Bruner, E. M. (1996). Tourism in Ghana: The representation of slavery and the return of the black diaspora. *American Anthropologist, 98*(2), 290–304.

Buckley, P. J., and Witt, S. F. (1987). Tourism in difficult areas II: Case studies of Calderdale, Leeds, Manchester and Scunthorpe. *Tourism Management, 10*(2), 138–152.

Buhalis, D., and Amaranggana, A. (2013). Smart tourism destinations. In Z. Xiang and I Tussyadiah (Eds.), *Information and communication technologies in tourism 2014.* Cham, Switzerland: Springer.

Butler, R. W. (1974). The social implications of tourist developments. *Annals of Tourism Research, 2*(2), 100–111.

Caldeira, A. M., and Kastenholz, E. (2018). Tourists' spatial behaviour in urban destinations: The effect of prior destination experience. *Journal of Vacation Marketing, 24*(3), 247–260.

Cater, C. (2010). Any closer and you'd be lunch. Interspecies interactions as nature tourism at marine aquaria. *Journal of Ecotourism, 9*(2), 133–148.

Chacko, H. E., and Marcell, M. H. (2008). Repositioning a tourism destination. *Journal of Travel & Tourism Marketing, 23*(2–4), 223–235.

Chang, H. H., and Lai, T.-Y. (2009). The Taipei MRT (mass rapid transit): Tourism attraction analysis from the inbound tourists' perspectives. *Journal of Travel & Tourism Marketing, 26*, 445–461.

Chigora, F., and Hoque, M. (2019). Places to cities: Comprehending brands as personalities for Zimbabwean urban tourism vibrancy. *African Journal of Hospitality, Tourism and Leisure, 8*(1), 1–9.

Choi, Y. H., Song, H., Wang, J.-H., and Hwang, J. (2019). Residents' perceptions of the impacts of a casino-based integrated resort development and their consequences: The case of the Incheon area in South Korea. *Journal of Destination Marketing & Management, 14*, 100390.

Coca-Stefaniak, J. A., and Morrison, A. M. (2018). City tourism destinations and terrorism: A worrying trend for now, but could it get worse? *International Journal of Tourism Cities, 4*(4), 409–412.

Condé Nast Traveler. (2017). The world's 20 best cities for architecture lovers. www.cntraveler.com/galleries/2013-07-07/cities-architecture-design-lovers-photos.

Conforti, J. M. (1996). Ghettos as tourism attractions. *Annals of Tourism Research, 23*(4), 830–842.

Dai, T., Hein, C., and Zhang, T. (2019). Understanding how Amsterdam City tourism marketing addresses cruise tourists' motivations regarding culture. *Tourism Management Perspectives, 29*, 157–165.

Daskalopoulou, I., and Petrou, A. (2009). Urban tourism competitiveness: Networks and the regional asset base. *Urban Studies, 46*(4), 779–801.

Destination Marketing Association International. (2005). *Standard DMO Performance Reporting: A Handbook for Destination Marketing Organizations (DMO)*. Washington, DC: DMAI.

Destination Marketing Association International. (2012). *DMO Uniform System of Accounts: Financial Reporting Practices for DMOs*. Washington, DC: DMAI.

Destination Marketing Association International. (2012). *Standard DMO Performance Reporting: A Handbook for Destination Marketing Organizations (DMO)*. Washington, DC: DMAI.

Draper, J., Thomas, L. Y., and Fenich, G. G. (2018). Event management research over the past 12 years: What are the current trends in research methods, data collection, data analysis procedures, and event types? *Journal of Convention & Event Tourism, 19*(1), 3–24.

du Rand, G. E., and Heath, E. (2006). Towards a framework for food tourism as an element of destination marketing. *Current Issues in Tourism, 9*(3), 206–234.

Duignan, M. B. (2019). London's local Olympic legacy. *Journal of Place Management and Development, 12*(2), 142–163.

Dupré, K. (2019). Trends and gaps in place-making in the context of urban development and tourism. *Journal of Place Management and Development, 12*(1), 102–120.

Dürr, E. (2012). Urban poverty, spatial representation and mobility: Touring a slum in Mexico. *International Journal of Urban and Regional Research, 36*(4), 706–724.

Edwards, D., Griffin, T., and Hayllar, B. (2008). Urban tourism research: Developing an agenda. *Annals of Tourism Research, 35*(4), 1032–1052.

Erkuş-Öztürk, H., and Terhorst, P. (2018). Economic diversification of a single-asset tourism city: Evidence from Antalya. *Current Issues in Tourism, 21*(4), 422–439.

Fenich, G. G. (2012). *Meetings, expositions, events, and conventions: An introduction to the industry* (3rd ed.). Boston, MA: Prentice Hall.

Fenich, G. G., Vitiello, K. L., Lancaster, K. F., and Hashimoto, K. (2015). Incentive travel: A view from the top. *Journal of Convention & Event Tourism, 16*(2), 145–158.

Franck, K. A. (2005). The city as dining room, market and farm. *Architectural Design, 75*(3), 5–10.

Frenzel, F., and Koens, K. (2012). Slum tourism: Developments in a young field of interdisciplinary tourism research. *Tourism Geographies, 14*(2), 195–212.

Freytag, T., and Bauder, M. (2018). Bottom-up touristification and urban transformations in Paris. *Tourism Geographies, 20*(3), 443–460.

Frost, W. (Ed.). (2011). *Zoos and tourism: Conservation, education, entertainment?* Bristol: Channel View Publications.

Gartrell, R. (1988). *Destination marketing for convention and visitor bureaus*. Dubuque, IA: Kendall Hunt Publishing.

Gertner, D. (2011). Unfolding and configuring two decades of research and publications on place marketing and place branding. *Place Branding and Public Diplomacy, 7*(2), 91–106.

Getz, D. (2008). Event tourism: Definition, evolution, and research. *Tourism Management, 29*, 403–428.

Gibson, H. J. (1998). Sport tourism: A critical analysis of research. *Sport Management Review, 1*(1), 45–76.

Gladstone, D. L., and Fanstein, S. S. (2001). Tourism in US global cities: A comparison of New York and Los Angeles. *Journal of Urban Affairs, 23*(1), 23–40.

Gotham, K. F. (2002). Marketing Mardi Gras: Commodification, spectacle and the political economy of tourism in New Orleans. *Urban Studies, 39*(10), 1735–1756.

Gotham, K. F. (2005a). Theorizing urban spectacles. *City, 9*(2), 225–246.

Gotham, K. F. (2005b). Tourism from above and below: Globalization, localization and New Orleans's Mardi Gras. *International Journal of Urban and Regional Research, 29*(2), 309–326.

Gratton, C., Shibli, S., and Coleman, R. (2005). Sport and economic regeneration in cities. *Urban Studies, 42*(5/6), 985–999.

Gravari-Barbas, M., and Guinand, S. (2017). *Tourism and gentrification in contemporary metropolises: International perspectives*. London: Routledge.

Gretzel, U., Sigala, M., Xiang, Z., and Koo, C. (2015). Smart tourism: foundations and developments. *Electronic Markets, 25*(3), 179–188.

Gretzel, U., Zhong, L., and Koo, C. (2016). Application of smart tourism to cities. *International Journal of Tourism Cities, 2*(2).

Griffin, T., and Hayllar, B. (2009). Urban tourism precincts and the experience of place. *Journal of Hospitality Marketing & Management*, *18*(2–3), 127–153.

Gunn, C. A. (1977). Industry pragmatism vs tourism planning. *Leisure Sciences*, *1*(1), 85–94.

Gursoy, D., and Kendall, K. W. (2006). Hosting mega events: Modeling locals' support. *Annals of Tourism Research*, *33*(3), 603–623.

Hale, P., and MacDonald, S. (2005). The Sydney Opera House: An evolving icon. *Journal of Architectural Conservation*, *11*(2), 7–22.

Hankinson, G. (2010). Place branding research: A cross-disciplinary agenda and the views of practitioners. *Place Branding and Public Diplomacy*, *6*(4), 300–315.

Hartal, G. (2019). Gay tourism to Tel-Aviv: Producing urban value? *Urban Studies*, *56*(6), 1148–1164.

Haywood, K. M., and Muller, T. E. (1988). The urban tourist experience: Evaluating satisfaction. *Journal of Hospitality & Tourism Research*, *12*(2), 453–459.

Heath, E., and Kruger, E. (2010). Branding and positioning an African capital city: The case of Tshwane in South Africa. In R. Maitland and B. W. Ritchie (Eds.), *City tourism: National capital perspectives*. Wallingford, England: CABI.

Henderson, J. C. (2009). Food tourism reviewed. *British Food Journal*, *111*(4), 317–326.

Henderson, J. C. (2014). Bidding for world heritage: Singapore's botanic gardens. *Tourism, Culture and Communication*, *14*(2), 63–75.

Hinch, T. D. (1996). Urban tourism: Perspectives on sustainability. *Journal of Sustainable Tourism*, *4*(2), 95–110.

Hoffman, L. M. (2003). The marketing of diversity in the inner city: Tourism and regulation in Harlem. *International Journal of Urban and Regional Research*, *27*(2), 286–299.

Hritz, N., and Ross, C. (2010). The perceived impacts of sport tourism: An urban host community perspective. *Journal of Sport Management*, *24*, 119–138.

Hughes, H. (2003). Marketing gay tourism in Manchester: New market for urban tourism or destruction of "gay space"? *Journal of Vacation Marketing*, *9*(2), 152–163.

Husbands, W., and Thompson, S. (1990). The host society and the consequences of tourism in Livingstone, Zambia. *International Journal of Urban and Regional Research*, *14*(3), 490–513.

Ignatov, E., and Smith, S. (2006). Segmenting Canadian culinary tourists. *Current Issues in Tourism*, *9*(3), 235–255.

Irimiás, A. (2012). The Chinese diaspora in Budapest: A new potential for tourism. *Tourism Review*, *67*(1), 23–33.

Jackson, L. A. (2008). Residents' perceptions of the impacts of special event tourism. *Journal of Place Management and Development*, *1*(3), 240–255.

Jansen-Verbeke, M., and van Rekom, J. (1996). Scanning museum visitors: Urban tourism marketing. *Annals of Tourism Research*, *23*(2), 364–375.

Jin, N. P., Lee, S., and Lee, H. (2015). The effect of experience quality on perceived value, satisfaction, image and behavioral intention of water park patrons. *International Journal of Tourism Research*, *17*(1), 82–95.

Kapera, I. (2018). Sustainable tourism development efforts by local governments in Poland. *Sustainable Cities and Society*, *40*, 581–588.

Kavaratzis, M. (2017). The participatory place branding process for tourism: Linking visitors and residents through the city brand. In N. Bellini and C. Pasquinelli (Eds.), *Tourism in the city*. Cham, Switzerland: Springer.

Kivela, J., and Crotts, J. C. (2006). Tourism and gastronomy: Gastronomy's influence on how tourists experience a destination. *Journal of Hospitality & Tourism Research*, *30*(3), 354–377.

Knapp, D., and Sherwin, G. (2005). *Destination BrandScience*. Washington, DC: Destinations International.

Kozak, M., and Baloglu, S. (2011). *Managing and marketing tourist destinations: Strategies to gain a competitive edge*. New York: Taylor & Francis.

Kurtzman, J. (2005). Economic impact: Sport tourism and the city. *Journal of Sport & Tourism*, *10*(1), 47–71.

Lahav, T., Mansfeld, Y., and Avraham, E. (2013). Inducing national media coverage for tourism in rural versus urban areas: The role of public relations. *Journal of Travel & Tourism Marketing*, *30*(4), 291–307.

Lai, S., and Ooi, C.-S. (2015). Branded as a World Heritage city: The politics afterwards. *Place Branding and Public Diplomacy*, *11*, 276–292.

Lasansky, D. M., and McLaren, B. (Eds.). (2004). *Architecture and tourism: Perception, performance and place*. Oxford: Berg.

Laslo, D. H., and Judd, D. R. (2005). Convention center wars and the decline of local democracy. *Journal of Convention & Event Tourism, 6*(1/2), 81–98.

Law, C. M. (1985). Urban tourism: Selected British case studies (Urban Tourism Research Project Working Paper). Salford, England: University of Salford.

Laws, E. (1995). *Tourist destination management: Issues, analysis and policies.* New York: Routledge.

Lawton, G. R., and Page, S. J. (1997). Analysing the promotion, product and visitor expectations of urban tourism: Auckland, New Zealand as a case study. *Journal of Trawl & Tourism Marketing, 6*(3–4), 123–142.

Lee, M. K., Yoon, H. Y., and Park, H. W. (2017). From online via offline to online: How online visibility of tourism information shapes and is shaped by offline visits. *Journal of Travel & Tourism Marketing, 34*(9), 1143–1154.

Le-Klähn, D.-T., and Hall, C. M. (2015). Tourist use of public transport at destinations: A review. *Current Issues in Tourism, 18*(8), 785–803.

Leslie, D., and Craig, C. (2000). The "all-suite" hotel: A study of the market in Scotland. *Journal of Vacation Marketing, 6*(3), 221–235.

Lim, C. C., and Bendle, L. J. (2012). Arts tourism in Seoul: Tourist orientated performing arts as a sustainable niche market. *Journal of Sustainable Tourism, 20*(5), 667–682.

Lipardi, V. (2013). The evolution and worldwide expansion of science centres. In A. M. Bruyas and M. Riccio (Eds.), *Science centres and science events: A science communication handbook.* Milan: Springer, 49–61.

Lohmann, G., Albers, S., Koch, B., and Pavlovich, K. (2009). From hub to tourist destination: An explorative study of Singapore and Dubai's aviation-based transformation. *Journal of Air Transport Management, 15*, 205–211.

Loi, K.-I., and Kim, W. G. (2010). Macao's casino industry: Reinventing Las Vegas in Asia. *Cornell Hospitality Quarterly, 51*(2), 268–283.

Long, J. G. (2005). The real cost of public funding for major league sports facilities. *Journal of Sports Economics, 6*(2), 119–143.

Long, P. (2017). The parallel worlds of tourism destination management and the creative industries: Exchanging knowledge, theory and practice. *Journal of Policy Research in Tourism, Leisure and Events, 9*(3), 331–340.

Long, P., and Morpeth, N. D. (2016). *Tourism and the creative industries: Theories, policies and practice.* London: Routledge.

Lopez de Avila, A. (2015). Smart destinations: XXI century tourism. Presented at the ENTER 2015 Conference on Information and Communication Technologies in Tourism, Lugano, Switzerland, 4–6 February 2015.

Lorentzen, A. (2009). Cities in the experience economy. *European Planning Studies, 17*(6), 829–845.

Lovatt, A., and O'Connor, J. (1995). Cities and the night-time economy. *Planning Practice and Research, 10*(2), 127–134.

Luo, J. M., Qiu, H., Goh, C., and Wang, D. (2016). An analysis of tourism development in China from urbanization perspective. *Journal of Quality Assurance in Hospitality & Tourism, 17*(1), 24–44.

McKercher, B., Okumus, F., and Okumus, B. (2008). Food tourism as a viable market segment: It's all how you cook the numbers! *Journal of Travel & Tourism Marketing, 25*(2), 137–148.

Mair, J. (2005). Incentive travel: A theoretical perspective. *Event Management, 19*, 543–552.

Martínez-Ruiz, M. P., Martínez-Caraballo, N., and Amatulli, C. (2010). Tourist destinations and luxury commerce: Business opportunities. *Journal of Place Management and Development, 3*(3), 205–220.

Mason, P. (2000). Zoo tourism: The need for more research. *Journal of Sustainable Tourism, 8*(4), 333–339.

Maxim, C. (2019). Challenges faced by world tourism cities: London's perspective. *Current Issues in Tourism, 22*(9), 1006–1024.

Meschkank, J. (2011). Investigations into slum tourism in Mumbai: Poverty tourism and the tensions between different constructions of reality. *GeoJournal, 76*(1), 47–62.

Mescon, T. S., and Vozikis, G. S. (1985). The economic impact of tourism at the port of Miami. *Annals of Tourism Research, 12*(4), 515–528.

Milman, A. (2009). Evaluating the guest experience at theme parks: An empirical investigation of key attributes. *International Journal of Tourism Research, 11*, 373–387.

Molinillo, S., Anaya-Sánchez, R., Morrison, A. M., and Coca-Stefaniak, J. A. (2019). Smart city communication via social media: Analysing residents' and visitors' engagement. *Cities, 94*, 247–255.

Morgan, N., Prirchard, A., and Pride, R. (2004). *Destination branding.* Oxford: Elsevier.

Morrison, A. M. (2019). *Marketing and managing tourism destinations*. London: Routledge.

Mullins, P. (1991). Tourism urbanization. *International Journal of Urban and Regional Research, 15*(3), 326–342.

O'Sullivan, D., and Jackson, M. J. (2002). Festival tourism: A contributor to local sustainable economic development? *Journal of Sustainable Tourism, 10*(4), 325–342.

Pagliara, F., La Pietra, A., Gomez, J., and Vassallo, J. M. (2015). High speed rail and the tourism market: Evidence from the Madrid case study. *Transport Policy, 37,* 187–194.

Paskaleva-Shapira, K. A. (2007). New paradigms in city tourism management: Redefining destination promotion. *Journal of Travel Research, 46,* 108–114.

Paulauskaite, D., Powell, R., Coca-Stefaniak, J. A., and Morrison, A. M. (2017). Living like a local: Authentic tourism experiences and the sharing economy. *International Journal of Tourism Research, 19*(6), 619–628.

Pearce, D. G. (2001). An integrative framework for urban tourism research. *Annals of Tourism Research, 28*(4), 926–946.

Pearce, D. G. (2007). Capital city tourism. *Journal of Travel & Tourism Marketing, 22*(3–4), 7–20.

Pike, S. (2005). *Destination marketing organisations: Bridging theory and practice.* Boston, MA: Elsevier Science.

Pike, S. (2008). *Destination marketing: An integrated communication approach.* Oxford: Butterworth-Heinemann.

Pinke-Sziva, I., Smith, M., Olt, G., and Berezvai, Z. (2019). Overtourism and the night-time economy: A case study of Budapest. *International Journal of Tourism Cities, 5*(1), 1–16.

Plaza, B. (2006). The return on investment in the Guggenheim Museum Bilbao. *International Journal of Urban and Regional Research, 30*(2), 452–467.

Postma, A., Buda, D.-M., and Gugerell, K. (2017). The future of city tourism. *Journal of Tourism Futures, 3*(2), 95–101.

Preuss, H. (2007). The conceptualisation and measurement of mega sport event legacies. *Journal of Sport & Tourism, 12*(3–4), 207–228.

Qu, H., and Zhang, H. Q. (1997). Hong Kong: A major urban tourism destination in South-East Asia. *Journal of Vacation Marketing, 3*(4), 363–372.

Quan, S., and Wang, N. (2004). Towards a structural model of the tourist experience: An illustration from food experiences in tourism. *Tourism Management, 25,* 297–305.

Quinn, B. (2006) Problematising "festival tourism": Arts festivals and sustainable development in Ireland. *Journal of Sustainable Tourism, 14*(3), 288–306.

Quinn, F. J. (1967). Revitalizing a tourism area: Saratoga's performing arts center. *Cornell Hotel & Restaurant Administration Quarterly, 7*(4), 98–101.

Rabbiosi, C. (2015). Renewing a historical legacy: Tourism, leisure shopping and urban branding in Paris. *Cities, 42,* 195–203.

Rath, J. (Ed.). (2007). *Tourism, ethnic diversity and the city.* New York: Routledge.

Ricci, P. R., and Holland, S. M. (1992). Incentive travel: Recreation as a motivational medium. *Tourism Management, 13*(3), 288–296.

Richards, G. (2001). The experience industry and the creation of attractions. In G. Richard (Ed.), *Cultural attractions and European tourism.* Wallingford, England: CABI.

Richards, G. (2011). Creativity and tourism: The state of the art. *Annals of Tourism Research, 38*(4), 1225–1253.

Richards, G., and Palmer, R. (2010). *Eventful cities: Cultural management and urban revitalization.* Oxford: Butterworth-Heinemann.

Riley, M. (1984). Hotels and group identity. *Tourism Management, 5*(2), 102–109.

Ritzer, G., and Stillman, T. (2001). The modern Las Vegas casino-hotel: The paradigmatic new means of consumption. *Management, 4*(3), 83–99.

Roberts, M. (2001). Contemporary issues in the public realm. In M. Roberts and C. Greed, C. (Eds.), *Approaching urban design: The design process.* Harlow: Longman.

Roberts, M. (2006). From "creative city" to "no-go areas": The expansion of the night-time economy in British town and city centres. *Cities, 23*(5), 331–338.

Rogerson, C. M. (2006). Creative industries and urban tourism: South African perspectives. *Urban Forum, 17*(2), 149–166.

Rogerson, C. M., and Rogerson, J. M. (2017). City tourism in South Africa: Diversity and change. *Tourism Review International, 21*(2), 193–211.

Rouleau, J. (2017). Every (nocturnal) tourist leaves a trace: Urban tourism, nighttime landscape, and public places in Ciutat Vella, Barcelona. *Imaginations Journal*, 7(2), 58–71.

Roult, R., Adjizian, J.-M., and Auger, D. (2016). Tourism conversion and place branding: The case of the Olympic Park in Montreal. *International Journal of Tourism Cities*, 2(1), 77–93.

Russo, P., and Scarnato, A. (2018). "Barcelona in common": A new urban regime for the 21st-century tourist city? *Journal of Urban Affairs*, 40(4), 455–474.

Sahin, S., and Baloglu, S. (2014). City branding: Investigating a brand advocacy model for distinct segments. *Journal of Hospitality Marketing & Management*, 23(3), 239–265.

Sanders, H. T. (2014). *Convention center follies: Politics, power, and public investment in American Cities*. Philadelphia, PA: University of Pennsylvania Press.

Sant, S.-L., Mason, D. S., and Hinch, T. D. (2013). Conceptualising Olympic tourism legacy: Destination marketing organisations and Vancouver 2010. *Journal of Sport and Tourism*, 18(4), 287–312.

Scheyvens, R. (2007). Poor cousins no more: Valuing the development potential of domestic and diaspora tourism. *Progress in Development Studies*, 7(4), 307–325.

Schmallegger, D., and Carson, D. (2010). Whose tourism city is it? The role of government in tourism in Darwin, Northern Territory. *Tourism and Hospitality Planning & Development*, 7(2), 111–129.

Selby, M. (2004). Consuming the city: Conceptualizing and researching urban tourist knowledge. *Tourism Geographies*, 6(2), 186–207.

Séraphin, H., Sheeran, P., and Pilato, M. (2018). Over-tourism and the fall of Venice as a destination. *Journal of Destination Marketing & Management*, 9, 374–376.

Sheehan, L., Ritchie, J. R. B., and Hudson, S. (2007). The destination promotion triad: Understanding asymmetric stakeholder interdependencies among the city, hotels, and DMO. *Journal of Travel Research*, 46(1), 64–74.

Sheldon, P. J. (1995). The demand for incentive ravel: An empirical study. *Journal of Travel Research*, 33(4), 21–28.

Shinew, K. J., and Backman, S. J. (1995). Incentive travel: An attractive option. *Tourism Management*, 16(4), 285–293.

Shoval, N. (2018). Urban planning and tourism in European cities. *Tourism Geographies*, 20(3), 371–376.

Silberberg, T. (1995). Cultural tourism and business opportunities for museums and heritage sites. *Tourism Management*, 16(5), 361–365.

Smith, S. L. J. (1985). Location patterns of urban restaurants. *Annals of Tourism Research*, 12(4), 581–602.

Specht, J. (2014). *Architectural tourism: Building for urban travel destinations*. Wiesbaden: Springer Gabler.

Spotts, D. M. (1997). Regional analysis of tourism resources for marketing purposes. *Journal of Travel Research*, 35(3), 3–15.

Stone, R. (2018). Lisbon's overtourism lesson: Living like a local is not enough. *Skift*, 31 May. https://skift.com/2018/05/31/lisbons-overtourism-lesson-living-like-a-local-is-not-enough/.

Su, L. J., Hsu, M. K., and Marshall, K. P. (2014). Understanding the relationship of service fairness, emotions, trust, and tourist behavioral intentions at a city destination in China. *Journal of Travel & Tourism Marketing*, 31(8), 1018–1038.

Tang, C., Weaver, D., and Lawton, L. (2017). Can stopovers be induced to revisit transit hubs as stayovers? A new perspective on the relationship between air transportation and tourism. *Journal of Air Transport Management*, 62, 54–64.

Thurnell-Read, T. (2011). Off the leash and out of control: Masculinities and embodiment in Eastern European stag tourism. *Sociology*, 45(6), 977–991.

Thurnell-Read, T. (2012). Tourism place and space. British stag tourism in Poland. *Annals of Tourism Research*, 39(2), 801–819.

Timothy, D. J. (2005). *Shopping tourism, retailing and leisure*. Clevedon, England: Channel View Publications.

Timur, S., and Getz, D. (2009). Sustainable tourism development: How do destination stakeholders perceive sustainable urban tourism? *Sustainable Development*, 17(4), 220–232.

Tomlinson, A. (2014). Olympic legacies: Recurrent rhetoric and harsh realities. *Contemporary Social Science*, 9(2), 137–158.

Tosun, C., Temizkan, S. P., Timothy, D. J., and Fyall, A. (2007). Tourist shopping experiences and satisfaction. *International Journal of Tourism Research*, 9, 87–102.

Travel Channel. (2019). World's top architecture cities. www.travelchannel.com/interests/arts-and-culture/photos/worlds-top-architecture-cities.

Travel Michigan. (2016). *Manual for Michigan assessment districts*. Lansing, MI: Michigan Economic Development Corporation.

Turner, R. S., and Rosentraub, M. S. (2002). Tourism, sports and the centrality of cities. *Journal of Urban Affairs, 24*(5), 487–492.

Um, S., and Crompton, J. L. (1987). Measuring resident's attachment levels in a host community. *Journal of Travel Research, 26*(1), 27–29.

UNWTO. (2008). *A practical guide to tourism destination management.* Madrid: World Tourism Organization.

UNWTO, and ETC. (2009). *Handbook on tourism destinations branding.* Madrid: World Tourism Organization; Brussels: European Travel Commission.

UNWTO, and ETC. (2011). *Handbook on tourism product development.* Madrid: World Tourism Organization; Brussels: European Travel Commission.

Uysal, Ü. E. (2013). Branding Istanbul: Representations of religion in promoting tourism. *Place Branding and Public Diplomacy, 9*(4), 223–235.

Valls, J.-F., Sureda, J., and Valls-Tuñon, G. (2014). Attractiveness analysis of European tourist cities. *Journal of Travel & Tourism Marketing, 31*(2), 178–194.

Vandermey, A. (1984). Assessing the importance of urban tourism: Conceptual and measurement issues. *Tourism Management, 5*(2), 123–135.

van Limburg, B. (1998). City marketing: A multi-attribute approach. *Tourism Management, 19*(5), 415–417.

Vesey, C., and Dimanche, F. (2003). From Storyville to Bourbon Street: Vice, nostalgia and tourism. *Journal of Tourism and Cultural Change, 1*(1), 54–70.

Wang, Y., and Pizam, A. (2011). *Destination marketing and management: Theories and applications.* Wallingford, England: CABI.

Whitson, D., and MacIntosh, D. (1996). The global circus: International sport, tourism and the marketing of cities. *Journal of Sport and Social Issues, 20*(3), 278–295.

Wiley, J. A., and Walker, D. M. (2011). Casino revenues and retail property values: The Detroit case. *The Journal of Real Estate Finance and Economics, 42*(1), 99–114.

Wu, C., Funck, C., and Hayashi, Y. (2014). The impact of host community on destination (re)branding: A case study of Hiroshima. *International Journal of Tourism Research, 16*(6), 546–555.

Wu, S.-T., and Chen, Y.-S. (2015). The social, economic, and environmental impacts of casino gambling on the residents of Macau and Singapore. *Tourism Management, 48*, 285–298.

Zaeimdar, M., and Bahmanpour, H. (2014). Environmental pollution in tourism area (case study: noise pollution). *Journal of Applied Science and Agriculture, 9*(2), 741–746.

10

THE EMERGENCE OF THE BUSINESS TOURISM CITY

Rob Davidson

Introduction

Most seminal works on urban tourism identify business tourism as one of the components of tourism in cities (Ashworth, 1989; Judd and Fainstein, 1999; Law, 2002; Hall and Page, 2003). Yet, despite its considerable importance to cities worldwide, business tourism has received comparatively scant attention in academic circles, compared with most leisure-related forms of urban tourism. By way of redressing the balance somewhat, this chapter examines the emergence of the global market for business tourism in cities, with particular focus on the development of the various segments of demand for business tourism and the process by which cities have equipped themselves with the physical and human infrastructure necessary to compete as destinations for this form of tourist activity.

Terminological issues

Investigation into any aspect of the business tourism industry rapidly and inevitably encounters challenges relating to the terminology in this field. Alternative terms widely used synonymously for business tourism include "business events" and "the meetings industry" (Davidson and Hyde, 2014; Locke, 2010). To these may be added the various acronyms that have been used in recent times to define the set of different activities that comprise business tourism. Marques and Santos (2017) identify some of these as MECE (Meetings, Events, Conventions, Exhibitions), MCE (Meetings, Conventions, Exhibitions), CEMI (Conventions, Exhibitions, Meetings, Incentives), MC & IT (Meetings, Conventions & Incentive Travel) and MICE (Meetings, Incentives, Conventions, Exhibitions). The MICE acronym holds particular appeal as a memorable, shorthand expression, but its use is beset with problems. For example, there is no universally agreed definition of what MICE stands for. As well as the version cited by Marques and Santos, MICE has been variously defined as Meetings, Incentives, Conferences and Exhibitions; and Meetings, Incentives, Conferences and Events, with the term "Congresses" occasionally being substituted for "Conferences". However, the main problem with this term is that it is barely recognised at all by anyone except those practitioners and academics who are active in this field. Moreover, the term is very rarely used in the US, the world's largest market for business tourism. However, the MICE acronym has been widely adopted elsewhere – in particular

162

in those regions in which the market is growing rapidly, such as the Middle East and South-East Asia. In China, for example, the title of the leading trade publication is "MICE China" and the annual event to celebrate this industry is named "World MICE Day" (Davidson, 2019). By way of contrast, the term "business tourism" is well-established in academic circles, where tourism educators and researchers have made a useful if limited contribution to our understanding of this industry.

But regardless of the term used to denote this industry as a whole, there is general agreement that overall it comprises work-related travel to events that "bring together colleagues from similar industries, professions or interest groups to connect with each other in order to share ideas and information, to make decisions, or simply to enjoy and celebrate their work-related achievements" (Davidson, 2019: 2). Most commentators distinguish between business tourism and individual business travel (Swarbrooke and Horner, 2012; Davidson and Cope, 2003), where the latter comprises the routine trips made by individual employees, generally travelling alone, in order to carry out their day-to-day duties or to visit clients in order to close deals, for example. Business tourism, by way of contrast, generally involves larger numbers of people converging temporarily in a particular destination for the purpose of attending more *occasional* types of work-related events, such as conferences, trade shows or exhibitions, and incentive trips. These elements will be taken to comprise the form of business tourism that is the focus of this chapter.

Business tourism as a catalyst for urban development

Authors frequently highlight the urban context within which most business tourism takes place (McCabe et al., 2000; Weber and Chon, 2002; Rogers, 2013). Cities are widely acknowledged as the most commonly used business tourism destinations as they tend to offer the supporting infrastructure required for the hosting of conferences and exhibitions – notably, venues, a wide range of accommodation and hospitality suppliers, as well as multiple transport connections. But there are exceptions, and business tourism is by no means exclusively confined to urban centres.

> Incentive trips, for example, may take place in resorts and spas or in wilderness areas offering nature-based activities for participants. And, for security reasons, high-profile political conferences such as the gatherings of the G20 group of nations have occasionally been held in remote communities such as mountain villages or on small islands, where the participants can be more easily protected from disturbances created by protestors. These examples are, however, fairly rare exceptions to the general rule that business events are predominantly urban-based.
>
> *(Davidson, 2019: 2)*

Given its propensity to be largely a city-focused phenomenon, the business tourism market is often regarded as a viable strategy for urban development. The reasons that usually underlie the choice of business tourism as a motor of urban development are closely linked to the various benefits that this form of tourist activity can bring to the destinations in which it takes place. Traditionally, the majority of these benefits have been considered in terms of the positive economic impacts that business tourism can yield wherever it takes place, generated for the main part by the spending of the organisers of conferences, exhibitions and incentive trips, as well as by the business tourists themselves. It is clear that most of that spending directly benefits the tourism and hospitality sectors in the host destinations. However, in recent years, academics and practitioners have increasingly highlighted the "beyond-tourism" impacts of conferences and exhibitions. Both types of benefits will now be outlined.

Tourism benefits

Business tourists are generally acknowledged to be high-yield visitors, and their average contribution to destinations' tourism economies has often been demonstrated to be higher than that of other tourists (Shone, 1998; Dwyer, 2002: 25; Wan, 2011: 130). For this reason, the benefits of business tourism are often measured in terms of visitor numbers, daily expenditure and the number of nights spent in the destination. Moreover, the visits of business tourists are not as bound to peak seasons as are those of leisure tourists. As business tourism generally takes place throughout the year, it can serve to smooth out seasonality problems for travel and hospitality companies operating in tourism destinations (Rogers, 2013).

A further direct contribution of business tourism to the wider tourism industry arises from the practice of some business visitors to extend their trip in order to enjoy some leisure time before and/or after the actual business event they are attending. The potential for such events to generate additional leisure tourism for cities in this way has been widely acknowledged (Davidson, 2002; Lichy and McLeay, 2018) as an additional benefit of this industry, with the ability to act as a catalyst for the short-break leisure market (Haven-Tang et al., 2007). Indeed, Kerr at al. (2012) suggest that some business tourists may even use their trips to a new destination to assess its potential for future leisure travel.

"Beyond-tourism" benefits

While the contribution that business visitors can make to the tourism economies of cities are clearly significant, growing attention has latterly focused upon the other "non-tourism" – or "beyond-tourism" impacts of business tourism, notably the important benefits that it may offer for knowledge transfer, knowledge creation and building the economic base in host destinations, making it an important driver for the intellectual development of those living and working in the cities in which it takes place. According to the World Tourism Organization (UNWTO):

> meetings and conventions essentially take place for the purposes of business, professional and scientific development as well as sharing knowledge and expertise. Therefore, both the delegates and the events themselves have a lot to offer to the host community … [Business] events in any area of discipline – particularly major national or international events – often attract literally the very best expertise in the world, which means local access to a high level of knowledge transfer and international exposure for local professionals. In areas like medical practice, this can have huge implications for how local skills develop – which creates, in turn, big benefits for the quality of service in the community. All these factors combine to create a strong and diverse return on an investment in the business events sector.
>
> *(UNWTO, 2014: 21)*

Other sources have focused upon the links between business tourism and the modern "knowledge economy". For example, Rogerson observes that conferences are

> fundamental to constructing "the network society" as well as for the functioning of knowledge-based economies. This leads to an increased need for updates and

knowledge transfer, most effectively and efficiently undertaken through the organising of meetings or the hosting of conferences.

(Rogerson, 2015: 184)

Deery and Jago (2010) also elaborate upon the role of business tourism in the knowledge creation and dissemination processes, emphasising its valuable role in improving business performance and underpinning innovation. While acknowledging that such impacts are longer-term and more difficult to quantify than tourism benefits, they nevertheless make a convincing case that business tourism can bring many beneficial outcomes leading to the creation and dissemination of innovative practices and the enhancement of individual and organisational performance in destinations.

The development of the business tourism market

Demand

The practice of human beings congregating for the purpose of conferring on affairs of state and discussing trade and commerce finds its roots far back in the ancient world, in locations such as the Agora of Athens and the Roman Forum. Throughout the succeeding centuries, such gatherings became an essential element of cultural, political and commercial life, directly contributing to the progress made by societies throughout the developed world (Rogers and Davidson, 2016).

In the modern era, Spiller (2002) notes that during the late nineteenth and early twentieth centuries, as industrialisation spread through the US and Western Europe, the need for meetings between business leaders and other entrepreneurs materialised, adding to the numbers of those already meeting to discuss and exchange ideas on political, religious, literary, recreational and other varied topics. Advances in transport technology during the same period, combined with increasing levels of prosperity for the growing middle classes and the rise of the professions in many countries, created a more mobile society in which travel to conferences became an increasingly frequent activity for many. Inextricably linked to the emergence of the professions at this time was the creation of the professional associations, regarded by Friedman and Phillips (2004: 187) as "an essential component of professionalism", as they provided the recognition necessary for workers to gain standing in their field. Beginning during the Industrial Revolution, a vast wave of new trade associations was established as new occupations were created by the impact of industrialisation on working practices. Later, as the industrialised nations became increasingly urbanised, many more professional associations were founded, including those representing medicine, law and accounting, which were created specifically to develop and enforce common standards of practice (Davidson, 2019). As such, associations were required by their own regulations to hold regular conferences (usually on an annual basis), they provided a considerable boost to the volume of demand for business tourism.

Running in parallel with the development of demand for meetings and conferences, came the materialisation of what would eventually become the exhibitions and trade fairs component of business tourism. The practice of creating fairs as temporary markets where buyers and sellers gathered to transact business originated in the ancient world and continued through medieval times when, according to Gopalakrishna and Lilien (2012), artisans and farmers exhibited their wares at local fairs. Such fairs were a convenient way for local producers to gain access to large numbers of potential buyers who came to attend the events from neighbouring towns and villages. In the modern era, the first industrial-scale exhibitions appeared in the form of "world fairs", such as the Crystal Palace Exhibition of 1851 (officially called the "Great Exhibition of the Works of Industry of All Nations") in London's Hyde Park. Opened by Queen Victoria, the first international trade fair was accessible by the general public as well as industrialists, and its commercial success generated many others. Those included the New York World Fair of 1853 and the Exposition Universelle in Paris, which was visited by over 32 million people in 1889 and by over 48 million in 1990 (Dee, 2011). By the early twentieth century, exhibitions were firmly established as a regular fixture in the business calendars of the industrialised nations, and the practice of affixing trade shows to the conference programmes of professional and trade associations was spreading.

By way of contrast, the practice of motivating and rewarding high-performing employees through the use of incentive travel is much more recent. The first example of the use of incentive trips as a motivational technique in the workplace is believed to have occurred in 1906 when the National Cash Register Company awarded its top salespeople from around the US diamond-studded pins and a free trip to the company's headquarters in Dayton, Ohio. Five years later, the winners were awarded a free trip to New York (Davidson, 1994).

In the second half of the twentieth century, a number of new factors facilitated the rapid expansion of the business tourism market. Lawson (2000) identifies these as:

- the expansion of government and quasi-government organisations, together with an increasing need for meetings between the public and private sectors;
- the growth of multinational corporations and pan-national agencies, necessitating more interdepartmental and inter-regional meetings;
- developments in associations, cooperatives, professional groups and pressure groups;
- changes in sales techniques, the use of product launches and sales promotion meetings;
- the need to update information and methods through in-company management training, continuing professional development and attendance at ad hoc or scheduled meetings;
- the development of subject specialisation, with conferences acting as a means of experts passing on information.

Most of these trends have continued into the present century, fuelling greater demand for business tourism-related activities in cities worldwide.

Supply

The supply of facilities specifically built by cities for the hosting of business tourism-related events is a relatively modern phenomenon. The earliest meetings and fairs were simply held in the marketplaces of towns and villages. Regarding gatherings for the purpose of discussion and debate, Rogers and Davidson (2016: 2), note that "The vast majority of such meetings were held locally, in locations such as public spaces, theatres and hotels. Later, the first purpose-built meeting venues, such as the elegant eighteenth-century assembly rooms ... in many British cities were constructed".

It was in the second half of the last century that, in order to reap the benefits of the burgeoning demand for the types of gatherings described above, a growing number of urban governments in the developed and the developing world made efforts to capitalise upon their potential as business tourism destinations. This targeting of the business tourism market took place in the context of what commentators have characterised as global economic restructuring and the decline of traditional manufacturing activities in many cities of the developed world (Law, 1996; Telfer, 2002; Rogerson, 2004). For these cities, the development of facilities and services for business tourism (as well as leisure tourism) became a vital element in their economic regeneration strategies, which were designed to reap the types of benefits that were discussed earlier in this chapter. Accordingly, the development of inner-city leisure spaces, waterfront developments, festival marketplaces, casinos, museums, conference centres and sports stadia were the physical manifestations of a major wave of local economic development initiatives surrounding urban tourism and economic regeneration (Rogerson, 2004).

Davidson and Hyde highlight the links between business tourism and the economic development of the destinations in which it takes place, noting that many cities have invested in meetings facilities, such as flagship conference centres, as an element of plans to regenerate urban areas in need of re-development:

> From Glasgow and Philadelphia to Cape Town and Dublin, large-scale meeting facilities have been built as a means of bringing prosperity and animation back into previously neglected parts of those cities. More intangibly, the fact of hosting a conference, in particular an international event, can be a source of pride and prestige for the city or country where the conference takes place, as well as a means of creating an image or brand for itself in the international community of nations.
>
> *(Davidson and Hyde, 2014: 3)*

The concept of purpose-built conference and exhibition venues was first conceived in the US in the early 1960s and the trend then spread to Europe in the following decade. For example, London's Wembley Conference Centre, which opened in 1977, was one of the first purpose-built conference centres in the UK. In its opening year, it hosted more than 350,000 people attending 300 events (Rogers and Davidson, 2016). The rate of growth of such venues has been extraordinary, particularly in the US. For example, in 1970, only 15 cities in the US had a facility to host a trade show for 20,000 people. By 1985, the number of cities that could accommodate such large numbers of attendees had increased ten-fold (Frieden and Sagalyn, 1989).

But the real value of investing enormous public resources in the construction of large-scale conference and exhibition centres has increasingly been called into question. A substantial body of literature focusing on the allocation of public resources to investment in business tourism venues in the hope of generating ancillary investment, high employment multipliers in the hospitality and retail sectors and local tax revenues suggests that such expectations are widely misplaced. For example, Eisinger's (2000: 327) research leads him to the conclusion that, in the US at least, projections of the potential economic benefits of investing public funds in conference centres are almost always exaggerated by the proponents of such developments, who generally include the consultants employed by local authorities to calculate the likely returns on such investments:

> The source of fantastic or exaggerated expectations is not simply the rhetoric of persuasion by local political and development elites bent on gaining public support for building a … convention center. The rhetoric itself is usually based on consultant projections about the economic impact of these amenities that use inflated spending and employment multipliers. Although the application of multipliers by neutral economic analysts is generally informed by empirical research or by reference to U.S. Department of Commerce standards, consultants hired by project proponents often seem to pull their multipliers out of thin air.

Further forthright criticism of the use of public funds for the construction of conference and exhibition venues was published in the form of a controversial report from the Brookings

Institution, entitled, *Space Available: The Realities of Convention Centers as Economic Development Strategy* (Sanders, 2005). Although the study focuses on the situation in the US, it raises a number of questions that business tourism cities in any country ought to consider. The report's author reached a similar conclusion to that of Eisinger (2000) – that despite the immense amount of public funding invested in new venues in the US, the expected benefits in terms of increased numbers of visitors and revitalisation of the city centre through the stimulation of new private investment and development has not always occurred in business tourism destinations. Sanders argued that in the US, the market for large-scale business tourism events was actually in decline (and had been even prior to the disruptions of 9/11) and therefore simply could not support the ongoing proliferation of US venue development in the early years of this century. Although the study focuses on exhibitions rather than conferences, there are serious implications for the owners and operators of conference centres if its findings are valid, since in the US these are the venues primarily used for such events.

Sanders (2005: 1) noted that, despite a perceived decline in the demand for conference centres,

> Nonetheless, localities, sometimes with state assistance, have continued a type of arms race with competing cities to hold these events, investing massive amounts of capital in new convention center construction and expansion of existing facilities. Over the past decade alone, public capital spending on convention centres has doubled to $2.4 billion annually, increasing convention space by over 50% since 1990. Nation-wide, 44 new or expanded convention centers are now in planning or construction.

In response to the publication of the Brookings study, several business tourism associations, including the Center for Exhibition Industry Research and the International Association for Exhibition Management, immediately issued statements rejecting its findings. Both associations questioned Sanders' conclusion that the market was in decline, claiming that, in limiting his data to the 200 largest trade shows in the US and to conference centres, he had overlooked the expanding market for conferences held in hotels, where many fledgling meetings incubate (Minton, 2005). Similarly, Hazinski and Detlefsen (2005) also question whether Sanders paints an accurate picture of the state of the business tourism industry, claiming that in focusing on trade shows, he relies on a small and unrepresentative sample of events that excludes consumer shows and other events that may also be held in conference centres.

Nevertheless, given the ongoing business tourism boosterism demonstrated by a growing number of cities worldwide, it is reasonable to suggest that the actual returns on the investment of public funds in conference and exhibition centres should be more systematically scrutinised and that externalities arising from these developments should also be taken into account. Morgan and Condliffe (2007) note that such externalities can include traffic congestion, loss of amenity for local residents or negative environmental impacts such as pollution from traffic. Eisinger (2000) adds to this list the opportunity costs produced by the diverting of public funds into such large-scale infrastructural developments for business tourism.

Intermediaries – business tourism city marketing

Responsible for bringing together supply and demand in the business tourism industry are those specialised intermediaries who market cities as destinations for conferences, exhibitions and incentive trips: convention bureaux. Harrill (2005) describes convention bureaux as "umbrella" marketing or promotional agencies, working on behalf of the extensive collection of local businesses that serve the business tourism industry. This system may be taken as an example of the "consolidated" approach to marketing, following the principle that consolidated efforts provide greater strength and unity and therefore enhanced results; while, by way of contrast, segmented, fragmented individual marketing programmes by individual operators generally yield less impact and success.

The origins of business tourism destination marketing have been traced back to an article by the American journalist Milton J. Carmichael, who, in February 1896, wrote in the *Detroit Journal*:

> During the past few years Detroit has built up a name as a convention city, delegates coming from hundreds of miles, manufacturers holding their yearly consultations around our hotels, and all without any effort on the part of the citizens, or any special attention paid to them after they got here. They have simply come to Detroit because they wanted to.... Can Detroit, by making an effort, this year secure the holding of 200 or 300 of these national conventions during the year of '97. It will mean the bringing here of thousands and thousands of men from every city in the union ... and they will expend millions of dollars.
>
> *(Ford and Peeper, 2007: 1107)*

In the same article, Carmichael argued that local businesses should join forces to begin a formal and organised promotion of Detroit as a desirable convention destination, in order to attract more business events. As a result, less than two weeks later, on 19 February 1896, members of the Chamber of Commerce joined with the Manufacturers Club to form a new organisation, the Detroit Convention and Businessmen's League, which had the objective of promoting that city as "a desirable convention destination" (Ford and Peeper, 2007: 1107).

In the modern era, convention bureaux have proliferated as a growing number of destinations have begun to target the business tourism market. And although the marketing functions of convention bureaux may be undertaken at geographical levels ranging from an entire country, through regions within a country, to individual cities or towns, "the vast majority of convention bureaux operate at the level of the individual city or town" (Rogers and Davidson, 2016: 12). Within destinations, convention bureaux have evolved as collaborative ventures between public and private entities that share the main objective of increasing the visibility of the local area as a viable destination for business tourism (Wang, 2008). They are, generally speaking, partnership entities in which the convention bureau's partners – conference centres and other venues, hotels and other forms of accommodation, transport operators, destination management companies, translators, etc. – pay fees in order to be represented in the marketing activities of the convention bureau. In return for such fees, the partners expect to gain exposure in the business tourism market and to receive help with submitting a proposal (bid) for a new event to a meeting planner or commissioning body (Carlsen et al., 2001).

According to Vallee (2008: 162), a convention bureau's tasks most commonly include working to:

- solicit, qualify and confirm groups to hold meetings, conventions and trade shows in the area it represents;
- assist meeting groups that have confirmed through attendance building and convention servicing;
- manage the destination brand through awareness building and customer relationship management;

- market [the destination] through targeted promotional and sales activities;
- facilitate relationships between meeting manager and travel trade buyers and sellers, with sellers generally composed of local businesses offering products and services;
- service visitors, including convention delegates, in the destination to encourage them to stay longer and see more of the area.

In order to carry out these responsibilities, convention bureaux employ a wide range of marketing tools, which Davidson (2019) itemises as: destination branding and positioning; advertising in trade publications aimed at meeting planners and other decision-makers; public relations; familiarisation trips and press trips; exhibiting at specialist business tourism trade shows; and ambassador programmes – identifying members of the community who have the potential to lobby to bring conferences to their cities through for example their membership of national or international associations.

Looking ahead – a new approach to business tourism destination marketing

In recent years, a new approach to marketing destinations for business tourism has emerged, as a growing number of convention bureaux have understood that, in addition to important factors such as price, accessibility and having the appropriate infrastructure for the hosting of meetings, the existence of relevant local economic and scientific expertise can also deliver an important competitive advantage when attempting to attract association and academic conferences. Davidson (2016) observes that in order to strengthen their bids for business events, convention bureaux are seeking to differentiate their destinations by developing strategic alliances with their local knowledge industries – such as universities and research institutions – and local businesses, to promote their cities as destinations for the meetings and conferences of associations representing those industry sectors that are active in their particular city. This harnessing of cities' local industrial and research expertise represents a major shift, from simply marketing the destination's meeting infrastructure hardware to promoting its intellectual software as well. It is a new concept in destination marketing that depends upon convention bureaux adopting a strong partnership approach to winning business events for their cities. Most commonly, they have to work with their colleagues in economic development, inward investment and education departments in order to reach out to local companies, organisations and research centres and involve them in bidding for conferences.

Examples abound of how this approach has been successful for cities. For example, the Convention Bureau of Hamburg, a city with a strong presence of companies operating in the logistics and transportation sectors, began to leverage the existence of those companies in order to promote the destination to meeting planners seeking destinations for large-scale business events in the global transportation industry. An early success of the Hamburg Convention Bureau, working in partnership with local transportation companies, was the winning of the IATA World Passenger Symposium in October 2015 (DMAI, 2016). Similarly, the Visit Denver convention bureau, in collaboration with the Denver Metro Chamber of Commerce, identified growth industries where Denver has a strong presence, such as aviation, aerospace, bioscience, energy, telecommunications, health care and information technology. Visit Denver is now targeting these industry segments, and one example is the close alliance it has developed with Anschutz Medical Campus in Denver, home of the University of Colorado Health Science Division, to identify and promote new sources of local expertise to international groups. This collaboration between the convention bureau, the local economic development team and the medical community resulted in bringing the World Conference on Lung Cancer to Denver (Brost, 2015).

It is argued that one reason why this approach is so effective in winning conferences is because conference planners are attracted by the idea that, when the presence of their industry is significant in the destination, local experts can be invited as speakers at their conferences; local specialists working in the field of the conference topic can boost delegate numbers; and local research centres, laboratories or factories can provide interesting sites for conference excursions (Davidson, 2019).

It would appear that this technique of harnessing cities' local industrial and research expertise is likely to become a widely established supplement to the marketing toolboxes of a growing number of convention bureaux in the future. To take the example of one country, the German Convention Bureau publishes an annual *Meeting & Event Barometer* that provides details of developments in the business tourism sector of that country. The 2016 edition showed that, for meeting planners, the presence of local expertise in their sector was becoming a critical factor to be taken into account in the choice of destination for their business events. Of the organisers surveyed, 73.7% believed that partnering their events with local companies and research organisations in their sector was becoming increasingly important (GCB, 2016). The same report showed that city convention bureaux in Germany were increasingly geared to this approach. Accordingly, in 2012, one-third of them had integrated local fields of competence into their marketing campaigns, while these figures rose to 44.4% in 2015, demonstrating that the cooperation of German convention bureaux with established scientific and business institutions in their cities has increased substantially in recent years (GCB, 2016).

Summary

This chapter has reviewed the role of business tourism as a catalyst for urban development, highlighting the tourism-related benefits of cities' hosting of conferences, exhibitions and incentive trips, as well as their "beyond-tourism" benefits. The development of the market for business tourism was outlined in terms of the origins and growth of demand for meetings and exhibitions and the corresponding expansion in the supply of facilities and services to cater to this demand. Finally, the role of convention bureaux in marketing cities as business tourism destinations was examined, with reference to the expanding range of techniques and tools that may be employed to attract conferences, exhibitions and incentive trips.

References

Ashworth, G. J., 1989. Urban tourism: An imbalance in attention. *Progress in Tourism, Recreation and Hospitality Management*, 1, 33–54.

Brost, K., 2015. *NextGen CVBs: The Changing Role of Destination Marketing Organizations*. Available at: www.themeetingmagazines.com/cit/cvbs-and-dmos/.

Carlsen, J., Getz, D., and Soutar, G., 2001. Event evaluation research. *Event Management*, 6(4), 247–257.

Davidson, R., 1994. *Business Travel*. Harlow: Pearson Education.

Davidson, R., 2002. Leisure extensions to business trips. *Travel and Tourism Analyst*, (5), 2–17.

Davidson, R., 2016. Harnessing local expertise and knowledge: A new concept in business tourism destination marketing. *Proceedings SITCON 2016 – Singidunum International Tourism Conference*. Belgrade: Singidunum University.

Davidson, R., 2019. *Business Events*. 2nd ed. Abingdon: Routledge.

Davidson, R., and Cope, B., 2003. *Business Travel: Conferences, Incentive Travel, Exhibitions, Corporate Hospitality and Corporate Travel*. Harlow: Pearson Education.

Davidson, R., and Hyde, A., 2014. *Winning Meetings and Events for Your Venue*. Oxford: Goodfellow Publishers.

Dee, R., 2011. *Sweet Peas, Suffragettes and Showmen: Events that Changed the World in the RHS Halls*. Andover: Phillimore & Co.

Deery, M., and Jago, L. K., 2010. *Delivering Innovation, Knowledge and Performance: The Role of Business Events*. Business Events Council of Australia.

DMAI, 2016. *The Evolving Role of DMOs in a Shifting Marketplace*. Destination Marketing Association International.

Dwyer, L., 2002. Economic contribution of convention tourism: Conceptual and empirical issues. *In*: K. Weber and K. Chon, eds. *Convention Tourism: International Research and Industry Perspectives*. Binghamton, NY: The Haworth Hospitality Press, 21–36.

Eisinger, P., 2000. The politics of bread and circuses: Building the city for the visitor class. *Urban Affairs Review*, 35(3), 316–333.

Ford, R. C., and Peeper, W. C., 2007. The past as prologue: Predicting the future of the convention and visitor bureau industry on the basis of its history. *Tourism Management*, 28(4), 1104–1114.

Frieden, B., and Sagalyn, L. 1989. *Downtown, Inc*. Cambridge, MA: The MIT Press.

Friedman, A., and Phillips, M., 2004. Balancing strategy and accountability: A model for the governance of professional associations. *Nonprofit Management & Leadership*, 15, 187–204.

GCB, 2016. *Meeting & Event Barometer Report*. German Convention Bureau.

Gopalakrishna, S., and Lilien, G. L., 2012. Trade shows in the business marketing communications mix. *In*: G. L. Lilien and R. Grewal, eds. *Handbook on Business-to-Business Marketing*. Cheltenham: Edward Elgar Publishing, 226–245.

Hall, C. M., and Page, S., 2003. *Managing Urban Tourism*. Harlow: Prentice Hall.

Harrill, R., ed., 2005. *Fundamentals of Destination Management and Marketing*. Washington, DC: International Association of Convention and Visitors Bureaus.

Haven-Tang, C., Jones, E., and Webb, C., 2007. Critical success factors for business tourism destinations. *Journal of Travel & Tourism Marketing*, 22(3/4), 109–120.

Hazinski, T., and Detlefsen, H., 2005. Is the sky falling on the convention center industry? Available at: www.hotel-online.com/News/PR2005_2nd/May05_ConventionBiz.html.

Judd, D. R., and Fainstein, S. S., eds., 1999. *The Tourist City*. New Haven, CT: Yale University Press.

Kerr, G., Cliff, K., and Dolnicar, S., 2012. Harvesting the "business test trip": Converting business travellers to holidaymakers. *Journal of Travel and Tourism Marketing*, 29, 405–415.

Law, C. M., 1996. Introduction. *In*: C. M. Law, ed. *Tourism in Major Cities*. London: International Thomson Business Press, 1–22.

Law, C. M., 2002. *Urban Tourism: The Visitor Economy and the Growth of Large Cities*. London: Cengage Learning EMEA.

Lawson, F., 2000. *Congress, Convention and Exhibition Facilities: Planning, Design and Management*. New York: Architectural Press.

Lichy, J., and McLeay, F., 2018. Bleisure: Motivations and typologies. *Journal of Travel & Tourism Marketing*, 35(4), 517–530.

Locke, M., 2010. A framework for conducting a situational analysis of the meetings, incentives, conventions, and exhibitions sector. *Journal of Convention & Event Tourism*, 11, 209–233.

McCabe, V., Poole, B., Weeks, P., and Leiper, N., 2000. *The Business and Management of Conventions*. Milton, QLD: John Wiley and Sons.

Marques, J., and Santos, N., 2017. Tourism development strategies for business tourism destinations: Case study in the central region of Portugal. *Tourism*, 65(4), 437–449.

Minton, E., 2005. What planners really want. *The Meeting Professional*, June.

Morgan, A., and Condliffe, S., 2007. Measuring the economic impacts of convention centers and event tourism: A discussion of the key issues. *Journal of Convention & Event Tourism*, 8(4), 81–100.

Rogers, T., 2013. *Conferences and Conventions: A Global Industry*. 3rd ed. Abingdon: Routledge.

Rogers, T., and Davidson, R., 2016. *Marketing Destinations and Venues for Conferences, Conventions and Business Events*. 2nd ed. Abingdon: Routledge.

Rogerson, C. M., 2004. Urban tourism and small tourism enterprise development in Johannesburg: The case of township tourism. *GeoJournal*, 60(3), 249–257.

Rogerson, C. M., 2015. The uneven geography of business tourism in South Africa. *South African Geographical Journal*, 97(2), 183–202.

Sanders, H., 2005. *Space Available: The Realities of Convention Centers as Economic Development Strategy*. Washington, DC: The Brookings Institution.

Shone, A., 1998. *The Business of Conferences*. Oxford: Butterworth-Heinemann.

Spiller, J., 2002. History of convention tourism. *In*: K. Weber and K. Chon, eds. *Convention Tourism: International Research and Industry Perspectives*. Binghamton, NY: The Haworth Hospitality Press, 3–20.

Swarbrooke, J., and Horner, S., 2012. *Business Travel and Tourism*. London: Routledge.

Telfer, D. J., 2002. Tourism and regional development issues. *In*: R. Sharpley and D. J. Telfer, eds. *Tourism and Development: Concepts and Issues*. Bristol: Channel View Publications.

UNWTO, 2014. *AM Reports, Volume Seven – Global Report on the Meetings Industry*. World Tourism Organization.

Vallee, P., 2008. Convention and visitors bureaus: Partnering with meeting managers for success. *In:* G. C. Ramsborg, B. Miller, D. Breiter, B. J. Reed and A. Rushing, eds. *Professional Meeting Management, Comprehensive Strategies for Meetings, Conventions and Events*. Chicago, IL: Kendall/Hunt, 161–178.

Wan, Y. K. P., 2011. Assessing the strengths and weaknesses of Macao as an attractive meeting and convention destination: Perspectives of key informants. *Journal of Convention & Event Tourism*, 12, 129–151.

Wang, Y., 2008. Collaborative destination marketing: Roles and strategies of convention and visitors bureaus. *Journal of Vacation Marketing*, 14(3), 191–209.

Weber, K., and Chon, K. S., 2002. *Convention Tourism: International Research and Industry Perspectives*. Binghamton, NY: Haworth Hospitality Press.

11

AN EVENTFUL TOURISM CITY

Hosting major international exhibitions in Melbourne

Valentina Gorchakova and Vladimir Antchak

Introduction

It is a truism that nowadays cities are using events as a tool to generate and maintain a range of benefits, including income generation, community pride, cultural vibrancy and image enhancement. Events have become an inseparable part of any urban policy agenda.

Staging ad hoc events is not sufficient for a city to achieve its wider objectives in tourism and to maximise the benefits of the hosted events; rather, events should be integrated in the physical space, culture and identity of the host city (Richards & Palmer, 2010). This integrative approach to events as a part of a city's wider ambitions is embraced by the term "eventfulness". As a concept, eventfulness is sometimes presented as a panacea for the urban problems caused by neoliberalism. Contemporary city orientation towards commercialisation, theming and entertainment raises a question of artificial and themed urbanity where places have been transformed into entertainment productions and stages for an ever-growing number of performances and events (e.g. Smith, 2016). "Eventfulness" implies that a city thinks holistically about events, that it elaborates event calendars, being mindful of how events affect each other, city residents and the various sectors and stakeholders involved.

Truly eventful cities are not just responding to market trends, but are among market leaders that are provoking publics (Richards, 2015). Richards and Duif (2019, p. 21) argue that in order to compete in the economy of attraction and fascination (Schmid, 2006) and to maximise positive effects from staged events, cities require programming, which is "a coherent series of strategic actions that are developed over time". Such activities can lead to the development of a sustainable and balanced portfolio of events (Antchak, 2017; Getz, 2017), where events complement each other and contribute to the development of a unique place identity.

This chapter explores how a portfolio of major international exhibitions has contributed to the development of eventfulness in Melbourne, Australia. The city of Melbourne has constructed an ambitious and competitive programme of major cultural events that has positively affected the brand of the destination and the overall attractiveness of the place.

Major cultural exhibitions are shown as an effective tool for enhancing Melbourne's vibrancy and making the city attractive for tourists and residents alike. In our study, these effects were found to be dependent upon the alignment of the exhibitions with regard to the city's context, objectives and experiences provided. The chapter draws together the idea of an "eventful" city with the concepts of storytelling and place identity and examines events as their key agents.

Place branding, storytelling and events

The global competition for capital and human resources (Hall, 2006), as well as endeavours by cities to be ranked as "best places to live and do business in", has prompted urban destinations to think about tourism, and the visitor economy in general, as a means to achieve wider long-term goals. Therefore, places are attempting to redefine themselves as distinctive identities through enhancing their local characteristics and communicating attractive brands (Ashworth & Page, 2011). Recent research by Oxford Economics (2014) shows that investment in destination marketing and promotion not only generates dividends by attracting visitors, but also raises the quality of life, develops transportation networks, builds a place's profile and deepens connections through staging events.

A place brand represents the complexity of the tangible and intangible assets that distinguish this place from others and are difficult to replicate. Kavaratzis and Ashworth (2006, p. 189) maintain that,

> A place needs to be differentiated through a unique brand identity if it wants to be, first, recognised as existing, secondly, perceived in the minds of place customers as possessing qualities superior to those of competitors and, thirdly, consumed in a manner commensurate with the objectives of the place.

Pike (2004, 2008) suggests that a city brand represents an identity for the brand's producer, and an image for the consumer. Brand identity is aspirational and reflects the desired perceptions associated with a brand, while brand image is understood as existing perceptions of a brand (Aaker, 1996). Brand identity may be conceptualised in terms of the vision and culture of the brand, which are at the core of how the brand is positioned, what relationships it generates and what personality it possesses (de Chernatony, 1999). Therefore, the uniqueness of a place should be identified in order to emphasise its points of difference. Events can highlight the distinctive features of a destination, which are critical for differentiating the place in the minds of potential visitors (Echtner & Ritchie, 1993). In many instances, however, the search for place identity is often marked by pursuing similar strategies, such as waterfront developments, creation of entertainment and shopping districts, construction of convention bureaus and stadiums and urban beautification (Spirou, 2011).

To solve the problem of sameness and load a brand with meaning, cities need to develop and distribute authentic stories (Richards & Duif, 2019). Storytelling has arguably become one of the most in-demand strategies in city branding and destination marketing. A real or fictitious story can showcase and communicate unique competitive advantages of a place and attract desirable attention. A good story about a place consists of many dimensions linked to different facets of the city. It consolidates multiple stakeholders and inspires attractiveness and visitation and overall meaningfulness of the place experience for both local residents and tourists. For instance, Copenhagen has developed five strategic core stories about the city. These are: design and architecture; gastronomy; sustainability; cultural heritage; tolerance and diversity (Wonderful Copenhagen, 2017).

The elements of a place story can then be communicated in terms of images, core messages and signs (Richards & Duif, 2019). Govers (2018, p. 59) argues that storytelling is a prerogative of "admired" or "imaginative" communities with a strong sense of belonging. Such communities enchant their audiences buy creating "amazing imaginative stuff" and captivating national and international attention.

In many cities, events have been employed to communicate positive stories about places, increase "visitability", as well as catalyse urban development and transformation. Events can be used to amplify visitation, diversify tourist products, regulate seasonality, rejuvenate destinations, consolidate local assets and bolster a destination's authenticity (Ziakas, 2014). Edinburgh has successfully positioned itself as the world leading festival city (Edinburgh Festival City, n.d.). One of the unique features of Edinburgh's events scene is that there is a formal consortium of 11 independent festivals, working collaboratively in a mutually beneficial manner, contributing to the creation of a "festival ecosystem" in the city (Antchak et al., 2019). Annually, Edinburgh delivers "over 3,000 events reaching audiences of more than 4.5 million and creating the equivalent of approximately 6,000 full time jobs" (BOP Consulting, 2018, p. 8).

Rotterdam has expended efforts to develop a programme of cultural events to improve the cultural image of the city (Richards & Wilson, 2004); Auckland uses events to become the most liveable city in the world (Antchak et al., 2019); Macau has recently focused its strategic global marketing campaign on showcasing the variety of festivals and events, under the brand "Experience Macao Event Style" (Couto, 2019). On the front of business events, Manchester has demonstrated a quality-driven approach by targeting and successfully hosting large-scale conferences of international calibre that are aligned with the city's brand and key industry sectors (Vokacova & Antchak, 2019). Consistent support of major cultural exhibitions in Canberra has led to the development of a varied and balanced portfolio of events that meet the needs of both local residents and tourists (Gorchakova, 2019).

Cities, however, should not only employ events to market themselves to the existing or potential visitors and residents but also use events as part of "making the place", both socially and culturally, through "developing a special 'atmosphere' unique to a place that cannot be found elsewhere" (Richards & Palmer, 2010, p. 419). This is particularly important as tourists seek to participate in the life of local communities more, enhancing their role in co-creation of the individual tourist experience (Jelincic & Senkic, 2019). The capability of events to change cities might be of particular relevance for places lacking an iconic landmark or a "character" – while observable and measurable characteristics of a place contribute to a functional impression that a tourist gets, the destination's atmosphere directly affects the tourist's psychological impressions (Guerreiro & Mendes, 2014). Events can add vivid atmosphere and bring in new people and ideas; they can consolidate tangible and intangible resources available to a city and present a "contemporary cultural vibrancy" (Prentice & Andersen, 2003, p. 8), while "weaving narratives to link resources and meanings into a coherent story" (Richards & Duif, 2019, p. 17).

Creativity appears to have become a leading notion in place-making and marketing. It is the strategy for a broader instrumentalisation of a place's cultural assets in a search for growth and innovation (Richards, 2011). The term "creative city" is often used when a place demonstrates its intention to be creative in terms of urban planning and design, as well as through its events, especially when they showcase the creative industries (Evans, 2017). Events as "creative spectacles" (Richards & Wilson, 2006) provide a direct link between creativity and tourism, build networks and enhance the feeling of festivity. Creativity challenges the serial reproduction of culture and places through the creation of new cultural forms and innovative cultural products (Richards & Wilson, 2006). A creative city becomes a place where "the surprise of discoveries and encounters is the major potential, the main productive force" (Levy, 2011, p. 33).

Eventfulness

Sporadically organised events are not capable of delivering long-term positive outcomes or of maintaining attention and interest in a destination. Events should be managed holistically as a part of the city's wider ambitions. This integrative approach to events is embraced by the term "eventful" city (Richards & Palmer, 2010).

Richards (2017) identifies three different strategies of eventfulness. All three strategies are focused on delivering certain benefits to a city, making it an attractive destination and a liveable place. Event-Centric Eventfulness focuses on the development of event policies, programmes and overall management of events in a city. However, the narrow focus of this approach does not guarantee a balanced distribution of event-led benefits into other sectors of the local economy or city life in general. A Sector-Centric Eventfulness is based on the understanding that events can generate more positive outcomes beyond the immediate direct impacts. This strategy values events as a platform for particular economic or socio-cultural activities in a city; it helps to unite stakeholders and consolidate sectoral resources. Finally, a Network-Centric Eventfulness provides an opportunity for a city to join an international network and to make the city a hub for different activities related or non-related to events. For example, by linking tourism, international events and creative industries, a city can overcome its limitations, reach new markets and use international networks to leverage added value.

Overall, the successful eventful city puts the "cultural ecology of a city" first (Richards & Palmer, 2010, p. 415). Events should be embedded in the city context, reflect its culture, diversity and identity, and be more concerned with the quality of events, rather than their quantity. This vision can be achieved by implementing diverse portfolio strategies.

The effective realisation of a balanced and well-planned event portfolio can bring significant attention to a destination. A series of events included into a portfolio can enhance particular features or competitive advantages of a destination, and hence develop its attractiveness. As a policy tool, an event portfolio can play a role of a catalyst for urban development, consolidating and stimulating different stakeholders, sustainably using available soft and hard resources and building confidence in the community. The literature suggests different approaches to event portfolio programming (e.g. Antchak, 2017; Antchak & Pernecky, 2017; Dragin-Jensen et al., 2016; Getz, 2013). The common argument is that an event portfolio should contain an array of different genres and types of events of local, national and international significance. Destinations can bid for events, create events or invest into already existing local events. To diversify the content of an event portfolio, some destinations apply creativity and include in their portfolios not only sporting competitions and large-scale festivals, but such cultural projects as touring exhibitions (Gorchakova, 2019), which are footloose events and usually do not happen in the same place twice (Getz, 2013).

Introducing touring exhibitions

Exhibitions designed to be exhibited in several venues for limited duration are described as "touring" or "travelling" (Belcher, 1991). Some major touring exhibitions often include original art works and artefacts, rarely seen together, while others present high-quality replicas. Such exhibitions are one-off occurrences and may be organised by museums, art galleries, fashion houses, film studios or private companies, and therefore the origin and nature of the exhibits vary significantly. The large financial and time resources involved make it important for these exhibitions to be shown in different museums and destinations, that is, to travel (Bradburne, 2001). These cultural projects are usually widely advertised in the media, generating a sensational attention to the event and its host destination (Sepulveda dos Santos, 2001).

Touring exhibitions can arguably be considered as part of the creative tourism phenomenon, place-making and strategic storytelling. Previous research (Carmichael, 2002; Mihalik & Wing-Vogelbacher, 1993) demonstrates that touring exhibitions are capable of attracting tourists, generating income for local business and facilitating cities' competitiveness and profile alongside other large-scale city events. Such events have a major power to draw visitors (Getz, 2013) and deliver once-in-a-lifetime, memorable experiences for the audience (Axelsen, 2006a; Kotler et al., 2008). A "true blockbuster" may attract an audience composed of 50% out-of-town visitors (Lord & Piacente, 2014).

The potential of major international cultural exhibitions in destination marketing remains a relatively neglected area in tourism and events studies. Only a few researchers have explored it so far: visitors' motivations were investigated by Axelsen and Arcodia (2004) and Axelsen (2006a, 2006b, 2007); exhibitions' potential for the tourism sector of a city were analysed by Mihalik and Wing-Vogelbacher (1993), Carmichael (2002) and, recently, by Calinao and Lin (2016); while Arnaud, Soldo and Keramidas (2012) focused on the effects of such cultural events on territorial governance and stakeholder management.

There is a clear demand to explore the phenomenon of major touring exhibitions further and understand the ways such events can be employed by destinations to build an international profile and capitalise on the financial and socio-cultural benefits from such events.

Research context: Melbourne, Australia

Melbourne is located in southeast Australia and is the capital of the state of Victoria. In 2018, the population of the city reached 5 million people (Population Australia, n.d.), and it is currently the fastest growing city in Australia. Melbourne may become the largest Australian city by 2028 (Longbottom & Knight, 2018).

The Victorian Major Events Company (VMEC) was established in 1991 with the aim of securing major events for Melbourne. Since 2016, the responsibilities of VMEC and Tourism Victoria, the state's previous tourism body, were brought together under a single entity of Visit Victoria, the state of Victoria's destination marketing organisation – "one body, one clear plan", as it was announced (Premier of Victoria, n.d.). Therefore, at the time of writing this chapter, Visit Victoria was an organisation whose remit covered both attracting major events to the city and developing tourism in the region. In Melbourne itself, there also exists a Greater Melbourne tourism organisation, Destination Melbourne, that markets the city within the state of Victoria and helps to develop the visitor industry.

In Melbourne, there are more than 100 art galleries, museums and other spaces operating in the cultural and creative industries. The National Gallery of Victoria is the oldest and – located across two galleries – one of the biggest in Australia. The Australian Centre for the Moving Image is a unique institution – it is Australia's only national museum of the moving image in all its forms – film, television, games and digital culture. The Melbourne Museum is a natural and cultural history museum that showcases the country's social history, Indigenous cultures, as well as science and natural environment. The Australian Centre for Contemporary Art, Centre for Contemporary Photography, Heide Museum of Modern Art and a number of independent galleries and artist-run spaces are located in Melbourne.

A research study by the Boston Consulting Group (2015) suggests that Melbourne has a compelling and diverse creative and cultural offering. In 2015, the city attracted more than ten million national and international visitors. In total 69% of the international visitors who participated in the research agreed that Melbourne is a creative city that offers many attractions, events and experiences. However, Melbourne's position as a leading cultural and creative destination in Australia has been challenged as other cities in the country are investing heavily in cultural infrastructure. In such a competitive environment, the strategic priority for Melbourne is to make culture a key part of the city proposition, optimise and expand the current offer, develop and communicate clear creative messages, protect current and invest in new cultural infrastructure, and enhance overall governance and collaboration.

Methodology

This chapter is based on a qualitative case study conducted in 2015–2016 as part of a larger research project (Gorchakova, 2017) that investigated the role of touring "blockbuster" exhibitions in the marketing of cities in Australia and New Zealand. The data were collected in a form of semi-structured interviews with destination managers, event planners, representatives of museums and other cultural institutions. The city and state policies, strategies and regulations in tourism, events and culture were sourced and analysed, along with statistical data, exhibition reports and publications in mass media. In total, six interviews and 15 documents were thematically analysed applying both manual coding and the qualitative analysis software, NVivo 12. Several quotes from the interviews are used in this chapter to support the main thesis.

Thematic analysis revealed an interplay of three key constructs that affect the realisation of the city's place-making and marketing strategies and also provide an innovative way to utilise events as key storytelling agents. These are "Creativity and major events", "Collaboration" and "Place representation through events". The following sections of this chapter will explore each of the constructs in detail.

Creativity and major events in Melbourne

Melbourne is a city that cannot be associated with one particular world-famous, well-recognised location or a landmark; therefore, the city has chosen to focus on something that it can control and what can significantly contribute to its competitive advantage – special events. Arguably, many cities around the world have used special events as tourism drivers, but their focus has overwhelmingly been on sport events (e.g. Antchak, 2017). Melbourne was no exception prior to the early 2000s. Around that time, VMEC began to think creatively in the events realm. In particular, it was recognised that in the cultural, lifestyle and creative space, there were possible opportunities that could also prove hugely popular with tourists. Thus, cultural exhibitions emerged on the radar of VMEC and they continue to be regarded as major city events today.

As a result of this new strategic direction, a special annual series of international "blockbuster" exhibitions, Melbourne Winter Masterpieces (MWM), has been organised in the city since 2004 to promote tourism and the city brand image. The exhibitions are deliberately scheduled for the winter season to mitigate the overall downtime in the city's visitor economy sector. Because these events take place indoors, they are not dependent upon the weather, as many sport events are. Therefore, an annual exhibition programme is able to fill the gap in the events calendar – and it delivers. *The Impressionists: Masterpieces from the Musée d'Orsay* was organised in June–September 2004. Its brand and content were strong enough to attract some 380,000 visitors – a record attendance for any art exhibition held in Australia before that time (Council of Trustees of the National Gallery of Victoria, 2005).

Under this programme, the National Gallery of Victoria has presented some of the most remarkable exhibitions ever hosted in Australia, featuring the works of Degas, Picasso, Dali, Van Gogh and masterpieces from the world's largest museums, such as the Hermitage, the Prado Museum, the Musée d'Orsay and the Museum of Modern Art. The Australian Centre for the

Moving Image has also hosted a number of major exhibitions. Among the most well-known are *David Bowie Is*, *Tim Burton: The Exhibition* and *Pixar: 20 Years of Animation*. The "blockbusters" at the Melbourne Museum include *Tutankhamun and the Golden Age of the Pharaohs* and *A Day in Pompeii*. Overall, the 25 exhibitions organised in Melbourne since 2004 as part of the MWM have attracted nearly 6.5 million visitors (Table 11.1).

Melbourne, as a host city, was able to secure exclusive rights in Australia for all MWM exhibitions. Previously, major exhibitions, when coming to Australia, toured a number of cities; however, VMEC suggested that Melbourne should have exclusive rights, and later the requirement transcended to include not only Australia but also New Zealand. As innovative as it was, the rationale was also sensible and therefore was welcomed by the exhibitions' lenders – their exhibits would not be subject to venue changes and freight risks, the logistics would be easier to handle with only one host institution involved, and all the organisational and insurance costs would be reduced. As one of the museum managers in Melbourne mentioned: "So the idea that they could sit in one very highly regarded gallery, rather than being on the road, being freighted around the country, was very appealing … it was a lightbulb moment". At the same time, VMEC understood that they would need to attract more visitors to cover the costs; therefore, an exhibition would have to be strong and appealing enough to draw audiences from outside of Melbourne and the state of Victoria, and VMEC would have to be more creative in marketing in order to get higher national visibility to secure those audiences.

Table 11.1 Melbourne Winter Masterpieces Series – attendance (2004–2018)

2018	Masterworks from Museum of Modern Art, New York	404,000
2018	Wonderland	179,000
2017	Wallace & Gromit and Friends: The Magic of Aardman	110,000
2017	Van Gogh and the Seasons	462,000
2016	Degas: A New Vision	198,000
2015	Masterpieces from the Hermitage	172,000
2015	David Bowie Is	199,000
2014	Italian Masterpieces	153,000
2014	DreamWorks Animation	220,000
2013	Monet's Garden	343,000
2013	Hollywood Costumes	204,000
2012	Game Masters	103,000
2012	Napoleon: Revolution to Empire	189,000
2011	Tutankhamun	796,000
2011	Vienna: Art and Design	172,000
2010	European Masters	200,000
2010	Tim Burton: The Exhibition	276,000
2009	A Day in Pompeii	333,000
2009	Salvador Dali: Liquid Desire	333,000
2008	Art Deco 1910–1939	241,000
2007	Guggenheim	180,000
2007	Pixar – 2007	147,000
2006	Picasso	224,000
2005	Dutch Masters	219,000
2004	The Impressionists	371,000
	Total	6,428,000

Source: Creative Victoria (2019).

Special programming of a number of smaller events around a "blockbuster" exhibition is part of this creative approach. Events include live music concerts, curatorial talks and family events and activities, including the setting up of photo booths or spaces where children can become creative themselves. When marketing major exhibitions, a sense of occasion is communicated – that there is something special, a must-see in town, and it is here only for a limited period of time. Such a message relays a sense of urgency and uniqueness and inspires visitors to come.

On its initiation, this was a different type of event portfolio model to try. The goal was to create a distinctive cultural brand and re-shape perspectives of the cultural institutions and other partners – including the overseas ones – around the way exhibitions could be organised as major events and contribute to brand identity and city attractiveness. Adding high-profile exhibitions to the city's events portfolio signified an ambitious and creative approach to events by the state's decision-makers and tourism and event planners, who, along with the cultural institutions, demonstrated an "appetite for new projects" and willingness to take risks. As a result, Melbourne has arguably got one of the most "stellar" arts event portfolios in Australasia. This may not have been achieved in the absence of productive stakeholder relationships.

Collaboration in Melbourne

The level of collaboration and partnership is advanced in Melbourne. The stakeholders do not work in silos, but instead are able to work together to organise and deliver an event. Naturally, the more that people work with one another, the more experience they gain and the higher are the chances of success of the projects. As one of the interviewees mentioned, obtaining initial experience was critical for the development of stakeholders' expertise, industry capacity and the reputation "of being a city that does events and exhibitions really well … that we are capable of hosting excellent events".

The term "Team Melbourne" was used by another interviewee to describe the high level of collaboration between the stakeholders in the city. The first MWM exhibition was claimed to be a result of the "assiduous" work of the NGV, VMEC and Art Exhibitions Australia, with indemnity support from the federal government, and the "vital role" of Tourism Victoria "in connecting Melbourne's interstate and international visitors with this important show" (Council of Trustees of the National Gallery of Victoria, 2005, p. 15).

Bringing stakeholders together at an early stage of the project is one of the main objectives of VMEC – "a very broad range of people – often 20 or 30 people around a table getting briefed on the project and putting their contributions in around … and so it kind of happens organically". Among those involved are the cultural institution that hosts an exhibition, the state and city authorities, tourism bodies, key sponsors, media partners, airlines and hotels. Through the committee forums and meetings, the participants would share objectives and find opportunities to work together, including through in-kind or monetary sponsorships, joint competitions, tourism packages or other activities. Collaborations create a conducive environment for joint work, events and projects in the future. The credibility of stakeholders is critical when looking for new opportunities, evaluating the feasibility of large-scale projects and bidding on major events.

Major cultural events are regarded as being "repeat business" in terms of building on the relationships with national and international partners so that there is a continuous nurturing of the connections established in Melbourne and the state of Victoria. As one of the interviewees acknowledged, "all of those relationships and all of those successful hostings do wonders for its opportunity to do the next deal, to make the next connection" – which reflects a forward-looking, proactive approach with a longer-term vision than reaping immediate financial gains.

Eventually, the track record of events organised in a city contributes to the development of industry capability and enhances its reputation as an events destination locally as well as worldwide. The high level of engagement and collaboration generates early awareness of what events are being planned, and when and where they will be organised. This knowledge allows the stakeholders to share the audiences, rather than splitting them, thus allowing them to work on the overall tourist experience together instead of creating and promoting different events separately. The Team Melbourne approach has proved to be one of the most significant factors in creating and enhancing the tourism and events industry in the city.

Place representation and events in Melbourne

One of the main marketing objectives of destination marketers is to make Melbourne known as the world's most "visitable" city. "Visitability" is a vision that builds on the concept of *liveability* – something that Melbourne is known for, having been recognised as the world's most liveable city for seven years in a row between 2011 and 2017, according to The Economist Intelligence Unit's (EIU) Liveability Index (*The Economist*, n.d.). *Visitability* implies the city possesses certain attributes and qualities that make it a desirable place to visit. Success in this area requires not only collaboration between various actors working in events, tourism and hospitality, but also the integrated promotion of the city, that is when marketing communications go out from a similar platform. This includes the coordination of the message delivered to potential tourists and actual city visitors at all points of contact, ensuring that the story about the city is consistent and memorable, which is a prerequisite for a strong brand image.

The Melbourne story revolves around its lifestyle, atmosphere and the quality of experiences that need to be discovered and explored: "There's intimacy to Melbourne … it's a city of small experiences, between bars and cafes, to the laneways, to the small shopping districts, to the things that are tucked away". The buzzing effect of events is appreciated by the tourism marketers – the city being "busy" – with people and events – is something that contributes to the "endless possibilities" in Melbourne. In the absence of a single outstanding attraction that will draw tourists into ticking that box on their must-see lists, the tourism planners aspire to offer visitors a range of experiences that will allow them to explore Melbourne's culture and subcultures, attend events and see various places of interest, that, combined, tell the story of the city. However, to learn this story for themselves, tourists need to make a decision to come.

"Blockbuster" exhibitions, along with other major events, have been used effectively in Melbourne as pull drivers to motivate people to come. Major cultural exhibitions offer an opportunity to see something special, rare and unique. To become part of the city's story, the topic of an exhibition needs to be of interest to Melbournians and be in line with the city's positioning and brand values. Other priorities include the quality of the exhibits, the exclusivity rights and the exhibition's brand, which needs to be strong enough to encourage domestic tourism. Over the years, MWM has gained a reputation of having a "stamp of quality". Its brand equity has become such that "the Australian market people sort of know that in winter in Melbourne there's a big show to go and see … it really works [now] as a tourism-driver".

Melbourne is often described in terms that include "cultural", "stylish", "warm and welcoming", "sophisticated" and "creative". The Winter Masterpieces series enhances the image of Melbourne as a "destination for cultural events", which are conceived as an integral part of the city's brand and positioning. A major exhibition, therefore, can be a potent instrument to foster the city's *visitability*.

Discussion and conclusion

In Melbourne, eventfulness is seen as part of the city's DNA, an innate quality of the city, of "who we are". Arguably, the city has now gone through all three stages of eventfulness: event-, sector- and network-centric (Richards, 2017). The comprehensive approach to events, along with pioneering and long-term thinking, and the collaboration across networks, has led to the re-conceptualisation of the city's event portfolio in a way that it now organically incorporates not only major international sporting events, but also major international cultural events.

Creativity in the context of place-making and tourism implies that an idea, or activity, is novel; that there is a room for the stakeholders to think in a different and innovative way; and that the idea is valuable and useful (Bilton, 2007). Event and tourism decision-makers have planned, organised, marketed and delivered exhibitions as major city events – an innovative approach to what had long been considered as predominantly museum and art gallery territory. Stakeholders have managed to create a sense of immediacy that is intrinsic to a special event, along with a must-see status, which together have resulted in significant, beyond usual, attendance, and contributed to the visitor economy during the otherwise low and relatively uneventful tourism season of winter.

The first blockbuster exhibition within the MWM series at NGV demonstrated that the city got the pieces of its "event puzzle" right, and the consecutive exhibitions proved just that. The quality and exclusivity of the exhibitions, the consistency of their organisation – the city hosting at least one every winter – and the associations with Melbourne, winter and something extraordinary (Melbourne Winter Masterpieces), have generated a "visiting habit" among tourists and enhanced the city's position as a cultural and creative capital. The level of momentum around the exhibition programme makes it possible for cultural agencies and tourism bodies to pursue international exhibitions as part of its wider events, tourism and destination marketing agenda. Importantly, the interests of the local residents are taken care of, too – Melburnians, and Victorians, gain access to some of the most well-known and rare collections in the world, learn new things, socialise and build rapport with the cultural institutions that host these exhibitions.

Major touring exhibitions have become embedded not only in the event portfolio but also in the rhythms of Melbourne, introducing new senses and meanings of the city and defining its "melodic uniqueness" (Antchak, 2018, p. 53). Melbourne Winter Masterpieces stands out as an example of an eventful city, whose ambition to create a unique event, the skill to co-brand it with the city, and the willingness to take risks and try new formats and ways of bringing major events, have made a genuine difference to the city's image and in the tourism domain.

This research has revealed the important role of major touring exhibitions in designing an attractive and fascinating brand identity in Melbourne. The analysis and discussion of the results show that creativity in the form of new approaches to cultural events, available soft and hard infrastructure such as collaborative networks and venues, and authentic meanings produced through place branding and storytelling (Richards & Duif, 2019) have provided the city with a solid platform for reputation and image development. The positive impacts of major cultural exhibitions were found to be dependent upon the alignment of these exhibitions with the city's context, objectives and experiences, and hinge upon the collaborative planning and shared vision of stakeholders.

Several areas for future research can be suggested. There have been structural changes in Melbourne in the past few years. In particular, VMEC as a separate entity, ceased to exist in 2016, and its responsibilities and duties were reallocated to Visit Victoria, the state's tourism body. Echoing one of the premises of an eventful city, Visit Victoria Event's webpage states that "the key to a successful event legacy is the lasting connection between the event, the host city and the community" (Visit Victoria, n.d.). It would be interesting to see if and how the change might have affected the connection between Melbourne Winter Masterpieces, the city and its community. What have been the dynamics in the stakeholder relationships since the restructuring took place? Have there been any innovative ideas or activities in the area of major events? Where does Melbourne stand with regard to its "visitability" nowadays?

Major international cultural exhibitions remain a novel topic in the events and tourism literature, notwithstanding the longevity of the phenomenon (Gorchakova, 2017). It is worth exploring the ways other cities in the world have been approaching this type of event. In particular, if there are any effects touring cultural exhibitions have had on the host city's event portfolios and cultural agenda? How incorporated are they in the wider urban and tourism policies? Have they contributed to the enhancement of a place's desirable image and supported its identity?

Overall, the planning and management of international touring exhibitions proves to be efficient once stakeholders from tourism, events and cultural sectors work together and approach these events holistically. The strategic decision-making around scheduling may have significant effects on event tourism in a city as indoor cultural events help address the issue of seasonality and add variety to the overall event offer in a destination. The brand of a major exhibition can be used efficiently in strategic storytelling, combining elements of place identity with its "blockbuster" status.

References

Aaker, D. A. (1996). *Building strong brands*. New York, NY: Free Press.

Antchak, V. (2017). Portfolio of major events in Auckland: Characteristics, perspectives and issues. *Journal of Policy Research in Tourism, Leisure and Events*. doi:10.1080/19407963.2017.1312421.

Antchak, V. (2018). City rhythms and events. *Annals of tourism research, 68*, 52–54. doi: https://doi.org/10.1016/j.annals.2017.11.006.

Antchak, V., & Pernecky, T. (2017). Major events programming in a city: Comparing three approaches to portfolio design. *Event Management, 21*(5), 545–561. doi:10.3727/152599517X15053272359013.

Antchak, V., Ziakas, V., & Getz, D. (2019). *Event portfolio management: Theory and methods for event management and tourism*. Oxford: Goodfellow Publishers.

Arnaud, C., Soldo, E., & Keramidas, O. (2012). Renewal of territorial governance through cultural events: Case study of the Picasso-Aix 2009 cultural season. *International Journal of Arts Management, 15*(1), 4–17.

Ashworth, G., & Page, S. J. (2011). Urban tourism research: Recent progress and current paradoxes. *Tourism Management, 32*, 1–15. doi:10.1016/j.tourman.2010.02.002.

Axelsen, M. (2006a). Defining special events in galleries from a visitor perspective. *Journal of Convention & Event Tourism, 8*(3), 21–43. doi:10.1300/J452v08n03_02.

Axelsen, M. (2006b). Using special events to motivate visitors to attend art galleries. *Museum Management and Curatorship, 21*(3), 205–221. doi:10.1080/09647770600302103.

Axelsen, M. (2007). Visitors' motivations to attend special events at art galleries: An exploratory study. *Visitor Studies, 10*(2), 192–204. doi:10.1080/10645570701585285.

Axelsen, M., & Arcodia, C. (2004). Conceptualising art exhibitions as special events: A review of the literature. *Journal of Convention & Event Tourism, 6*(3), 63–80. doi:10.1300/J452v06n03-05.

Belcher, M. (1991). *Exhibitions in museums*. Leicester: Leicester University Press.

Bilton, C. (2007). *Management and creativity: From creative industries to creative management*. Oxford: Blackwell.

BOP Consulting. (2018). *Edinburgh Festivals: The network effect*. Retrieved from www.edinburghfestivalcity.com/assets/000/003/791/The_Network_Effect__July_2018__original.pdf?1531301203.

Boston Consulting Group. (2015). Melbourne as a global cultural destination. Retrieved from https://creative.vic.gov.au/__data/assets/pdf_file/0006/115764/BCG-Melbourne-as-a-Global-Cultural-Destination-Summary-for-CV-website.pdf.

Bradburne, J. M. (2001). A new strategic approach to the museum and its relationship to society. *Museum Management and Curatorship, 19*(1), 75–84. doi:10.1016/S0260-4779(01)00025-5.

Calinao, D. J., & Lin, H.-W. (2016). The cultural tourism potential of a fashion-related exhibition – the case of Alexander McQueen: Savage Beauty at the Victoria and Albert Museum. *Journal of Heritage Tourism*, 1–14. doi:10.1080/1743873X.2016.1206550.

Carmichael, B. A. (2002). Global competitiveness and special events in cultural tourism: The example of the Barnes Exhibit at the Art Gallery of Ontario, Toronto. *Canadian Geographer/Le Géographe canadien, 46*(4), 310–324. doi:10.1111/j.1541-0064.2002.tb00753.x.

Council of Trustees of the National Gallery of Victoria. (2005). *Year in review: NGV 04/05 annual report.* Retrieved from www.ngv.vic.gov.au/wp-content/uploads/2014/09/ngv_corp_annualreport_2004_05_1.pdf.

Couto, U. (2019). The eventful city in a complex economic, social and political environment: The case of Macao. In J. Mair (Ed.), *The Routledge handbook of festivals* (pp. 194–203). London: Routledge.

de Chernatony, L. (1999). Brand management through narrowing the gap between brand identity and brand reputation. *Journal of Marketing Management, 15*(1), 157–179. doi:10.1362/026725799784870432.

Dragin-Jensen, C., Schnittka, O., & Arkil, C. (2016). More options do not always create perceived variety in life: Attracting new residents with quality- vs. quantity-oriented event portfolios. *Cities, 56*, 55–62.

Echtner, C. M., & Ritchie, J. R. B. (1993). The measurement of destination image: An empirical assessment. *Journal of Travel Research, 31*(4), 3–13. doi:10.1177/004728759303100402.

Edinburgh Festival City. (n.d.). Retrieved from www.edinburghfestivalcity.com/.

Evans, G. (2017). Creative cities: An international perspective. In J. Hannigan & G. Richards (Eds.), *The Sage handbook of new urban studies* (pp. 310–329). Los Angeles, CA: Sage.

Getz, D. (2013). *Event tourism: Concepts, international case studies, and research.* New York, NY: Cognizant Communication Corporation.

Getz, D. (2017). Developing a framework for sustainable event cities. *Event Management, 21*(5), 575–591.

Gorchakova, V. (2017). *Touring blockbuster exhibitions: Their contribution to the marketing of a city to tourists.* PhD Doctoral dissertation. AUT University, Auckland, New Zealand.

Gorchakova, V. (2019). Event portfolios and cultural exhibitions in Canberra and Melbourne. In V. Antchak, V. Ziakas, & D. Getz, *Event portfolio management: Theory and methods for event management and tourism* (pp. 125–243). Oxford: Goodfellow Publishers.

Govers, R. (2018). *Imaginative communities: Admired cities, regions and countries.* Antwerp: Reputo Press.

Guerreiro, M., & Mendes, J. (2014). Experiencing the tourist city: The European capital of culture in re-designing city routes. *Journal of Spatial and Organizational Dynamics, 2*(4), 288–306.

Hall, P. (2006). Seven types of capital city. In D. L. A. Gordon (Ed.), *Planning twentieth century capital cities* (pp. 8–14). Abingdon: Routledge.

Jelincic, D. A., & Senkic, M. (2019). The value of experience in culture and tourism: The power of emotions. In N. Duxbury & R. Greg (Eds.), *A research agenda for creative tourism* (pp. 41–53). Cheltenham: Edward Elgar.

Kavaratzis, M., & Ashworth, G. (2006). City branding: An effective assertion of identity or a transitory marketing trick? *Place Branding, 2*(3), 183–194. doi:10.1057/palgrave.pb.5990056.

Kotler, N., Kotler, P., & Kotler, W. I. (2008). *Museum marketing and strategy: Designing missions, building audiences, generating revenue and resources,* 2nd ed. San Francisco, CA: Jossey-Bass.

Levy, J. (2011). The city is back (in our minds). In H. Schmid, W.-D. Sahr, & J. Urry (Eds.), *City and fascination: Beyond the surplus of meaning* (pp. 33–48). Farnham: Ashgate.

Longbottom, J., & Knight, B. (2018). Melbourne's population explosion threatens to create a "Bangkok situation". *ABC News*, 16 October. Retrieved from www.abc.net.au/news/2018-10-15/melbourne-will-be-australias-biggest-city-which-party-has-policy/10358988.

Lord, B., & Piacente, M. (2014). *Manual of museum exhibitions,* 2nd ed. Lanham, MD: Rowman & Littlefield.

Mihalik, B. J., & Wing-Vogelbacher, A. (1993). Traveling art expositions as a tourism event: A market research analysis for Ramesses the Great. *Journal of Travel & Tourism Marketing, 1*(3), 25–42. doi:10.1300/J073v01n03_02.

Oxford Economics. (2014). *Destination promotion: An engine for economic development.* Retrieved from https://mktg.destinationsinternational.org/acton/attachment/9856/f-0732/1/-/Dest_Intl_2014_Destination_Promotion_An_Engine_of_Economic_Development_-_Full_Report.pdf.

Pike, S. (2004). *Destination marketing organisations.* Amsterdam: Elsevier.

Pike, S. (2008). *Destination marketing: An integrated marketing communication approach.* Oxford: Butterworth-Heinemann.

Population Australia. (n.d.). Melbourne population 2019. Retrieved from www.population.net.au/melbourne-population/.

Premier of Victoria. (n.d.). Visit Victoria: A new era for tourism and major events. Retrieved from www.premier.vic.gov.au/visit-victoria-a-new-era-for-tourism-and-major-events/.

Prentice, R., & Andersen, V. (2003). Festival as creative destination. *Annals of Tourism Research, 30*(1), 7–30. doi:10.1016/S0160-7383(02)00034-8.

Richards, G. (2011). Creativity and tourism: The state of the art. *Annals of Tourism Research, 38*(4), 1225–1253. doi:10.1016/j.annals.2011.07.008.

Richards, G. (2015). Developing the eventful city: Time, space and urban identity. In S. Mushatat & M. Al Muhairi (Eds.), *Planning for event cities* (pp. 37–46). Ajman, United Arab Emirates: Municipality and Planning Dept. of Ajman.

Richards, G. (2017). Emerging models of the eventful city. *Event Management, 21*(5), 533–543.

Richards, G., & Duif, L. (2019). *Small cities with big dreams: Creative placemaking and branding strategies.* New York, NY: Routledge.

Richards, G., & Palmer, R. (2010). *Eventful cities: Cultural management and urban revitalisation.* Oxford: Butterworth-Heinemann.

Richards, G., & Wilson, J. (2004). The impact of cultural events on city image: Rotterdam, cultural capital of Europe 2001. *Urban Studies, 41*(10), 1931–1951. doi:10.1080/0042098042000256323.

Richards, G., & Wilson, J. (2006). Developing creativity in tourist experiences: A solution to the serial reproduction of culture? *Tourism Management, 27*, 1209–1223. doi:10.1016/j.tourman.2005.06.002.

Schmid, H. (2006). Economy of fascination: Dubai and Las Vegas as examples of a thematic production of urban landscapes. *Erdkunde, 60*(4), 346–336.

Sepulveda dos Santos, M. (2001). The new dynamic of blockbuster exhibitions: The case of Brazilian museums. *Bulletin of Latin American Research, 20*(1), 29–45. doi:10.1111/1470-9856.00003.

Smith, A. (2016). *Events in the city: Using public spaces as event venues.* Abingdon: Routledge.

Spirou, C. (2011). *Urban tourism and urban change: Cities in a global economy.* New York, NY: Routledge.

The Economist. (n.d.). The Global Liveability Index. Retrieved from www.eiu.com/topic/liveability.

Visit Victoria. (n.d.). Why events love Melbourne. Retrieved from https://corporate.visitvictoria.com/events/major-events.

Vokacova, Z., & Antchak, V. (2019). Sector-focused approach to business events in Manchester. In V. Antchak, V. Ziakas, & D. Getz, *Event portfolio management: Theory and methods for event management and tourism* (pp. 144–154). Oxford: Goodfellow Publishers.

Wonderful Copenhagen. (2017). *The end of tourism as we know it.* Retrieved from http://localhood.wonderfulcopenhagen.dk/wonderful-copenhagen-strategy-2020.pdf.

Ziakas, V. (2014). *Event portfolio planning and management: A holistic approach.* Abingdon: Routledge.

12

HOW CREDIBLE IS VANCOUVER'S NEW DESTINATION BRAND?

An analysis of a destination's brand promise and the tourist's brand experience

Rami K. Isaac and Jacqueline Wichnewski

Introduction

Within today's changing business environment and the growth of the tourism industry, destinations are facing more competition than ever before. New tourism spots and substitutable products and services increase the destination choice for tourists (Dinnie, 2016; García et al., 2011). In this regard, with the change in demand, the preference of destinations these days is defined by the experience, rather than the destination itself. Hence, besides the physical attributes which tourists are seeking, destinations increasingly have to satisfy visitors' emotional needs (UNWTO & ETC, 2009).

Brand experience is empirically distinct from other concepts such as destination image, destination brand personality and destination brand identity (García et al., 2011; Usakli & Baloglu, 2011; UNWTO & ETC, 2009). From a management point of view, an experience is a lasting impression that is formed in the mind of the customers through the encounter with the holistic offer of a brand (Klaus & Maklan, 2007). In this regard, Brakus et al. (2009) introduced the conceptualisation of brand experience, which evaluates the sensory, affective, intellectual and behavioural dimensions of a brand. Depending on how many of these dimensions are stimulated and to what degree, the resulting brand experience can be more or less intense (Zarantonello & Schmitt, 2010). Moreover, brand experience can positively affect customer satisfaction and brand loyalty (Barnes et al., 2014). In this context, a large part of the branding literature appears to allege that emotional and cognitive experiences are likely to be associated key drivers (Oliver, 1994; Phillips & Baumgartner, 2002; Bigné et al., 2005). This means, whether a brand promise is in line with the tourist experience, can be directly correlated to customer satisfaction, the willingness to recommend the destination to others, as well as a tourist's intention to revisit the place (Barnes et al., 2014).

The tourism industry of the City of Vancouver is still developing. The seaport city is located on the Western coast of British Columbia in Canada and is renowned for its natural beauty and cultural diversity (Destination BC Corp., 2019). In 2017, the city received 10.3 million overnight visitors, representing the fourth consecutive record-breaking year (The Metro Vancouver Convention and Visitors Bureau, 2018a). However, in order to remain successful in the future and to live up to the change in consumer demand, the city introduced their first established destination brand in June 2018. Concerning this matter, the city's new brand promise is communicated via the following statement: "Vancouver is a place that connects people and inspires them to live with passion". Thus, a clear focus is placed on the psychological attributes the city has to offer, more precisely on the experience visitors can obtain by engaging with the people living in the city. Furthermore, in order to educate the stakeholders and incorporate the brand within the destination, a document with associated brand guidelines has been released. Although the tourism board has been supported by other professional agencies, Tourism Vancouver represents the official destination marketing and management organisation for the City of Vancouver (The Metro Vancouver Convention and Visitors Bureau, 2018b).

Several researchers have already examined various concepts regarding the visitors' perspective on destination branding (García et al., 2011; Usakli & Baloglu, 2011; UNWTO & ETC, 2009). However, only limited research has been conducted comparing the brand identity, resulting in the brand promise, with the actual brand experience of a destination's visitors. This study contributes by filling this gap in the literature. Therefore, this chapter aims to identify whether the communicated brand promise of Vancouver's new established destination brand is in line with the brand experience perceived by the tourists visiting the city, in order to determine possible discrepancies.

Literature review

Destination branding

Although the historical roots of branding can be traced back to the late nineteenth century, branding within the context of tourism destinations, such as countries, cities or regions, is a relatively recent phenomenon (Blain et al., 2005). Dinnie (2016) and UNWTO and ETC (2009) claim that the increase of destination branding is derived from a globally growing competition due to more emerging tourism destinations and a consequently increasing destination choice for tourists. Moreover, in line with the increased focus on the psychological and emotional elements of brands, the preference of destinations nowadays is defined by the experience, rather than solely the tourism product (UNWTO & ETC, 2009). More precisely, branding a destination implies the obligation of taking into account the perspectives and needs of a range of different stakeholders, in order to capture the *identity* of a destination. In this way, it requires a long-term effort, is less controllable and appears to be more complex than the branding of products and services (Morrison, 2019; Dinnie, 2016).

Some examples of city destination branding are Porto: Brand "Porto". It is visualised by icons and illustrations that depict the city's unique gastronomic, architectural, cultural and geographical elements, and relevant focus on stakeholder involvement (authenticity). Another example is San Francisco, which is one of the most cosmopolitan cities of the USA, the financial centre and cultural hub of the West Coast – like Vancouver for Canada. The Golden Gate Bridge is a relevant icon for the city, which is represented in their logo. In addition, they recently introduced their new positioning statement "For those who embrace the bold and seek the unexpected". San Francisco's optimistic spirit is a constant celebration of individuality and the belief that here, in the most beautiful city in the world, all things are possible.

Brand identity

Creating a strong and unique brand identity can be regarded as the first and most important step within the process of destination branding. "Identity" is defined by the Oxford Dictionary (2019) as "the fact of being who or what a person or thing is", including an extended definition as "the characteristics determining this".

In this regard, it is relevant to clarify that the destination brand identity is based on the overall identity of the destination (DNA). More precisely, since the identity comprises extensive information about a destination's history, culture and people, the brand identity is solely built upon a limited range of all the consistent parts of the identity. This means, those engaged in constructing a destination's brand identity – in this case the destination management organisations (DMOs) – need to be selective in identifying the elements of the overall identity, which can usefully serve to reach the objectives of the branding strategy (Dinnie, 2016). More importantly, it is relevant to emphasise that the creators of the destination brand decide upon the identity that will be communicated (Morgan et al., 2011) and therefore, on the way the brand will be perceived by tourists visiting the respective destination. This entails, that a brand identity is often considered as a *preferred* or *desired identity* (Pike, 2009). In this context, Ooi (2011) states that cities, regions or countries tend to present only the positive aspects of the destination, which entails that many aspects are ignored, as these are not considered attractive by the branding authorities.

Aaker (1996) provides an aspect of brand identity and demonstrates how it comprises both, the core characteristics of the destination, but also how these can be extended to contain value adding elements. He argues that brand identity should not be regarded as static but be open to change if necessary. The brand identity consists of two integral parts; the core identity and the extended identity (Aaker, 1996). The core identity represents the brand's soul and should include the fundamental characteristics, which make the brand valuable and unique. In contrast to the core identity, the extended identity is less persistent to change, which means that, if needed, it can and should be altered. The extended identity varies per organisation and can contain elements such as mission, vision, values, personality, distinguishing preferences and emotional benefits (Ghodeswar, 2008; Kapferer, 2008). In this way, it provides the brand with consistency and completeness (Aaker, 1996). Hence, this part should be considered as equally important as the core identity.

The Identity Structure Model presented by Aaker (1996) is introduced to highlight the two different dimensions of a brand's identity and will be used in this chapter in order to identify the brand identity of the city of Vancouver. Besides this and based on the agreement of Kapferer (2008) and Morrison (2019), the third dimension of identity will be identified, more precisely, the visual and unique design of the brand, which normally contains a logo, as well as some words or phrases and symbols.

Brand promise

Researchers have repeatedly indicated that a brand represents a certain *promise* to the customers. As according to Light and Kiddon (2009), this so-called "brand promise" can be defined as "a statement describing the intended promised experience of a brand" (p. 78), which is delivered to every customer, at any time and at any place. It answers the question: "what kind of brand experience do we wish to promise and deliver to every customer every time?" (Light & Kiddon, 2009, p. 27). Based on general agreement, this *statement* derives from the essence of both emotional and functional benefits, reflecting the heart, soul and spirit of the brand that customers can expect to obtain from experiencing a brand's product and services (Knapp, 2000). In other words, it derives from the core brand identity (Morrison, 2019; Light & Kiddon, 2009; Ghodeswar, 2008). Therefore, the promise of the brand is just as important for destinations as it is for other service organisations.

Prior to the experience within the destination, the respective brand promise is conveyed through online and offline marketing communications. In this context, the brand promise will shape all of an organisation's subsequent brand marketing messages. Hence, brand promise could also be referred to as the expectations that customers have of the brand, based upon what they have seen in these marketing communications and the way the communications are delivered (Morrison, 2019; UNWTO & ETC, 2009). In regard to destination branding, the respective DMO is in charge of leading, guiding and coordinating a destination's online and offline "critical promise points", in other words, all those interactions in the virtual and material world when the destination's brand promise is encountered and evaluated by its (potential) visitors (Morgan et al., 2011; Baker, 2007). Therefore, Blain et al. (2005) suggest that the marketing activities, executed by the DMO, should convey a promise of a memorable travel experience that is uniquely associated with the respective destination. Also, it is relevant to emphasise that unless the destination establishes an emotional relationship with its potential visitors, it will be difficult for the destination to gain a foothold in their imagination and project the promise convincingly (UNWTO & ETC, 2009).

The previously indicated complexity of destination branding is reflected in the fulfilment of the brand promise. Due to the variable nature of tourism products, such a complex experience is difficult to guarantee every time. In fact, all the elements of the destination experience are not under the control and partly not even under the direct influence of the DMOs (Baker, 2007; Blain et al., 2005).

Brand experience

The concept of brand experience is empirically distinct from other concepts, such as destination image or destination brand personality (García et al., 2011; Usakli & Baloglu, 2011; UNWTO & ETC, 2009) and is increasingly raising the attention of marketing scholars.

As already indicated previously, marketers are actively focusing on the brand experiences, as consumers seek a memorable experience which involves pleasing interaction at every brand contact (Morgan et al., 2011; Blain et al., 2005). From a management point of view, an experience is a lasting impression that is formed in the minds of customers through the encounter with the holistic offer of a brand (Klaus & Maklan, 2007). In this context, Light and Kiddon (2009) add that the experience of a brand comprises both, functional and emotional elements, which together define the distinctiveness of a brand. Brakus et al. (2009) introduced the conceptualisation of brand experience, which evaluates a customer's sensory, affective, intellectual and behavioural responses induced by brand-related stimuli.

The dimensions of brand experience

In accordance with Schmitt (1999) and Brakus et al. (2009), the different dimensions of the concept of brand experience can be explained as follows:

- The sensory dimension is linked to aspects of the brand that generate physical experiences based on the consumer's five senses – sight, smell, taste, touch and hearing. This, for example, could be the sound of traditional music, the taste of the food, the beauty of the city's skyline or the smell of the sea.

- In contrast, the affective dimension considers feelings, sentiments and emotions evoked by a brand, such as the feeling of being welcome in the hotel or the love of the city's architecture. Thus, it also involves social interaction.
- Further, the intellectual dimension represents the stimulation of thought, curiosity and self-reflection when experiencing the brand. This might occur during a thought-provoking guided tour or museum exhibit.
- Lastly, the behavioural dimension refers to physical actions and behaviours evoked by interactions with the brand, such as being encouraged to go surfing in the ocean. It can also entail changing behaviours, lifestyles and habits of a consumer alternatively, for example, to start recycling waste.

Depending on how many of these dimensions are stimulated and to what degree, the resulting brand experience can be more or less intense (Zarantonello & Schmitt, 2010).

The relationship between brand promise and brand experience

Taking into account the different concepts discussed above, it can be concluded that the ultimate aim of a destination brand is to ensure that visitors' experiences of the destination – from the moment they are exposed to the destination's marketing communications until their return home – are positive, which encourages them to book trips, and return as an advocate of the respective destination by developing emotional and long-lasting relationships. This is initiated by enticing people with a brand promise – derived from the destination's core identity – about what they will experience at the destination. The brand should then be reflected in the way the destination is presented to its visitors, by all stakeholders involved, from visitor centres and hoteliers to residents. The visitor's brand experience at the destination must be in line with the expectations generated by the marketing activities of the DMO: the delivery of the *brand experience* and the *brand promise* must be brand-compliant – entailing that the essence of the brand is confirmed in the visitor's eyes.

Proposed framework

The following conceptual framework (Figure 12.1) of this chapter's research was developed based on the literature review. It illustrates the perspectives of both the destination marketing organisation and the visitors, more precisely, the relationships between brand identity, brand promise and brand experience of a destination, using the city of Vancouver as an example. In regard to the perspective of the destination marketing organisation, the brand identity of Vancouver's destination brand is expressed in the brand promise, which is communicated to (potential) visitors through marketing activities. At the destination, visitors then experience the brand on four different levels: sensory, affective, intellectual and behavioural – which vary in intensity, but together determine the overall destination brand experience and thus, the visitor's perspective. If promise and experience differ from each other, a gap occurs, which is likely to have a negative effect on the brand. When brand promise and experience are in parallel, this generates consistency, which represents a critical factor for successful tourism destination brands. Thus, the goal of this chapter is to compare Vancouver's destination brand identity, expressed in the brand promise, with the brand experience as perceived by tourists visiting the city.

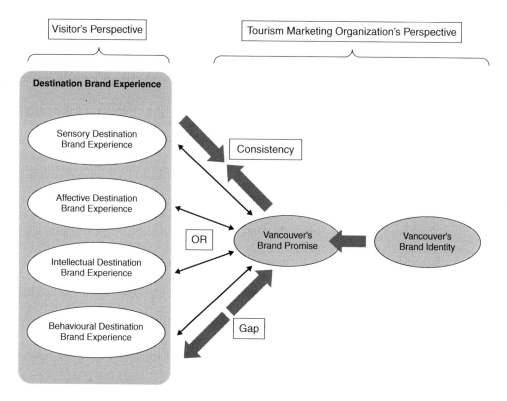

Figure 12.1 Conceptual framework.

Methodology

A focus was placed on carrying out qualitative field research, as the aim was to obtain new insights instead of gathering a quantity of statistically accurate data. The qualitative research has been conducted in two steps. First, the brand identity of Vancouver as a tourism destination and its expression in the brand promise were investigated. In the second step, using the respective knowledge obtained, the brand experience of the city's visitors was examined and evaluated. The research methods that have been applied to investigate each perspective were semi-structured personal interviews and structured e-mail interviews. The qualitative field research took place within the city of Vancouver in British Columbia, Canada and was conducted between 30 March and 4 May 2019.

Expert interviews are conducted in order to obtain profound knowledge about a particular subject in a specific professional field. Tourism Vancouver has been identified as the official destination marketing and management organisation for the city of Vancouver and thus, the responsible destination branding authority. Accordingly, the organisation is also responsible for the development of the new brand identity and corresponding brand promise that is communicated to (potential) visitors. In order to complement the obtained secondary data, an expert interview was conducted with Mr Stephen Pearce, the Vice President of Marketing, which took place on 11 April 2019.

In order to generate insight into the brand experience of the tourists visiting the city of Vancouver, the semi-structured interview method was also selected. Thus, the use of more general questions and topics intended to offer the participants freedom to respond. Moreover, it enabled the researcher to receive unbiased, in-depth knowledge regarding the participants' perspectives on their brand experience, as expressed in their own words. Apart from the investigation of tourists' overall experiences, the interviews were constructed to receive detailed insights into the four different dimensions of brand experience: sensory, affective, intellectual and behavioural (Brakus et al., 2009).

Sample

Fourteen participants were interviewed, from which five filled in the questionnaire by e-mail. The interviewees were approached in three different locations within the Downtown area – namely, at Waterfront, English Bay and the hostel HI Vancouver Central – representing strategic locations that are popular among tourists.

Approaching the participants, a random sampling method was applied. This means each member of a population of interest – in this case Vancouver's tourists – had an equal probability of being chosen. This is significant, as the brand promise must be in line with the brand experience of *any* visitor coming to the city – irrespective of whether they are among the target markets of the organisation. The interviews covered an average time frame of 25 to 35 minutes. The interviews were audio-recorded and transcribed.

Data analysis

The transcribed interviews were analysed by means of a qualitative content analysis. In order to categorise and organise the relevant data, codes were developed separately for both the expert interview and the tourist interviews. For the expert interview a focus was placed on the different parts of the identity, while the codes for the tourist interviews resulted from the different dimensions of the brand experience concept. Subsequently, results were extracted in a neutral manner, discussed and interpreted towards the research question of this study.

Findings: Vancouver's brand promise

Vancouver's brand strategy

After a two-year process and extensive research, Tourism Vancouver introduced the city's first established destination brand in Vancouver's history, in June 2018. Before, Tourism Vancouver only had a corporate brand, which entailed the development of a new strategy and visual identity, in order to reach their vision of manifesting Vancouver "as the most exciting, attractive and welcoming city destination in North America, and a must-visit year-round destination" (Pearce, 2019; The Metro Vancouver Convention and Visitors Bureau, 2019). In this context, it is relevant to point out that the brand has specifically been designed to target tourists, which entails that it needs to be separated from other city brands, such as the economic development – or municipal city brand.

Tourism Vancouver segments their key target markets based on the meetings and conventions business, the travel trade and the fully independent consumer market. In regard to the originating countries, Vancouver places a focus on core geographic markets including Canada and the United States, but also on five further relevant markets, namely, China, Mexico, Australia, Germany and the United Kingdom (MMGY Global, 2017)

Vancouver's brand identity

The core identity of Vancouver's new destination brand, more precisely the leading brand essence and the basis for all other brand attributes, is summarised in the phrase "See the beauty in everything" (Tourism Vancouver, 2018, p. 10). As explained by Mr Pearce (2019), the organisation's Vice President of Marketing, this notion derives from the fact that Vancouver, at first sight, has nothing special – "there is no icon, we do not have an Eiffel Tower, no Disney-land". Therefore, the newly established brand identity was created based on the inspiration from the multi-dimensional nature of the city and its inhabitants, embracing nature, culture, food and art. Although many of the assets Vancouver possesses, including mountains, the ocean, forests and a diverse community, can be found anywhere in the world, the organisation claims that in Vancouver "it comes together in a way that [they] think is particularly intimate, particularly special". More precisely, it is reputed to create a "vibe" – a feeling of rejuvenation and a sense of health (Pearce, 2019; Tourism Vancouver, 2018).

This sense of health and wellness should be related to the optimism and the natural beauty that is present in Vancouver. In this context, the organisation describes the "beauty" that can be experienced in the city, on two levels – the inside and the outside. The "beautiful outside" is described to be reflected by the harmony between urban and nature, including the mountains that cradle the city, the water that surrounds it, the city's fresh air and clean streets, as well as its Stanley Park, which is one of the largest urban parks in North America. In this context, even though Vancouver is known for its rain, which might be anticipated negatively, the organisation promotes it as a refreshing and renewing character of the city. Furthermore, in regard to the "beautiful inside", the organisation refers to the city's inclusive and multicultural community, characterised by a fusion of various personalities and lifestyles, which makes visitors feel welcome and allows them to be themselves. In this regard, they promote: it is about "acceptance, not tolerance" (Tourism Vancouver, 2018, p. 10).

In addition, as indicated by Mr Pearce (2019), in Vancouver the people "do not look at the outdoors; they celebrate the outdoors" and "take their playtime very seriously", which can again be related to the notion of health and wellness. This work to live mentality implies the community's active engagement in the great number of outdoor activities the city has to offer, such as hiking, biking, scuba diving, sailing or skiing, which together represent a relevant part of Vancouver's tourism product. Furthermore, besides being physically and socially active, the city also claims to be environmentally active (Tourism Vancouver, 2018). This means, when talking about food, for example, "it is all about organic, all about homegrown", as well as about "staying away from species that are harmed" (Pearce, 2019). Thus, a focus is placed on sustainability in order to maintain the natural beauty of the place.

Vancouver's visual identity

The establishment of Vancouver's new destination brand also entailed the creation of a new visual identity. According to van den Bosch and de Jong (2005), a visual identity, if implemented and managed correctly and consistently, is a useful tool to further influence the reputation of an organisation through its design. Moreover, it represents one of the first and most significant impressions tourists have when they are planning a trip to a destination. In this context, the most important component of the visual identity is the brand logo, which identifies and distinguishes the brand in consumers' minds.

Vancouver's new destination brand focuses on a simple word-based logo that reads the textual mark "VANCOUVER". Inspired by a compass, the contrasting coloured part of the

Figure 12.2   Vancouver's brand logo.

Source: Tourism Vancouver (2018).

logo (Figure 12.2) illustrates "the collective impact (circle of the compass) of multiple perspectives (the compass points above and below the 'O') that connects visitors to the experience they desire" (Tourism Vancouver, 2018, p. 13). This means, the organisation wants to emphasise the diversity of the city, which is reflected in its culture, nature, food and activities. Hence, due to the multiple perspectives that are embraced and welcomed within the same destination, visitors are expected to be able to experience Vancouver in regard to their individual perspectives and in this way, satisfy their individual needs. In order to reach this goal and to experience Vancouver at its fullest, visitors should not be tied to any plans and expectations, instead they should "let go and allow the city to guide [them]" (Tourism Vancouver, 2018, p. 13). Mr Pearce (2019) puts this in his own words, explaining: "if you open yourself to new experiences and to be a little more whimsical in terms of how you engage with the destination, you may, in fact, find little treasures around the corner that were completely unexpected". Furthermore, in line with the brand's essence to "see the beauty in everything", the compass motif should represent the visitor's exploration of both the physical "beauty" of the city itself and its surroundings as well as the "beauty" inside the diverse group of inhabitants and their characteristics (Tourism Vancouver, 2018).

When it comes to the use of the logo, in order to ensure consistency and correctness, the organisation proposes two different versions of the logo – the stacked and the horizontal version – as presented In Figure 12.3 below.

The stacked logo represents the primary and preferred option of the two, as it provides more visual impact (Tourism Vancouver, 2018). In this regard, as explained by Mr Pearce (2019), the organisation, for example, is planning "to have a little sculpture outside that will actually allow people to use it as an Instagram moment". The secondary option, which is the horizontal version of the logo, was created for areas of application that do not allow for a lot of space. Additionally, highlighting the compass within the logo, three main colours were chosen building

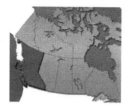

Figure 12.3 Brand logo versions.

Sources: Vancouver Tourism (2018); Province British Columbia [Image] (2018).

the new colour palette: blue, red and yellow. The use of varying colours should also represent the different perspectives. In regard to the typography, the organisation selected Proxima Soft and Proxima Nova. As the city's marketing communications are global, these typefaces are both modern and welcoming, as well as clearly legible and functional, which entails that they can be easily adapted to various languages (Pearce, 2019; Tourism Vancouver, 2018).

Referring to the "beautiful outside", the harmony between city and nature should be represented, capturing its vibrant, refreshing nature and optimistic energy. This might, for example, show Stanley Park's seawall at dawn or the skyline of the city photographed from a boat. Concerning this matter, it is relevant for the organisation to "inspire viewers and make them realize that Vancouver is at one with nature and not just surrounded by it" (Tourism Vancouver, 2018, p. 40). Moreover, regarding the "beautiful inside", it is about capturing moments of connection by featuring intimate and authentic moments of single or small groups of people experiencing Vancouver. In this regard, the photographs used should evoke a feeling "allowing the viewer to see themselves being a part of whatever they're seeing at any particular moment" (Tourism Vancouver, 2018, p. 13).

When it comes to the communication style that should be used in promotional activities, the organisation developed the so-called "voice" of the brand, which can also be referred to as brand personality. The voice reflects the brand's core values and embraces the following personality traits: unpretentious, sophisticated, welcoming, refreshing, genuine and optimistic (Tourism Vancouver, 2018). In this regard, the notion of being unpretentious refers to a simple and inclusive brand tone, which is welcoming everyone. Equally, it complies with the direction of presenting authentic, genuine or "real" material that does not lead to misconceptions about the city. Moreover, sophisticated should not be regarded from a luxury perspective, instead it refers to a sense of aspiration among the city and its people, or as described by Mr Pearce (2019): "being a little more aware and being able to relate to a whole variety of ways and means". Lastly, refreshing again relates to the notion of rejuvenation and health.

Brand Vancouver – visualised by a brand wheel

Based on the information provided above, the brand wheel has been used as a tool to summarise and visualise Vancouver's new destination brand (Figure 12.4). A brand wheel is a templated approach that helps in breaking down a brand into all its building blocks (Element Three, 2018). Equally, it represents the identity structure, more precisely the core identity and the extended identity, according to Aaker (1996). In this chapter, Vancouver's new destination brand is broken down into seven building blocks: embracing the core identity or the essence, the extended identity is composed of the mission, vision, values, voice, distinguishing features and emotional benefits.

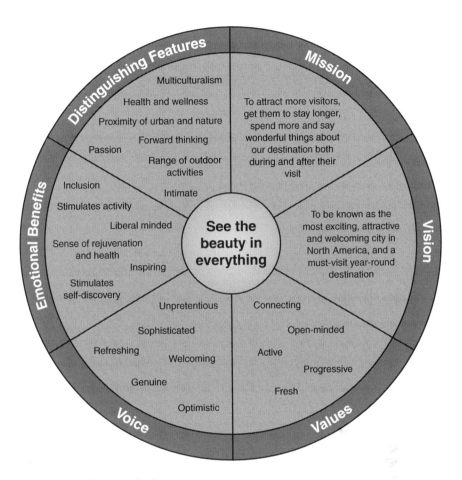

Figure 12.4 Vancouver's brand wheel.

Source: Authors.

The brand is reflected in all of the tourism marketing organisation's activities, including the brand video, social media communication and printed material. In order to ensure correctness and consistency within the industry, as well as to position the new destination brand successfully, Tourism Vancouver is focused on educating and informing the various industry players. These efforts of Tourism Vancouver imply their direction that the brand "cannot be seen as a brand owned by an organization; it has to be seen as a destination brand owned by the community" (Pearce, 2019).

Vancouver's brand promise

Tourism Vancouver's brand promise states that "Vancouver is a place that connects people and inspires them to live with passion" (Tourism Vancouver, 2018, p. 10). In this context, it is important to point out that this promise is not used as a slogan. More importantly, it drives the organisation's thinking and equally should drive the industry's thinking, relative to what is communicated and to the connections they are trying to make with the consumer (Pearce, 2019). According to Mr Pearce (2019), it means that "this destination in all the way, shape and forms, it touches people in a way that is uplifting, that is healing … that touches people's soul". This means, (potential) visitors can expect to experience a feeling of rejuvenation, including moments of self-reflection, encouraged by the city's opportunities, as well as the optimism and open mindset of its people.

Communication of the brand promise

As the brand promise is not communicated as a slogan, the way Vancouver's destination brand is communicated visually and verbally in the organisations marketing activities represents a significant part of the promise that is made to the city's (potential) visitors. Since the internet represents the main source of information for today's travellers, relevant online promotional material has been analysed. This includes but is not limited to campaigns and images, like promotional videos, the organisation's website and social media appearances.

Brand video

In line with the launch of Vancouver's new destination brand and logo in June 2018, a new promotional brand video was produced (see www.youtube.com/watch?v=vRPlS02Yc20). Overall, the brand video is showcasing the city's natural scenery and recreation-based activities, creating expectations of various natural landscapes and countless sport opportunities. Moreover, it clearly communicates all the building blocks of Vancouver's brand identity. A focus is placed on presenting people within these sceneries, who are engaging in the different activities and in this way telling the story of the city. Such activities include jogging or biking through Stanley Park, paddle boarding at the waterfront, having a communal dinner with a group of friends and so on. The showcasing of people in their intimate moments, individually or in a group, evokes different feelings such as excitement, motivation, warmth or intimacy and allows the viewers to immerse themselves into the experience portrayed. Likewise, it represents the promise of the experiences and feelings (potential) visitors can expect from their stay in Vancouver.

Furthermore, the compass motif, which is the key element of the new brand logo representing "letting go and allowing the city to guide you" (Tourism Vancouver, 2018, p. 13), is literally guiding the viewer through the video incorporating the storytelling.

Website

Tourism Vancouver provides four different official websites in the languages English, Chinese, Spanish and Japanese, which equally are adapted to the preferences of the respective target markets. The function of the website is to inform (potential) visitors about the city and what it has to offer. When opening the website, the first thing that catches the attention of the viewer is the new brand logo with a big moving banner in the background that zooms in on the skyline of Vancouver. Hence, the first impression of Vancouver already shows the proximity of the city to nature, being amidst water and mountains. Further down on the homepage, an interactive map of the city's neighbourhoods is displayed, numbering the 14 neighbourhoods. Navigating to each number, an image and short description of the respective neighbourhood appears. The map provides an introduction to the variety of the city, in terms of cultures and sights. Moreover, it demonstrates that the city is easy to explore, which can be related to its intimacy.

Navigating to the sub-page under the title "activities", several categories are displayed, featuring activities related to nature, culture, food and art, such as "BC Wine Country", "Nightlife", "Skiing & Snowboarding", "Indigenous Tourism", "Golfing", "Sporting Events" and so on. A further section on the website emphasises the city's focus on health and wellness, summarising the information about the city's walkability, its spa and yoga offerings and recreation activities in the phrase: "Vancouver has plenty of options that are good for the mind and the body".

Social media

Vancouver is actively promoted on Facebook, Instagram and Twitter via the accounts of the organisation Tourism Vancouver. The newly established brand logo constitutes the profile picture of all accounts. Further, the responsible marketers are consistently posting to keep their followers up to date. Posts appear very frequently, even multiple times per day.

The official Facebook account is called *Inside Vancouver*, counting approximately 250,000 followers. In regard to the content, a mix of reshared posts of the organisation's official blog and photos taken by visitors can be identified, which are appealing to both residents and visitors. The combination of informative and user-generated content represents a good strategy to keep their followers engaged. The blog posts, for example, provide inspiration for activities and places to visit or announce upcoming festivals. In this way, the organisation is not only informing (potential) visitors about the countless activities and events taking place in Vancouver that bring people together, it also creates excitement in them to visit the city. Moreover, resharing photos taken by other travellers represents a form of online word-of-mouth (WOM), which serves as an excellent social proof and may influence a person's decision to visit the destination. The photos show the various landscapes of the city from different perspectives, for example from a plane, from under a tree or at night-time and should present the beauty of the city creating a positive feeling towards the destination and inspire (potential) travellers.

On Instagram, the official account of Tourism Vancouver is also called *Inside Vancouver*, counting over 200,000 followers. In line with the visual identity, bright and vibrant colours are used, adding energy and drama to the professional photographs. It is striking that almost all of the pictures are user-generated content, which is emphasised by acknowledging the account of the photographer in the description line of the posts. Besides inspiration for (potential) travellers, this approach especially shows authenticity, which represents a significant part of the brand identity. The "highlights" section presents the different neighbourhoods of Vancouver through a collection of stories, which mainly present links to associated inspiring articles, such as "things to do", as published on the organisation's website. This again underlines the active character that can be expected of the city (Inside Vancouver [@inside_vancouver], n.d.).

Blogs

Beside the visual incentives provided through Instagram, Tourism Vancouver's official multi-author blog called *Inside Vancouver*, represents the organisation's most important tool for the creation of inspiring, updated and informative content about the city. As the blog posts are written by Vancouverites, the notion of receiving "insider" information can be related to the intimate atmosphere in Vancouver that is part of the brand identity and thus creates a *connection* to (potential) visitors prior to their stay. In this context, the use of multiple authors provides different genuine perspectives, stories and opinions shared by locals about various topics and activities within nature, culture, food and art (www.insidevancouver.ca/). In this way, the city may be appealing to various types of tourists with differing interests and perspectives, which reinforces the emotional connection to the city. Hence, the blog also represents variety, which the city embraces.

To conclude, it can be stated that the brand promise of Vancouver's newly established destination brand is based on a well thought-out and coherent brand strategy, aiming to position the city as an open-minded and modern destination, featuring a unique combination of varied landscapes and a diverse community that creates a feeling of rejuvenation and a sense of health.

Visitor's brand experience

This section displays the findings of the qualitative field research that has been conducted in order to identify the brand experience of tourists visiting the destination Vancouver.

Background of the interview participants

Fourteen tourists visiting Vancouver were interviewed for the purpose of this chapter. Out of these 14 respondents, eight were female and six were male. The majority of the respondents were aged between 21 and 27 years. This can be related to the fact that young people were much more willing to participate in the study than older people. Further, most of the respondents had their accommodation located in Downtown (6) or South Vancouver (5), while two of them stayed in West Vancouver and one even outside the city in Burnaby. The visitors interviewed came from seven different countries (Spain, Ireland, Germany, New Zealand, Canada, UK, the Netherlands), which indicates a high geographic coverage. In this context, the majority, however, had German nationality (6).

In regard to their travel behaviour, the majority of the respondents visited the city on their own (7), while five of them were accompanied by their significant other and two by a friend. The length of stay varied significantly, from only three days to up to two-and-a-half years. In this regard, it is relevant to mention that six of these travellers were on work visas, which are valid for one or two years, depending on the nationality. Hence, for most of them Vancouver represented the starting point of their journeys. This entails that their purpose of visit was to explore the Canadian country and culture. Moreover, Vancouver was also selected due to the destination's mild climate and accessibility. For the remaining respondents, who were not on work visas, the main purpose of visiting the city was vacation and relaxation.

Expectations

Tourists visiting Vancouver, and thus the respondents of this research, encounter the brand promise, which is communicated visually and verbally in Tourism Vancouver's marketing activities, during their information search prior to their visit. This entails the creation of associated expectations.

It was found that respondent expectations clearly reflected the brand promise, which was defined in the previous section. The majority of the respondents anticipated Vancouver as a city in *nature*, surrounded by *beautiful* and *varied landscapes*. In this regard, almost half of the respondents indicated that they expected to be offered *various activities* and *things to do*, especially in the *outdoors*. These expectations especially reflect the imagery and video material used to promote the city. Moreover, the majority of the respondents expected to encounter *very friendly* and *welcoming people*, which some of them related to a widespread stereotype – "man hört ja immer, dass die Kanadier so sehr freundlich sind" [one often hears that the Canadians are very friendly]

(Respondent 5). Additionally, three respondents also pictured Vancouver as a *clean city*, which is in line with the organisation's claim of being environmentally active. Lastly, as well as being indicated as a reason for visiting the city in the first place, three further respondents expected Vancouver to be the *mildest city in Canada*. Hence, this emphasises the idea of promoting Vancouver as a year-round destination.

Sensory experience

Concerning the outcome of the interviews, it was found that the respondents frequently indicated attributes that are in line with the brand identity. Hence, the city was commonly described as *beautiful*, reflecting the essence of the brand. This was especially related to the expanse of varied nature comprising mountains, forests, parks, lakes and the ocean – combining *urban and nature*. This sensory experience of being able to be "in the middle of the city and walk in nature without leaving it" (Respondent 9) has commonly been described in response to the question of what makes Vancouver unique. In this context, the city's *Stanley Park* has been emphasised several times, which equally is considered as a unique feature in the brand identity. Moreover, in accordance with the brand's value of being *fresh*, most of the respondents, in fact, experienced the city as *very clean* with a smell of *fresh air* and *nature* throughout the place. The latter was especially regarded as impressive by the respondents, as they have not experienced this in other bigger cities. Further comparing Vancouver to other cities, even though several respondents experienced the city life as *active* and *busy*, they did not consider it as noisy. Respondent 4 summarised this in the following sentence: "Vancouver seems busy without being loud". Regarding Stanley Park, a few respondents even emphasised that they enjoyed the quietness and silence while being surrounded by nature.

These very positive sensory experiences, however, have been accompanied by significantly negative ones. Almost all the respondents mentioned the outstanding number of *homeless people*, crowded together in certain areas – especially East Hastings, which, in this multitude, they had never seen before in one place. Accordingly, these areas were not as clean as and were noisier than the rest of the city. A further aspect, which has been associated more negatively for some and less for others, is the *smell of cannabis* throughout the city, representing a "huge contrast" (Respondent 1) to the *fresh air*. In this context, it is relevant to point out that the consumption of non-medical cannabis was legalised in October 2018 (Government of British Columbia, 2018). Hence, some respondents may not have been aware of this, as demonstrated by Respondent 5: "Man musste sich erst einmal daran gewöhnen das offen auf der Straße zu sehen" [One first had to get used to seeing this on an open street].

Affective experience

As described by Pearce (2019), "a brand is ultimately an emotive construct", aimed at establishing a meaningful relationship with the customers, which drives them to "pay a little more sometimes". Hence, the affective experience of Vancouver's destination brand represents the most important dimension for the organisation.

Regarding the outcome of the interviews and in connection to their sensory experience, most of the respondents felt *amazed* by the variety of landscapes the city lies within, as well as by "having a forest basically inside the city" (Respondent 1), referring to Stanley Park. In this context, Respondent 3 describes his experience like "Ich musste viele Male stehen bleiben oder einfach sitzen und genießen" [I had to remain standing or sitting several times and simply enjoying]. Furthermore, during the activities in the city, especially in the outdoors – for example, while strolling through Stanley Park or laying on the beach at English Bay – the majority of the respondents felt *relaxed* and *in peace* or "being in the moment" as described by Respondent 2. The latter has particularly been related to the *silence* and *quietness* experienced in certain areas of the city. In this context, many respondents also indicated that they felt *free* in the city, which they referred to both the *range of activities* offered and the ability "einfach das zu machen wo ich gerade Lust drauf habe" [simply to do whatever I feel like] (Respondent 7), as well as the *accessibility* of the city – since "being able to just walk and bike everywhere and catch trains and buses … is making it so easy" (Respondent 4). Thus, the feelings described clearly reflect what the organisation claims as the destination's unique feature, namely, to create a sense of health and a feeling of rejuvenation.

When it comes to social interactions within Vancouver, all of the respondents felt overall *welcome* in the city, clearly reflecting the brand personality (voice). Some of them also indicated that, even though they were travelling on their own, they did not feel alone. This was, first of all, related to the city's *multicultural* character, which provides the ability "to meet people from all over the world" (Respondent 2). Moreover, they described the local people as being very *friendly*, *open* and especially *accepting*, which made it easy for them to *fit in* and *meet people* with the same mindset. "I feel really myself here", as mentioned by Respondent 4. This reflects the *unpretentious* personality of the brand. Further, besides the multiple cultures that *connect* people, the respondents also emphasised the wide range of activities offered that bring people together. Another aspect that corresponds to the *welcoming* character of the city is based on the fact that a few of the respondents experienced the locals as being curious and showing great interest in them as people. In this regard, Respondent 7 explained: "Man wird ständig angesprochen, wenn gemerkt wird, dass man nicht von hier kommt" [People constantly come up and talk to you, when they notice that you are not from here]. Overall, it is relevant to point out that deviations in the respondent experiences with locals have been determined based on their nationalities. This means German respondents who come from a culture which is known for being more formal and serious, were more positive than, for example, the Irish respondent who comes from a very open-minded society. The latter, thus, rather described the locals as *shy*, which made it hard to make local friends.

Although the respondents felt generally *happy* and *safe* in the city, they emphasised a rather negative affective experience in regard to the areas with a higher density of the *homeless population*. Contrary to their expectations, seeing such an outstanding number of them or walking through the respective neighbourhoods made the respondents feel *shocked* and more *unsafe*. "It was a different world", as stated by Respondent 6. Further, a few of the respondents even experienced them being rough with other people or themselves – "their behaviour can sometimes be intimidating" (Respondent 2). Hence, these two different faces of Vancouver constitute complete opposites.

Behavioural experience

Based on the outcome of the interviews, taking into account the physical actions that the respondents performed during their stays in Vancouver, it became clear that they commonly engaged in *outdoor activities*. This, among many other options, includes walking, biking, hiking, snowboarding, camping and being on the beach. Concerning this matter, it has again been emphasised how easy it is to get around in the city. Summarised by Respondent 6: "it is part of the vibe of the city – being fit and doing a lot of outdoor activities". Likewise, this responds to the brand's value of being *active*. In line with the *various landscapes* and the presence of *Stanley Park*, the respondents were very positive and *enjoyed* immersing themselves in *nature* – "it was nice getting to disconnect from technology and just explore" (Respondent 2). This, again, refers to the notion of *wellness* and *rejuvenation* as part of the brand identity.

When it comes to their behaviour, almost all the respondents – except those who visited for only three days – did not follow a certain plan, instead they engaged in activities based on the *inspiration* and *recommendations* they received from other travellers and locals. As explained by Respondent 6 "usually you just meet someone, who does something new and then you just end up doing what they do". Others simply relied on their intuition. Hence, in accordance with the destination brand, more precisely the meaning of the compass, they *let themselves go* and *allowed the city to guide them*.

Some of the respondents stated that they have become more *active*, in terms of taking part in more *outdoor activities*, since they arrived. This may also be related to the wide range of different opportunities the city has to offer. Furthermore, four of the respondents indicated that they became more *conscious of their environment* "due to the eco-friendly behaviour of the city and its people, and their approach to draw attention to it" (Respondent 11), which resulted in improved waste management, such as in the form of using reusable water bottles. For Respondent 7, this change "hat die Aktivitäten natürlich noch viel mehr bereichert" [had enriched her behavioural experience even more]. Taking into account the different nationalities of the respondents, one of the Spanish interviewees stated an interesting behavioural change. More precisely, based on his realisation that people in Vancouver usually pay for the bus, even though it is not controlled, he explained that "in my country, people do not tend to do that that often" (Respondent 1). Thus, this *honest* and *respectful* behaviour encouraged him to follow their example.

Intellectual experience

The majority of the respondents focused their physical actions based on the *inspirations* and *recommendations* which they had received from other travellers and locals. This clearly related to the brand's promise that Vancouver *inspires* people. Furthermore, most of the respondents considered Vancouver as a *progressive* city, which is in line with the values of the brand. More precisely, they describe the city as very *environmentally friendly*, based on, for instance, the city's focus on *preservation* of nature – Stanley Park – and *animal rights*. The latter can especially be experienced in the Aquarium. Thus, Respondent 9 pointed out: "I think that they are way farther than other countries when it comes to taking care of the environment". Their self-reflection led to behavioural changes, as explained in the previous section.

Furthermore, regarding the cultural differences of the respondents, it is striking that especially the German and Swiss-German respondents emphasised the *tolerant* character of the city, which relates especially to its *multiculturalism*, but also to the focus on *gender equality*. Respondent 3 explained: "es ist nicht so wie in der Schweiz, hier ist es sehr viel toleranter … Man akzeptiert die Menschen so wie sie sind". [It is not like in Switzerland, here it is much more tolerant … People are accepted the way they are]. Thus, in line with the brand promise, visitors recognise that the city is not just *physically active*, but also *environmentally* and *socially*.

Summarising the findings in regard to the four dimensions of brand experience, it can be discerned that the overall brand experience of the respondents was very positive. This has been emphasised by a frequent use of expressions, such as "I like" or "I enjoyed".

As indicated earlier, depending on how many of these four dimensions of brand experience are stimulated and to what degree, the resulting brand experience can be more or less intense (Zarantonello & Schmitt, 2010). Hence, it was found that the length of stay of the respondents clearly influenced the intensity of their experience. The sensory experience represented the most dominant dimension, followed in importance by the affective and behavioural experiences.

Hence, the intellectual experience was stimulated to a lesser degree than the other dimensions, which as a result is suggested to be less significant in determining satisfaction and intention to recommend. This outcome is likely to be due to the type of destination, namely, an outdoor-activity-oriented destination, while a murder-mystery weekend in a dark tourism destination is more likely to stimulate intellectual experiences.

Conclusions

The purpose of this chapter was to identify whether the communicated brand promise of Vancouver's newly established destination brand is in line with the brand experience perceived by the tourists visiting the city. The findings demonstrate that Vancouver's new destination brand is based on extensive research and thoughtful creation of a new brand identity and logo. This entails that the visitor's brand experience was found to be, to a large extent, in line with the brand promise that is conveyed via the organisation's marketing communications, reflecting the destination's natural heritage, social capital and progressive activity. Relating the attributes of Vancouver's brand promise to the concept of brand experience, it can be determined that the brand promise comprises all of the dimensions of the brand experience, with an emphasis on the sensory and affective levels. In addition, Vancouver was overall experienced very positively by the respondents. Accordingly, it becomes clear that the imagery and video material used to promote the city significantly influence the creation of associated expectations. In regard to the different dimensions of brand experience, the findings suggest that the visitor outcomes are mostly driven by how they feel in a physiological way, as their senses encounter rich stimuli from the destination such as the beauty of the landscapes or the fresh air. Also, the affective and behavioural experiences of the visitors represented a key factor in their overall experience, describing that they felt relaxed, in peace or active, which all related to the brand's claim of creating a special "vibe" of rejuvenation and sense of health. In regard to the intellectual dimension, the visitors clearly felt inspired by the encountered population.

Furthermore, it is relevant to take into account Tourism Vancouver's promise statement: "Vancouver is a place that connects people and inspires them to live with passion" (Tourism Vancouver, 2018, p. 10), which summarises the brand's promise. Based on the respondent's description of the city as being welcoming, open-minded and friendly, offering a wide range of activities, the results imply that the destination clearly provides the social capital and opportunities that could connect people. It can also be argued that the process of *connecting* requires time, which would thus not apply to short-term visitors. In addition, as shown in the results, this phenomenon also depends on perceptions related to different nationalities. Furthermore, regarding the part that the city "inspires them to live with passion", the findings do not provide an adequate assessment of this part of the promise. Even though it was found that the visitors felt inspired by the people they encountered, none of them indicated a "life-changing" event. In other words, *inspiring people to live with passion* sounds very permanent and carries the connotation of not having a fulfilled life, when not living in the city.

Overall, it was found that the brand promise is, to a large extent, in line with visitors' brand experiences. Since there is no literature available on this research, this chapter adds a new dimension to the research scope of destination branding. Thus, it can be stated that the newly established destination brand of Vancouver represents a good example for future research and serves as a best practice in aligning brand promise – reflecting the brand identity – with brand experience.

Recommendations and future research

The organisation's promise statement "Vancouver is a place that connects people and inspires them to live with passion" (Tourism Vancouver, 2018, p. 10), only partially reflects the respondents' brand experience. It is advised that the second part, namely "inspires people to live with passion", be rephrased. A suggested rephrasing might include "living in the moment", which was mentioned by one of the respondents. This phrase has the connotation of enjoying every moment during a stay in Vancouver and every opportunity the city has to offer, as well as, in this way, seeing the beauty in everything.

In this context, since Tourism Vancouver's promise statement is not used as a slogan – in other words, it is not supposed to constitute an entertaining phrase that captures attention and persists in the minds of the consumers – the organisation is advised to consider developing a slogan that reflects the promise of the destination brand. As proven by best practices, such as "I Amsterdam" or "What happens in Vegas, stays in Vegas", slogans can have a significant impact on the destination's reputation and awareness. They can be a useful tool to establish an emotional connection between the consumer and the destination, which, equally, is the aim of Tourism Vancouver.

Instagram nowadays is regarded as the most important social media channel for tourism destinations and was also found to be a relevant information source for the respondents in this study. It is recommended that the content published should be adapted increasingly to the brand promise. This means, since the promise statement is not communicated as a slogan to (potential) visitors, it should be reflected in all of Tourism Vancouver's marketing activities. The analysis of the brand promise has shown that the content on Instagram is focused on showcasing the variety of landscapes that can be experienced in the city. However, in regard of the visual identity, it is lacking imagery displaying people that reflect both the connecting and active character of the city and enable the viewer to emotionally connect to the destination, prior to their visit.

For future research, we would suggest that a focus can be placed on the varied types of visitors. The outcome of this research study indicates that cultural background has an influence on the brand experience at the destination. Therefore, we would expect the brand experience to vary by visitor's gender, age, purpose or nationality. The outcome might be relevant in terms of emphasising the importance of tailoring a brand in regard to different visitor profiles. In addition, we recommend testing the research model, more precisely the compliance of brand promise and brand experience, in a wider variety of destination brands and in different types of destinations.

References

Aaker, D. A. 1996. *Building strong brands.* New York: Free Press.

Baker, B. 2007. *Destination branding for small cities: Essentials for successful place branding.* London: Creative Leap Books.

Barnes, S. J., Mattsson, J., & Sørensen, F. 2014. Destination brand experience and visitor behaviour: Testing a scale in the tourism context. *Annals of Tourism Research*, 48, 121–139.

Bigné, J. E., Andreu, L., & Gnoth, J. 2005. The theme park experience: An analysis of pleasure, arousal and satisfaction. *Tourism Management*, 26, 833–844.

Blain, C., Levy, S. E., & Ritchie, J. R. B. 2005. Destination branding: Insights and practices from destination management organizations. *Journal of Travel Research*, 43 (4), 328–338.

Brakus, J., Schmitt, B. H., & Zarantonello, L. 2009. Brand experience: What is it? How is it measured? Does it affect loyalty? *Journal of Marketing*, 73, 52–68.

Destination BC Corp. 2019. Vancouver. Available at www.hellobc.com/places-to-go/vancouver/ (Accessed 17 March 2019).

Dinnie, K. 2016. *Nation branding: Concepts, issues, practice.* New York: Routledge.

Element Three. 2018. Defining your brand identity: Creating a brand wheel. Available at https://element-three.com/blog/defining-brand-creating-brand-wheel/ (Accessed 23 April 2019).

García, J. A., Rico, M. G., & Collado, A. M. 2011. A destination-branding model: An empirical analysis based on stakeholders. *Tourism Management*, 33, 646–661.

Ghodeswar, B. M. 2008. Building brand identity in competitive markets: A concept model. *Journal of Product & Brand Management*, 17 (1), 4–12.

Government of British Columbia. 2018. Cannabis. Available at www2.gov.bc.ca/gov/content/safety/public-safety/cannabis (Accessed 3 May 2019).

Inside Vancouver [@inside_vancouver]. n.d. Posts [Instagram profile]. Available at www.instagram.com/inside_vancouver/ (Accessed 28 April 2019).

Inside Vancouver [@insidevancouver]. n.d. Posts [Facebook profile]. Available at www.facebook.com/insidevancouver/ (Accessed 28 April 2019).

Kapferer, J. 2008. *The new strategic brand management: Creating and sustaining brand equity long term*, 4th edition. Cornwall: Kogan Page Limited.

Klaus, P., & Maklan, S. 2007. The role of brands in a service-dominated world. *Journal of Brand Management*, 15 (2), 115–122.

Knapp, D. E. 2000. *The brand mindset.* New York: McGraw-Hill.

Light, L., & Kiddon, J. 2009. *Six rules for brand revitalization: Learn how companies like McDonald's can reenergize their brands.* Upper Saddle River, NJ: Pearson Education.

MMGY Global. 2017. *Request for proposal: Bringing the destination brand to life.* Vancouver, BC: MMGY Global.

Morgan, N., Pritchard, A., & Pride, R. 2011. *Destination brands: Managing place reputation*, 3rd edition. Oxford: Elsevier.

Morrison, A. M. 2019. *Marketing and managing tourism destinations*, 2nd edition. New York: Routledge.

Oliver, R. L. 1994. Conceptual issues in the structural analysis of consumption emotion, satisfaction and quality: Evidence in a service setting. *Advances in Consumer Research*, 21, 16–22.

Ooi, C.-S. 2011. Paradoxes of city branding and societal changes. In K. Dinnie (Ed.), *City branding: Theory and cases.* Basingstoke: Palgrave Macmillan, pp. 54–61.

Oxford Dictionary. 2019. Definition of identity in English. Available at https://en.oxforddictionaries.com/definition/identity (Accessed 15 March 2019).

Pearce, S. 2019. 11 April. Personal interview.

Phillips, D., & Baumgartner, H. 2002. Consumption emotions in the satisfaction response. *Journal of Consumer Psychology*, 12, 243–252.

Pike, S. 2009. Destination brand positions of a competitive set of near-home destinations. *Tourism Management*, 30, 857–866.

Schmitt, B. 1999. Experiential marketing. *Journal of Marketing Management*, 15 (1–3), 53–67.

The Metro Vancouver Convention and Visitors Bureau. 2018a. New record: 10.3 million Vancouver visitors in 2017. Available at www.tourismvancouver.com/media/articles/post/new-record-103-million-vancouver-visitors-in-2017/# (Accessed 11 November 2018).

The Metro Vancouver Convention and Visitors Bureau. 2018b. Introducing Vancouver's destination brand. Available at www.tourismvancouver.com/meetings/articles/post/introducing-vancouvers-destination-brand/ (Accessed 11 November 2018).

The Metro Vancouver Convention and Visitors Bureau. 2019. Vancouver's health and wellness culture. Available at www.tourismvancouver.com/vancouver/vancouvers-health-and-wellness-culture/ (Accessed 27 April 2019).

Tourism Vancouver. 2018. Tourism Vancouver brand guidelines. Available at https://tourismvancouver.app.box.com/s/mxukkodellnjpwrqkwrqm7p8p7miklo0 (Accessed 20 April 2020).

United Nations World Tourism Organization (UNWTO) & European Travel Commission (ETC). 2009. *Handbook on tourism destinations branding*: With an introduction by Simon Anholt. Madrid: World Tourism Organization.

Usakli, A., & Baloglu, S. 2011. Brand personality of tourist destinations: An application of self-congruity theory. *Tourism Management*, 32, 114–127.

van den Bosch, A., & de Jong, M. 2005. How corporate visual identity supports reputation. *Corporate Communications: An International Journal*, 10 (2), 225–316.

Zarantonello, L. & Schmitt, B. H. 2010. Using the brand experience scale to profile consumers and predict consumer behaviour. *Journal of Brand Management*, 17, 532–540.

13

FAMILY TOURISM

Past, present and opportunities

Xinran Y. Lehto, Jun Chen and Uyen Le

Introduction

Although its definition can vary among disciplines, cultures and times, family is commonly understood as "two or more persons living together and related by blood, marriage, or adoption" (US Census Bureau, 2019). As contemporary families travel, experience providers respond with family-centric programmes and activities. For example, the city of Charleston presents "Family Day at Marion Square" as a tradition of its annual Piccolo Spoleto festival. Science Centre Ontario provides a permanent exhibit called KidSpark for children under eight. The Children's Museum of Indianapolis envisions itself to be the global leader among all museums in serving children and families. It is estimated that family tourism accounts for a third of the world's leisure travel, and this will remain true in the future (Morrison et al., 2018). Take the US for example, where research shows that there is much interest from families to go on family vacations (88%), and that shorter vacations appear to be the most common with summer being the peak travelling season for this traveller market, according to a national survey conducted by New York University (NYU) and the Family Travel Association (Minnaert, 2017). Theme parks, beach vacations and family road trips remain the most preferred family vacation types, although cruises and all-inclusive resorts are among the favourite choices for the future. In the same survey, however, family tourists' evaluations of how well the travel industry has accommodated their needs are underwhelming. Some of the underperforming areas include affordability, and the quality of packaged options, accommodations for large families, online assistance, and travel assistance for families with special needs (young or old). The classic image of a travelling family has always been full of cheerfulness and optimism as one envisions the picture of happy parents with a few boisterous children in tow. However, not all elements of family travel are as rosy. Travelling with infants or toddlers, for example, means lap seats on the plane. When young children are on the road, there are a lot more details that parents need to plan for and manage. The NYU study attests to the promise of this travel market and also points to unaddressed traveller needs and research opportunities in this area.

Family tourism commands our attention for a number of reasons. The sheer market size of this travelling population is reason number one. The prevalence of this form of travel makes it by default an important area for researchers to investigate. A second rationale is that family travel involves small group dynamics that possess unique characteristics and exhibit needs that are differentiated from other types of travel companionship. Knowledge of the specific needs and tendencies of family travellers can inform family tourism practitioners, allowing them to act upon research insights in their offerings, and thereby better serve this consumer segment. A third reason is that the concept of family itself is continuously evolving. In addition to the classic nuclear families of two parents with children, various other forms of family have taken shape, resulting in more varieties of family travel composition. The changing family structure must command our continuous attention. Furthermore, in the context of today's increasing digitalisation of all facets of life and consumers' tendency towards physical inactivity, the importance of family tourism as a mechanism for providing meaningful family time and enhancing family well-being must not be overlooked (Lehto & Lehto, 2019). The significance of family tourism research in this sense goes beyond business management relevancy into the realm of tourism's social responsibility.

Aim and approach

This study aims to (1) provide an overview of the family tourism research landscape – including a timeline, authorship, study subjects, research topicality, theories and methodological practices; and (2) discuss missing links and future research opportunities in family tourism.

The first step in our study was to conduct an extensive search of published research on family tourism. Several research databases were searched, including the ABI/INFORM Global, Business Source Complete, ScienceDirect, Hospitality and Tourism Complete and Google Scholar to identify relevant literature on this topic. The search terms included "family tourism", "family travel", "family trip", "family tourists/travellers/visitors", "family holiday(ing)", "family vacation", "family leisure", as well as a combination of words that include family, children, tourism, experience, management and marketing. Through our research, we identified a total of 159 articles, produced by 184 authors, published in a wide number of academic journals over a 45-year period (1975–2019). This collection of literature may not be exhaustive but we deemed it to be a reasonable data foundation to allow our analyses to be conducted, patterns to be derived and conclusions to be made.

The first step in our analysis was to perform content analysis of the 159 studies on family tourism using both quantitative and qualitative approaches. This included construction of a database with information categories pertaining to authorship, authorship affiliation, topic, theory, research method, sample origin, travel destination and publication year. Research themes were also identified using researcher judgement, which was then cross-validated through continuous discussions among the authors to address discrepancies.

An overview of family tourism research

Family tourism studies the dynamics of the family unit "on the road". A baseline question is what constitutes family tourism? Family tourism, family travel, family vacation, family holiday(ing) are a few terms seemingly used interchangeably to refer to this phenomenon. For this research, we chose to use the term "family tourism" as the overarching construct. There are some variations of definitions. For example, Schänzel et al. (2005) suggested that "family holiday" involves leisure travel away from home for more than one day undertaken by a family group with at least one child. This definition takes into account length of stay and the presence of children. The researchers later, however, moved away from these qualifying parameters, defining family tourism at a more conceptual level – referring to it as "a purposive time spent together, as a family group, doing activities different from normal routines that are fun but that may involve compromise and conflict at times". This later definition specifically conceptualises family holidays as possessing characteristics of "fun", "non-routine", "compromise" and/or "conflict" but did not specify mobility. How far does a family need to travel for their activity to be conceptualised as family holiday is unspecified? In that sense, family staycation can be a form of family tourism. The qualifier of activities being non-routine does not necessarily suggest that families must travel to environments outside a family's everyday living community. For this study, we therefore defined family tourism as a domain of research that focuses on travel that involves the family unit without including other qualifiers such as distance, etc.

Family tourism has received research attention as early as in the mid-1970s. Since then there has been a surge in academic interest in family tourism in the past five years as evidenced by the observation that almost half of the studies gathered were published between 2015 and 2019 ($N = 73$). In fact, the overwhelming majority of family tourism-related studies were produced from 2005 onward. This shows that research focusing on family tourism is a relatively recent phenomenon, although families have been travelling for much longer and family travel accounts for a major form of travel companionship. This trend is clearly evident from Figure 13.1 which presents a timeline of the annual frequency of publications from 1975 to 2019.

Researchers

Where are the family tourism researchers based? The geographical distribution of researchers' work institution affiliation spans seven regions of the world, namely, Africa, Arctic regions, Central and South America, Australia and New Zealand, Asia and South Pacific, USA and Canada, and Europe (Figure 13.2). When breaking down by countries, it seems there is a high concentration of family tourism researchers from the USA (29.35%) and UK (11.96%), followed by researchers from Australia (7.61%), New Zealand (7.07%) and China (6.52%). Researchers from the rest of the world account for about 37%, including researchers from Canada, Denmark, Spain, Malaysia, Iran, Ghana, Brazil, Italy, Iran, Ireland, Serbia, South Africa, Portugal, Hong Kong, Netherlands, Israel, Turkey, Finland, Croatia, Greece, Bosnia and Herzegovina, Thailand, Austria, Indonesia, Norway, Switzerland, Sweden and Mexico. When it comes to individual authors, a CiteSpace analysis shows that authorship networks in family tourism research appear to be highly concentrated, indicating that family tourism has yet to draw interest from a broader group of researchers (Figure 13.3).

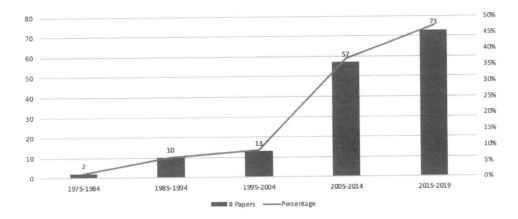

Figure 13.1 A timeline of family tourism publications (1975–2019).

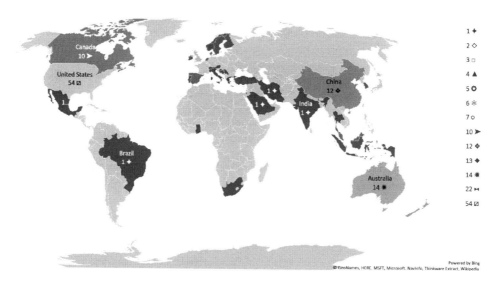

Figure 13.2   Geographic distribution of researcher affiliations.

Research subjects

Although not all study subjects in the reviewed articles had an articulated travel destination, for those who did, the top travelled destinations were the USA (25%), UK (12.5%), China (8.93%) and Malaysia (5.36%). These five destinations account for about 60% of all travel destinations to which the studied sample populations had taken trips. Where do the research samples come from? A closer examination of the study subjects' geographic origins shows that the family tourists being studied come from Central and South America, Australia and New Zealand, Asia, South Pacific, USA and Canada, and Europe, with no sample representation from Africa or Arctic Regions. Although there was a wide representation of study subject origins, the data show a highly concentrated geographic distribution when breaking the samples down by country (Table 13.1). More than half of the studied samples come from five countries: the USA, UK, China, Australia and New Zealand. Other more visible sampled visitor country origins include Denmark, Canada, Germany, Malaysia, the Netherlands and Spain.

Figure 13.4 presents a historical and regional view of the study subjects. Traditionally, it is apparent that there has been a persistent research interest in family visitors from Western cul-

Table 13.1 Top five traveller origins by country

Traveller country origin	Articles and % of total
USA	19 (17.59%)
UK	12 (11.11%)
China	10 (9.26%)
Australia	8 (7.41%)
New Zealand	8 (7.41%)
Others	51 (47.22%)

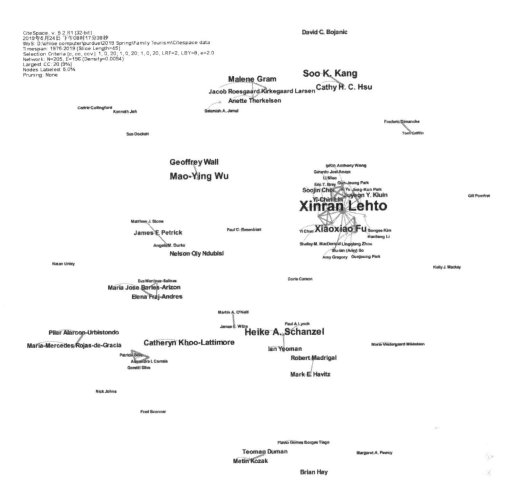

Figure 13.3 Authorship networks.

tures. However, researchers have paid much greater attention to family visitors from Asia from 2005 onwards. Much less research utilised samples from Central and South America. Additionally, family tourists from Africa and Arctic Regions have received scant attention from researchers.

Research methods, themes and theory lenses

Based on our analysis, family tourism researchers have embraced both qualitative (46%) and quantitative (46%) research methods in their investigations. About 8% of research in this field utilised a mixed method approach. The significant utilisation of various qualitative methods may reflect the complexity of understanding the contextual nuances of the family system in tourism and the challenges associated with uncovering the perspectives of multiple related members in the system and the views of (young) children in particular. When examining methodological trends from a historical angle, we note an interesting upward swing for the use of qualitative

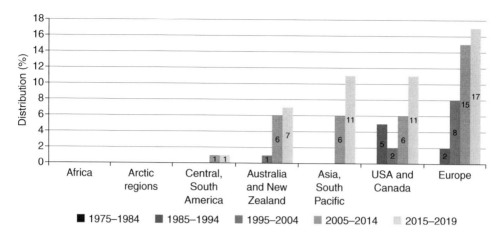

Figure 13.4 Regional distribution of travellers' origin based on periods.

approaches, and a downward swing in the applications of quantitative methods (Figure 13.5). In fact, during the period of 2015–2019, the number of studies following the qualitative tradition surpassed those of a quantitative nature.

Another interesting result is that our analysis shows that family tourism research largely addresses a few broad themes, namely, children, family role, destination activities, travel companionship and experiences. Children have attracted considerable attention in recent years as indicated in the total number of publications (23%). With regard to this theme, researchers have focused on topics such as decision roles, special needs children, young children, information search, learning, children as study subjects, views on travel and perceptions of specific types of travel experience. The second most popular theme was family role corresponding to 16% of the studies, followed by destination activities (14%), travel companionship (9%) and specific experience evaluations (6%). Other topics (23%) addressed include family function, travel benefits and motivation, and travel assistance.

A closer examination of research themes over time shows an evolving pattern (Figure 13.6). During the period 1975–1984, there were only two themes that the researchers focused on: adverse effects of family tourism and family role influences. The number of research topics

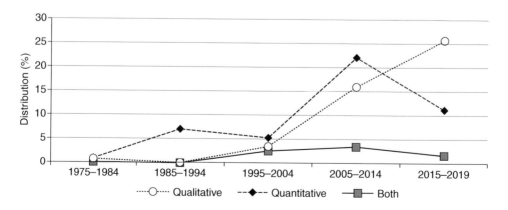

Figure 13.5 A longitudinal view of family research methodology.

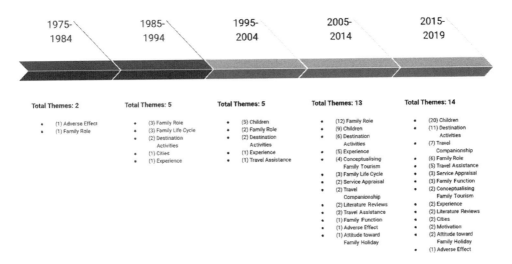

Figure 13.6   Timeline distribution of themes in family tourism research.

appears to broaden over the past five decades, with a narrow set of topics prior to 2005 and a much larger set of topics post 2005. This correlates with the much greater number of publications since 2015.

Another finding is that researchers have attempted to couch their research in various theoretical lenses from social psychology, psychology, sociology and family studies. Table 13.2

Table 13.2 Theoretical lenses

Theories	Examples of application
Pull–push factor theory	Shavanddasht (2018)
Experiential learning theory	Stone & Petrick (2017)
Associative network theory	Yang & Lau (2019)
Reminiscence bump	Tung & Ritchie (2011)
Consumer socialisation theory	Watne et al. (2014)
Ecocultural theory	Mactavish et al. (2007)
Family life cycle	Hong et al. (2005)
Social exchange	Singh & Nayak (2014)
Goffman's dramaturgical sociology	Gram et al. (2018)
Reversal theory	Pomfret (2018)
Justice theory	Park et al. (2008)
Leisure constraints theory	Agate et al. (2015)
Performance theory	Drenten (2018)
Spill over theory	Sthapit & Björk (2017)
Personal construct theory	Schänzel & Yeoman (2015a, 2015b)
Piaget's cognition development theory	Li et al. (2017)
Regulatory focus theory	Ram et al. (2014)
Resource theory	Kang & Hsu (2005)
Social judgement theory	Madrigal et al. (1992)
Social-emotional selectivity theory	Lee & Graefe (2010)
The Circumplex Model of Marital and Family Systems	Lehto et al. (2009)

represents a sample of theoretical concepts and theories utilised in family tourism research. For the most part, these theories have been used to serve as a theoretical backdrop, guide hypothesis development or support the validity of empirical discoveries.

Topicality of family tourism

Although the significance of family travel has always been highly valued due to both its market size and its importance for families, it must be recognised that compared to other forms of travel, family travel is inherently more complex and can come with challenges – from budgeting, to details of trip preparation, special service needs, family members varying interests and needs when on site in destinations and increased safety concerns to name just a few. How can research in family tourism inform us?

Family role influence

Despite the shifts in travel trends and changes in family structure, family as a tourism consumption unit remains as important as ever. As family represents a system, how family roles influence family travel elements have been a popular topic throughout the past 45 years of research. Since family members in the system have distinct household roles, their needs, expectations and interpretations of such consumptive scenarios can vary.

Family role influence on family vacation decision-making has been a classic topic, perhaps due to its direct tie to consumptive choices. Although most of the studies on family roles focus on how parents as a unit make vacation decisions, some have explored the differing characteristics and perspectives of wives versus husbands. It is widely acknowledged that joint decision-making is the most prevalent decision model applied by spouses (e.g. Hsu & Kang, 2003; Nanda et al., 2007; Nichols & Snepenger, 1988; Ritchie & Filiatrault, 1980; Rojas-de-Gracia & Alarcón-Urbistondo, 2017; Srnec et al., 2016). However, some decisions tend to be dominated by either partner. Rojas-de-Gracia and Alarcón-Urbistondo (2017) indicated that searching for information was normally conducted autonomously by either the wife or the husband. The study of Ritchie and Filiatrault (1980) showed that decisions concerning the length of the vacation, the amount of money to spend and the actual dates of the vacation were more husband-dominated.

When it comes to parenthood, Ram, Uriely and Malach-Pines (2014) revealed that parents showed less authoritative "parenting style" when they were on vacation with children. Parents can also experience role reversal in some situations where children can teach parents skills (e.g. recreational skills, online navigation skills) (e.g. Lehto et al., 2009; Yu et al., 2018). Mothers view family vacations as an opportunity to relax and release parental control. For example, Mottiar and Quinn's research (2012) indicating that self-catering holidays were an "escape" from the daily life for females, but women and mothers' genderised roles such as cleaning clothes and preparing food that were normally performed by female respondents at home are often maintained. Interestingly, from the father's perspective, Schänzel and Jenkins (2017) noted that sharing leisure-based holidays with their children were special times for fathers to have fun, to (re)build and maintain family relations, and to experience fatherhood. It was also noted that parents attach higher regard for family vacations, especially in their beneficial functions for the family, compared to the perspectives of children (Fu et al., 2014).

Family travel assistance

According to UNWTO (2019), among the more than one billion international travellers, a high percentage is composed of "families with young children and/or seniors, and persons with disabilities or other special needs". Travel is not equally easy for everyone. Families with special needs members in particular require specialised travel assistance and products. All of us will need accessible environments, products and services at some point of our lives due to conditions resulting from a disability, illness, pregnancy, or young or advanced age. Accessible tourism within the framework of family tourism bears particular relevance, simply because travellers with special needs usually travel with their families and caretakers. This is an important area of research, albeit it has seen limited attention from researchers (e.g. Yu et al., 2018; Mintz, 2018).

Aside from the more prominent financial unaffordability these families may face, mobility can be a prominent issue for family caregivers. Mactavish, Mackay, Iwasaki and Betteridge (2007), for example, investigated the meaning of life quality for family caregivers (i.e. biological and adoptive parents, and adults with siblings) of individuals with intellectual disability and noted that vacations have become an opportunity to escape caregiving responsibility. Such mobility is an essential respite for family caregivers. However, taking a vacation alone for respite time appears to be a luxury inconceivable to achieve for many caregivers, for financial reasons, time commitments and mobility constraints. For caretakers of a member with extended illness, hospital trips, constant private home care needs and other daily necessities nullify the possibility of travelling for restorative time. This disadvantaged traveller population has attracted attention from the media (e.g. Brown, 2019), but not nearly enough attention from family tourism researchers. Brown (2019) suggests that a vacation for a caregiver is a necessary form of self-care, which both can and needs to happen. He suggests that "local parks", "community festivals", "play tourist in your own city" (visit local sites and places that are not previously travelled to), and "taking photos" (of sights and sounds when running errands) are less glamorous but practical forms of travel for caregivers. Another means of travel is what Brown termed "armchair travel" with disabled family members. For example, online browsing and virtual travel to places are some of the suggested ways to take a journey in one's mind. As family members, care-receivers have significant meaning and also play important roles in the entire family. Their needs should not be ignored. The importance of relationality between the care-receivers and family caregivers in family mobility is another topic that should be further studied from sociological and psychological perspectives.

When families that have members who possess one or more forms of disabilities do travel, travel can sometimes be more of a challenge than a positive experience. A major barrier is that it is more difficult for caregivers to follow their ordinary caregiving routines during a journey. The result is that the affected families may travel less frequently and staying at home often becomes the most reassuring solution. The complexity of the issue is revealed by recent research focusing on families of children with autism (Sedgley et al., 2017). Sedgley discusses how tourism can provide wellbeing benefits to individuals with developmental difficulties such as children with autism and revealed the picture of such families on vacation through their lived experiences. Their research describes such a vacation as a journey of mixed emotions for the mother – bringing attention to the stressful caregiving role of mothers and their emotional labour during such a holiday.

Further perspectives on this issue are provided by multiple research studies that have consistently noted that families' travel motivations are multi-faceted. Important elements include family bonding, family socialising, parent relaxation and escape, as well as building children's intellectual competence and physical competency mastery via travel. Given the fact that children with disabilities in general require constant adult companionship, it makes this family population different from families with typically developing children. For this population, the parents' well-being is much more closely tied to the children. Whereas the typically developing families appear to have family, self and children-oriented motivations, this travel group is centred much more on the children.

Individuals with impairment usually travel with caregivers and these individuals need special travel assistance. Hence, developing and maturing facilities catering for different special needs is an area that deserves attention for both researchers and industry practitioners (Yau et al., 2004). The needs of the care-receivers were illustrated by Mintz (2018). Mintz's study depicts how immature designs and facilities for this travelling population are insufficient. Using Disney as a case in point, insufficient stations in Disney Parks for recharging motorised wheelchairs created inconvenience to groups with special needs, Mintz (2018) argued that the standard of equality should not be measured as simple equality which focuses on equal waiting time for everyone. Instead it should be measured in terms of whether an equal access to opportunities is provided that allows tourists with disabilities to enjoy the attractions and their Disney experience as well as the non-disabled tourists. Ethically, the question of how to measure and provide equality for families with special care needs has offered public policymakers and tourism service providers an important perspective on how to improve public wellbeing. Another is on how best to facilitate travellers with babies, toddlers and small children? This is an area that has received very limited attention from researchers.

Our review shows that some of the special needs-related literature has focused on families with infants and young children. Whittle (2019), for example, explored the impact of sling use on the travel experience of families with babies and young children. This study suggests that connecting with young babies makes parents feel a sense of wholeness but long-distance travelling with high requirements for individuals' walkability would be a tough task for young children. As a form of assistance, slings to some extent made it more convenient for parents to take young babies along with them and improved the whole experience of parents during the family journey.

Children, children

Families will continue to embrace travelling with children going forward as travel continues to be regarded as a unique opportunity for strengthening family togetherness and bonding. For example, close to a quarter of US family vacationers travel internationally with their children. Children have received increasing attention from researchers. Children from various age groups are a visible element in family tourism research – infants, toddlers, pre-teens, teenagers and adult children, albeit research in this area is still rather insufficient. Topics related to children in family tourism research are diversifying from the classic decision-making angle. In addition to children's role in the decision-making process, other popular topics include children's special needs, communication and negotiation with parents, learning processes and outcomes during and after a journey, and perception of travel experience from children's perspectives.

Children in decision-making: Children's role in family vacation decision-making is a prominent area of interest (Gram, 2007). It is noted that children in general have gained an increasing influence in the travel decision-making processes of families. Although parents are known to have the decisive votes at final decision stages, children's preferences and wishes, especially those of the teenagers, are taken into high consideration by parents (e.g. Curtale, 2018; Dunne, 1999; Gram, 2007; Thornton et al., 1997). The study of Thornton et al. (1997) shows that the physical needs of children (e.g. arrangement of mealtimes, need for sleep) and their ability to negotiate with parents have great influence on the behaviours of family tourists, and that their influence power and methods vary by age phases.

Children's age plays a role in that their input is being considered incrementally more as they progress into adolescence and become more experienced. Gram's (2007) research, for example, notes that in the family holiday process, children can play the roles of "pestering" (e.g. I want to go to Legoland!), "blackmailing" (e.g. if we are not going to Italy, I am not coming!), "leveraging" peer power (e.g. my friends are going there) and "negotiating" (e.g. what to do or where to eat when on site at a destination). Another interesting observation in Gram's research is that the data suggest that German children are perceived as having more power than their counterparts from Danish families, showcasing the need for cross-cultural understanding of family role influence.

Family decisions are normally not made at one point in time but are an ongoing process with constant negotiating over the course of the holiday. Decision-making involving children is also characterised by occasional conflicts and compromises on the part of family members to ensure a harmonious family holiday. Conflict resolution strategies seem to be correlated to the age phases of children and different communication patterns between children and parents. Therkelsen (2010), for example, showed that whereas socio-oriented communication patterns where parents control and monitor their children's consumption behaviour may be prevalent in families with young children, as the children grow older, these seem to be supplemented by more concept-oriented communication patterns where children are to form their own opinions on consumption issues through discussion and weighing alternatives and thus children's consumer skills and competences are developed. Singh and Nayak (2016) also argued that adolescents express more in concept-oriented families and use more various types of resolution strategies compared with socio-oriented families.

Child-centric services and programmes: These have gained recent attention from family tourism researchers. A number of studies in this area represent market-driven attempts such as analysing parents' views of children's programmes and amenities in accommodation (e.g. Agate et al., 2015; Birchler, 2012; Gaines et al., 2004; Khoo-Lattimore et al., 2018; Kowisuth, 2015). Studies have noted that security and hygiene are the most important elements in the selection of programmes in hotels. Child-friendly interactions between children and hotel employees are another important factor for parents (e.g. Birchler, 2012; Gaines et al., 2004; Khoo-Lattimore et al., 2018; Kowisuth, 2015). Special amenities including bassinets, bottle warmers, stationary for children to write and draw were suggested to be offered in family-oriented and child-friendly accommodations (Khoo-Lattimore et al., 2018). A number of studies look into summer camp programmes for children and youth. Dresner and Gill (1994) note that summer nature camps benefit children's interpersonal relationships, feelings of connection with the natural world and, importantly, campers' self-esteem. Paris (2001; 2008) chronicles the history of the American summer camps and highlights the importance of summer camps in shaping children's development. From the lens of cultural comparison, research (e.g. Khoo-Lattimore et al., 2018; Lehto et al., 2017a) has also noted that while sharing similar perceptions with their Western counterparts about the role of summer camps for children, Chinese parents have unique preferences and motives when it comes to summer overseas camps for children. Khoo-Lattimore et al. (2018) suggest that the parents' preferences for activity programming can vary from culture to culture. Western families tend to prefer their children to participate in physical activities related to the children's interests such as fishing, swimming, dancing and playing games, whereas Asian parents tend to seek more explicit learning-oriented activities and programmes. Lehto et al. (2017a) note that Chinese parents place particular high value on a camp programme that facilitates visits to well-known universities – a pragmatic element that perhaps reflects Chinese parents' tendency to link overseas summer camps with potential future college choices for their children. These differing preferences may be explainable by the Confucian values of travel as a means for education and acquisition of knowledge (Lehto et al., 2017b) versus the Western travel value association with hedonic pleasure, creativity, freedom and excitement (Khoo-Lattimore et al., 2018).

Crying babies on board: Children have their own special needs and behave differently from adults. Small children, for example, may not have the ability to self-regulate or express emotions and feelings. By investigating public online news sites and discussion boards about the debate over crying babies on planes, Small and Harris (2014) suggest that service providers (in their study, airlines) should engage in "customer compatibility management". These researchers advocate for targeting homogenous consumers and managing customer-to-customer inter-actions. Their study has led to thoughtful debates in family tourism on such issues as social inclu-sion/exclusion and justice issues referring to the rights and responsibilities of different groups of passengers including babies.

Children's own voice: As future tourists, children's experiences from their own perspectives also draw great academic interest (e.g. Cullingford, 1995; Wu et al., 2019; Rhoden et al., 2016; Fu et al., 2014; Hay, 2017; Drenten, 2018). Wu et al. (2019) suggest that a memorable family travel experience to children was centred upon family togetherness and physical activities. These observations coincide with what is noted by Rhoden et al. (2016) which revealed that children prefer to be physically active. Nonetheless, although children enjoy the togetherness with parents that help build closer emotional bonds, they also recognise the opportunities to connect with their siblings and to socialise with children from outside the family group (Hay, 2017). Research also shows that adult destination preferences and choices can be influenced signifi-cantly by their childhood travel experiences. Cullingford (1995) notes that children are gullible to different sources of information. This tendency can lead them to imbibe stereotypes and prejudices toward certain countries, places or people. Overall, however, the meaning of family travel for children is underrepresented (Carr, 2011).

Learning: Learning is another emphasised topic in children-related research of family tourism. Travel is seen as an excellent educational tool, especially for museums and heritage attractions – they are regarded as natural play-based learning environments for children (Dockett et al., 2011; Frost & Laing, 2017; Sanford, 2010; Yang & Lau, 2019). Yang and Lau (2019) for example, investigated the influence of motivation for children's experiential learning on engage-ment at World Heritage Sites, and the subsequent influence on children's learning outcomes and educational benefits. Interestingly, they found that the brand awareness of World Heritage Sites has impacts on learning effectiveness. Well-known World Heritage Sites bring more effective experiential learning for children, compared with less well-known World Heritage Sites. In the study of Lehto et al. (2017a, 2017b), experiential learning for children is identified as a specific travel benefit sought by Chinese parents. Expanding children's horizons, extending children's knowledge and their learning about culture, history and people are some of the spe-cific aspects that parents sought to accomplish through family travel. Parents' desire for impart-ing children's learning orientation can vary by culture in its explicitness and magnitude, as noted by Khoo-Lattimore et al. (2018). In the case of Chinese family travellers it seems that experien-tial learning needs of children take precedence over family unit needs (togetherness and shared experiences) and those of parental individual oriented needs (such as relaxation). Parents extend teaching and learning to the travel setting in a very explicit way.

Families in motion: consumptive experiences

Family travellers are unique in that they travel as a small group dynamic but the group membership in the dynamic is permanent – long term as opposed to that of other group travellers where their social interaction and membership are liminal in nature. How do families interact with their travel environments and settings and yet interact among themselves? What are the shared aspects of experiences – how does the "I" and "me time" interface with the "We" and "Us" within the family when experiencing places and consuming products? What is the co-created experience like? What do family travellers do? How do they engage themselves with activities, programmes, attraction sites and sounds? How do they engage each other while visiting a destination? How do they interact with the environment they are in? These elements of family tourism should be a core focus as what they engage themselves in onsite is the core element of a family tourism experience. This is why families travel to certain places. Specific types of attractions, products and experiences represent important research contexts in family tourism. The existing literature has examined various attractions that include nature-based special interest tourism with family members (e.g. camping tourism), activity-related family tourism (e.g. skiing tourism), travelling and attending special events with family members (e.g. weddings, conferences) and attractions targeted specifically at families (e.g. theme parks and children's museums).

Research on nature-based family tourism spans the topics related to family experience (e.g. Goodenough et al., 2015; Mikkelsen & Blichfeldt, 2015), motivation and satisfaction (e.g. Lee & Graefe, 2010) and destination choice (e.g. Mikulić et al., 2017). Mikkelsen and Blichfeldt's (2015) research assessed the family travel experience by examining how caravanning could provide family members with balance between interdependence and autonomy. Caravanning seems to offer "real, quality family time" where adults are allowed more "own time" and "spouse time", and children also have extraordinary opportunities to engage in "own time". Goodenough et al. (2015) focused on the aspect of family togetherness and noted that leisure programmes based on woodland activities seemed to alleviate parental pressure derived from their children's disconnection from nature and engender parents' self-confidence as "competent parents" during the process of guiding their children to accomplish outdoor activities. Lee and Graefe (2010) explored the relationship between demographic characteristics and the motivation for family recreation in a natural tourism destination and found that married visitors and visitors with children were more motivated and the consequent satisfaction was higher for families having children. Mikulić et al. (2017) uncovered that infrastructure-related campsite attributes (e.g. sanitary facilities, water and electricity), safety and ecological standards were recognised as the most important attributes for both campsite choice and the onsite experience and suggested these campsite attributes be focused upon as basic requirements for camping businesses to build a strong and sustainable competitive position.

Other studies include the work of Schänzel, Hull and Velvin (2017) who identified the effects of global factors, including climate change, environmental/cultural conflict, energy and economic shocks, increasing competition with sun destinations, changes in demography and changing consumers, that impact future scenarios for families engaged in skiing tourism. As a special travel type, the accompanying partners' conference travel increases the intimacy in the couple's relationship and allows opportunities for conferees to communicate with their spouses both emotionally and intellectually and thus help facilitate conferees' serendipity and their pursuit of personal interests (Yoo et al., 2016).

Motivational factors seem to be a study focus in theme park research for family tourism (e.g. Bakir & Baxter, 2011; Johns & Gyimóthy, 2003; McClung, 1991). McClung (1991) unveiled that the most important factors influencing visitors' park attendance included climate, children's desire to attend and cost, and suggested theme parks direct their marketing activities to capitalise on children's influence as a primary factor for attracting more visitors.

Clearly destination activity – what families do at a travel destination and how they create experiences together – is an area of research that has drawn attention. Activities together as a family have been noted to promote family cohesion – the affection, support, helpfulness and caring among family members (Lehto et al., 2012). Lehto et al.'s research identified a seven-factor structure of family destination activities, covering a wide spectrum of leisure activities that families enjoy on a trip. These seven factors are city interests (e.g. museums, zoos/aquariums, city sightseeing), active nature pursuits (hiking, camping), shopping (e.g. for books, toys, arts and crafts), dining and entertaining (e.g. fine dining, local specialties and delicacies), family social events (e.g. VFR, festivals, sports events), outdoor sports (e.g. swimming, skiing) and farm-based activities (e.g. farmers market, visiting farms). In the context of families travelling with children with disabilities, Kim and Lehto (2012) identified a set of activities that are specifically relevant to this group of family travellers. Although some preferred activities are similar with the activities of the typical family travellers, their study also noted that wellness activities such as yoga, health spa and wellness classes are of particular interest to parents travelling with children with disabilities, especially for the parents. This again suggests a desire for respite experiences for such parents.

In the context of family reunion travel, it is noted that family reunion travellers seem to participate in both recreational/leisure activities as well as a special set of reunion specific activities such as board/table games, family photo-sharing, extended family photo-taking, family videos, family talent shows, making family trees, family history research and visits to sites significant to family history (Kluin & Lehto, 2012).

Beyond the nuclear family

Researchers have explored family scenarios beyond the traditional nuclear family composition. Travelling with extended family members, for example, has received academic attention. Research related to extended family tourism covers travels with various family compositions. Examples include: (1) Extended family travel/multi-generational travel (e.g. Kluin & Lehto, 2012; Pearlman, 2018; Yun & Lehto, 2009); (2) visiting friends and relatives (VFR) (e.g. Lehto et al., 2001; Backer, 2012; Griffin & Dimanche, 2017; Kashiwagi et al., 2018; Stepchenkova et al., 2015); (3) grandparents–children travel (Grandtravel) (Shavanddasht, 2018); (4) aunts–children travel (professional aunt no kids – PANKs) (e.g. Camargo & Tamez, 2015); and (5) LGBT family (e.g. Lucena et al., 2015; Trussell et al., 2015).

Travel motivation and activities are common topical areas threading throughout research on these various forms of travel that consider travel member composition. Take family reunion travel, for example. This topic has been researched as a sub-segment of multiple generational travel, referring to "gatherings of multiple family units composed of at least three generations on a recurring basis, not for special events such as weddings or funerals" (Yun and Lehto, 2009). Such travel is seen as a form of family ritual and understood in the tourism context as a form of interactional mobilisation to sustain family structure. From this perspective, Kluin and Lehto (2012) developed a four-factor family reunion travel motivational scale that includes: (1) maintaining family history and togetherness; (2) maintaining immediate family cohesion; (3) family communication; and (4) family adaptation. The order of importance of these four factors appears to follow a reverse pyramid structure with extended family needs outweighing those of immediate family needs and personal/individual needs.

VFR is another extended family-related area of research interest. The VFR travel market has received research attention from the late 1990s, as destinations tap this segment for growth strategy, as VFR is seen to be an economic contributor for communities and is less vulnerable to market forces such as recessions (Backer, 2012). Kashiwagi et al. (2018) suggested that international students are important resources for tourism practitioners for they can trigger international VFR visits by their friends and relatives.

Grandtravel is another area of focus due to its increase in popularity in recent years for reasons related to increased health, wealth, desire for more shared time with grandparents. Shavanddasht (2018) further segments Grandtravel into three clusters including "family lovers" who enjoyed spending time with their grandkids and emphasised the basic values of family togetherness; "multi-purpose seekers" who expected "quality", "enjoyment"; and "historical/natural attractions" and "knowledge hunters" who preferred to enhance their knowledge by visiting new places.

Minority configurations of family that do not fit the heteronormative model have also received recent albeit limited attention from the family tourism literature. A recent paper by Lucena, Jarvis, and Weeden (2015), for example, called for research to understand travel motivation, decision strategy and destination choices of lesbian and gay parented families. They noted a need for research to understand how same-sex families navigate their sexuality while on holiday and whether their social interaction can be impaired in heteronormative travel spaces.

Discussion: research opportunities

When individuals travel, they are likely to travel with family. This is the case not only with leisure travellers but can be true for *bleisure* travellers (e.g. conference travellers) as they increasingly attempt to balance work with family time and leisure. This trend will continue as families are increasingly looking for opportunities for quality family time. Millennials, for example, do not want to wait until their children are older; they travel with children under the age of five (Resonance, 2018). Given the prevalence of family travel, the importance of research of families in tourism cannot be overstated. This research provides a delineation of the existing family tourism literature. A look back at the research landscape of this area can serve not only as a synopsis of what it has become, but also as an invitation for researchers to actively partake in building a futurist research agenda for family tourism.

While our analysis reveals an uptick in research effort in this area, especially the studies over the past decade, we have noted that research in this area is insufficient both in its scope and depth. Given that family travellers are front and centre in the traveller landscape, this topic deserves much more research attention than it has been given. Our study shows that research in family tourism is highly centred on family travellers from a very small number of countries and is conducted by highly concentrated networks of researchers. Much more research is needed especially in the context of developing nations. Our analysis suggests that opportunities abound for researchers.

Supplier-side topics

As the existing studies on family tourism have a distinct consumer orientation, data from the supply side are sorely needed. How do family tourism suppliers provide consumers with family-oriented or family-friendly products? What are the design considerations? For example, it is noted that ridesharing platforms (e.g. Uber, Lyft) have begun to work on providing customers with the list of family-friendly vehicles that offer car seats to infants/toddlers/young children. It would be interesting to examine how the sharing economy as a whole systematically integrates family-friendly elements within the P2P business model. There is an increasingly urgent call for the hospitality and tourism industries to use wellness parameters to design experiences and services and therefore play an increased role in safeguarding consumer wellness (Lehto & Lehto, 2019). This raises questions such as how well have family wellness needs been factored into tourism products, experiences and services? How well do they perform? How do services cater to family travellers with special needs such as the very old, the very young, the sick, the pregnant or the disabled? How has the principle of universal accessibility been implemented? Accessibility and equitable participation by all are three principles advocated by UNWTO as quality and competitiveness measures for tourism destinations (UNWTO, 2019). Research on accessible environments, products and services are called for. Another supply-side topic that has seen little attention is ChildSafe tourism (UNICEF, 2018). The exploitation of children as part of tourism is an area of concern. How have policies, standards and procedures been implemented to protect children's physical and emotional wellbeing and prevent children from being exploited as part of tourism offerings?

There is also a need to understand marketing in family tourism. How do destinations market to family travellers? How do they engage family travellers, especially on the newer platforms and forms of communication such as social media and other online social networks and applications? How effective are they? Most of the current advocating voices for family tourism appear to have come from online travel magazines, travel bloggers and social influencers. A recent phenomenon is the multitude of ratings and recommendations by these family travel influencers (e.g. "Top 10 family-friendly places you must see") – how valid are their recommendations? What evaluative criteria do they use? Or are they simply the new form of advertising?

Demand-side topics

Opportunities also exist in the intersection between family tourism and digital technology. An example would be "how is digital technology changing the classic family road trip experience?" Dating back to the1960s, road trips with children have been a common travel experience for American families. "Dad at the wheel, Mom reading the map, siblings playing games in the back seat" (Ratay, 2018) is a nostalgic memory of the baby boomers. In contrast, in the digital era, Generation X and Millennials have grown to expect shorter vacations and more instant gratification (Rugh, 2008). Devices like audiobooks/iPads/smart phones have become a constant entertainment companion for family road trips. While these contemporary technology devices help children psychologically shorten the travel distance and manage boredom, how has this digital element of a family trip altered the family tourism experience itself? Smartphones and tablet-based digital applications have become indispensable items for contemporary family travellers. These digital platforms are now used by some parents as pacifiers for car/bus/train/airplane rides, long waits and meals out. Increasingly, digital applications are designed to assist family vacation experiences – from information delivery, social interactions and interactions with the environments surrounding them, to experiencing digitalised attractions (e.g. virtual reality and augmented reality). How do these family-friendly digital technologies affect family travellers? How do they influence family travellers' decision-making processes? How do they influence their shared vacation experiences? How do they impact family functioning on the road? How do they affect children's experiences and their wellbeing? Are screen time and digital contents healthy for family travellers? How much might be too much?

Despite the recent research interest in understanding family tourism experiences, the work that has been done is far from sufficient. There is a great need for more research in this area. For example, how do family travellers interface with specific types of tourism destinations (e.g. cities), tourism sectors (e.g. cruises, aquariums, resorts), attractions and programmes (e.g. events, festivals, museums), and services and facilities (e.g. restaurants, airlines, shops and lodging facilities). Such issues are especially important to urban destinations that rely heavily on family travellers. However, there has been very little research as to how families consume a travel product such as streetscape, museums, architecture, musicals and restaurants.

To address these issues a children's lens is particularly needed. There has been increased effort in examining children's perspectives. However, children's evaluations of existing family-friendly tourism products and services need to be much better understood. A better understanding of children's perspectives may be able to spur the tourism and hospitality industries to be more creative and innovative in their experience offerings. The benefits of travel to children is another understudied area. Although family leisure time and togetherness are positively correlated with childhood socialisation and development, little research has been dedicated to empirically understanding whether travel inherently benefits children (Durko & Petrick, 2016).

Another issue is that current research provides insufficient insights into how families co-consume tourism products and co-create experiences – not enough attention is paid to the shared consumption of specific types of attractions and experiences. There is a need to consider the family from an interactive point of view, not just from a child's lens or the mother's lens. Although the latter lenses are important, where is the relational lens in the picture? How are children related to the adults and how do they interact with the adults and the setting/environment?

Additionally, how do families interface with sustainability? How do they make choices and behave based on sustainability principles? Further, a better cultural lens is much needed. For example, how do cultural dynamics interact with family dynamics? How do cultural values influence family travel behaviour and consumptive experience? For example, how does cultural distance between the travellers' home origin and a destination influence the family consumptive experience? How does cultural distance influence family travellers in their decision-making, trip behaviour and experiences?

Methodological opportunities

Methodologically, much more work is needed that considers the fact that a family represents a relational group dynamic. Research involving multiple families' members is necessary because it can better advance our understanding of family tourism behaviour, experiences, perspectives and emotions beyond perspectives of individuals alone. Despite the call for data from multiple member perspectives, there has been underuse of this approach for data and analytical approaches.

Special methodological considerations are needed for acquiring and appropriately analysing data from dyads and families because family data are interdependent and require analytical techniques that can accommodate this non-independency. Data collection from children remains particularly challenging despite efforts using innovative methods to understand children's perspectives (Poria & Timothy, 2014).

Additionally, longitudinal research designs appear to be non-existent. It would be of great value for such research design if researchers attempted to understand how family travellers and their behaviours change over time. For example, how does family travel correlate with children's development? Future travel styles? Career choices? A longitudinal lens would be appropriate for such investigations.

Conclusion

Despite an increased body of research, family tourism is still an under-researched area. This chapter sought to establish a better understanding of this domain of research. We conducted a systematic search of family tourism literature using key words including family vacation and family holiday. We identified 159 articles, produced by 184 authors and published over a five-decade period. Based on a meticulous content analysis of these studies, we developed an overview of family tourism as a field, what it has become in theoretical development, topicality and themes, and methodological advances. We suggest ways to move the family research field from a descriptive delineation to a more mature stage built upon theory and methodological knowledge.

Our analysis of the existing research indicates that while family tourism has gained a healthy dose of attention from researchers, it is far from being studied sufficiently. There is so much yet to be systematically examined. In a nutshell, future research in family tourism calls for much finer views of segments within the family travel market, finer views of impacts of family travel on the family dynamic, especially on the children, finer views of travel moments and experiences, and finer views of service providers.

Family tourism research does not bear ramifications only for the practice and destination management. It adds to our understanding and generates insights for this baseline societal unit in the travel space. How travellers function as a family on the road completes and compliments the broader literature of family studies. The sustained and growing demand for family tourism calls for a more energised focus on research in this area.

References

Agate, J. R., Agate, S. T., & Birchler, K. (2015). A vacation within a vacation: children's day programs and parental satisfaction. Tourism Culture & Communication, 15(1), 21–32. https://doi-org.ezproxy.lib.purdue.edu/10.3727/109830415X14339495039379.

Backer, E. (2012). VFR travel: why marketing to aunt Betty matters. In E. Backer, H. Schänzel, & I. Yeoman (Eds.). Family Tourism: Multidisciplinary Perspectives. Bristol: Channel View Publications, 81–92.

Backer, E., & Schänzel, H. (2013). Family holidays: vacation or obli-cation? Tourism Recreation Research, 38(2), 159–173.

Bakir, A., & Baxter, S. (2011). "Touristic fun": motivational factors for visiting Legoland Windsor Theme Park. Journal of Hospitality Marketing & Management, 20(3–4), 407–424.

Birchler, K. M. (2012). A vacation within a vacation: An examination of how child participation in day programs during family vacations influences parental satisfaction with the vacation. Thesis. Paper 848.

Brown, F. (2019). "Travel" still possible for caregivers. channel 3000, April 29. Retrieved from www.channel3000.com/travel-still-possible-for-caregivers.

Camargo, B. F., & Tamez, M. (2015). Professional aunt, no kids: an unexplored segment of family tourism. E-Review of Tourism Research, 12(3–4), 161–171.

Carr, N. (2011). Children's and Families' Holiday Experiences. London: Routledge.

Cullingford, C. (1995). Children's attitudes to holidays overseas. Tourism Management, 16(2), 121–127.

Curtale, R. (2018). Analyzing children's impact on parents' tourist choices. Young Consumers, 19(2), 172–184.

Dockett, S., Main, S., & Kelly, L. (2011). Consulting young children: experiences from a museum. Visitor Studies, 14(1), 13–33.

Drenten, J. (2018). When kids are the last to know: embodied tensions in surprising children with family vacations. Young Consumers, 19(2), 199–217.

Dresner, M., & Gill, M. (1994). Environmental education at summer nature camp. The Journal of Environmental Education, 25(3), 35–41.

Dunne, M. (1999). The role and influence of children in family holiday decision making. International Journal of Advertising and Marketing to Children, 1(3), 181–191.

Durko, A. M., & Petrick, J. F. (2016). Family and relationship benefits of travel experiences: a literature review. Journal of Travel Research, 52(6), 720–730. https://doi.org/10.1177/0047287513496478.

Frost, W., & Laing, J. (2017). Children, families and heritage. Journal of Heritage Tourism, 12(1), 1–6.

Fu, X., Lehto, X., & Park, O. (2014). What does vacation do to our family? Contrasting the perspectives of parents and children. Journal of Travel & Tourism Marketing, 31(4), 461–475.

Gaines, B., Hubbard, S., Witte, J., & O'Neill, M. (2004). An analysis of children's programs in the hotel and resort industry market segment. International Journal of Hospitality & Tourism Administration, 5(4), 85–99.

Goodenough, A., Waite, S., & Bartlett, J. (2015). Families in the forest: guilt trips, bonding moments and potential springboards. Annals of Leisure Research, 18(3), 1–20.

Gram, M. (2007). Children as co-decision makers in the family? The case of family holidays. Young Consumers, 8(1), 19–28.

Gram, M., Therkelsen, A., & Larsen, J. (2018). Family bliss or blitz? Parents' and children's mixed emotions towards family holidays. Young Consumers, 19(2), 185–198.

Griffin, T., & Dimanche, F. (2017). Urban tourism: the growing role of VFR and immigration. Journal of Tourism Futures, 3(2), 103–113. doi:10.1108/jtf-10-2016-0036.

Hay, B. (2017). Missing voices: Australian children's insights and perceptions of family holidays. Hospitality & Society, 7(2), 133–155. https://doi.org/10.1386/hosp.7.2.133_1.

Hong, G., Fan, J., Palmer, L., & Bhargava, V. (2005). Leisure travel expenditure patterns by family life cycle stages. Journal of Travel & Tourism Marketing, 18(2), 15–30.

Hsu, C. H. C., & Kang, S. K. (2003). Profiling Asian and Western family independent travelers (fits): an exploratory study. Asia Pacific Journal of Tourism Research, 8(1), 58. https://doi-org.ezproxy.lib.purdue.edu/10.1080/10941660308725456.

Johns, N., & Gyimóthy, S. (2003). Postmodern family tourism at Legoland. Scandinavian Journal of Hospitality and Tourism, 3(1), 3–23.

Kang, S. & Hsu, C. (2005). Dyadic consensus on family vacation destination selection. Tourism Management, 26(4), 571–582.

Kashiwagi, S., Nagai, H., & Furutani, T. (2018). VFR travel generated by international students: the case of Japanese students in Australia. Turizam, 66(1), 89–103.

Khoo-Lattimore, C., del Chiappa, G., & Yang, M. J. (2018). A family for the holidays: delineating the hospitality needs of European parents with young children. Young Consumers, 19(2), 159–171. doi:10.1108/yc-08-2017-00730.

Kim, S., & Lehto X. Y. (2012). Leisure travel of families of children with disabilities: motivation and activities. Tourism Management, 37, 13–24.

Kluin, J. & Lehto, X. (2012). Measuring family reunion travel motivations. Annals of Tourism Research, 39(2), 820–841.

Kowisuth, P. (2015). Factors influence the hotel selection of tourists travelling with children to Phuket. Thesis submitted to Price of Songkla University.

Lee, B., & Graefe, A. (2010). Promotion of family recreation for a nature-based destination. Journal of China Tourism Research, 6(1), 50–60.

Lehto, X. Y., Choi, S., Lin, Y.-C., & MacDermid, S. (2009). Vacation and family functioning. Annals of Tourism Research, 36(3), 459–479.

Lehto, X. Y., Fu, X., Kirillova, K., & Bi, C. (2017a). What do parents look for in an overseas youth summer camp? Perspectives of Chinese parents. Journal of China Tourism Research, 1–22.

Lehto, X. Y., Fu, X., Li, H., & Zhou, L. (2017b). Vacation benefits and activities: understanding Chinese family travelers. Journal of Hospitality and Tourism Research, 41(3), 301–328.

Lehto, X. Y., & Lehto, M. R. (2019). Vacation as a public health resource: towards a wellness-centered tourism design approach. Journal of Hospitality and Tourism Research, 43 (7), 935–960.

Lehto, X., Lin, Y., Chen, Y., & Choi, S. (2012). Family vacation activities and family cohesion. Journal of Travel & Tourism Marketing, 29(8), 835–850.

Lehto, X. Y., Morrison, A. M., & O'Leary, J. T. (2001). Does the visiting friends and relatives' typology make a difference? A study of the international VFR market to the United States. Journal of Travel Research, 40(2), 201–212.

Li, M., Wang, D., Xu, W., & Mao, Z. (2017). Motivation for family vacations with young children: anecdotes from the Internet. Journal of Travel & Tourism Marketing, 34(8), 1047–1057. https://doi-org.ezproxy.lib.purdue.edu/10.1080/10548408.2016.1276007.

Lucena, R., Jarvis, N., & Weeden, C. (2015). A review of gay and lesbian parented families' travel motivations and destination choices: gaps in research and future directions. Annals of Leisure Research, 18(2), 1–18.

McClung, G. (1991). Theme park selection: factors influencing attendance. Tourism Management, 12(2), 132–140.

Mactavish, J. B., Mackay, K. J., Iwasaki, Y., & Betteridge, D. (2007). Family caregivers of individuals with intellectual disability: perspectives on life quality and the role of vacations. Journal of Leisure Research, 39(1), 127–155.

Madrigal, R., Havitz, M., & Howard, D. (1992). Married couples' involvement with family vacations. Leisure Sciences, 14(4), 287–301.

Mikkelsen, M., & Blichfeldt, B. S. (2015). "We have not seen the kids for hours": the case of family holidays and free-range children. Annals of Leisure Research, 18(2), 1–20.

Mikulić, J., Prebežac, D., Šerić, M., & Krešić, D. (2017). Campsite choice and the camping tourism experience: investigating decisive campsite attributes using relevance-determinance analysis. Tourism Management, 59, 226–233.

Minnaert, L. (2017). U.S. family travel survey 2017. NYU School of Professional Studies and Family Travel Association.

Mintz, K. (2018). Is Disney disabling? Disability & Society, 33(8), 1366–1371.

Morrison, A. M., Lehto, X. Y., & Day, J. (2018). The Tourism System. Dubuque, IA: Kendall Hunt Publishing.

Mottiar, Z., & Quinn, D. (2012). Is a self-catering holiday with the family really a holiday for mothers? Examining the balance of household responsibilities while on holiday from a female perspective. Hospitality & Society, 2(2), 197–214. https://doi.org/10.1386/hosp.2.2.197_1.

Nanda, D., Hu, C., & Bai, B. (2007). Exploring family roles in purchasing decisions during vacation planning. Journal of Travel & Tourism Marketing, 20(3–4), 107–125.

Nichols, C. M., & Snepenger, D. J. (1988). Family decision making and tourism behavior and attitudes. Journal of Travel Research, 26(4), 2–6. https://doi.org/10.1177/004728758802600401.

Paris, L. (2001). The adventures of Peanut and Bo: summer camps and early-twentieth-century American girlhood. Journal of Women's History, 12(4), 47–76.

Paris, L. (2008). Children's Nature: The Rise of the American Summer Camp. New York: New York University Press.

Park, O.-J., Lehto, X., & Park, J.-K. (2008). Service failures and complaints in the family travel market: a justice dimension approach. Journal of Services Marketing, 22(7), 520–532.

Pearlman, D. (2018). Characteristics of family reunion travelers. Journal of Convention & Event Tourism, 19(2), 99–119.

Pomfret, G. (2018). Conceptualising family adventure tourist motivations, experiences and benefits. Revista Turismo & Desenvolvimento (RT&D)/Journal of Tourism & Development, (27/28), 463–465. Retrieved from http://search.ebscohost.com.ezproxy.lib.purdue.edu/login.aspx?direct=true&db=hjh&AN=130540933&site=ehost-live.

Poria, Y., & Timothy, D. J. (2014). Where are the children in tourism research? Annals of Tourism Research, 47, 93–95.

Ram, Y., Uriely, N., & Malach-Pines, A. (2014). Releasing control: parents on vacation. International Journal of Tourism Research, 16(3), 232–240. https://doi-org.ezproxy.lib.purdue.edu/10.1002/jtr.1921.

Ratay, R. (2018). Don't Make Me Pull Over! An Informal History of the Family Road Trip. New York: Scribner.

Resonance (2018). The future of U.S. Millennial travel. http://media.resonanceco.com/uploads/2018/08/Resonance-2018-Future-of-US-Millennial-Travel-Report-1.0.pdf, accessed 30 April 2020.

Rhoden, S., Hunter-Jones, P., & Miller, A. (2016). Tourism experiences through the eyes of a child. Annals of Leisure Research, 19(4), 424–443.

Ritchie, J., & Filiatrault, P. (1980). Family vacation decision-making: a replication and extension. Journal of Travel Research, 18(4), 3–14.

Rojas-de-Gracia, M.-M., & Alarcón-Urbistondo, P. (2017). Couple roles in subdecisions on family vacations. Cornell Hospitality Quarterly, 59(2), 160–173.

Rugh, S. S. (2008). Are We There Yet? The Golden Age of American Family Vacations. Lawrence, KS: University Press of Kansas.

Sanford, C. (2010). Evaluating family interactions to inform exhibit design: comparing three different learning behaviors in a museum setting. Visitor Studies, 13(1), 67–89.

Schänzel, H. A., Hull, J. S., & Velvin, J. (2017). Family tourism and the ski experience at Aun Peaks Resort, Canada. In C. Lee, S. Filep, J. N. Albrecht, W. J. L. Coetzee (eds) CAUTHE 2017: Time for Big Ideas? Re-thinking the Field for Tomorrow. Dunedin, New Zealand: Department of Tourism, University of Otago.

Schänzel, H., & Jenkins, J. (2017). Non-resident fathers' holidays alone with their children: experiences, meanings and fatherhood. World Leisure Journal, 59(2), 156–173.

Schänzel, H. A., Smith, K. A., & Weaver, A. (2005). Family holidays: A research review and application to New Zealand. Annals of Leisure Research, 8(2–3), 105–123.

Schänzel, H. A., & Yeoman, I. (2015a). Trends in family tourism. Journal of Tourism Futures, 1(2), 141–147. doi:10.1108/jtf-12-2014-0006.

Schänzel, H., & Yeoman, I. (2015b). The future of family tourism. Tourism Recreation Research, 39, 343–360. doi:10.1080/02508281.2014.11087005.

Sedgley, D., Pritchard, A., Morgan, N., & Hanna, P. (2017). Tourism and autism: journeys of mixed emotions. Annals of Tourism Research, 66(C), 14–25.

Shavanddasht, M. (2018). Grandparent's segmentation by the tourism motivation: travelling with or without grandchildren. Young Consumers, 19(2), 141–158.

Singh, R., & Nayak, J. (2016). Parent–adolescent conflict and choice of conflict resolution strategy. International Journal of Conflict Management, 27(1), 88–115.

Small, J., & Harris, C. (2014). Crying babies on planes: aeromobility and parenting. Annals of Tourism Research, 48, 27–41.

Srnec, T., Lončarić, D., & Prodan, M. P. (2016). Family vacation decision making process: evidence from Croatia. Faculty of Tourism & Hospitality Management in Opatija. Biennial International Congress. Tourism & Hospitality Industry, 432–445. Retrieved from http://search.ebscohost.com.ezproxy.lib.purdue.edu/login.aspx?direct=true&db=hjh&AN=117824423&site=ehost-live.

Stepchenkova, S., Shichkova, E., Kim, H., Pennington-Gray, L., & Rykhtik, M. (2015). Segmenting the "visiting friends and relatives" travel market to a large urban destination: the case of Nizhni Novgorod, Russia. Journal of Destination Marketing & Management, 4(4), 235–247.

Sthapit, E., & Björk, P. (2017). Activity participation home and away: examining the spillover theory among families on holiday. Anatolia: An International Journal of Tourism & Hospitality Research, 28(2), 209–223. https://doi-org.ezproxy.lib.purdue.edu/10.1080/13032917.2017.1311272.

Stone, M. J., & Petrick, J. F. (2017). Exploring learning outcomes of domestic travel experiences through mothers' voices. Tourism Review International, 21(1), 17–30. https://doi-org.ezproxy.lib.purdue.edu/10.3727/154427217X14858894687478.

Therkelsen, A. (2010). Deciding on family holidays: role distribution and strategies in use. Journal of Travel & Tourism Marketing, 27(8), 765–779. https://doi.org/10.1080/10548408.2010.526895.

Thornton, P. R., Shaw, G., & Williams, A. M. (1997). Tourist group holiday decision-making and behaviour: the influence of children. Tourism Management, 18(5), 287–297.

Trussell, D., Xing, T., & Oswald, A. (2015). Family leisure and the coming out process for LGB young people and their parents. Annals of Leisure Research, 18(3), 323–341.

Tung, V., & Ritchie, J. (2011). Investigating the memorable experiences of the senior travel market: an examination of the reminiscence bump. Journal of Travel & Tourism Marketing, 28(3), 331–343.

UNICEF (2018). Research on sexual exploitation of children in travel and tourism. www.unicef-irc.org/research/research-on-sexual-exploitation-of-children-in-travel-and-tourism/.

UNWTO (2019). Accessibility. Retrieved from www.unwto.org/accessibility, accessed 30 April, 2020.

US Census Bureau (2019). America's families and living arrangements: 2019. Retrieved from www.census.gov/data/tables/2019/demo/families/cps-2019.html, accessed 28 April 2020.

Watne, T., Brennan, L., & Winchester, T. (2014). Consumer socialization agency: implications for family decision-making about holidays. Journal of Travel & Tourism Marketing, 31(6), 681–696.

Whittle, R. (2019). Baby on board: The impact of sling use on experiences of family mobility with babies and young children. Mobilities, 14(2), 137–157.

Wu, M., Wall, G., Zu, Y., & Ying, T. (2019). Chinese children's family tourism experiences. Tourism Management Perspectives, 29, 166–175.

Yang, F. X., & Lau, V. M.-C. (2019). Experiential learning for children at World Heritage Sites: the joint moderating effect of brand awareness and generation of Chinese family travelers. Tourism Management, 72, 1–11.

Yau, M. K.-S., McKercher, B., & Packer, T. L. (2004). Traveling with a disability: More than an access issue. Annals of Tourism Research, 31(4), 946–960.

Yoo, H., McIntosh, A., & Cockburn-Wootten, C. (2016). Time for me and time for us: conference travel as alternative family leisure. Annals of Leisure Research, 19(4), 444–460.

Yu, X., Anaya, G., Miao, L., Lehto, X., & Wong, I. (2018). The impact of smartphones on the family vacation experience. Journal of Travel Research, 57(5), 579–596.

Yun, J., & Lehto, X. Y. (2009). Motives and patterns of family reunion travel. Journal of Quality Assurance in Hospitality & Tourism, 10(4), 279–300.

14

THE IMPACT OF MILLENNIALS ON URBAN TOURISM

Dae-Young Kim and Ye-Jin Lee

Introduction

"Whenever and wherever I want to travel, I go". This notion sounds idealistic but hardly practical. However, a new generation – Millennials – is putting this mindset into action. The United Nations estimates that nearly 200 million travellers fall into this age cohort (22–37 years old). Millennials generate more than $180 billion in tourism revenue annually and are rapidly becoming the most important demographic to the hospitality and tourism industries (UNWTO, 2012). They desire something differentiated and personalised when it comes to travel, particularly in the context of urban tourism.

Millennials, also termed Generation Y, tend to be more independent in their religious and political views, more entrepreneurial, less likely to be married, more distrustful of authority, better educated, more likely to be in debt, more closely linked to their peers, and more travel-oriented than any generation before them. In the context of tourism, Millennials attempt to meet their own needs and wants by emphasising intangible experiences, including travel, whereas older generations view travel as an extravagance, rather focusing on work and performance.

To date, several studies describing Millennials and proposing future implications from these descriptions have been conducted by the mass media, commercial consultants and independent research companies. Despite a consensus about the characteristics of Millennials, information about this generational group often reveals contradictory claims that are leading to incongruent conclusions and recommendations in different industry sectors (Benckendorff et al., 2010). Likewise, there is a lack of comprehensive managerial implications and suggestions in the field of urban tourism. This chapter discusses the overall patterns of Millennial travellers in how they search for travel information, behavioural characteristics on trips, and trip activities to the specific context of urban tourism. The goal of this chapter is to provide urban tourism scholars and practitioners with a better understanding of how marketing and programming related to travel can be tailored to Millennial audiences.

Who are Millennials?

There has been a profusion of terms describing the cohort of individuals born between 1980 and 1994 (Howe & Strauss, 2000). Termed Millennials, or generation Y, individuals in this group make up roughly 25% of the overall US population. This number is far greater than that of the previous generations of Baby Boomers (born between 1944 and 1964) and Generation X (born between 1965 and 1979). The Brookings Institution estimates that Millennials will make up about 75% of the workforce by 2025 (Winograd & Hais, 2014).

Millennials are said to be different from other generations in their way of thinking, behaving and learning due to their extensive exposure to modern technology (Bennett & Maton, 2010). Many studies and market reports have revealed that Millennials are more educated, racially diverse and comfortable with technology as a result of having been exposed to the internet from their formative years than other age cohorts. They are said to have a higher level of comfort using technology and possess more fluent information search skills than members of previous generations (e.g. Costello et al., 2004; Oblinger et al., 2005). The era of Web 2.0 allowed this generation easier access to information and the ability to share their opinions and thoughts publicly. Furthermore, Millennials are closely linked to their peers on social media platforms such as Facebook, Twitter and Instagram. Although non-Millennials tend to place more value on personal connections, Millennials (i.e. "Digital natives") are connected to others in real time regardless of physical distance via technology. Social media contributes to this trend by allowing Millennials to broaden their peer community network. They do not delineate their world between online and offline – these two different spheres are seamlessly blended.

Moreover, Millennials are more entrepreneurial and join companies that are in sync with their values. Previous generations used to follow the prescribed path of graduating from high school, going to college, getting married and having children. However, there are no such rules for Millennials. With their skilled experiences, they find opportunities everywhere and explore the world proactively. As mentioned above, Millennials are more travel-oriented than any other previous generation. They report a desire to visit every continent in their life more often than non-Millennials (70% vs 48%) and travel abroad as much as possible (75% vs 52%). Due in part to their life stage, Millennials express an overwhelming enthusiasm for adventure, pursuing opportunities as if they are keeping a scorecard and checking off experiences as they go. Their general sense of adventurousness also manifests itself in interests in other cultures, exotic foods and novel activities. There are multiple streams of evidence that distinguish this cohort from other generations (Table 14.1).

The influence of Millennials is already changing the landscape of the tourism and hospitality industries. Notably, tourism reflects a dynamic environment that is entangled with other social transformations. Globalisation, the development of transportation and the ubiquity of communication tools via information technology have all had a significant impact on the development of innovative tourism practices (Kim et al., 2014). Ever-changing trends in the tourism industry can be properly understood only through the lens of an overall, worldwide society (Monaco, 2018). Distinctive generational travel patterns can be recognised: Millennials, as the travellers of the future, will take about 320 million international trips in 2020 (Wyse Travel Confederation, 2015).

Urban tourism and Millennials

Tourism constitutes a central component in the economy, social life and geography of many cities worldwide. It is, therefore, a key element in urban development. Because urban destinations offer a broad range of cultural, architectural, technological, social and natural experiences and products for leisure and business, urban tourism can be represented as a driving force in terms of the development of many cities and countries.

The inclination of Millennials has resulted in different travel-related expectations from their older counterparts. From the viewpoint of Millennials, spending a whole week in Paris is more

Table 14.1 General characteristics of different generations

	Millennials	Generation X	Baby Boomers
Birth	Born 1980 to 1994	Born 1965 to 1979	Born 1944 to 1964
Ages in 2018	24–38	39–53	54–74
Size	95 million (in the US)	82 million (in the US)	76 million (in the US)
Media consumption	• Prefer cord-cutting in favour of streaming services • Comfortable with mobile devices • Multiple social media accounts	• Consume both traditional media and digital	• Traditional media (i.e. TV, magazines and radio)
Banking habits	• Little patience for efficiency • Trust brands with superior product history (e.g. Apple and Google) • Seek digital tools to manage their debt	• Use online but prefer transactions in person	• Go in branch to do transactions
Shaping events	• The Great Recession • The technological explosion of the internet and social media • 9/11	• End of the Cold War • Rise of personal computing	• Post WW2 optimism • Vietnam War • The social and environmental movement
What is on the financial horizon	• Entering the workforce with high amounts of student debt • Prefer access over ownership	• Raise a family, pay off student debt and take care of aging parents.	• Largest growing demographic for student loan debt • Balancing children's success with retirement goals

Note
Adapted from https://communityrising.kasasa.com/gen-x-gen-y-gen-z/. Copyright (2018) by KASASA.

meaningful than simply taking a photo in front of the Eiffel Tower, for instance. They pursue more engagement with cities such as San Francisco or New York rather than just stopping to shoot a picture at the Golden Gate Bridge or Times Square. This particularly complex self-consciousness holds multiple layers of implication but ultimately will significantly impact how and why Millennials make many of their travel decisions. In order to meet their needs, it is essential to design city experiences by acknowledging their characteristics. Several questions arise when pondering the core values of Millennials: What are their favoured tourist practices? What are the emerging needs of this generation that the tourist industry must consider? What are the beneficial tools to support Millennials' tourist choices?

Millennials crave an authentic experience

The concept of "authentic" travel is hardly new. However, looking at Millennials' demand for authenticity reveals a complex portrait that is nevertheless surprisingly precise. According to market reports, the top five preferred activities of Millennials when exploring a new city was described as follows: (a) discovering local heritage and traditions, (b) visiting small shops with locally made items, (c) discovering local youth culture and trends, (d) visiting a food market and (e) visiting a local festival (Ruspini & Melotti, 2016). These activities are all related to an "authentic" experience and soaking up the very essence of a destination through local experiences that are activity-based. They are more willing to pay the costs for a worthwhile experience such as cultural/educational, culinary, voluntourism and adventure tourism (i.e., backpacking/glamping tourism) to be connected with locals (GlobalData, 2019).

Moreover, a survey conducted by Oxford Economics (2018) mentioned that almost 85% of Asian Millennials responded that they are interested in "living like a local" while travelling. This finding demonstrates that authenticity is a key motivator that brings Millennials to experience the true colour of destinations, as well as supporting their perceptions of value by fostering independence and finding "hidden gems". Likewise, far from obsessing over touristic places, having "fireside" moments with local "insider" knowledge is more appreciated by Millennials. Similarly, many emphasise trying local food is very important, which often outranks visiting museums, monuments or the beach. If that is the case, "insider information" coming from locals will make their experience more valuable. Not every Millennial differs significantly from their non-Millennial counterparts in terms of how much they claim to prioritise the authentic culture of the places they visit. Nonetheless, this generation has created its authenticity through e-word of mouth, such as hashtag-enabled local meet-ups.

Millennials look for personalisation

Another hallmark of travelling Millennials is personalisation. Every Millennial has different tastes: some into camping, others enjoy staying at boutique hotels or getting pampered at spas, and others prefer large beach resorts. Some Millennials like to discover the culinary culture of a country in Michelin-starred restaurants, while others prefer street food (Fromm, 2018). They expect personalisation that will take them to a different world than what has been seen in the past; they are not banking on a strict itinerary. Millennials are well informed about their intended destinations and have already researched where they are going. As the "digital native" generation, Millennials are more receptive to personalisation techniques (McGee, 2017). Interestingly, McGinnis (2016) noted in the *State of the Connected Consumer* report that 63% of Millennial consumers are willing to share data with companies in exchange for personalised offers and discounts. In other words, these individuals do not solely pursue brand loyalty; a brand that capitalises their desire for new experiences, and a taste for adventure by differentiating its offering will do well with Millennials.

They use technologies to support their tourist choices

The development of Web 2.0 made it possible to connect with like-minded peers via the internet. "Digital natives" consider social media to be an integral part of their travel experiences. As grown-ups with profound connections to the world via social media, they enrich their travel-related experiences by posting, sharing and commenting in cyberspace. According to an Expedia survey, 42% of Millennials prefer to record photos on social media rather than physical albums or personal computers (Expedia, 2016). Likewise, more than 40% of travellers aged 18–33 indicate that "Instagrammability" comes first when choosing their next holiday spot (Hayhurst, 2017). That is, the picture-worthiness of a particular place beats other factors.

Another result of using technologies among Millennials has prompted the exponential growth of online travel agencies (OTAs). Numerous OTAs (e.g. TripAdvisor, Yelp, Expedia and Priceline.com) are leading companies in the tourism industry that work as platforms for travel-related reviews, metasearch, last-minute bookings and group deals available for every country. Also, Millennial travellers prefer making travel reservations on their smartphones or tablets. More than half of Millennials (51%) favoured this option compared with only 36% of travellers over the age of 60 (Berrigan, 2018). Moreover, technology is impacting different areas of services such as hotel check-in, restaurant bookings and airport kiosks. Millennials know what they want in terms of technology that encompasses speed, convenience, value and security. Online platforms also spark the sharing economy – well-known examples include Airbnb, Uber and Lyft. Google maps are also invaluable for obtaining directions while travelling, particularly in foreign countries.

Millennials do not decide without consulting their friends

Millennials are more likely to make choices after consulting with a wide range of their peers than non-Millennials. These peers include not only family and friends but also a much wider net of contacts via social media due to their high level of media consumption compared with previous generations. About 40% of Millennials acknowledge their peer's influence on a day-to-day basis. One explanation for this situation is that social media travel suggestions, peer-generated ratings and reviews are simply more genuine than brand-sponsored commercials, which is not much of a surprise. Recommendations from peers represent a way to mitigate the risk of booking a reservation at an unknown, independent hotel without a major brand attached to it (Arnold, 2018; Oates, 2014). Hotels seeking to attract Millennials need to be more considerate of this cohort and their increased levels of communication relative to their older counterparts.

Millennials seek out affordability and spend selectively

When Millennials are searching for travel destinations, they look for overall value and affordability. Low-cost airlines are a good example. Millennials might hesitate to travel if transportation costs are expensive. However, when a price becomes reasonable, the trip is a go (Fromm, 2018).

A collaborative form of consumption brings unbeatable benefits to Millennials as well. For example, Airbnb offers a cheaper option for Millennials for finding accommodations, given the more extensive choice of possibilities and access to locals. Whenever travellers have a hard time using transportation, Uber and Lyft are always there. Millennials spend their money in a curated

Table 14.2 Comparison of Millennials' and others' tourism characteristics

	Millennials	*Previous generations*
Information source	Mostly via the internet: Peers, Social media, Online Travel Agency (OTA)	Traditional media: Travel agency, Travel book, Friends
Travel price	Prioritize affordability (e.g. Sharing economy, Low-Cost Carrier)	Prioritize brand (e.g. airline, hotel)
Preferred travel type	Like a local: flexibility	Bundled package
Travel activities	Authentic, Genuine, Experiential	Popular, Famous destinations
Risk-taking	More likely	Less likely; comfortable taking
Main purpose	Diverse (e.g. Food, Shopping)	Mostly sightseeing

way that sends a message about themselves and their opinions of brands. This trend drove the growth of the sharing economy (e.g. Airbnb, Couchsurfing, Uber, EatWith and Withlocals). Using a product or service temporarily and paying less as a result certainly beats the price of ownership, and Millennials agree. This mindset helps the sustainability of the tourism industry in terms of balancing demand and supply.

Significant global urban tourism destinations

Different cities have a unique combination of nature and human-featured images that have distinct characteristics. It makes cities, in particular, world-class cities more important as tourist attractions. As tourism brings economic wealth to the cities, destinations work hard to attract as many travellers as possible, to become the best travel destinations through focusing their tourism resources. Sometimes, this phenomenon leads the cities themselves to be changed as increasing travel demands bring the necessity of building new infrastructure. Below are several examples of successful urban tourism destinations which have displayed a significant rise in visitor arrivals.

Hong Kong

Since the late 1980s and early 1990s, the tourism industry of Hong Kong (Figure 14.1) has been an essential part of the economy as it shifted gear toward the service sector. Accordingly, there has been a sharp increase in domestic tourists from Mainland China. Hong Kong is truly a destination like no other. While the city attracts travellers of all ages, Millennials love its rare combination of Western influence, traditional culture, tasty cuisine, easy transportation and family-friendliness. From modern skyscrapers to traditional rooflines, Hong Kong is a feast for the eyes, perfect for self-taught Millennial photographers (Dan, 2018).

What is more, most people speak fluent English, and all government signs are in English; this makes getting around even more accessible. Because it is a relatively secure city in comparison to other destinations, most Millennials enjoy the relaxed, easy-going atmosphere of Hong Kong without fearing for their safety. In 2017, the Hong Kong government released the "Smart City Blueprint" with a vision to build Hong Kong into a world-class smart city. This project maps out development plans, aiming to (a) enhance the effectiveness of city management and improve people's quality of living, (b) increase Hong Kong's attractiveness toward global businesses and sustainability by (c) making use of innovation and technology from six different domains: smart

Figure 14.1 Night-time view of Hong Kong Island shoreline.

Source: Photo: https://pixabay.com/photos/hong-kong-peek-night-long-exposure-1791067/.

mobility, smart living, smart environment, smart people, smart government and smart economy.

Because of the benefits mentioned above, Hong Kong visitor arrivals have been steadily increasing for the past few decades (*Trading Economics*, 2019). However, its recent political situation has badly affected its overall economy, including travel industries. The number of visitors has declined, flights and reservations have been cancelled and people search more on safety-related keywords when looking up information on the destination (BBC News, 2019). Considering Millennials as the largest market, it is assumed that this cohort would also become wary, which is expected to hurt the destination image. Though some say this effect would be only temporary, and not matter in the long term, the outcome should not be underestimated. It is highly recommended for major tourism stakeholders such as the government, locals and tourism companies to work closely in order to resolve this issue and minimise the negative impact properly.

Seoul, South Korea

Seoul (Figure 14.2) is a dynamic city with twice the population density of New York City. Recently, the city introduced a new urban development project named "Smart Seoul", which reflects the people-oriented atmosphere (Hwang & Choe, 2013; Seoul Metropolitan Government, 2016). This project is composed of three domains: (a) integrating modes of transportation with higher accessibility, (b) making mobility affordable for all, and (c) transforming the urban landscape and removing the unhealthy, car-centric infrastructure. For South Korea, like other well-known destinations, Millennials are a huge market as they make up to 45% of the overall

Figure 14.2   Lotte World Tower, Seoul.

Source: Photo: https://pixabay.com/photos/lotte-world-tower-seoul-1791802/.

tourist market. This ratio is growing at a fast pace as people in their twenties accounted for 16.5% of total tourists in 2008 and represent 25% of total visitors in 2018 – showing a 50% increase.

One of the explanations of this phenomenon is the "K-wave" (Korean wave) spurred by popular Korean entertainment such as K-pop music, K-drama, Korean beauty and fashion. Seemingly, one can easily find examples from social media and broadcast platforms. Recently, the song "Gangnam Style" by Psy kindled this movement, followed by K-pop boy group BTS, whose "Idol" music video reached the fastest 100 million views in 2018 (Herman, 2018). In terms of drama, during airing of the hit K-drama "My Love from Another Star", the main actress' fashion items, like lipsticks and shoes, sold out quickly in several places, including China, Paris and New York. Likewise, K-beauty items can be found in Sephora, and without any difficulty, which would not have been imaginable five years ago.

Barcelona, Spain

Another example of a major smart city is Barcelona, Spain (Figure 14.3), which has been utilising innovative technologies to change the daily life of its citizens. This city is particularly interesting because it has reinvented itself over the past 30 years after the economy collapsed in the 1980s, due to stagnation and severe unemployment. By investing in the Internet of Things (IoT) for urban systems, the project to transform the overall economy and social profile toward modern-city tourism was launched under the 22@ umbrella. It includes urban transformation and innovation development with district heating, telecommunications and pneumatic garbage-collection systems. This movement surely profits the city by reducing congestion and emissions,

Figure 14.3 Parc Güell, Barcelona.

Source: Photo: https://pixabay.com/photos/parc-guell-gaud%C3%AD-barcelona-spain-332390/.

saving water and power costs. In turn, the city's commitment toward economic development for smarter urban infrastructure brings better living conditions to all residents, workers and visitors. Through these improvements, Barcelona estimates that IoT systems have helped save $58 million on water, increased parking revenues by $50 million per year and generated 47,000 new jobs. Through smart lighting, the city reports saving an additional $37 million annually (Adler, 2016).

Discussion and implications

Having good content

Millennials are always looking for new experiences. Although individuals in this generation have less disposable income to spend than previous generations, they are becoming increasingly affluent as many come of working age. As discussed above, Millennials view travel as an integral part of their self-identity and expression (Airbnb, 2016). They like to explore cities to find unusual and unique activities that distinguish themselves from others. When authenticity is combined with locals on-site, typical activities bring about more enjoyment. Importantly, good content is vital. Serving these needs for new destinations, personalised trips and local activities is more attractive to new consumers (GlobalData, 2019). In other words, Millennials seek out interactions with the destination and locals. Sometimes, tourism/destination marketers feel there are not enough resources or exciting activities to induce Millennial tourists to their place. Even in this case, it becomes no problem at all. Having a story to their destination is an excellent alternative to tourism operators. With historical, cultural, supernatural and other aspects that capture the attention and imagination, everything can be combined as storytelling (see Alton, 2018; Goh et al., 2013; Hsiao et al., 2013).

The task of creating interesting content that grabs the attention of Millennials via social interactions will be the responsibility of the cities. A city's tourism can be revived by addressing Millennials' expectations. In general, destinations should provide experiential offerings by reflecting the fact that Millennials pursue a bit of everything (Charles, 2018). Millennials want to uncover places, have a drink at a local bar, eat local foods, go to music festivals, have culinary experiences and take a dance class. At the same time, they also want to see the Mona Lisa at the Louvre and go to the British Museum or another classic tourist destination. Though there seems to be a lack of consistency in terms of Millennials' desires, these interests do have something in common: the quality of the content.

Millennials are an ever-changing market

Who would have thought about companies such as Airbnb or Uber a decade ago? As Millennials set the ground for new business models, the overall size of the tourism pie focused on Millennials is steadily increasing via the addition of previously non-existent markets. These needs are becoming universal – Millennials no longer represent a direct business target.

In this vein, several international companies (e.g. WeWork, Regus) are now leading new types of travel by offering global travellers the opportunity to mix business travel with leisure time ("bleisure") (Lichy & McLeay, 2018). This type of travelling combines a meeting or work commitment with spending non-working time exploring nearby places. Indeed, more business travellers are taking leisure time while on their work trips (Oates, 2014). Ninety-four percent of Millennial travellers under the age of 35 conveyed interest in bleisure (Skift, 2014). Leisure time does not have to be too fancy; Millennials agree. With its full scope, this type of travel concept captures both business and leisure travellers (Lichy & McLeay, 2018). For instance, it involves spending a single or a couple of days at an intermediate location before/after flying to a work-related destination. If a car is necessary for a work trip, visiting nearby cities for a few days while not working is another alternative. Short adventures are also possible: set the alarm earlier than usual or block off a free hour during the day to check out a local attraction or wander around a street. Furthermore, this cohort is bringing people with them on their travels: more than half of bleisure travellers reported that they travel with their families (Oates, 2014).

Artificial Intelligence (AI) is also bound to have a significant effect on the travel-related decision-making process (Stalidis et al., 2015; Zsarnoczky, 2017). This technology not only allows travellers to personalise and optimise their trip itinerary and overall costs – it also helps hospitality and tourism service providers deliver more individualised and personalised experiences based on travellers' temperaments, tastes and real-time moods. Information technology (IT) has become a key to bringing this generational characteristic into tourism, and it will continue to play a significant role in the experiences of both travellers and service providers (Xiang et al., 2015).

Sustainable tourism is the key to long-term development

Sustainability is another developing trend that merits attention. A study conducted by Deloitte (2017) found that Millennials are very willing to pay for sustainable and socially responsible products and services. In terms of tourism, their social awareness results in more ethical travel that is in sync with pursuing a local experience. For example, a lot of Millennials are more interested in having a coffee or tea at a small café, which benefits the local economy, than stopping at a large-scale franchise store.

Given developments in technology, gentrification is creating social strains as well. Sharing economy platforms such as Airbnb and Uber is forcing local governments to create regulations. For example, Airbnb rentals are increasing prices in popular destination cities such as San Francisco and Paris, and 24/7 noise is disturbing local neighbours (Wachsmuth & Weisler, 2018). Also, Uber is provoking a conflict with the taxi industry because permit prices for taxi owners are plummeting (Henley, 2017). Overtourism, no matter how sensitive, can destroy the fragile balance of local life. It is accordingly time for governments and local authorities to implement regulation plans. Minimising the negative impacts that tourists have on environments via heritage sites can better balance the pros and cons of tourists visiting destinations. Popular destinations such as Venice, Italy, Angkor, Cambodia, and Boracay, Philippines, are already suffering serious problems due to unsustainable tourism (Fox, 2019; *The Telegraph*, 2018). Best practices are becoming the norm and – with an understanding of the issues and a desire to enact change for the better – solutions will come. Seasoned travellers have also become more aware of their impacts and are helping to preserve the places they are visiting. Millennials have transformed how the travel world operates, and we are happy to say, so far, it seems for the best.

The important role of social media

Millennials respect and rely on the opinions of their peers. If they find something cool in a friend's post, they tend to give it more credit than they would from other sources (e.g. magazine advertisement, sponsored blog post). In this overwhelming, information-oriented society, Millennials trust something genuine that comes from their peers. City marketers can accordingly attract Millennials to post and share their experiences on social media, creating a compelling narrative for other Millennials to follow their experiences. For example, peer-to-peer marketing is a good motivator to influence other people to be inspired by their behaviour and follow. Over 97% of Millennial travellers (and 60% of overall travellers) share their travel photos while on a trip (Fromm, 2018).

User-generated content has an impact on every aspect of travel behaviour from planning a trip to during a trip to after a trip (Cox et al., 2009; Fotis et al., 2012). With photos shared in real-time, Millennials' peers are more inspired to visit new destinations. Sharing pictures after a trip on social media causes Millennials to become excited about planning their next adventures. Peers can be connected via online by following hashtags that they are interested in, thereby encouraging newcomers to travel as well.

More smart cities will be developed

"Smart city" is an umbrella term referring to how information and communication technology (ICT) helps to improve the efficiency of a city's operations. This term also relates to promoting the local economy and citizens' quality of life. Interestingly, branding a city as a "smart city" often yields better results in terms of external identity and image than implementing its initiatives alone (Gavaldà & Ribera-Fumaz, 2012). Based on the example of Barcelona, public image and marketing are just as crucial as just being itself. Interestingly, Barcelona's smart–city marketing strategy has proven to be even more successful than the city's strategy itself (Gascó et al., 2016).

Sharing a clear vision and goals and ensuring the active participation of responsible stakeholders is important for a smart city's development and its sustainability. No smart city can view its citizens as only recipients of interventions; they are instead partners who define the city they want to be a part of. The High Line in Manhattan, a place where many international tourists visit to enjoy its impact on the urban and cultural landscape, is portrayed not only as a symbol of post-modern urban beautification but also as a dazzling district with unique value (Rainey, 2014). Likewise, Millennials have profoundly transformed cities, and the revolution is still ongoing as their needs evolve. Destinations like Barcelona, Seoul and Hong Kong are enacting urban and tourist policies to attract more Millennials. These policies can help cities become friendlier to tourists and stand out from the urban competition.

New technology infrastructure can provide the benefits of Wi-Fi hotspots, electric vehicle charging facilities, improved traffic flow and car parking information. With the Internet of Things (IoT) sensors, real-time information related to weather, air quality and temperature can be relayed. Ultimately, the real-time collected data can benefit cities by helping managers and decision-makers to better understand smart-city initiatives (McKinsey, 2018).

References

Adler, L. (2016) How smart city Barcelona brought the internet of things to life. *Data-Smart City Solutions*. [Online] Available from: https://datasmart.ash.harvard.edu/news/article/how-smart-city-barcelona-brought-the-internet-of-things-to-life-789 [Accessed 26 March 2019].

Airbnb. (2016) Airbnb and the rise of Millennial travel. [PDF file] Available from: www.airbnbcitizen.com/wp-content/uploads/2016/08/MillennialReport.pdf/ [Accessed 10 October 2019].

Alton, L. (2018) What Irish tourism marketing can tell us about the value of authentic storytelling. *Skyword*. [Online] Available from: www.skyword.com/contentstandard/creativity/what-irish-tourism-marketing-can-tell-us-about-the-value-of-authentic-storytelling/ [Accessed 18 October 2019].

Arnold, A. (2018) Here's how much Instagram likes influence Millennials' choice of travel destinations. *Forbes*. [Online] Available from: www.forbes.com/sites/andrewarnold/2018/01/24/heres-how-much-instagram-likes-influence-millennials-choice-of-travel-destinations/#4ee501644eba [Accessed 27 March 2019].

BBC News. (2019) Hong Kong protests: how badly has tourism been affected? [Online] Available from: www.bbc.com/news/world-asia-china-49276259/ [Accessed 10 October 2019].

Benckendorff, P., Moscardo, G. & Pendergast, D. (Eds.). (2010) *Tourism and Generation Y*. Wallingford: CABI.

Bennett, S. & Maton, K. (2010) Beyond the "digital natives" debate: Towards a more nuanced understanding of students' technology experiences. *Journal of Computer Assisted Learning*, 26(5), 321–331.

Berrigan, T. (2018) China international travel monitor 2018 sees millennials driving market changes and preferences. *Inbound*. [Online] Available from: www.inboundreport.com/2018/08/21/china-international-travel-monitor-2018-sees-millennials-driving-market-changes-preferences/ [Accessed 4 March 2019].

Charles, M. (2018) Millennial travellers and how they've changed travel for the better. *Fiz*. [Online] Available from: www.fiz.com/blog/travel-trends/millennial-travellers/ [Accessed 25 March 2019].

Costello, B., Lenholt, R. & Stryker, J. (2004) Using Blackboard in library instruction: Addressing the learning styles of Generations X and Y. *The Journal of Academic Librarianship*, 30(6), 452–460.

Cox, C., Burgess, S., Sellitto, C. & Buultjens, J. (2009) The role of user-generated content in tourists' travel planning behavior. *Journal of Hospitality Marketing & Management*, 18(8), 743–764.

Dan. (2018) How is Hong Kong attracting millennial travelers? *G'day World*. [Online] Available from: www.gdayworld.com/travel-tips/how-is-hong-kong-attracting-millennial-travelers/ [Accessed 29 March 2019].

Deloitte. (2017) *The 2017 Deloitte Millennial Survey*. [PDF file] Available from: www2.deloitte.com/content/dam/Deloitte/lt/Documents/human-capital/lt-deloitte-millennial-survey-2017-executive-summary.pdf/ [Accessed 4 February 2019].

Expedia. (2016) *Millennial Traveler Report*. [PDF file] Available from: www.foresightfactory.co/wp-content/uploads/2016/11/Expedia-Millennial-Traveller-Report-Final.pdf/ [Accessed 24 February 2019].

Fotis, J. N., Buhalis, D. & Rossides, N. (2012) Social media use and impact during the holiday travel planning process. In M. Fuchs, F. Ricci & L. Cantoni (Eds.) *Information and Communication Technologies in Tourism 2012: Proceedings of the International Conference in Helsingborg, Sweden, January 25–27, 2012* (pp. 13–24). Vienna: Springer.

Fox, K. (2019) Venice becomes the front line in the battle against Overtourism. *CNN Travel.* [Online] Available from: www.cnn.com/travel/article/venice-tourism-overcrowding-intl/index.html/ [Accessed 14 October 2019].

Fromm. J. (2018) How are Millennials using travel technology? *Forbes.* [Online] Available from: www.forbes.com/sites/jefffromm/2018/07/31/how-are-millennials-using-travel-technology/ [Accessed 25 March 2019].

Gascó, M., Trivellato, B. & Cavenago, D. (2016) How do southern European cities foster innovation? Lessons from the experience of the smart city approaches of Barcelona and Milan. In J. R. Gil-Garcia, T. A. Pardo & T. Nam (Eds.) *Smarter as the New Urban Agenda* (pp. 191–206). Cham: Springer.

Gavaldà, J. & Ribera-Fumaz, R. (2012) Barcelona 5.0: from knowledge to smartness? IN3 Working Paper Series. [Accessed 24 March 2019].

GlobalData. (2019) *Trends in Global Millennial Travel.* [PDF file] Available from: www.aaaa.org/wp-content/uploads/2019/04/GlobalData-Trends-in-Global-Millennial-Travel-2019-03.pdf/ [Accessed 10 October 2019].

Goh, K. Y., Heng, C. S. & Lin, Z. (2013) Social media brand community and consumer behavior: Quantifying the relative impact of user-and marketer-generated content. *Information Systems Research,* 24(1), 88–107.

Hayhurst, L. (2017) Survey highlights Instagram as key factor in destination choice among millennials. [Online] Available from: www.travolution.com/articles/102216/survey-highlights-instagram-as-key-factor-in-destination-choice-among-millennials/ [Accessed 4 March 2019].

Henley, J. (2017) Uber clashes with regulators in cities around the world. *The Guardian.* [Online] Available from: www.theguardian.com/business/2017/sep/29/uber-clashes-with-regulators-in-cities-around-the-world/ [Accessed 30 March 2019].

Herman, T. (2018) BTS' "Idol" music video is fastest to reach 100M views in 2018. *Billboard.* [Online] Available from: www.billboard.com/articles/columns/pop/8472934/bts-idol-music-video-fastest-to-reach-100m-views-in-2018/ [Accessed 2 March 2019].

Howe, N. & Strauss, W. (2000) *Millennials Rising: The Next Great Generation.* New York: Vintage.

Hsiao, K. L., Lu, H. P. & Lan, W. C. (2013) The influence of the components of storytelling blogs on readers' travel intentions. *Internet Research,* 23(2), 160–182.

Hwang, J. S. & Choe, Y. H. (2013) Smart cities – Seoul: A case study. ITU-T Technology Watch Report, Geneva. [Online] Available from: www.itu.int/en/ITU-T/techwatch/Pages/smart-city-Seoul.aspx/ [Accessed 18 March 2019].

KASASA. (2018) General characteristic of different generations [Table]. Available from: https://community rising.kasasa.com/gen-x-gen-y-gen-z/ [Accessed 30 March 2019].

Kim, S., Kim, D.-Y. & Wise, K. (2014) The effect of searching and surfing on cognitive responses to destination image on Facebook pages. *Computers in Human Behavior,* 30, 813–823.

Lichy, J. & McLeay, F. (2018) Bleisure: Motivations and typologies. *Journal of Travel & Tourism Marketing,* 35(4), 517–530.

McGee, T. (2017) How millennials are changing retail patterns. *Forbes.* [Online] Available from: www.forbes.com/sites/tommcgee/2017/01/23/the-rise-of-the-millennial/#4abda24f5f74/ [Accessed 4 March 2019].

McGinnis, D. (2016) Please take my data: Why consumers want more personalized marketing. *Salesforce blog.* [Online] Available from: www.salesforce.com/blog/2016/12/consumers-want-more-personalized-marketing.html/ [Accessed 29 March 2019].

McKinsey. (2018) Smart cities: Digital solutions for a more livable future. [Online] Available from: www.mckinsey.com/industries/capital-projects-and-infrastructure/our-insights/smart-cities-digital-solutions-for-a-more-livable-future/ [Accessed 14 October 2019].

Monaco, S. (2018) Tourism and the new generations: Emerging trends and social implications in Italy. *Journal of Tourism Futures,* 4(1), 7–15.

Oates, G. (2014) Why the power of peer reviews is different for millennial travelers. *Skift.* [Online] Available from: https://skift.com/2014/02/17/why-the-power-of-peer-reviews-is-different-for-millennial-travelers/ [Accessed 28 March 2019].

Oblinger, D., Oblinger, J. L. & Lippincott, J. K. (2005) *Educating the Next Generation.* Boulder, CO: Educause.

Oxford Economics. (2018) Data & digital platforms: Driving tourism growth in Asia Pacific. [PDF file] Available from: https://s3.amazonaws.com/tourism-economics/craft/Latest-Research/TE_APAC-Data-Digital-Platforms-2018.pdf/ [Accessed 28 March 2019].

Rainey, J. (2014) New York's High Line Park: An example of successful economic development. [PDF file] Available from: www.greenplayllc.com/wp-content/uploads/2014/11/Highline.pdf/ [Accessed 14 October 2019].

Ruspini, E. & Melotti, M. (2016) How is the millennial generation reshaping cities and urban tourism. In *Proceedings of the Conference "Urban planning and tourism consumption"* (pp. 1–3).

Seoul Metropolitan Government. (2016) Mayor Park Won Soon declares "Solution to world's economic crisis is social economy" at Global Social Economy Forum (GSEF) assembly. [Online] Available from: http://english.seoul.go.kr/mayor-park-won-soondeclares-solution-worlds-economic-crisis-social-economy-global-social-economy-forum-gsef-assembly/? cat=46/ [Accessed 29 March 2019].

Skift. (2014) *The Bleisure Report 2014.* [PDF file] Available from: https://skift.com/wp-content/uploads/2014/10/BGH-Bleisure-Report-2014.pdf/ [Accessed 30 March 2019].

Stalidis, G., Karapistolis, D. & Vafeiadis, A. (2015) Marketing decision support using Artificial Intelligence and Knowledge Modeling: Application to tourist destination management. *Procedia: Social and Behavioral Sciences*, 175, 106–113.

The Telegraph. (2018) What's happening in Boracay, the island paradise ruined by tourism? [Online] Available from: www.telegraph.co.uk/travel/destinations/asia/philippines/articles/boracay-closure-when-will-island-reopen/ [Accessed 14 October 2019].

Trading Economics. (2019) Hong Kong visitor arrivals. [Online] Available from: https://tradingeconomics.com/hong-kong/tourist-arrivals/ [Accessed 10 October 2019].

UNWTO. (2012) UNWTO and WYSE travel confederation launch global declaration to promote youth travel. [Online] Available from: http://media.unwto.org/en/press-release/2012-09-25/unwto-and-wyse-travel-confederation-launch-global-declaration-promote-youth [Accessed 31 March 2019].

Wachsmuth, D. & Weisler, A. (2018) Airbnb and the rent gap: Gentrification through the sharing economy. *Environment and Planning A: Economy and Space*, 50(6), 1147–1170.

Winograd, M. & Hais, M. (2014) How millennials could upend Wall Street and corporate America. Brookings Institution.

Wyse Travel Confederation. (2015) Millennial traveler report. [Online] Available from: www.wysetc.org/2014/11/new-research-download-our-new-millennial-traveller-report-for-a-unique-insight-into-this-influential-generation-of-travellers/ [Accessed 19 March 2019].

Xiang, Z., Magnini, V. P. & Fesenmaier, D. R. (2015) Information technology and consumer behavior in travel and tourism: Insights from travel planning using the internet. *Journal of Retailing and Consumer Services*, 22, 244–249.

Zsarnoczky, M. (2017) How does Artificial Intelligence affect the tourism industry? *VADYBA*, 31(2), 85–90.

15

VISITING OLDER FRIENDS AND RELATIVES

Opportunities for tourism cities

Elisa Backer (Zentveld)

Introduction

Visiting Friends and Relatives (VFR) is a major form of travel in most countries around the world. As destinations grow, so too do VFR numbers. This is irrespective of how attractive the city is because the attraction (not entirely but largely) lies with the people. Tourism cities face particular challenges with accommodating visitors because they must provide infrastructure to cope with the growing population, the needs of the tourists and the needs of VFR travellers who may be attracted to the city because of the people, the city attractions or possibly a combination of both.

While many aspects about growing populations challenge tourism cities, this chapter discusses the aging population and quality of life. Such a discussion offers the reader a unique lens in which to consider VFR. VFR travel has historically been considered from an economic perspective (Asiedu, 2008; Backer, 2007, 2010, 2012; Backer & Hay, 2015; Hu & Morrison, 2002; Lee et al., 2005; McKercher, 1996, 1995; Yousuf & Backer, 2015). Such an approach is understandable given that tourism is considered a sub-section of the library Dewey catalogue system. Although, often tourism has little to do with economics. VFR travel has more recently started to be considered from a social perspective (Capistrano, 2013; Griffin, 2013) and in particular new research has commenced to consider the role of the host of VFRs (Backer, 2007, 2008; Backer & Laesser, 2011; Schänzel & Brocx, 2013; Shani & Uriely, 2012; Yousuf & Backer, 2017). A small number of VFR articles have started to consider the relationship between VFR and quality of life (Backer, 2019; Backer & Weiler, 2018).

This chapter takes a unique approach to discussing VFR travel by focusing on a segment of VFR travel not previously considered: that of Visiting Older Friends and Relatives (VOFR). In discussing the segment of VOFR, this chapter does so through the lens of quality of life. Quality of life is an important health measure, and critical for the aging population. As will be discussed in this chapter, quality of life encompasses many aspects that are highly relevant to understanding and planning cities with aging populations. Since people are living longer, society's housing needs are changing, along with their medical, transport and social needs. VFRs form an integral part of the social needs of all people but especially the elderly, who can easily suffer from social isolation and loneliness.

Elderly people are often at risk of feeling isolated and lonely. Once people retire, their sense of purpose can alter, sense of self-esteem, as well as social connections. As has been outlined in recent research (Backer, 2014; Backer & King, 2017; Backer & Weiler, 2018), the process of aging can result in visits from friends and relatives becoming especially important for quality of life. However, while people are living longer, the health and mobility of the elderly can vary enormously. Some elderly people can have mobility constraints while others can be healthy and mobile. Aging can also affect behaviour. The elderly can become self-absorbed and set in their ways and perhaps even be aggressive. This poses a number of complex issues for family and friends who intend visiting, and naturally can exacerbate social isolation if behaviour becomes too challenging for friends and relatives to accept and their volume of visits reduce. Thus, understanding all quality of life components including behavioural is important in understanding best practices for VOFR.

This chapter not only explores quality of life for older people, it also discusses housing and transport trends related to the aging in tourism cities, as these things impact on quality of life and naturally have direct implications for planning and managing tourism cities. This chapter concludes by offering tourism city planners a number of insights in which commercial accommodation may be appropriate for city planning.

Cities are already needing to plan to cater for the aging population such as more single storey accommodation, almshouses, improved pavements, more regular public transport stops, more shading and more time to negotiate pedestrian crossings. However, as will be discussed in this chapter, it may also be helpful to consider forms of commercial accommodation in those development areas to enable options for visitors who are travelling to visit their aging friend or relative. This may also be helpful in encouraging VFR trips where aging friends/relatives are challenging to visit. VFRs staying in commercial accommodation can retain some personal space while being geographically close during visits, and that process can be beneficial, enhance quality of life and as a result encourage visitation. The simple aspect of being able to have individual accommodation to provide relief when friends or relatives visit aging people may assist greatly in ensuring those trips do not impact negatively on well-being, as well as to reduce conflict and stress. Such actions may in turn ensure that visits are not only more enjoyable but can be more frequent. These concepts will become increasingly important in light of the increasing life expectancy rates among tourism cities.

VFR travel

One of the largest and most significant forms of tourism is Visiting Friends and Relatives (VFR) travel. VFR travel is recognised as being a sizable form of travel worldwide. The size of this form of travel has been reported as comprising around half of the US pleasure traveller market (Braunlich & Nadkarni, 1995; Hu & Morrison, 2002) and over half of Australia's domestic travel market (Backer, 2012, 2015).

Despite the size of VFR travel, it tends to be overlooked as a market segment and rarely appears in the marketing plans of destination marketing organisations and tourism businesses. Part of this reason is due to a lack of research into the relationship of VFR travellers with industries, and as such an assumption that VFR travellers do not participate in tourism activities and inject negligible funds into local economies (Backer, 2007). In addition, the relationship that the host has in shaping VFR trips has been under-researched and not fully explored as a marketing opportunity.

Interest in VFR travel seems to go in cycles. Jackson's seminal article (1990), which was 13 years later reprinted in the same journal (2003), stimulated considerable interest by suggesting

that this type of travel was much bigger than official estimates suggested. Jackson (1990) identified that people can quite easily be visiting friends and relatives, but self-classify themselves as
holidaymakers, therefore underestimating the official data regarding the size of VFR travel. As
such, VFR travel is likely to be of greater value to the host economy than tourism industries
may understand. After this suggestion was raised by Jackson (1990), interest in the field developed
for a number of years, leading to a wave of research through the mid-1990s (for example Braunlich & Nadkarni, 1995; Hay, 1996; King, 1996; McKercher, 1994, 1995; Morrison et al., 1995;
Seaton, 1994; Seaton & Palmer, 1997; Seaton & Tagg, 1995; Yaman, 1996). Interest also led to
a special issue of an international journal (*Journal of Tourism Studies*, 1995, 6 (1)) and an international conference (VFR Tourism: Issues and Implications, 1996) dedicated to this area of
tourism.

Morrison et al. (2000) and Hu and Morrison (2002) led the next surge of interest in the early
years of the 2000s, with focus on trip behaviour and guest motivations. They also claimed that
the segment was under-appreciated by tourism marketers. Thirdly, Backer (2007, 2008) and
Young et al. (2007) revived interest with renewed evidence on the economic importance of the
VFR segment, including a focus on the hosts in addition to the guests, as well as consideration
of how to define and model VFR travel (Backer, 2007, 2012). This latter interest provided an
attempt to put forward a marketing communication strategy using the hosts as a channel (Young
et al., 2007). This wave of interest resulted in highlighting the economic stability effect that
marketing to VFR travellers can have in economic recessions (BBC World News, 2009). Most
recently, focus has been given to the social aspects of hosting (for example Yousuf & Backer,
2017), migration considerations (for example Huang et al., 2017; Provenzano & Baggio, 2017;
Rogerson, 2017), and new broader aspects of how VFR impacts on other areas such as disaster
recovery (Backer & Ritchie, 2017), analysis through the context of literature and culture
(Seaton, 2017), social tourism (Backer & King, 2017) and the differences between friends and
relatives (Backer et al., 2017). Significantly, recent additions to VFR travel have included the
first scholarly book on the subject (Backer & King, 2015) as well as a special issue in a leading
tourism journal dedicated to VFR travel (Backer & Morrison, 2017).

The aspect raised in this chapter – considering commercial accommodation for VOFR –
may at first seem at odds with the concept of VFR travel. This stems from a fundamental
misunderstanding of VFR travel behaviour. Some early definitions of VFR travel that have
classified VFR in terms of behaviour have resulted in furthering the misperception that VFR
travel is where friends or relatives stay with the person they are visiting. For example, King
(1994) stated that VFR travel is categorising visitors by the type of accommodation that they
use. Boyne, Carswell and Hall (2002) proposed that

> a VFR tourism trip is a trip to stay temporarily with a friend or relative away from the
> guest's normal place of residence, that is, in another settlement or, for travel within a
> continuous settlement, over 15 km one-way from the guests' home.
>
> *(p. 246)*

They admitted that their definition "largely avoids rather than confronts some of the key conceptual issues" (pp. 246–247). Similarly, Kotler, Bowen and Makens (2006) state that "VFR, as
the name suggests, are people that stay in the homes of friends and relatives" (p. 748). These
suggestions reinforce the implied notion that VFR travellers do not stay in commercial accommodation. In fact, according to Navarro and Turco (2004), the perception that VFR travellers
make little use of commercial accommodation and do not tend to frequent restaurants, cafes,
pubs and clubs is why VFR travel has not been clearly defined.

A definition was put forward by Backer (2007) that "VFR travel is a form of travel involving a visit whereby either (or both) the purpose of the trip or the type of accommodation involves visiting friends and/or relatives" (p. 369). This has since been used as the basis for a definitional model to visually highlight that there are in fact three distinct VFR types, and by measuring VFR by purpose of visit *or* accommodation type, only two of the three groups will be included (Backer, 2012). Accordingly, the discussion in this book chapter concerning planning for VOFR through consideration of commercial accommodation is a focus on one of those three core types of VFR traveller groups, that of CVFR (Commercial Visiting Friends and Relatives; that is, those VFRs that travel to a destination specifically to visit a friend or relative but stay in commercial accommodation). Research undertaken by Backer (2010) indicates that the CVFR segment is around 26% of all VFR travel.

Despite increasing sophistication in approaches to market segmentation, VFR travel has remained a stalwart of tourism data classification schemes, along with holiday/leisure and business. VFR travel can include the young or old, can be motivated by family visits, other leisure, sport or business activities or exploitation of accommodation and transport. Significantly, VFR travel can represent an important form of travel for the elderly. While the elderly may lose confidence or interest in leisure tourism, especially if they are the sole survivor, VFR can give them a reason to travel. The elderly may be motivated to visit a new grandchild, a christening, wedding, significant milestone birthday or simply social reconnections with family. Accordingly, VFR is a vital topic for examination with respect to tourism cities and the focus on the elderly with respect to planning seems important. As the elderly become a stronger segment due to increasing life expectancy rates, the importance of the topic will only grow.

Life expectancy rates

Society is seeing a demographic trend towards an aging population. Increasingly OECD countries are realising that their elderly populations are a fast-growing segment, and for some countries the elderly are the fastest growing segment within their population groups (Rosenberg & Everitt, 2001). As people live longer, this creates pressures on town planners and taxpayers. Increasing consideration is required for infrastructure such as housing as well as medical facilities. Concerns relating to the sustainability of health as well as social care services have been raised in relation to the aging population (Lilburn et al., 2018). Other considerations include transport links such as public transport, shading, park benches, timing on pedestrian crossings and other mobility issues. Countries that have high life expectancy rates are those most needing to consider the needs of their aging populations.

According to the World Health Organization (WHO), the average life expectancy rate for the world in 2016 was 72.0 years (World Health Organization, 2016). The top three countries for life expectancy have rates that exceed 84 years with Hong Kong having the highest life expectancy, followed by Macau, then Japan. The global average life expectancy rate is weighed down by those countries with low life expectancy rates such as Nigeria (54.8 years), Central African Republic (53.8 years), Chad (53.6 years) and Sierra Leone (52.7 years) (World Population Review, 2019). In fact, 64 countries have life expectancy rates below 70 years (World Population Review, 2019). Fifteen countries have life expectancy rates below 60 years of age (World Population Review, 2019). The top ten countries for life expectancy rates are outlined in Table 15.1.

It is also useful to note the disparity in life expectancy between men and women (Table 15.1). For the first three countries, the difference is considerable (for example almost six years difference between men and women in Japan). Taking into account that it is common for

Table 15.1 Top ten countries for life expectancy

Ranking	Country	Life expectancy (years)	Life expectancy males (years)	Life expectancy females (years)
1	Hong Kong	84.308	81.370	87.292
2	Macau	84.188	81.220	87.120
3	Japan	84.118	80.870	87.322
4	Switzerland	83.706	81.780	85.540
5	Spain	83.500	80.748	86.180
6	Singapore	83.468	81.444	85.404
7	Italy	83.416	81.230	85.506
8	Australia	83.314	81.478	85.164
9	Iceland	83.152	81.762	84.542
10	France	82.946	80.058	85.814

Source: World Population Review (2019).

heterosexual couples to have an age gap, most commonly with the female being several years younger than the male, heterosexual females may find themselves living alone for a decade or more. Further, it is not uncommon to have significant age gaps between couples, and in some African countries around 30% of relationships involve a large age gap (Karantzas, 2018). Thus, women present as a group at particular risk of social isolation and loneliness as they may be living alone for a decade or more.

Key aspects

There are two key aspects that are important to highlight for researchers and industry practitioners concerned with future development for tourism cities relating to managing an aging population. Those aspects are quality of life and infrastructure. This section will discuss each aspect separately.

Quality of life

Quality of life is now recognised as a key health indicator. While there are a range of definitions that exist for quality of life, it "is broadly defined as an individual's happiness or satisfaction with life and environment including needs and desires, aspirations, lifestyle preferences, and other tangible and intangible factors which determine overall well-being" (Cutter, 1985, p. 1). There is some confusion regarding quality of life, perhaps because the term is used academically as well as through casual conversation. There is also some confusion between quality of life and well-being, where some may feel that the terms are synonyms (Uysal et al., 2016). However, while there is certainly a close relationship between quality of life and well-being, the terms are not identical. Well-being is affected by quality of life; it is focused on those positive components of a person's life such as happiness and other emotions that are positive (Backer, 2019).

The impact to our lives of emotions, positive or negative, is significant. Emotions can be considered as "affective responses governing our daily lives" (Ramer et al., 2019, p. 1). With increasing focus on quality of life as a key health construct, understanding the impact of emotions on our quality of life, and treating quality of life as a preventive health measure seems important. While quality of life has been studied for many decades in numerous disciplines it is relatively new in tourism (Backer, 2019). However, a small number of articles have examined the relationship between emotions and travel experiences (Lin et al., 2014; Nawijn et al., 2013). Understanding the role of quality of life has consistently been revealed as being important and recently was shown to have a significant relationship with impacting mortality (Alimujiang et al., 2019). Further, examining the impact of VFR trips on the emotions of the entire family dynamics during VFR experiences would be beneficial to guide future research (Ramer et al., 2019).

The relationship between VFR travel and quality of life is not yet well understood, but while it can create additional stress and negative emotions in many situations, it can also enhance quality of life (Backer, 2019). Understanding what variables make VFR trips more stressful and what variables tend to enhance quality of life experiences would be of immense value. This is especially important given that reducing loneliness has been shown to improve the quality of life for older people (Zhu et al., 2018) and can be used for preventive healthcare (Backer & King, 2017).

Infrastructure

An aging population requires careful consideration for city planners. It is accordingly a major concern for transport routes and transport provision (Mercado et al., 2010). An aging population also tends to result in an increase in demand for health-related services, aged care facilities and social security services (Carbonaro et al., 2018), which places pressure on other aspects of infrastructure for planners and government.

Of note, while there does tend to be a focus on mobility aspects for the aging population regarding city planning, many members of the aging population are very able and healthy and therefore may have differing needs. Thus, the traditional focus on mobility aspects may need reconsideration in light of meeting the needs of a diverse demographic group who have greatly different preferences, health levels, physical abilities and resources (Mercado et al., 2010). By way of an example, the number of elderly people in Ontario, Canada, who still hold a driver's licence almost doubled between 1988 and 2006 (Mercado et al., 2010).

Certainly, the continuation of mobility (through driving) can be very important to aging people but with the aging population comes an increased risk of car crashes. The serious injury and fatality rates per distance travelled are higher for the group of elderly drivers than middle-aged drivers (Oxley et al., 2010). Therefore, consideration of transport and traffic systems with older drivers in mind will be critical if car accident rates are to be managed in the future.

And while the increasing proportion of the elderly will also put additional strain on public transport, technological changes regarding motor vehicles is likely to result in new transport options. One example could be an "Uber"-style self-drive option where a person does not own a motor vehicle but can call for a driverless car to arrive at their house, from where they independently get transported to their destination. Certainly self-driving cars represent a potential market for all demographics, and companies can be expected to aggressively compete for leadership in this emerging domain (Hancock et al., 2019). While the elderly may initially struggle with feeling comfortable to trial self-driving motor vehicles, friends and relatives may assist in demonstrating the effectiveness of the technology to help elderly feel confident and regain independence otherwise lost. This may be especially beneficial with managing quality of life levels and also reducing strain on relatives through offering care and support to their elderly kin.

Conclusions

This chapter has raised several aspects relevant to tourism city planning. Consideration of planning for and aspects peculiar to an aging population such as social care needs, transport requirements and healthcare needs have been discussed in the literature previously. These issues are not new to the literature. What is new to the literature is the notion of planning in tourism cities for visiting older friends and relatives. Thus, this chapter has made a significant contribution to future research and development by raising a new tourism segment and a new opportunity for tourism city planning – that of planning commercial accommodation to support and encourage visiting older friends and relatives. As discussed previously, VOFR is a new segment that has not been previously mentioned in literature or practice.

Interestingly, while there is general acknowledgement of the interdependence of mobility and old age, the focus is generally on how aging impacts mobility as opposed to how mobility impacts age (Schwanen & Páez, 2010). Naturally, altered mobility in later life can also impact on social connections with family and friends (Rosenbloom, 2010). Thoughtful planning in tourism cities can consider how these social connections may be served through planning. As this chapter has outlined, one component of that network of planning could be commercial accommodation. While planning for "tourism" accommodation tends to be based at CBDs (central business districts) and near tourism attractions, there may also be purpose in considering commercial accommodation options for outlying suburbs that are perhaps purposefully designed for clusters of elderly. That is, it seems sensible to consider a network of needs for the elderly such as suburbs with pedestrian lights that last a little longer, more park benches for stopping frequently on the walk to bus stops, and shading. Those clusters could also have suitable commercial accommodation for catering for VFRs. That accommodation may be self-contained apartments rather than motels and hotels, or even accommodation through groups such as HomeAway. Such aspects with a VFR mindset for connecting the elderly could encourage more visitation which in turn may contribute positively to quality of life and serve as a preventive health measure. Further research into this area would be helpful, with a focus on those larger tourism cities that are witnessing significant improvements in life expectancy. It is felt that consideration of developing commercial accommodation to support VOFR travellers could be important to encourage visitation, which may reduce loneliness, social exclusion and improve quality of life for the elderly. Such aspects could in turn potentially hold preventive healthcare benefits, which would assist in reducing the burden placed on taxpayers already occurring through an aging population.

References

Alimujiang, A., Wiensch, A., Boss, J., Fleischer, N. L., Mondul, A. M., McLean, K., … Pearce, C. L. (2019). Association between life purpose and mortality among US adults older than 50 years. *JAMA Network Open, 2*(5), e194270. https://doi.org/10.1001/jamanetworkopen.2019.4270.

Asiedu, A. (2008). Participants' characteristics and economic benefits of visiting friends and relatives (VFR) tourism: An international survey of the literature with implications for Ghana. *International Journal of Tourism Research, 10*(6), 609–621.

Backer, E. (2007). VFR travel: An examination of the expenditures of VFR travellers and their hosts. *Current Issues in Tourism, 10*(4). https://doi.org/10.2167/cit277.0.

Backer, E. (2008). VFR travellers: Visiting the destination or visiting the hosts? *Asian Journal of Tourism and Hospitality Research, 2*(April), 60–70.

Backer, E. (2010). Opportunities for commercial accommodation in VFR travel. *International Journal of Tourism Research, 12*(4), 334–354. https://doi.org/10.1002/jtr.

Backer, E. (2012). VFR travel: It is underestimated. *Tourism Management, 33*(1), 74–79. https://doi.org/10.1016/j.tourman.2011.01.027.

Backer, E. (2014). The relationship between VFR travel and social tourism in Australia. In C. Ryan (Ed.), *The New Zealand Tourism and Hospitality Research Conference. Tourism in the Asia Pacific Region, 9–12 December 2014* (pp. 54–69). Hamilton, NZ: The University of Waikato Management School.

Backer, E. (2015). VFR travel: Its true dimensions. In E. Backer & B. King (Eds.), *VFR Travel Research: International Perspectives* (pp. 59–72). Bristol: Channel View Publications.

Backer, E. (2019). VFR travel: Do visits improve or reduce our quality of life? *Journal of Hospitality and Tourism Management, 38*(March), 161–167. https://doi.org/10.1016/j.jhtm.2018.04.004.

Backer, E., & Hay, B. (2015). Implementing VFR travel strategies. In E. Backer & B. King (Eds.), *Visiting Friends and Relations: Exploring the VFR Phenomenon*. Bristol: Channel View Publications.

Backer, E., & King, B. (Eds.). (2015). *VFR Travel Research: International Perspectives*. Bristol, UK: Channel View Publications.

Backer, E., & King, B. (2017). VFR traveller demographics: The social tourism dimension. *Journal of Vacation Marketing, 23*(3), 191–204. https://doi.org/10.1177/1356766716665439.

Backer, E., & Laesser, C. (2011). VFR purpose of trip does not always mean staying with hosts: predicting VFR purpose of trip and choice of accommodation. In *CAUTHE 2011 National Conference: Tourism: Creating a Brilliant Blend* (pp. 913–917). Adelaide: University of South Australia.

Backer, E., Leisch, F., & Dolnicar, S. (2017). Visiting friends or relatives? *Tourism Management, 60*, 56–64. https://doi.org/10.1016/j.tourman.2016.11.007.

Backer, E., & Morrison, A. M. (2017). VFR Travel Special Issue. *International Journal of Tourism Research*.

Backer, E., & Ritchie, B. (2017). VFR travel: A viable market for tourism crisis and disaster recovery? *International Journal of Tourism Research, 19*(4), 400–411.

Backer, E., & Weiler, B. (2018). Travel and quality of life: Where do socio-economically disadvantaged individuals fit in? *Journal of Vacation Marketing, 24*(2), 159–171. https://doi.org/10.1177/135676 6717690575.

BBC World News. (2009). Family traffic: Fast track program. Retrieved from http://news.bbc.co.uk/player/nol/newsid_8040000/newsid_8040900/8040921.stm?bw=bb&mp=wm&news=1&nol_storyid=8040921&bbcws=1#.

Boyne, S., Carswell, F., & Hall, D. (2002). Tourism and migration: New relationships between production and consumption. In C. H. A. M. Williams (Ed.), *Tourism and Migration: New Relationships between Production and Consumption* (pp. 241–256). Dordrecht: Kluwer.

Braunlich, C., & Nadkarni, N. (1995). The importance of the VFR market to the hotel industry. *The Journal of Tourism Studies, 6*(1), 38–47.

Capistrano, R. C. G. (2013). Visiting friends and relatives travel, host–guest interactions and qualitative research: methodological and ethical implications. *Asia-Pacific Journal of Innovation in Hospitality and Tourism, 2*(1), 87–100.

Carbonaro, G., Leanza, E., McCann, P., & Medda, F. (2018). Demographic decline, population aging, and modern financial approaches to urban policy. *International Regional Science Review, 41*(2), 210–232. https://doi.org/10.1177/0160017616675916.

Cutter, S. (1985). *A Geographer's View on Quality of Life*. Pennsylvania, PA: Commercial Printing.

Griffin, T. (2013). A paradigmatic discussion for the study of immigrant hosts. *Current Issues in Tourism*, (June), 1–12. https://doi.org/10.1080/13683500.2012.755157.

Hancock, P. A., Nourbakhsh, I., & Stewart, J. (2019). On the future of transportation in an era of automated and autonomous vehicles. *Proceedings of the National Academy of Sciences, 116*(16), 7684–7691. https://doi.org/10.1073/pnas.1805770115.

Hay, B. (1996). An insight into the European experience: A case study on domestic VFR tourism within the UK. In H. R. Yaman (Ed.), *VFR Tourism: Issues and Implications* (pp. 52–66). Melbourne: Victoria University of Technology.

Hu, B., & Morrison, A. M. (2002). Tripography: Can destination use patterns enhance understanding of the VFR market? *Journal of Vacation Marketing, 8*(3), 201–220. https://doi.org/10.1177/13567667 0200800301.

Huang, W.-J., King, B., & Suntikul, W. (2017). VFR tourism and the tourist gaze: Overseas migrant perceptions of home. *International Journal of Tourism Research, 19*(4), 421–434.

Jackson, R. T. (1990). VFR tourism: Is it underestimated? *Journal of Tourism Studies, 1*(2), 10–17.

Karantzas, G. (2018). Mind the gap: Does age difference in relationships matter? Retrieved 13 March 2019, from https://theconversation.com/mind-the-gap-does-age-difference-in-relationships-matter-94132.

King, B. (1994). What is ethnic tourism? An Australian perspective. *Tourism Management, 15*(3), 173–176.

King, B. (1996). VFR tourism: A future research agenda. In H. R. Yaman (Ed.), *VFR Tourism: Issues and Implications* (pp. 85–89). Melbourne: Victoria University of Technology.

Kotler, P., Bowen, J., & Makens, J. (2006). *Marketing for Hospitality and Tourism* (4th ed.). New Jersey: Pearson Education.

Lee, G., Morrison, A. M., Lehto, X., Webb, J., & Reid, J. (2005). VFR: Is it really marginal? A financial consideration of French overseas travellers. *Journal of Vacation Marketing, 11*(4), 340–356. https://doi.org/10.1177/1356766705056630.

Lilburn, L., Brehency, M., & Pond, R. (2018). "You're not really a visitor, you're just a friend": How older volunteers navigate home visiting. *Ageing and Society, 38*(4), 817–838. https://doi.org/10.1017/s0144686x16001380.

Lin, Y., Kerstetter, D., Nawijn, J., & Mitas, O. (2014). Changes in emotions and their interactions with personality in a vacation context. *Tourism Management, 40*, 416–424. https://doi.org/10.1016/j.tourman.2013.07.013.

McKercher, B. (1994). *Report on a Study of Host Involvement in VFR Travel to Albury Wodonga*. Albury: Charles Sturt University.

McKercher, B. (1995). An examination of host involvement in VFR travel. In R. Shaw (Ed.), *Proceedings from the National Tourism and Hospitality Conference, 14–17 February 1995. Council for Australian University Tourism and Hospitality Education* (pp. 246–255). Canberra, ACT: Bureau of Tourism Research.

McKercher, B. (1996). Host involvement in VFR travel. *Annals of Tourism Research, 23*(3), 701–703.

Mercado, R., Páez, A., & Newbold, K. B. (2010). Transport policy and the provision of mobility options in an aging society: A case study of Ontario, Canada. *Journal of Transport Geography, 18*(5), 649–661. https://doi.org/10.1016/j.jtrangeo.2010.03.017.

Morrison, A. M., Hsieh, S., & O'Leary, J.T. (1995). Segmenting the VFR market by holiday activity participation. *Journal of Tourism Studies, 6*(1), 48–63.

Morrison, A. M., Woods, B., Pearce, P. L., Moscardo, G. M., & Sung, H. H. (2000). Marketing to the VFR market: An international analysis. *Journal of Vacation Marketing, 6*(2), 102–118.

Navarro, R., & Turco, D. (2004). Segmentation of the visiting friends and relatives travel market. *Visions in Leisure and Business, 13*(1), 4–16.

Nawijn, J., Mitas, O., Lin, Y., & Kerstetter, D. (2013). How do we feel on vacation? A closer look at how emotions change over the course of a trip. *Journal of Travel Research, 52*(2), 265–274. https://doi.org/10.1177/0047287512465961.

Oxley, J., Langford, J., & Charlton, J. (2010). The safe mobility of older drivers: A challenge for urban road designers. *Journal of Transport Geography, 18*(5), 642–648. https://doi.org/10.1016/j.jtrangeo.2010.04.005.

Provenzano, D., & Baggio, R. (2017). The contribution of human migration to tourism: The VFR travel between the EU28 member states. *International Journal of Tourism Research, 19*(4), 412–420.

Ramer, S. I., Zorotovich, J., Roberson, P. N. E., Flanigan, N., & Gao, J. (2019). Effects of pre-existing family dynamics on emerging adult college students' emotions over the course of fall break. *Journal of Destination Marketing and Management*, (January). https://doi.org/10.1016/j.jdmm.2019.01.004.

Rogerson, C. M. (2017). Unpacking directions and spatial patterns of VFR travel mobilities in the Global South: Insights from South Africa. *International Journal of Tourism Research, 19*(4), 466–475.

Rosenberg, M., & Everitt, J. (2001). Planning for aging populations: Inside or outside the walls. *Progress in Planning, 56*(3), 119–168. https://doi.org/10.1016/S0305-9006(01)00014-9.

Rosenbloom, S. (2010). How adult children in the UK and the US view the driving cessation of their parents: Is a policy window opening? *Journal of Transport Geography, 18*(5), 634–641. https://doi.org/10.1016/j.jtrangeo.2010.05.003.

Schänzel, H., & Brocx, M. (2013). VFR travel: The host experience of the Polynesian community in Auckland. In *Proceedings to the CAUTHE 2013 Conference, Lincoln, New Zealand, 11–14 February*.

Schwanen, T., & Páez, A. (2010). The mobility of older people: An introduction. *Journal of Transport Geography, 18*(5), 591–595. https://doi.org/10.1016/j.jtrangeo.2010.06.001.

Seaton, A. V. (1994). Are relatives friends? Reassessing the VFR category in segmenting tourism markets. In A. V. Seaton, C. L. Jenkins, R. C. Wood, P. U. C. Dieke, M. M. Bennett, L. R. MacLellan, & R. Smith (eds) *Tourism: The State of the Art* (pp. 316–321). Chichester. Wiley.

Seaton, A. (2017). Qualitative approaches to the phenomenology of VFR travel: The use of literary and cultural texts as resources. *International Journal of Tourism Research, 19*(4), 455–465.

Seaton, A. V., & Palmer, C. (1997). Understanding VFR tourism behaviour: The first five years of the United Kingdom Tourism Survey. *Tourism Management, 18*(6), 345–355.

Seaton, A. V., & Tagg, S. (1995). Disaggregating friends and relatives in VFR tourism research: The Northern Ireland evidence 1991–1993. *Journal of Tourism Studies, 6*(1), 6–18.

Shani, A., & Uriely, N. (2012). VFR tourism: The host experience. *Tourism Management, 39*(1), 421–440.

Uysal, M., Sirgy, M. J., Woo, E., & Kim, H. (2016). Quality of life (QOL) and well-being research in tourism. *Tourism Management, 53*, 244–261.

World Health Organization. (2016). Global Health Observatory data. Retrieved 21 February 2019, from www.who.int/gho/mortality_burden_disease/life_tables/situation_trends/en/.

World Population Review. (2019). Life expectancy by country 2019. Retrieved 24 April 2019, from http://worldpopulationreview.com/countries/life-expectancy/.

Yaman, H. (ed.) (1996). *VFR Tourism: Issues and Implications*. Melbourne: Victoria University of Technology.

Young, C. A., Corsun, D. L., & Baloglu, S. (2007). A taxonomy of hosts visiting friends and relatives. *Annals of Tourism Research, 34*(2), 497–516.

Yousuf, M., & Backer, E. (2015). A content analysis of Visiting Friends and Relatives (VFR) travel research. *Journal of Hospitality and Tourism Management, 25*, 1–10. https://doi.org/10.1016/j.jhtm.2015.07.003.

Yousuf, M., & Backer, E. (2017). Hosting friends versus hosting relatives: Is blood thicker than water? *International Journal of Tourism Research, 19*(4), 435–446.

Zhu, Y., Liu, J., Qu, B., & Yi, Z. (2018). Quality of life, loneliness and health-related characteristics among older people in Liaoning province, China: A cross-sectional study. *BMJ Open, 8*(11), 1–7. https://doi.org/10.1136/bmjopen-2018-021822.

16

DARK TOURISM AND CITIES

John J. Lennon

The phenomenon of dark tourism was identified as such and categorised by Lennon and Foley (1996, 2000). For many years humans have been attracted to sites and events that are associated with death, disaster, suffering, violence and killing. Indeed, such sites appear to exert a dark fascination for visitors (Sharpley and Stone, 2009). The area has seen growth in coverage including work in a number of fields: interpretation (Lennon, 2009); selective commemoration (Lennon, 2009; Lennon and Wight, 2007; Lennon and Smith, 2004); cross disciplinary research in the fields of death studies (Lennon and Mitchell, 2007); criminology (Botterill and Jones, 2010); literature (Skinner, 2012); problematic heritage (Ashworth, 1996; Tunbridge and Ashworth, 1995) and in the architectural legacy of dark sites (Philpott, 2016). Conversely, death and acts of mass killing are a major deterrent for the development of certain destinations and yet such acts can become the primary purpose of visitation in others.

As early as 1993 Rojek had referred to "Black spots" and "Fatal Attractions" to highlight sites of fatality which he identified as a feature of the post-modern condition (Rojek, 1993, p. 136). However, such a definitional framework was considered too narrow and Lennon and Foley (2000) hypothesised that there are aspects of the ancient, modern and post-modern to be identified within the spectrum of dark tourism. The phenomena developed included:

- Visits to death sites and disaster scenes
- Visits to sites of mass or individual death
- Visits to sites of incarceration
- Visits to representations or simulations associated with death
- Visits to re-enactments and human interpretation of death

The linkages between cities and dark tourism can be made through geography and history. Cities and urban visitation have been at the forefront of tourism development and growth. As nodal centres, gateways, historical trading sites and concentrations of population, cities are geographically important. Visitors are attracted to cities for a variety of reasons (Peterson, 2018; Page and Hall, 2002) and dark sites are now accepted as visitor attractions, elements of the tourist experience and a motivator of travel (Sharpley and Stone, 2009). However, the linkage with cities and urban locations is not straightforward and this is part of a wider relationship with heritage. The interpretation of heritage is the result of complex interactions and pressures between stakeholders and interest groups. This pressure is acute in the case of "dark" or disso-nant sites (Ashworth and Hartmann, 2005). Heritage is a contested concept and is invariably impacted by competing ideologies, interpretation, funding and a host of other factors. Lowenthal (1998) highlighted that merely defining heritage, let alone agreeing acceptable and verifiable truth(s), will remain elusive. In city tourist attraction sites, visitor centres and those locations considered as dark sites, such issues are confronted. Commemoration, history and its problem-atic and contrasting representation in heritage centres, historic buildings and a variety of "dark" sites is the result of complex interactions of contrasting perceptions, ideologies and interests.

Dark tourism has generated much more than purely academic interest. The term has entered the mainstream and is a popular subject of media attention. The enduring appeal has been reinforced in many urban settings including for example: New York, Washington, DC, Dublin, Belfast, Londonderry/Derry, Paris, Berlin, Munich, Prague, Moscow and Pripyat. These cities are all examples of places where the shared trauma of dark events and heritage has been com-memorated in museums, memorials, interpretation or burial sites. This has formed the basis of much of the academic exploration to date which has had a focus on nations and sites (sometimes in urbanised centres) as the sample of cities examined in Table 16.1 indicates.

The hypothesis that dark sites are historical and cities as centres of urban population are at the centre of history, while logical is only a partial interpretation of this confused landscape. Tragic heritage sites are found dispersed in rural, highland and island locations, yet there is a clear correlation between such sites, urban centres and visitation. The relationship is heightened by tourism in cities, which in many cases will far outperform non-urban locations (Peterson, 2018). The success of urban tourism has of course helped urban attractions and they tend to dominate as most visited attractions at a national level. For example, in the UK there has been a consistent dominance of city-based attractions over recent decades (ALVA, 2019).

Education, memorialisation, learning and preservation of historical records are frequently used to justify and explain motivation for development and visitation to such sites. Such sites in cities and elsewhere are defined by a heritage that has changed over time. Many factors imbue the meanings and content of urban places. This is a function of a plethora of competing influences and agendas: political, economic, cultural, demographic and historical. For example, Berlin is populated by many sites related to the Nazi regime and the Communist regime of the DDR (former East Germany). As the Reich capital this is to be expected but the land-scape of Germany is also peppered with evidence of the darker elements of this nation's past. The selection, interpretation and conservation of elements of the past of Berlin and Germany are critical in understanding what is considered and how it is represented (Ashworth, 2008). The sites listed in the Table 16.1, for Berlin, provide multiple narratives around World War Two, the DDR and the Cold War. They exhibit a range of content driven by influences as diverse as simple commercial gain to the complex interaction of political, economic and ideo-logical agendas (Gegner and Ziino, 2012). In such city sites, interpretation is used to articulate heritage through objects, artefacts, audio recordings, place, imagery or entirely created experi-ences. These sites have the potential to re-represent the past, as evidence, experience and as a critical element of the tourism product. What helps in this debate is exploration of what is interpreted and conserved and what is ignored. Cities such as Berlin clearly define and display their difficult past for residents and tourists; however, other towns and cities choose not to commemorate but rather overlook or selectively appraise those darker elements. This is an important part of the dark tourism narrative evidenced in destinations as diverse as Phnom Penh (Cambodia) and Lety (Czech Republic). A short discussion of contrasting approaches in these locations will illustrate the fundamental challenge for urban locations of a dark or dis-sonant heritage.

Table 16.1 Selected cities with highlighted dark tourism sites sampled

City location (nation)	Locations with tourist appeal and related infrastructure
New York (USA)	Ground Zero and 9/11 Memorial and Museum 9/11 Tribute Centre Ellis Island Museum
Washington, DC (USA)	Arlington National Cemetery Pentagon 9/11 Monument Vietnam Veterans Memorial Korean Veterans Memorial Smithsonian Museum of Air and Space (Missiles and weapons of mass destruction and Enola Gay) US Holocaust Memorial Museum
Dublin (Ireland)	General Post Office Kilmainham Gaol Glasnevin Cemetery
Belfast (Northern Ireland)	Tour of the "Troubles" Murals of the "Troubles" Crumlin Road Gaol Titanic Experience Titanic Quarter Former Sites of Incarceration
Paris (France)	Père Lachaise Graveyard Catacombs Shoah Memorial Napoleon's Tomb Drancy Memorial Deportation Memorial Immigration Museum Pont de l'Alma (Death site Diana, Princess of Wales)
Berlin (Germany)	Berlin Wall Sachsenhausen Plotzensee Memorial Palace of Tears Topography of Terror, Bebelplatz book-burning site and monument German Resistance Memorial Site House of Wannsee Conference Stasi Prison Berlin Story Bunker Holocaust Museum Jewish Museum
Prague (Czech Republic)	Jewish Quarter Jewish Graveyard KGB Museum Memorial to the Victims of Communism Museum of Communism Jan Palach Sculpture Nuclear Bunker
Moscow (Russia)	Lenin Mausoleum Novodevichiy Cemetery Lubyanka (site of KGB headquarters) Gulag History Museum Stalin Bunker Sakharov Centre Muzeon Sculpture Park
Pripyat (Ukraine)	Evacuated town within Chernobyl exclusion zone

Dark tourism and the city of Phnom Penh

Phnom Penh and Cambodia are, for many, synonymous with the period of the Khmer Rouge (1975–1979). On 17 April 1975 Khmer Rouge soldiers entered Phnom Penh to begin a period of rule that was barbaric even by the standards of the twentieth century. The Khmer Rouge obliterated the hierarchical political and social culture of the former regime and reconstructed Cambodian society from ground zero. They rejected modern technology, the marketplace and the division of labour. The small Khmer Rouge force of 60,000 dominated this nation of over 7 million. However, the lack of trained and educated personnel among the Khmer Cadre meant the remnants of the previous regime were considered a "threat" to the revolution. As a consequence, a large proportion of the old elite, skilled labourers, educated professionals and urban dwellers were targeted for extermination, leaving peasant workers and revolutionary soldiers to dominate. The adopted doctrine of self-reliance left many to starve or die in a country bereft of medicine, food and pesticide (Kiernan, 1996).

The world reacted with shock, disbelief and incredulity as the Khmer Rouge emptied cities, obliterated evidence of a consumer society, destroyed books and libraries, ended diplomatic relations, abolished money, markets, foreign exchange, private property and established state control over all foreign and domestic trade. The only external links of any consequence maintained were with China and North Korea (Ponchaud, 1978 and Bizot, 2003). Revolutions and civil wars frequently produce killings and initial periods of revenge attacks, yet under the Khmer Rouge, killing was a systematic programme that rose in amplitude throughout the four-year period of rule. The leadership were intent on a radical agricultural revolution built upon international isolation and punishing levels of independence (Jackson, 1989). In practice, this took the form of extreme attacks on the fabric of the existing society. In the new agricultural communes, families were segregated by gender resulting in the dilution of family and parental influence. Thus, the family and religion, the pillars upon which traditional Khmer society had been built, were completely undermined. Large communal farms were forcibly developed, and individual ownership of land was abolished (Quinn, 1976).

After the Khmer Rouge defeat in 1979, Cambodia was devastated. Approximately one-third of the population were dead (Dunlop, 2005). Mass graves from this period populate the country; some 19,440 mass graves and 167 former security offices/prisons are recorded (Jackson, 1989). These graves contain the bodies of those deliberately executed. They do not contain or record the young, the old or the sick who died in forced evacuations, nor those who died from malnutrition, forced labour, paucity of medicines or other causes. During the period of the Khmer Rouge rule it is widely acknowledged that between 1.5 and 2 million people died. The exact figure may never be known, and accounts differ (see Jackson, 1989; CIA, 1980; Vickery, 1984). Yet, as a percentage of population (estimated at 7.3 million in 1975) this was arguably one of the worst genocides ever perpetrated.

At a national level, the city of Phnom Penh is the location of the only two major visitor sites providing evidential narrative and interpretation of this regime. These sites located in the Cambodian capital are S-21 (the Tuol Sleng Genocide Museum) and Choeung Ek (the "Killing Fields" site). Tuol Sleng is a former secondary school that became a prison (known as S-21) to detain and torture individuals accused of opposing the regime (Chhang and Kosal, 2005). The prisoners and victims of the S-21 facility, were estimated at approximately 10,500 including over 2,000 children, who were also killed here (Chhang and Kosal, 2005). Other authors suggest that detailed review of documentation held at S-21 indicates a figure closer to 20,000 executions at this site (Dunlop, 2005; Chandler, 1999). The regime and the nature of incarceration were brutal, and the shackles and torture instruments used by the guards are now exhibited. In 1979, after capture by the invading North Vietnamese army, S21 (like many of the incarceration sites) was found to contain definitive documentary evidence of execution and torture (Hawk, 1989).

Admission and entry to the site is pitifully low and entrance charges are state controlled. Indeed, since 1979, the museum has faced continual financial challenges and is not allowed to generate revenues from admission, retail or other sources. The government justification for this policy of low-cost admission charge is to ensure access for Cambodian nationals. The narrative of this site is far from complete and the meticulously documented confessions, extracted under duress by the Khmer Rouge, are decaying. For a visitor the enormity of the genocide perpetrated is hard to appreciate or fully understand in a museum starved of funds. City-based heritage attractions, such as S-21, provide the signposts and records to document and preserve the bloody record of the Khmer Rouge. However, the deterioration so clear at S-21 cannot be viewed neutrally. Habermas (1970) has successfully theorised on the impossibility of non-ideological interpretation and it is accepted that both neutrality and accuracy remain elusive concepts from a curatorial perspective. Heritage and the interpretation of history will always be in a socio-political environment which will invariably influence the construction and nature of the interpretation or its absence. Despite some chilling displays, the overall investment in conservation and interpretation at S-21 has been minimal. The building is deteriorating and exhibits of furniture, photographs of victims and documentary evidence are all at risk (Hughes, 2003). Investment by the government of Cambodia has not been delivered to this important part of the city's heritage. Furthermore, S21 has changed and altered over time. Borders and parameters of the site have been lost, building use has changed. As such changes occur the site authenticity and heritage are eroded, yet it remains an essential tourist stop for visitors. However, in its present form it is not a place where those visitors will fully appreciate the tragic past of this nation. The site and its contents suggest a passive remembrance rather than calls for justice or understanding.

The Choeung Ek "Killing Fields" site is perhaps the most famous attraction related to the Khmer Rouge period. It is enshrined in popular consciousness as the site synonymous with the 1984 film, *The Killing Fields*. Annual visitation to the site is in excess of 250,000 and there is a predominance of international visitors. After the Angkor Wat temple complex in Siem Reap, it constitutes the most significant site of tourist visitation in the nation. It is reached by an arterial route south of the capital and many tour buses and coaches now make the 20-minute trip to Choeung Ek from the city centre.

The site centrepiece is the Stupa containing the skulls of over 8,000 victims and the wider site is full of mass graves over which the visitor traverses. The interpretation of such graves is by hand-painted signs in Khmer and poorly translated English. With each monsoon season more topsoil is eroded and skeletal remains, clothing, shoes and other evidence of this tragic place come to the surface. In places, larger bones have been collected and throughout the site tourists film and photograph this spectacle. The site has many non-exhumed graves and the history and archaeology of this place is only partly documented. For a visitor, hoping to understand the complexity of how the Khmer Rouge gained and retained power, created this regime of terror and perpetrated such a large genocide, this site will not provide that understanding. Interpretation remains weak, poorly translated and selective. The site offers a spectacle where tragedy is partially labelled rather than explained, where history is overlooked, and conservation is largely ignored.

These are the only two developed sites, located in or near the city that deal with this past. One killing field site of 19,440 and one prison of 167 nationally. Most have been lost, demolished or overgrown by vegetation. Choeung Ek and S-21 have been effectively starved of funds with admission charges "capped" unlike, for example, the Angor Wat complex, which charges international visitors a more realistic price with a strong emphasis on use of funds for conservation.

Angor Wat, by comparison, represents a very positive period in Cambodian history (and its imagery is reflected in national icons, flags, coins and paper currency); while S-21, and Choeung Ek, represent the other end of that historic continuum. These sites associated with Phnom Penh, challenge the presentation of Cambodia as a "gentle land" with an idealised view of a people locked in the embrace of Theravada Buddhism's peaceful doctrines. Such a selective and partial approach at a city level is not unconnected to how this nation has developed post 1979. The Kingdom of Cambodia's failure to punish, or even remove from positions of authority, those responsible for acts of genocide remains a shadow on the nation. The delayed response of the Cambodian government and the international community to the overwhelming evidence in the form of photographs and documentation is tragic. When, in October 2004, after seven years of negotiation the Cambodian government agreed with the UN to bring to trial the surviving leaders of the Khmer Rouge, just three men – Nuon Chea, Kang Kek Ieu and Khieu Samphan – were finally sentenced to life in prison in 2017. This followed a trial that commenced in 2006 and cost almost US$300 million (Mydans, 2017a). The Cambodian prime minister since 1979, Hun Sen, frequently criticised the legal process and vociferously argued against any further trials of perpetrators (Mydans, 2017b). Prime Minister Hun Sen, as a former Khmer Rouge Divisional Commander, has also not faced legal redress and in 2017, was responsible for the abolition of the main opposition party and Cambodian democracy (Holmes, 2017). How a city and a nation deal with a difficult and tragic past is fundamental to its development as a democratic and civilised society. Tourism sites in such capitals provide a lens through which we can examine the success or otherwise of this development.

Lety and the "Gypsy" camp

The persecution of the Roma and Sinti peoples (hereafter referred to as Roma) has continued throughout European history, particularly during World War Two. The Roma were considered by the Nazi authorities to be an inferior race and were pursued relentlessly. These events remain however "one of the most neglected chapters in the history of the Nazi regime".

These events have been described as a forgotten holocaust (Fraser, 1995). Yet, in the region of 250,000 European Roma and Sinti were killed during this period, although some sources place the figure much higher, at around 500,000 (Fraser, 1995; Guy, 2001). This level of loss was clearly horrendous and proportionately, the Roma suffered greater than any other classification group of victims, except Jews.

During World War Two, persecution was evident in the Czech lands with Roma who were transported to the work camps that had been established at the town of Lety, in Bohemia, and at Hodonin, in Moravia. Deportation came about by edicts issued during 1942 and 1943, partly aimed at the prevention of criminality and the arrest of persons categorised as anti-social (Necas, 1982, 1999). The camp at the town of Lety was originally a forced labour camp, yet excessive work, hunger and overcrowding contributed to the prisoners' sickness, which claimed the lives of many men, women and children. During 1943 and 1944, transports of prisoners were arranged from Lety (and the so-called "Gypsy" camp) to Auschwitz (Necas, 1982, 1999). It is important to note that both the Lety and Hodonin "Gypsy" camps were run solely by Czech personnel with the police authorities having access to detailed personal information concerning Roma that provided a documentary basis for arranging arrest, deportation and extermination (Necas, 1982; Gabal, 1997; Polansky, 1998; Lewy, 2000). Following liberation of Lety and Hodonin, only 600

Roma men and women returned to their homes in the then Czechoslovakia. Clearly, there is little to be proud of in this narrative of Czech nationals involved in arrest, incarceration and deportation of Roma over this period. The interpretation and commemoration clearly reflect a past that many find difficult to accommodate.

The current urban settlement of Lety is situated in Southern Bohemia, just off the main arterial route south of the Czech Republic capital of Prague. It comprises a collection of red roofed houses, a hotel/restaurant and a small factory. To the east of the urban settlement of Lety, the location of the "Agpi" pig and pork processing factory can be found. The factory is located in a tranquil position next to a small lake and is surrounded by forest and low-lying hills. It was originally developed during the Communist period of rule in the then Czechoslovakia. It was privatised in 1994 and the factory processes 13,000–15,000 pigs and employs approximately 20 people (Trojan, 1999). It is on this site, during World War Two, that the Lety "Gypsy" camp was situated. Objections to the location and factory operation have been recorded for decades. The controversy surrounding the non-commemoration of the Lety camp increased from 1994, following an American historian's investigation and discovery of many documents relating to the operation of the camp with clear evidence of Czech nationals' perpetration of atrocities on this site (Polansky, 1998). However, the historical interest in the location has done little to bring about the closure of the pig farm (Gabal, 1997; Polansky, 1998; Mbabuike and Evans, 2000). However, in 1996, the Czech government did provide support for the erection of a single information board and the placing of a contemporary stone memorial at two locations near the site. Notably, Lety contains no signage to the nearby site of Roma incarceration. Indeed, visitors struggle to locate the site, which enjoys only a single signpost in Czech. The locations of the interpretive board and commemorative sculpture are untended and overgrown. At this point the erected interpretation board is not visible. A walk along the track to the left reveals the single interpretation board.

A monument is contained nearly in the shape of a fragmented sphere. It contains only the words in Czech:

Obetem cikanskeho tabora v Letech 1942–1943. Nezapomente. Me Bisteren.

[The victims of the Gypsy Camp 1942–1943. Remember Me]

This comprises the complete interpretation to be found at this site. Pape and Polansky (1998) have argued that the Czech state and public have intentionally suppressed and distorted the history of this camp to the disadvantage of the Roma. They suggest important facts about the camp are concealed and that post-war governments have had little interest in presenting an accurate historical picture. Disputes continue concerning: the extent of the atrocities that took place; responsibility for the atrocities; and the operation of the pork processing factory on the site. Since the erection of the single interpretive board in 1996 no additional investment or development has occurred at the site. Such inactivity as regards this tragic site at Lety must also be considered within the context of the ongoing prejudice against the Roma population. Lety ignores its past, failing to present its tragic history and attracts few visitors. This town's dark past will remain consequently hidden and its evidence obliterated by the presence of this large-scale industrial pig farm.

Understanding the dark tourism agenda in cities and urban spaces

In the case of historical events such as the holocaust and mass killing such as was experienced in Cambodia, collective trauma of the national population is frequently cited as a reason for the inability to reconcile the past with the present. This creates an historical discontinuity as current residents and nationals seek to distance themselves from that difficult past (Roth et al., 2017). Such attitudes, whether evident in museum interpretation, conservation or even omission can be seen as a response to the period of repression and fear. Collective trauma is a possible explanation following the cataclysmic events that impacted greatly on the fabric of many parts of Cambodia. As Hirschberger (2018) records, aside from the horrific loss of life and impact on survivors, collective trauma is also a crisis of meaning. This author delineates a journey or process that commences with collective trauma and which then transforms to a collective memory. The outcome is a system of meaning and understanding of the past that allows individuals and groups to redefine their collective identity. This collective memory of difficult heritage, for example in the incarceration of the Roma in the former Czechoslovakia, contrasts with individual memory. These dark sites as places of evidence persist beyond the survivors. At a human level, whether in Phnom Penh or Lety, Kalinowska (2012) identified defensive elements in the collective psyche that provide a stabilising context for a society or nation's identity and new self-image. Thus, what is commemorated and conserved and what is ignored, destroyed or redeployed is crucial to understanding such responses in cities and nations generally.

The disregard for such heritage, its selective interpretation and the limited conservation can be seen as a historical repression of this period, as government and citizens focus on the future. The challenges of conservation and interpretation of dark sites is an issue not unique to urban locations. However, such sites if developed and conserved can offer learning, and provide evidential heritage of the difficult and sometimes dark past. However, in many cities non-commemoration, deterioration and loss of sites has occurred. This is not simply an issue of ideologically driven selectivity. Other factors such as ownership of historical narratives, historiography and, at an operational level, conservation skills and locally translated economic priorities are also important. The range of factors reaffirms that objects and sites, in cities or elsewhere, do not exist in isolation and are imbued with meaning. The sensitive interpretation of objects, buildings and locations can help local populations comprehend elements of their shared history, which may at first be irreconcilable with their current existence.

Heritage remains a contested terrain and the pursuit of historical "accuracy" is invariably compromised. Partial or selective narratives have been defined as a process of creating multiple constructions of the past whereby history is never an objective recall, but is rather a partial interpretation, based on the way in which we view ourselves in the present. The two short case studies examined are defined by a dark heritage. Maintaining the narrative of such sites in an evidence-based, transparent and politically neutral way is fundamentally important since such sites constitute for many visitors and tourists the major (and in many cases only) learning experience related to such dark and troubled history.

References

ALVA (2019) Latest Visitor Figures 2018, www.alva.org.uk/details.cfm?p=423.

Ashworth, G. (1996) Holocaust Tourism and Jewish Culture: The Lessons of Krakow–Kazimierz, in Robinson, M. Evans, N. and Callaghan, P. (Eds.), *Tourism and Culture: Towards the 21st Century*, Athenaeum Press.

Ashworth, G. (2008) The Memorialisation of Violence and Tragedy: Human Trauma as Heritage, in Graham, B. and Howard, P. (Eds.), *The Ashgate Research Companion to Heritage and Identity*, Farnham, Ashgate.

Ashworth, G. and Hartmann, R. (2005) *Horror and Human Tragedy Revisited: The Management of Sites of Atrocities for Tourism*, New York, Cognizant Communications Corporation.

Bizot, F. (2003) *The Gate*, London, Harvill Press.

Botterill, D. and Jones, T. (2010) *Tourism and Crime*, Oxford, Goodfellow Publishing.

Central Intelligence Agency (CIA) (1980) *Kampuchea: A Demographic Catastrophe*, Washington, DC, US Government Printing Office.

Chandler, D. P. (1999) *Brother Number One: A Political Biography of Pol Pot* (Revised Edition), Thailand, Silkworm Books.

Chhang, Y. and Kosal, P. (2005) *Genocide Museum Tuol Sleng (Former Khmer Rouge S-21 Prison)*, Phnom Penh, Government of Cambodia, Tuol Sleng Museum and Documentation Centre of Cambodia.

Dunlop, N. (2005) *The Lost Executioner*, London, Bloomsbury.

Fraser, A. (1995) *The Gypsies* (2nd Edition), Oxford and Malden, MA, Blackwell.

Gabal, I. (1997) Confronting a Roma Killing Field in the Czech Lands: The East Must Evaluate Its History Based on New Truths, *Transitions Magazine*, 4 (1).

Gegner, M. and Ziino, B. (Eds.) (2012) *The Heritage of War*, Routledge, London.

Guy, W. (2001) The Czech Lands and Slovakia: Another False Dawn?, in Guy, W. (Ed.), *Between Past and Future*, Hatfield, University of Hertfordshire Press.

Habermas, J. (1970) *Towards a Rational Society*, London, Heinemann Educational Books.

Hawk, D. (1989) The Photographic Record, in Jackson, K. D. (Ed.), *Cambodia 1975–78 Rendezvous with Death*, Sussex, Princeton University Press.

Hirschberger, G. (2018) Collective Trauma and the Social Construction of Meaning, *Frontiers in Psychology*, 9, www.ncbi.nlm.nih.gov/pmc/articles/PMC6095989/.

Holmes, O. (2017) Death of Democracy in Cambodia as Court Dissolves Opposition, *The Guardian*, 16 November, 5.

Hughes, R. (2003) *The Abject Artefacts of Memory: Photographs from Cambodia's Genocide*, London and Delhi, Media, Culture and Society Publications.

Jackson, K. D. (Ed.) (1989) *Cambodia 1975–78 Rendezvous with Death*, Sussex, Princeton University Press.

Kalinowska, M. (2012) Monuments of Memory: Defensive Mechanisms of the Collective Psyche and Their Manifestation in the Memorialization Process, *Journal of Analytical Psychology*, 57 (4), 425–444.

Kiernan, B. (1996) *The Pol Pot Regime*, Thailand, Silkworm Books.

Lennon, J. (2009) Tragedy and Heritage in Peril: The Case of Cambodia, *Tourism Recreational Research*, 3 (2), 116–123.

Lennon, J. J. and Foley, M. (1996) Editorial: Heart of Darkness, *International Journal of Heritage Studies*, 2 (1), 195–197.

Lennon, J. J. and Foley, M. (2000) *Dark Tourism: The Attraction of Death and Disaster*, London, Continuum.

Lennon, J. J. and Mitchell, M. (2007) Dark Tourism: The Role of Sites of Death in Tourism, in Mitchell, M. (Ed.) *Remember Me: Constructing Immortality – Beliefs on Immortality, Life and Death*, London, Routledge.

Lennon, J. J. and Smith, H. (2004) A Tale of Two Camps: Contrasting Approaches to Interpretation and Commemoration in the Sites at Terezin and Lety, Czech Republic, *Tourism Recreation Research*, 29 (1), 15–25.

Lennon, J. J. and Wight, C. (2007) Selective Interpretation and Eclectic Human Heritage in Lithuania, *Tourism Management*, 28 (1), 519–529.

Lewy, G. (2000) *The Nazi Persecution of the Gypsies*, New York, Oxford University Press.

Lowenthal, D. (1998) *The Heritage Crusade and the Spoils of History*, Cambridge, Cambridge University Press.

Mbabuike, M. C. and Evans, A. M. (2000) Others Victims of the Holocaust, *Dialectical Anthropology*, 25, 1–25.

Mydans, S. (2017a) Khmer Rouge Trial, Perhaps the Last, Nears End in Cambodia, *The The New York Times*, 23 June, 4.

Mydans, S. (2017b) 11 Years, $300 Million and 3 Convictions: Was the Khmer Rouge Tribunal Worth It? *The New York Times*, 10 April, 6.

Necas, C. (1982) The Czech Gypsies during the Nazi Occupation, *Journal of the Gypsy Lore Society*, 4 (2).

Necas, C. (1999) Bohemia and Moravia: Two Internment Camps for the Gypsies in Czech Lands, in Kenrick, D. (Ed.), *In the Shadow of the Swastika: The Gypsies during the Second World War*, Hatfield, University of Hertfordshire Press.

Page, M. and Hall, C. M. (2002) *Managing Urban Tourism*, London, Prentice Hall.

Peterson, D. (Ed.) (2018) *Urban Tourism in the 21st Century*, London, Willford Press.

Philpot, G. (2016) *Relics of the Reich*, Barnsley, Pen and Sword.

Polansky, P. (1998) *Black Silence*, New York, Cross-Cultural Communications.

Ponchaud, F. (1978) *Cambodia Year Zero*, Harmondsworth, Penguin.

Quinn, K. M. (1976) Political Change in Wartime: The Khmer Krahom Revolution in Southern Cambodia 1070–74, *Naval College Review*, 28 (Spring).

Rojek, C. (1993) *Ways of Seeing Modern Transformations in Leisure and Travel*, London, Macmillan.

Roth, J., Huber, M., Juenger, A. and Liu, J. H. (2017) It's about Valence: Historical Continuity or Historical Discontinuity as a Threat to Social Identity, *Journal of Social and Political Psychology*, 5, 320–341.

Sharpley, R. and Stone, P. (Eds.) (2009) *The Darker Side of Travel*, Bristol, Channel View Publications.

Skinner, J. (2012) *Writing the Dark Side of Travel*, London, Berghahn Books.

Trojan, V. (1999) That Shameful Pig Farm, *The New Presence Magazine*, in *The Patrin Web Journal*, www.geocities.com/Paris/5121/lety.htm, accessed 10 June 2002.

Tunbridge, J. E. and Ashworth, G. J. (1995) *Dissonant Heritage: The Management of the Past as a Resource in Conflict*, Chichester, John Wiley.

Vickery, M. (1984) Democratic Kampuchea: Themes and Variations, in Chandler, D. (Ed.) *Cambodia 1975–1982* (178–198), Boston, MA, South End Press.

17

GREEN CITIES AS BIRD WATCHING DESTINATIONS

David Newsome and Greg Simpson

Introduction and background

Birds have always fascinated people, because of their visibility, colour, song and behaviour. Bird watching is one of the most popular of natural history activities that people from all walks of life pursue (Connell 2009; Kjølsrød 2019; Moss 2009; Ma et al. 2013). The presence of native birds in the urban fabric helps to shape the attractiveness of a city, especially when those birds can be appreciated in association with an abundance of urban green space (Hails 1992; Hails and Kavanagh 2013; Parker and Simpson 2018a). Tourism is among the largest and most rapidly growing industries on the planet and the nature-based tourism (NBT) market segment represents 20% of the global tourism industry (Center for Responsible Travel 2016; United Nations World Tourism Organization (UNWTO) n.d.). With a conservative estimate being that hundreds of thousands of tourists travel to participate in bird watching every year, avitourism is an increasingly important sub-niche of the rapidly growing NBT market segment (Cong et al. 2017; Moss 2009; Newsome 2015; Steven et al. 2015; Steven and Newsome 2020). There are clear economic benefits for international tourism cities that promote urban bird watching as a means of achieving the UN Sustainable Development Goals (Marasinghe et al. in press; Steven et al. 2013; Newsome 2015). Economic development, however, needs to be balanced against the potential disturbance and resultant reduced fitness of bird species targeted by urban avitourism and potential impacts on the environmental values of green spaces that support those species (Barter et al. 2008; Newsome 2017; Marasinghe et al. in press).

With the general trend towards ever more people living in and visiting cities, increased urban population densities are, however, placing pressure on existing green spaces such as city parks and remnants of native vegetation (Simpson and Newsome 2017; Simpson and Parker 2018a). To counter this, the installation of innovative forms of green infrastructure (GI) such as living green-skins planted on the walls and roofs of buildings has been promoted (Hails 1992; Hails and Kavanagh 2013; Parker and Simpson 2018a; Simpson and Parker 2018b). These urban green spaces can provide habitat for wild birds, which, as a combined experience, adds to the aesthetic appeal of cityscapes by softening urban vistas and adding distinctiveness and appeal to urban environments (Salminen et al. 2012; Jones and Newsome 2015; Parker and Simpson 2018b; Shanahan et al. 2015a;). Furthermore, several authors report that urban green space and its associated wildlife in cities is becoming more important as a means of offering rest, relaxation and quieter spaces for urbanised human populations. City greening programmes are now well recognised as providing for better living conditions and healthier urban populations (Bratman et al. 2015; Fuller et al. 2007; Keniger et al. 2013; Parker and Zingoni de Baro 2019; Shanahan et al. 2015b; Simpson and Parker 2018a; World Health Organisation 2019). This GI and its attendant wildlife now extend well into the tourism sphere, with specific parks, urban nature reserves and greenscaped buildings (i.e. green roofs and walls) being marketed in their own right as appealing and interesting tourism features (e.g. Kolczak 2017; Milne 2006; Ryan et al. 2012; Simpson and Newsome 2017; Wild Bird Society of Taipei 2005).

Underpinning the review on GI and city liveability conducted by Parker and Simpson (2018a), in an earlier review of the importance of green roofs for birds, Fernández-Cañero and González-Redondo (2010) made the point that the practice of setting up green roofs and roof gardens is not new but is centuries old. Moreover, Fernández-Cañero and González-Redondo (2010) posited that green roofs comprise an important habitat for birds and the "quality" of a green roof is vital for wildlife and especially important when it provides food resources, a water supply, and cover from predators and for safe conditions breeding (Figure 17.1). The green roof phenomenon, in the modern context, has been particularly noticeable since the late 1990s (Peck et al. 1999). Furthermore, the use of green roofs by birds has been documented across the world. For example, London in the UK (Johnston and Newton 2004; Grant 2006), Berne in Switzerland (Baumann

Figure 17.1　Roof top display garden at the University of Greenwich. The Green Roofs and Living Walls Centre (https://greenroofslivingwalls.org/) conducts research and provides information to the community and local schools. The photograph depicts a roof-top garden containing a range of native species of plant and a pond which can provide food, cover and water for birds.

2006; Cantor 2008), Singapore (Wang et al. 2017; Belcher et al. 2019) and New York in the USA (Partridge and Clark 2018).

Partridge and Clark (2018) reiterate the work of previously mentioned authors in regard to the importance of green space in urban environments for wildlife. Strengthening the call of Simpson and Newsome (2017), Partridge and Clark (2018) also caution that the expansion of urban areas permanently replaces bird habitats and that city greening is becoming more important in the provision of new habitats for birds. This is especially so in the context of a recent report from BirdLife International (BI 2018) indicating a continuing decline in the world's birds. Furthermore, there is a more serious issue at the heart of things here because as much as 40% of the world's bird populations are currently in decline (BI 2018). Admittedly, many of these species occur in degraded and fragmented rainforest environments and on islands (e.g. Alwis et al. 2016; Azman et al. 2011; Mansor and Sah 2012; Marasinghe et al. in press; Perera et al. 2017; Zakaria and Rajpar 2010), but cities have a role to play in raising awareness about the plight of birds.

Urban green spaces, as bird habitats with their component birds can therefore serve as indirect and direct recreational assets for local urban populations, and as tourism attractions for transit and dedicated visitors in the domain of international tourism cities (Parker and Simpson 2018a; Simpson and Newsome 2017). Milne (2006) lists 61 cities in his *Where to Watch Birds in World Cities* guide and includes destinations such as Cairo, Hong Kong, Taipei, Budapest, London, Boston and Sao Paulo. Tall buildings, public parks, botanical gardens, sewage works, wetlands and coastal environments are listed places to watch birds in urban environments. Following on from this, the remainder of this chapter provides an account of four selected cities with green spaces that readily operate as bird watching destinations, illustrates further the significance of urban green space for birds, and highlights an important trend and tourism product for international tourism cities into the future.

Singapore, Republic of Singapore

Entering the twenty-first century, there was concern about the loss of nature in the city-state of Singapore (Wee and Hale 2008). The foreword to the *Birds of Singapore* (Hails and Jarvis 1987) noted that "a mark of civilisation is the importance which its people give to wild creatures". Since that time, however, significant attention has been paid to the retention of remaining natural vegetation and promotion of its values (Hails and Kavanagh 2013). The creation of additional green space has led to Singapore being identified as the "Garden City" (Kolczak 2017). At the same time that urban nature was being lost, there was increasing attention being given to the connection between nature conservation and city liveability. In Singapore this dates back to the 1980s (Hails 1985). Today the promotion of Singapore's green spaces and wildlife as liveability features of the island-state are seen to considerably enhance Singapore's standing as a garden city destination, further increasing its status as an international tourism city (TripAdvisor 2017; 2019a).

Two government reports published in 1985 and 1987 established the policy environment that set the scene for a strategy to *Bring Back the Birds* to Singapore (Hails 1985; 1987; Hails and Kavanagh 2013). Subsequent work and positive results drove an extended greening of the city with wildlife conservation and wild bird populations in mind. Accordingly, the government of Singapore initiated a programme that aimed to protect existing remnants of native vegetation and restore native trees and natural habitats for the purpose of increasing biodiversity and amenity values for Singaporeans (Hails 1985; Hails and Kavanagh 2013).

By employing ornithologists to advise on the specific habitat requirements of different groups of birds, consideration was given to the layout and design of parks, the importance of connecting habitats to one another and the selection of species for planting (Hails and Kavanagh 2013). The programme involved identifying and planning for suitable habitats that contained a good diversity of species and variable plant structural complexity, with a combination of large, medium and small trees along with low vegetation and ground cover (Hails and Kavanagh 2013). Open habitats were also deemed important to increase the diversity of birds along with the inclusion of ponds and streams, provision of nest boxes and the retention of large dead trees with holes used as nests by certain species. Careful consideration was also given to the use of pesticides in parks, to minimise the impact on insect species eaten by the wild birds. The overall strategy aimed to provide for a wide range of habitats for different species. Hails and Kavanagh (2013) report on the importance of providing plants that provide a range of habitats and food resources for birds. One such group in tropical environments are the figs. For example, Hails and Kavanagh (2013) observe that *Ficus microcarpa*, besides providing fruit and associated insects for many birds, has many dangling aerial roots and the trunk of the tree has many convolutions, which increase both micro and macro habitat diversity for birds.

At the start of the twenty-first century, Singapore had a checklist of about 300 species of birds (Hails and Jarvis 1987), but the list is increasing under the initiatives outlined above and the *Bring Back the Birds* programme (Hails and Kavanagh 2013). An example is the return of the Oriental Pied Hornbill (*Anthracoceros albirostris convexus*). The National Parks Board of Singapore initiated a captive breeding programme for the Oriental Pied Hornbill (Figure 17.2) and with

Figure 17.2　Oriental pied hornbill (*Anthracoceros albirostris*) occurs in Singapore. The bird was extinct as a breeding bird in Singapore in 1885, but in more recent times was recorded sporadically visiting Singapore from nearby Malaysia. It was seen regularly in the 1990s in mangroves at Pulau Ubin, an island situated on the northern side of Singapore. The re-establishment of this species as a breeding bird was assisted by captive breeding and release and a conservation project to provide nesting boxes which increased the availability of potential nest sites. The estimated population today is about 60–100 birds.

Source: www.nparks.gov.sg/mygreenspace/issue-15-vol-4-2012/conservation/hornbills-in-the-lion-city.

the aid of nest boxes birds released back into the wild have bred and the hornbills have re-established themselves in Singapore (Koh 2009). In addition, Singapore is now probably the best place in the world for bird watchers to see the endangered Straw Headed Bulbul (*Pycnonotus zeylanicus*), whose populations have been decimated in other parts of Asia by collecting for the cage bird trade (Yong et al. 2018).

While enhancing the liveability of Singapore, this ecological approach to city planning and design has created the conditions for Singapore to become a bird watching destination in its own right. Singapore now has a full range of bird watching sites available to the international tourism city visitor (Milne 2006). TripAdvisor, which has been shown to be a valid method of rating NBT experiences (Prakash et al. 2019), ranks visiting wild bird watching sites high among the 900 Singapore-based tourist experiences reviewed (Kinstler 2018; TripAdvisor 2019a). Noted bird watching sites include the Botanic Gardens (#4 of 900), Central Catchment (MacRitchie) Nature Reserve (#10 of 900), Sungai Buloh Nature Park (#28 of 900), Bukit Timah Nature Reserve (#63 of 900), Pasir Ris Park (#117 of 900) and Bukit Batok Nature Park (#178 of 900). Moreover, Sungai Buloh Nature Park and adjacent reserves in Singapore play an important role in educating residents and tourists alike about birds. An educative component is a key element in authentic NBT experiences (Newsome et al. 2013; Patroni et al. 2018; 2019).

Accordingly, the conservation successes and interest in urban bird watching detailed in this case study help to consolidate the *birds in the city* tourism agenda. Urban green spaces that support avitourism thus provide much scope to increase the diversity of city experiences and tourist engagement through deep learning and the appreciation of nature in an international tourism city.

Dubai, United Arab Emirates

Despite the widespread loss of nature, frequently due to the expansion of cities to accommodate urban development and/or an ever-growing human population, there remains hope and there are positive developments taking place in many urban environments (Parker and Simpson 2018a; Simpson and Parker 2018b). Dubai is one such example, where effective conservation and restoration has taken place within the central part in a large city. This has led to the creation of a bird watching destination that attracts the attention of international tourists who are passing through or holidaying in Dubai (Ryan et al. 2012; Newsome and Rodger 2013).

The Ras Al Khor Wildlife Sanctuary did not exist before 1980, until a decision was made to replant mangroves in order to restore and protect the site as a bird reserve (Ryan et al. 2012; Simpson and Newsome 2017). The previously dredged Dubai Creek has become important regarding the conservation of Greater Flamingos (*Phoenicopterus roseus*) that arrive in Dubai as a stopover and overwintering site during their migration through the Middle East. Since the original restoration efforts, and following protection from tree cutting and hunting, Ras Al Khor has become an important inner-city nature reserve. The site now comprises 6.2 km^2 of mangroves, mudflats, shallow waters with fringing vegetation, and a buffer zone that prohibits development impacting the reserve. The flamingo population is currently estimated to be at 3,000–3,500 overwintering birds, 500 of which have become permanent residents at Ras Al Khor, which includes an increasing number of breeding pairs (Figure 17.3).

Ras Al Khor has therefore become globally recognised as an important wetland bird conservation site in the Middle East and is additionally protected under the Ramsar Convention because of its importance for flamingo conservation (Ryan et al. 2012). The site now attracts 150 species of birds, in combination with a structured feeding programme to encourage the

Figure 17.3   Greater flamingos (*Phoenicopterus roseus*), Ras Al Khor Wildlife Sanctuary, Dubai.

flamingos to remain and breed (Ryan et al. 2012; Newsome and Rodger 2013). Site modification to enhance habitat conditions for birds includes refuge areas for resting birds, platforms and artificial perches, mud dams and an artificial island. The daily feeding of flamingos is entirely conducted as a management activity and in association with the location of a viewing hide facilitates close observation and photography (Newsome and Rodger 2013). There are two main viewing hides, which contain binoculars and information on bird identification for visitors. Access to the reserve is supervised by onsite wardens.

A visitor survey administered by Ryan et al. (2012) revealed that visitors to Ras Al Khor appreciated the "beauty of the flamingos" and the relative solitude that the wetland afforded. In 2019, TripAdvisor (2019b) ranked visiting Ras Al Khor Sanctuary as the #1 wildlife tourism attraction in Dubai and at #64 of the 390 "Things to Do in Dubai". The only nature-related tourism attractions that TripAdvisor ranked higher than visiting Ras Al Khor where manufactured experiences consisting of two artificial aquariums (#15 and #16), a Dolphinarium (#20) and the artificial Green Planet Biosphere (#57).

Such findings indicate the importance of wetland bird habitat restoration (see also Perth case study) and the protection of wildlife in cities, thus increasing the potential for bird watching as an additional activity for residents and tourists alike.

New York City, USA

An iconic global destination, New York City (NYC) is renowned as the city that is so good, they named it twice and is a benchmark for the liveability of all other cities on the planet (The

Economist Intelligence Unit 2016; Kenny 1978; Parker 2017). The "MillionTreesNYC" and "Forever Wild" programmes of planting native trees and protecting urban nature spaces have enhanced the environment of New York, increased biodiversity, and have provided habitats for birds and other wildlife (NYC Parks n.d.a; n.d.b). There are 51 parks and reserves in the city that contain Forever Wild sites that conserve and protect ecologically important remnant and restored patches of natural vegetation (NYC Parks n.d.b). The focus of enhancing biodiversity in the city, and NYC's position along the migration route for waterfowl and many birds of prey and warblers that fly along the Atlantic seaboard during the northern spring and autumn seasons, sets the scene for New York to be an international bird watching destination. The metropolitan area of New York has a checklist of about 400 species of resident and migratory birds (Milne 2006). Many of these species can be seen in the heart of the city at locations such as Central Park (Manhattan), Van Cortlandt Park (north of The Bronx), Pelhau Bay Park (The Bronx), Prospect Park (Brooklyn), Brooklyn Marine Park (shore of Jamaica Bay) and on Jamaica Bay itself (NYC Parks n.d.c). Central Park is a long-established iconic tourism destination and despite its inner-city location, offers the bird watcher the possibility of seeing 120 species in a day (Milne 2006; TripAdvisor 2018). Central Park also provides the bird watching visitor with the prospect of viewing as many as 20 species of migrating warbler during optimum weather conditions (Milne 2006).

New York also serves as an excellent example of the current trend to establish living green-skin elements, such as rooftop gardens and nature spaces, planted on city buildings (e.g. Greenroofs.com 2019; NYC Parks n.d.d; Parker and Simpson 2018a). Partridge and Clark (2018) compared bird utilisation and arthropod abundance between green roofs with conventional rooftops in New York. It is relevant to point out that the green rooftops, in their study sites in Manhattan, The Bronx and Brooklyn, varied in composition. Vegetation composition varied from sedum (a succulent utilised because it can cope with a range of environmental conditions), sedum mixed with grasses and herbs, urban vegetable farms, through to a green roof comprising a mixture of shrubs, fruit trees, grasses and sedum. Partridge and Clark (2018) found 41 species of bird utilising green roofs as compared to 14 on conventional roofs. Most of the birds recorded on green roofs were insectivorous species, reflecting the importance of arthropod presence on green roofs. Generally speaking, the more complex and species-rich the green roof vegetation was, the more likely it will host a greater number of bird species. Partridge and Clark (2018) concluded that green roofs are functioning as stopover habitats during the spring bird migration.

The NYC study (Partridge and Clark 2018), and other work conducted in Singapore (Belcher et al. 2019), strongly indicates that green roofs, as well as enhancing the quality and liveability of the urban environment, add value to the quality of bird habitats in cities. As suggested by Parker and Simpson (2018a), the combination of green public open space (POS), conservation of biodiversity and the provision of urban biodiversity using innovative GI such as green roofs contribute to a city being a viable and sustainable bird watching destination that will attract birds and the avitourists who seek to see them.

Perth, Western Australia

Perth has been celebrated as one of the world's most liveable cities (Jones and Newsome 2015; Parker 2017; Parker and Simpson 2018b). As for most state capital cities in Australia, planners in Perth have accounted for green areas such as parks, gardens and the retention of patches of remnant vegetation (Stephenson and Hepburn 1955 cited in Grose 2007; Western Australian Planning Commission 2004). From the earliest days of British colonisation in 1829, the development of Perth as a metropolitan area has incorporated large agricultural and recreational green spaces as prominent features of the city (Simpson and Newsome 2017). Those areas persist today in locations such as Kings Park, Bold Park and Herdsman Lake (Jones and Newsome 2015; Parker and Simpson 2018b). Many resident birds have survived in the city, because compared to many cities around the world, parks and remnant nature spaces in Perth are relatively large and have good connectivity between different patches of habitat (Patorniti et al. 2016; Parker and Simpson 2018b). Testimony to a city that can function as a bird watching destination was an activity that involved more than 130 citizen scientists who volunteered their time surveying bird watching sites in Perth in 1985. This was done in order to contribute to the first edition of a bird watching guide to Perth (Van Delft 1988). Those citizen scientists identified 174 species from a range of sites. Well-recognised bird watching sites within the metropolitan area include the Swan River at Ashfield, Lake Monger, Kings Park, Bold Park, Alfred Cove, Herdsman Lake, Thomson's Lake, Lake Forrestdale and Lake Claremont (Van Delft 1988; 1997). That work demonstrated Perth as a viable bird watching destination some 30 years ago and the city remains a bird watching destination where visitors can appreciate native and endemic species today (Rykers 2017; Simpson and Newsome 2017).

Lake Claremont presents an interesting case in that the water body and adjacent vegetation demonstrate a complex history of post-colonisation human use, degradation and recovery of a lake that now presents as an important urban green POS and urban tourism resource (Simpson and Newsome 2017). The importance of Lake Claremont (Figure 17.4) as an urban bird watching destination dates back to at least the second half of the twentieth century (Simpson and Newsome 2017). That is in spite of the fact that from early in its post-colonisation history, the

Figure 17.4   Lake Claremont, Perth, Western Australia.

lake has been modified for agricultural production, polluted with chemicals, parts filled in with construction waste and domestic rubbish, its fringing vegetation modified and the component wildlife over-hunted, which resulted in a totally degraded wetland ecosystem by the 1960s (Simpson and Newsome 2017). Lake Claremont acts an Australian example, akin to many urban wetlands around the world that have been degraded or even lost (van Asselen et al. 2013; Davidson 2014). In the opening decades of the twenty-first century wetlands are now increasingly recognised as vitally important ecological, urban liveability and tourism resources for cities such as Perth (Department of Conservation and Land Management 2005; Dooley et al. 2004; 2006; Parker and Simpson 2018a; Simpson and Newsome 2017).

The story of the restoration and recovery of Lake Claremont is important because it acts as an exemplar of a negative environmental history that has been *turned around* to refocus the values of the lake in a modern context. Simpson and Newsome (2017) document these changes, which include weed control and restoration of native vegetation in the riparian zone of the lake. Of particular importance is the conversion of a golf course, which originally replaced natural vegetation and degraded land near the lake, to natural vegetation. A reflection of the way the community values wetlands today, this rehabilitation of the landscape was, largely, initiated and actioned by a *Friends of Lake Claremont* community-based environmental group with support from the Town of Claremont local government authority. The lake is now officially designated as "Bush Forever" site and "conservation category wetland" of the Swan Coastal Plain (Simpson and Newsome 2017).

Lake Claremont is highly valued by the local, and increasingly by the national and international, bird watching community. The years of restoration efforts had resulted in the Lake Claremont bird list increasing from 54 to 96 species since the closure of the golf course in 2005 and maturation of the restored natural vegetation (Simpson and Newsome 2017). A 2015 survey of visitors to Lake Claremont, which included intrastate, interstate and international tourists, found that 10% of survey participants were at the lake for the purposes of watching the birdlife (Parker 2017; Parker et al. 2018). In 2019, TripAdvisor (2019c) rated a visit to Lake Claremont as the "Top Attraction" of the Claremont local government area and two-thirds of the reviews made mention of the local birdlife.

This case study demonstrates how the combination of enlightened policy settings and environmentally focused community values and commitment can rapidly restore degraded wetlands and nature spaces to encourage the re-establishment of bird species that were locally extinct in the urban fabric. As highlighted by Simpson and Newsome (2017), Partridge and Clark (2018) and Parker and Zingoni de Baro (2019), it is, however, essential that cautious management is exercised for urban green POS such as Lake Claremont, or else those gains could just as rapidly be reversed, with the associated loss of urban biodiversity and opportunities for watching birds in the city. In contrast, enhancing the environmental values and diversity of bird species at such urban green POS enhances the liveability and attractiveness of the urban fabric for both residents and tourists alike. Such an approach goes forward to improve the environmental, social and economic sustainability and the competitiveness of international tourism cities.

Challenges for the future and recommendations for further research and policy development

Lepczyk et al. (2017) maintain that we still have a lot to learn about birds and cities and there is a need for standardised monitoring systems and further research. Future challenges include human population growth and the resultant pressure to clear undeveloped natural areas and existing bird habitats for living space. Climate change, such as the drying climate being experienced by Perth, Australia, is likely to result in more wildfires that have the capacity to alter species diversity and the structure of vegetation in at risk cities. This in turn will impact on food availability, the amount of cover and the quality of nesting sites for birds.

Shwartz et al. (2014) caution that a lot of work remains to be done in attempting to understand the complex relationship between people and urban nature. These relationships include the mechanisms that affect urban biodiversity, ecosystem services provided by urban nature, the viability of urban wildlife, how urban populations perceive biodiversity and associated ecosystem services and the means by which people engage with nature and wildlife in cities. Shwartz et al. (2014) go on to provide recommendations for practitioners and urban planners. The first of these involves the enhancement of urban biodiversity experiences for city dwellers so that people will see the benefits of urban nature which can lead to greater awareness and support for green city initiatives. The second point made by Shwartz et al. (2014) emphasises the need for whole urban matrix planning. This is because cities comprise a wide range of green areas under a combination of public and private ownership. For example, city authorities could make it easier for private landowners to adopt green building design, native vegetation retention and facilitate private citizen management of remnant vegetation. The final recommendation refers to the need for further research that can inform policy decision-making processes.

Despite some of the challenges considered earlier, there is much room for optimism and the prospect for a rich research agenda that aims to maintain and increase the presence of wild birds in our cities. To that end, other cities not specifically mentioned earlier that have developed an urban bird watching image, and have recognised urban bird watching sites of international interest, include Malaga (e.g. Guadalhorce Nature Reserve, Spain), Barcelona (e.g. Delta del Llobregat, Spain), Berlin (Spandauer Fost, Germany), Geneva (Geneva Harbour, Switzerland), Buenos Aries (Costanera Sur Reserve, Argentina) and Entebbe (Entebbe Botanical Gardens, Uganda).

Conclusion

The sheer number of bird watchers today and their growing presence as tourists in cities around the world punctuates the importance of diverse offerings in urban green spaces that support a high diversity of bird species to deliver urban wildlife tourism experiences. This chapter highlights that cities of all types potentially have the capacity to offer the international traveller the opportunity to experience new species and nature encounters without leaving the wider environs of a tourism city. Visitors can thus engage in a pleasant and convenient stopover if just transiting through to another destination or to make the city a destination of choice for avitourists. Parks, gardens, nature reserves and innovative GI, such as green roofs, provide bird habitats and with adequate conservation, birds will be present and additional species can become established. Green infrastructure and wild birds are important aspects of city tourism in terms of the joy and health benefits that birds and urban nature bring to people, thus enhancing the quality of life in city environments. What could be more powerful than residents and tourists being able to see that birds can survive in the city, providing they have suitable protection via conservation of wetlands, remnant forests and other green infrastructure such as green roofs and green building design? Bird watching tourism has a strong future alongside the global push for urban greening and liveability agendas that innovative international tourism cities are pursuing.

References

Alwis, N., Perera, P. and Dayawansa, N.P., 2016. Response of tropical avifauna to visitor recreational disturbances: A case study from the Sinharaja World Heritage Forest, Sri Lanka. *Avian Research*, 7, 15. https://doi.org/10.1186/s40657-016-0050-5.

Azman, N.M, Latip, N.S.A., Sah, S.A.M., Akil, M.A.M.M., Shafie, N.J. and Khairuddin N.L., 2011. Avian diversity and feeding guilds in a secondary forest, an oil palm plantation and a paddy field in riparian areas of the Kerian River Basin, Perak, Malaysia. *Tropical Life Sciences Research*, 22 (2), 45–64.

Barter, M., Newsome, D. and Calver, M., 2008. Preliminary quantitative data on behavioural responses of Australian Pelican (Pelecanus conspicillatus) to human approach on Penguin Island, Western Australia. *Journal of Ecotourism*, 7 (2–3), 197–212.

Baumann, N., 2006. Ground-nesting birds on green roofs in Switzerland: Preliminary observations. *Urban Habitats*, 4 (1), 37–50.

Belcher, R.N., Sadanandan, K.R., Goh, E.R., Chan, J.Y., Menz, S. and Schroepfer, T., 2019. Vegetation on and around large-scale buildings positively influences native tropical bird abundance and bird species richness. *Urban Ecosystems*, 22 (2), 213–225.

BirdLife International (BI), 2018. *State of the World's Birds: Taking the Pulse of the Planet*. Cambridge: BirdLife International. Available from: www.birdlife.org/sites/default/files/attachments/BL_Report-ENG_V11_spreads.pdf [Accessed 19 May 2019].

Bratman, G.N., Daily, G.C., Levy, B.J. and Gross, J.J., 2015. The benefits of nature experience: Improved affect and cognition. *Landscape and Urban Planning*, 138, 41–50.

Cantor, S.L., 2008. *Green Roofs in Sustainable Landscape Design*. New York: WW Norton & Company.

Center for Responsible Travel, 2016. *The Case for Responsible Travel: Trends & Statistics 2016*. Available from: www.responsibletravel.org/whatWeDo/The_Case_for_Responsible_Travel_2016_Final.pdf [Accessed 18 May 2019].

Cong, L., Newsome, D., Wu, B. and Morrison, A.M., 2017. Wildlife tourism in China: A review of the Chinese research literature. *Current Issues in Tourism*, 20 (11), 1116–1139.

Connell, J., 2009. Birdwatching, twitching and tourism: Towards an Australian perspective. *Australian Geographer*, 40 (2), 203–217.

Davidson, N., 2014. How much wetland has the world lost? Long-term and recent trends in global wetland area. *Marine and Freshwater Research*, 65, 934–941.

Department of Conservation and Land Management, 2005. *Thomson's Lake Nature Reserve Management Plan*. Perth: Conservation Commission of Western Australia. Available from: www.dpaw.wa.gov.au/images/documents/parks/management-plans/decarchive/thomsons_lake.pdf [Accessed 19 May 2019].

Dooley, B., Bowra, T., Hohloch, S., Loughton, B. and Rajah, R., 2004. *Herdsman Lake Regional Park Management Plan*. Perth: Conservation Commission. Available from: www.dpaw.wa.gov.au/images/documents/parks/management-plans/decarchive/herdsman_lake.pdf [Accessed 19 May 2019].

Dooley, B., Bowra, T., Strano, P., Davis, I., Murray, R. and McGowan, J., 2006. *Beeliar Regional Park Management Plan*. Perth: Conservation Commission. Available from: www.dpaw.wa.gov.au/images/documents/parks/management-plans/decarchive/beeliar_management_plan_18_10_2006.pdf [Accessed 19 May 2019].

Fernández-Cañero, R. and González-Redondo, P., 2010. Green roofs as a habitat for birds: A review. *Journal of Animal and Veterinary Advances*, 9 (15), 2041–2052.

Fuller, R.A., Irvine, K.N., Devine-Wright, P., Warren, P.H. and Gaston, K.J., 2007. Psychological benefits of greenspace increase with biodiversity. *Biology Letters*, 3 (4), 390–394.

Grant, G., 2006. Extensive green roofs in London. *Urban Habitats*, 4 (1), 51–65.

Greenroofs.com, 2019. Jacob K Javits Convention Center. Available at: www.greenroofs.com/projects/jacob-k-javits-convention-center/ [accessed 30 April 2019].

Grose, M.J., 2007. Perth's Stephenson–Hepburn Plan of 1955: 10% POS, and housing then and now. *Australian Planner*, 44 (4), 20–21.

Hails, C., 1985. *Studies of the Habitat Requirement and Management of Wild Birds in Singapore*. Unpublished report. Singapore: Ministry of National Development.

Hails, C., 1987. *Improving the Quality of Life in Singapore by Creating and Conserving Wildlife Habitats*. Unpublished report. Singapore: Ministry of National Development.

Hails, C., 1992. Improving the quality of life in Singapore by creating and conserving wildlife habitats. In: C. Ben-Huat, and N. Edwards, eds. *Public Space: Design, Use and Management*. Singapore: Singapore University Press, 138–158.

Hails, C. and Jarvis, F., 1987. *Birds of Singapore*. Singapore: Times Editions.

Hails, C. and Kavanagh, M., 2013. Bring back the birds! Planning for trees and other plants to support Southeast Asian wildlife in urban areas. *Raffles Bulletin of Zoology*, 29, 243–258.

Johnstone, J. and Newton, J., 2004. *Building Green: A Guide for Using Plants on Roofs, Walls and Pavement*. London: Greater London Authority. Available from: https://brightonandhovebuildinggreen.files. wordpress.com/2017/07/johnstone-and-newton-building-green.pdf [Accessed 19 May 2019].

Jones, C. and Newsome, D., 2015. Perth (Australia) as one of the world's most liveable cities: A perspective on society, sustainability and environment. *International Journal of Tourism Cities*, 1 (1), 18–35.

Keniger, L.E., Gaston, K.J., Irvine, K.N. and Fuller, R.A., 2013. What are the benefits of interacting with nature? *International Journal of Environmental Research and Public Health*, 10 (3) 913–935.

Kenny, G., 1978. *New York, New York (So Good They Named It Twice)*. Gerard Kenny: D & J Arlon Enterprises Ltd.

Kinstler, L., 2018. How TripAdvisor changed travel. *Guardian*, 17 August. Available at: www.theguardian. com/news/2018/aug/17/how-tripadvisor-changed-travel [Accessed 19 May 2019].

Kjølsrød, L., 2019. You can really start birdwatching in your backyard, and from there the sky's the limit. In: L. Kjølsrød, ed. *Leisure as Source of Knowledge, Social Resilience and Public Commitment: Specialised Play*. London: Palgrave Macmillan, 145–168.

Koh, S.K., 2009. My greenspace: Return of the hornbills. Singapore: National Parks Board. Available from: www.nparks.gov.sg/mygreenspace/issue-02-vol-2-2009/conservation/return-of-the-hornbills [Accessed 29 April 2019].

Kolczak, A., 2017. Urban innovator: This city aims to be the world's greenest. *National Geographic*, 28 February. Available from: www.nationalgeographic.com/environment/urban-expeditions/green-buildings/green-urban-landscape-cities-Singapore/ [Accessed 11 April 2019].

Lepczyk, C.A., La Sorte, F.A., Aronson, M.F., Goddard, M.A., MacGregor-Fors, I., Nilon, C.H. and Warren, P.S., 2017. Global patterns and drivers of urban bird diversity. In E. Murgui and M. Hedblom, eds. *Ecology and Conservation of Birds in Urban Environments*. Cham: Springer, 13–33.

Ma, Z., Cheng, Y., Wang, J. and Fu, X., 2013. The rapid development of birdwatching in mainland China: A new force for bird study and conservation. *Bird Conservation International*, 23 (2), 259–269.

Mansor, M.S. and Sah, S.A.M., 2012. The influence of habitat structure on bird species composition in lowland Malaysian rain forests. *Tropical Life Sciences Research*, 23 (1): 1–14.

Marasinghe, S.S., Simpson, G.D., Perera, P.K.P. and Newsome, D., in press. Scoping recreational disturbance of shorebirds to inform the agenda for research and management in tropical Asia. *Tropical Life Sciences Research*.

Milne, P., 2006. *Where to Watch Birds: World Cities*. New Haven, CT: Yale University Press.

Moss, S., 2009. Birding past, present and future: A global view. In: J. del Hoyo, A. Elliott and D.A. Christie, eds. *Handbook of the Birds of the World*, Vol. 14: *Bush-shrikes to Old World Sparrows*. Barcelona: Lynx Ediciones.

Newsome, D., 2015. Conflicts between cultural attitudes, development and ecotourism: The case of bird watching tours in Papua New Guinea. In: K. Markwell, ed. *Animals and Tourism: Understanding Diverse Relationships*. Bristol: Channel View Publications, 94–210.

Newsome, D., 2017. A brief consideration of the nature of wildlife tourism. In: J.K. Fatima, ed. *Wilderness of Wildlife Tourism*. Oakville, Ontario: Apple Academic Press, 21–26.

Newsome, D., Moore, S.A. and Dowling, R.K., 2013. *Natural Area Tourism: Ecology, Impacts and Management*. 2nd ed. Bristol: Channel View Publications.

Newsome, D. and Rodger, K., 2013. Feeding of wildlife: An acceptable practice in ecotourism? In: R. Ballantyne and J. Packer, eds. *International Handbook on Ecotourism*. Cheltenham: Edward Elgar, 436–451.

New York City (NYC) Parks, n.d.a. *MillionTreesNYC*. Available from: www.nycgovparks.org/trees/milliontreesnyc [Accessed 19 May 2019].

New York City (NYC) Parks, n.d.b. *Forever Wild: Nature in New York City*. Available from: www.nyc govparks.org/greening/nature-preserves [Accessed 19 May 2019].

New York City (NYC) Parks, n.d.c. *Nature in New York City?!* Available from: www.nycgovparks.org/pagefiles/50/forever_wild_brochure.pdf [Accessed 30 April 2019].

New York City (NYC) Parks, n.d.d. *Green Roofs*. Available from: www.nycgovparks.org/greening/sustainable-parks/green-roofs [Accessed 30 April 2019].

Parker, J., 2017. *A Survey of Park User Perception in the Context of Green Space and City Liveability: Lake Claremont, Western Australia*. Thesis (Master's). Murdoch University. Available at: https://research repository.murdoch.edu.au/id/eprint/40856/ [Accessed 20 May 2019].

Parker, J. and Simpson, G.D., 2018a. Public green infrastructure contributes to city livability: A systematic quantitative review. *Land*, 7 (4), 161. https://doi.org/10.3390/land7040161.

Parker, J. and Simpson, G., 2018b. Visitor satisfaction with a public green infrastructure and urban nature space in Perth, Western Australia. *Land*, 7 (4), 159. https://doi.org/10.3390/land7040159.

Parker, J., Simpson, G.D. and Newsome, D., 2018. Who visits the Lake Claremont public open space and what are those visitors doing? Murdoch University. Available from: www.murdoch.edu.au/School-of-Veterinary-and-Life-Sciences/_document/Research-Bulletins/5.16-Lake-Claremont_HR.pdf [Accessed 20 May 2019].

Parker, J. and Zingoni de Baro, M.E., 2019. Green infrastructure in the urban environment: A systematic quantitative review. *Sustainability*, 11(11), 3182. https://doi.org/10.3390/su11113182.

Partridge, D.R. and Clark, J.A., 2018. Urban green roofs provide habitat for migrating and breeding birds and their arthropod prey. *PLoS ONE*, 13 (8), 0202298. https://doi.org/10.1371/journal.pone.0202298.

Patorniti, G., Brennan, K., Hanly, P., Prideaux, C., Olejnik, C., Debono, S., Fitzgerald, B. and MacGregor, E., 2016. *Swan Coastal Plain South: Management Plan 85.* Kensington: Department of Parks and Wildlife. Available from: www.dpaw.wa.gov.au/images/documents/parks/management-plans/swan_coastal_plain_south_management_plan.pdf [Accessed 19 May 2019].

Patroni, J., Day, A., Lee, D., Chan, L., Kim, J., Kerr, D., Newsome, D. and Simpson, G.D., 2018. Looking for evidence that place of residence influenced visitor attitudes to feeding wild dolphins. *Tourism and Hospitality Management*, 24(1), 87–105.

Patroni, J., Newsome, D., Kerr, D., Sumanapala, D.P. and Simpson, G.D., 2019. Reflecting on the human dimensions of wild dolphin tourism in marine environments. *Tourism and Hospitality Management*, 25 (1), 1–20.

Peck, S.W., Callaghan, C., Kuhn, M.E. and Bass, B., 1999. *Greenbacks from Green Roofs: Forging a New Industry in Canada.* Ottawa: Canada Mortgage and Housing Corporation. Available from: http://citeseerx.ist.psu.edu/viewdoc/download?doi=10.1.1.196.7020&rep=rep1&type=pdf [Accessed 19 May 2019].

Perera, P., Wijesinghe, S., Dayawansa, N., Marasinghe, S. and Wickramarachchi, C., 2017. Response of tropical birds to habitat modifications in fragmented forest patches: A case from a tropical lowland rainforest in south-west Sri Lanka. *Community Ecology*, 18 (2), 175–183. https://doi.org/10.1556/168.2017.18.2.7.

Prakash, S.L., Perera, P., Newsome, D., Kusuminda, T. and Walker, O., 2019. Reasons for visitor dissatisfaction with wildlife tourism experiences at highly visited national parks in Sri Lanka. *Journal of Outdoor Recreation and Tourism*, 25, 102–112.

Ryan, C., Ninov, I. and Aziz, H., 2012. Ras Al Khor – Eco-tourism in constructed wetlands: Post modernity in the modernity of the Dubai landscape. *Tourism Management Perspectives*, 4, 185–197.

Rykers, E., 2017. Urban birdwatching guide to Perth. *National Geographic*, 11 October. Available at: www.australiangeographic.com.au/topics/wildlife/2017/10/urban-birdwatching-guide-to-perth/ [Accessed 30 April 2019].

Salminen, O., Ahponen, H., Valkama, P., Vessman, T., Rantakokko, K., Vaahtera, E., Taylor., A., Vasander, H. and Nikinmaa, E., 2012. Benefits of green infrastructure: Socio-economic importance of constructed urban wetlands (Nummela, Finland). In: M. Kettunen, P. Vihervaara, S.D. D'Amato, T. Badura, M. Argimon and P. Ten Brink, eds. *Socio-economic Importance of Ecosystem Services in the Nordic Countries: Synthesis in the Context of The Economics of Ecosystems and Biodiversity (TEEB)*. Copenhagen: Nordic Council of Ministers, 245–254. https://doi.org/10.6027/tn2012-559.

Shanahan, D.F., Lin, B.B., Gaston, K.J., Bushand, R. and Fuller, R.A., 2015a. What is the role of trees and remnant vegetation in attracting people to urban parks? *Landscape Ecology*, 30 (1), 153–165.

Shanahan, D.F., Lin, B.B., Bush, R., Gaston, K.J., Dean, J.H., Barber, E. and Fuller, R.A., 2015b. Toward improved public health outcomes from urban nature. *American Journal of Public Health*, 105 (3), 470–477.

Shwartz, A., Turbé, A., Julliard, R., Simon, L. and Prévot, A.C., 2014. Outstanding challenges for urban conservation research and action. *Global Environmental Change*, 28, 39–49.

Simpson, G. and Newsome, D., 2017. Environmental history of an urban wetland: From degraded colonial resource to nature conservation area. *Geo: Geography and Environment*, 4 (1), e00030. doi:10.1002/geo2.30.

Simpson, G.D. and Parker, J., 2018a. Data for an Importance–Performance Analysis (IPA) of a public green infrastructure and urban nature space in Perth, Western Australia. *Data*, 3 (4), 69. https://doi.org/10.3390/data3040069.

Simpson, G.D. and Parker, J., 2018b. Data on peer-reviewed papers about green infrastructure, urban nature, and city liveability. *Data*, 3 (4), 51. https://doi.org/10.3390/data3040051.

Steven, R., Castley, J.G. and Buckley, R., 2013. Tourism revenue as a conservation tool for threatened birds in protected areas. *PLoS ONE*, 8 (5), p.e62598. https://doi.org/10.1371/journal.pone.0062598.

Steven, R., Morrison, C. and Castley, J.G., 2015. Birdwatching and avitourism: A global review of research into its participant markets, distribution and impacts, highlighting future research priorities to inform sustainable avitourism management. *Journal of Sustainable Tourism*, 23 (8–9), 1257–1276.

Steven, R. and Newsome, D., 2020. Birdwatchers as a consumer tribe in nature-based tourism. In: C. Pforr, M. Volgger and R. Dowling, eds. *Consumer Tribes in Tourism: Contemporary Perspectives on Special Interest Tourism*. New York: Springer.

The Economist Intelligence Unit, 2016. *A Summary of the Liveability Ranking and Overview: August 2016*. London: The Economist Intelligence Unit Limited. Available from: www.tagesschau.de/ausland/ranking-101~_origin-fa2e3e36-d44d-4633-bff7-c8b471e645f6.pdf [Accessed 2019].

TripAdvisor, 2017. *Gardens by the Bay*. Available from: www.tripadvisor.com.au/ShowUserReviews-g294265-d2149128-r481291718-Gardens_By_The_Bay-Singapore.html [Accessed 18 May 2019].

TripAdvisor, 2018. *Bird Watching in Central Park*. Available from: www.tripadvisor.com.au/ShowTopic-g60763-i5-k11516862-Bird_watching_in_Central_Park-New_York_City_New_York.html [Accessed 18 May 2019].

TripAdvisor, 2019a. *Top Attractions in Singapore 2019*. Available from: www.tripadvisor.com.au/Attractions-g294265-Activities-Singapore.html [Accessed 18 May 2019].

TripAdvisor, 2019b. *Top Attractions in Dubai 2019*. Available from: www.tripadvisor.com.au/Attractions-g295424-Activities-Dubai_Emirate_of_Dubai.html [Accessed 18 May 2019].

TripAdvisor, 2019c. *Lake Claremont*. Available from: www.tripadvisor.com/Attraction_Review-g950978-d8449318-Reviews-Lake_Claremont-Claremont_Greater_Perth_Western_Australia.html [Accessed 1 May 2020].

United Nations World Tourism Organization (UNWTO), n.d. *Why Tourism? Tourism – an Economic and Social Phenomenon*. Available from: www2.unwto.org/content/why-tourism [Accessed 18 May 2019].

van Asselen, S., Verburg, P.H., Vermaat, J.E. and Janse, J.H., 2013. Drivers of wetland conversion: A global meta-analysis. *PLoS ONE*, 8 (11), e81292. https://doi.org/10.1371/journal.pone.0081292.

Van Delft, R. 1988. *Birding Sites around Perth*. Perth: University of Western Australia Publishing.

Van Delft, R. 1997. *Birding Sites around Perth*. 2nd ed. Perth: University of Western Australia Publishing.

Wang, J.W., Poh, C.H., Tan, C.Y.T., Lee, V.N., Jain, A. and Webb, E.L., 2017. Building biodiversity: Drivers of bird and butterfly diversity on tropical urban roof gardens. *Ecosphere*, 8 (9), e01905. https://doi.org/10.1002/ecs2.1905.

Wee, Y.C. and Hale, R., 2008. The Nature Society (Singapore) and the struggle to conserve Singapore's nature areas. *Nature in Singapore*, 1, 41–49.

Western Australian Planning Commission (WAPC), 2004. Metropolitan Region Scheme MRS Amendment No 1082/33. Bush Forever and related lands. Perth: Government of Western Australia.

Wild Bird Society of Taipei, 2005. *Birdwatching in Taiwan*. Taipei: Wild Bird Society of Taipei.

World Health Organisation, 2019. Urban green spaces. Available from: www.who.int/sustainable-development/cities/health-risks/urban-green-space/en/ [Accessed 19 May 2019].

Yong, D.L., Lim, K.S., Lim, K.C., Tan, T., Teo, S. and Ho, H.C., 2018. Significance of the globally threatened Straw-headed Bulbul *Pycnonotus zeylanicus* populations in Singapore: A last straw for the species? *Bird Conservation International*, 28 (1), 133–144.

Zakaria, M. and Rajpar, M.N., 2010. Bird species composition and feeding guilds based on point count and mist netting methods at the Paya Indah Wetland Reserve, Peninsular Malaysia. *Tropical Life Sciences Research*, 21 (2), 7–26.

PART III

Product and technology developments for tourism cities

Part III takes a supply-side view on tourism in urban areas. The goal is to review key aspects of product and technology developments. It begins by considering some of the significant products often associated with cities including culture and heritage, and city markets. Coastal and harbour cities offer distinctive opportunities for tourism and two examples are discussed in Part III in Sydney, Australia and Port Elizabeth, South Africa. Cities are known for having portfolios of diverse attractions and Part III also discusses these urban "drawing cards". Another feature of many cities is that they have old and new parts, and there is a chapter in Part III that reviews how this "ambidexterity" works for city tourism. Then there is walking in cities, which is such a great way to explore these fascinating places.

As hinted to in the Introduction to the *Handbook*, technology is the source of many of the trends occurring in tourism supply and demand. Part III delves into these trends in greater detail with coverage of smart tourism, e-tourism and social media.

A brief description of the 11 chapters in Part III follows.

Stephen Boyd first provides a narrative over the decades regarding the extent of the size of the research field on cultural and heritage tourism, and how scholarly engagement has evolved and matured over time. Using literature reviewing as the main data source, he follows with a number of sections that explore certain facets of that relationship more closely, namely clustering of heritage and cultural supply within the inner-city region and the development of new heritage and cultural spaces in the form of trails and routes. The author concludes by providing a conceptual model of cultural and heritage tourism in a city context.

Joan Henderson discusses the characteristics and roles of open-air and covered markets in urban tourist destinations where they often serve as a visitor attraction and amenity. The attributes of various types of market and their appeal to tourists are considered as well as contrasts around the world, illustrated by specific examples from popular cities, including Bangkok, Hong Kong, Kuala Lumpur, London, New York and Singapore. She reviews general and speciality markets, food markets and cooked food markets using secondary data. The results have practical implications for urban planners and policy makers together with the tourism industry of private enterprise and public agencies which have responsibilities for the approval, operation and advancement of markets as a resident and visitor resource.

Ece Kaya and **Deborah Edwards** use the case of Darling Harbour in Sydney, Australia to pinpoint the key challenges with such creative, waterfront developments in major cities. They argue that urban tourism spaces can be more than what Darling Harbour offers now. Urban tourism spaces can be used in more creative and productive ways by both residents and tourists. Drawing on theoretical perspectives this chapter discusses what those creative and productive uses might be and seeks to address the question of what Darling Harbour can offer to please both the residents of Sydney and tourists.

Sello Samuel Nthebe and **Magdalena Petronella (Nellie) Swart** outline current perspectives on tourist attractions in terms of how they shape urban destination development and specifically through the lens of business tourists. They explore how historical and intellectual developments associated with tourist attractions have contributed to the development of urban tourism destinations. An outline is given of the different types of tourist attractions and impacts, to support the main claims and developmental stages. Four factors are highlighted: (i) the range of tourist attractions, (ii) security at those attractions, (iii) their authenticity and (iv) the location of the tourist's accommodation.

Hugues Séraphin, with a focus on France, shows that the ambidextrous model of tourism cities contributes to the good performance of those destinations. The author conceptually and critically discusses ambidextrous management in tourism in general, and then its application to cities' design. This chapter is written based on secondary research. Ambidexterity means that cities have old and historical parts (e.g. *le vieux* Lyon; *le vieux* Strasbourg; *le vieux* Carcassonne) and new and modern parts (e.g. *le nouveau* Lyon; *le nouveau* Strasbourg; *le nouveau* Carcassonne). The chapter provides evidence to support the fact that as an approach, and/or concept, ambidexterity is suitable to the tourism industry.

Cina van Zyl presents a practical approach to, and a solution for, marketing and branding tourism in urban areas. The author starts by introducing key definitions and concepts, providing a theoretical base on which to brand the city. City branding constraints are emphasised. Then city attractors are introduced and applied to the case of Port Elizabeth, South Africa to illustrate the city's branding possibilities. The Port Elizabeth case study is used to explore the realities faced by tourism cities and determine how to manage this activity. A SWOT analysis illustrates the city's current position, some lessons learnt and several future focus opportunities.

J. Andres Coca-Stefaniak and **Gildo Seisdedos** review some of the key parallels between the concepts of smart cities and smart tourism destinations. Evidence of new trends in the smart concept as well as the widening of the smart tourism destination concept to neighbouring regions of established smart tourism cities are discussed with reference to examples from practice in Europe and China. A new typology is proposed for smart tourism destinations that encompasses tourism cities as well as their wider region.

Sebastian Molinillo, **Rafael Anaya-Sánchez** and **Antonio Guevara-Plaza** discuss critically, based on a review of recent literature on the topic, the challenges faced by urban tourism destinations with regard to e-tourism and new technologies. Publications are grouped into categories and qualitative analysis is applied to identify the main topics and challenges for urban tourism destinations. Based on this analysis, the chapter discusses tourist experience, value co-creation, mobile technologies, the sharing economy, social media, governance, information management and management of the destination offer. The theoretical contribution of this chapter revolves around a comprehensive literature review of e-tourism and urban destinations. Practical implications are also discussed, with an emphasis on the potential implementation of technological developments.

Ulrike Gretzel discusses the many ways in which social media shape city tourism experiences and influence the way tourism cities can be managed and marketed. The author draws attention to the mediatisation of urban spaces and its consequences for tourists, residents and tourism providers. She argues that social media enable new forms of urban tourism that enhance tourism experiences but can also lead to overtourism and negative impacts on the city and its residents. The chapter concludes that more research is needed to deeply understand the diverse and complex roles social media play in city tourism.

Claire Papaix and J. Andres Coca–Stefaniak explore the role of transport in tourism cities beyond its functional value of connecting people and places. Drawing from literature in tourism and transport studies, this chapter argues the case for an experiential approach to the design of transport systems in urban tourism destinations with a special emphasis on elements of wellbeing co-creation that involve local residents as well as tourists.

Blake Morris discusses how the artistic medium of walking can contribute to reflexive modes of tourism that encourage critical engagement with the everyday, rather than an escape from it. The author describes three different modes of artistic walking practices that are available to tourists. Then, he moves on to specific examples of artistic practices, including audio–walks, psychogeographic drifts and mass participation walks. These include the site-specific, such as the Loiterer Resistance Movement's First Sunday Walks in Manchester and Deveron Projects' annual Slow Marathon in Scotland, as well as those that can be done anywhere in the world, as is the case with Jennie Savage's *A Guide to Getting Lost* (2014). The chapter explains how these walks differ from more traditional walking tourism activities and argues they offer modes of critical engagement with the city that encourages a reflexive tourism.

18

CULTURAL AND HERITAGE TOURISM IN CONTEMPORARY CITIES

Stephen W. Boyd

Introduction

Cultural and heritage tourism are well recognised as two of the oldest forms of tourism. Scholarly literature has debated how both have been defined, their importance to destinations, the impacts they generate and the challenges they face (du Cros and McKercher, 2015; Timothy and Boyd, 2003; Richards, 2007; Smith and Robinson, 2006). Both are umbrella terms that define a product category involving cultural and/or heritage products that are built, intangible and which in many cases have been modified over time. Combined they represent the most important segments of global tourism; this is not surprising given that all destinations have culture and a heritage that has been passed down. Both types of tourism have a strong experiential dimension attached to them and are popular across all age ranges and destinations. The focus of this chapter is to examine both within the specific context of the city. Cities have played a major role in the shaping of society over time; they are important places because of the heritage and culture they contain. However, the city has not been a setting that has evoked much debate by scholars; the number of major texts on urban tourism has been rather limited over the years (see for example Page, 1995 and Page and Hall, 2003). This is somewhat surprising as cities are places that strongly associate with heritage and culture; however, interest in both by visitors has varied over time.

It is, however, possible to argue that the relationship between cities and their appeal to visitors regarding their culture and heritage has been cyclical over time. The appeal of the early Grand Tour travellers was to the cities because of their heritage and culture and not coastlines, unless cities were located on coasts (Towner, 1996). The advent of beach holidays and the rise of mass tourism to coastal areas, either adapted or deliberately designed for tourism, however, resulted in the loss of appeal for cities as places to visit exclusively. Even with the development of special interest forms of tourism including heritage and cultural, many destinations promoted regions peripheral to their cities and wider environs. The heritage and culture of destinations was often not located within a city region context. Only in the past few decades have tourists rediscovered the city as a place to visit. That interest has been driven by a number of factors. First, the rise of short break holidays to principally cities as destinations in their own right. Second, the route networks of low-cost airlines to both cities and other secondary urban centres have connected large markets to many city destinations; this is certainly the case in Europe. Third, heritage and cultural capital have been important in reshaping many post-industrial economies of cities, and cities have become interesting places again to visit because of their mix of visitor attractions. However, the popularity of cities as destinations in their own right has created unforeseen problems around the emerging concept of overtourism, and that for some cities they are facing the problems that other tourism regions have faced in the past and that within them are the seeds of their own possible demise.

Cities have played a key role as destinations which encapsulate attractions and experiences that may be broadly grouped under cultural and heritage tourism, including the arts, festivals, events, cultural quarters linked to past histories and individuals, customs and traditions, food and drink, monumental built heritage, religion, indigenous culture, creative industries and dissonant heritage of place (e.g. Ashworth and Tunbridge, 2000; Timothy and Boyd, 2003; du Cros and McKercher, 2015). While both types of tourism (cultural and heritage) can take place in different geographic contexts, urban, rural, peripheral and isolated, the urban/city setting is often where many of these attraction types are found and consumed. What is of interest in this chapter is the role that the city space has played as host to these opportunities, attractions and experiences. From a market perspective, both types of tourism appeal across the spectrum from those that are classed as "serious", "casual" or "serendipitous".

The history of tourism will reveal that cities have played host in the past to major events and festivals (carnivals being a classic example of this), they are where different cultural groups would coalesce and maintain their traditions and cultures (modern-day events and festivals help showcase this today). Cities have been those key settlements where monumental heritage got "built" and "added to" over the years (cathedrals, museums, art galleries, monuments, castles, city halls, places associated with government and governance) and that much of this in terms of city planning saw it clustered within a city's historic core; often representative of the original city. The extent to which tourism scholars interested in examining heritage and cultural tourism have done so within the specific city setting is what is at the heart of this chapter.

In order to do so, the chapter is structured as follows. The first section provides a narrative over the decades regarding the extent of the size of the research field, and how scholarly engagement has evolved and matured over time. This is followed by a number of sections that explore certain facets of that relationship more closely. It is impossible in the confines of a single chapter to do credit to all possible areas of research; rather a selection of themes and issues are addressed that the author deems to be of interest and value to the reader wanting to understand the relationship that cultural and heritage tourism has had with the city as a geographic entity, namely clustering of heritage and cultural supply within the inner-city region, the development of new heritage and cultural spaces in the form of trails and routes, the value of brands and accolades to cities, in particular European Capital of Culture status, and the risks associated with cities if their heritage and cultural offer is overconsumed and number exceed the carrying capacity of cities; the latter of which is addressed in the conclusion. Attention now shifts to provide a narrative of the engagement tourism scholars have had with the city as their case study where the focus has been on heritage and culture.

Relationship between cultural and heritage tourism and cities over time

When it comes to what is of interest to tourism researchers, it is fair to say that focus has not been with cities and urban spaces; instead it has been with coasts and beaches (3 S [sun, sea and sand] product development) as well as natural and peripheral spaces (nature tourism, ecotourism), and that the rise of interest over the years on scholarship regarding cities, especially with reference to culture and heritage has been one of slow adoption followed by clear evidence of traction and an established body of research over time, especially since the turn of the millennium. Table 18.1 provides a snapshot of the themes this body of research has covered over the years based on

Table 18.1 Research activity on heritage, culture and cities over time

Period	Number of publications	Key themes and authors
1971–1980	2	Cities as places to be managed (Bosselman, 1978)
1981–1990	8	Urban regeneration (McNulty, 1985); elements of inner-city tourism (Jansen-Verbeke, 1986)
1991–2000	69	Heritage tourism product in cities (Caffyn and Lutz, 1999); cultural production in cities (Richards, 1996); heritage management in cities (van der Borg et al., 1996); urban heritage tourism development (Chang et al., 1996; Jansen-Verbeke and Lievois, 1999); heritage planning in cities (Ashworth and Tunbridge, 1999); culture and regeneration (Shaw and Macleod, 2000); role of the cultural sector and sustainable urban tourism (Russo and van der Borg, 2000; Laws and Le Pelley, 2000); impact case studies (Glasson, 1994; Curtis, 1998; Ashworth, 2000); role of heritage in commodification of cities (Meethan, 1996); modelling the tourist-historic city (Ashworth and Tunbridge, 2000)
2001–2010	182	Relationship between tourism, cities and visitor economy (Law, 2002); heritage within a destination lifecycle (Russo, 2002); built heritage in former colonial cities (Henderson, 2002a, 2002b); capital city function as linked to heritage and culture (Hall, 2002); sustainable tourism development in the context of heritage cities (van der Borg, 2004); regeneration (as linked to cultural development – Smith, 2006; creative turn – Richards and Wilson, 2006a; former soviet cities – Rátz et al., 2008; post-oil Gulf cities – Alraouf, 2010; heritage precinct development – Wheeler et al., 2009); management of multi-heritage site cities (Pria et al., 2007); popular culture and heritage (Frost, 2008)
2011–2019	615	State of the art review on creativity and tourism (Richards, 2011); ECoC focused research (e.g. Lui, 2014a, 2014b, 2014c); capital cities as open air museums (Saidi, 2012); cultural urban walking routes (Ferreira et al., 2012); *Routledge Handbook of Cultural Tourism* (Smith and Richards, 2013); cultural experience and authenticity (Wickens, 2017; Ram et al., 2016); Special issues: *Journal of Tourism and Cultural Change*, 2016 – contested heritage in former Yugoslavia; *International Journal of Tourism Cities*, 2017 – communist legacies, heritage and cities; *Cultural Tourism and Sustainability* – impacts on historic city centres (García-Hernández, et al., 2017); and *Tourism Geographies*, 2018 – overtourism, placemaking and heritage.

Source: Authors own work using CABI Leisure and Tourism Abstract Database (1978–2019).

a review of the Leisure and Tourism Abstracts CABI database. The author undertook an advanced search between "cultural and heritage tourism" and "cities" and examined this within specific time periods, commencing in the 1970s when the first research was undertaken. Of note here is the work by Bosselman (1978). Although this early study did not specifically focus on cities exclusively, it was the first noticeable study that drew scholarly attention to the challenges that planned and unplanned development creates for land-use settings, where the environments that were examined included cities. However, research that connects heritage and culture within a city context remained in its infancy with only a few publications across the 1981–1990 period that focused on the potential that heritage and culture offered cities from the perspective of regeneration of past industrial space (see McNulty, 1985). An exception was the work of Jansen-Verbeke (1986) who was the first to establish what were the major elements that made up inner-city tourism; a simple study but one that emerged to be of major importance in shaping scholarly thinking on cities, their functions and the types of tourists to whom cities could appeal.

Tourism scholarship over the years has been shaped by seminal works; in the case of resort/destination development, Butler's (1980) Tourism Area Life Cycle (TALC) is a case in point. In the context of research on the urban landscape, an early seminal work was that by Jansen-Verbeke, previously mentioned; a study that helped to position both cultural and heritage features as key structural elements of inner-city tourism. Her research suggested that the heritage and culture within a city must be part of what she termed as the primary elements of urban/city tourism when examined from a supply-side perspective. According to Jansen-Verbeke, this was composed of both what were termed "activity place" (cultural and heritage elements here included cultural facilities, events, exhibitions) and "leisure setting", which comprised both physical setting (e.g. monuments, buildings, art objects, waterfronts) and the cultural characteristics within that leisure setting (e.g. language, local customs, folklore).

While Jansen-Verbeke considered shopping and markets to be secondary elements, today both can be positioned as part of the primary elements of urban/city tourism as both shopping precinct and city farmers' markets help to showcase the heritage and culture of a place. What Jansen-Verbeke's work achieved was to firmly position heritage and culture as critical components of any understanding of tourism that takes place within cities; it offered the building blocks for future research. In terms of understanding the morphology of the tourist city, Burtenshaw et al. (1991) would later suggest that this comprised many regions of activity or city type; as such they matched up the users of cities with city resources to create distinctive regions of activity or city types such as the historic city and the cultural city. In short, this helped link heritage and cultural features of cities to distinct precincts and tourism opportunities.

By the mid-1990s, a number of important studies emerged that helped to showcase the potential that heritage and culture could offer cities. Richards (1996) illustrated the cultural capital that many European cities offered within their historic centres, that cultural capital helped to fuel interest in heritage tourism, and that many European capital cities were in competition for the heritage tourist. Research followed that focused on how best to manage tourism within European heritage cities (see van der Borg et al., 1996) but also how to plan cities from a heritage perspective as well as noting the links between heritage and commodification (see Meethan, 1996; Ashworth and Tunbridge, 1999). Other scholars published research that saw the connection between culture and regeneration, especially the opportunity that culture and heritage offered to promote sustainable urban tourism and offset the negative impacts associated with the economic restructuring of many cities (Chang et al., 1996; Shaw and MacLeod, 2000; Russo and van der Borg, 2000). Research here ranged from single case studies such as Glasson's (1994) work on Oxford and Curtis' (1998) study of Canterbury, England, to comparative case examination such as the work by Chang et al. (1996), who compared Montreal and Singapore.

At the end of this time period (1991–2000), a major text was produced by Ashworth and Tunbridge (2000) that focused specifically on the tourist-historic city. A tourist-historic city model illustrated the evolution of change from original city, to historic city core to tourist-historic city. This major work not only looked to map this evolutionary change within case examples but also focused on the challenges of managing the tourist-historic city against modernisation and change. It was the latter aspect that Ashworth (2000) further developed when he cautioned if heritage tourism was something that could be in harmony or in conflict with urban environments, examining this relationship for both tourist-historic cities and multifunctional cities that had distinct heritage precincts.

Against this backcloth of scholarly endeavour, the 2001–2010 era witnessed a "take-off" of research that connected heritage and cultural tourism to the urban/city setting. As revealed in Table 18.1, a number of themes emerged over this decade, driven by a number of key scholars. For instance, scholars had recognised that cities had started to take on the latter stages of Butler's life cycle, especially those that offered an attraction space around their heritage and culture. Russo (2002) argued that many heritage cities faced what he termed a "vicious cycle", which was the self-feeding linkage between excursionist tourists attracted to cities in their latter stages of their destination lifecycle and the decline in that city's attractiveness. Other scholars recognised the opportunities that heritage tourism offered cities where their heritage had political and socio-cultural significance as well as economic value for its residents (Henderson, 2002a, 2002b), and the benefits that could be derived from heritage and culturally related tourism for capital cities given the uniqueness of how these cities function (Hall, 2002). The relationship between tourism, culture and regeneration is returned to in the edited book by Smith (2006). This edited work examined the impact of tourism's relationship to urban regeneration and cultural development from a local people's point of view of their sense of place, heritage and identity. A key contribution in this edited work was that of Richards and Wilson (2006a) who argued for what they termed as the need for "the creative turn" in achieving successful urban regeneration; this involved creativity that combined creative spaces, events and tourism.

The scholarly field by this time (2001–2010) had expanded to include research that covered a broad church of issues such as understanding tourist motivations to visit specific sites in multi-heritage site cities; the work by Poria et al. (2007) on Jerusalem is a case in point. Others focused on a specific type of heritage product; for example, Frost (2008) examined how Melbourne, Australia, looked to music, specifically the work of hard rock band AC/DC to create a unique space and heritage of the city. In a similar vein, the work of Rátz et al. (2008) illustrated how in the case of the city of Budapest tourism and cultural regeneration helped to transform old spaces into new spaces. For example, the repacking of itineraries around a set of attractions making up a distinct urban trail. Others examined how heritage and culture products must be presented as authentic and not help create a hyper tradition and authentic fake to facilitate tourists' consumption of place (see for example the work of Alraouf (2010) on Bahrain).

Over the current decade (2011–2019), it would be a fair assumption to state that the field of research on cultural and heritage tourism in relation to cities is one resembling maturity. Cities have become very popular as the venue in which research has been undertaken; the venues being favoured as predominantly single case studies. There has been limited research beyond the case study; but it is within that context that research has focused on particular projects such as the European Capital of Culture (ECoC) and World Heritage Sites of culture situated within cities, to name but two. Beyond the empirical case studies, there have been a few conceptual papers that challenged some conventional thinking. In particular, Richards (2011) offered a state-of-the-art paper where he argued that cities needed to consider the opportunities to promote creative tourism as opposed to cultural tourism. He argued that most cities have major cultural and heritage attractions, churches, museums, heritage centres and that as such cultural tourism has become a form of mass tourism where the product offer suffers from what he termed "serial reproduction". Page and Hall (2003) had earlier referred to the serial monotony facing urban tourism as all urban places were starting to resemble sameness and be viewed as having some degree of placeless-ness where they risked becoming a "nonplace". Richards advocated the importance that residents could offer in transforming cities as places of culture to places of creativity, in both the attractions they provide, the events they host and that through being creative the authenticity of experience is safeguarded.

Evidence of the maturing of research in sub-fields was the edited handbook on cultural tourism by Smith and Richards (2013) where the context of the city featured across many parts of its structure; some of this will be addressed in later sections of the chapter. Another notable development with an established field of academic inquiry was the emergence of a number of special issues within scholarly journals. Examples here include that edited by Naef and Pioner (2016) where the relationship between tourism, conflict and contested heritage in former Yugoslavia was examined. Here scholars looked at a number of issues, including how heritage is interpreted, the role that memory has played in cities, the rise of the martyred city and the extent to which dissonant heritage is a prominent feature within cities that have been impacted by conflict. Iankova (2017) guest edited a special issue on the communist legacies within cities and the tourism opportunities this offers. Lastly, Lew (2018) guest edited a special issue that examined European cultural cities from the perspective of placemaking, heritage and overtourism; the latter being a concept that will be returned to at the end of the chapter.

Against such a backcloth of research as revealed in Table 18.1, it is hard to justify what areas deserve more in-depth attention. A number are set out below as they aim to illustrate what dimensions of cities and strategies cities have embarked on to be popular as destinations for tourists that are culturally and heritage minded.

Urban design and linkage to cultural and heritage tourism products

Place and space are important concepts and attributes of tourism (Crouch, 2013), and cities historically have offered tourism a unique and confined space within which a clear heritage and cultural product has evolved or been deliberately designed (Lew, 2017). The physical realm of the historic core with its tangible built heritage – often where the seat of decision-making (city hall), spiritual values (cathedral, main church) and trade (town square and market) are found in close proximity to one another – created an opportunity to cluster key heritage and cultural attractions. This is reflected in the work of Ashworth and Tunbridge (2000) who argued in their model of the tourist-historic city that this would overlap between the historic city and the central business district of the modern city; that key attractions would be those that receive high levels of visitation and be highly clustered in terms of their locale. Over time it would create a relationship between tourism and historic cores where the core became subject to "tourismification" and change with more cultural tourism features added to the core products (Jansen-Verbeke, 1998). An urban-centred spectrum opportunity model called the urban tourism opportunity spectrum (UTOS) highlighted the extent to which products, attractions and related services were clustered across cities, around key attractions; it was also a model that saw a real-life application across European cities (Jansen-Verbeke and Lievois, 1999). According to Lew (2017) cities have long used their tangible heritage (parks and gardens, pedestrianised areas, courtyards, tree-lined avenues, historic cores with their built heritage) alongside that which is intangible (creativity, stories, customs) to facilitate placemaking to both city residents as well as incoming tourists. However, clustering of key attractions in the same geographic space created over time potential dangers of over-visitation and congestion in parts of the city.

This became very noticeable in the work of Russo (2002) – his research on Venice revealed how historic city centres can suffer from what he referred to as the "vicious cycle" of cultural tourism experience. A majority of visitors can be attracted to a few key sites that over time are at such risk of congestion with declining quality of visit experience offered to the visitor that they choose an alternative destination. This "serious cultural and heritage tourism" market is then replaced with a lower-spend day-trip excursionist "casual" and "serendipitous" market, and in the case of Venice much of this market is cruise-sector driven. The "sameness" of place or placelessness is reflected in the research by Richards and Wilson (2006b) who suggested that cities could suffer from the "serial reproduction of cultural tourism" with a cultural and heritage offering that was not that dissimilar from others. It is not a difficult argument to make that most cities have a cathedral, flagship heritage centres, town hall, main square and piazza, monuments, museums and art galleries. Perhaps cultural and heritage tourism in cities has been the victim of its own success and this has led to the shift toward thinking around creative tourism (Richards, 2013). This involves greater engagement between visitors and residents that taps into local heritage, culture and tradition by creating new spaces, events that are community-driven and linked, and where co-creation between locals and tourists is to the benefit of local people over traditional tourism providers. If cities were to retain their appeal, other researchers point to the association between place attachment and authenticity (Ram et al., 2016). They argued that those tourists who are culturally and heritage minded are drawn to cities because of the heritage and iconic value they perceive from visiting major flagship cultural and heritage attractions (Ram et al., 2016).

Cities, regeneration and change

A major strand of research has been linked to the relationship of cultural and heritage tourism and the regeneration of certain spaces in the contemporary city. Researchers have looked at this in a number of broad dimensions: how cultural and heritage tourists have created specific markets for cities (Law, 1993; Chang et al., 1996; Timothy and Boyd, 2003; du Cros and McKercher, 2015); the revitalisation of post-industrial spaces within cities (Garcia, 2004; Carruthers, 2013); the relationship between regeneration in cities and the cultural development by residents across a range of spaces within different cities; those that would familiarise themselves with tourism and those that do not (Smith, 2006); the regeneration of specific spaces in cities, such as cultural precincts and quarters (Roadhouse, 2010; Hall, 2013); and the role of ethnic quarters in cities as cultural tourism products (Shaw, 2013). With regard to cultural quarters, Hall (2013) questions the benefits they offer as many are copies and therefore lack authenticity; often within these spaces are found cultural products such as museums, theatres that add to the serial reproduction of cultural tourism opportunity and limit the extent to which they contribute to the wider goals of regeneration, making places distinctive and competitive. The same argument can be applied to ethnic quarters in cities; where is their distinctiveness and for which markets are they catering? Shaw (2013) notes that on the one hand they are part of a wider strategy to create cosmopolitan space for a selective market that needs to be entertained, whereas on the other hand they offer opportunity for cross-fertilisation of ideas between different ethnic residential groups and the tourists that visit. Although previously discussed, the focus of scholars on regeneration and change has engendered a wider debate of the need to shift away from traditional cultural tourism product development toward creative forms of tourism that tap into bottom-up, community initiatives and opportunities.

Trails and routes in cities

Another emerging area of scholarly engagement is the increase in research on linear spaces within cities; research that examines city routes and trails over research that focuses on single point attractions such as heritage centres and museums (Timothy and Boyd, 2015). Routes have been in our landscape for centuries, either as organic features that have evolved from an initial use into a tourism use or purposive and designed for tourism. When urban/city landscapes are considered, routes often take on a purposive nature and are used by many city marketing bodies as the most cost-effective way to connect points of similar interest (for example, the Gaudi Trail, Barcelona; Boston Freedom Trail). Within the context of cities, research on urban trails has highlighted the importance of theming as well as substitution; in the case of the latter the development of the Cultural Avenue Project, Budapest, was designed to divert the visitor away from the well-visited attractions to less known or popular attractions creating multiple sightseeing routes (Rátz et al., 2008). Theming within trails creates scope for creativity on the part of the tourist as well as providing the opportunity to be more experiential in how they are designed, tapping into a range of experiences that fall into the realms of education, entertainment, aesthetics and escapism and making greater use of the senses (MacLeod, 2013; Timothy and Boyd, 2015).

Cities and European Capital of Culture

Another major strand of research by scholars has been to examine the importance placed on city governments bidding to achieve the accolade of European Capital of Culture (ECoC). The programme was initiated in 1985 with two precise goals of promoting dialogue among European citizens and developing a sense of belonging to the cause of the European Community. Since 1985, when Athens was awarded with the first title, there have been over 50 cities that have been given the ECoC title and have hosted a year-long series of community participative events that have aimed to highlight each recipient city's strong European cultural dimension. In more recent years, bidding cities have had to also demonstrate a forward-looking legacy of the development of the city beyond being named as ECoC. Over time, a number of strands of research on culture and heritage and its links with ECoC have emerged. The first strand of scholarly research has been that associated with identifying what are the critical success factors involved in a winning bid – in particular the role of the arts in any cultural regeneration strategy put in place for bidding cities, in particular Liverpool which was awarded the title for 2008 (Favre, 2004; Avery, 2006). A second broad research strand has been to focus on the impacts of success, whether this has been the economic and socio-cultural impacts of major events (Liu, 2014a), improved city governance (Paskaleva et al., 2009), improved awareness of local identity by city residents (Platt, 2011) or the role of events in cultural tourism development and wider city image (Liu, 2014b, 2014c). A third strand of research has looked specifically at events, especially cities bidding for major events as part of their wider year-long programme of events and the costs involved if not successful (Richards and Marques, 2016), the wider effects of event monitoring and evaluation (Richards, 2015), and the opportunity to foster co-creation of events between local residents and wider cultural stakeholders (Åkerlund and Müller, 2012). A final strand of research has been to examine the legacy of being a successful city (Quinn, 2010), and in particular the benefits it has brought to tourism in increased overnight stays in the follow-up year (Falk and Hagsten, 2017), the extent to which it has increased the "smartness" of cities (Bernardino et al., 2018), as well as the political effects of hosting ECoC in terms of engendering regime change of the host city (Žilič-Fišer and Erjavec, 2017).

Conclusion

The focus of this chapter has been on understanding the city in the context of cultural and heritage tourism, first from the perspective of a chronological narrative of the extent to which tourism scholars have been interested in cities for cultural and heritage tourism, and second, by offering a more in-depth discussion around a number of selected themes. Figure 18.1 schematises that discussion as it illustrates the complexity of culture and heritage as a city's primary elements, the nature of demand for cultural and heritage tourism, and the types of cities that cater to that broad spectrum of demand, and the outcomes that may result for cities.

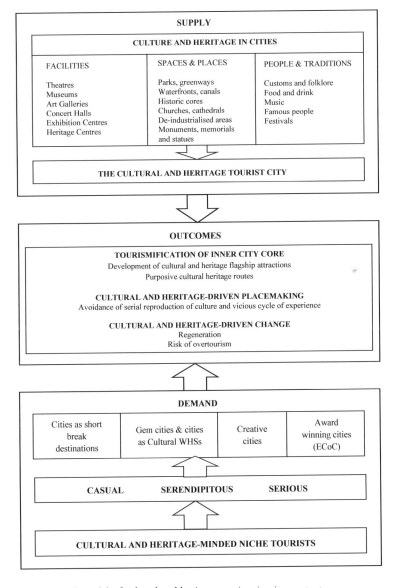

Figure 18.1 Conceptual model of cultural and heritage tourism in city context.

One particular outcome illustrated in the figure is that of overtourism. To what extent is this new phenomenon linked to tourism (particularly the heritage and cultural tourists who actually use cities)? What areas of the city do tourists visit (Ashworth and Page, 2011)? And what is the extent to which city through placemaking emphasises both their cultural and heritage appeal (Lew, 2017)? What is emerging is a distinct unrest in city residents in major European cities (Séraphin et al., 2018). However, what is not clear are the factors behind this – changing perception of the relationship between residents and visitors or greater social activism among city residents? The emerging research on understanding overtourism will be vital if cities are to remain as both popular and accessible for the cultural and heritage tourist.

References

Åkerlund, U. and Müller, D.K. (2012) "Implementing tourism events: the discourses of Umea's bid for European Capital of Culture 2014". *Scandinavian Journal of Hospitality and Tourism*, 12(2), 164–180.

Alraouf, A.A. (2010) "Regenerating urban traditions in Bahrain. Learning from the Bab-Al-Bahrain: the authentic fake". *Journal of Tourism and Cultural Change*, 8(1/2), 50–68.

Ashworth, G.J. (2000) "Heritage tourism and urban environments: conflict or harmony?". In: Briassoulis, H. and van der Straaten, J. (eds) *Tourism and the Environment: Regional, Economic, Cultural and Policy Issues* (pp. 283–304). Dordrecht, Netherlands: Kluwer Academic Publishers.

Ashworth, G.J. and Page, S.J. (2011) "Urban tourism research: recent progress and current paradoxes". *Tourism Management*, 32(1), 1–15.

Ashworth, G.J. and Tunbridge, J.E. (1999) "Old cities, new pasts: heritage planning in selected cities of Central Europe". *GeoJournal*, 49(1), 105–116.

Ashworth, G.J. and Tunbridge, J.E. (2000) *The Tourist-Historic City: Retrospect and Prospect of Managing the Heritage City*. New York: Pergamon.

Avery, P. (2006) "Born again: from dock cities to cities of culture". In: Smith, M. (ed.) *Tourism, Culture and Regeneration* (pp. 151–162). Wallingford: CABI.

Bernardino, S., Santos, J.F. and Ribeiro, J.C. (2018) "The legacy of European Capitals of Culture to the 'smartness' of cities: the case of Guimarães 2012". *Journal of Convention and Event Tourism*, 19(2), 138–166.

Bosselman, F.P. (1978) *In the Wake of the Tourist*. Washington, DC: The Conservation Foundation.

Burtenshaw, D., Bateman, M. and Ashworth, G.J. (1991) *The European City: A Western Perspective*. London: David Fulton Publishers.

Butler, R.W. (1980) "The concept of a tourist area cycle of evolution and implications for management of resources". *Canadian Geographer*, 24(1), 5–12.

Caffyn, A. and Lutz, J. (1999) "Developing the heritage tourism product in multi-ethnic cities". *Tourism Management*, 20(2), 213–221.

Carruthers, C. (2013) "Cultural tourism development in the post-industrial city: developing strategies and critical reflection". In: Smith, M. and Richards, G. (eds) *The Routledge Handbook of Cultural Tourism* (pp. 311–316). Oxford: Routledge.

Chang, T.C., Milne, S., Fallon, D. and Pohlmann, C. (1996) "Urban heritage tourism". *Annals of Tourism Research*, 23(2), 284–305.

Crouch, D. (2013) "Space and place-making: space, culture and tourism". In: Smith, M. and Richards, G. (eds) *The Routledge Handbook of Cultural Tourism* (pp. 247–251). Oxford: Routledge.

Curtis, S. (1998) "Visitor management in small historic cities". *Travel and Tourism Analyst*, 3, 75–89.

du Cros, H. and McKercher, B. (2015) *Cultural Tourism*, 2nd ed. Oxford: Routledge.

Falk, M. and Hagsten, E. (2017) "Measuring the impact of the European Capital of Culture programme on overnight stays: evidence from the last two decades". *European Planning Studies*, 25(12), 2175–2191.

Favre, C. (2004) "Liverpool: winning and sustaining the European Capital of Culture title". *Hospitality Review*, 6(3), 5–13.

Ferreira, L., Aguiar, L. and Pinto, J.R. (2012) "Cultural tourism, tourist routes and impacts on destinations". *Cultur: Revista de Cultura e Turismo*, 6(2), 109–126.

Frost, W. (2008) "Popular culture as a different type of heritage: the making of AC/DC lane". *Journal of Heritage Tourism*, 3(3), 176–184.

Garcia, B. (2004) "Cultural policy and urban regeneration in Western European Cities: lessons learned from experience, prospects for the future". *Local Economy*, 19(4), 312–326.

García-Hernández, M., De la Calle-Vaquero and Yubero, C. (2017) "Cultural heritage and urban tourism: Historic city centres under pressure". *Sustainability*, 9(8), 1346.

Glasson, J. (1994) "Oxford: a heritage city under pressure, visitor impacts and management responses". *Tourism Management*, 15(2), 137–144.

Hall, C.M. (2002) "Tourism in capital cities". *Tourism* (Zagreb), 50(3), 235–248.

Hall, C.M. (2013) "Regeneration and cultural quarters: changing urban cultural space". In: Smith, M. and Richards, G. (eds) *The Routledge Handbook of Cultural Tourism* (pp. 332–338). Oxford: Routledge.

Henderson, J.C. (2002a) "Heritage attractions and tourism development in Asia: a comparative study of Hong Kong and Singapore". *International Journal of Tourism Research*, 4(5), 337–344.

Henderson, J.C. (2002b) "Built heritage and colonial cities". *Annals of Tourism Research*, 29(1), 254–257.

Iankova, K. (ed.) (2017) "Special Issue: Communist legacies and cities: dimensions and tourism opportunities". *International Journal of Tourism Cities*, 3(3), 205–320.

Jansen-Verbeke, M. (1986) "Inner-city tourism: resources, tourists and promoters". *Annals of Tourism Research*, 13, 79–100.

Jansen-Verbeke, M. (1998) "Tourismification of historic cities". *Annals of Tourism Research*, 25(4), 739–742.

Jansen-Verbeke, M. and Lievois, E. (1999) "Analysing heritage resources for urban tourism in European cities". In: Pearce, D.G. and Butler, R.W. (eds) *Contemporary Issues in Tourism Development* (pp. 81–107). London: Routledge.

Law, C.M. (1993) *Urban Tourism: Attracting Visitors to Large Cities*. London: Mansell.

Law, C.M. (2002) *Urban Tourism: The Visitor Economy and the Growth of Large Cities*, 2nd ed. London: Continuum.

Laws, E. and Le Pelley, B. (2000) "Managing complexity and change in tourism: the case of a historic city". *International Journal of Tourism Research*, 2(4), 229–246.

Lew, A.A. (2017) "Tourism planning and place making: place-making or placemaking?". *Tourism Geographies*, 19(3), 448–466.

Lew, A.A. (ed.) (2018) "Special issue: Tourism and urban planning in European cities: overtourism, place-making and heritage". *Tourism Geographies*, 20(3), 371–572.

Liu Y. (2014a) "Socio-cultural impacts of major event: evidence from the 2008 European Capital of Culture, Liverpool". *Social Indicators Research*, 115(3), 983–998.

Liu Y. (2014b) "Cultural events and cultural tourism development: lessons from the European capitals of Culture". *European Planning Studies*, 22(3), 498–514.

Liu Y. (2014c) "The impact of cultural event on city image: an evaluation of the 2008 European Capital of Culture, Liverpool". *International Journal of Leisure and Tourism Marketing*, 4(1), 19–30.

MacLeod, N. (2013) "Cultural routes, trails and the experience of place". In: Smith, M. and Richards, G. (eds) *The Routledge Handbook of Cultural Tourism* (pp. 369–374). Oxford: Routledge.

McNulty, R.H. (1985) "Revitalizing industrial cities through cultural tourism". *International Journal of Environmental Studies A*, 25(4), 225–228.

Meethan, K. (1996) "Consuming (in) the civilised city". *Annals of Tourism Research*, 23(2), 322–340.

Naef, P. and Pioner, J. (eds) (2016) "Special Issue: Tourism, conflict and contested heritage in former Yugoslavia". *Journal of Tourism and Cultural Change*, 14(3), 181–290.

Page, S.J. (1995) *Urban Tourism*. Oxford: Routledge.

Page, S.J. and Hall, C.M. (2003) *Managing Urban Tourism*. Harlow: Prentice Hall.

Paskaleva, K., Besson, E. and Sunderland, M. (2009) "Tourism and European Capitals of Culture: the role of destination competitiveness governance". *International Journal of Tourism Policy*, 2(1/2), 107–123.

Platt, L. (2011) "Liverpool 08 and the performativity of identity". *Journal of Policy Research in Tourism, Leisure and Events*, 3(1), 31–43.

Poria, Y., Biran, A. and Reichel, A. (2007) "Different Jerusalem's for different tourists: capital cities – the management of multi-heritage site cities". *Journal of Travel and Tourism Marketing*, 22 (3/4), 121–138.

Quinn, B. (2010) "The European capital culture initiative and cultural legacy: an analysis of the cultural sector in the aftermath of Cork 2005". *Event Management*, 13(4), 249–264.

Ram, Y., Bjork, P. and Weidenfeld, A. (2016) "Authenticity and place attachment of major visitor attractions". *Tourism Management*, 52, 110–122.

Rátz, T., Smith, M. and Michalkó, G. (2008) "New places in old spaces: mapping tourism and regeneration in Budapest". *Tourism Geographies*, 10(4), 429–451.

Richards, G. (1996) "Production and consumption of European cultural tourism". *Annals of Tourism Research*, 23(2), 261–283.

Richards, G, (2007). *Cultural Tourism: Global and Local Perspectives*. Binghamton, NY: Haworth Press.

Richards, G. (2011) "Creativity and tourism: the state of the art". *Annals of Tourism Research*, 38(4), 1225–1253.

Richards, G. (2013) "Tourism development trajectories: from culture to creativity?". In: Smith, M.K. and Richards, G. (eds) *Routledge Handbook of Cultural Tourism* (pp. 297–303). Oxford: Routledge.

Richards, G. (2015) "Evaluating the European capital of culture that never was: the case of BrabantStad 2018". *Journal of Policy Research in Tourism, Leisure and Events*, 7(2), 118–133.

Richards, G. and Marques, L. (2016) "Bidding for success? Impacts of the European Capital of Culture bid". *Scandinavian Journal of Hospitality and Tourism*, 16(2), 180–195.

Richards, G. and Wilson, J. (2006a) "The creative turn in regeneration: creative spaces, spectacles and tourism in cities". In: Smith, M.K. (ed.) *Tourism, Culture and Regeneration* (pp. 12–24). Wallingford: CABI.

Richards, G. and Wilson, J. (2006b) "Developing creativity in tourist experiences: a solution to the serial reproduction of culture?" *Tourism Management*, 27, 1209–1223.

Roadhouse, S. (ed.) (2010) *Cultural Quarters: Principles and Practice*, 2nd ed. Bristol: Intellect.

Russo, A.P. (2002) "The 'vicious circle' of tourism development in heritage cities". *Annals of Tourism Research*, 29(1), 165–182.

Russo, A.P. and van der Borg, J. (2000) "The strategic importance of the cultural sector for sustainable urban tourism". In: Fossati, A. and Panella, G. (eds) *Tourism and Sustainable Economic Development* (pp. 71–98). New York: Kluwer Academic Publishers.

Saidi, H. (2012) "Capital cities as open-air museums: a look at Québec City and Tunis". *Current Issues in Tourism*, 15(1/2), 75–88.

Séraphin, H., Sheeran, P. and Pilato, M. (2018) "Over-tourism and the fall of Venice". *Journal of Destination Marketing and Management*, 9, 374–376.

Shaw, S. (2013) "Ethnic quarters: exotic islands or trans-national hotbeds of innovation". In: Smith, M. and Richards, G. (eds) *The Routledge Handbook of Cultural Tourism* (pp. 339–345). Oxford: Routledge.

Shaw, S.J. and MacLeod, N.E. (2000) "Creativity and conflict: cultural tourism in London's city fringe". *Tourism, Culture and Communication*, 2(3), 165–176.

Smith, M.K. (ed.) (2006) *Tourism, Culture and Regeneration*. Wallingford: CABI.

Smith, M.K. and Richards, G. (eds) (2013) *The Routledge Handbook of Cultural Tourism*. Oxford: Routledge.

Smith, M.K. and Robinson, M. (eds.) (2006) *Cultural Tourism in a Changing World: Politics, Participation and (Re)presentation*. Clevedon: Channel View Publications.

Timothy, D.J. and Boyd, S.W. (2003) *Heritage Tourism*. Harlow: Prentice Hall.

Timothy, D.J. and Boyd, S.W. (2015) *Tourism and Trails: Cultural, Ecological and Management Issues*. Bristol: Channel View Publications.

Towner, J. (1996) *An Historical Geography of Recreation and Tourism in the Western World: 1540–1940*. Chichester: Wiley.

van der Borg, J. (2004) "Tourism management and carrying capacity in heritage cities and sites". In: Coccossis, H. and Mexa, A. (eds) *The Challenge of Tourism Capacity Assessment: Theory and Practice* (pp. 163–179). Aldershot: Ashgate.

van der Borg, J., Costa, P. and Gotti, G. (1996) "Tourism in European cities". *Annals of Tourism Research*, 23(2), 306–321.

Wheeler, F., Reeves, K., Laing, J. and Frost, W. (2009) "Niche strategies for small regional cities: a case study of the Bendigo Chinese Precinct plan". *Tourism Recreation Research*, 34(3), 295–306.

Wickens, E. (2017) "The consumption of cultural experiences in city tourism". *Tourism and Hospitality Research*, 17(3), 264–271.

Žilič-Fišer, S. and Erjavec, K. (2017) "The political impact of the European Capital of Culture: 'Maribor 2012 gave us the power to change the regime'". *International Journal of Cultural Policy*, 23(5), 581–596.

19

OUTDOOR AND INDOOR MARKETS IN TOURISM CITIES

Joan C. Henderson

Introduction

The chapter is concerned with the characteristics and roles of open–air and covered markets in urban tourist destinations where they often serve as a visitor attraction and amenity. The attributes of various types of market and their appeal to tourists are considered as well as contrasts around the world, illustrated by specific examples from popular cities. Conditions related to the economy, society and culture, and climate all influence markets which are often seen as an embodiment of place by destination marketing organisations and visitors and valued for their perceived authenticity. The sights, sounds and smells of busy marketplaces add to their attractiveness and they provide opportunities to purchase goods and souvenirs of differing sorts, quality and price. While tourists can be a significant source of business for stallholders and markets are regularly highlighted in promotion, there are possible threats from tourism which are also discussed in the chapter. Uncertainties about future prospects are acknowledged, although there is evidence of sustained interest and a rise in popularity of certain forms of markets. They are found to be evolving in accordance with wider social and economic changes, indicative of their dynamism and vibrancy which can enliven cities for the benefit of both local communities and tourists.

Tourism cities and shopping

Cities, by their nature, draw visitors for many reasons and might therefore all be termed tourism cities. However, some are clearly more visited than others and size and status are determinant factors with capital cities often the busiest and having a mix of business and leisure travellers. Smaller cities possessing distinctive features may also act as magnets for tourists such as historic centres with a wealth of heritage. Various rankings exist based on international arrival figures, although data are not always reliable and overlook domestic visitors who cannot be accurately measured. Nevertheless, listings by agencies such as Euromonitor yield useful insights into leading cities which were identified as Hong Kong, Bangkok, London, Singapore, Macau, Dubai, Paris, New York, Shenzhen and Kuala Lumpur in 2016 (Euromonitor International 2017). Circumstances in these cities with regard to the supply of indoor and outdoor markets and representations for tourists are returned to in a subsequent section.

Whatever a city's scale or special qualities, the retail sector is likely to be an important constituent of the tourism industry in those which are tourist hubs (Page and Hall 2003). For larger cities, retail offerings can be a key selling point and competitive advantage. Shopping of multiple sorts and in diverse venues is a regular tourist pastime and accounts for a significant proportion of expenditure (Timothy 2005). Holidaymakers are a potentially lucrative group of customers given their propensity to spend more than usual. Some will be looking for goods that they cannot find at home and are particular to the place yet may be disappointed in an era of globalisation in which the same merchandise is available across the world. The retail malls ubiquitous in major cities of Asia, for example, contain a similar mix of shops selling universally recognised branded goods which are often Western in origin. Browsing in familiar stores in new environments and the possibility of extended choices at lower prices due to advantageous currency exchange rates may motivate some shopping tourists, demonstrated by the enthusiasm for luxury European brands shown by certain Chinese travellers. However, others will be interested in searching out settings and products which are localised rather than globalised. Indoor and outdoor markets can satisfy these requirements and hence contribute to the visitor retail and overall experience in a manner explored in this chapter. At the same time, modes of shopping are not mutually exclusive, and tourists may enjoy both high-end malls of an internationalised style and informal street bazaars. The ability to proffer a range of retailing would thus seem to be an asset for most city destinations.

Urban markets

There is no standard definition of a market which is commonly understood to be a place where buyers and sellers gather to do business. Traditional retail markets are public spaces comprising stallholders who occupy indoor or outdoor sites, and many are well established. Long histories are exemplified by instances in Europe (Guardia *et al.* 2010; URBACT 2015) and the term market town is still employed in the UK to describe settlements which have had a market at their heart since the mediaeval period. The souks which are characteristic of the Arab world have been traced back further (Gharipour 2012) and still operate in the old quarters of cities in the Middle East and North Africa alongside modern versions. The eighteenth-century Grand Bazaar in Istanbul is among the world's largest and oldest covered markets with 61 streets and over 4,000 shops. Some historic urban marketplaces have disappeared as a consequence of planning strategies or neglect whereas others survive, albeit evolving over the years and continuing to do so in response to local and wider conditions and trends. A degree of modernisation has occurred extending to technological innovation such as the use of social media for advertising, cashless payment which has been adopted by many food vendors in Chinese cities, and market management software. Market trading, however, is conducted mainly in cash and certain technologies may be unavailable, unfamiliar and too costly for traders and management.

Again, there is no formal classification of types and general markets sell an assortment of goods whereas others are more specialised; for example, antiques, arts and crafts, and flea markets. Food is often a component, but some focus on this exclusively and farmers' markets have become prominent. A distinction can be made between open and covered markets, although they may be combined, and wares are similar. Days and hours of operation also vary, and some are weekend, weekly or more infrequent affairs. In addition, they can be linked to events or seasons such as those celebrating religious occasions and accompanying arts festivals, sometimes termed fairs, which can be classed as speciality markets. Night markets are a feature of several cities, especially in Asia. Many markets remain in the hands of the public sector, which rents out pitches, but they can be private initiatives, and both are officially regulated. Nevertheless, a few belong to the informal economy and the policing of illegal street trading is a challenging task for municipal administrations.

Urban markets are agreed to have potential positive economic, socio-cultural and environmental impacts generally. Revenue and jobs are generated, and trade is boosted for other shops in the vicinity and local suppliers (Hallsworth *et al.* 2015; London Development Agency 2010; URBACT 2015). Goods in neighbourhood markets are usually affordable for those on low incomes and markets are sites of social inclusion and exchange, contributing to community wellbeing and sustainability (Morales 2009). They are manifestations of cultural heritage and place identity, embracing aspects of contemporary society which are increasingly multicultural (Gilli and Ferrari 2018). Towns and cities are animated and improved by the presence of market trading, adding to their liveability (Guardia *et al.* 2010).

Despite their many merits and a growth in speciality markets, traditional markets appear to be in decline in parts of the world because of competition from alternative types of retailing. Authorities have a part to play in supporting markets and these trends have policy implications. There is an appreciation of the economic and socio-cultural losses attendant on market closures, yet other forms of urban land use might be deemed more appropriate and productive. Regulation can be a challenge and there are possible damaging environmental impacts to consider. Governments have sometimes been reluctant to invest resources in public markets and urban policies may be unsympathetic. In some European cities, there are criticisms of gentrification whereby markets have transformed themselves to cater to the more affluent and thereby alienated less privileged former customers (Gonzalez and Waley 2012). The so-called gentry are assumed to be citizens, but tourists too can be seen as an outcome of and vehicle for the gentrification process and the cause of undesirable changes. Not everyone concurs about the disadvantageous effects of tourism and its relationship with markets is discussed below. Whatever the customer mix, the busiest markets are potential terrorist targets as demonstrated by the attacks on Borough Market and nearby London Bridge in 2017 and on a Christmas market in Berlin in 2016.

The tourist appeal of urban markets

Markets around the world succeed in drawing tourists and their spending, confirming that they are an attraction and amenity of actual and prospective importance. As such, they are interesting research arenas for the exploration of contemporary tourism issues pertaining to the tourist experience, tourist–resident interactions, sharing economy providers, impacts and overtourism, and management. Shopping opportunities are provided, including the purchase of locally produced goods, and there is often a promise of bargains. Some tourists are willing to enter into and enjoy bargaining in markets where this is customary (Correia and Kozak 2016) such as Beijing's Silk Market (Wu *et al.* 2014). Market visits are a chance to interact with residents (Kikuchi and Ryan 2007), creating a feeling of participating in everyday life. They can encapsulate a destination and sometimes its exoticism as in the souks of Morocco and Tunisia, which are depicted as must-see sights. Insights are gained into the culture and possibly history of the location, resulting in a better understanding. Some markets are events with an entertainment function, occasionally hosting performances, and enhancing the visitor experience. These characteristics partly explain why markets are used in place making to cultivate positive images, conveying notions of vibrancy and authenticity (Pappalepore *et al.* 2014). Authenticity is, however, an elusive concept which is personally and socially constructed. It may be more imaginary than real at markets (Beer *et al.* 2012) and other sites where tourists are hoping to encounter destination life.

While bringing new business, popularity among tourists can have drawbacks for markets traditionally dealing with residents who may be deterred by excessive numbers of sightseers.

Stalls might shift to selling souvenirs and other items more oriented towards visitors, further discouraging regular customers. There is a risk of the loss of the enticing sense of place (URBACT 2015) and "touristification" (Gilli and Ferrari 2018, p. 146), possibly ending in the eventual abandonment by both residents and tourists. Balancing the demands of the two groups is therefore a critical undertaking if the distinctive personalities and ambience of markets are to be preserved and their sustainability increased (Crespi-Vallbona *et al.* 2017), although this is not always necessary. Urban markets are more likely to be shared by resident and visitor populations than those in beach resorts catering to holidaymakers, for example, but some exist primarily to serve tourists. These and certain speciality markets are not rooted in the locale so that there is no supposed authenticity or community space to be endangered. Similar observations apply to the aforementioned threat of gentrification and it must be recalled that markets may be designed from the beginning for the more affluent, even if such signs of income inequalities among city dwellers are unpalatable to some.

Additional disadvantages of markets busy with tourists that have implications for authorities include exacerbation of difficulties such as littering and waste disposal, congestion of pedestrian and vehicular traffic, parking pressures and policing imperatives. Some markets are known for counterfeit merchandise (Correia and Kozak 2016) and sales to tourists help perpetuate criminality. Visitors may also be less vigilant and vulnerable to the pickpockets who are lured by the gatherings of people which typify major markets. Problems and opportunities overall and with specific reference to tourism depend partly on the market and selected key forms are now considered.

General and speciality markets

General markets sell an array of goods such as clothing and accessories, household articles, toys, flowers and food. Stalls can be mixed or organised into separate zones, which is usually the case for foodstuffs. Products on offer and the tangible and intangible features of the market are shaped by economic, political, socio-cultural and environmental factors. These dynamics also influence speciality markets which concentrate on certain sorts of items or themes which were indicated earlier. Regional differences are disclosed in the examples of Christmas markets in the West and Asian night markets. The former has acquired popularity in more European cities in recent years, selling Christmas-related products and food and drink during Advent. They are a way of stimulating tourism in the off-season and encouraging spending, especially on cold winter days and evenings. Many patrons are locals, but the largest markets are visited by excursionists and tourists from overseas and can be a basis for short package tours. Shopping is not necessarily the main reason for attendance and people are motivated by the idea of the market as a cultural and recreational occasion (Brida *et al.* 2017). They anticipate an authentic experience which adds to the value allotted to markets by them, although expectations are not always met, and disappointment ensues (Casteran and Roedere 2013).

A Christmas market can be termed a night market, but obviously unlike those in many Asian cities. The event has been dated back to Imperial China (Yu 2004) and is found in Chinatowns worldwide, including those in North American cities. The popularity of night markets is apparent across East Asia and one instance is the Taiwanese capital of Taipei where permanent market sites for eating and shopping are heavily promoted to tourists. Some were introduced only in the 1980s and are intended to boost local economies, supply evening entertainment and legalise illicit trading (Hsieh and Chang 2006). Visitors are seeking to immerse themselves in the native culture, especially by sampling Taiwanese delicacies, and buy low-priced goods in a bustling and colourful atmosphere (Lee *et al.* 2008). Engaging in banter with stallholders is another agreeable dimension of the visit (Chuang *et al.* 2014) but constrained by language barriers which can inhibit transactions by foreign visitors in all markets.

Street vending is a very strong tradition in South East Asia where night markets of assorted scales are regular occurrences and embedded in daily life. Some open during the day and into the night and can also be labelled general markets according to their offerings. Pasar Malam is the name given in Indonesia, Malaysia and Singapore to those organised in neighbourhoods after dark which comprise itinerant traders (Ibrahim and Leng 2003; Zakariya and Ware 2010) and there are temporary markets marking cultural and religious festivals. The weather perhaps augments the appeal of these markets for visitors from a temperate climate who can enjoy exploring in the balmy evening air after the intense heat of the day, even though heavy tropical rain can be disruptive for outdoor buyers and sellers.

Food markets

As previously stated, food is commonly an element of general markets while a recurring theme of speciality markets and its pre-eminence warrants separate consideration. Many speciality markets in developed countries are devoted to expensive fine or gourmet foods, demand for which has increased in tandem with the rise of a so-called foodie culture of people knowledge-able about food and prepared to spend substantial amounts on quality produce (Getz *at al.* 2014). These trends are further demonstrated by the food tourism movement (UNWTO 2017) in which food and its appreciation is a primary reason for travel and determinant of trip satisfaction (Ellis *et al.* 2018). City food markets are wide ranging from a small collection of temporary street stalls to extensive covered markets with a long history and an often-pleasing architectural style. Some rely heavily on visitors and encompass restaurants, cafes, events and entertainments. Products sold by certain traders may be tailored to the needs of tourists who are likely to favour non-perishable foodstuffs which are easy to carry as souvenirs. Other markets remain centred on the local clientele, even if seeing a significant number of outsiders. Alterations in order to accommodate tourism, as well as calls for more epicurean goods from wealthier citizens, are being resisted in several European capitals because of worries about the erosion of market identity and exclusion of the less well-off (Gilli and Ferrari 2018).

Farmers' markets have existed for centuries, but a new generation has emerged which have been described as postmodern (Beer *et al.* 2012). Originally largely confined to rural settlements, they are now run in many cities and particularly those of North America, Europe and Australia/New Zealand (Frost *et al.* 2016). Operating models are both non-profit making and commercial and community groups may be involved (Zittlau and Gorman 2012). A laudable example is the Souk El Tayeb in the Lebanese capital of Beirut which was launched by an individual keen to provide a means of distribution for contemporary small-scale farmers while retaining something of the traditional souk atmosphere. It has an additional purpose as an expression of "peace and social harmony" (URBACT 2015, p. 28) in a city once renowned for its violent divisions. Emphasis in farmers' markets as a whole is on freshness and quality, rather than inexpensive price, and some trade in organic foods in line with healthier eating advocacy.

Ensuring that vendors are bona-fide enterprises is necessary and there are efforts at regulation and certification to prevent misrepresentations (Page and Hall 2003) and the taking over of market space by non-farming and non-local ventures (Beer *at al.* 2012). Buyers at the best farmers' markets gain from having access to locally sourced produce and farmers are able to reach customers directly. Markets are enjoyable, educational and social and there is an interesting meeting of the urban and rural when they are held in city centres. Tourists have a chance to learn about and sample specialities of the area and wider region, engaging with the destination in a way deemed meaningful. There is an implication and expectation of authenticity (Hall *et al.* 2008), although the food sold is not necessarily traditional and may be informed by other food cultures. The hosting of farmers' markets is especially useful for cities positioning themselves as centres of food tourism, bolstering images of places where food of the highest standards is readily available from the immediate surroundings.

While beyond the scope of this chapter, it should be recognised that wholesale food markets can also be a curiosity for tourists as seen in Japan (Japan Guide 2018). A notable illustration is Tokyo's fish market at Tsukiji prior to its relocation. It was included in tourist itineraries and guidebooks, despite the early morning start for visitors, and one of the biggest and oldest such markets. Indeed, the steady stream of onlookers provoked complaints about interference in the work of dealers and restrictions were imposed (Bloomberg 2017).

Cooked food markets

Cooked food outlets are frequently incorporated into markets selling perishable and non-perishable foodstuffs but can be conceived of as a discrete type of market when a number are gathered together (Kowalczyk 2014). These often have individual or communal dining areas and sell what is widely known as street food which takes numerous forms around the world (de Cassia Vieira Cardoso *et al.* 2014). Traders are usually associated with less developed nations where they are relied upon by the urban poor for affordable meals, yet customers can be from all classes of society. Notions of street food are also undergoing revision and it now embraces sophisticated and expensive dishes of international and fusion cuisines. Several cities are famed for their traditional street food sold by scattered stalls and carts and collections along streets or in open-air and undercover markets. Best street food city rankings are published which appear highly subjective, but similar selections imply a degree of consensus. Frommer's (2017) is fairly typical and lists Bangkok, Hong Kong, Istanbul, Kuala Lumpur, Marrakech, Mexico City, Mumbai, Paris, Rio de Janeiro, Singapore, Tel Aviv and Tokyo. The inclusion of street food for the first time in the recently inaugurated Michelin Guides for the Asian cities of Bangkok, Hong Kong and Singapore (Henderson 2017) signifies its importance there as well as Michelin endeavours to remain relevant and accurately report provision.

Street food has attributes of novelty and authenticity and is capable of animating urban environments (Newman and Burnett 2013). These are appealing to tourists and cooked food markets are lively spaces of colours, sounds and smells where locals and their eating habits can be observed. Akin to markets generally, they communicate the essence of a place and its people and act as a theatre in which tourists are both performers and audience members. However, some outsiders may be worried about unfamiliar dishes and ingredients and ways of cooking alongside hygiene (Choi *et al.* 2013). Cooked food poses particular regulatory dilemmas because of the serious consequences of lapses in hygiene, resulting in schemes to try and impose minimum standards. Street traders as a whole are at risk in many cities where they do not fit with muni-cipal physical and economic plans (Oz and Elder 2012; Yatmo 2008). Informal trading is especially targeted, and vendors are constantly threatened with losing their livelihood. Moves against them have been reported in the Vietnamese cities of Ho Chi Minh City and Hanoi (Dao Truong 2018; Kim 2015) and Jakarta in Indonesia, symptomatic of the tensions between development and traditional practices in rapidly modernising South East Asian cities. The sustainability of street food and some markets is thus uncertain, and more examples are cited in the next section.

Markets in leading city destinations

As already suggested, destination marketing organisations utilise markets in their promotion and this is evidenced by the websites for the aforementioned ten most popular cities (Euromonitor International 2017). Hong Kong boasts of its street markets and 20 of particular interest are identified. Specialisations include antiques, Chinese medicine, jade, electronic items, flowers and fashion. Ladies Market in Kowloon is possibly the most visited general market, dealing mainly with clothing and accessories as well as souvenirs, and tourists can "practice their haggling skills" at the 100 stalls stretching for a kilometre (Hong Kong Tourism Board 2018). Shopping opportunities are more confined in neighbouring Macau, but a large outdoor market selling food, flowers and clothing and two flea markets are advertised (Macao Government Tourism Office 2018). Shenzhen's nomination as a leading tourist city is rather surprising and perhaps due to the high volume of business travellers to the Chinese industrial and commercial centre which is near to Hong Kong. TripAdvisor (2018) records 41 flea and street markets, but these encompass wholesalers and no official online English language material could be found.

Bazaars and night markets are a shopping category in advertising for the Malaysian capital of Kuala Lumpur and the largest is Petaling Street in Chinatown which sells clothes, food, electronic goods and fresh produce. Nearby Central Market occupies a 1928 building and is given over to handicrafts. Markets serving Indian and Malay communities are highlighted and the city has numerous other night and flea markets (Visit KL 2018). These and a multicultural population are also characteristics of Singapore which has over 100 officially run markets and hawker centres. The historic districts of Chinatown, Little India and Kampong Glam have street stalls and the first two have larger permanent markets. While some trading is directed at the specific ethnic group, souvenirs are available and dominate in certain sections of Chinatown, which is one of the city state's top unpaid attractions. Temporary markets are organised across the island for Chinese New Year, Deepavali and Ramadan and the most extensive in the traditional ethnic quarters are accompanied by illuminated street decorations and presented as cultural events by the Tourism Board (Singapore Tourism Board 2018). Bugis Market in the city centre has a wide range of items and regular flea and arts/design markets are held elsewhere.

Street food is integral to daily life in Singapore, Kuala Lumpur, Hong Kong and Macau, which are famous for their hawkers working in roadside markets and indoors as well as from single stalls. It is portrayed by marketers as an expression of cultural heritage and racial diversity in the cases of Kuala Lumpur and Singapore. Hawkers are a tourist resource and the award of a Michelin star to two from Singapore in 2016 garnered worldwide publicity, turning them into visitor attractions. Despite its interest to visitors and widespread patronage by residents, food hawking is a job shunned by many citizens because of the physical demands and comparatively low pay. There are thus doubts about the prospects of conventional hawker food and centres in Singapore especially, although young entrants who are entrepreneurial and part of a "hip" hawker culture and the upgrading of centres are grounds for optimism (Henderson 2017). Concerns about the survival of street food have also been voiced in Hong Kong where urban planning policies are making it difficult for traders to continue (BBC 2013) and in Bangkok.

The Thai capital is rated among the best cities for street food globally, but there is extensive illegal business, which authorities are struggling to regulate while appreciative of the social benefits of the sector. There are professed anxieties about safety due to congested pavements and roads, unsanitary conditions and criminality (Boonjubun 2017; Tangworamongkon 2014) with Singapore often held up as a model to emulate because of its licensing regime which incorporates strict hygiene rules, inspection, grading and centralisation of activity. News about plans to severely confine Bangkok's street trade prompted protests at home and overseas, leading to a statement by the Tourism Authority of Thailand. It sought to reassure that traditional food would still be obtainable subject to the law, but more curbs are expected. Bangkok is additionally known for its night markets (Tourism Authority of Thailand 2018) selling other products besides food, exemplified by Chatuchak Weekend Market which covers 35 acres. Floating markets of boats on the city's river and canals are also a distinctive and picturesque means of retailing, albeit several now are primarily tourist sights rather than fully functioning markets (Bangkok.com 2018).

Street and market trading conducted informally and its linkages to poverty are fewer pressing issues in the developed world, but this is not to deny the existence of urban deprivation and violation of laws in cities such as London. There were an estimated 162 markets in Greater London in 2010 (London Development Agency 2010), many serving local communities and without a tourism role. The official visitor guide contains information about 34 markets which are described as street (18), food (8), clothes (3), crafts (2), antiques (2) and books (1). A few are well known visitor attractions such as Camden, Covent Garden, Portobello Road and Borough Markets. Some have a long history while subject to reinvention in a bid to satisfy a modern consumer society. Most are located in prosperous areas of central London (Visit London 2018), certain of which have been gentrified relatively recently. Paris is another European capital which recognises the tourist predilection for markets. Seven of its flea markets offering "all types of objects for all budgets" and numerous second-hand markets which "liven up different parts of the city every weekend" are showcased on the destination management organisation's website. Also appearing are 50 out of the 80 food and specialist covered and open-air neighbourhood markets and numerous Christmas markets (Paris Tourist Office 2018).

Among the top ten tourist cities, New York advertises the highest number of markets. A search of a city website (NYC & Company 2018) yielded details of over 455 street markets, although some seem fairly small and ephemeral. Private blogs also indicate large volumes and great variety with artisan, farmers, food and flea markets (Markets of New York City 2018). Finally, and in a contrasting instance from the Middle East, tourists in the United Arab Emirates can "travel back in time" at the perfume, gold, spice and textile souks rich in sensory pleasure in the "heart of old Dubai". They can also visit twenty-first-century iterations of the souk concept, one of which has a "heritage theme", in air-conditioned comfort. A "market boom" is claimed to be occurring in the emirate comprising "community meeting places" selling local food, arts and crafts, and fashion and accessories. There is a flea market and, somewhat unexpectedly, a farmers' market in "the shadows of the Burj Khalifa" (Visit Dubai 2018), the futuristic tower which has come to symbolise modern Dubai.

Conclusion

Markets emerge as an important feature of many urban areas, the presence of which can add to overall liveability and tourist attractiveness. They are deemed worthy of mention in destination advertising, including that of the world's most visited cities. Taking diverse forms, markets cannot be divorced from their settings and wider conditions which gives rise to marked contrasts by location. Nevertheless, some common patterns are discernible worldwide and the appeal of markets to tourists appears to transcend international boundaries. Details of the merchandise are referred to in marketing, but equal attention is given to intangibles such as atmosphere. There are often allusions to tantalising sounds, sights and smells and the crowds who generate an exciting bustle. The marketplace is depicted as a social and cultural as well as economic space, acting as a representation of a destination and place of recreation and entertainment. At the same time, tourists can pose a threat to the character of some markets by driving away locals and paradoxically eroding visitation motivation.

Markets thus deserve study in a tourism context and reveal aspects of tourist behaviour and destination marketing strategies, suggesting what makes cities interesting to visitors and illuminating attempts to communicate favourable attributes in promotion. They also are sites of impacts where both the positive and negative consequences of tourism are on display, generating debate about how best to manage visitors and encourage sustainability. These are all areas for future research, identifying general principles and practices and exploring differences according to specific national and local dynamics. Market appeal and variations depending upon factors such as age, nationality and lifestyle are of special significance in terms of their survival and suitable marketing. The results of such work therefore have practical implications for urban planners and policy makers together with the tourism industry of private enterprise and public agencies which have responsibilities for the approval, operation and advancement of markets as a resident and visitor resource.

References

Bangkok.com, 2018. Bangkok markets: Where to find Thai markets in Bangkok. [online]. Available from: www.bangkok.com/shopping-market/ [Accessed 13 May 2018].

BBC, 2013. Hong Kong hawkers' survival battle. [online]. Available from: www.bbc.com/news/av/business-23967722/hong-kong-hawkers-survival-battle [Accessed 17 May 2018].

Beer, S., Murphy, A. and Shepherd, R., 2012. Food and farmers markets. *In*: C. McIntyre, ed. *Tourism and retail: The psychogeography of liminal consumption*. New York: Routledge, 111–142.

Bloomberg, 2017. Visit Tsukiji, a "great wonder of the world" while you still can. [online]. Available from: www.bloomberg.com/news/features/2017-06-08/how-to-visit-tsukiji-fish-market-in-tokyo-before-it-closes [Accessed 20 May 2018].

Boonjubun, C., 2017. Conflicts over streets: The eviction of Bangkok street vendors. *Cities*, 70, 22–31.

Brida, J.G., Meleddu, M. and Tokarchuk, O., 2017. Use value of cultural events: The case of the Christmas markets. *Tourism Management*, 59, 67–75.

Casteran, H. and Roedere, C., 2013. Does authenticity really affect behaviour? The case of the Strasbourg Christmas market. *Tourism Management*, 36, 153–163.

Choi, J., Lee, A. and Ok, C., 2013. The effect of consumers' perceived risk and benefit on attitude and behavioural intention: A study of street food. *Journal of Travel and Tourism Marketing*, 30, 222–237.

Chuang, Y.F., Hwang, S.N., Wong, J.Y. and Chen, C.D., 2014. The attractiveness of tourist night markets in Taiwan: A supply-side view. *International Journal of Culture, Tourism and Hospitality Research*, 8 (3), 333–334.

Correia, A. and Kozak, M., 2016. Tourists' shopping experiences at street markets: Cross-country research. *Tourism Management*, 56, 85–95.

Crespi-Vallbona, M., Perez, M.D. and Miro, O.M., 2017. Urban food markets and their sustainability: The compatibility of traditional and tourist uses. *Current Issues in Tourism*. [online]. Available from: https://doi.org/10.1080/13683500.2017.1401983 [Accessed 17 May 2018].

Dao Truong, V. 2018. Tourism, poverty alleviation and the informal economy: The street vendors of Hanoi, Vietnam. *Tourism Recreation Research*, 43 (1), 52–67.

de Cassia Vieira Cardosa, R., Companion, M. and Marras, S.R. eds., 2014. *Street food: Culture, economy, health and governance*. London: Routledge.

Ellis, A., Park, E., Kim, S. and Yeoman, I., 2018. What is food tourism? *Tourism Management*, 68, 250–263.

Euromonitor International, 2017. *Top 100 city destination ranking*. London: Euromonitor International.

Frommer's, 2017. The world's best street food: 12 top cities. [online]. Available from: www.frommers.com/slideshows/818551-the-world-s-best-street-food-12-top-cities [Accessed 15 May 2018].

Frost, W., Laing, J., Williams, K., Strickland, P. and Lade, C., 2016. *Gastronomy tourism and the media*. Bristol: Channel View Publications.

Getz, D., Robinson, R., Andersson, T. and Vujicic, S., 2014. *Foodies and food tourism*. Oxford: Goodfellow.

Gharipour, M., 2012. The culture and politics of commerce. *In*: M. Gharipour, ed. *The bazaar in the Islamic city: Design, culture and history*. New York: University of Cairo Press, 1–50.

Gilli, M. and Ferrari, S., 2018. Tourism in multi-ethnic districts: The case of Porta Palazzo market in Torino. *Leisure Studies*, 37 (2), 146–157.

Gonzalez, S. and Waley, P., 2012. Traditional retail markets: The new gentrification frontier? *Antipode*, 45 (4), 965–983.

Guardia, M., Fava, N. and Oyon, J.L., 2010. Retailing and proximity in a liveable city: The case of Barcelona public markets system. *In*: M. Schrenk, V.V. Popovich and P. Zeile, eds. *REAL CORP 2010 Proceedings, Vienna 18–20 May*, 619–628.

Hall, C.M., Mitchell, R., Scott, D. and Sharples, L., 2008. The authentic market experience of farmers' markets. *In*: C.M. Hall and L. Sharples, eds. *Food and wine festivals and events around the world: Development, management and markets*. Oxford: Butterworth-Heinemann, 198–230.

Hallsworth, A., Ntounis, N., Parker, C. and Quin, S., 2015. *Markets matter: Reviewing the evidence and detecting the market effect*. Manchester: The Institute of Place Management.

Henderson, J.C., 2017. Street food, hawkers and the Michelin Guide in Singapore. *British Food Journal*, 119 (4), 790–802.

Hong Kong Tourism Board, 2018. Street markets and shopping streets. [online]. Available from: www.discoverhongkong.com/seasai/shop/where-to-shop/street-markets-and-shopping-streets [Accessed 12 May 2018].

Hsieh, A.T. and Chang, J., 2006. Shopping and night markets in Taiwan. *Tourism Management*, 27, 138–145.

Ibrahim, M.F. and Leng, S.K., 2003. Shoppers' perceptions of retail development: Suburban shopping centres and night markets. *Journal of Retail and Leisure Property*, 3 (2), 176–189.

Japan Guide, 2018. Markets. [online]. Available from: www.japan-guide.com/e/e2452.html [Accessed 18 May 2018].

Kikuchi, A. and Ryan, C., 2007. Street markets as tourist attractions: Victoria Market, Auckland, New Zealand. *International Journal of Tourism Research*, 9, 297–300.

Kim, A.M., 2015. *Sidewalk city: Remapping public space in Ho Chi Minh City*. Chicago, IL: University of Chicago Press.

Kowalczyk, A., 2014. From street food to food districts: Gastronomy services and culinary tourism in an urban space. *Turystyka Kuturowa*, 9, 2136–160.

Lee, S.H., Chang, S.C., Hou, J.S. and Lin, C.H., 2008. Night market experience and image of temporary residents and foreign visitors. *International Journal of Culture, Tourism and Hospitality Research*, 2 (3), 217–233.

London Development Agency, 2010. *London's retail street markets: Final draft report*. London: Regeneris Consulting.

Macao Government Tourist Office, 2018. Shopping by category. [online]. Available from: http://en.macaotourism.gov.mo/shopping/shopping_detail.php?c=3&id=65 [Accessed 13 May 2018].

Markets of New York City, 2018. A guide to the best artisan, farmer, food and flea markets. [online]. Available from: www.marketsofnewyork.com [Accessed 15 May 2018].

Morales, A., 2009. Public markets as community development tools. *Journal of Planning, Education and Research*, 28 (4), 426–440.

Newman, L.L. and Burnett, K., 2013. Street food and vibrant urban spaces: Lessons from Portland, Oregon. *Local Environment*, 18 (2), 233–248.

NYC & Company, 2018. Shopping. [online]. Available from: www.nyco.com/search/default_collection/bWFya2v02cw [Accessed 15 May 2018].

Oz, O. and Elder, M., 2012. Rendering Istanbul's periodic bazaars invisible: Reflections on urban transformation and contested space. *International Journal of Urban and Regional Research*, 36 (2), 297–314.

Page, S.J. and Hall, C.M., 2003. *Managing urban tourism*. Harlow: Pearson.

Pappalepore, I., Maitland, R. and Smith, A., 2014. Prosuming creative urban areas: Evidence from East London. *Annals of Tourism Research*, 44, 227–240.

Paris Tourist Office, 2018. Shopping. [online]. Available from: https://en.parisinfo.com/shopping [Accessed 13 May 2018].

Singapore Tourism Board, 2018. Singapore's annual cultural events. [online]. Available from: www.visitsingapore.com/editorials/singapores-annual-cultural-events/#festivals-events-singapore [Accessed 12 May 2018].

Tangworamongkon, C., 2014. *Street vending in Bangkok: Legal and policy frameworks, livelihood challenges and collective responses*. Cambridge, MA: WIEGO Law and Informality Resources.

Timothy, D.J., 2005. *Shopping tourism, retailing and leisure*. Bristol: Channel View Publications.

Tourism Authority of Thailand, 2018. Bangkok. [online]. Available from: www.tourismthailand.org/About-Thailand/Destination/Bangkok [Accessed 15 May 2018].

TripAdvisor, 2018. Flea and street markets in Shenzhen. [online]. Available from: www.tripadvisor.com.sg/Atractions-g297415-ctivities-c26-t142-Shenzhen_Guangdong.html [Accessed 13 May 2018].

UNWTO, 2017. *Second global report on gastronomy tourism.* Madrid: United Nations World Tourism Organization.

URBACT, 2015. *Urban markets: Heart, soul and motor of cities.* Barcelona: City of Barcelona.

Visit Dubai, 2018. Discover all that's possible in Dubai. [online]. Available from: www.visitdubai.com/en/shop-dine-relax/shopping [Accessed 12 May 2018].

Visit KL, 2018. Bazaars and night markets. [online]. Available from: www.visitkl.gov.my/visitklv2/index.php?r=column/ctwo&id=40 [Accessed 13 May 2018].

Visit London, 2018. London street markets. [online]. Available from: www.visitlondon.com/things-to-do/shopping/market/street-market/ [Accessed 12 May 2018].

Wu, M.Y., Wall, G. and Pearce, P.L., 2014. Shopping experiences: International tourists in Beijing's Silk Market. *Tourism Management,* 41, 96–106.

Yatmo, Y.A., 2008. Street vendors as "out of place" urban elements. *Journal of Urban Design,* 13 (3), 387–402.

Yu, S.D., 2004. Hot and noisy: Taiwan's night market culture. *In*: D.K. Jordan, A.D. Morris and M.L. Moskowitz, eds. *The minor arts of daily life: Popular culture in Taiwan.* Honolulu, HI: University of Hawai'i Press, 129–149.

Zakariya, K. and Ware, S.A., 2010. Walking through night markets: A study on experiencing everyday urban culture. Paper presented at the 11th International Joint World Cultural Tourism Conference 2010 in Hangzhou, China, 12–14 November.

Zittlau, J. and Gorman, C., 2012. Farmers markets as an authentic tourist experience: The case of Dublin. [online]. Available from: https://arrow.dit.ie/cgi/viewcontent.cgi?article=1002&context=dgs [Accessed 11 May 2018].

20

SOMETHING FOR EVERYONE?

The challenge of touristic urban spaces

Ece Kaya and Deborah Edwards

Introduction

There is no question that urban tourism is considered a valued component of a city's economy (Edwards et al. 2008, Ashworth and Page 2011). This is evident in the latter part of the twentieth century, by cities transforming industrial spaces into tourism precincts as they sought to provide leisure opportunities for visitors. With the advent of the Festival Market model of the 1980s (Edwards et al. 2008), taking inspiration from the early approaches to waterfront developments that occurred in Boston, Baltimore, Seattle, San Francisco and London (Florio and Brownill 2000, Bruttomesso 2004, Gospodini 2006, Jones 2007, Fainstein 2008, Smith and Ferrari 2012), Sydney transformed a disused working port – Darling Harbour – into an attractive tourism precinct. The precinct is continually being revitalised with the most recent change a redevelopment of the Sydney Convention and Exhibition Centre site into the new International Convention Centre Sydney.

Darling Harbour's main attractions are now the convention centre and numerous hotels, restaurants and cafes where people can eat, drink and stay and, at times, enjoy festival activities. As a tourism precinct Darling Harbour has been successful receiving approximately 4.6 million domestic and international visitors in year end March 2018 (Destination NSW 2018). The major motivations of these visitors were eating out, sightseeing and shopping (Destination NSW 2018). Although these developments are meant to be based on the public interest and improving the livelihood of the city centre, Darling Harbour's focus on the construction of hotels, restaurants, cafes, bars and tourism shopping has created an attraction space for middle- and upper-class residents and tourists (Kaya 2018) shifting it from a neighbourhood to a commodity (Edwards et al. 2010).

It is evident that urban tourism creates considerable income and a significant number of jobs in those facilities. However, what we have in Darling Harbour is not very different than many other waterfront precincts (Kaya 2020). In this chapter we argue that urban tourism spaces can be more than what Darling Harbour offers now, more than cafes, restaurants and hotels that represent a mode of consumption. Urban tourism spaces such as this can be used in more creative and productive ways by both residents and tourists. Drawing on theoretical perspectives this chapter discusses what those creative and productive uses might be and seeks to address the question of what Darling Harbour can offer to please both the residents of Sydney and tourists.

The shift in the urban waterfronts

The decline of working waterfronts coupled with demand for urban transformation has seen widespread recognition since the later 1960s (Hoyle 2000). The development of new transport models, technological improvements in containerisation and changes in shipping industries created the need for deindustrialisation, and port cities have faced urban and economic decline since the second half of the twentieth century. Cities that were once functionally dependent upon their ports for economic growth developed beyond that point, with global trends towards their transformation for other uses. Waterfront areas have therefore emerged as a distinctive type of tourist district/precinct (Griffin and Hayllar 2006). The goal of these projects and transformations is a new production of the city, with an embraceable planned approach creating commercial and recreational areas that would contribute to a healthier economic and urban life.

In the 1980s, the idea of creativity was developed with terms of culture, arts, cultural resources and cultural planning, and in the mid-1990s, the cultural industries became creative industries and the creative economy (Landry 2012). Harvey (1987 cited in Urry 2002, p. 13) argues that "display and spectacle are the symbols" of a society and that every city has to present itself as exciting and innovative, creative and as a safe place in which to live and consume. In this process, city governments encouraged strategies to increase the attractiveness of cities and they focused on the creation of consumption spaces such as "nouvelle cuisine restaurants", art galleries, coffee bars, hotels and real estate developments for middle-class elites (Zukin 1998, p. 825).

The phenomenon of waterfront redevelopment is a highly visible example of contemporary urban restructuring enabling the emergence of creative quarters which construct the link between cultural consumption and cultural production (Evans 2009b). In many cities, efforts have been made and are currently being made, to renew the strengths of the waterfront through large-scale renewal projects (West Melbourne Waterfront, Australia; False Creek South in Vancouver, Canada; Hammarby Lake City, Stockholm). These changes dramatically alter the original character and function of the port area from a site of production to a landscape more readily associated with "post-Fordist" consumption practices. A landscape which has been redeveloped and transformed in a local context, concomitantly takes on a global orientation, mimicking other "post-modern" urban waterfront destinations. However, this global orientation which inspired governments of big cities to create spaces for urban festivals and harbourside shopping malls can result in the standardisation of consumer culture and the detachment of cities from their histories (Zukin 2009). Sydney's urban waterfront was also looking for a place in the global world, for the creation of a new identity and for the marketing of this new identity in the process (Hoyle 2000). With an intention of creating a place for people and tourism Darling Harbour's redevelopment was regarded as a mega-project; a tool for urban renewal (Oakley 2014).

Urban renewal has become an instrument to establish continuity between the past and the future in cities that have lost their identity and become standardised. Redevelopment has been offered as a solution to form new areas of investment and consumption when cities have started to look for marketing niches – through the "manufacturing" of new identities in order to increase their attraction, or by means of new investments in some high-potential areas of the city that had lost their economic priorities (Harvey 1989, Smith 2002, Kaya 2020). Cities have striven to reclaim, through urban tourism, the economic resources lost in the deindustrialisation process.

Creative space and challenges of urban tourism

Urban tourism regulated by local authorities tried to create a new identity and articulate the urban space to global and regional markets. Therefore, urban tourism has become a strategy to promote cities in the urban competition that is played out among the world's cities within the new order of global capitalism. Urban tourism not only brought economic promotion, but also led to socio-cultural transformations of cities (Kaya 2020).

The World Travel and Tourism Council (WTTC) (2019, p. 5) reports that in 2018 tourism accounted for 10.4% of global GDP and 319 million jobs or 10% of total employment in 2018. It is not surprising then that urban tourism is considered a major economic contributor of the twenty-first century with waterfronts as the gathering areas between maritime and urban environments (Ashworth and Page 2011). Urban tourism generally gets its legitimacy from the rhetoric that city investments create new investment areas in different sectors, in turn increasing economic returns and distributing them in the interest of all parts of the community that create a workforce, protect and sustain historical zones (Edwards et al. 2008). Urban tourism produces reinterpretations of locality to promote the district to the international tourism market by means of a more local discourse (Kaya 2020, Selby 2004). However, the priorities of tourism-led transformation lead to increased demand for tourism, for goods and services, resulting in the increase of local prices and the creation of a high level of inflation (Kreag 2001) thereby ignoring the living, housing and other needs of residents. This also means that tourism facilities managed and owned by outsiders, or by a few local elites, can lead to fewer benefits for local people (Inskeep 1991).

Almost every city in the world has been inclined to manufacture itself into a well-groomed identity from the landmarks of the past, and then to introduce this recreated identity to the world by means of advertisements. Because of the "city myths" created, only an image appears, and the multi-identity structure of the cities is somewhat diminished (Urry 2002). This has become an important point of criticism on how cities sustain both their visitors and residents, by what they offer.

In the beginning of the twenty-first century, culture and creativity became core objectives to successful cities searching for new foundations in city development (Kunzmann 2004). Creativity in tourism has also become a way of "developing tourism products and experiences, revitalising existing products, valorising cultural and creative assets, providing economic spin-offs for creative development, using creative techniques to enhance the tourism experience and adding buzz and atmosphere to places" (Richards 2014, p. 120). Art and creativity create cultural innovations that develop cultural places and clusters through the process of social interaction and communication (Frey 2009). This is highly evident in the redesign of urban spaces, for example Hudson Yards, New York and the Millennium Park, Chicago.

A creative city constitutes cultural and physical infrastructure for investing in the arts, culture and creative sectors. The creative class, also called knowledge workers, has hard and soft location preferences determined by economic forces. Quality of university milieu, leisure facilities, hedonistic environments, lifestyle environments and accessibility attract scientists, engineers and creative professionals (Baum et al. 2007). Creative cities also contain quality of place that supports public welfare and creativity that inspires change (Frey 2009). In this context, urban waterfronts have become spaces of opportunity for the creative class rather than being problem spaces (Desfor and Laidley 2010). Rotterdam Harbour's old docks, Tobacco Warehouse at Stanley Dock, Manhattan's High Line and Vancouver's Granville Island are some examples of the creative use of redundant waterfront and obsolete infrastructure for hosting local food markets, theatres, arts and craft studios, urban parks and graffiti walls for local artists.

Creative space is a space of hybridity and constitutes a link between local community and artists by fostering inspiration (Santagata 2002). Linkages between community life and vibrant cultural agendas; between work and leisure, between working space and private residential space in a sort of 24/7 lifestyle; between knowledge-oriented or creative-oriented new facilities represent distinctive characteristics of creative spaces. Creative spaces, which provide art activities and flexible use of public space, can be defined as a combination of cultural life and local creative ecosystems (Kunzmann 2004). Therefore, connectivity between creativity and culture in spaces provides community participation in arts, crafts and group activities (Evans 2009a). In this respect, the notion of "successful placemaking" has been provided as an approach that brings the physical characteristics of place and the process of social interaction together. This approach relies on the idea that the physical characteristics of a place have influence on how its users interact with it (Ferrari et al. 2012). Project for Public Spaces (2006) argues that 80% of the success of public space depends on the activities they offer.

Placemaking and design principles for waterfronts

Creative city-inspired placemaking ideas have been incorporated in cities' development strategies. Placemaking converts public spaces into places that connect with their residents; that provide rich experiences and sense of belonging; that have meaning and identity (Shaw and Montana 2016). Placemaking is the concept that aims to establish a positive image, and coherent space between activities and locales. In this section we consider waterfronts in the context of micro-, meso- and macro-scales and we discuss the attributes required to make successful places.

Scales to assess placemaking

Placemaking in waterfronts has been discussed within three scales: the micro-, meso- and macro-scale. The micro-scale considers the waterfront at a human scale, recognising the physical qualities, visual and social realm of the place. The meso-scale evaluates the waterfront development area's connections to the surrounding city and provides design guidance. The macro-scale places the waterfront in the perspective of a political-economy focusing on marketing approaches within wider regional, national and international contexts (Ferrari et al. 2012).

While the planning of waterfronts should consider the interests of various social groups they are often places that result in a halo effect increasing the prices of nearby residences and lead to gentrification. This is why, planning of urban waterfronts needs to have the social dimension in consideration which means the waterfront should offer different experiences as well as understand the qualities of the community (Sairinen and Kumpulainen 2006). The domination of commercial-tourist functions over residential ones can be problematic as it can limit the use of these areas to a few hours a day. Diversifying activities for residents and encouraging 24-hour activity will be favourable for waterfronts to thrive throughout the year.

In the urban design that aims to create stimulating city spaces, waterfronts offer a great opportunity for public enjoyment and add value to public and private development. Public parks, promenades and streetscape improvements can provide coherent, publicly accessible pedestrian space systems. This space system also needs to be supported by multiple modes of transportation which provide people with different modes of access depending on their needs. In this respect, mixed-use developments are beneficial in creating opportunities for social interaction, socially diverse communities, greater urban vitality and street life, better access to facilities, greater feelings of safety through more "eyes on the street" and more consumer choice of lifestyle (Yeang 2000). Having squares and gathering spaces surrounded by commercial land uses integrated with promenades, foreshores and docks contributes to an active space design (Moughtin 2003). To envision waterfronts as lively public destinations for people to revisit and enjoy places public goals at the centre of development.

Encouraging community engagement and local ownership can establish a pride and a more long-term financial expediency by shaping the waterfronts regarding the community's needs and desires. This is why adoption of a shared community vision rather than a master plan will build enthusiasm among the public, will avoid prescribed project implementations and will encourage stakeholders to become committed to the waterfront's success (Project for Public Spaces 2009). "Have Your Say" created by NSW Department of Planning, Industry and Environment aims to receive community feedback on local government area plans and state planning initiatives such as open spaces and parklands, improving energy affordability and increasing housing choice. It is not specifically designed for waterfronts, but it is a positive initiative which emphasises community contributions in decision making.

Another effective principle of creating successful waterfronts is to create multiple destinations within a destination by using the "Power of Ten". A community defines ten uses and activities that nearby residents, businesses, community organisations and other stakeholders can enjoy (Project for Public Spaces 2009). Hence, ten destinations need to be incorporated as a whole waterfront vision to achieve continuity in the public space. Helsinki's Esplanade is a great example that connects the city centre and the waterfront for people to reach by foot or bike. Establishing a strong connection to water or types of water such as fountains or spray play areas also optimises the public space (Project for Public Spaces 2009). Project for Public Spaces (2007) considers Circular Quay, The Rocks and the Botanical Garden as places that offer a unique waterfront atmosphere. On the other hand, Darling Harbour has been argued as a waterfront location that cut the community–city–water connection due to the construction of the Western Distributor (Kaya 2020, Farrelly 2002, Freestone 2000) with its large concrete structures.

Kop van Zuid Project in Rotterdam, the IJ-oevers Project in Amsterdam and the Hafencity Project in Hamburg and the regeneration projects of the Scheldt quays and the Islet in Antwerp are some of the case studies examined by Erkök (2009) within five criteria: urban space/recreation, housing, cultural environment, land use pattern and infrastructure/mobility (Table 20.1). The case studies revealed that water contributes to the development of collective space and creates a sense of belonging, that establishing human–water connections by applying innovative design solutions like water squares enhances the quality of urban life, and that using multiple factors of the urban space produces original spaces as seen in the combination of leisure and flood safety implications in Hafencity, Hamburg. Erkök (2009) argues that housing on waterfronts leads to a more equal day–night use of space and using architecturally innovative designs encourages attraction as seen in Amsterdam and Hamburg. Landmarks on waterfront spaces also establish a greater cultural environment and create vibrancy as can be found in the Music building in Amsterdam (Erkök 2009) or the Australian Centre for the Moving Image (ACMI) in Melbourne. Newly added public transport infrastructure enables better connections with the city centre and increases accessibility (Erkök 2009).

Place attributes

In *Death and Life of Great American Cities*, Jane Jacobs (1961) defined successful urban place as a place with a mixed variety of activities such as tea houses and cafes, ethnic grocery stores, cake shops, cinemas, galleries, pubs and clubs and delicatessens; a place where these activities are accessible. In this context, Montgomery (1998) illustrated the qualities of successful urban places

Table 20.1 Waterfront regeneration projects

City	Rotterdam		Amsterdam	Hamburg	Antwerp	
	Kop van Zuid	Rotterdam 2035 and Water squares	IJ-oevers	Hafencity	't Eilandje	Quay redesign
Project						
Programme	Residential offices, education, leisure, culture, tourism	Residential, leisure, sports	Residential, leisure, culture, tourism	Residential offices, retail, leisure, culture, tourism, flood protection	Residential, offices, leisure, culture	Leisure, flood protection
Landmarks	Erasmus Bridge, Hotel New York		Whale, Silodam, Music building, Filmmuseum	Elbphilharmonie Concert Hall	MAS (Museum by the Stream)	
Added infrastructure	Erasmus Bridge, metro stop, tramline extension	More water use for public transport	Tunnel, NorthSouth metro line, U-tram	New U4 underground line	Tram	
Driving force for regeneration or development	Poor image of Rott-south, empty port sites, need for new. attractive residences	Water management problems, demand for more water-related living	Reintroducing the city to river, reutilizing old port areas, creating attractive homes close to the city centre	Reutilising old port areas and expanding the city centre by 40%, solving the occasional flooding problem	Weak relationship of the city and Scheldt, empty port sites	Idle quay, flooding problem, broken contact of the city with water
Spaces of interaction with water	High-quality design and lively waterfronts, terraces with panoramas of Maas and the city	Water squares, water roofs, homes on water	Man-made islands on the IJ, quays, bridges, beaches on IJ	Promenades, quays, plazas, waterfront terraces, floating pontoons	Boardwalks, floating platforms, breakwaters and ground levels along the waterfront	Promenades, raised platforms providing perception of the river as opportunity for city dwellers
Public space qualities	High-quality and walkable public realm, use of public art	Innovative multifunctional water squares, flexible dykes combined with city parks	High-quality open spaces accompanying residential use at waterfronts	Vibrant and high-quality open spaces by and on the water, flood protection combined with public–private spaces	Inner water, surface as a public square, qualitative redevelopment public domain + new functions for docks	Quays as multifunctional public spaces, complete public use, quays as spaces in between the river and city
Culture initiators	Luxor theatre, museums, outdoor culture events		Westergasfabriek, MDSM yard, Music building	Elbphilharmonie Concert Hall, museums, university	MAS (Museum by the Stream) several museums	
Diversity	Variety of residential styles by different architects working on each block	Water-related housing, water squares tailor made for different neighbourhood contexts	Variety of residential types and styles	Diversity in functions, forms in the new city centre	Accentuating unique mix, island character and lively urban neighbourhoods	Using new dam as a tool and combining it with a variety of public spaces
Housing qualities	Mixture of high- and low-income housing for a wider social mix	New typologies of floating homes, buildings on poles	A mix of social, middle-income and higher-income housing, good-quality and high-quantity housing	A variety of possibilities for the mix of offices and dwellings	Luxurious housing along the quays of 'Willemdok; "living by the water"	

Source: Reprinted [adapted] from Erkök (2009).

by providing conditions to fit the built form to activity and image. He argued that those conditions should support each other. For him, activity is the product that incorporates vitality and diversity.

Vitality is how we differentiate successful places from others. It exemplifies the pedestrian flow during different times of the day, the number of cultural events and celebrations over the year, current active street life and the uptake of facilities (Montgomery 1998). Diversity is associated with the number of primary land uses, including: residential; the extent of local or independent businesses and shops; patterns in the existence of evening and night-time activity; the presence and size of street markets; the availability of cinemas, theatres, wine bars, cafes, pubs, restaurants and other cultural and meeting places offering different services at varying prices and degrees of quality; the availability of gardens, squares and corners to enable people-watching; the availability of various unit sizes of property at different degrees of cost that enable small businesses to stay in business without being affected by sudden rises in rent and/or property taxes; the degree of innovation and confidence in new architecture, so that where possible there should be a variety of building types, styles and design; and the existence of an active street life (Montgomery 2008).

Image is how people perceive a place. In this sense, visitors use landmarks to construct their knowledge about a place, to orient themselves and to understand a city and its image. People's individual values and ideas derived from their cultural identities shape their perceptual expectations of a place. The image of place is their set of feelings and impressions about that place (Montgomery 1998). These sentiments are captured in Figure 20.1 the Place Diagram (Project for Public Spaces 2013).

The Place Diagram (Project for Public Spaces 2013) is a framework for developing urban places culturally, economically and environmentally. Drawing on Montgomery's (1998) conditions for achieving an urban context it outlines four key attributes required to make great places: access and linkage, comfort and image, uses and activities, and sociability (Figure 20.1). First places need to be accessible and well connected to their surroundings. An accessible city allows people from all ages, backgrounds and capabilities to engage with the activities, resources, services and information they need. Second they should be comfortable and have a good image. In other words, residents should have positive perceptions of safety, cleanliness and places to sit and relax. Third is that people of different ages, demographics and life stages can participate in a range of activities, and fourth is that places create environments that encourage socialisation and revisitation. At the core of the Place Diagram framework is that the success of cities occurs at the level of human interaction. Ferrari et al. (2012) state that "*a successful place should fulfil people's emotional needs and even influence their mood*" (p. 160).

Taken together, it is evident that the success of a waterfront development is closely related to what it offers as non-cliché designs which produce a creative public space allowing all levels of society to participate in and enjoy. It has long been argued that Darling Harbour missed this opportunity when it was first redeveloped, but it still has the potential to become a more dynamic and creative urban space.

Darling Harbour – is it something for everyone?

Darling Harbour comprises four precincts; King Street Wharf, Cockle Bay, Darling Quarter and Darling Square (Figure 20.2). Darling Harbour sits between the Central Business District (CBD) to the east, Chinatown to the south and Pyrmont and the residential areas to the west.

The development of Darling Harbour was led by the construction of separate developments which were aimed at transforming the public domain of Darling Harbour and generating better

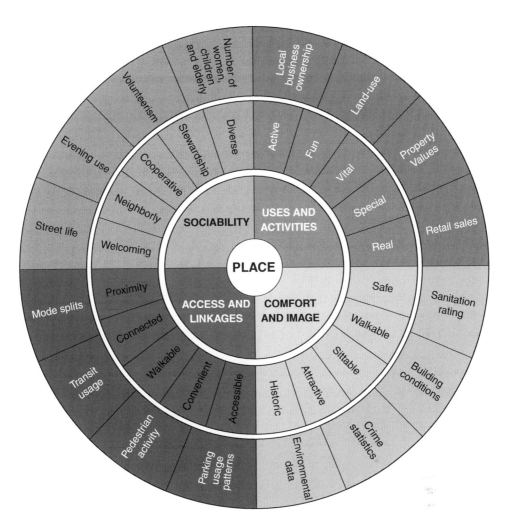

Figure 20.1 The Place Diagram.

Source: Project for Public Spaces (2013).

connectivity between Darling Harbour and Sydney CBD. The urban space/recreation, housing, cultural environment, land use pattern and infrastructure/mobility of King Street Wharf, Cockle Bay, Darling Quarter and Darling Square is represented in Table 20.2. These projects and transformations aim at a "new production of the city", with an embraceable planned approach to providing a healthier economic and urban life (Kaya 2020).

At the micro-scale, the purpose of regeneration was to create an urban neighbourhood by providing a childcare centre, a new library (Darling Square) and apartments occupied by families. An old ice-making factory and disued rail line were incorporated to generate a sense of place (Maddox 2019). In a way, it is good to see that this regeneration learned from the past and now values place attachment, but it is still evident that tourists do not visit Darling Harbour either for its heritage or its historical significance (Kaya 2020). The Urban Institute's Arts and Culture Indicators Project (ACIP) (Jackson et al. 2006) in the United States indicates that arts

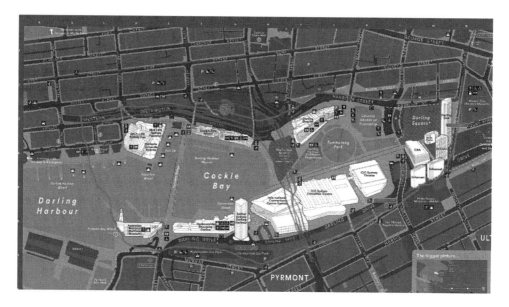

Figure 20.2 Map of Darling Harbour.

Source: Reprinted [adapted] from https://darlingharbour.com/media/2660/darling_harbour_map_2018.pdf.

and culture, and opportunities for creative expression enable healthy places. There is a demand for robust public support and engagement with creative endeavour, for precincts and infrastructure that are connected and integrated for stronger cultural participation (City of Sydney Council 2014). Darling Harbour has the potential to offer more frequent cultural experiences and creative participation.

At the meso-scale, Darling Harbour has a better connectivity to its surroundings. Especially, Darling Quarter provided a new pedestrian street which connects Darling Harbour South to the City and a pedestrian boulevard which runs adjacent to dining spaces, children's theatre, grassed areas, a water feature and a playground. The development of Darling Quarter and Darling Square included integrated transport links to buses, trains and light rail with more efficient walkways and cycleways. Likewise, Cockle Bay, now has a 220-metre-wide boulevard with continuity between Barangaroo Reserve and Circular Quay, Cockle Bay and Chinatown and a connected walkway to Central Station.

At the macro-scale, the developments have generated recognition in both regional and international contexts. The number of international visitors to Darling Harbour in March 2018 was 2.5 million, up 3% compared to March 2017 and there were 2.1 million domestic visitors to Darling Harbour in the year ended March 2018 (Destination NSW 2018).

The redevelopment of King Street Wharf, Cockle Bay, Darling Quarter and Darling Square precincts constitute key placemaking attributes of the Place Diagram (Project for Public Spaces 2013) – sociability, uses and activities, comfort and image, access and linkage. We employ these attributes to evaluate Darling Harbour.

Darling Harbour attracts visitors from all ages and offers a comfortable, safe and clean environment (Hayllar et al. 2010). Night-time building lighting and illumination of the playground enable 24-hour use. The precinct provides open green areas, rugs, deck chairs, table tennis tables, free WIFI along with the nearby shops, cafes and bars. However, local business ownership, local

Table 20.2 The urban space/recreation, housing, cultural environment, land use pattern and infrastructure/mobility of Darling Harbour

Project	King Street Wharf	Cockle Bay	Darling Quarter	Darling Square
Program	Residential, commercial, retail, dining, leisure	Retail, dining, leisure and commercial facility with new park facilities	Commercial, public open spaces, mixed use, retail, community, recreation	Retail, dining, residential, education and cultural institutions
Landmarks	Sea Life Sydney Aquarium, Madame Tussauds, Wildlife Sydney Zoo	Pyrmont Bridge, The International Convention Centre Sydney (ICC Sydney), Australian National Maritime Museum, IMAX (reopens in 2020), Harbourside Shopping Centre, Kingpin Bowling	Children's Playground Village Green, OPEN, and the Monkey Baa Theatre, Chinese Garden of Friendship	Commonwealth Bank Place, Darling Square Library, Steam Mill Lane, Little Hay Street, The Exchange Laneway bars, craft breweries and gin distilleries, artisanal bakeries and fusion food pop ups
Added infrastructure	Light rail. New city pedestrian connections	Light rail. New city pedestrian connections	Light rail. New city pedestrian and cycle path connections	Light rail. New city pedestrian and cycle path connections
Driving force for regeneration or development	Waterfront redevelopment. Regeneration of precinct	Demolition of existing site improvements	Improving the public domain that was first developed in 1980s. Revitalisation of the western edge of the Sydney CBD	Revitalisation of the site between Darling Harbour, Haymarket, Pyrmont and the city
Spaces of interaction with water	Waterfront promenade. Ferry, charter and cruise ship services	Waterfront promenade. Water spiral fountain	Playground with interactive water play. Water park	
Public space qualities	Walkable public realm	Walkable public realm	Innovative, high-quality and walkable public realm, community garden, open area for people to picnic and relax and features oversized rugs, deck chairs, table tennis tables and free WIFI and board games	High-quality and walkable public realm, open space, green space
Culture initiators		Powerhouse Museum	300-seat children's theatre. The water play area designed to reference the site's industrial history through the use of pumps and water wheels.	Darling Square Library, Paddy's Market, Chinatown
Diversity	Mixed-use precinct: residential and commercial development	Mixed-use precinct: public domain and commercial development	Mixed-use precinct: public domain and commercial development	Mixed-use precinct: residential and commercial development
Housing qualities	High-income housing	High-income housing	Middle- and high-income housing	Middle- and high-income housing

markets, Indigenous representation, its historic connection to the past are not strongly evident in Darling Harbour.

The new landmarks of Darling Harbour now include the reconstructed International Convention Centre Sydney (ICC Sydney), ICC Sydney Exhibition Centre, ICC Sydney Theatre and the 590-room Sofitel hotel. Additional landmarks due to open in 2020 are the 20-storey W Hotel and an IMAX theatre. However, these new developments encroach on the precinct's meaningful open spaces and connections. There are wide concrete walkways without purposes, few seating arrangements or artwork engagements on the front promenade of the Harbourside Shopping Centre and along Cockle Bay. The use of the water is not significant and there are no water-related activities.

Darling Quarter offers pre-schoolers and their carers free yoga sessions on Fridays led by a qualified yoga instructor who is also a nanny and an actress. Little Creators provides art and music workshops and story time on a daily basis in specific months, especially during school holidays. Father's Day Markets, Melbourne Cup Day at various venues, Matsuri Japan Festival and New Beginnings Spring Festival are some of temporary events happening in Darling Harbour. Seasonally, Christmas decorations and fireworks or Easter celebrations also attract tourists and bring the Sydney community together, but the participative learning in the arts, authentic experiences connected to the history or special character of the place do not reflect permanence.

According to Darling Harbour's Visitor Profile (2018), the major motivations of both domestic and international tourists who visited Darling Harbour were eating out at cafes and restaurants, going to pubs, clubs and discos, sightseeing and going shopping (Destination NSW 2018). Darling Harbour was considered an "essential part of visiting Sydney", with tourists stating that they did not think that Darling Harbour represented the real life of Sydneysiders (Griffin and Edwards 2012). They were unsure about the precinct being a place where tourists could interact with Sydney people and with other tourists. Therefore, tourists did not think that the precinct offered different and interesting sounds. They were undecided as to whether the precinct enabled them to easily find their way to other places in Sydney or offer activities for extended periods of time (Griffin and Edwards 2012).

Another Australian case study found that one of the most popular activities for both domestic and international visitors was Darling Harbour. It was stated that Darling Harbour was visited by 82% of respondents (Edwards et al. 2009). Nevertheless, the Aquarium (one of the main attractions) and other sites within the Darling Harbour precinct were visited by 29.7% of visitors or less (Edwards et al. 2009). The Powerhouse Museum and the Australian Maritime Museum were more attractive to domestic visitors rather than international visitors (Edwards et al. 2009).

As Darling Harbour has been purposefully developed, managed and marketed for tourism, one of our colleagues asked an interesting question about Darling Harbour while considering its constant redevelopment: Does Sydney do anything other than eating and sleeping? We work in a university situated in the heart of Sydney that has Darling Harbour as its backdrop. This provides us with an intimacy of this urban space. Similar to the sentiment of our colleague we too have noticed that through Darling Harbour's successive redevelopments (from working port to leisure precinct) Darling Harbour seems to have a narrow set of activities – tourist activities, eating and accommodation. Certainly, the space is shared by tourists, local visitors, city workers and those who live nearby. But this sharing generally occurs on the way to and from the abundant eating venues in the area. Really there is not much else on offer – except for the paid activities such as Sea Life Sydney Aquarium, Madame Tussauds and Wildlife Sydney Zoo. Although Darling Harbour represents the application of successful placemaking criteria with a range of successful attributes and shows an example of a produced place for everyone, a lack of creative space to have a more participatory city life and to offer connectivity between residents and tourists still remains.

New avenues for Darling Harbour

Building a distinctive creative precinct is a cultural policy component of the City of Sydney Council. Sydney Your Say Open Forum is an initiative that aims to engage the community with Sydney's future plans. Participants in Sydney Your Say Open Forum stated that they expect to feel surprised in their city, they want to see unique events in unlikely places, they want to visit museums that are open late or have more food festivals or street markets like they have in Bangkok or in Melbourne (City of Sydney Council 2014). Sydney residents appreciate the presence of events and festivals, but they require consistency; they expect more everyday, creative and outdoor activities spread throughout the year.

Hanoi residents in Vietnam go to Hoan Kiem Lake every Sunday to play and relax. Couples go there every night and dance with different genres of music. Such local food and produce and art markets could become a permanent part of Darling Harbour similar to The Rocks Markets which are open every Friday, Saturday and Sunday. Live music, creative workshops, cooking classes and artist talks could foster a stronger connection between people and place. The representation of Darling Harbour's industrial history and identity via interactive digital wall installations similar to the WALL project created by the Museum of Copenhagen could tell the story of place and reinforce its local character.

Aboriginal history, art and sculpture also require greater attention in Darling Harbour. Pop-up open-air museums like the pop-up museum for the National Trust Heritage Festival created by the Addison Road Community Centre could encourage a collective heritage experience.

Increasing creative expression in the public realm is also linked to the creation of distinctive precincts. Permanent and temporary activities through sculpture, digital history walls, live music, dance, film, writing, murals and street art would bring creative life to the urban space. Arranging legal walls for graffiti and murals that could be repainted by artists and children could be encouraged. Street art and "legal graffiti" offer new paths to the creative space (McAuliffe 2012) such as in the Union Lane Street Art Project in Melbourne or the Alley Project on Grafton Street in Cairns. Designing workshops to teach these types of arts to both residents and tourists could not only create a distinct identity for the Darling Harbour precinct but enable income opportunity to local artists.

Concluding remarks

In this chapter, we discussed the role of placemaking principles in creating successful urban waterfronts. Creative precincts provide opportunities for cultural and local expressions and contribute to the quality of public space, encourage public participation and allow connection between residents and tourists. In this way creative precincts build positive images at micro-, meso- and macro-scales. Redeveloped urban waterfronts can then become the catalysts for inclusive lively communities.

Darling Harbour's redevelopment into a safe and clean public space was more in keeping with a "Disney theme park", which means the creation of a safe and clean public space in which people who do not know each other can trust each other and have a good time (Zukin 1998, p. 832). Eventually, the space has evolved with tourism development, business services, education, property and cultural industries. It has become a place where tourists feel both comfortable and some connection to Sydney people because here, unlike in the streets of the CBD, both tourist and local are in a similar, playful state of mind (Edwards et al. 2008). The perceptions that Darling Harbour catered to a diverse range of people, and moreover that everyone was welcome and had equal licence to "play" were prominent positive components of tourist experiences (Edwards et al. 2008). In this sense, creative placemaking ideas and creative cultural policy implications would allow Darling Harbour to become a more unique public urban space.

References

Ashworth, G. and Page, S. J. 2011. Urban Tourism Research: Recent Progress and Current Paradoxes. *Tourism Management*, 32, 1–15.

Baum, S., Yigitcanlar, T., Horton, S., Velibeyoglu, K. and Gleeson, B. 2007. The Role of Community and Lifestyle in the Making of a Knowledge City. Griffith University, Brisbane.

Bruttomesso, R. 2004. Complexity on the Urban Waterfront. In R. Marshall, ed. *Waterfronts in Post-Industrial Cities*. London: Taylor & Francis, 39–50.

City of Sydney Council 2014, *Creative City: Cultural Policy and Action Plan 2014–2024*. Available from: www.Cityofsydney.Nsw.Gov.Au/__Data/Assets/pdf_File/0011/213986/11418-Finalisation-Of-Cultural-Policy-Document-July-2016.pdf [Accessed 17 October 2019].

Desfor, G. and Laidley, J. 2010. Introduction: Fixity and Flow of Urban Waterfront Change. In G. Desfor, J. Laidley, Q. Stevens and D. Schubert, eds. *Transforming Urban Waterfronts*. New York: Routledge, 1–14.

Destination NSW 2018. Darling Harbour Visitor Profile Year Ended March 2018. Available from: www.destinationnsw.com.au/wp-content/uploads/2018/10/darling-harbour-visitor-profile-ye-march-2018.pdf?x15361 [Accessed 7 October 2019].

Edwards, D., Griffin, T. and Hayllar, B. 2010. Darling Harbour: Looking Back and Moving Forward. In B. Hayllar, T. Griffin and D. Edwards, eds. *City Spaces – Tourist Places*. Oxford: Butterworth-Heinemann, 275–294.

Edwards, D., Griffin, T., Hayllar, B., Dickson, T. and Schweinsberg, S. 2009. *Understanding Tourist "Experiences" and "Behaviour" in Cities: An Australian Case Study*. CRC for Sustainable Tourism Pty Ltd.

Edwards, D., Small, K., Griffin, T. and Hayllar, B. 2008. Sites of Experience: The Functions of Urban Tourism Precincts. *CAUTHE 2008: Tourism and Hospitality Research, Training and Practice*, 803.

Erkök, F. 2009. Waterfronts: Potentials for Improving the Quality of Urban Life. *A | Z ITU Journal of the Faculty of Architecture*, 6, 126–145.

Evans, G. 2009a. Creative Spaces and the Art of Urban Living. In T. Edensor, D. Leslie, S. Millington and N. Rantisi, eds. *Spaces of Vernacular Creativity*. London: Routledge, 19–32.

Evans, G. 2009b. *From Cultural Quarters to Creative Clusters: Creative Spaces in the New City Economy*. Stockholm: Institute of Urban History.

Fainstein, S. S. 2008. Mega-Projects in New York, London and Amsterdam. *International Journal of Urban and Regional Research*, 32, 768–785.

Farrelly, E. 2002. Opening up the Cahill Expressway Won't Be a Dynamic Change. *The Sydney Morning Herald*, 3 December. Available from: www.Smh.Com.Au/Articles/2002/12/02/1038712881320.Html [Accessed 13 October 2015].

Ferrari, M. S. G., Jenkins, P. and Smith, H., 2012. Successful Placemaking on the Waterfront. In H. Smith and M. S. Garcia Ferrari, eds. *Waterfront Regeneration: Experiences in City-Building*. London: Routledge, 153–175.

Florio, S. and Brownill, S. 2000. Whatever Happened To Criticism? Interpreting the London Docklands Development Corporation's Obituary. *City*, 4, 53–64.

Freestone, R. 2000. Planning Sydney: Historical Trajectories and Contemporary Debates. In J. Connell, ed. *Sydney: The Emergence of a World City*. Melbourne: Oxford University Press, 119–143.

Frey, O. 2009. Creativity of Places as a Resource for Cultural Tourism. In G. Maciocco and S. Serreli, eds. *Enhancing the City*. Dordrecht: Springer, 135–154.

Gospodini, A. 2006. Portraying, Classifying and Understanding the Emerging Landscapes in the Post-Industrial City. *Cities*, 23, 311–330.

Griffin, T. and Edwards, D. 2012. Importance–Performance Analysis as a Diagnostic Tool for Urban Destination Managers. *Anatolia*, 23 (1), 32–48.

Griffin, T. and Hayllar, B. 2006. Historic Waterfronts as Tourism Precincts: An Experiential Perspective. *Tourism and Hospitality Research*, 7, 3–16.

Harvey, D. 1987. Flexible Accumulation through Urbanization: Reflections on "Post-Modernism" in the American City. *Antipode*, 19, 260–286.

Harvey, D. 1989. From Managerialism to Entrepreneurialism: The Transformation in Urban Governance in Late Capitalism. *Geografiska Annaler: Series B, Human Geography*, 71, 3–17.

Hayllar, B., Griffin, T. and Edwards, D. 2010. *City Spaces–Tourist Places*. Oxford: Butterworth-Heinemann.

Hoyle, B. 2000. Global and Local Change on the Port-City Waterfront. *Geographical Review*, 90, 395–417.

Inskeep, E. 1991. *Tourism Planning: An Integrated and Sustainable Development Approach*. New York: Wiley.

Jackson, M. R., Kabwasa-Green, F. and Herranz, J. 2006. *Cultural Vitality in Communities: Interpretation and Indicators*. Washington, DC: The Urban Institute.

Jacobs, J. 1961. *The Death and Life of Great American Cities*. New York: Vintage.

Jones, A. L. 2007. On the Water's Edge: Developing Cultural Regeneration Paradigms for Urban Waterfronts. In M. K. Smith, ed. *Tourism, Culture and Regeneration*. Wallingford: CABI, 143–150.

Kaya, E. 2018. Touristification of Industrial Waterfronts: The Rocks and Darling Harbour. *International Journal of Humanities and Social Sciences*, 5 (2).

Kaya, E. 2020. Post-Industrial Waterfront of Sydney: Place from Production to Consumption. In *Transformation of Sydney's Industrial Historic Waterfront*. Singapore: Springer.

Kreag, G. 2001. *The Impacts of Tourism*. Minnesota, MN: Sea Grant.

Kunzmann, K. 2004. Culture, Creativity and Spatial Planning. *Town Planning Review*, 75, 383–404.

Landry, C. 2012. *The Creative City: A Toolkit for Urban Innovators*. Abingdon: Routledge.

McAuliffe, C. 2012. Graffiti or Street Art? Negotiating the Moral Geographies of the Creative City. *Journal of Urban Affairs*, 34, 189–206.

Maddox, G. 2019. Population of a Small Town: The Transformation of Darling Harbour. *The Sydney Morning Herald*, 7 January. Available from: www.smh.com.au/national/nsw/population-of-a-small-town-the-transformation-of-darling-harbour-20181112-p50fgo.html [Accessed 10 October 2019].

Montgomery, J. 1998. Making a City: Urbanity, Vitality and Urban Design. *Journal of Urban Design*, 3, 93–116.

Montgomery, J. 2008. Manners Maketh the City. Reflections on Behaviour and Manners in Urban Places. *Journal of Urban Design*, 13, 159–162.

Moughtin, C. 2003. *Urban Design: Street and Square*, 3rd ed. Oxford: Architectural Press.

Oakley, S. 2014. A Lefebvrian Analysis of Redeveloping Derelict Urban Docklands for High-Density Consumption Living, Australia. *Housing Studies*, 29, 235–250.

Project for Public Spaces 2006. *Eleven Principles for Creating Great Community Places*. Available from: www.pps.org/article/11steps [Accessed 9 October 2019].

Project for Public Spaces 2007. *What Is Placemaking?* Available from: www.pps.org/article/what-is-placemaking [Accessed 9 October 2019].

Project for Public Spaces 2009. *How to Transform a Waterfront?* Available from: www.pps.org/article/turn-waterfrontaround [Accessed 9 October 2019].

Project for Public Spaces 2013. *What Makes a Successful Place?* Available from: www.pps.org/reference/grplacefeat/ [Accessed 9 October 2019].

Richards, G. 2014. Creativity and Tourism in the City. *Current Issues in Tourism*, 17, 119–144.

Sairinen, R. and Kumpulainen, S. 2006. Assessing Social Impacts in Urban Waterfront Regeneration. *Environmental Impact Assessment Review*, 26, 120–135.

Santagata, W. 2002. Cultural Districts, Property Rights and Sustainable Economic Growth. *International Journal of Urban and Regional Research*, 26, 9–23.

Selby, M. 2004. *Understanding Urban Tourism: Image, Culture and Experience*. London: I.B. Tauris.

Shaw, K. and Montana, G. 2016. Place-Making in Megaprojects in Melbourne. *Urban Policy and Research*, 34, 166–189.

Smith, H. and Ferrari, M. S. G. 2012. *Waterfront Regeneration: Experiences in City-Building*. London: Routledge.

Smith, N. 2002. New Globalism, New Urbanism: Gentrification as Global Urban Strategy. *Antipode*, 34, 427–450.

Urry, J. 2002. *Consuming Places*. London: Routledge.

World Travel and Tourism Council (WTTC) 2019. The Economic Impact of Travel and Tourism. Available from: www.wttc.org/-/Media/Files/Reports/Economic-Impact-Research/Regions-2019/World2019.pdf [Accessed 22 October 2019].

Yeang, L. D. 2000. *Urban Design Compendium*. London: English Partnerships/Housing Corporation.

Zukin, S. 1998. Urban Lifestyles: Diversity and Standardisation in Spaces of Consumption. *Urban Studies*, 35, 825–839.

Zukin, S. 2009. *Naked City: The Death and Life of Authentic Urban Places*. Oxford: Oxford University Press.

21

SHAPING URBAN DESTINATIONS

Perspectives on tourist attractions

Sello Samuel Nthebe and Magdalena Petronella (Nellie) Swart

Introduction

Tourist attractions have shaped the development of urban tourist cities since the earliest developments in the field of tourism. The history of tourist attractions is captured in ancient archaeological sites, which are known for their economic contribution to modern-day cities. Tourist attractions stimulate the interest and curiosity of travellers and are often highlighted as "must-see" or "bucket-list" experiences. The aim of this chapter is to outline current perspectives on tourist attractions in terms of how they shape urban destination development, and to do so specifically through the lens of business tourists. We explore how historical and intellectual developments associated with tourist attractions have contributed (and continue to contribute) to the development of urban tourism destinations. An outline is given of the different types of tourist attractions and impacts, to support the main claims and developmental stages. Four factors are highlighted here, namely: (i) the range of tourist attractions, (ii) security at those attractions, (iii) their authenticity and (iv) the location of the tourist's accommodation. These factors do not only impact people's desires to visit tourist attractions, but have also proven to be crucial elements of destination attractiveness (Dimitrov, Stankova, Vasenska & Uzunova, 2017; Lee & Huang, 2014; Owusu-Frimpong, Nwankwo, Blankson & Tarnanidis, 2013) and destination competitiveness, except for the location of the tourist's accommodation, (Bianchi, 2018). In addition, these factors impact leisure and bleisure experiences at a destination (The Economist Intelligence Unit, 2019). This is clarified by aligning those factors with the principal contributions of, and major criticisms associated with, the development of tourist attractions at urban destinations. The chapter concludes with four proposals outlining how future research on tourist attractions can continue to spur sustainable development at urban tourism destinations.

Historical and intellectual developments related to tourist attractions

The history of tourist attractions can be traced from Egypt, where archaeologists found examples of visitors' graffiti dating to 1244 BC, carved in the pyramid of Giza (Weaver & Lawton, 2009). It is nonetheless challenging to accurately trace the history of tourist attractions, mainly due to: (i) difficulties in determining whether a site was a tourist attraction during its early beginnings and (ii) the complexities presented by determining the purpose behind a visit to the site

(Swarbrooke, 1995). Figure 21.1 illustrates the historical development of tourist attractions by highlighting how visits to certain sites were the sole prerogative of specific elites/groups and determining the purpose of their visits.

From Figure 21.1, it is apparent that from the "earliest beginnings" until "the Renaissance", visits to religious shrines and cultural sites were activities deemed exclusive to specific elites or groups. The resourceful Greeks and Romans utilised their resources to visit architectural and artistic sites during the period described as the "earliest beginnings" (Lubbe, 2003). Visits to religious shrines were a mandatory activity solely for pilgrims who went on pilgrimages during the "Medieval period" (Swarbrooke, 1995). "The Renaissance" elite comprised traders who travelled for the purpose of establishing trade routes, but ultimately ended up developing a desire to learn about other cultures and partake of some sightseeing in the process (Cook, Yale & Marqua, 2010). As Figure 21.1 indicates, in the "17th and 18th centuries", the popularity of health-related benefits associated with spas and beaches meant that, for the majority of people, leisure was not their main motivation for visiting such sites. Visiting specific locations for leisure purposes was a luxury enjoyed only by the Greeks, Romans and "Renaissance" elite.

From the nineteenth century onwards, that reality was changed due to the broadening of the range of sites that could be visited for leisure purposes (Swarbrooke, 1995). This was a result of, among others, people of the nineteenth century recognising mountains as being ideally suited to mountaineering, and spas and beaches as ideal for leisure, with events gaining popularity in the twentieth century (Swarbrooke, 1995). The broadening of the range of sites that could be visited for leisure purposes grew significantly after World War II and resulted in recognition of the economic benefits of tourism (see Figure 21.1). Notably, prior to this realisation, this range of sites could have been referred to as "visitor attractions" as they seemed to attract visitors, rather than specifically tourists (Rosendahl, 2009; Swarbrooke, 2002). The evident economic benefits which tourism yielded post-World War II through to the millennium, spurred efforts to formulate a definition for a tourist attraction that would encapsulate a wide variety of sites/ activities which lure visitors (see Lawton, 2005; Rosendahl, 2009; Weidenfeld, Butler & Williams, 2010). This posed a challenge, because "due to the complexity and diversity of the attractions sector, there is no accepted definition which embraces all attractions" (Swarbrooke, 2002: 4). According to Weidenfeld *et al.* (2010: 2), a tourist attraction is "a sole component, geographical area or independent locality which, based on a single primary element, is considered an attraction by tourists or visitors". Alternatively, tourist attractions can be referred to as "features of a destination which influence a tourist's tourism activities at the destination and the motivation of potential tourists" (Middleton & Clarke, 2001: 349). Tourist attractions include a destination's natural or man-made attributes that attract tourists (Lawton, 2005). These definitions are consistent with Wall's (1997) notion that a tourist attraction has three characteristics, namely a place/location, an attractive image and tourists who frequent it.

Tourist attractions form a prominent sector of the tourism industry – even more so in recent years (Amir, Osman, Bachok & Ibrahim, 2015b; Deloitte, 2018; Human Sciences Research Council for South African Tourism and Department of Environmental Affairs and Tourism, 2001 (hereafter HSRC); Middleton & Clark, 2001). Around the early 2000s, tourists' spending on tourist attractions was the fourth highest component of tourism expenditure (Amir *et al.*, 2015b; HSRC, 2001) – and this has since increased to become the third largest component (Deloitte, 2018). It therefore should come as no surprise that tourism research highlights tourist attractions as one of the significant features motivating tourists to visit a destination (Hieda, 2015; Lepp & Gibson, 2011; Lo & Qu, 2014; Tanford, Montgomery & Nelson, 2012). In fact, tourism-related research acknowledges the availability of tourist attractions as a positive factor in contributing to a destination's image (Jalivand, Samiei, Dini & Manzari, 2012; Ramkissoon,

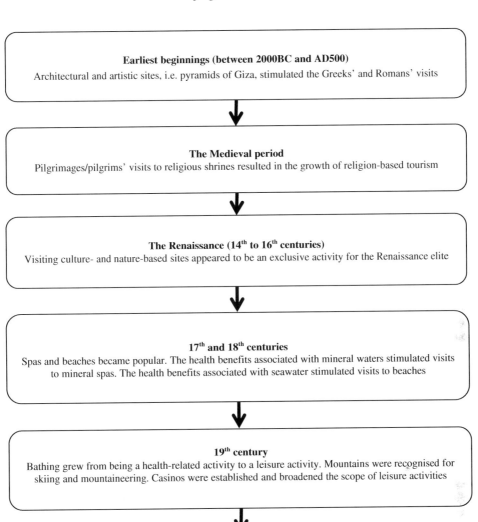

Earliest beginnings (between 2000BC and AD500)
Architectural and artistic sites, i.e. pyramids of Giza, stimulated the Greeks' and Romans' visits

The Medieval period
Pilgrimages/pilgrims' visits to religious shrines resulted in the growth of religion-based tourism

The Renaissance (14th to 16th centuries)
Visiting culture- and nature-based sites appeared to be an exclusive activity for the Renaissance elite

17th and 18th centuries
Spas and beaches became popular. The health benefits associated with mineral waters stimulated visits to mineral spas. The health benefits associated with seawater stimulated visits to beaches

19th century
Bathing grew from being a health-related activity to a leisure activity. Mountains were recognised for skiing and mountaineering. Casinos were established and broadened the scope of leisure activities

20th century
Events became popular during this era

Post-World War II
The significant growth in the variety of tourist attractions resulted in tourism being recognised for its economic role

Figure 21.1 History of tourist attractions.

Uysal & Brown, 2011). In this regard, Beppu City in Japan (Hieda, 2015), Cape Town (City of Cape Town, 2013), Hong Kong (Lo & Qu, 2014), Macau (McCartney, 2008; Wan, 2011) and Mauritius (Ramkissoon *et al.*, 2011) are among the destinations that regard the availability of tourist attractions as a pull factor for tourists.

Major claims and developments

In this section, the major claims and developments relating to tourist attractions are discussed in the context of the types of attractions and the impact these have on the destination.

Types of tourist attractions

Existing definitions (see above) permit almost anything to be deemed a tourist attraction. These definitions are consistent with Lew's (1987) view that tourist attractions include both sites/facilities and services. The classification of tourist attractions received particular attention from the 1980s, when destinations aimed to pinpoint and organise their portfolios of tourist attractions (Kušen, 2010). This was due to the responsible entities at those destinations realising the importance of tourist attractions to tourism (Lew, 1987). Even now, this much is clear: an absence of tourist attractions at a specific location results in an absence of tourists and thus there is no tourism to speak of (Ngwira & Kankhuni, 2018). Lew (1987) was among the first scholars who made an effort to classify tourist attractions, arguing that three categories exist, from an ideographic, cognitive and organisational perspective. According to Lew (1987), the ideographic perspective focuses on the unique elements of a tourist attraction; the cognitive pertains to tourists' experiences at such an attraction, and the organisational concerns the location, capacity and temporary/permanent nature of said attraction. Lew's (1987) classification was, however, a response to the need to establish tourist attraction categories to suit particular research orientations: the cognitive perspective serves studies which are oriented towards desired experiences, the idiographic aids studies oriented towards distinguishing between nature, nature–human and human-based tourist attractions, and the organisational serves studies which are oriented towards the attraction's location, capacity and temporary/permanent nature.

Following Lew (1987), Swarbrooke (1995) introduced the following four types of tourist attractions:

- Natural, i.e. wildlife;
- Man-made, but not originally designed primarily to attract visitors, e.g. churches;
- Man-made and purpose-built to attract tourists, e.g. museums; and
- Special events, e.g. festivals.

In addition to these four types, tourism-related facilities such as renowned restaurants and accommodation establishments can be regarded as tourist attractions (Swarbrooke, 1995). Swarbrooke's (1995) typology can be linked to Lew's (1987) idiographic perspective/category, which distinguishes tourist attractions on the basis of whether they are natural, combine natural and human elements or mainly comprise a human element. Special events can be linked to Lew's (1987) organisational perspective/category, which distinguishes tourist attractions on the basis of, among others, whether they are temporary or permanent. Unlike Lew's (1987) classification, which aligns tourist attractions with diverse research orientations, Swarbrooke's (1995) typology is oriented towards the management of those attractions.

Taking into account Lew's (1987) classification and Swarbrooke's (1995) typology, Wall (1997) introduced a rather unique classification that is limited to tourist attractions whose economic potential and vulnerability due to excessive use are of concern. Wall (1997) suggests that these attractions be categorised – based on their physical attributes – into the following three types:

- Point, which refers to tourist attractions that require a small location/area, i.e. waterfalls, to be visited by a significant number of individuals;
- Linear, which refers to tourist attractions such as trails, rivers and coastlines;
- Areas, which refer to places such as parks which are attractive to many and have the capacity to attract even more people.

Of the three approaches discussed here, Lew's (1987) idiographic and organisational perspective/ categories and Swarbrooke's (1995) typology in particular are endorsed by a number of scholars (Deng, King & Bauer, 2002; Kušen, 2010; Pearce & Benkendorff, 2006). Based on physical characteristics, Deng *et al.* (2002) highlight waterways, scenery and wildlife as categories resorting under natural tourist attractions. Similarly, Kušen's (2010) list of 16 categories used for managing the functionality of tourist attractions, includes flora, spas, trails and events. Pearce and Benkendorff's (2006) classification resulted in 18 categories of various types of tourist attractions, which include casinos, farms, theme parks and museums.

Impact of tourist attractions on the destination

Tourists' motivation for visiting a specific destination is likely to include reasons such as the desire to visit its tourist attractions (Bar-Kołelis & Wiskulski, 2012; Reisinger, Mavondo & Crotts, 2009; Tanford *et al.*, 2012; Yousefi & Marzuki, 2012). Notably, tourists from different cultural backgrounds prefer different tourist attractions (Reisinger *et al.*, 2009). This is consistent with Bar-Kołelis and Wiskulski's (2012) observation that tourists' needs and interests play a significant role. This implies that destinations that seek to attract a significant number of tourists need to respond to both their needs and their interests (Bar-Kołelis & Wiskulski, 2012), and cater for people from diverse cultural backgrounds (Reisinger *et al.*, 2009). Thus, a destination with a variety of tourist attractions is likely to be perceived as attractive (Dimitrov *et al.*, 2017; Lee & Huang, 2014) and competitive (Bianchi, 2018; Edward & George, 2008). The significance of leading leisure, business and bleisure destinations, e.g. Tokyo and Sydney, is attributed to having a variety of tourist attractions (The Economist Intelligence Unit, 2019).

For decades, tourist attractions have predominantly been associated with leisure tourists (Crompton & McKay, 1997; Jun, Vogt & MacKay, 2007; Kim & Brown, 2012; Swarbrooke, 1995; Yousefi & Marzuki, 2012). From the early 1990s, however, tourist attractions became associated with, among others, business tourists (Davidson, 2003; McCartney, 2008; Nelson & Rys, 2000; Rittichainuwat & Mair, 2012; Robinson & Callan, 2002; Witt, Gammon & White, 1992). This prompted Davidson's (2003) call for leisure activities to be included in business tourism – a call received with enthusiasm by a number of tourism scholars (i.e. McCartney, 2008; Rittichainuwat & Mair, 2012; Tanford *et al.*, 2012; Wan, 2011). In 2008, the possibility of Macau becoming a business tourism destination was explored (McCartney, 2008), but a lack of sufficient tourist attractions was found to be among the factors impeding it from becoming a business tourism destination (McCartney, 2008; Wan, 2011). In the United Kingdom (UK), Robinson and Callan (2005) found that the availability of tourist attractions enhances the attractiveness of the location of an events venue. Rittichainuwat and Mair (2012) and Tanford *et al.*

(2012) concluded that the availability of tourist attractions motivates business tourists to visit a destination. Despite these findings, tourists' cultural backgrounds have an impact on their preferences when it comes to what passes for a tourist attraction (Bar-Kołelis & Wiskulski, 2012; Reisinger *et al.*, 2009): tourists from Northern Europe prefer natural attractions, those from the United States of America (USA) prefer heritage attractions and those from Asia prefer entertainment (Reisinger *et al.*, 2009). Lastly, shopping seems to be the preferred leisure activity of Ukrainians and Russians (Bar-Kołelis & Wiskulski, 2012).

Principal contributions

Leisure tourists' significant spending on tourist attractions is well recognised in this industry (Amir *et al.*, 2015b; HSRC, 2001). Although business tourists have displayed a willingness to spend money on tourist attractions (Amir *et al.*, 2015b; Jones & Li, 2015), certain factors determine their interest in visiting such sites, and consequently spending their money there. Such factors include (i) the range of tourist attractions (Elston & Draper, 2012; Shin, 2009; Whitfield & Webber, 2011), (ii) security at those sites (Rittichainuwat & Chakraborty, 2012), (iii) their authenticity (Fawzy, 2010; Shin, 2009; Yankholmes & McKercher, 2015) and (iv) the location of their hotel/accommodation (Fawzy, 2010; Visser, 2007; Zhou, Ye, Pearce & Wu, 2014). It is worthwhile investigating what principally contributes to business tourists' interest in tourist attractions, in the context of the impact of the aforementioned four factors.

Range of tourist attractions

The availability of a range of tourist attractions is crucial for destinations wishing to attract business tourists (Crouch & Louviere, 2004; Elston & Draper, 2012). Destinations that lack a range of tourist attractions should consider investing in such attractions, in order to draw business tourists (Wan, 2011). Table 21.1 provides examples of various types of attractions that are of interest to business tourists, with the attractions classified by type on the basis of their physical attributes (first column, Table 21.1). The second column provides examples of categories and the third column references studies that revealed business tourists' interest in the respective tourist attractions (see Table 21.1).

Nthebe (2017) adopted the outline in Table 21.1 to investigate domestic business tourists' interest in a range of tourist attractions in Pretoria, the capital city of South Africa. That study

Table 21.1 Attractions of interest to business tourists

Tourist attraction types	Categories	Sources
Museums	Historical	Chiang *et al.*, 2016; Fawzy, 2010; Shin, 2009; Nelson & Rys, 2000; Ramkissoon *et al.*, 2011
Sports	Golf courses	Elston & Draper, 2012; Whitfield, 2009
Nature	Wildlife	Elston & Draper, 2012; Terzi *et al.*, 2013
Entertainment	Nightlife	Nelson & Rys, 2000
	Events	Lin *et al.*, 2010
	Sports events	Donaldson & Ferreira, 2009
Shopping	Shopping malls	Luo & Lu, 2011; Xue & Cox, 2008
Famous restaurants	Franchises	Visser, 2007

Source: Authors' own compilation.

revealed that the majority of domestic business tourists are more likely to visit shopping malls and view wildlife, and less likely to visit sports-related attractions such as golf courses.

Security at tourist attractions

Just as security is vital for the success of a destination (Amir, Ismail & See, 2015a; George, 2003), so it is for the success of tourist attractions (Boakye, 2010; George & Mawby, 2013). Leisure and business tourists alike are unlikely to visit tourist attractions when they perceive the destination to be unsafe (Amir *et al.*, 2015a). Following George's (2003) findings on the impact which a perceived low level of security has on leisure and business tourists' freedom to explore Cape Town, the City of Cape Town (2013) announced that it had taken measures to address safety and security. As far as could be determined, no studies have been conducted to determine whether or not the measures taken in fact improved tourists' perceptions in this regard. Additional impacts of security on the destination are highlighted in destination attractiveness research (Lee & Huang, 2014) and destination competitiveness research (Bianchi, 2018). Tourists' perceptions regarding a destination as a safe destination boosts destination's competitiveness, e.g. Taiwan's (Lee & Huang, 2014), and destination's attractiveness, e.g. Chile's (Bianchi, 2018). Furthermore, Singapore has proven to be very attractive to leisure, business and bleisure tourists due to its perceived high level of security (The Economist Intelligence Unit, 2019).

Concerning tourist attractions, the crowds of tourists who are attracted to specific attractions are largely exposed to crimes such as cell phone theft (Boakye, 2010). The security concerns which (over)crowding at tourist attractions raises may be addressed by employing more staff (Jin & Pearce, 2011). Rittichainuwat and Chakraborty (2012) add that the visibility of security personnel at tourist attractions is important for business tourists. Their stance is supported by Nthebe (2017), who found that the presence of security personnel at tourist attractions was important for domestic business tourists in Pretoria.

Authenticity

The importance of managing and protecting a destination's heritage tourist attractions has long been acknowledged by tourism planners, as a measure of safeguarding the destination's competitiveness (Jamieson & Jamieson, 2014). This is because heritage tourist attractions are expected to exhibit the destination's real history and culture (Chhabra, 2012). While a destination's history is mainly exhibited by museums (Cohen & Cohen, 2012; Lacher, Oh, Jodice & Norman, 2013), its culture is embodied in its communities (Ashworth & Page, 2011; Brown, 2013). In summary, heritage is constituted by both the destination's history and culture (Meskell & Scheermeyer, 2008).

To achieve authenticity, heritage tourist attractions should depict the destination's real heritage (Bryce, Curran, O'Gorman & Taheri, 2015; Chhabra, 2010; Cohen, 1988; Meskell & Scheermeyer, 2008; Taylor, 2001). This will satisfy tourists whose motivation to travel is a chance to experience something of the heritage and history of the destination (Hieda, 2015). This desire motivates leisure tourists to travel (Ramkissoon & Uysal, 2011; Yousefi & Marzuki, 2012), but it is a secondary motivation for business tourists (Fawzy, 2010; Shin, 2009; Yankholmes & McKercher, 2015). Fawzy (2010) found that the availability of cultural tourist attractions in the proximity of their hotel was important to business tourists, whereas Shin (2009) highlights enjoying heritage as one of business tourists' motives for visiting a destination. Yankholmes and McKercher (2015) confirm that business tourists seeking to experience a destination's heritage will be attracted to heritage sites. Thus, the availability of heritage sites within a destination enhances the destination's attractiveness, e.g. Ghana's (Owusu-Frimpong *et al.*, 2013) and destination's competitiveness, e.g. Chile's (Bianchi, 2018). The lack of heritage sites can be attributed to Adelaide and Brisbane's inability to surpass leading destinations, e.g. Sydney and Melbourne, in attracting significant numbers of leisure, business and bleisure tourists (The Economist Intelligence Unit, 2019).

Hotel locations

The impact of a hotel's location on tourists' movement within a destination has attracted research attention over the years (Lew & McKercher, 2006; Shoval, McKercher, Ng & Birenboim, 2011; Visser, 2007). This is attributed to tourists' consideration of the distance between the location of their hotel and a tourist attraction, when deciding whether or not to visit that site (Lew & McKercher, 2006). For example, Visser (2007) highlights that tourist attractions that are not located within the hotel's proximity are unlikely to be visited by leisure tourists. Shoval *et al.* (2011) concur, noting that unless tourist attractions located outside the hotel location's proximity are prominent features of a destination, they are unlikely to merit a visit. Nthebe (2017) confirms this finding in respect of domestic business tourists in Pretoria. Thus, it can be argued that the attractiveness of the location where a tourist intends to stay is enhanced by the availability of nearby attractions (Fawzy, 2010; Yang, Tang, Luo & Law, 2015; Xue & Cox, 2008). However, the option for bleisure (Expedia Group, 2018), leisure and business tourists to utilise public transport facilities (George, 2003) makes the accessibility of tourist attractions beyond the proximity of the hotel more convenient (Lew & McKercher, 2006). After all, accessibility is not only determined by distance, but also by the availability of public transport (Lockwood & Medlik, 2001; Witt *et al.* 1992). Given the benefits of such facilities, Issahaku and Amuquandoh (2013), Xue and Cox (2008) and Yang *et al.* (2015) found that the availability of public transport facilities near the hotel's location is a crucial requirement for business tourists. Consequently, hotels that are located within the proximity of tourist attractions and public transport facilities further enhance the destination's attractiveness (Owusu-Frimpong *et al.*, 2013).

Main criticism

As discussed, business tourists are willing to spend money and time on tourist attractions. Their willingness to increase this spending is, however, affected by: (i) a range of tourist attractions, (ii) security at those sites, (iii) the authenticity of an attraction and (iv) their hotel's location. Although tourism-related literature highlights how these factors impact business tourists' interest in visiting tourist attractions, there are research gaps that still need to be explored. The main criticism relating to business tourists' interest in tourist attractions is discussed in the context of these four factors.

Range of tourist attractions

Although research by Fawzy (2010), Nelson and Rys (2000) and Nthebe (2017) shows that business tourists are drawn to different attractions, these results need to be interpreted with caution. In a study exploring the factors influencing the UK's attractiveness as a business tourism destination, Nelson and Rys (2000) found that the availability of golf courses, historical attractions and nightlife were important. Nelson and Rys (2000) did not explore a broader range of tourist attractions, or the impact of other attractions such as wildlife sanctuaries and shopping malls. Fawzy (2010), who investigated the importance of a hotel's location for business tourists in Cairo, highlights the availability of nearby historical attractions as an important attribute, but did not look into a broader range of attractions (e.g. shopping malls and golf courses). Nthebe (2017) explored the likelihood of domestic business tourists visiting attractions in Pretoria and found that the majority were very likely to visit shopping malls and wildlife sanctuaries or parks. The unavailability of a casino prior to 2017, in Pretoria and not the City of Tshwane, made it less likely that domestic business tourists in Pretoria would visit such venues, therefore this was excluded from this chapter (Nthebe, 2017).

Based on the above, it appears that there is a lack of research exploring business tourists' interest in a broader range of tourist attractions. This can be attributed to, among others, the destination's portfolio of such attractions (Nthebe & Swart, 2017).

Security at tourist attractions

Given business tourists' likelihood of visiting attractions, ensuring a high level of security at those locations is important (Boakye, 2012). A study by George (2003) highlights the impact which security has on business tourists. In Cape Town, George (2003) found that business tourists are unlikely to visit tourist attractions, irrespective of whether it is during the day or at night, when they perceive a destination to be unsafe. Thus, ensuring a high level of security at tourist attractions can contribute towards creating a safe destination for tourists (George & Swart, 2012). Security measures such as ensuring the presence of security personnel may result in business tourists perceiving a particular tourist attraction as safe (Rittichainuwat & Chakraborty, 2012). Nthebe (2017) found that the same applies to domestic business tourists in Pretoria. In the context of leisure tourism, the presence of (i) public policing personnel in the streets and (ii) security personnel at major tourist attractions resulted in tourists feeling safe in Cape Town during the 2010 FIFA World Cup (George & Swart, 2012). However, creating a perception of a safe tourist attraction should be pursued with caution, as a high number of security personnel can either (i) make tourists feel safe or (ii) raise concerns about the tourist attraction's level of safety (Amir *et al.*, 2015a).

Nthebe (2017) and Rittichainuwat and Chakraborty (2012) investigated the importance business tourists attach to security personnel being present at tourist attractions, but not whether their presence raised concerns about safety levels – an issue which George and Swart (2012) also neglected to investigate. As far as could be determined, there is a lack of research into business tourists' concerns about a high number of security personnel at a site.

Authenticity

Business tourists have shown an interest in visiting heritage attractions (Fawzy, 2010; Nelson & Rys, 2000; Shin, 2009; Yankholmes & McKercher, 2015). Nonetheless, the representation of a destination's genuine heritage has been the subject of numerous debates around authenticity (Bryce *et al.*, 2015; Chhabra, 2010; Cohen, 1988; Cohen & Cohen, 2012; Meskell & Scheermeyer, 2008). Despite these debates, there is consensus that a crucial attribute of authenticity is the representation of genuine heritage (Bjerregaard, 2015; Steiner & Reisinger, 2006; Taylor, 2001).

A destination's heritage is exhibited by historical tourist attractions such as museums, as well as tours of cultural attractions such as townships (also known as informal settlements). Fawzy (2010) highlighted the availability of nearby historical tourist attractions as an important determinant when business tourists chose the location of a hotel in Cairo. Nelson and Rys (2000) found that the availability of historical tourist attractions had an influence on the UK's attractiveness as a business tourism destination, but their study did not explore the influence of cultural attractions. In Gwangju, Korea, Shin (2009) found that business tourists are motivated by the opportunity to explore the destination's history and culture. Nthebe (2017) concurs, having found that domestic business tourists were interested in experiencing the history and culture of Pretoria. The studies by Fawzy (2010), Nelson and Rys (2000) and Nthebe (2017) did not investigate business tourists' perceptions of the degree of authenticity required of a heritage attraction's representation of the destination's history or culture. Arguably, that can only be investigated through in-depth research at a heritage tourist attraction (Bjerregaard, 2015; Chhabra, 2010).

Hotel locations

The location of a hotel/tourist accommodation will determine a business tourist's decision whether or not to visit certain attractions (Lew & McKercher, 2006). This is attributed to the proximity of the hotel to tourist attraction(s) (Fawzy, 2010; Xue & Cox, 2008; Zhou *et al.*, 2014) and public transport facilities (Issahaku & Amuquandoh, 2013; Xue & Cox, 2008; Yang *et al.*, 2015). It has been established that business tourists prefer hotels that are situated close to attractions (Fawzy, 2010; Zhou *et al.*, 2014), but as far as could be determined, no studies have sought to determine whether business tourists in fact visit nearby tourist attractions. Shoval *et al.* (2011) confirm that most leisure tourists in Hong Kong visited nearby tourist attractions, rather than sites situated further afield.

Furthermore, while business tourists' likelihood of using public transport facilities to visit tourist attractions is acknowledged (George, 2003; Lew & McKercher, 2006), the literature is largely silent on whether business tourists actually make use of public transport facilities to visit such attractions. Gutiérrez and Miravet (2016) found that both business and leisure tourists utilised public transport facilities while visiting Costa Daurada, but that study did not indicate whether business tourists used public transport to visit tourist attractions in particular.

Importance of area/perspective and anticipated future development

Business tourists' interest in visiting tourist attractions is crucial for destinations that seek to attract visitors to their shores and retain them (City of Cape Town, 2013; McCartney, 2008; Wan, 2011). Destinations such as Cape Town (City of Cape Town, 2013) and Macau (McCartney, 2008; Wan, 2011) have identified the lack of sufficient tourist attractions as an impediment to attracting business tourists. The present chapter has discussed which factors have an impact on business tourists' interest in visiting tourist attractions, and while these factors have attracted the attention of researchers, further research is required to lay a foundation for developing strategies that will result in destinations becoming more attractive to this group of tourists. Below, four future developments are anticipated and outlined.

First, with destinations' continuing investment in developing a broader portfolio of tourist attractions (City of Cape Town, 2013; Wan, 2011), it is anticipated that the lack of research exploring business tourists' interests in a broader range of tourist attractions will be addressed. Second, future studies are likely to address the paucity of research exploring whether the presence of a high number of security personnel raises business tourists' concerns about a site's security. Third, future studies are expected to address the dearth of research into business tourists' perceptions of the degree of authenticity of a heritage tourist attraction's representation of history or culture. Finally, it is anticipated that future studies will investigate whether business tourists prefer to visit nearby tourist attractions and/or use public transport facilities to visit such sites.

Conclusion

The aim of this chapter was to outline current perspectives on tourist attractions, in shaping urban destination development. This was mainly done from the perspective of business tourists. Evidence was provided on how historical and intellectual developments over centuries supported the development of urban tourism destinations. The different types of tourist attractions were confirmed as major stimuli in the development of tourism destinations in urban areas. Principal contributions to, and major criticism associated with, the development of tourist attractions at urban destinations were supported by four factors: investment in tourist attractions at urban destinations, the presence or absence of security personnel as custodians of safety at the destination, the degree of authenticity of a heritage tourist attraction's representation of history or culture, and the preference of business tourists to visit attractions by making use of public transport – all of which provide an array of future research opportunities.

References

Amir, F., Ismail, N. and See, P., 2015a. Sustainable tourist environment: Perception of international women travelers on safety and security in Kuala Lumpur. *Procedia – Social and Behavioral Sciences*, 168, pp. 123–133.

Amir, S., Osman, M., Bachok, S. and Ibrahim, M., 2015b. Understanding domestic and international tourists' expenditure patterns in Melaka, Malaysia: Result of CHAID analysis. *Procedia – Social and Behavioral Sciences*, 172, pp. 390–397.

Ashworth, G. and Page, S.J., 2011. Urban tourism research: Recent progress and current paradoxes. *Tourism Management*, 32(1), pp. 1–15.

Bar-Kołelis, D. and Wiskulski, T., 2012. Cross-border shopping at Polish borders: Tri City and the Russian tourists. *GeoJournal of Tourism and Geosites*, 1(9), pp. 43–51.

Bianchi, C., 2018. Exploring the attractiveness of Chile as a vacation for international tourists. *Tourism Analysis*, 23, pp. 351–364.

Bjerregaard, P., 2015. Dissolving objects: Museums, atmosphere and the creation of presence. *Emotion, Space and Society*, 15, pp. 74–81.

Boakye, K.A., 2010. Studying tourists' suitability as crime targets. *Annals of Tourism Research*, 37(3), pp. 727–743.

Boakye, K.A., 2012. Tourists' views on safety and vulnerability. A study of some selected towns in Ghana. *Tourism Management*, 33(2), pp. 327–333.

Brown, L., 2013. Tourism: A catalyst for existential authenticity. *Annals of Tourism Research*, 40, pp. 176–190.

Bryce, D., Curran, R., O'Gorman, K. and Taheri, B., 2015. Visitors' engagement and authenticity: Japanese heritage consumption. *Tourism Management*, 46, pp. 571–581.

Chhabra, D., 2010. Back to the past: A sub-segment of Generation Y's perceptions of authenticity. *Journal of Sustainable Tourism*, 18(6), pp. 793–809.

Chhabra, D., 2012. Authenticity of the objectively authentic. *Annals of Tourism Research*, 39(1), pp. 499–502.

Chiang, H.H., Tsaih, R.-H. and Han, T.S., 2016. Measurement development of service quality for museum websites displaying artefacts. In R.H. Tsaih, T.S. Han (Eds.), *Managing Innovation and Cultural Management in the Digital Era: The Case of the National Palace Museum*. Abingdon: Routledge.

City of Cape Town, 2013. *Draft Tourism Development Framework: 2013 to 2017*. [Online]. Available: www.houtbayheritage.org.za/Draft%20HYS_Tourism_Development_Framework_Background_Eng.pdf [Accessed 25 July 2014].

Cohen, E., 1988. Authenticity and commoditization in tourism. *Annals of Tourism Research*, 15(3), pp. 371–386.

Cohen, E. and Cohen, S.A., 2012. Authentication: Hot and cool. *Annals of Tourism Research*, 39(3), pp. 1295–1314.

Cook, R., Yale, R. and Marqua, J., 2010. *Tourism: The Business of Travel*. 4th ed. Upper Saddle River, NJ: Pearson.

Crompton, J.L. and McKay, S.L., 1997. Motives of visitors attending festival events. *Annals of Tourism Research*, 24(2), pp. 425–439.

Crouch, G.I. and Louviere, J.J., 2004. The determinants of convention site selection: A logistic choice model from experimental data. *Journal of Travel Research*, 43(2), pp. 118–130.

Davidson, R., 2003. Adding pleasure to business: Conventions and tourism. *Journal of Convention & Exhibition Management*, 5(1), pp. 29–39.

Deloitte, 2018. *Travel and Hospitality Outlook*. [Online]. Available: www2.deloitte.com/content/dam/Deloitte/us/Documents/consumer-business/us-cb-2018-travel-hospitality-industry-outlook.pdf [Accessed 23 February 2019].

Deng, J., King, B. and Bauer, T., 2002. Evaluating natural attractions for tourism. *Annals of Tourism Research*, 14, pp. 553–575.

Dimitrov, P., Stankova, Z., Vasenska, I. and Uzunova, D., 2017. Increasing attractiveness and image recognition of Bulgaria as a tourism destination. *Tourism & Management Studies*, 13(3), pp. 39–47.

Donaldson, R. and Ferreira, S., 2009. (Re-)creating urban destination image: Opinions of foreign visitors to South Africa on safety and security? *Urban Forum*, 20(1), pp. 1–18.

Edward, M. and George, B., 2008. Tourism development in the state of Kerala, India: A study of destination attractiveness. *European Journal of Tourism Research*, 1(1), pp. 16–38.

Elston, K. and Draper, J., 2012. A review of meeting planner site selection criteria research. *Journal of Convention & Event Tourism*, 13(3), pp. 203–220.

Expedia Group, 2018. *Unpacking Bleisure Travel Trends*. [Online]. Available: https://info.advertising.expedia.com/hubfs/Content_Docs/Rebrand2018/Bleisure%20Select%20Meeting_June%202018.pdf?hsCtaTracking=b78cb326-ba0a-4192-a0e7-1f3cfcf995cd%7Ccf9f079a-da2b-4cae-ac5f-e58277edd18a [Accessed 9 October 2019].

Fawzy, A., 2010. Business travelers' accommodation selection: A comparative study of two international hotels in Cairo. *International Journal of Hospitality & Tourism Administration*, 11(2), pp. 138–156.

George, R., 2003. Tourist's perceptions of safety and security while visiting Cape Town. *Tourism Management*, 24, pp. 575–585.

George, R. and Mawby, R., 2013. Security at the 2012 London Olympics: Spectators' perceptions of London as a safe city. *Security Journal*, 37, pp. 1–11.

George, R. and Swart, K., 2012. International tourists' perceptions of crime risk and their future travel intentions during the 2010 FIFA World Cup™ in South Africa. *Journal of Sport & Tourism*, 17(3), pp. 201–223.

Gutiérrez, A. and Miravet, V., 2016. The determinants of tourist use of public transport at the destination. *Sustainability*, 8, pp. 1–16.

Hieda, M., 2015. Motivations for travel: A study of tourists in Beppu, Japan. *Journal of Hospitality & Tourism*, 13(2), pp. 10–22.

Human Sciences Research Council for South African Tourism and Department of Environmental Affairs and Tourism, 2001. *South African Domestic Tourism Survey: Marketing the Provinces*. Available: www.researchgate.net › profile › Oumar_Bouare › publication › links. [Accessed 15 July 2015].

Issahaku, A. and Amuquandoh, F., 2013. Dimensions of hotel location in the Kumasi Metropolis, Ghana. *Tourism Management Perspectives*, 8, pp. 1–8.

Jalivand, R.M., Samiei, N., Dini, B. and Manzari, P.Y., 2012. Examining the structural relationships of electronic word of mouth, destination image, tourist attitude toward destination and travel intention: An integrated approach. *Journal of Destination Marketing & Management*, 1, pp. 134–143.

Jamieson, W. and Jamieson, M., 2014. The complementary role of urban design and tourism planning in tourism destination management and development. *Journal of Hospitality & Tourism*, 12(1), pp. 70–87.

Jin, Q. and Pearce, P., 2011. Tourist perception of crowding and management approaches at tourism sites in Xi'an. *Asia Pacific Journal of Tourism Research*, 16(3), pp. 325–338.

Jones, C. and Li, S., 2015. The economic importance of meetings and conferences: A satellite account approach. *Annals of Tourism Research*, 52, pp. 117–133.

Jun, S., Vogt, C. and MacKay, K., 2007. Relationships between travel information search and travel product purchase in pretrip contexts. *Journal of Travel Research*, 45, pp. 266–274.

Kim, A.K. and Brown, G., 2012. Understanding the relationships between perceived travel experiences, overall satisfaction, and destination loyalty. *Anatolia: An International Journal of Tourism & Hospitality Research*, 23(3), pp. 328–347.

Kušen, E., 2010. A system of tourism attractions. *Tourism Review*, 58(4), pp. 409–424.

Lacher, G.R., Oh, C., Jodice, L.W. and Norman, W.C., 2013. The role of heritage and cultural elements in coastal tourism destination preferences: A choice modelling-based analysis. *Journal of Travel Research*, 52(4), pp. 534–546.

Lawton, L.J., 2005. Resident perceptions of tourist attractions on the Gold Coast of Australia. *Journal of Travel Research*, 44(2), pp. 188–200.

Lee, C. and Huang, H., 2014. The attractiveness of Taiwan as a bicycle tourism destination: A supply-side approach. *Asia Pacific Journal of Tourism Research*, 19(3), pp. 273–299.

Lepp, A. and Gibson, H., 2011. Reimaging a nation: South Africa and the 2010 FIFA World Cup. *Journal of Sport & Tourism*, 16(3), pp. 211–230.

Lew, A., 1987. A framework of tourist attraction research. *Annals of Tourism Research*, 29(2), pp. 422–438.

Lew, A. and McKercher, B., 2006. Modeling tourist movements: A local destination analysis. *Annals of Tourism Research*, 33(2), pp. 403–423.

Lin, W., Sun, M., Poovendran, R. and and Zhang, Z., 2010. Group event detection with a varying number of group members for video surveillance. *IEEE Transactions on Circuits and Systems for Video Technology*, 20(8), pp. 1057–1067.

Lo, A. and Qu, H., 2014. A theoretical model of the impact of a bundle of determinants on tourists' visiting and shopping intentions: A case of mainland Chinese tourists. *Journal of Retailing and Consumer Services*, 4, pp. 969–989.

Lockwood, A. and Medlik, S., 2001. *Tourism and Hospitality in the 21st Century*. Oxford: Butterworth Heinemann.

Lubbe, B., 2003. *Tourism Management in Southern Africa*. Department of Tourism Management, University of Pretoria. Pretoria: Pearson Education South Africa.

Luo, A. and Lu, X., 2011. A study of inbound business tourists' shopping behavior and influencing factors: A case study of the Canton Fair in Guangzhou. *Journal of China Tourism Research*, 7, pp. 137–167.

McCartney, G., 2008. The CAT (Casino Tourism) and the MICE (Meetings, Incentives, Conventions, Exhibitions): Key development considerations for the convention and exhibition industry in Macau. *Journal of Convention & Event Tourism*, 9(4), pp. 293–308.

Meskell, L. and Scheermeyer, C., 2008. Heritage as therapy: Set pieces from the new South Africa. *Journal of Material Culture*, 13(2), pp. 153–173.

Middleton, V. and Clarke, J., 2001. *Marketing in Travel and Tourism*. 3rd ed. Oxford: Butterworth Heinemann.

Nelson, R. and Rys, S., 2000. Convention site selection criteria relevant to secondary convention destinations. *Journal of Convention & Exhibition Management*, 2(2/3), pp. 71–83.

Ngwira, C. and Kankhuni, Z., 2018. What attracts tourists to a destination? Is it attractions? *African Journal of Hospitality, Tourism and Leisure*, 7(1), pp. 1–19.

Nthebe, S., 2017. Hotel front office staff and interest in tourist attractions: Their influencing role in business tourists' visiting intentions. Unpublished Master's dissertation. Pretoria: University of South Africa.

Nthebe, S.S. and Swart, M.P., 2017. The mediating role of tourist attractions in the relationship between hotel employees and business tourists' intentions to visit tourist attractions in Pretoria, South Africa. *African Journal for Physical Activity and Health Sciences*, June (Supplement), pp. 133–145.

Owusu-Frimpong, N., Nwankwo, S., Blankson, C. and Tarnanidis, T., 2013. The effect of service quality and satisfaction on destination attractiveness of sub-Saharan African countries: The case of Ghana. *Current Issues in Tourism*, 16(7–8), pp. 627–646.

Pearce, P. and Benkendorff, P., 2006. Benchmarking, usable knowledge and tourist attractions. *Journal of Quality Assurance in Hospitality and Tourism*, 7(1/2), pp. 29–52.

Ramkissoon, H. and Uysal. M. S., 2011. The effects of perceived authenticity, information search behaviour, motivation and destination imagery on cultural behavioural intentions of tourists. *Current Issues in Tourism*, 14(6), pp. 537–562.

Ramkissoon, H., Uysal, M. and Brown, K., 2011. Relationship between destination image and behavioral intentions of tourists to consume cultural attractions. *Journal of Hospitality Marketing & Management*, 20(5), pp. 575–595.

Reisinger, Y., Mavondo, F.T. and Crotts, J.C., 2009. The importance of destination attributes: Western and Asian visitors. *Anatolia: An International Journal of Tourism & Hospitality Research*, 20(1), pp. 236–253.

Rittichainuwat, B. and Chakraborty, G., 2012. Perceptions of importance and what safety is enough. *Journal of Business Research*, 65, pp. 42–50.

Rittichainuwat, B. and Mair, J., 2012. Visitor attendance motivations at consumer travel exhibitions. *Tourism Management*, 33(5), pp. 1236–1244.

Robinson, L.S. and Callan, R.J., 2002. Professional UK conference organizers' perceptions of important selection and quality attributes of the meetings product. *Journal of Convention & Exhibition Management*, 4(1), pp. 1–17.

Robinson, L.S. and Callan, R.J., 2005. UK conference delegates' cognizance of the importance of venue selection attributes. *Journal of Convention & Event Tourism*, 7(1), pp. 77–91.

Rosendahl, T., 2009. Tourist attractions and their marketing communication: Off the peg or tailor-made? *Journal of Promotion Management*, 15(1), pp. 269–285.

Shin, Y., 2009. Examining the link between visitors' motivations and convention destination image. *Tourismos: An International Multidisciplinary Journal of Tourism*, 4(2), pp. 29–45.

Shoval, N., McKercher, B., Ng, E. and Birenboim, A., 2011. Hotel location and tourist activity in cities. *Annals of Tourism Research*, 38(4), pp. 1594–1612.

Steiner, C.J. and Reisinger, Y., 2006. Understanding existential authenticity. *Annals of Tourism Research*, 33(2), pp. 299–318.

Swarbrooke, J., 1995. *The Development and Management of Visitor Attractions*. London: Butterworth Heinemann.

Swarbrooke, J., 2002. *The Development and Management of Visitor Attractions*. 2nd ed. London: Butterworth Heinemann.

Tanford, S., Montgomery, R. and Nelson, K.B., 2012. Factors that influence attendance, satisfaction, and loyalty for conventions. *Journal of Convention & Event Tourism*, 13(4), pp. 290–318.

Taylor, J.P., 2001. Authenticity and sincerity in tourism. *Annals of Tourism Research*, 28(1), pp. 7–26.

Terzi, C.M., Sakas, D.P. and Seimenis, I., 2013. International events: The impact of the conference location. *Procedia – Social and Behavioral Sciences*, 73, pp. 363–372.

The Economist Intelligence Unit, 2019. *The 2019 Bleisure Barometer: Asia's Best Cities for Work and Recreation*. [Online]. Available: https://fivestarcities.economist.com/wp-content/uploads/ECO056-JP-ANA-Bleisure-Business-and-Leisure-5-ver20190212.pdf [Accessed 9 October 2019].

Visser, G., 2007. Urban tourism in Bloemfontein: Current dynamics, immediate challenges and future prospects. *Urban Forum*, 18(4), pp. 351–370.

Wall, G., 1997. Tourism attractions: Points, lines, and areas. *Research Notes and Reports*, 96, pp. 240–243.

Wan, Y.K.P., 2011. Assessing the strengths and weaknesses of Macao as an attractive meeting and convention destination: Perspectives of key informants. *Journal of Convention & Event Tourism*, 12(2), pp. 129–151.

Weaver, D. and Lawton, L., 2009. *Tourism Management*. Sydney: Wiley Australia.

Weidenfeld, A., Butler, R.W. and Williams, A.M., 2010. Clustering and compatibility between tourism attractions. *International Journal of Tourism Research*, 12(1), pp. 1–16.

Whitfield, J.E., 2009. Why and how UK visitor attractions diversify their product to offer conference and event facilities. *Journal of Convention & Event Tourism*, 10(1), pp. 72–88.

Whitfield, J. and Webber, J., 2011. Which exhibition attributes create repeat visitation? *International Journal of Hospitality Management*, 30, pp. 439–447.

Witt, S.F., Gammon, S. and White, J., 1992. Incentive travel: Overview and case study of Canada as a destination for the UK market. *Tourism Management*, 13(3), pp. 275–287.

Xue, X.H. and Cox, C., 2008. Hotel selection criteria and satisfaction levels of the Chinese business traveler. *Journal of China Tourism Research*, 4, pp. 261–281.

Yang, Y., Tang, J., Luo, H. and Law, R., 2015. Hotel location evaluation: A combination of machine learning tools and web GIS. *International Journal of Hospitality Management*, 47, pp. 14–24.

Yankholmes, A. and McKercher, B., 2015. Understanding visitors to slavery heritage sites in Ghana. *Tourism Management*, 51, pp. 22–32.

Yousefi, M. and Marzuki, A., 2012. Travel motivations and the influential factors: The case of Penang, Malaysia. *Anatolia: An International Journal of Tourism & Hospitality Research*, 23(2), pp. 169–176.

Zhou, L., Ye, S., Pearce, P. and Wu, M., 2014. Refreshing hotel satisfaction studies by reconfiguring customer review data. *International Journal of Hospitality Management*, 38, pp. 1–10.

22

"LE VIEUX" AND "LE NOUVEAU"

The ambidextrous model of French tourism cities

Hugues Séraphin

Introduction

Heritage (either tangible or intangible) tourism is using the past as a tourism product and service (Park, 2014).

> The fascination with experiencing and consuming the past continues to grow all across the globe. On an international scale heritage tourism experiences encourage people to encounter and appreciate diverse historic environments and cultural assets, thereby enhancing the understanding of different people and cultures.
>
> *(Park, 2014: 32)*

That said, this past (heritage) is very often re-appropriated, in other words, commodified for touristic consumption (Park, 2014). This quest for authenticity has seen many destinations in the world swamped with visitors (Rickly, 2019). Authenticity can therefore be seen as an asset (potential to attract visitors) but also an issue (over visitation) for destinations. This is to be related to the Janusian faced character of the tourism industry (Sanchez & Adams, 2008). Indeed, for every positive impact of the industry, there is a negative impact (Sanchez & Adams, 2008). This ambidextrous nature of the tourism industry is calling for an ambidextrous management approach of the industry.

This chapter focuses on the structure of French tourism cities that could be said to be structured following an ambidextrous model (old/new). The objective of the chapter is to discuss the benefits and limitations of this structure, with regard to overtourism (and related perverse impacts). In so doing, the chapter is structured around four main sections:

"Ambidexterity in tourism" introduces the concept of organisational ambidexterity or ambidextrous management. "Contextual framework" provides information on France as a destination, as the case study of this chapter is based on France. "Contextual and conceptual framework" shows how the concept of ambidextrous management applies to the way French cities are structured/organised. And "Discussion", highlights the positive aspects and limitations about the structure of most French cities.

From a methodological point of view, this conceptual chapter adopts a positivist approach, with the overall objective to conceptualise the way French cities are organised.

Ambidexterity in tourism

The Janus-faced character of the tourism industry

Originally called oppositional thinking, Janusian thinking is a reference to Janus, the Roman god with two faces, who looked in opposite directions simultaneously, and also played an essential role in the creation of the world (Rothenberg, 1971).

As a result, Janusian thinking is a process that involves conceiving and utilising multiple incompatible opposites or contradictory ideas, concepts, images or antitheses simultaneously (Rothenberg, 1971, 1996). Creativity, in other words, the development of new and valuable phenomena, is the main outcome of this process (Rothenberg, 1996). Janusian thinking is also closely associated to artistic creation and related domains (Rothenberg, 1971). The Janusian thinking process has influenced many other concepts such as the Yin and the Yang in Eastern culture; the Being and Becoming in Western culture; God and the Devil; Eros and Thanatos; etc., all of which convey an integration of opposites (Rothenberg, 1971). That said, it is important to mention the fact that

> there is a wide variety of types of opposition, ranging from strong opposition containing logical antithesis or contradiction to mild opposition consisting of simple contrast.... Strong opposition or logical antithesis has the greatest shock or surprise value; it conveys the greatest sense of novelty and may also convey the greatest truths.
>
> *(Rothenberg, 1971: 204)*

This range of oppositions is also reflected in organisational ambidexterity (Papachroni et al., 2015), with a range of tensions (spatial separation; temporal separation; moving beyond spatial and temporal separations) triggered by a variety of approaches to ambidexterity (structural, contextual, temporal ambidexterity; paradox theory). For Sanchez and Adams (2008), tourism is a Janus-faced character due to the fact the industry is simultaneously a quick fix to socioeconomic development, and also undermines destinations' national aspiration such as social equity. In a nutshell, they explained that for every positive impact, the tourism industry generates a negative one (Sanchez & Adams, 2008).

Organisational ambidexterity

In the business environment, organisational ambidexterity which is in essence the involvement of two polar opposites, namely exploitation vs exploration (Smith, 2017), could be considered as a practical application of the Janusian thinking process. Papachroni, Heracleous and Paroutis (2015), argue that in organisational ambidexterity the paradoxical poles (exploration and exploration) and the tensions coming out of the oppositions (old/new; capability/rigidity; continuity/change; chaos/inertia; leverage/stretch; efficiency/flexibility) interrelate over time. This interrelation leads to success (Smith, 2017). Organisational ambidexterity or ambidextrous management is a new and emerging management approach investigated in tourism academic research (Mihalache & Mihalache, 2016; Séraphin & Yallop, 2019), which has proven to lead to improvement of performance in tourism and non-tourism-related organisations (Séraphin & Butcher, 2018). At the moment, organisational ambidexterity (OA) in tourism has been used mainly in research on leadership and Human Resources Management (HRM); innovation and performance; and finally, in destination management and sustainability. In HRM, ambidextrous leadership style contributes to the improvement of staff performance, by reducing the perverse impacts of lack of motivation such as lateness and turnover (Bouzari & Karatepe, 2017). As for innovation and performance, Cheng, Tang, Shih and Wang (2016), and Tang (2014), provided evidence that hotels' market performance can be improved with an OA, by developing simultaneously the type and range of services on offer (exploration), while improving the quality of current services (exploitation). The combination of both continually contributes to create value for customers and subsequently market performance, as customers' willingness to return will be enhanced. Finally, in terms of destination management, Martínez-Pérez, García-Villaverde and Elche (2016) have argued that cultural tourism destinations should adopt an OA approach that would consist in bridging gaps between locals and outsiders. The final outcome would be the development of social capital between both groups.

Conceptual framework

- Janusian thinking and organisational ambidexterity are related as they both involve conceiving and utilising multiple incompatible opposites or contradictory ideas, concepts, images or antitheses simultaneously (exploration and exploitation; old/new; capability/rigidity; continuity/change; chaos/inertia; leverage/stretch; efficiency/flexibility) (Papachroni et al., 2015; Rothenberg, 1971; Séraphin & Yallop, 2019; Smith, 2017).
- Janusian thinking and/or organisational ambidexterity, contribute to creativity; innovation; improvement of performance; improvement of customers' experience (Cheng et al., 2016; Mihalache & Mihalache, 2016; Séraphin & Yallop, 2019; Smith, 2017; and Tang, 2014).

Figure 22.1 summarises the conceptual framework of this chapter.

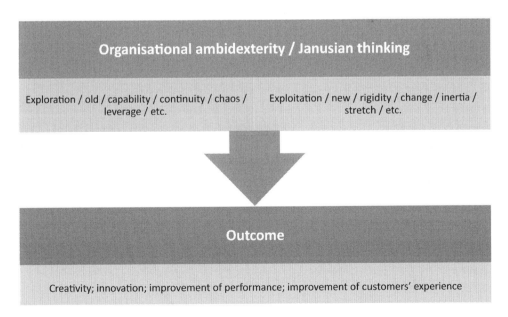

Organisational ambidexterity / Janusian thinking

Exploration / old / capability / continuity / chaos / leverage / etc.

Exploitation / new / rigidity / change / inertia / stretch / etc.

Outcome

Creativity; innovation; improvement of performance; improvement of customers' experience

Figure 22.1 Organisational ambidexterity/Janusian-thinking and outcomes.
Source: Author.

Contextual framework

France as a destination: key figures (French Ministry of Economy and Finance, 2019)

France is the leading destination in the world in terms of number of visitors. In 2017, France received 86.9 million visitors. Spain and the USA are the closest followers (Figure 22.2).

Table 22.1 International tourist arrivals by country of destination

Rank in 1980	Rank in 2017	Destination	Arrivals in 2017 (€ millions)	Change 2017/2016 (%)
1	**1**	**France**	**86.0**	**5.1**
3	2	Spain	81.9	8.8
2	3	United States	76.9	0.7
18	4	China	60.7	2.4
4	5	Italy	58.2	11.1
8	6	Mexico	39.3	12.0
7	7	United Kingdom	37.7	5.3
52	8	Turkey	37.6	24.1
9	9	Germany	37.5	5.3
27	10	Thailand	35.4	8.9
		Total World	**1,326.4**	**7.0**

Source: French Ministry of Economy and Finance (2018).

That said, in terms of income generated by the tourism industry, France is only the third destination, as preceded by the USA and Spain (Figure 22.3).

Table 22.2 International tourism revenue by country of destination

Rank in 1980	Rank in 2017	Destination	Arrivals in 2017 (€ millions)	Change 2017/2016 (%)
1	1	United States	186.0	1.9
4	2	Spain	60.3	10.3
2	**3**	**France**	**53.7**	**9.0**
27	4	Thailand	50.9	13.1
5	5	United Kingdom	45.3	12.2
3	6	Italy	39.2	7.7
24	7	Australia	36.9	9.3
6	8	Germany	35.3	4.2
N/A	9	Macao	31.5	17.6
N/A	10	Japan	30.1	14.4
		Total World	**1,186.3**	**5.5**

Source: French Ministry of Economy and Finance (2018).

Visitors generally choose rural destinations when visiting France (Figure 22.4).

Table 22.3 Tourism in metropolitan France by type of destination

	Breakdown by trips (%)	*Breakdown by nights (%)*	*Average length of stay (nights)*
Seashore destinations	**29.9**	**32.1**	**6.8**
Rural seashore	7.5	11.1	7.3
Urban seashore	15.5	20.9	6.6
Mountain destinations	**19.1**	**21.9**	**5.6**
Non-resorts	13.5	14.6	5.3
Resorts	5.6	7.3	6.4
Rural destinations	**21.8**	**18.6**	**4.2**
Urban destinations	**30.3**	**21.7**	**3.5**
Not reported	5.9	5.7	4.9
Metropolitan France	**100**	**100**	**4.9**

Source: French Ministry of Economy and Finance (2018).

France as a destination and revenue management

Based on Figures 22.2 and 22.3, it seems that when it comes to revenue management (RM), France is underperforming. The purpose of RM, also called yield management (Ivanov & Zhechev, 2012; Lacagnina & Provenzano, 2016) is to create demand by dividing customers into different segments in order to allocate them the right product or service, at the right time and at the right price in a way that maximises the revenue of an organisation (Cross, 1997; Ivanov & Zhechev, 2012; Vives et al., 2018). This view is also supported by (Lacagnina & Provenzano, 2016: 779) who argued that: "revenue management aims at improving the performance of an organisation by selling the right product/service to the right customer at the right time". With the development of ICT (information and communication technologies), especially with the emergence of Online Travel Agencies (e.g. Expedia), review websites (e.g. TripAdvisor), and social media (e.g. Facebook), the choice of "the right distribution channel" has become crucial (Tyrrell, 2017). Revenue management on destination and corporate levels works through the use of various pricing, non-pricing and combined revenue management tools. On corporate level, pricing techniques include dynamic pricing, price differentiation, price presentation and lowest price guarantee; non-pricing techniques include inventory management (capacity management, over contracting and overbooking, room-availability guarantee, length-of-stay control), and 100% satisfaction guarantee, while combined techniques consist of channel management and optimal room-rate allocation (Ivanov, 2014). The economic performance of a destination improves with its ability to attract the right types of visitors (Favre, 2017). The ideology developed by Favre (2017) is encouraging destinations to move from a mass tourism strategy to a special interest tourism (SIT) strategy. This form of tourism contributes to the happiness and quality of life of locals as they have an opportunity to showcase their culture/heritage (Park, 2014), but would also enable them to maintain long-term livelihoods (Rickly, 2019). Equally important, this strategy would enable the destination to retain existing visitors. "Destinations experiencing overtourism inspire some tourists to choose different holiday locations" (Rickly, 2019). As for locals, authenticity is associated with equity and sense of place and community wellbeing (Rickly, 2019). That said, this change of paradigm, if it happens, will take time as France as a tourist destination is quite conservative (Séraphin, 2017).

The most visited destinations in France

The top ten most visited destinations (cities) in France are (Locatour [Online]):

- Paris: 30 million tourists/year
- Lourdes/Lyon: 6 million tourists/year
- Toulouse: 5.6 million tourists/year
- Nice: 4.3 million tourists/year
- Larochelle: 4 million tourists/year
- Honfleur: 3.5 million tourists/year
- Carcassonne: 3.2 million tourists/year
- Strasbourg: 3.1 million tourists/year
- Le Mont Saint-Michel: 3 million tourists/year
- Bordeaux: 2.7 million tourists/year

All the above cities (and other cities in France) are built on the model of an historical ("le vieux") side, and modern side ("le nouveau") of the city. Indeed, "le nouveau" is assimilated to the dynamic and modern side of the city. This side of the city concentrates all the urban activities and is also characterised by modern architecture (Levy, 1983). As for the old side ("le vieux"), it is characterised by its very narrow streets and old style architecture (Levy, 1983). Cities in France are basically a combination of old and new; modern and history (Levy, 1983). The old side of the city is appearing as an anachronism (Levy, 1983).

Contextual and conceptual framework

Overview

Based on information collected so far in this paper, tourism cities in France could be said to be structured following an ambidextrous model (Figure 22.5).

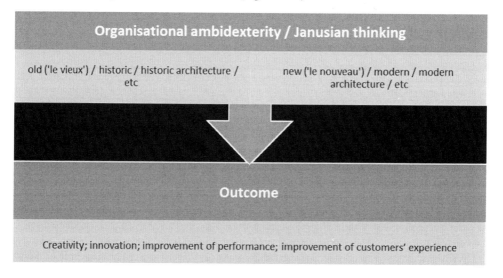

Figure 22.2 Organisational ambidexterity of French tourism cities.
Source: Author.

Carcassonne as a case study

Built over the ruins of a former Roman fortress, "La Cite" of Carcassonne was the theatre of the crusades against the heretics in the thirteenth century (Office du Tourisme Carcassonne). On the right bank of the River Aude is the medieval town, a UNESCO World Heritage Site. Within the medieval side of the city can be found a large number of shops and craftsmen. Approximately 120 people live in this part of the city (Office du Tourisme Carcassonne).On the left bank of the River Aude is the modern side of the city, also called "La Bastide St Louis" (Office du Tourisme Carcassonne). Figure 22.6 is quite symbolic.

On the left side is the historic city (past), and on the right side the modern city (present). In the middle is River Aude and "Le Pont vieux" bridging the gap between the past and the present. Carcassonne exemplifies the ambidextrous design of French cities.

Discussion

Overtourism and tourismphobia (including anti-tourism movement)

There are many existing definitions of overtourism. Among these are, the definition suggested by Milano et al. (2018 cited in Milano et al., 2019: 1):

> Overtourism is described as the excessive growth of visitors leading to overcrowding in areas where residents suffer the consequences of temporary and seasonal tourism peaks, which have caused permanent changes to their lifestyles, denied access to amenities and damaged their general well-being.

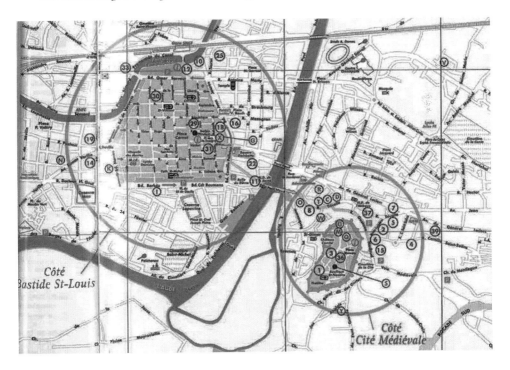

Figure 22.3 Map of Carcassonne.

Source: Office Municipale Carcassonne.

As for Goodwin (2017 cited in Milano et al., 2019: 91), he suggests the following definition: "a condition where hosts or guests, locals or visitors, feel that there are too many visitors and that the quality of life in the area or the quality of the experience has deteriorated unacceptably". The term "overcrowding" often associated with overtourism (Canada, 2019; Milano et al., 2019; Pecot & Ricaurte-Quijano, 2019), is defined as "the process and results of an intensified use of (semi-) public space, which are perceived as disruptive forces by multiple stakeholders" (Gerritsma, 2019: 125). This phenomenon is exacerbated by the intensification of touristification of places that are not originally tourism destinations or places, such as favelas, historically classified as marginal and dangerous (Da Cunha, 2019). Additionally, overtourism is associated with terms such as "saturation"; "European urban context"; "socio-ecological transformation"; "local wellbeing"; "quality of life" (Pecot & Ricaurte-Quijano, 2019); "conflicts"; "water over-exploitation"; "pollution" (Canada, 2013); "environmental degradation" (Canosa et al., 2019); "excess"; "discontents" (Milano et al., 2019).

The ambidextrous design of cities in France should in most cases prevent the development of tourismphobia (and anti-tourism movements) due to the fact that the historical quarters of the cities are the honey pot for tourists (not the modern part). This side of the city accommodates a very limited number of the population. To stop tourists disrupting the life of locals, many destinations have opted for the segregation of visitors from locals, and/or discouraging tourists to visit some areas. This is a strategy that Séraphin, Sheeran and Pilato (2018) called "trexit" or tourism exit. The way current French cities are designed is naturally implementing a "trexit" strategy.

Resilience of the local population

To some extent, what overtourism has done is to test the resilience of local populations with regard to external aggression, i.e. the influx of a large number of people visiting their place of living. In Lucerne (Switzerland), the Kleintheater, a small independent theatre company has for instance developed a satirical performance around overtourism issues in their city (Webber et al., 2019). Based on existing research (Croes et al., 2017; Getz, 2008; Haifeng et al., 2012; Hutton, 2016; Ozer et al., 2017), locals' attitude in an overtourism context change, and move from:

> Stage 1 – *Locals are helpless victims.* Despite the fact their life is disrupted by visitors, they do not protest or fight back because they are not supported by the local authorities. They mainly appear as victims of a system that supports the wellbeing and happiness of visitors above the wellbeing and happiness of locals.
>
> Stage 2 – *Peaceful activists.* Residents hold a positive attitude to tourism development until the number of visitors reaches a point that threatens their original lifestyle. Ant-tourism movements or resistant identities, which are voicing locals' concerns regarding the continuously growing number of tourists, are evidence of this change of attitude towards visitors.
>
> Stage 3 – *Vandals.* Locals are showing that they are neither passive nor powerless. At this stage, locals have chosen violence.
>
> Stage 4 – *Resilience.* Resilience is the ability to bounce back from stress and adversity or change that can emerge from the experience of trauma (Hutton, 2016). The Kleintheater theatre in Lucerne exemplifies both definitions.

In French cities designed following the ambidextrous model discussed in this chapter, locals' attitude toward tourists must be at stage 1, for those living in the tourist side of the city. For

Table 22.4 The co-integrative model of tourism development

STAGE 1: PLANNING CHARACTERISTICS

1.1. Early stage of post-
conflict and post-disaster
- Country has just come out of a conflict
- Country has just come out of a disaster
- Country is no longer visitor friendly
- No or hardly any tourists

1.2. Destination to be
- Economic development
- Poverty decreases
- Creation or reopening of a DMO
- Country is visited by a limited number of people

STAGE 2: HUBRIS CHARACTERISTICS

2.1. Social capital
development

Pre-commodification/Euphoria/Exploration:
- Small groups of tourists
- Visitor experiences are fun, inspirational, safe, open and accessible to all with no discrimination
- Good visitors/locals relationship (involving local residents in tourism development through working relationships, consultation and cooperation)
- No significant impact of tourism on the visited area (the quality of the urban, coastal and rural environment is maintained or enhanced without physical or visual degradation)
- Tourism industry is prosperous
- Locals are engaged and benefit from the industry (tourism activities strengthen and sustain the quality of life of the communities)

STAGE 3: NEMESIS AND CATHARSIS CHARACTERISTICS

3.1. Nemesis

Involvement/Early commodification/Apathy:
- Development of facilities for visitors
- Promotion of the area by DMO
- The number of visitors increases
- Increase of investments
- Land transformation and resource extraction

Development/Advanced commodification:
- Continuing growth of visitors
- Further effort (by DMO) to promote the destination
- The destination receives a large number of visitors

Consolidation/Pre-destruction:
- Number of visitors may exceed the number of inhabitants
- Destination now has a tourism-based economy
- Detachment between tourists and locals
- The industry continues to grow
- New types of businesses which were not part of the community offerings are introduced

3.2. Catharsis

Stagnation/Advanced destruction/Antagonism:
- Destinations have reached a record number of tourist arrivals
- The destination is no longer fashionable
- Community experiences backlash of tourism
- Destination suffers the strain of tourism (overtourism)
- Imposes burdens on local residents

Decline:
- Destination is unable to compete with rivals
- Change in the characteristics and attractions of the tourist area

(Continued)

Table 22.4 Continued

STAGE 4: BOOMERANG CHARACTERISTICS

4.1. Solutions	Agents of change:
	• Transformational leadership
	• Enterprise reform
	• Technology innovation
	• Corporate transparency
	• Stakeholder engagement
	• Social responsibility
	• Integrated value
4.2. Back to Hubris stage	Pre-commodification/Euphoria/Exploration/Social capital development:
	• Small groups of tourists
	• Visitor experiences are fun, inspirational, safe, open and accessible to all with no discrimination
	• Good visitors/locals relationship (involving local residents in tourism development through working relationships, consultation and cooperation)
	• No significant impact of tourism on the visited area (the quality of the urban, coastal and rural environment is maintained or enhanced without physical or visual degradation)
	• Tourism industry is prosperous
	• Locals are engaged and benefit from the industry (tourism activities strengthen and sustain the quality of life of the communities)

Source: Séraphin, Gowreesunkar, Zaman and Lorey (2019).

those living in the modern side of the city. Their view of tourism is probably different. The stages described above do not consider destinations not victims of overtourism. Séraphin, Gowreesunkar, Zaman and Lorey (2019), have developed a more complex and precise index (The Cointegrative Model of Tourism Development), to identify the stages of tourism development and their characteristics, including the evolution of the perception of locals (Table 22.1). Based on this index, most French cities, because of their ambidextrous designed model are at stage 2 (*Hubris*). At this stage we are including all the leading tourism destinations in the world that are not suffering from overtourism or other major issues related to tourism development. From our perspective, this is the best stage for a destination to be at and therefore to remain. Some aspects of this stage have already been described and discussed in already existing models. Leading destinations that are not built according to this model are at stage 3.1. This is the stage where destinations suffering from overtourism and other negative impacts of tourism are found. This is the stage for destinations to avoid being. If they are already at this stage, it is important for destination management organisations (DMOs) to come forward with solutions to emerge from this stage. That said, strategies applied so far have proven to be unsuccessful. Séraphin et al. (2018) have urged for an ambidextrous management of tourism. This approach consists in using existing strategies to mitigate the negative impacts of tourism (exploitation) to develop a new one tailored to meet the needs of the destinations (exploration).

Conclusion

Summary

French tourism cities are structured in an ambidextrous way. This structure presents some positive aspects, among these are the preservation of the quality of life of locals, while giving visitors opportunities to enjoy local heritage without impacting too much on the life of locals. In terms of limitation, this structure of the French cities does not give local DMOs any opportunities to test the resilience of the local residents. That said, the model: "Cointegrative Model of Tourism Development" could help them to locate the attitude and perception of locals toward tourism.

Theoretical contribution

Ambidextrous management is new to tourism literature. So far it has been used to discuss the performance of tourism organisations and destinations. It has never been used to discuss the way a city is structured. By doing this, this chapter is filling a gap by showing the extent to which the concept of ambidextrous management is broad and suitable to the tourism industry. This chapter has provided additional evidence to support the fact that as an approach and/or concept, ambidexterity is totally suitable to the tourism industry. Finally, this chapter could be said to have contributed to existing literature on overtourism as it is offering an alternative to overcome this issue. That said, in the case of French cities, this alternative is almost natural. Indeed, the way the cities are structured is not to deal with the issue of overtourism but happens to be quite useful.

Practical contributions

This chapter provides the managers of DMOs where tourismphobia is an issue an idea of how a city could potentially be structured to avoid anti-tourism movements. As one size does not fit all, it does not mean that the French model if applied to other destinations would work. That said, the French model provides evidence that having cities split between the tourism side and the non-tourism side reduces annoyance for the locals, without stopping interactions between locals and visitors.

Future research

Using France as an example has shown that the destination is not performing very well in terms of revenue management. Future research on tourism cities should be looking at products and services that could be developed by cities to yield the maximum income from tourists without having to increase numbers. Additionally, Beal, Séraphin, Modica, Pilato and Platania (2019) have provided evidence that heritage plays a meditating role between locals and visitors in French cities. That said, it would be interesting to investigate whether or not the modern side of cities ("Le nouveau") plays a mediation role between locals and visitors. If they do, how? Last but not least, it would be interesting to do a comparative study (using multi criteria decision analysis), between French cities built following an ambidextrous structure and French cities not following this pattern, in order to see which performs better. The result of such a study would enable the identification of which model is the most effective and sustainable.

References

Beal, L., Séraphin, H., Modica, G., Pilato, M., & Platania, M. (2019). Analysing the mediating effect of heritage between locals and visitors. An exploratory study using Mission Patrimoine as a Case Study. *Sustainability*, *11*, 1–15.

Bouzari, M., & Karatepe, O. M. (2017). Test of a mediation model of psychological capital among hotel salespeople. *International Journal of Contemporary Hospitality Management*, *29*(8), 2178–2197. https://doi.org/10.1108/IJCHM-01-2016-0022.

Canada, E. (2019). Responses to overtourism in Guanacaste (Costa Rica): A rural water conflict perspective. In Milano, C., Cheer, J. M., & Novelli, M. (Eds), *Overtourism: Excess, discontents and measures in travel and tourism*, Wallingford: CABI, pp. 107–124.

Canosa, A., Graham, A., & Wilson, E. (2019). Progressing a child-centred research agenda in tourism studies. *Tourism Analysis*, *24*, 95–100.

Cheng, J.-S., Tang, T.-W., Shih, H.-Y., & Wang, T.-C. (2016). Designing lifestyle hotels. *International Journal of Hospitality Management*, *58*, 95–106. https://doi.org/10.1016/j.ijhm.2016.06.010.

Croes, R., Rivera, M. A., Semrad, K., & Khalizadeh, J. (2017). Happiness and tourism: Evidence from Aruba, Institute for Tourism Studies. Retrieved from: http://stars.library.ucf.edu/cgi/viewcontent.cgi?article=1040&context=dickpope-pubs, Accessed 24 May 2019.

Cross, R. G. (1997). Launching the revenue rocket: How revenue management can work for your business. *Cornell Hotel and Restaurant Administration Quarterly*, *38*(2), 32–43.

Da Cunha, N. V. (2019). Public policies and tourist saturation in the Favelas of Rio de Janeiro. In Milano, C., Cheer, J. M., & Novelli, M. (Eds), *Overtourism: Excess, discontents and measures in travel and tourism*, Wallingford: CABI, pp. 152–166.

Favre, C. (2017). The Small2Mighty tourism academy: Growing business to grow women as a transformative strategy for emerging destinations. *Worldwide Hospitality and Tourism Themes*, *9*(5), 555–563.

French Ministry of Economy and Finance. (2019). Key facts on tourism in France. Retrieved from: www.entreprises.gouv.fr/files/files/directions_services/etudes-et-statistiques/Chiffres_cles/Tourisme/2019-04-key-facts-on-tourism-2018.pdf, Accessed on 17 June 2019.

Gerritsma, R. (2019). Overcrowded Amsterdam: Striving for a balance between trade, tolerance and tourism. In Milano, C., Cheer, J. M., & Novelli, M. (Eds), *Overtourism: Excess, discontents and measures in travel and tourism*, Wallingford: CABI, pp. 125–147.

Getz, D. (2008). *Event Studies: Theory, research and policy for planned events*. London: Elsevier.

Haifeng, Y., Jing, L., & Mu, Z. (2012). Rural community participation in scenic spot. A case study of Danxia Mountain of Guangdong, China. *Journal of Hospitality & Tourism*, *10*(1), 76–112.

Hutton, M. (2016). Neither passive not powerless: Reframing economic vulnerability via resilient pathways. *Journal of Marketing Management*, *32*(3–4), 252–274.

Ivanov, S. (2014). *Hotel revenue management: From theory to practice*. Varna: Zangador.

Ivanov, S. & Zhechev, V. (2012). Hotel revenue management: A critical literature review. *Tourism Review*, *60*(2), 175–197.

Lacagnina, V. & Provenzano, D. (2016). An integrated fuzzy-stochastic model for revenue management: The hospitality industry case. *Tourism Economics*, *22*(4), 779–792.

Levy, J. P. (1983). Quartiers anciens et centre ville: Nouveaux enjeux d'une politique de rehabilitation. *Revue geographiques des Pyrenees et du Sud-Ouest*, *54*(1), 101–126.

Locatour. Retrieved from: https://blog.locatour.com/top-10-villes-touristiques-france/, Accessed on 17 June 2019.

Martínez-Pérez, Á., García-Villaverde, P. M., & Elche, D. (2016). The mediating effect of ambidextrous knowledge strategy between social capital and innovation of cultural tourism clusters firms. *International Journal of Contemporary Hospitality Management*, *28*(7), 1484–1507. https://doi.org/10.1108/IJCHM-082014-0405.

Mihalache, M., & Mihalache, O. R. (2016). Organisational ambidexterity and sustained performance in the tourism industry. *Annals of Tourism Research*, *56*, 14–144.

Milano, C., Cheer, J. M., & Novelli, M. (2019). *Overtourism: Excess, discontents and measures in travel and tourism*. Wallingford: CABI.

Office du Tourisme Carcassonne. (2020). Carcassonne. Retrieved from: www.tourisme-carcassonne.fr/, Accessed 1 May 2020.

Ozer, S., Bertelsen, P., Singla, R., & Schwartz, S. J. (2017). Grab your culture and walk with the global: Ladakhi students' negotiation of cultural identity in the context of globalisation-based acculturation. *Journal of Cross-Cultural Psychology*, *48*(3), 294–318.

Papachroni, A., Heracleous, L., & Paroutis, S. (2015). Organisational ambidexterity through the lens of paradox theory: Building a novel research agenda. *The Journal of Applied Behavioural Science*, *5*(1), 71–93.

Park, H. Y. (2014). *Heritage Tourism*. Abingdon: Routledge.

Pecot, M., & Ricaurte-Quijano, C. (2019). Todos a Galapagos? Overtourism in wilderness areas of the global south. In Milano, C., Cheer, J. M., & Novelli, M. (Eds), *Overtourism: Excess, discontents and measures in travel and tourism*, Wallingford: CABI, pp. 70–85.

Rickly, J. M. (2019). Overtourism and authenticity. In Dodds, R. & Butler, R. (Eds), *Overtourism: Issues, realities and solutions*, Berlin: De Gruyter, pp. 46–61.

Rothenberg, A. (1971). The process of Janusian thinking in creativity. *Archives of General Psychiatry*, *24*(3), 195–205.

Rothenberg, A. (1996). The Janusian process in scientific creativity. *Creativity Research Journal*, *9*(2), 207–231.

Sanchez, P. M., & Adams, K. M. (2008). The Janus-faced character of tourism in Cuba. *Annals of Tourism Research*, *35*(1), 27–46.

Séraphin, H. (2017). Terrorism and tourism in France: The limitations of dark tourism. *Worldwide Hospitality and Tourism Themes*, *9*(2), 187–195.

Séraphin, H., & Butcher, J. (2018). Tourism management in the Caribbean: The case of Haiti. *Caribbean Quarterly*, *64*(2), 254–283.

Séraphin, H., Gowreesunkar, V., Zaman, M., & Lorey, T. (2019). Limitations of trexit (tourism exit) as a solution to overtourism. *Worldwide Hospitality and Tourism Themes*, *11*(5), 566–581.

Séraphin, H., Sheeran, P., & Pilato, M. (2018). Over-tourism and the fall of Venice as a destination. *Journal of Destination Marketing & Management*, *9*, 374–376.

Seraphin, H., & Yallop, A. (2019). Proposed framework for the management of resorts Mini Clubs: An ambidextrous approach. *Leisure Studies*, 1–13. https://doi.org/10.1080/02614367.2019.1581249.

Smith, S. M. (2017). Organisational ambidexterity: Welcome to paradox city. *Human Resources Management Digest*, *25*(1), 1–3.

Tang, T.-W. (2014). Becoming an ambidextrous hotel: The role of customer orientation. *International Journal of Hospitality Management*, *39*, 1–10. https://doi.org/10.1016/j.ijhm.2014.01.008.

Tyrrell, T. (2017). Revenue management. In Lowry, L.L. (Eds), *The Sage international encyclopedia of travel and tourism*. Thousand Oaks, CA: Sage, pp. 1030–1034.

Vives, A., Jacob, M., & Payeras, M. (2018). Revenue management and price optimisation techniques in the hotel sector: A critical literature review. *Tourism Economics*, *24*(6), 720–752.

Weber, F., Eggli, F., Ohnmacht, T., & Stettler, J. (2019). Lucerne and the impact of Asian group tours. In Dodds, R. & Butler, R. (Eds), *Overtourism: Issues, realities and solutions*, Berlin: De Gruyter, pp. 169–184.

23

THE TRANSITION OF A COASTAL INDUSTRIAL CITY INTO A WORLD-CLASS TOURISM AND MARITIME CITY

The case of Port Elizabeth

Cina Van Zyl

Introduction

There are destinations on any traveller's list: Paris, Venice, New York. There are others one hopes to visit if time and money were not a constraint: Tokyo, Sydney, Buenos Aires. Furthermore, some cities do not appear on a list, yet one ends up going there for work. Port Elizabeth (PE) in South Africa: how to keep a city competitive in a robust market when compared to cities in the same country, such as Cape Town or Durban?

This chapter presents a practical approach to, and a solution for, marketing and branding tourism in urban areas. An academic wrote the chapter with contribution from a tourism practitioner. The chapter starts by introducing key definitions and concepts, providing a theoretical base on which to brand the city. The city branding constraints to be mindful of are emphasised. Then city attractors are introduced and applied to the case of Port Elizabeth to illustrate the city's branding possibilities. The Port Elizabeth case study is used to explore the realities faced by tourism cities and determine how to manage this activity. A SWOT analysis illustrates the city's current position, some lessons learnt and several future focus opportunities.

Literature review

Destinations/cities and tourism

Tourists seek to fulfil their needs through complex patterns of using services and interacting with *destinations* (Prebensen *et al.* 2014, Pearce and Zare 2017). A destination can vary in scale as tourism researchers use it to include whole regions, even countries and in other cases, the term refers to a specific city, site or attraction (Crouch and Ritchie 2005, Grün and Dolnicar 2016, Jovicic 2016). Prayag and Ryan (2012) state a destination is a specific area, such as a country, a city or an island. In this chapter, a destination is seen as a city in a country, and the

346

terms are used interchangeably. Following on are various definitions used applicable to the context of the research.

Urban tourism is considered a relatively new area of research until recently often neglected by tourism researchers (Ashworth 1989, Law 2002, Maxim 2019, Page and Hall 2003). This lack of research on a relevant topic has changed, and more research is now being conducted due to the rapid growth sustained by this form of tourism (Maitland 2009, Maxim 2016) as well as the resulting policy issues associated with it (Pearce 2001).

Limited research on tourism cities is attributed to the complex nature of the phenomenon of urban tourism and the "multifunctional nature of cities" (D. Pearce 2011, p. 59). Tourism is less visible in cities where it represents only one activity among many others embedded in the economy of the city (Edwards *et al.* 2008, Maitland and Newman 2009). Despite the importance of research on tourism development in cities, Ashworth and Page (2011) note the gap in research (except for research by Maitland and Newman 2009). Cities are the main gateway for tourists visiting a country, and their success has a direct impact on the number of visitors it attracts. Cities are faced by challenges to update their appeal to visitors continuously and to maintain their distinctiveness (Maitland 2012, p. 1).

Researchers have classified cities into different typologies based on the particular characteristics they present (Maitland and Newman 2009, Page and Hall 2003). World tourism cities accommodate world-class attractions (Law 2002) and events. They "have become points where knowledge is transformed into productive activities" (Ashworth and Page 2011, p. 4), and are centres of business and cultural excellence that offer visitors some benefits such as easier accessibility through connected airports, scheduled tourism services, diverse accommodation facilities and a variety of entertainment options (Edwards *et al.* 2008). These city attractors are introduced in a follow-up section.

Port Elizabeth, the case study for this chapter, belongs to a coastal and marine area. The question to be asked is, is PE a world-class tourism city? The typology – a world tourism city, which refers to tourism occurring in world cities rather than cities that are dependent on tourism for their profile, such as Venice or Bath (Ashworth 2010) could provide an interesting angle. World tourism cities within the African context have not received much attention, which opens up areas of future research on the continent being the rationale for this chapter.

Coastal and marine tourism – although distinct forms of tourism – are very closely related, due to the water/sea element (Hall 2001). *Marine tourism* usually refers to an activity that takes place in the water or deep oceans – such as cruising, sailing and other water-based activities and nautical sports such as scuba diving, sailing, water-skiing, windsurfing and wildlife mammal watching (European Commission 2014, Orams 1998). *Coastal tourism* refers to the type of tourism which takes place at the seaside – the water/sea element is predominant and considered as the primary assist and advantage (Page and Connell 2006). According to Hall (2001), coastal tourism is closely related to marine (maritime) tourism (since it covers activities taking place at the coastal waters too) although it covers beach-based tourism and recreational activities such as swimming, sunbathing, coastal walks etc. (European Commission 2014, Diakomihalis 2007). *Marine and coastal tourism* are both among the oldest forms of tourism and represent the largest segment of the tourism industry (Papageorgiou 2016). Being one of the most significant segments of the maritime economic sectors, as well as the most significant component of the tourism industry, both coastal and maritime tourism are subjected to planning procedures to assure minimum impact on the natural ecosystem and on the local economy (fisheries, aquaculture etc.). Many people live and move to coastal cities, while tourists also choose to visit coastal destinations (Miller and Auyong 1991, Papageorgiou 2016). The so-called "blue-growth" – "smart, sustainable and inclusive economic and employment growth from the oceans, seas and coasts" – should play a vital role in the organisation of tourism development (European Commission 2014).

Tourists of today have become more knowledgeable and more sensitive (Correia and Kozak 2016). These tourists make use of online information provided by social media, blogs, applications, etc., rather than the use of offline information (Chung and Koo 2015), thus "smart" tourism (Gretzel *et al.* 2015) as a trend needs an introduction. Though beyond the scope of this research for cities to stay relevant, destination management organisations (DMO) need to familiarise themselves with the concept of *smart tourism* and a smart city going forward. Smart tourism is "a new buzzword applied to describe the increasing reliance of tourism destinations, their industries and their tourists on emerging forms of ICT that allow for massive amounts of data to be transformed into value propositions" (Gretzel *et al.* 2015). A smart city is thus a city that incorporates information and communication technology (ICT) to enhance the quality and performance of public services such as energy, transportation and utilities in order to reduce resource consumption, wastage and overall costs (Techopedia 2019). Comparative case studies implementing smart tourism are rare (Buonincontri and Micera 2016) and called upon in future research (Um and Chung 2019). The South African context requires further research.

Classic concepts of city branding/destination or city marketing

Many studies of the image, branding, marketing or competitiveness of destinations are reported as initial research on destinations (Um and Chung 2019). This section reports on the origins thereof and observations on branding products and cities to provide context.

A *city brand* is a construct – a structure which is often researched, designed and consciously put in place (Heely 2011). A destination brand image can be defined as the consumer's perception of a brand, represented by a network of associations linked to the brand name in the consumer's memory (Keller 1993). Anholt (2007) stated that it is not synonymous with image, identity or reputation of a city though academics frequently define it as such. A brand seeks to challenge and change received image, identity and reputation. Viewed in this way image/identity/reputation is a sociocultural construct, while branding is a business discipline dating back to the nineteenth century, businesses introduced it in order to advertise products and services. From the twentieth century, branding has arguably become a central organising principle of contemporary business organisation (Olins 2008). Apple, Orange, Coke and BMW illustrate how they shape the products, environments and communication processes of these companies, as well as the way they treat and relate to their employees and external stakeholders. The principles and practices underpinning branding of products are transferable to the branding of cities – it remains the case that the process of branding cities poses a more significant set of challenges than in the case of products.

The origins of city branding such as *I Love New York*, though difficult to date, probably started around the twentieth century. In 1977, the advertising agency Well, Rich and Green, launched a campaign to design a logo and a slogan for the state of New York to help project a positive image of the city region which had a reputation for civil disorder and uncleanliness (Heely 2011). Other destination branding initiatives in America, run by city convention bureaux, include, for example, in the case of Las Vegas, *What happens here stays here*. In Europe, city branding started in the 1980s with *I Amsterdam* city brand platform, still alive and well (Heely 2011, p. 125). The Dutch are known for pioneering city branding and city marketing (Braun 2008, Amsterdam City Council 2004, p. 10). The success story is documented in literature (Heely 2011), but in essence, a bottom-up approach is followed where every "Amsterdammer has to be a seller of his town". The residents act as frontline ambassadors of their city. Furthermore, a brand such as *I Amsterdam* is only likely to succeed where a city branding association (CBA) (Netherlands City Marketing Network) is created to act as custodian of the brand. In the early twenty-first century, a more robust and ambitious concept of city branding emerged in the form of an "umbrella".

Five key markets or audiences were to be addressed – the residents and businesses of a city, potential students and inward investors, and prospective tourists. Despite the topicality of city branding, many cities with strong images/reputations/identities do not have a formal city branding platform on which they promote themselves as tourism destinations or places in which to work, study and live, such as, for instance, London, Paris, Rome, Florence and Venice. Similarly, the case of Port Elizabeth. Some cities use an alternative approach to brand themselves on the back of significant events and mega attractions as in the case of the Barcelona Olympic Games. A few city brands pass the longevity test as the campaigns come and go with nothing memorable to remember. There is no formula for a successful city branding implementation, but in order to understand and apply city branding, one needs to be aware of its inherent limitations or constraints.

City branding constraints

Destination or city branding challenges have been well reported in literature by various authors (Heely 2011, Morrison 2019, Morgan *et al.* 2011, Page and Hall 2003, P.L. Pearce 2011) and for this chapter grouped into the following five main themes or constraints:

- The first constraint is the lack of influence and control over the sociocultural and political influences which shape that which city branding is ultimately seeking to transform, namely image, reputation and identity of a city.
- The second constraint is the weak relationship of city branding to other aspects of urban branding and the city economy and its governance generally. Brands and city advertising are subject to public discussion and criticism.
- The third constraint is considerable confusion about city branding among politicians, stakeholders, residents, businesses and media, which requires a team effort.
- The fourth constraint is inadequate or lack of sufficient funding to support city branding efforts, as city branding does not pay for itself.
- The fifth constraint is the difficulty in evaluating or measuring city branding projects, but it would be better to get some thoughts on how to achieve the best odds of successfully counting the things that count.

In most cases, limited information is available on correct tourism policies that local authorities use in the capital or cities (Maxim 2019, p. 1012). Tourism is a complex phenomenon, an experience good, that overlaps with other policy areas, and therefore strategies and plans which influence its development are rarely dedicated exclusively to the activity, in this case, city branding (Page and Hall 2003). P.L. Pearce (2011) enforced this view by arguing that in most cases, tourism is part of broader urban policies and does not have a separate strategy.

DMOs do not have total control over the destination product that is being branded. The multiplicity of attractions encompassed by any destination brand makes the process of branding destinations a challenge (Pike 2005). It is essential for city branding projects to be aware of these constraints and attempt to improve on how their respective cities are perceived and branded to the broader tourism audience.

City attractors or attributes

Attractions, attributes, resources of a destination/city are key elements or components of its appeal and play a vital role in the destination's success (Vinyals-Mirabent 2019). Tourists of today seek the consumption of holistic experiences far from the traditional sun-sea-and-sand experience (de San

Eugenio Vela 2011, Sack 1992). Various attributes appeal to visitors which could be used effectively in the city branding. According to the Travel and Tourism Competitiveness Index, *natural and cultural resources* are highlighted as "the principal reason to travel" to a destination (Schwab *et al.* 2017, p. 8).

Many studies reported and have identified the various components of tourism destination management and marketing. The interest of this chapter is to identify the components or attractors of a tourism destination which could make the city of PE unique. One should note that the service or motivation of tourism depends on the purpose of the visit to the destination (Hsu *et al.* 2010).

Research by Buhalis (2000) presented six types of *tourism destination component*. It is crucial for DMOs to manage the demand, supply and quality of the proposed six As (i.e. attractions, accessibility, amenities, available packages, activities and ancillary services) to maximise the competitiveness of the destination and the quality of services provided to tourists. Um and Chung (2019) added smart tourism technology and service level attributes to the six As framework by Buhalis (2000) to cater for the desired tourism experience of the new tourist. Among others, Kim (2014) identified ten key destination attributes – local culture, the varieties of activities, hospitality, infrastructure, environment management, accessibility, quality of service, physiography, place attachment and superstructure – that destination managers can use to manipulate memorable tourism experiences. The list of categories used in the case study is based on research by Vinyals-Mirabent (2019) which identified items from other research studies by Echtner and Ritchie (1993), Choi, Lehto and Morrison (2007) and Mazanec and Wöber (2010).

Destination managers need to work on and promote both the tangible attractors and the desired perception. These destination attractors are critical features making the brand unique and distinguishing it in consumers' minds towards the desired positioning (Echtner and Ritchie 1993, Qu *et al.* 2011, Stepchenkova and Morrison 2008). The multiplicity of attractions encompassed by any destination brand makes the process of branding destinations a challenge (Pike 2005).

The case of the city of Port Elizabeth

The Port Elizabeth case study illustrates how a coastal city with incredible natural resources and tourism assets has diversified its economic base and substantially reduced its dependency on the motor industry. The port of Port Elizabeth, with its proximity to heavily industrialised and intensively farmed areas, has facilities for the handling of all commodities – bulk, general and container cargo. The port imports large volumes of containerised components and raw materials as the centre of the country's motor vehicle manufacturing industry. Today employment in the metro is mainly in the automotive, auto component, food pharmaceutical, tourism, agriculture, textile and rubber industries. Port Elizabeth has been dubbed the "Detroit of South Africa" and is home to General Motors SA and Uitenhage is host to Volkswagen SA. The city's tourism potential was never developed until disinvestment in the 1980s, and the city was forced to consider all options for *diversifying its economic base* of which tourism was considered the best option (Nelson Mandela Bay Tourism Report (NMBT) 2019). In retrospect, tourism has revitalised the local economy of the city of Port Elizabeth, encouraging investment in tourism infrastructure.

The coastal city of Port Elizabeth is poised to take its place as an important tourism hub in South Africa. Nelson Mandela Bay, also referred to as Algoa Bay, is the only metropolitan city that was allowed by the former president to bear his name. By adopting his name, the metro strives to align itself with the "spirit of freedom" eminent in the life and philosophy of this Nobel Prize Winner. Nelson Mandela Bay Metropolitan City embraces the cities of Port Elizabeth, Uitenhage and Despatch. The estimated population of the metro comprises nearly 1.5 million people, making it South Africa's fifth-largest city in terms of population size and the second largest in terms of area (Statistics South Africa 2019).

In 1987, when Port Elizabeth was being written off as the "Ghost on the Coast", as a direct result of disinvestment during the apartheid era, the words of a derogatory song said, "The only thing that moved in Port Elizabeth was the wind and the sea" (NMBT 2019). However, wind and sea when combined make perfect weather conditions for sailing and Algoa Bay is considered by many to be, "*the water sports capital of South Africa*". All the coastal assets of Nelson Mandela Bay add up to a *world-class coastal tourism destination*. Table 23.1 introduces a list of main categories of attractors luring potential tourists to the city.

Based on the information collected from various SA Tourism, foreign and domestic tourism surveys and the experience of a local tourism practitioner in the area, the following two strategic priorities could contribute significantly to tourism growth in the Eastern Cape:

- The development of private game reserves together with the expansion of the Greater Addo Elephant National Park (GENP), the Baviaans Mega Reserve (BMR) including the proposed Greater Karoo Mega Reserve (GKMR), are likely to be the main drivers of eco-tourism growth in the Eastern Cape Province. These substantial natural assets should encourage more international tourists to visit the province. However, the provincial nature reserves should be positioned in the marketplace to attract domestic tourists many of whom cannot afford to stay in luxury game parks. Furthermore, domestic tourists can be attracted to visit provincial nature reserves during the shoulder periods, especially in winter, when local tourists like to take a short break not too far from home.
- The extension of the Port Elizabeth Airport runway and the upgrading of the East London and Mthatha regional airports would substantially improve airline access to the Eastern Cape Province. The purpose of upgrading should be to attract direct flights into the province from foreign destinations, including unscheduled flights (charters). Direct airline access to the Eastern Cape would facilitate the development of competitive package tours designed and priced to compensate for the increase in pre-paid expenses as a direct result of the global fuel crisis. The Eastern Cape should be able to offer the most competitive beach, bush and bird combination packages in South Africa.

Table 23.1 List of Port Elizabeth's categories of attributes or attractors

Categories of attractors	Short description	Examples in the city of PE
Accommodation	Lodging possibilities and their characteristics.	PE offers a variety of family-friendly accommodation.
Architecture, heritage and cultural attractions	Monuments, archaeological sites, buildings, historical events, shipwrecks, etc.	Amanzi Springs is a Palaeolithic site. Algoasaurus bauri is a dinosaur species uniquely discovered in Algoa Bay. The Nelson Mandela Bay Freedom Statue is situated in PE. The Prester John Memorial is located near the City Hall. Explorer Bartholomeus Dias arrived in Port Elizabeth (September 1488), looking for a trade route to India. Alexandria Coastal Dune field is the largest coastal dune field (mixed-use) in the Southern Hemisphere and was declared a World Heritage Site.

(Continued)

Table 23.1 Continued

Categories of attractors	Short description	Examples in the city of PE
Climate	General climate, temperature and weather.	Warm, dry summers and mild winter temperatures entice water sports enthusiasts throughout the year.
Events, festivals and fairs	Events which stand out from the general agenda.	PE is the bottlenose dolphin capital of the world (18 April 2016) celebrated with an annual Dolphin Festival.
		The city presents the annual South African Nippers Surf Lifesaving event.
		The World Boardsailing Championships take place at Hobie Beach.
		PE hosts sports events such as the 2010 Soccer World Cup, Rugby 7s World Championships and Ironman 70.3 World Championship.
		PE has an International Convention Centre.
Food and drink	Traditional gastronomy, typical food and beverages, restaurants, cellars, etc.	A variety of restaurants, catering for every need, are located in the Boardwalk and on the seashore.
Infrastructure and transportation	How to get to and move around the city: by train, plane, bus, boat, public transport, etc.	Various modes of transport are available in PE.
		The Port of Ngqura (2008). Third deep-water port in South Africa, with dedicated customs facilities and state-of-the-art container terminals and transport logistics.
		Port Elizabeth Airport to be expanded to international status.
		Coega Industrial Development Zone (IDZ) provides 11,500 hectares of land for development.
		The port of Port Elizabeth is the third most visited port in the country, after Cape Town and Durban.
Landscape and natural resources	Context about parks and gardens, landscape, green areas, riversides, views, etc.	The city is home to five biomes: thicket, grassland, Nama-Karoo, fynbos and forest.
		40 km of golden beaches along the Indian Ocean coastline; Pollock Beach is renowned among surfers for its "big" waves; Kings Beach is hailed as one of the country's safest beaches; Humewood Beach, Wells Estate and King's Beach have been awarded "blue flag" status.
		Algoa Bay is home to eight whale species, seven dolphin species, 25 shark species including the great white, five seawater ray species, penguins and a variety of sea birds.
		St Croix Island Marine Reserve is a breeding ground for the jackass penguin.

Categories of attractors	Short description	Examples in the city of PE
Leisure activities	Recreational activities and attractions such as zoos, swimming pools, casinos, theme parks, etc.	The Addo Elephant National Park is now one of the few "Big Seven" national parks in the world, offering game viewing of the "Big Five" land-based species as well as dolphin and whale watching. Situated near Schoenmakerskop, Sardinia Bay Nature Reserve provides opportunity for diving, surfing and scenic walks. Algoa Bay hosts one of the largest colonies of African Penguins in the world (in the region of 30,000 penguins). "Largest population estimate to date for bottlenose dolphins along the South African coast" (Reisinger and Karczmarski 2010). Swartkops Estuary is home to 6,000 bird species and known for its variety of water birds. Van Staden's Wildflower Reserve conserves a number of endemic, rare and threatened plants. Wind and sea when combined make perfect weather conditions for sailing, surfing, scuba diving, angling, snorkelling, kite-surfing and fly-fishing. Cruises, operated by Raggy Charters, are available for dolphin, whale and African penguin watching. A bucket list item of tourists worldwide is the experience to swim with dolphins. Various outdoor activities are available such as bird-watching, horse-riding, canoeing, quad biking and picnicking.
Local culture, history and lifestyle	Context about the history of the city, cultural legacy, and evolution over time. Character of the population, daily life, locals etc.	Perhaps it is the *coastal and marine tourism assets* of the area that could demonstrate to the world that humankind, wildlife, marine life, commerce and industry can all live in harmony in a unique coastal and marine tourism conservation estate unlike any other in the world. The area is known for first meetings of the Khoisan, British, Dutch, German and Xhosa. Today, Port Elizabeth is still a melting pot for *a myriad of races and nationalities – a true representative of the Rainbow Nation.* Diversity of cultures can be a powerful tool. Life in the township is very different to life in the suburbs. Experience life in the townships. Community pride for resident attending The Annual Dolphin Festival.

(*Continued*)

Table 23.1 Continued

Categories of attractors	Short description	Examples in the city of PE
Nightlife	Nightclubs, cocktail bars, night walks and other night activities.	Algoa Bay boasts a vibrant nightlife.
Political and economic factors	Information about the government, political stability of the country, money, currency, money exchange etc.	Give information on what a tourist needs to know about PE. Migration to coastal cities is taking place. The migration to the city has a positive impact on the economy of PE. The port of Port Elizabeth has the facilities to handle bulk, general and container cargo. The city is the centre of motor vehicle manufacturing, employment mainly in automotive, food, pharmaceutical, tourism, agriculture, textile and rubber industries. Coega Industrial Development Zone (IDZ) provides 11,500 hectares of land for development.
Safety/Service/ Shopping	Safety of the destination. Information about the service staff. Shopping areas, attractive shops, souvenir shopping etc.	Visitors are cautioned to be streetwise when travelling in the city.
Sports	Physical activities, outdoor sports activities, gyms and pavilions, popular sports activities, etc.	PE offers a variety of sports activities such as golf, rugby, cricket, road running, biathlon, cycling and water sport.
Tourism products and packages	Activities and packages prepared by the tourism office and other stakeholders: sightseeing, bus tours etc.	Examples include Raggy Charters, Nelson Mandela Bay Tourism, Adrenalin Addo Zipline, Heavenly Stables (horse riding), Bayworld Oceanarium, Sundays River ferry, Holmeleigh Farmyard, Thunzi Bush Lodge, Kragga Kamma Game Park, Bay West City Ice Rink, Sundays River sand boarding and the Seaview Game Predator Park. Safaris to private game reserves and the Addo Elephant National Park. Several advertisements and films are shot in PE promoting film tourism.
Wellness	Spas, wellness centres, ways to get away from the city stress, thermal facilities, etc.	A variety of spas and wellness centres offers diverse services. Swimming with bottlenose dolphins appeals to a worldwide audience. The interaction is known to release stress and have a positive impact on health and well-being.

SWOT analysis of Port Elizabeth city tourism

Table 23.2 describes some strengths, weaknesses, opportunities and threats (SWOT) of Port Elizabeth city tourism. The information presented in the SWOT analysis of Port Elizabeth city tourism is elements to consider by different stakeholders in improving the city brand. Port

Table 23.2 SWOT analysis of Port Elizabeth city tourism

Strengths	Weaknesses
• PE is uniquely located, offering coastal and wildlife experiences to travellers. • The city boasts the largest bottlenose dolphin population in the world. • The name Nelson Mandela is world renowned, and the city is associated with it and carries the name Nelson Mandela Metropole. • PE is known as the friendly city. • PE has favourable weather to accommodate tourists and visitors through all four seasons. • The city has sound infrastructure, an airport, hotels, restaurants, shopping malls, a conference centre, world-class sports stadiums for soccer, rugby and cricket. • The port of Port Elizabeth is the third most visited port in the country, after Cape Town and Durban.	• Inability to market PE by using digital distribution channels and social media. • There is a lack of a dedicated and well-resourced city branding association. • Scientific research marketing programmes to understand new trends and consumers' expectations regarding coastal and city tourism are deficient. • Cape Town and Johannesburg offer stiff competition as preferred destinations for foreign visitors. • The city is perceived to be a gateway destination and not the main destination.
Opportunities	Threats
• Ongoing development opportunities of PE as a destination for business, sporting events, festivals and ecotourism exist. • Culture and heritage tourism may express the social and cultural values of city residents. • Develop a sustainable township tourism value chain that includes small to medium businesses to produce high-quality service and deliver positive local tourism experience. • The city of Port Elizabeth can do more to keep their visitors in the city for more extended periods as this delivers sustained benefit to product owners and the city's economy at large. • The city needs to host tourism-generating events to reduce the seasonal gap in winter. • The opportunity exists to create a gastronomy destination. • Development of township tourism as a conduit for giving back or contributing to society. • The substantial natural assets (5 biomes) should encourage more international tourists to visit the province. • Community buy-in and involvement of the residents to act as a frontline ambassador of their city.	• Security concerns in some areas of the city and country are a significant consideration for international visitors. • Government policies do not support tourism to its fullest potential. • Entrepreneurs experience difficulty in developing and starting new business ventures in travel and tourism. • PE is a long-haul destination for international travellers and does not have an international airport.

Elizabeth tourism has developed over the last years regarding the change in focus of the economy, the community, marketing organisations, universities, restaurants, the marine industry, entrepreneurs, financial institutions and others that use marine and maritime as an instrument to position PE as a tourism hub in South Africa as a unique destination to visit for ecotourism, water sports, adventure and wildlife.

Concluding remarks

Regardless of limited research on city tourism and city branding reported in the African context, there is agreement on the simplified marketing message to be sent out as a means to market and brand a country, region or city. Based on the research done, what could be stronger than *We are Nelson Mandela Bay* as a slogan for the city – giving recognition and honouring Mandela in the metropolis. Port Elizabeth should capitalise on this brand name with which it is associated. The literature review provided a theoretical base and defined key concepts such as coastal and marine tourism in the context on which to brand the city. Branding a city is challenging, and cities should be cautious of the pitfalls in the industry. City attractors or attributes such as the only city in the world that borders on a "Big Seven" wildlife mega reserve or *containing elements of five biomes* – vegetation types in South Africa or the bottlenose dolphin capital of the world illustrate the uniqueness of the city of Port Elizabeth.

This chapter suggests that the concept of a smart city should be explored, and further research is called upon due to innovation and new communication technologies. As an effective destination marketing tool, future research should exert more efforts towards smart cities tourism websites, for Port Elizabeth and other cities in South Africa. This chapter can be used to assist destination or city managers seeking direction in city branding.

References

Amsterdam City Council, 2004. Amsterdam City. Amsterdam: Amsterdam City Council, p. 10.

Anholt, S., 2007. *Competitive identity: the new brand management for nations, cities and regions*. Basingstoke: Palgrave Macmillan.

Ashworth, G.J., 1989. Urban tourism: an imbalance in attention. *In*: C.P. Cooper, ed. *Progress in tourism, recreation and hospitality research*. London: Belhaven, 33–54.

Ashworth, G.J., 2010. Book review: World tourism cities: developing tourism off the beaten track. *Tourism Management*, 31 (5), 696–697.

Ashworth, G.J. and Page, S.J., 2011. Urban tourism research: recent progress and current paradoxes. *Tourism Management*, 32 (1), 1–15.

Braun, E., 2008. *City marketing: towards an integrated approach*. Thesis (PhD). Rotterdam: Erasmus University.

Buhalis, D., 2000. Marketing the competitive destination of the future. *Tourism Management*, 21 (1), 97–116.

Buonincontri, P. and Micera, R., 2016. The experience co-creation in smart tourism destinations: a multiple case analysis of European destinations. *Information Technology & Tourism*, 16 (3), 285–315.

Choi, S., Lehto, X.Y., and Morrison, A.M., 2007. Destination image representation on the web: content analysis of Macau travel related websites. *Tourism Management*, 28 (1), 118–129.

Chung, N. and Koo, C., 2015. The use of social media in travel information search. *Telematics and Informatics*, 32 (2), 215–229.

Correia, A. and Kozak, M., 2016. Tourists' shopping experiences at street markets: cross–country research. *Tourism Management*, 56 (1), 85–95.

Crouch, G.I. and Ritchie, J.B., 2005. Application of the analytic hierarchy process to tourism choice and decision making: a review and illustration applied to destination competitiveness. *Tourism Analysis*, 10 (1), 17–25.

de San Eugenio Vela, J., 2011. *Teoria i mètodes per a marques de territori*. Barcelona: Editorial UOC.

Diakomihalis, M., 2007. Greek maritime tourism: evolution, structures and prospects. *In*: A.A. Pallis, ed. *Maritime transport: the Greek paradigm, research in transportation economics*. London: Elsevier, 419–455. http://dx.doi.org/10.1016/S0739-8859(07)21013-3.

Echtner, C.M. and Ritchie, J.R.B., 1993. The measurement of destination image: an empirical assessment. *Journal of Travel Research*, 31 (4), 3–13.

Edwards, D., Griffin, T., and Hayllar, B., 2008. Urban tourism research: developing an agenda. *Annals of Tourism Research*, 35 (4), 1032–1052.

European Commission, 2014. *A European strategy for more growth and jobs in coastal and marine tourism*. Brussels: European Commission. EC – Com 86 (final).

Gretzel, U., Sigala, M., Xiang, Z, and Koo, C., 2015. Smart tourism: foundations and developments. *Electronic Markets*, 25 (3), 179–188.

Grün, B. and Dolnicar, S., 2016. Response style corrected market segmentation for ordinal data. *Marketing Letters*, 27 (4), 729–741.

Hall, M., 2001. Trends in ocean and coastal tourism: the end of the last frontier? *Journal of Ocean and Coastal Management*, 44 (9–10), 601–618. http://dx.doi.org/10.1016/S0964-5691(01)00071-0.

Heely, J., 2011. *Inside city tourism: a European perspective*. Bristol: Channel View Publications.

Hsu, C.H., Cai, L.A., and Li, M., 2010. Expectation, motivation, and attitude: a tourist behavioral model. *Journal of Travel Research*, 49 (3), 282–296.

Jovicic, D.Z., 2016. Key issues in the conceptualization of tourism destinations. *Tourism Geographies*, 18 (4), 445–457.

Keller, K.L., 1993. Conceptualizing, measuring, and managing customer-based brand equity. *Journal of Marketing*, 57 (1), 1–22.

Kim, J.H., 2014. The antecedents of memorable tourism experiences: the development of a scale to measure the destination attributes associated with memorable experiences. *Tourism Management*, 44 (5), 34–45.

Law, C.M., 2002. *Urban tourism: the visitor economy and the growth of large cities*. 2nd ed. London: Continuum.

Maitland, R., 2009. Introduction: national capitals and city tourism. *In*: R. Maitland and & B.W. Ritchie, eds. *City tourism: national capital perspectives*. Wallingford: CABI, 1–13.

Maitland, R., 2012. Global change and tourism in national capitals. *Current Issues in Tourism*, 15 (1–2).

Maitland, R. and Newman, P., eds. 2009. *World tourism cities: developing tourism off the beaten track*. New York: Routledge.

Maxim, C., 2016. Sustainable tourism implementation in urban areas: a case study of London. *Journal of Sustainable Tourism*, 24 (7), 971–989.

Maxim, C., 2019. Challenges faced by world tourism cities: London's perspective. *Current Issues in Tourism*, 22 (9), 1006–1024.

Mazanec, J.A. and Wöber, K.W., 2010. *Analysing international city tourism*. 2nd ed. Vienna: Springer.

Miller, M.L. and Auyong, J., 1991. Coastal zone tourism: a potent force affecting environment and society. *Marine Policy*, 15 (2), 75–99. http://dx.doi.org/10.1016/0308-597X(91)90008-Y.

Morgan, N., Pritchard, A., and Pride, R., 2011. *Destination brands: managing place reputation*. 3rd ed. Amsterdam: Elsevier.

Morrison, A.M., 2019. *Marketing and managing tourism destinations*. 2nd ed. London: Routledge.

Nelson Mandela Bay Tourism Report (NMBT), 2019. Official tourism information for Port Elizabeth. Available from: www.nmt.co.za [Accessed 21 October 2019].

Olins, W., 2008. *The brand handbook*. London: Thames and Hudson.

Orams, M., 1998. *Marine tourism: development, impacts and management*. London: Routledge.

Page, S.J. and Connell, J., 2006. *Tourism: a modern synthesis*. 3rd ed. Andover, MA: Cengage Learning EMEA.

Page, S.J. and Hall, C.M., 2003. *Managing urban tourism*. Harlow: Pearson Education.

Papageorgiou, M., 2016. Coastal and marine tourism: a challenging factor in Marine Spatial Planning. *Ocean & Coastal Management*, 129, 44–48.

Pearce, D., 2001. An integrative framework for urban tourism research. *Annals of Tourism Research*, 28 (4), 926–946.

Pearce, D., 2011. Tourism, trams and local government policy making in Christchurch: a longitudinal perspective. *In*: D. Dredge and J. Jenkins, eds. *Stories of practice: tourism policy and planning*. Farnham: Ashgate, 57–78.

Pearce, P.L., 2011. *Tourist behaviour and the contemporary world*. Bristol: Channel View Publications.

Pearce, P.L. and Zare, S., 2017. The orchestra model as the basis for teaching tourism experience design. *Journal of Hospitality and Tourism Management*, 30, 55–64.

Pike, S., 2005. Tourism destination branding complexity. *The Journal of Product and Brand Management*, 14 (4), 258–259.

Prayag, G. and Ryan, C., 2012. Antecedents of tourists' loyalty to Mauritius: the role and influence of destination image, place attachment, personal involvement, and satisfaction. *Journal of Travel Research*, 51 (3), 342–356.

Prebensen, N.K., Woo, E., and Uysal, M.S., 2014. Experience value: antecedents and consequences. *Current Issues in Tourism*, 17 (10), 910–928.

Qu, H., Kim, L.H., and Im, H.H., 2011. A model of destination branding: integrating the concepts of the branding and destination image. *Tourism Management*, 32 (3), 465–476.

Reisinger, R.R. and Karczmarski, L., 2010. Population size estimate of Indo-Pacific bottlenose dolphins in the Algoa Bay region, South Africa. *Marine Mammal Science*, 26 (1), 1748–7692.

Sack, R.D., 1992. *Place, modernity, and the consumer's world: a relational framework for geographical analysis.* Baltimore, MD: Johns Hopkins University Press.

Schwab, K., *et al.*, 2017. *Travel & tourism competitiveness report 2017.* Geneva: World Economic Forum.

Statistics South Africa, 2019. *Population, Port Elizabeth (Eastern Cape, South Africa).* Pretoria: Stats SA.

Stepchenkova, S. and Morrison, A.M., 2008. Russia's destination image among American pleasure travelers: revisiting Echtner and Ritchie. *Tourism Management*, 29 (3), 548–560.

Techopedia, 2019. *What is a Smart City?* Available from: www.technopedia.com [Accessed 21 October 2019].

Um, T. and Chung, N., 2019. Does smart tourism technology matter? Lessons from three smart tourism cities in South Korea. *Asia Pacific Journal of Tourism Research*. https://doi.org/10.1080/10941665.2019.1595691 [Accessed 21 October 2019].

Vinyals-Mirabent, S., 2019. European urban destinations' attractors at the frontier between competitiveness and a unique destination image: a benchmark study of communication practices. *Journal of Destination Marketing & Management*, 12, 37–45.

Further reading

Morrison, A.M., 2019. *Marketing and managing tourism destinations.* 2nd ed. London: Routledge.

24

SMART URBAN TOURISM DESTINATIONS AT A CROSSROADS

Being "smart" and urban are no longer enough

J. Andres Coca-Stefaniak and Gildo Seisdedos

Introduction

Concepts such as *smart* or *smartness* have evolved over time from rather narrow technological interpretations in the form of mobile devices to more nuanced applications involving geographical locations (e.g. smart cities, smart tourism destinations). As a result of this, *smart places* have arisen partly as a result of the widening impact of new and disruptive technologies on the spaces we live in, including cities, regions and countries (Hedlund, 2012; Zygiaris, 2013; Vanolo, 2014). Urban tourism destinations are not immune to these global trends, particularly as regards their strategic positioning (Buhalis and Amaranggana, 2014) to compete for a larger and/or higher value share of the tourism market, regardless of whether their priority is leisure or business. In line with this, the use of information and communication technologies (ICTs) has developed substantially over the last two decades to deliver new experiences for tourists and visitors, while supporting wider automatisation processes (Gretzel, 2011), which remain a common challenge for urban managers and tourism destination managers alike (Hughes and Moscardo, 2019). Key channels for ICTs today include social networks, big data analysis, artificial intelligence, the internet of things (Vicini *et al.*, 2012), sensor equipment and other monitoring and data processing systems (Haubensak, 2011).

This chapter will review some of the key parallels between the concepts of smart cities and smart tourism destinations. This review will also cast a critical perspective on the *smart* concept, which has been traditionally dominated by technology-based approaches, even if a new generation of smart initiatives is beginning to emerge with a more human–centred focus. Evidence of this new trend as well as the widening of the smart tourism destination concept to neighbouring regions of established smart tourism cities will be discussed with reference to examples from practice in Europe and China. In line with these developments and given the knowledge gap that appears to exist in scholarly research concerning this urban–regional interphase with regard to smart cities and smart tourism destinations, a new typology is proposed for smart tourism destinations that encompasses tourism cities as well as their wider region.

To conclude, this chapter argues that smart tourism destinations are at a strategic crossroads in their development, which needs to move beyond traditionally favoured technology-focused initiatives towards a new generation of smart tourism destinations that balance often conflicting global–local trends. These include, among others, overtourism, climate change, terrorism, gentrification, growing demands from local residents for more liveable cities, declining city centre shopping due to the digital retail revolution, and the search for authentic transformational experiences by new generations of tourists and visitors. It is argued that visionary tourism cities will adopt a new strategic positioning that revolves around urban sustainability as a holistic paradigm – urban living labs being a good example of this – which will lead to a new generation of smart tourism destinations – the sustainable smart tourism destination. A conceptual framework is offered for further research and practice in this field. This framework combines elements from existing sustainability and tourism frameworks by adopting a systems-based approach to the management of urban tourism destinations, with elements of smart innovation used as catalysts for tackling a wide range of factors affecting the sustainability of tourism destinations in a sphere termed the "acceptable change domain", which captures the global–local tensions alluded to earlier in the context of tourism destinations at different stages of their life cycle.

The *smart* revolution

Smart cities and smart tourism destinations cannot be viewed in isolation from the wider concept of "smartness" and the – often digital – "smart" revolution affecting everyday aspects in most industrialised countries – from urban infrastructure management and the ways in which people have come to interact with others to the security of financial transactions and key governance aspects in political processes such as general elections. So, is there such a thing as "smartness"? In many ways, smartness, like many related concepts, remains an elusive notion today as it is often domain dependent, referring to anything from smart TV sets and smart cars to smart systems and devices (Alter, 2019), urban energy management (Battarra *et al.*, 2016), the environmental sustainability of cities (Balducci and Ferrara, 2018) or cross-agency information-sharing for better decision-making (Gil-Garcia *et al.*, 2019), or urban governance (Gil-Garcia *et al.*, 2016), among others. Regardless, in essence, the concept has often been used to refer to the impact that ICTs have had on society and the economy (see, for instance, Dameri, 2017), with ICTs often used as an umbrella term to denote a wide array of technologies and advances in communication and connectivity (see, for instance, Rutherford, 2011; or, for a recent tourism-focused review on this topic, see Ivars-Baidal *et al.*, 2019). The speed of innovation in this field, often referred to as "disruptive technologies" – a term first coined by Bower and Christensen (1995) to denote technologies able to displace current incumbents due to their high level of innovation – has led some thinkers to claim that humanity is, in effect, facing a fourth industrial revolution (Schwab, 2017) exemplified by major advances in robotics, artificial intelligence, nanotechnology, quantum computing, biotechnology, the internet of things (IoT), the industrial internet of things (IIoT), decentralised consensus, fifth-generation wireless technologies (5G), 3D printing and fully autonomous vehicles (World Economic Forum, 2016), to mention but a few examples, and their huge impacts on the challenge of educating future generations (Peters, 2017).

This fourth industrial revolution and particularly the ICTs acting as facilitators and catalysts for change, carries major implications for urban management and liveability in towns and cities around the globe. These range from enhanced digital monitoring using sensors and external data sources, to improved control systems with embedded software, real-time optimisation of processes (e.g. crowd flow and management) using advanced algorithms, and even autonomous self-diagnosing systems able to combine tracking, monitoring and optimisation (Porter and Heppelmann, 2014). However, ultimately, perhaps one of the most widespread albeit contested manifestations (Greenfield, 2013) of this fourth industrial revolution in the context of urban environments is the emergence of the concept of the smart city. This concept was first coined in the United States by IBM and CISCO several decades ago. Since then, smart cities have consolidated largely as a form of visioning for improving local economies, enhancing mobility, delivering environmental sustainability, improving quality of life in cities and enabling better governance (e.g. Abella *et al.*, 2017; Angelidou, 2014; Caragliu *et al.*, 2011; Vanolo, 2014; Picon, 2015; Hajer and Dassen, 2014; Monitor Deloitte, 2015) and even living test beds for urban innovation (Sassen, 2011; Zygiaris, 2013) and engagement with visitors and residents (Molinillo *et al.*, 2019) even if the use of place branding and marketing techniques by smart cities and smart tourism destinations remain a major challenge (Coca-Stefaniak, 2019). In spite of this seemingly endless list of benefits smart cities have attracted criticism from scholars on historical and philosophical grounds as constructs serving primarily a financial elite (Cugurullo, 2018) through a form of market triumphalism (Gibbs *et al.*, 2013) that promotes a standardising approach to the design of urban futures (Sadowski and Bendor, 2019) with arguably opaque approaches to urban planning and development (Kitchin, 2015; Kummitha and Crutzen, 2017). Other scholars (e.g. Hollands, 2008) have gone even further by denouncing the self-congratulatory labelling of smart cities in what amounts to little else than a revamped version of a preceding concept – the entrepreneurial city. All in all, smart cities and their strategic focus continue to evolve subject to all these forces and have even developed offshoots, such as smart tourism destinations. This is explored in more detail next.

From smart cities to smart tourism: exploring parallels

The development of smart city research in what remains a nascent – though rapidly growing – field of knowledge in academia has spanned now nearly three decades. Although there exist a number of systematic reviews of the literature on smart cities (Ramaprasad *et al.*, 2017; Ruhlandt, 2018; Lytras and Visvizi, 2018; Ismagilova *et al.*, 2019), agreement on a single definition of the concept remains as elusive as the broadness of its remit. Cocchia (2014), for instance, carried out a review of the literature on smart cities and digital cities spanning 19 years from 1993 to 2012 and concluded that the smart city concept was associated in the literature with interpretations as diverse as wired city, virtual city, ubiquitous city, intelligent city, information city, digital city, smart community, knowledge city, learning city, sustainable and green city, among others. Crucially, this study also found an exponential growth in academic publications on smart cities between 1993 and 2012 with the most cited definitions of smart city during this period outlined in Table 24.1.

Indeed, there is evidence to suggest that contemporary interpretations of the smart city concept are increasingly evolving beyond initial – somewhat simplistic – technology-centred and rather homogenising approaches (Alizadeh, 2017) towards a focus on improving the quality of life of residents and communities (e.g. Albino *et al.*, 2015), while building on their specific idiosyncrasies to enhance their competitiveness. For instance, this situation becomes apparent in the context of new smart cities built entirely following smart principles of urbanisation and urban management. The experience of Masdar, Songdo IBD and Skolkovo suggest specific patterns of place-making along the lines of *smart urban labs* (Sengers *et al.*, 2018) with a focus on attracting only highly skilled and talented residents through a wide range of taxation facilities and subsidies facilitating their relocation (Kolotouchkina and Seisdedos, 2018) in a manner that echoes the creative class arguments of Richard Florida (2006) and other scholars (see Thite, 2011), even if the marketing and branding of smart tourism destinations on their own merits of *smartness* remain in their infancy (Molinillo *et al.*, 2019) and a rich vein for further research (Coca-Stefaniak, 2019).

Table 24.1 Definitions of smart city

Focus	Definition	Source
Governance	"A city to be smart when investments in human and social capital and traditional (transport) and modern (ICT) communication infrastructure fuel sustainable economic growth and a high quality of life, with a wise management of natural resources, through participatory governance."	Caragliu *et al.* (2011)
Technology	"Smart city is defined by IBM as the use of information and communication technology to sense, analyze and integrate the key information of core systems in running cities."	IBM (2010)
	"Smart City is the product of Digital City combined with the Internet of Things."	Su *et al.* (2011)
Environmental	"Smart City is a city in which it can combine technologies as diverse as water recycling, advanced energy grids and mobile communications in order to reduce environmental impact and to offer its citizens better lives."	Setis-Eu (2012)
Human capital	"Smart community – a community which makes a conscious decision to aggressively deploy technology as a catalyst to solving its social and business needs – will undoubtedly focus on building its high-speed broadband infrastructures, but the real opportunity is in rebuilding and renewing a sense of place, and in the process a sense of civic pride … Smart communities are not, at their core, exercises in the deployment and use of technology, but in the promotion of economic development, job growth, and an increased quality of life. In other words, technological propagation of smart communities isn't an end in itself, but only a means to reinventing cities for a new economy and society with clear and compelling community benefit."	Eger (2009)
Innovation and learning	"(Smart) cities as territories with high capacity for learning and innovation, which is built-in the creativity of their population, their institutions of knowledge creation, and their digital infrastructure for communication and knowledge management."	Komninos (2011)
Multidisciplinary	"Smart city is a high-tech intensive and advanced city that connects people, information and city elements using new technologies in order to create a sustainable, greener city, competitive and innovative commerce, and an increased life quality."	Bakıcı *et al.* (2012)
	"A smart city is understood as a certain intellectual ability that addresses several innovative socio-technical and socio-economic aspects of growth. These aspects lead to smart city conceptions as 'green' referring to urban infrastructure for environment protection and reduction of CO_2 emission, 'interconnected' related to revolution of broadband economy, 'intelligent' declaring the capacity to produce added value information from the processing of city's real-time data from sensors and activators, whereas the terms 'innovating', 'knowledge' cities interchangeably refer to the city's ability to raise innovation based on knowledgeable and creative human capital."	Zygiaris (2013)

Although a number of different conceptual frameworks exist to illustrate the smart city concept and synthesise its many definitions, Cohen's (2013) smart city wheel remains arguably an early attempt at acknowledging the holistic and interdisciplinary nature of this concept. This framework identifies six aspects of smartness in cities, namely smart governance (including issues of transparency of data and decision-making); smart environment (mainly related to energy use and the sustainable management of resources); smart mobility (positing a mixed-model approach to the use of transport, combining mass public transport with ICTs and the rental of e-bikes, for instance); smart economy (largely related to the implementation of ICTs in economic strategies); smart people (e.g. human capital); smart living (quality of life in terms of health, safety, cultural vibrancy and happiness) (Lim *et al.*, 2018). This framework has been largely adopted and adapted by the European Union, which classifies the new services offered by smart cities into categories such as smart environment, smart mobility, smart living, smart people, smart economy and smart governance (Manville *et al.*, 2014). Other smart city frameworks developed since appear to revolve around the same concepts, albeit with specific nuances in each case (for a review, see Govada *et al.*, 2017), even if, more recently, some scholars (Ahvenniemi *et al.*, 2017) have started to advocate the use of the term "smart sustainable cities" so as to combine the generalised socio-economic sustainability focus of smart city frameworks with the more environmentally skewed focus of sustainable city frameworks.

Against this backdrop of the far more established, if perhaps still somewhat fuzzy, concept of smart cities (or even smart sustainable cities), a parallel concept has started to emerge in tourism – the smart tourism destination. Professor Dimitrios Buhalis is arguably the forefather of this concept and first acknowledged its roots in the field of smart cities (Buhalis, 2000). Since then, smart tourism destinations have been interpreted in terms of their focus on the use of ICTs to enhance tourism processes (Wang *et al.*, 2013), using technology to address tourists' personal needs (Huang *et al.*, 2012) and, more recently, enhancing their experiences (Guo *et al.*, 2014; Zhu *et al.*, 2014; Buhalis and Amaranggana, 2015). It is this latter point, the emphasis of smart tourism destinations on the provision of experiences for visitors while attaining quality of life for residents that differentiates them from being merely smart cities, as illustrated in Table 24.2. This point was later succinctly argued by Boes *et al.* (2016), who also pioneered the first conceptual framework specific to smart tourism destinations using an ecosystem approach, even if part of the framework, namely its *smart innovations* element, is distinctively anchored in much earlier work on smart cities by Giffinger *et al.* (2007), which posited the relevance of factors such as smart mobility, smart government, smart people, smart economy, smart living and smart environment in this context. Inevitably, Cohen's (2013) smart city wheel framework, discussed earlier, also bears many resemblances to the work by Giffinger *et al.* (2007).

All in all, while further studies continue to explore the parallels between smart cities and smart tourism destinations (see, for instance, Jasrotia and Gangotia, 2018), a general consensus appears to be emerging among scholars on the importance for both concepts to be more human-centred (Giovannella and Rehm, 2015; Lara *et al.*, 2016; Johnson and Samakovlis, 2019) and even consider contested approaches such as degrowth (March, 2018) in order to achieve more sustainable futures.

The next section explores examples from practice in the management of smart tourism destinations and draws relevant parallels to the above discussion.

Table 24.2 Smart tourism destination definitions

Definitions of smart tourism destinations	Source
"Places utilizing the available technological tools and techniques to enable demand and supply to co-create value, pleasure, and experiences for the tourist and wealth, profit, and benefits for the organizations and the destination."	Boes *et al.* (2015)
"Bringing smartness into tourism destinations meaning that destinations need to interconnect multiple stakeholders through a dynamic platform mediated by ICT in order to support prompt information exchange regarding tourism activities through machine-to-machine learning algorithm which could enhance their decision-making process."	Buhalis and Amaranggana (2014)
"A tourism destination is said to be smart when it makes intensive use of the technological infrastructure provided by the smart city in order to: (1) enhance the tourism experience of visitors by personalizing and making them aware of both local and tourism services and products available to them at the destination and (2) by empowering destination management organizations, local institutions and tourism companies to make their decisions and take actions based upon the data produced within the destination, gathered, managed and processed by means of the technology infrastructure."	Lamsfus *et al.* (2015)
"An innovative tourist destination, built on an infrastructure of state-of-the-art technology guaranteeing the sustainable development of tourist areas, accessible to everyone, which facilitates the visitor's interaction with and integration into his or her surroundings, increases the quality of the experience at the destination, and improves residents' quality of life."	Lopez de Avila (2015)
"A tourism system that takes advantage of smart technology in creating, managing and delivering intelligent touristic services/experiences and is characterised by intensive information sharing and value co-creation."	Gretzel *et al.* (2015)

Source: Adapted from Koo *et al.* (2016).

Implementing the smart tourism destination concept – examples from practice

One of the most recent international initiatives to recognise the achievements of tourism cities in the sphere of smart tourism is the recently launched (July 2019) European Union's European Capital of Smart Tourism initiative (EU, 2019). Contrary to the widely used smart cities wheel framework (Cohen, 2013), this award programme identifies four areas of excellence specific to smart tourism destinations: accessibility, digitalisation, sustainability, cultural heritage and creativity. Somewhat refreshingly, and in line with issues discussed in the previous section, this framework places a higher emphasis on the contribution of smart tourism destinations to sustainability, culture and creativity, possibly inspired partly by Hawkes' (2001) four pillar sustainability framework.

Accessibility is interpreted through a wide spectrum of issues ranging from physical accessibility for visitors with disabilities, to digital accessibility, city cards or signage. Digitalisation acknowledges initiatives that facilitate the dissemination of information to specific target groups, collecting information for smarter management of tourism cities and granting physical and psychological accessibility through innovation. Sustainability, on the other hand, is grouped into three groups of best practice categories. The first one focuses on how tourism cities combat climate change or adapt to it; the second one revolves around the preservation and enhancement of the natural environment; while the third one focuses on initiatives tackling the seasonality of tourism and encouraging the spread of tourist flows away from major urban tourism cities and to surrounding areas within the region to alleviate pressure on resources and local communities in tourism cities. Finally, the smart management of cultural heritage and creativity in smart tourism destinations are encouraged by rewarding practices that revive traditions and cultural heritage sustainably, building capacity and reach through community infrastructures and using cultural heritage for new creative initiatives that support the wider strategy of smart tourism cities. Table 24.3 outlines details of the 2019 winners for the European Capital of Smart Tourism initiative.

Table 24.3 European Capital of Smart Tourism 2019 winners

City	Summary of best practices
Helsinki (Finland)	The city's smart public transport system has enjoyed a rise in user satisfaction over the last two years. Additionally, an "Uber boat" system is currently being considered with driverless buses being trialled on open streets. Helsinki has been ranked second at the Accessible City Awards in 2015. Also, multilingual "Helsinki Helpers" are stationed at main attractions to offer assistance to visitors. Helsinki has plans in place (including143 specific measures) to become carbon neutral by 2035. The Helsinki Road Map prevents overcrowding and supports local business as it guides tourists around the city, while 75% of hotel rooms are certified environmentally friendly. Helsinki is also increasing the share of cycling, walking and electric cars and trains.
	Powered by its open approach to public data – available free for all since 2009 – Helsinki has become a hotbed of software innovation, including the ad-free MyHelsinki.fi website, featuring recommendations from local residents. Helsinki's traditional saunas feature a wide array of environmentally friendly options using sustainable wood and powered by water, solar heating and wind.
Lyon (France)	Lyon has won several accolades for accessibility, including the 2017 Access City Award. Visitors with disabilities and reduced mobility can move around the city with complete autonomy, taking advantage of a completely adapted transport network and smart signage. Lyon's museums offer adapted tours – those with hearing impairments are allowed to touch works of art – and many restaurants provide speaking menus. In 2019, 40,000 visitors to the city experienced the benefits of the Lyon City Card, which provides users with various discounts, free public transport and entrance to 23 museums and other attractions. In the future, visitors will be able to take advantage of the ONLYLYON Experience, receiving live geo-located tourist information direct to their smartphones to reduce congestion.
	Lyon-Saint-Exupery is one of 25 airports in just nine countries to be classed as carbon neutral, and sustainable development is one of the city's main priorities. An example of this is the "Lyon, Ville Equitable et Durable" label which identifies companies, shops, producers and events encouraging responsible consumption. Artists taking part in the Festival of Lights, meanwhile, are rewarded for taking a sustainable approach to their installations.

Source: EU (2019).

Other cities also received awards in European Smart Tourism 2019. These included Ljubljana (Slovenia) in the category of sustainability; Malaga (Spain) in the category of accessibility; Copenhagen (Denmark) in the category of digitalisation; Linz (Austria) in the category of cultural heritage and creativity.

There is, of course, a wide range of examples of smart tourism destinations beyond Europe. Due to the diversity of their strategic priorities, it would seem appropriate to create a typology of smart tourism destinations, as shown in Figure 24.1. The matrix-type typology suggested here is defined by two axes: urban character and relevance of tourists as stakeholders. Every quadrant of this matrix represents a different underlying concept of smart destination, with different policy development requirements. The urban character axis is governed by population density to express how intensely urban the destination is, which has important implications for resource use. This typology also encompasses a growing phenomenon of integrated regional approaches to tourism, which are often centred on one or more urban tourism destinations, which tend to act as hubs for the region with tourists often visiting other areas either as day-trippers or with longer-term stays. This non-urban (hybrid) category in our typology captures not only these wider – often inter-urban – regions but also tourism constructs (e.g. costa, riviera) and smaller attractions of a non-urban nature, such as natural parks and islands. Similarly, the typology offered here captures a much-neglected aspect of smart tourism destinations and smart cities: the rural sphere. Indeed, smart cities as a concept would appear to represent almost an oxymoron to rural locations. Yet, it is these more isolated – and low population density – places that are increasingly most in need of innovative solutions (see, for instance, Bock (2016) for applications of social innovation to rural areas), which instead tend to be restricted almost exclusively to their more cosmopolitan neighbours. This rural–urban paradox became quite apparent when the European Union adopted the notion "smart" in its new ten-year growth strategy, Europe 2020, stating that Europe should become a smart economy, though with little guidance – or even strategic vision – with regard to how this smart economy concept should apply to the same rural regions (Naldi *et al.*, 2015) that are now suffering from a phenomenon that could be described somewhat naïvely as "brain-drain" (Carr and Kefalas, 2009) or, perhaps more realistically, as "depopulation" (Viñas, 2019) due to the lure of better jobs and standard of

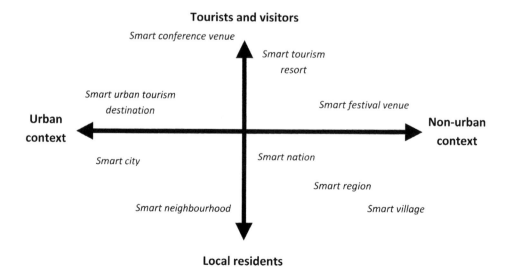

Figure 24.1 Urban and non–urban smart tourism destination typology.

living offered by cities, even when, from a purely tourism-based perspective, it is these rural locations that remain the main (resigned?) curators of the very authenticity that new generations of tourists crave (Sims, 2009; Jyotsna and Maurya, 2019).

In the lower part of this matrix, the specificities of tourists as stakeholders – including their priorities and behaviour – can be easily incorporated under the smart city umbrella. So, when looking for global best practices on smart tourism destinations, the upper part of this matrix is the right place to look for cutting-edge policies and developments.

Macau (China) is a good illustrative example for this typology as a global tourism city with 31 million visitors annually, which effectively amounts to 48 times the city's population. Given Macau's rather restricted geographical land area (30.3 km²), these large visitor numbers place a major strain on the city's resources and the environment with a detrimental impact on the well-being of its local residents. Macau is also the world's largest gambling destination (Shenga and Gub, 2018), with this sector delivering a gross revenue approximately seven times larger than that of Las Vegas in the United States and representing over 45% of the city's GDP. In order to deal with these major challenges to the city's resources, the Macao Government Tourist Office (MGTO) launched in March 2019 three "smart tourism" projects where cloud computing plays a vital role in delivering better services for visitors and residents alike as well as supporting the tourism sector. In collaboration with AliCloud of the Alibaba Group, the three smart tourism initiatives launched by the MGTO include a tourism data exchange platform, a visitor observation application and a smart visitor flow application. The tourism data exchange platform represents the foundation of these smart initiatives and is hosted by the government of Macau's cloud computing network capturing a variety of data related to tourism in the territory. On the other hand, the visitor observation app looks at basic attributes related to the behaviour of visitors, their preferences as well as their travel patterns. Lastly, the smart visitor flow app delivers predictions – four hours in advance – of the density of visitor flow in several tourist attractions on a 24-hour basis, seven days a week, making it easier to organise visitor itineraries. The smart visitor flow app currently covers 20 of the city's most visited tourism attractions, including several located in Macau's historic centre, which holds UNESCO World Heritage Site status.

At the other end of the spectrum, on a regional level or perhaps rather a tourism construct level, the Smart Costa del Sol (Spain) is a good example of the challenges faced by tourism destinations that do not conform strictly to the urban tourism typology, even when tourists remain clearly key stakeholders in its policy making and delivery of services. The Costa del Sol is neither a city nor a province or a region. Instead, it is formed by a heterogeneous group of villages around a key smart city and smart tourism destination for southern Spain: Malaga. Malaga is the sixth largest city in Spain with a population in excess of 500,000 with smaller towns nearby such as Torremolinos, Benalmádena, Fuengirola, Mijas, Marbella, Estepona, Casares and Nerja, which are also important tourism destinations regionally. The Costa del Sol is home to a population of 1.5 million and hosts annually 12 million visitors with 26 million overnight stays and a combined income from tourism of 11.1 million euros (Turismo y Planificación Costa del Sol, 2017). The Smart Costa del Sol project has been launched as part of the first round of smart city proposals in the "A Way to Build Europe" initiative of the European Regional Development Fund (ERDF). The Smart Costa del Sol initiative involves a partnership of 13 municipalities in the province of Málaga: Alhaurín de la Torre, Antequera, Benalmádena, Estepona, Fuengirola, Málaga, Marbella, Mijas, Nerja, Rincón de la Victoria, Ronda, Torremolinos and Vélez-Málaga. Its main objective is to build more efficient and sustainable cities through smart resource management to increase the social and economic wellbeing of residents and visitors alike. Using a public–private partnership approach that includes IDOM and Wellness Telecom, these municipalities have teamed up to deliver a digital transformation that will inject smartness into current tourism management processes and decision-making. The programme is structured into three major components, namely: a smart city platform that will connect various local initiatives and enable data sharing between municipalities; an open data portal that will publish project data in a way that residents and other organisations can access and understand; and a smart irrigation system that will monitor water consumption and manage the irrigation of parks and public spaces.

Future challenges for smart urban tourism cities

Following on from the above discussions, two issues are beginning to become rather apparent. On the one hand, smart urban tourism destinations can no longer adopt a city-centric approach to smartness. Instead, as smaller peripheral destinations begin to increasingly emerge as viable options to "decongest" overcrowded global tourism cities and reduce the pressure on their resources, smart tourism initiatives will need to become more regional in their approach and less focused specifically on the metropolises that have dominated this concept from its outset. This chapter has suggested a smart tourism typology (see Figure 24.1), which may go some way in influencing on-going practice and future academic research in this field. Secondly, smart tourism destination initiatives promoted by the European Union and other funding bodies elsewhere in the world are increasingly beginning to focus on aspects of sustainability and sustainable urban management, including the development of smarter human capital. This is perhaps a factor that was initially somewhat overlooked by smart tourism destinations, which mirrored themselves largely in the – now almost obsolete – techno-centric approach of the first smart city pilots. Instead, a growing understanding of the need for cities and other tourism destinations to adapt to environmental changes by developing resilience strategies and, at the same time, providing leadership with regard to innovative urban management solutions often referred to as urban living labs will lead to a shift in policy making whereby sustainability will need to be at the heart of smart solutions for urban tourism destinations and their wider regions. In essence, while some smart urban tourism destinations will continue to revel in their techno-centric initiatives, the more forward-looking ones will pursue instead a new paradigm: the sustainable smart tourism destination.

In line with this, tourism cities will increasingly see themselves as ecosystems of stakeholders and, by doing so, adopt a systems-based approach to their development (Morrison *et al.*, 2018; Bosak, 2019), the management of their resources and the wider environment. At the same time, they will look to capitalise on innovations rooted in the smart city paradigm (see Giffinger *et al.*, 2007) that can deliver positive impacts on the way these tourism cities are managed (Boes *et al.*, 2015) as well as their longer-term governance processes (e.g. policy-development envelope). However, at the core of the tourism destination's ecosystem will remain the same attractors and resources (see Figure 24.2) that have traditionally contributed to its authenticity and, by default, its unique competitive positioning and socio-economic sustainability, as stipulated by Crouch and Ritchie (1999) in their Calgary Model.

Sandwiched between the destination management sphere and its core (attractors and resources) a key domain will continue to develop in the future and attract scrutiny in line with society's growing awareness of environmental sustainability issues. This domain, referred to in Figure 24.2 as the Acceptable Change Domain, adopts the sustainability doughnut principle first posited by Raworth (2012) and is linked conceptually to earlier research on the limits of acceptable change first used in the conservation of wilderness areas (Stankey *et al.*, 1985) and later applied to tourism destinations (see, for instance, Ahn *et al.*, 2002; or Frauman and Banks, 2011). The Acceptable Change Domain, which also incorporates elements from Wang and Pizam's

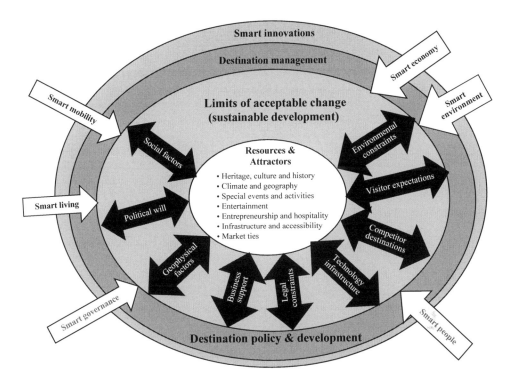

Figure 24.2 Conceptual framework for the management of sustainable smart tourism destinations.

Source: Coca-Stefaniak (2019).

(2011) destination marketing and management conceptual framework, will be the main battle-field for the sustainable smart tourism destination of the future, particularly in the case of tourism cities. This domain will host issues likely to shape tourism cities in the future, including, overtourism, terrorism, climate change, town centre retail businesses struggling to compete with online retail, political conflict, gentrification, changes in visitor behaviour and expectations, and the future proofing of technological solutions, to mention but a few. The acceptability of changes in this domain and, by default, the ability of destination management to expand its influence in tourism cities will grow in a vis-à-vis fashion influenced largely by the solutions that the smart innovations domain will be able to offer on a number of fronts, including environmental, governance, quality of life, local economy, human capital and mobility. It is the positively symbiotic co-existence of these two domains – the acceptable change domain and the smart innovations domain – that will deliver the resilience of a new generation of tourism destinations – the sustainable smart tourism destination.

References

Abella, A., Ortiz-de-Urbina-Criado, M. and de-Pablos-Heredero, C. (2017), "A model for the analysis of data-driven innovation and value generation in smart cities' ecosystems", *Cities*, *64*, pp. 47–53.

Ahn, B., Lee, B. and Shafer, C.S. (2002), "Operationalizing sustainability in regional tourism planning: an application of the limits of acceptable change framework", *Tourism Management*, *23*(1), pp. 1–15.

Ahvenniemi, H., Huovila, A., Pinto-Seppä, I. and Airaksinen, M. (2017), "What are the differences between sustainable and smart cities?", *Cities*, *60*, pp. 234–245.

Al Nuaimi, E., Al Neyadi, H., Mohamed, N. and Al-Jaroodi, J. (2015), "Applications of big data to smart cities", *Journal of Internet Services and Applications*, 6(1), pp. 1–5.

Albino, V., Berardi, U. and Dangelico, R.M. (2015), "Smart cities: Definitions, dimensions, performance and initiatives", *Journal of Urban Technology*, 22(1), pp. 3–21.

Alizadeh, T. (2017), "An investigation of IBM's Smarter Cites Challenge: What do participating cities want?", *Cities*, 63, pp. 70–80.

Alter, S. (2019), "Making sense of smartness in the context of smart devices and smart systems", *Information Systems Frontiers*, pp. 1–13.

Angelidou, M. (2014), "Smart city policies: A spatial approach", *Cities*, 41, pp. S3–S11.

Bakıcı, T., Almirall, E. and Wareham, J. (2012), "A smart city initiative: The case of Barcelona", *Journal of the Knowledge Economy*, 4, pp. 135–148.

Balducci, F. and Ferrara, A. (2018), "Using urban environmental policy data to understand the domains of smartness: An analysis of spatial autocorrelation for all the Italian chief towns", *Ecological Indicators*, 89, pp. 386–396.

Battarra, R., Fistola, R. and La Rocca, R.A. (2016), City smartNESS: The energy dimension of the urban system. In: R. Papa and R. Fistola (eds) *Smart energy in the smart city: Green energy and technology*, Cham: Springer, pp. 1–23.

Bock, B.B. (2016), "Rural marginalisation and the role of social innovation; a turn towards nexogenous development and rural reconnection", *Sociologia Ruralis*, 56(4), pp. 552–573.

Boes, K., Buhalis, D. and Inversini, A. (2016), "Smart tourism destinations: ecosystems for tourism destination competitiveness", *International Journal of Tourism Cities*, 2(2), pp. 108–124.

Bosak, K. (2019), The tourism system. In: S.F. McCool and K. Bosak (eds) *A research agenda for sustainable tourism*, Cheltenham: Edward Elgar, pp. 14–20.

Bower, J.L. and Christensen, C.M. (1995), "Disruptive technologies: Catching the wave", *Harvard Business Review*, 73(1), pp. 43–53.

Buhalis, D. (2000), "Marketing the competitive destination of the future", *Tourism Management*, 21, pp. 97–116.

Buhalis, D. and Amaranggana, A. (2014), Smart tourism destinations. In: Z. Xiang and I. Tussyadiah (eds), *Information and communication technologies in tourism 2014*, Cham: Springer, pp. 553–564.

Buhalis, D. and Amaranggana, A. (2015), Smart tourism destinations enhancing tourism experience through personalisation of services. In: I. Tussyadiah and A. Inversini (eds) *Information and communication technologies in tourism 2015*, Cham: Springer, pp. 377–389.

Caragliu, A., Del Bo, C. and Nijkamp, P. (2011), "Smart cities in Europe", *Journal of Urban Technology*, 18(2), pp. 65–82.

Carr, P.J. and Kefalas, M.J. (2009), *Hollowing out the middle: The rural brain drain and what it means for America*, New York: Beacon Press.

Coca-Stefaniak, J.A. (2019), "Marketing smart tourism cities: a strategic dilemma", *International Journal of Tourism Cities*, 5(4), pp. 513–518.

Cocchia, A. (2014), Smart and digital city: A systematic literature review. In: R.P. Dameri and C. Rosenthal-Sabroux (eds) *Smart city: How to create public and economic value with high technology in urban space*, New York: Springer, pp. 13–43.

Cohen, B. (2013), *Smart Cities Wheel*, www.smart-circle.org/smartcity/blog/boyd-cohen-the-smart-city-wheel/ (accessed 15 September 2019).

Crouch, G.I. and Ritchie, J.B. (1999), "Tourism, competitiveness, and societal prosperity", *Journal of Business Research*, 44(3), pp. 137–152.

Cugurullo, F. (2018), The origin of the smart city imaginary. In: C. Lindner and M. Meissner (eds) *The Routledge companion to urban imaginaries*, London: Routledge, pp. 113–124.

Dameri, R.P. (2017), Using ICT in smart city. In: R.P. Dameri, *Smart city implementation*. Progress in IS series. Cham: Springer, pp. 45–65.

Eger, J.M. (2009), "Smart growth, smart cities and the crisis at the pump: A worldwide phenomenon", *I-Ways*, 32(1), pp. 47–53.

European Union (2019), *European Capital of Smart Tourism*, https://smarttourismcapital.eu/ (accessed 23 October 2019).

Florida, R. (2006), "The flight of the creative class: The new global competition for talent", *Liberal Education*, 92(3), pp. 22–29.

Frauman, E. and Banks, S. (2011), "Gateway community resident perceptions of tourism development: Incorporating importance–performance analysis into a limits of acceptable change framework", *Tourism Management*, 32(1), pp. 128–140.

Gibbs, D., Krueger, R. and MacLeod, G. (2013), "Grappling with smart city politics in an era of market triumphalism", *Urban Studies*, *50*(11), pp. 2151–2157.

Giffinger, R., Fertner, C., Kramar, H., Kalasek, R., Pichler-Milanović, N. and Meijers, E. (2007), *Smart cities – ranking of European medium-sized cities*, Vienna University of Technology, http://curis.ku.dk/ws/files/37640170/smart_cities_final_report.pdf (accessed 21 October 2019).

Gil-Garcia, J.R., Pardo, T.A. and De Tuya, M. (2019), "Information sharing as a dimension of smartness: Understanding benefits and challenges in two megacities", *Urban Affairs Review*, https://doi.org/10.1177/1078087419843190.

Gil-Garcia, J.R., Zhang, J. and Puron-Cid, G. (2016), "Conceptualizing smartness in government: An integrative and multi-dimensional view", *Government Information Quarterly*, *33*(3), pp. 524–534.

Giovannella, C. and Rehm, M. (2015), "A critical approach to ICT to support participatory development of people centered smart learning ecosystems and territories", *Aarhus Series on Human Centered Computing*, *1*(1), p. 2.

Govada, S.S., Spruijt, W. and Rodgers, T. (2017), Smart city concept and framework. In: T.M.V. Kumar (ed.) *Smart Economy in Smart Cities*, Singapore: Springer, pp. 187–198.

Greenfield, A. (2013), *Against the smart city*, New York: Do Publ.

Gretzel, U. (2011), "Intelligent systems in tourism: A social science perspective", *Annals of Tourism Research*, *38*(3), pp. 757–779.

Gretzel, U., Werthner, H., Koo, C. and Lamsfus, C. (2015), "Conceptual foundations for understanding smart tourism ecosystems", *Computers in Human Behavior*, *50*, pp. 558–563.

Guo, Y., Liu, H. and Chai, Y. (2014), "The embedding convergence of smart cities and tourism internet of things in China: An advance perspective", *Advances in Hospitality and Tourism Research*, *2*(1), pp. 54–69.

Hajer, M.A. and Dassen, T. (2014), *Smart about cities: Visualising the challenge for 21st century urbanism*, Rotterdam: PBL Pub.

Haubensak, O. (2011), *Smart cities and the Internet of things*, Zurich: ETH.

Hawkes, J. (2001), *The fourth pillar of sustainability: Culture's essential role in public planning*, Melbourne: Common Ground.

Hedlund, J. (2012), *Smart city 2020: Technology and society in the modern city*, Microsoft Services.

Hollands, R.G. (2008), "Will the real smart city please stand up?", *City*, *12*(3), pp. 303–320.

Huang, X.K., Yuan, J.Z. and Shi, M.Y. (2012), Condition and key issues analysis on the smarter tourism construction in China. In: F.L. Wang, J. Lei, R.W.H. Lau and J. Zhang (eds) *Multimedia and signal processing*, Berlin: Springer, pp. 444–450.

Hughes, K. and Moscardo, G. (2019), "ICT and the future of tourist management", *Journal of Tourism Futures*, doi:10.1108/JTF-12–2018–0072.

IBM (2010), "Smarter thinking for a smarter planet", www.ibm.com/ibm/history/ibm100/us/en/icons/smarterplanet/ (accessed 22 October 2019).

Ismagilova, E., Hughes, L., Dwivedi, Y.K. and Raman, K.R. (2019), "Smart cities: Advances in research—An information systems perspective", *International Journal of Information Management*, *47*, pp. 88–100.

Ivars-Baidal, J.A., Celdrán-Bernabeu, M.A., Mazón, J.N. and Perles-Ivars, Á.F. (2019), "Smart destinations and the evolution of ICTs: A new scenario for destination management?", *Current Issues in Tourism*, *22*(13), pp. 1581–1600.

Jasrotia, A. and Gangotia, A. (2018), "Smart cities to smart tourism destinations: A review paper", *Journal of Tourism Intelligence and Smartness*, *1*(1), pp. 47–56.

Johnson, A. and Samakovlis, I. (2019), "A bibliometric analysis of knowledge development in smart tourism research", *Journal of Hospitality and Tourism Technology*, https://doi.org/10.1108/JHTT-07-2018-0065.

Jyotsna, J.H. and Maurya, U.K. (2019), "Experiencing the real village: A netnographic examination of perceived authenticity in rural tourism consumption", *Asia Pacific Journal of Tourism Research*, *24*(8), pp. 750–762.

Kitchin, R. (2015), "Making sense of smart cities: Addressing present shortcomings", *Cambridge Journal of Regions, Economy and Society*, *8*, pp. 131–136.

Kolotouchkina, O. and Seisdedos, G. (2018), "Place branding strategies in the context of new smart cities: Songdo IBD, Masdar and Skolkovo", *Place Branding and Public Diplomacy*, *14*(2), pp. 115–124.

Komninos, N. (2011), "Intelligent cities: Variable geometries of spatial intelligence", *Intelligent Buildings International*, *3*(3), pp. 172–188.

Kummitha, R.K.R. and Crutzen, N. (2017), "How do we understand smart cities? An evolutionary perspective", *Cities*, *67*, pp. 43–52.

Lamsfus, C., Wang, D. and Alzua-Sorzabal, A. (2015), "Going mobile: Defining context for on-the-go travelers", *Journal of Travel Research*, *54*(6), 691–701.

Lara, A.P., Da Costa, E.M., Furlani, T.Z. and Yigitcanla, T. (2016), "Smartness that matters: Towards a comprehensive and human-centred characterisation of smart cities", *Journal of Open Innovation: Technology, Market, and Complexity*, *2*(2), article 8.

Lim, C., Kim, K. Maglio, P. (2018), "Smart cities with big data: References models, challenges and considerations", *Cities*, 82, pp. 86–99.

Lopez de Avila, A. (2015), "Smart destinations: XXI century tourism". Presented at the ENTER2015 Conference on Information and Communication Technologies in Tourism, Lugano, Switzerland, 4–6 February.

Lytras, M. and Visvizi, A. (2018), "Who uses smart city services and what to make of it: Toward interdisciplinary smart cities research", *Sustainability*, *10*(6), 1998.

Macao Government Tourism Office (2019), www.gov.mo/en/news/110041/ (accessed 23 October 2019).

Manville, C., Cochrane, G., Cave, J., Millard, J., Pederson, J.K., Thaarup, J.K., Liebe, A., Wisner, M., Massink, R. and Kotterink, B. (2014), *Mapping smart cities in the EU*, European Parliament, Directorate General for Internal Policies, Policy Department A: Economic and Scientific Policy, www.europarl. europa.eu/RegData/etudes/etudes/join/2014/507480/IPOL-ITRE_ET(2014)507480_EN.pdf (accessed 21 October, 2019).

March, H. (2018), "The smart city and other ICT-led techno-imaginaries: Any room for dialogue with degrowth?", *Journal of Cleaner Production*, *197*, pp. 1694–1703.

Meijer, A. and Bolívar, M.P.R. (2016), "Governing the smart city: A review of the literature on smart urban governance", *International Review of Administrative Sciences*, *82*(2), pp. 392–408.

Molinillo, S., Anaya-Sánchez, R., Morrison, A.M. and Coca-Stefaniak, J.A. (2019), "Smart city communication via social media: Analysing residents' and visitors' engagement", *Cities*, *94*, pp. 247–255.

Monitor Deloitte (2015), *Smart cities: Not just the sum of its parts*, www2.deloitte.com/content/dam/ Deloitte/xe/Documents/strategy/me_deloitte-monitor_smart-cities.pdf (accessed 19 September 2019).

Morrison, A.M., Lehto, X.Y., Day, J.G. and Mill, R.C. (2018), *The tourism system*, New York: Kendall Hunt Publ.

Naldi, L., Nilsson, P., Westlund, H. and Wixe, S. (2015), "What is smart rural development?", *Journal of Rural Studies*, *40*, pp. 90–101.

Peters, M.A. (2017), "Technological unemployment: Educating for the fourth industrial revolution", *Journal of Self-Governance and Management Economics*, *5*(1), pp. 25–33.

Picon, A. (2015), *Smart cities: A spatialized Intelligence*, New York: Palgrave Macmillan.

Porter, M.E. and Heppelmann, J.E. (2014), "How smart, connected products are transforming competition", *Harvard Business Review*, *92*(11), pp. 64–88.

Ramaprasad, A., Sánchez-Ortiz, A. and Syn, T. (2017), A unified definition of a smart city. In: *International Conference on Electronic Government*, London: Springer, pp. 13–24.

Raworth, K. (2012), *A safe and just space for humanity: Can we live within the doughnut?*, Oxford: Oxfam GB, https://oxfamilibrary.openrepository.com/bitstream/handle/10546/210490/dp-a-safe-and-just-space-for-humanity-130212-en.pdf?sequence=13 (accessed 22 October 2019).

Ruhlandt, R.W.S. (2018), "The governance of smart cities: A systematic literature review", *Cities*, *81*, pp. 1–23.

Rutherford, J. (2011), "Rethinking the relational socio-technical materialities of cities and ICTs", *Journal of Urban Technology*, *18*(1), pp. 21–33.

Sadowski, J. and Bendor, R. (2019), "Selling smartness: Corporate narratives and the smart city as a socio-technical imaginary", *Science, Technology, & Human Values*, *44*(3), pp. 540–563.

Sassen, S. (2011), *Talking back to your intelligent city*, http://voices.mckinseyonsociety.com/talking-back-to-your-intelligent-city/ (accessed 20 July 2016).

Schwab, K. (2017), *The fourth industrial revolution*, New York: Crown Business.

Sengers, F., Späth, P. and Raven, R. (2018), Smart city construction: Towards an analytical framework for smart urban living labs. In: S. Marvin, H. Bulkeley, Q. Mai, K. McCormick and P. Voytenko (eds) *Urban living labs*, London: Routledge, pp. 74–88.

Setis-Eu (2012), *European Initiative on Smart Cities*, https://setis.ec.europa.eu/set-plan-implementation/ technology-roadmaps/european-initiative-smart-cities (accessed 22 October 2019).

Shenga, M. and Gub, C. (2018), "Economic growth and development in Macau (1999–2016): The role of the booming gaming industry", *Cities*, 75, pp. 72–80.

Sims, R. (2009), "Food, place and authenticity: Local food and the sustainable tourism experience", *Journal of Sustainable Tourism, 17*(3), pp. 321–336.

Stankey, G.H., Cole, D.N., Lucas, R.C., Petersen, M.E. and Frissell, S.S. (1985), *The limits of acceptable change (LAC) system for wilderness planning,* (INT-176).

Su, K., Li, J. and Fu, H. (2011), "Smart city and the applications", *IEEE International Conference on Electronics,* Communications and Control (ICECC), pp. 1028–1031.

Thite, M. (2011), "Smart cities: Implications of urban planning for human resource development", *Human Resource Development International, 14*(5), pp. 623–631.

Turismo y Planificación Costa del Sol (2017), Media Pack 2017, www.visitcostadelsol.com (accessed 23 October 2019).

Vanolo, A. (2014), "Smartmentality: The smart city as disciplinary strategy", *Urban Studies, 51*(5), pp. 883–898.

Vicini, S., Bellini, S. and Sanna, A. (2012), "How to co-create Internet of Things-enabled services for smarter cities", *SMART 2012: The First International Conference on Smart Systems, Devices and Technologies,* pp. 55–61.

Viñas, C.D. (2019), "Depopulation processes in European rural areas: A case study of Cantabria (Spain)", *European Countryside, 11*(3), pp. 341–369.

Wang, D., Li, X. and Li, Y. (2013), "China's 'smart tourism destination' initiative: A taste of the service-dominant logic", *Journal of Destination Marketing & Management, 2*(2), pp. 59–61.

Wang, Y. and Pizam, A. (2011), *Destination marketing and management: Theories and applications,* London: CABI.

World Economic Forum (2016), *The future of jobs: Employment, skills and workforce strategy for the fourth industrial revolution.* Global Challenge Insight Report. Geneva: World Economic Forum.

Zhu, W., Zhang, L. and Li, N. (2014), Challenges, function changing of government and enterprises in Chinese smart tourism. In: Z. Xiang and L. Tussyadiah (eds.), *Information and communication technologies in tourism,* Cham: Springer.

Zygiaris, S. (2013), "Smart city reference model: Assisting planners to conceptualize the building of smart city innovation ecosystems", *Journal of the Knowledge Economy, 4*(2), pp. 217–231.

25

eTOURISM CHALLENGES FOR URBAN TOURISM DESTINATIONS

Sebastian Molinillo, Rafael Anaya-Sánchez and Antonio Guevara-Plaza

Introduction

The term eTourism refers to the use of information and communication technologies (ICTs) in tourism and reflects the digitalisation of all processes and the value chain in the tourism industry, as well as in strategic relationships with stakeholders (Buhalis and O'Connor 2005). In the last two decades the emergence and expansion of ICTs (Figure 25.1), such as the web, online booking systems, destination management systems, online marketing platforms, mobile devices and social networks, among other developments, have attracted the attention of many researchers in the tourism field (Gretzel *et al.* 2016, Uysal *et al.* 2016, Kim *et al.* 2018b). These technologies have brought new challenges, risks and dilemmas, in aspects as diverse as information processing (see Mariani *et al.* 2018), the adoption of knowledge management infrastructures (Ávila-Robinson and Wakabayashi 2018), the reinterpretation of destination brands (de Rosa and Dryjanska 2017) and the redefinition of the role of the tourist (Xiang 2018). As a consequence, despite the possibilities offered by ICTs, companies in the sector do not use them to their full potential, or even always appropriately (Whyte 2018).

ICTs play a critical role in the competitiveness of destinations (Buhalis and Law 2008, Boes *et al.* 2016, Gretzel *et al.* 2016). These new technologies help travellers save time, facilitate the personalisation of services, allow them to enjoy social experiences and transform the way that customers interact with brands and destinations (Bloomberg Media Group 2019, Powell 2019). For urban destinations the technological challenge is an important strategic element in value creation and differentiation in the global context of increasing competition (Mariani *et al.* 2018). To better understand tourists and to adapt to their needs and preferences, it is important that technological advances are understood; these include procurement platforms, data analysis and relationship management systems (Bowen and Whalen 2017, Ávila-Robinson and Wakabayashi 2018, Sigala 2018). Interconnectivity, data capture and knowledge creation are fundamental for decision-making in this context (Sigala 2018).

On the one hand, technologies allow the replacement of human interactions (Bowen and Whalen 2017) while providing similar or even higher levels of satisfaction (Sheivachman 2017a). On the other, they encourage the establishment of more interactive relationships among companies, tourists and destinations, and among tourists themselves, affecting the processes of destination image and visit intention creation (Ávila-Robinson and Wakabayashi 2018, Mariani *et al.* 2018).

Figure 25.1   Information and communication technologies.

ICTs allow tourists to co-create their tourism experiences (Buhalis and Foerste 2015) through interaction with other tourists, residents, businesses, destinations and interconnected devices, affecting their satisfaction, well-being and future behaviours (Dickinson *et al.* 2016, Lin *et al.* 2017, Lund *et al.* 2018, Rihova *et al.* 2018). Technological advances re-engineer the whole process of development, management and marketing of destinations (Buhalis and Law 2008) and empower the actors in tourism (Sigala 2018) to the point that they build destination image through informal conversations outside the control of the destination (Lund *et al.* 2018).

Moreover, in the near future, one of the greatest challenges for destinations will be related to the opportunities generated by the constant growth in the availability of data and information (Xiang 2018). This phenomenon will require destinations to apply greater interdisciplinary understanding than hitherto to ensure they adapt their strategies to changes driven by technological advances (Sigala 2018). These developments mean that there has never been such an exciting time for the travel and tourism industry (Bowen and Whalen 2017).

The purpose of this chapter is to discuss critically, based on a review of recent literature on the topic, the challenges faced by urban tourism destinations with regards to eTourism and new technologies. Publications are grouped into categories and qualitative analysis is applied to identify the main topics and challenges for urban tourism destinations. Based on this analysis, the chapter discusses tourist experience, value co-creation, mobile technologies, sharing economies, social media, governance, information management and management of the destination offer. The theoretical contribution of this chapter, thus, revolves around a comprehensive literature review of eTourism and urban destinations. Practical implications are also discussed, with a particular emphasis on the potential implementation of technological developments.

Technology developments in the tourism environment

International tourism has experienced major continuous growth in the last decade, with an increase not only in the number of international tourist arrivals, but also in the number of destinations that have been opened up to tourism worldwide (UNWTO 2018). This growth calls attention to issues that affect cities and their stakeholders, such as sustainability, competitiveness, technological developments, financing, innovation, demand-led behaviour and development of the offer, among others. The expansion of ICTs is perhaps one of the most important issues because it has given rise to a new context for both tourists and businesses, destinations and local residents (Ávila-Robinson and Wakabayashi 2018, Kim et al. 2018a). Technologies have taken tourism management from a static position, where managers and tourists used them only as accessories, to a dynamic position where destinations and stakeholders interact in real time to co-create knowledge and tourist experiences (Sigala 2018, Xiang 2018, Buhalis and Sinarta 2019).

Among recent technological developments, researchers have highlighted three particular areas of interest for tourism: (1) the effect of tools such as smartphones, GPS and virtual reality on tourist behaviour and data collection (Thimm and Seepold 2016); (2) the creation of collaborative experiences with the tourist; and (3) destination management and governance (Gretzel et al. 2016, Ávila-Robinson and Wakabayashi 2018). Bowen and Whalen (2017) emphasised technology, with a focus on robotics and artificial intelligence, big data analytics, social media and the sharing economy, while Kim et al. (2018a) highlighted the advantages of information technology management. From an integrative approach, Sigala (2018) argued that ICTs have created a complex socio-technical smart-tourism ecosystem, characterised by the greater generation of intelligence, transformation of the user's experience and the more efficient management of destinations' marketing strategies and decision-making processes. According to Xiang (2018), in tourism in the last two decades, ICTs have evolved from marketing-driven tools to knowledge-creation tools – but great challenges will be faced in the immediate future "related to the interactions between human and data, networks and machine intelligence" (p. 149).

Technological challenges for the management of urban destinations

Tourist experience

In the tourism context, the tourist experience is defined as "a constant flow of thoughts and feelings during moments of consciousness which occur through highly complex psychological, sociological, and cognitive interaction processes" (Kang and Gretzel 2012, p. 442). Two of the most significant advances in the area of tourist experiences are the increase in the degree of co-creation and the integration of ICTs (Neuhofer et al. 2014).

As tourists now seek to "live" experiences rather than simply buy products (Seaton 2017), cities are being challenged to adapt to this environment and offer them new, memorable experiences (Park and Santos 2017) (Figure 25.2). Destination management organisations (DMOs) compete for travellers' attention; technologies allow them to maintain communication with the travellers during their visits and send useful and timely information to improve their experiences (Sorrells 2019). These technologies combine with the interpersonal interactions that take place between tourists and private company/DMO employees to generate an interactive and unique experience. These interactions, whether face-to-face or through ICTs, are at the core of this industry (Powell 2019). Wearing and Foley (2017) emphasised that creative and interactive processes create a better understanding of current tourist urban experiences. Of the experiences that tourists have while planning and during their trips, those that are unique and unexpected are the most memorable. These experiences are shared during and after the journey through social media, especially when positive emotions and experiences have been generated (Kim and Fesenmaier 2017). In this context, service providers help to

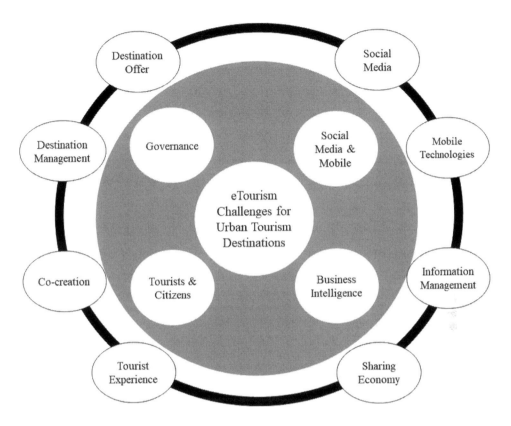

Figure 25.2 Technological challenges for the management of urban destinations.

generate memorable and authentic experiences (Zatori *et al.* 2018), knowing that the effect of positive experiences is very small compared to the effect of negative experiences (Jackson 2019). In this regard, perhaps destinations should pay special attention to their ability to react in real time. According to Buhalis and Sinarta (2019) and Boes *et al.* (2016), ICTs help destinations enhance the tourist experience through co-creation by engaging them through real-time service based on dynamic big data mining, artificial intelligence and contextualisation.

In conclusion, destinations must offer experiences tailored to the tourists' needs. As Fromm (2019) commented,

> the expectations of Icelandic tourists aged over 60 and those of Chinese Millennials, for example, are clearly going to be very different so it's crucial for travel brands to invest in truly understanding a diverse range of expectations and motivations to be able to offer the most suitable and personalized tourism product.

To ensure that the experience is personalised, this adaptation to the consumer's needs must involve the redesign of DMOs' official websites. This is the case with the city of Miami and the island of Puerto Rico; their new websites offer hyper-personalised services using cookies and different types of content, adapted to each stage of the tourist experience.

Co-creation

Studies focusing on quality of life and well-being in tourism point out that their effects are mainly focused on the relationship between two groups, local residents and tourists. Tourist activities affect residents in their social life, leisure, cultural offerings, infrastructure saturation, pollution, etc. (Uysal *et al.* 2016). Rivera *et al.* (2016) analysed the evaluation that residents make about the well-being they derive from tourism (e.g. income, quality of life), and concluded that tourism development influences happiness, especially through aspects not directly related to income. For example, the quality of life of local residents is closely related to the sustainability of the destination (Mathew and Sreejesh 2017). It is also important to avoid conflicts between residents and tourists, because should these increase this can affect the environment (e.g. environmental degradation, reduction of local resources) and impact on local lifestyle, culture and saturation of common spaces and public services (Romão *et al.* 2018).

Woo *et al.* (2015) showed that the value that local residents attribute to tourism positively affects their satisfaction, both material and immaterial, which can lead them to support tourism development. When residents perceive economic and sociocultural benefits, they increase their collaboration in the co-creation of the value of the tourism offer. Tourists participate more when they feel welcome (Lin *et al.* 2017). The co-creation of experiences by tourists improves their evaluation of their journeys, makes them more loyal and increases their satisfaction (Mathis *et al.* 2016). ICTs encourage interrelationships between stakeholders and destinations and value co-creation (Buhalis and Amaranggana 2013, Gretzel *et al.* 2015). Therefore, destinations should create the structures necessary to support suppliers in the implementation of strategies to foster greater cooperation between residents and tourists, which will positively affect value co-creation. Participation increases social value and improves the relationship between hosts and guests and between tourists and receiving communities (Rihova *et al.* 2018).

Conversely, where residents feel hostile towards tourism, due to its negative effects on their communities (Pinke-Sziva *et al.* 2019), this can degenerate into value-destruction processes (Lin *et al.* 2017). Similarly, when negative emotions are generated in the tourist by incongruous or unsatisfactory practices, this, in turn, creates feelings of frustration or unhappiness which can cause value destruction (Malone *et al.* 2018).

Mobile technologies

Mobile technologies, in particular the smartphone, have changed and personalised the tourist experience (Neuhofer *et al.* 2015, Cacho *et al.* 2016, Wang *et al.* 2014, 2016). The use of the smartphone before, during and after a stay at a destination "is shaped by complex interactions between contextual factors, cognitive beliefs, previous experiences and everyday use" (Wang *et al.* 2014, p. 11). Not only can tourists use mobile devices to search for information and make reservations about services (e.g. hotels, planes, cars) at any time and place, they can also access services that provide micro-experiences, such as accessing flight-status information or estimating the waiting time for public transport (Wang and Wang 2010). The combination of applications that offer geolocation, ubiquity, access to information, social networks, entertainment and other useful tools for tourists, promote the use of smartphones in tourism (Vallespín *et al.* 2017). In addition, mobile devices encourage tourists to share their experiences with family and friends, other visitors and local residents and to interact with the destination and improve the tourist experience (Neuhofer *et al.* 2012, Cacho *et al.* 2016). For these reasons, destinations such as Japan, Singapore and India have begun to offer free Wi-Fi connections, or provide their visitors with SIM cards, so that they can remain connected; these can include free apps that allow access to multiple services.

The needs of tourists while actually on their travels increase the importance and effectiveness of mobile technology in supporting their real-time decision-making (Wang *et al.* 2012). Thus, to contribute to positive tourist experiences, destinations should create Internet access for mobile devices, particularly in places such as airports, railway stations, hotels and at tourist attractions (da Costa Liberato *et al.* 2018). In addition, destinations can use a growing number of mobile technologies to increase their attractiveness and satisfaction. For example, Buhalis and Foerste (2015) suggested that SoCoMo marketing can provide the tourist with dynamic and personalised recommendations for each destination, due to a combination of social media, context-aware marketing (location-based) and mobile devices. Other mobile technologies that can help DMOs improve the tourist experience are mobile apps (Gupta *et al.* 2018), gamification (Xu *et al.* 2017), augmented reality (Paulo *et al.* 2018) and virtual reality (Kim *et al.* 2018b), among others. For example, the Swiss city of Laax, focused on winter sports, has developed a gamification-based app for skiers that combines useful services, information and a loyalty programme (Sorrells 2019). The province of Malaga has developed an augmented reality app that provides data (including contact information) about tourist attractions and other important services, such as banks and petrol/gas stations. Similarly, Helsinki has developed a mobile platform for the benefit of its Chinese tourists. Specifically, they collaborated with Tencent and Idean, which have great penetration in this market, and integrated the services of Lonely Planet, Yelp, Google Translator, Uber and an e-wallet. In addition, tourist guides now use apps that go beyond the provision of simple descriptions, to include mini games, such as puzzles, riddles and hunts for virtual items in real-world locations (Constine 2017). Along similar lines, Oslo has developed an augmented reality app in which one of its most famous playwrights acts as a guide.

Therefore, mobile devices help to improve the tourist experience, even in an affective, emotional sense, although an obsessive dependency on the smartphone can generate negative feelings (Lalicic and Weismayer 2018). In fact, although digital dysconnectivity limits access to tourist services (Tanti and Buhalis 2017), a large percentage of tourists want to digitally disconnect when they are on vacation (Dickinson *et al.* 2016), which has given rise to tourist destinations that expressly advertise mobile phone dead zones (Gretzel 2014).

Sharing economy

The "sharing economy", "peer to peer economy" or "collaborative consumption" are terms that refer to the practice of sharing (or exchanging) access to underutilised goods and services through digital platforms. Although the practice of sharing with others is nothing new in social relationships, the forms of sharing that have emerged in recent years are characterised by their ability to facilitate interactions between strangers, a strong reliance on digital technologies and the participation of high cultural capital consumers (Schor and Fitzmaurice 2015).

The collaborative economy has a wide field of action in the tourism world and has created a large number of marketplaces and businesses that have provided opportunities for professionals and small companies. In addition, it offers new ways to travel due to the possibility of sharing resources and experiences (Fundación Orange 2016). Thus, in the tourism field, the "sharing economy" creates challenges for individuals, businesses, destinations and tourism services. According to Cheng (2016, Cheng and Foley 2018), the main topics hitherto covered by the literature are related specifically to new business models in the accommodation sector (e.g. Airbnb, Domio) and their effects on duration and type of stay, the tourist experience and competition with the hotel sector. Airbnb is seen as a disruptive model that has radically modified the accommodation sector, allowed urban tourists access to informal accommodation, reduced travel costs and offer a more authentic experience at the destination (Guttentag 2015). Its success has caused rapid expansion into large cities and rural, beach and mountain destinations (Adamiak 2018).

"Sharing economy" models have more recently emerged in other sectors. In urban transport, one can hire vehicles with drivers, for example Uber and Cabify, and car-share, for example, using BlaBlacar, and share bicycles and skateboards using, for example Mobike, OFO and Wind. In the catering sector there are food delivery business models, for example Deliveroo and Eat-Street, "ghost kitchens", for example UberEats and DoorDash. In social eating and meal sharing, where one shares one's table and home, there are EatWith and Meal Sharing.

The growth of the "sharing economy", although it allows destinations to offer alternative services to those of the traditional economy (Juul 2015), raises dilemmas and challenges for the affected sectors in terms of regulation and taxation (Williams and Horodnic 2017) and for the coexistence and sustainability of destinations (McLaren and Agyeman 2015). These dilemmas should lead destinations to improve their coexistence and sustainability, which is being demanded by both residents and different tourist profiles (Fromm 2019). Some companies have already begun efforts to reduce their negative effect on destinations. Airbnb has launched a new fund in Europe, called the Community Tourism Programme, to finance projects to foster local customs and traditions.

Social media

The Internet, and especially social media, influences how tourists look for information, and create and share content on destinations. DMOs must adapt their marketing strategies to this channel, combining actions under their direct control (e.g. promotions, institutional campaigns) with others in which tourists actively participate. For example, some US destinations are, on the one hand, investing heavily in social media advertisements, sending out calls-to-action and offering tool-planning trips, and on the other they are seeking to generate interactions with, and between, users, and to take advantage of the content (images, reviews, etc.) that these generate. In this sense, Jansson (2018) suggested that when tourists view and share content through social media this affects the behaviour of other tourists, because this content, whether positive or negative, is used as an information source by these tourists.

Similarly, Lund *et al.* (2018) noted that content shared by tourists via social networks has greater impact than DMO strategies on destination brands, as tourists are more trusting of their peers than commercially motivated DMOs. Stojanovic *et al.* (2018) emphasised the importance of social networks in the relationship between destinations and tourists, as they create greater brand awareness and brand equity than traditional media. Thus, DMOs and local retailers should seek ways to encourage tourists to engage with social media, that is, to persuade them to generate and share content about the destination (e.g. Mariani *et al.* 2018). Therefore, destinations should analyse which mechanisms (e.g. website features, content, interactivity) lead the tourist to engage with the destination's social media website, as these have been shown to have positive effects on loyalty and trust in the destination, among other outcomes. In particular, DMOs must adapt their messages to the channel, because destination image and visit intention are influenced differently based on content and the websites used (e.g. YouTube, Instagram, Facebook) (Molinillo *et al.* 2018).

In addition, social media have empowered the tourist in the creation of destination image. For some tourist profiles it is not enough to live the experience, they also want to share it with others through their social media, thereby projecting an image and lifestyle (Seaton 2017). DMOs have to understand the processes that lead tourists to influence destination image. So *et al.* (2018) noted that the social visibility of the consumption of the tourist product leads to cognitive, affective and evaluative identification with the destination, which prompts tourists to share their experiences on social media. Lund *et al.* (2018) argued that this poses a challenge for DMOs as they must cope with a reality where destination brands are formed mainly as a result of experiences shared by tourists. These authors suggested that DMOs must recognise that content generated by tourists is an ally in the creation of destination image and, through story-telling, break down the barriers between online and offline environments. In this regard, de Rosa *et al.* (2019) showed the difficulty destinations face in achieving convergence between the brand identity of a city (projected image) and its brand image (perceived image). Similarly, destinations face difficulties in eliminating or reducing stereotypical views and negative aspects of their image, for which they must select an appropriate audience, adjust their message and use trusted sources (Avraham 2018).

Governance

Governance refers to "the interactions among structures, processes and traditions that determine how power and responsibilities are exercised, how decisions are taken, and how citizens or other stakeholders have their say" (Graham *et al.* 2003, pp. 2–3). Tourist destinations embody complex systems due to the convergence of diverse interests of multiple stakeholders which are not always aligned. Governance is a means of resolving conflicts between parties (Laws *et al.* 2011) and to achieve sustainable development that minimises adverse impacts and maximises benefits for local communities (de Bruyn and Alonso 2012). Destinations have to be attractive for visitors and residents, as well as sustainable (economically, socially and environmentally), which involves coexistence among all stakeholders (Romão *et al.* 2018).

The changes in the sector brought about by ICTs and new players require destinations to modernise their governance procedures through the incorporation of decision-making, participation, transparency, creation and knowledge sharing (with stakeholders) systems, in line with what Buhalis and Amarangana (2013) called smart-tourism destinations. Using ICTs, DMOs can expand their influence on local governance beyond just serving tourist interests, by integrating local infrastructure and services to balance stakeholder interests, by promoting social innovation and building relationships of trust (Go and Trunfio 2011). However, the varying perceptions and interests of the different stakeholders can hinder the adoption of destination management systems (Sigala 2013). The introduction of networking capabilities that include all stakeholders will be positive and successful for destinations and give greater power and acceptance to the DMO within the destination (Volgger and Pechlaner 2014). Along with leadership, Boes *et al.* (2015) argued that innovation, social capital, human capital and entrepreneurs are the key elements in leveraging ICTs for the co-creation of value and tourist experiences and the competitiveness of destinations. Other authors have added different elements, such as the knowledge transfer process between all the organisations in a destination and even between destinations (Werner *et al.* 2015, Hardy *et al.* 2018) and, more recently, adaptive co-management, understood as a dynamic approach to governance that aims to create sustainable development in respect of natural resources, stakeholder participation in management and decision-making, and adaptive learning (i.e. learning-by-doing) (Islam *et al.* 2018). For example, to prepare for a massive growth in visitor numbers, the Omani government plans to implement an advanced visitor processing information system to connect data and services in different functional areas to make the entry process into the country efficient, flexible and safe.

Information management

ICTs have given destinations access to an unprecedented amount of aggregated data (big data), from very diverse sources; these can be grouped into three categories: (1) online user-generated content (UGC) (e.g. text, photos, videos); (2) device data from mobile technologies and the Internet of Things (IoT) (mobile roaming data, GPS data, Bluetooth data, Wi-Fi, RFID, sensors); and (3) transaction data by operations (e.g. web search data, webpage visiting data, online booking data) (Li *et al.* 2018).

DMOs should not be limited simply to collecting this data, but should also conduct analyses, identify key elements and prepare/obtain and publish reports that can assist decision-making processes (UNWTO 2019). UGC has transformed the tourist's decision-making processes, experience and relationship with the destination (Law *et al.* 2014, Navío-Marco *et al.* 2018). The analysis of UGC can help destinations in their decision-making, based on a better knowledge of the tourist, and to design marketing strategies with more personalised messages, transparency and trust in their stakeholder relationships, and to improve the tourist experience and value co-creation (Del Vecchio *et al.* 2018). Similarly, it can help reduce the negative effects of over-tourism on local communities by exposing the causes of problems and proposing solutions (Sheivachman 2017b). Through the IoT, using sensors, computational cores and telecommunication systems, destinations can gather real-time information on issues as diverse as traffic congestion, pollution, energy consumption and waste management, among others; in addition, through mobile devices (e.g. smartphones, tablets, smartwatches) they can identify and geolocate their targets, understand their spatial behaviour, make recommendations based on the environment and provide IoT-linked services (Zanella *et al.* 2014). Some hotel companies, such as Marriott and Hilton, already use the IoT to improve their clients' experiences (Powell 2019). Finally, the destination can use transaction data to learn more about tourist behaviour, design marketing strategies, make predictions and ensure search engine optimisation (SEO) (Li *et al.* 2018). In this sense, brands and destinations should ensure tourists receive increased value in exchange for the information they provide (Fromm 2019). They must also understand that their role is not simply to attract tourists to destinations, but to create better experiences for them to enjoy during their stays (Sorrells 2019). For example, Amsterdam processes large volumes of data to analyse waiting times at major attractions, then sends mobile notifications to advise tourists about queues and suggest alternative venues, thus allowing them to better manage their time.

Big data poses destinations a real challenge, not for data collection, but for their real-time, efficient analysis and interpretation. All data types provide different information, which require various analytical techniques to solve probably dissimilar problems. To achieve this, some authors have suggested implementing technologies such as machine-learning tools and artificial intelligence (Del Vecchio *et al.* 2018, Allam and Dhunny 2019).

Integrated destination management systems

Developments in ICTs, the different online business models, the collaborative economy, content co-creation and data analysis tools, among others, raise questions about DMOs' marketing management. Tourists employ these tools to access services and plan trips in an integrated way, using multiple devices (Sheivachman 2017a). At present, many DMOs promote their regions in traditional ways, attending fairs, making fam trips, advertising in traditional and online media, promoting their websites on social media and creating mobile applications. These actions provide destination information to residents, tourists and companies, but do not constitute proper knowledge management.

On the contrary, tourism distribution and social media companies have been much faster than DMOs in collecting, processing and managing the information that exists on the Internet about tourists and destinations. These companies have developed participatory processes to create multiple-information channels and to promote, distribute and market tourism products and services, outside the knowledge of the destination management entities.

Online search engines have today great ability to redirect user requests to the following types of intermediary:

- *Aggregator sites*, also known as metasearch engines. These enable comparisons of different tourist services and products. They usually belong to the largest distribution companies (e.g. Expedia, Priceline). These platforms redirect customers to the actual service provider, online travel agencies or product consolidators. They also allow price comparisons of products/services of companies in their own or other distribution groups. These systems, depending on product type, carry out a specific marketing function.
- *Online travel agencies* (OTAs). Most of these depend on technology suppliers to produce their catalogue of services; therefore, their offer is led by the objectives of the big distribution companies, because they need to be part of a larger distribution group, or use the services of one of them, to make direct sales. This category can also include *Internet Distribution Systems* (IDS), which allow group marketing of particular types of products, such as rural houses, hotels, etc. (e.g. *booking.com*); this business model is based on the customer making a direct payment to the hotel, instead of the agency.
- *Global distribution systems* (GDSs). These use highly integrated technologies and hold large product inventories; the aforementioned groups hire their services for their marketing. These three categories (i.e. aggregators, OTAs and GDSs) represent super-intermediation in the distribution sector.
- *Sharing economy*. Technologies have facilitated the proliferation of platforms that promote experiences, rental of tourist accommodation, etc., which not only impact on destinations in ways previously discussed, but even prevent DMOs from knowing how many tourists will be in their destination on any given date, which, in some cases, could cause infrastructure overload, overcrowding of services and damage to the destination's global image. This category of intermediary is problematic for destination management, because ever more services are being integrated into specific areas (e.g. accommodation, catering, transport) without consultation with DMOs.

Faced with this system of distribution and tourism hybridisation, DMOs are gradually beginning to attempt to use ICTs to improve, among other things, their knowledge of visitors, tourist flows, destination load capacity, return on advertising, etc. DMOs, therefore, need to implement integrated destination management systems (IDMS) to serve at both technical and operational levels (Figure 25.3). The current problem for most destinations lies in the lack of plans to introduce appropriate technological systems. If DMOs continue to use systems which are not fit for purpose, they will not be able to coordinate their actions. Thus, while they will be able to gather some information, they will have to continue to pay technology companies for data acquisition and processing.

An IDMS must have a technological architecture based on different subsystems that allow internal and external interoperability. The main subsystems of the IDMS will be information, marketing, decision-making support and integration (Guevara Plaza 2014). Each of these should be understood as a set of discrete modules that participate in different subsystems at the same time. IDMSs will enable efficient destination management, analysis of the traceability and

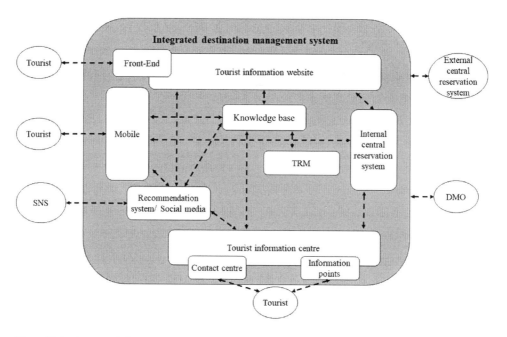

Figure 25.3 Integrated destination management system.

management of tourist flows, identification of demand profile (Leiva *et al.* 2013) and, above all, provide knowledge to guide tourism policy. To implement IDMSs, DMOs must top-down and bottom-up standardise their governance processes; there must be public–private and public–public alignment and collaboration, as only thus will it be possible to obtain the knowledge needed to manage the destination efficiently.

Conclusions

This chapter reviews recent literature in the field of eTourism to identify the main trends and challenges facing urban destinations. ICTs allow DMOs to generate knowledge through the collection and analysis of large amounts of data, and support their decision-making, governance, destination marketing, interaction with stakeholders and their real-time adaptation to their needs. In turn, tourists use technologies to improve their experiences, participate in the process of destination-image creation and interact socially. Thus, DMOs are faced with the great challenge of adopting and implementing technologies in a way that positively contributes to the creation of value for all destination stakeholders, in the context of today's continuous technological innovation.

In addition, DMOs also face a great challenge related not only to the direct use they make of technologies, but also in terms of how stakeholders use them. The new sharing economy business models pose just some of the many challenges that destinations will most probably have to address in the near future. Ensuring that disruptive business models are compatible with sustainable development, and that the needs and rights of all stakeholders are respected, will be one of the future policy cornerstones of urban destinations.

This chapter, which does not purport to be a thorough review of the new technologies available to destinations, outlines the technological context and some of the main challenges facing

DMOs. It contributes by providing information that can help researchers and practitioners to establish research agendas and guide sectoral policies and design action plans to improve the competitiveness and development of urban destinations.

References

Adamiak, C., 2018. Mapping Airbnb supply in European cities. *Annals of Tourism Research*, 71, 67–71.

Allam, Z. and Dhunny, Z.A., 2019. On big data, artificial intelligence and smart cities. *Cities*, 89, 80–91.

Ávila-Robinson, A. and Wakabayashi, N., 2018. Changes in the structures and directions of destination management and marketing research: A bibliometric mapping study, 2005–2016. *Journal of Destination Marketing & Management*, 10, 101–111.

Avraham, E., 2018. Nation branding and marketing strategies for combatting tourism crises and stereotypes toward destinations. *Journal of Business Research*, In Press. https://doi.org/10.1016/j.jbusres.2018.02.036.

Bloomberg Media Group, 2019. *World, Transformed Megatrends and Their Implications for Travel & Tourism.* Available from www.wttc.org/-/media/files/reports/2019/megatrends-2019-world-transformed.pdf [Accessed 30 September 2019].

Boes, K., Buhalis, D., and Inversini, A., 2015. Conceptualising smart tourism destination dimensions. *In: Information and Communication Technologies in Tourism 2015* (pp. 391–403). Springer, Cham.

Boes, K., Buhalis, D., and Inversini, A., 2016. Smart tourism destinations: Ecosystems for tourism destination competitiveness. *International Journal of Tourism Cities*, 2 (2), 108–124.

Bowen, J. and Whalen, E., 2017. Trends that are changing travel and tourism. *Worldwide Hospitality and Tourism Themes*, 9 (6), 592–602.

Buhalis, D. and Amaranggana, A., 2013. Smart tourism destinations. *In:* Z. Xiang and L. Tussyadiah, eds., *Information and Communication Technologies in Tourism 2014.* Zurich: Springer International Publishing, 553–564.

Buhalis, D. and Foerste, M., 2015. SoCoMo marketing for travel and tourism: Empowering co-creation of value. *Journal of Destination Marketing & Management*, 4, 151–161.

Buhalis, D. and Law, R., 2008. Progress in information technology and tourism management: 20 years on and 10 years after the Internet – the state of eTourism research. *Tourism Management*, 29 (4), 609–623.

Buhalis, D. and O'Connor, P., 2005. Information communication technology revolutionizing tourism. *Tourism Recreation Research*, 30 (3), 7–16.

Buhalis, D. and Sinarta, Y., 2019. Real-time co-creation and nowness service: Lessons from tourism and hospitality. *Journal of Travel & Tourism Marketing*, 36 (5), 563–582.

Cacho, A., Mendes-Filho, L., Estaregue, D., Moura, B., Cacho, N., Lopes, F., and Alves, C., 2016. Mobile tourist guide supporting a smart city initiative: A Brazilian case study. *International Journal of Tourism Cities*, 2 (2), 164–183.

Cheng, M., 2016. Sharing economy: A review and agenda for future research. *International Journal of Hospitality Management*, 57, 60–70.

Cheng, M. and Foley, C., 2018. The sharing economy and digital discrimination: The case of Airbnb. *International Journal of Hospitality Management*, 70, 95–98.

Constine, J. 2017. Nexto replaces boring audio guides with tourism games. *TechCrunch*, 4 DEC. Available from https://techcrunch.com/2017/12/04/nexto/ [Accessed 2 October 2019].

da Costa Liberato, P.M., Alén-González, E., and de Azevedo Liberato, D.F.V., 2018. Digital technology in a smart tourist destination: The case of Porto. *Journal of Urban Technology*, 25 (1), 75–97.

de Bruyn, C. and Alonso, A.F., 2012. Tourism destination governance: Guidelines for implementation. *In:* E. Fayos-Sola, J.A.M. da Silva, and J. Jafari, eds., *Knowledge Management in Tourism: Policy and Governance Applications.* Bingley: Emerald Group Publishing, 221–242.

de Rosa, A.S., Bocci, E., and Dryjanska, L., 2019. Social representations of the European capitals and destination e-branding via multi-channel web communication. *Journal of Destination Marketing & Management*, 11, 150–165.

de Rosa, A.S. and Dryjanska, L., 2017. Visiting Warsaw for the first time: Imagined and experienced urban places. *International Journal of Culture, Tourism and Hospitality Research*, 11(3), 321–340.

Del Vecchio, P., Mele, G., Ndou, V., and Secundo, G., 2018. Creating value from social big data: Implications for smart tourism destinations. *Information Processing and Management*, 54, 847–860.

Dickinson, J.E., Hibbert, J.F., and Filimonau, V., 2016. Mobile technology and the tourist experience: (Dis)connection at the campsite. *Tourism Management*, 57, 193–201.

Fromm, J., 2019. Megatrends of the tourism and travel industry. *Forbes*, 28 March. Available from www. forbes.com/sites/jefffromm/2019/03/28/megatrends-of-the-tourism-and-travel-industry/#6bbe4244c18c [Accessed 2 October 2019].

Fundación Orange, 2016. *La transformación digital en el sector turístico*. Available from www.fundacionorange. es/wp-content/uploads/2016/05/eE_La_transformacion_digital_del_sector_turistico.pdf [Accessed 1 October 2019].

Go, F.M. and Trunfio, M., 2011. E-services governance in public and private sectors: A destination management organization perspective. *In*: A. D'Atri, M. Ferrara, George, J.F., and Spagnoletti, P., eds., *Information Technology and Innovation Trends in Organizations*. Heidelberg: Physica-Verlag, 11–19.

Graham, J., Amos, B., and Plumptre, T., 2003. Governance principles for protected areas in the 21st century. Available from www.files.ethz.ch/isn/122197/pa_governance2.pdf [Accessed 25 February 2019].

Gretzel, U., 2014. Travel unplugged: The case of Lord Howe Island, Australia. *In*: K. Mackay, ed., *Proceedings of the TTRA Canada Annual Conference*. Yellowknife, Canada.

Gretzel, U., Sigala, M., Xiang, Z., and Koo, C., 2015. Smart tourism: Foundations and developments. *Electronic Markets*, 25 (3), 179–188.

Gretzel, U., Zhong, L., and Koo, C., 2016. Application of smart tourism to cities. *International Journal of Tourism Cities*, 2 (2).

Guevara Plaza, A. and Rossi Jiménez, C., 2014. Las tics aplicadas a la gestión de destinos turísticos. *In:* D. Flores Ruiz, ed., *Manual de gestión de destinos turísticos*. Valencia: Tirant Humanidades, 243–273.

Gupta, A., Dogra, N., and George, B., 2018. What determines tourist adoption of smartphone apps? An analysis based on the UTAUT-2 framework. *Journal of Hospitality and Tourism Technology*, 9 (1), 50–64.

Guttentag, D., 2015. Airbnb: Disruptive innovation and the rise of an informal tourism accommodation sector. *Current Issues in Tourism*, 18 (12), 1192–1217.

Hardy, A., Vorobjovas-Pinta, O., and Eccleston, R., 2018. Enhancing knowledge transfer in tourism: An Elaboration Likelihood Model approach. *Journal of Hospitality and Tourism Management*, 37, 33–41.

Islam, M.W., Ruhanen, L., and Ritchie, B.W., 2018. Adaptive co-management: A novel approach to tourism destination governance? *Journal of Hospitality and Tourism Management*, 37, 97–106.

Jackson, M., 2019. Utilizing attribution theory to develop new insights into tourism experiences. *Journal of Hospitality and Tourism Management*, 38, 176–183.

Jansson, A., 2018. Rethinking post-tourism in the age of social media. *Annals of Tourism Research*, 69, 101–110.

Juul, M., 2015. The sharing economy and tourism. Available from www.europarl.europa.eu/RegData/etudes/BRIE/2015/568345/EPRSBRI(2015)568345EN.pdf [Accessed 27 February 2019].

Kang, M. and Gretzel, U., 2012. Effects of podcast tours on tourist experiences in a national park. *Tourism Management*, 33 (2), 440–455.

Kim, C.S., Bai, B.H., Kim, P.B., and Chon, K., 2018a. Review of reviews: A systematic analysis of review papers in the hospitality and tourism literature. *International Journal of Hospitality Management*, 70, 49–58.

Kim, J. and Fesenmaier, D.R., 2017. Sharing tourism experiences: The posttrip experience. *Journal of Travel Research*, 56 (1), 28–40.

Kim, M.J., Lee, C.-K., and Jung, T., 2018b. Exploring consumer behavior in virtual reality tourism using an extended stimulus-organism-response model. *Journal of Travel Research*, 1–21. https://doi.org/10.1177/0047287518818915.

Lalicic, L. and Weismayer, C., 2018. Being passionate about the mobile while travelling. *Current Issues in Tourism*, 21 (8), 950–963.

Law, R., Buhalis, D., and Cobanoglu, C., 2014. Progress on information and communication technologies in hospitality and tourism. *International Journal of Contemporary Hospitality Management*, 26 (5), 727–750.

Laws, E., Agrusa, J., Scott, N., and Richins, H., 2011. Foreword: Tourist destination governance: Practice, theory and issues. *In*: E. Laws, H. Richins, J. Agrusa, and N. Scott, eds., *Tourist Destination Governance: Practice, Theory and Issues*. Wallingford, UK and Cambridge, MA: CABI, 1–16.

Leiva, J.L., Enciso, M., Rossi, C., Cordero, P., Mora-Bonilla, A., and Guevara, A., 2013. Context-aware recommendation using fuzzy formal concept analysis, *ICSOFT 2013*, 617–623.

Li, J., Xu, L., Tang, L., Wang, S., and Li, L., 2018. Big data in tourism research: A literature review. *Tourism Management*, 68, 301–323.

Lin, Z., Chen, Y., and Filieri, R., 2017. Resident-tourist value co-creation: The role of residents' perceived tourism impacts and life satisfaction. *Tourism Management*, 61, 436–442.

Lund, N.F., Cohen, S.A., and Scarles, C., 2018. The power of social media storytelling in destination branding. *Journal of Destination Marketing & Management*, 8, 271–280.

McLaren, D. and Agyeman, J., 2015. *Sharing Cities: A Case for Truly Smart and Sustainable Cities*. Cambridge, MA: The MIT Press.

Malone, S., McKechnie, S., and Tynan, C., 2018. Tourists' emotions as a resource for customer value creation, cocreation, and destruction: A customer-grounded understanding. *Journal of Travel Research*, 57 (7), 843–855.

Mariani, M.M., Mura, M., and Di Felice, M., 2018. The determinants of Facebook social engagement for national tourism organizations' Facebook pages: A quantitative approach. *Journal of Destination Marketing & Management*, 8, 312–325.

Mathew, P.V. and Sreejesh, S., 2017. Impact of responsible tourism on destination sustainability and quality of life of community in tourism destinations. *Journal of Hospitality and Tourism Management*, 31, 83–89.

Mathis, E.F., Kim, H., Uysal, M., Sirgy, J.M., and Prebenden, N.K., 2016. The effect of co-creation experience on outcome variable. *Annals of Tourism Research*, 57, 62–75.

Molinillo, M., Liébana-Cabanillas, F., Anaya-Sánchez, R., and Buhalis, D., 2018. DMO online platforms: Image and intention to visit. *Tourism Management*, 65, 116–130.

Navío-Marco, J., Ruiz-Gómez, L.M., and Sevilla-Sevilla, C., 2018. Progress in information technology and tourism management: 30 years on and 20 years after the internet. Revisiting Buhalis & Law's landmark study about eTourism. *Tourism Management*, 69, 460–470.

Neuhofer, B., Buhalis, D., and Ladkin, A., 2012. Conceptualising technology enhanced destination experiences. *Journal of Destination Marketing Management*, 1 (1), 36–46.

Neuhofer, B., Buhalis, D., and Ladkin, A., 2014. A typology of technology-enhanced tourism experiences. *International Journal of Tourism Research*, 16 (4), 340–350.

Neuhofer, B., Buhalis, D., and Ladkin, A., 2015. Smart technologies for personalized experiences: A case study in the hospitality domain. *Electronic Markets*, 25 (3), 243–254.

Park, S. and Santos, C.A., 2017. Exploring the tourist experience: A sequential approach. *Journal of Travel Research*, 56 (1), 16–27.

Paulo, M.M., Rita, P., Oliveira, T., and Moro, S., 2018. Understanding mobile augmented reality adoption in a consumer context. *Journal of Hospitality and Tourism Technology*, 9 (2), 142–157.

Pinke-Sziva, I., Smith, M., Olt, G., and Berezvai, Z., 2019. Overtourism and the night-time economy: A case study of Budapest. *International Journal of Tourism Cities*, 5 (1), 1–16.

Powell, L. 2019. How luxury hospitality can use technology to stay human. *Skift*, 26 February. Available from https://skift.com/2019/02/26/how-luxury-hospitality-can-use-technology-to-stay-human/ [Accessed 2 October 2019].

Rihova, I., Buhalis, D., Gouthro, M.B., and Moital, M., 2018. Customer-to-customer co-creation practices in tourism: Lessons from customer-dominant logic. *Tourism Management*, 67, 362–375.

Rivera, M., Croes, R., and Lee, S.H., 2016. Tourism development and happiness: A residents' perspective. *Journal of Destination Marketing & Management*, 5, 5–15.

Romão, J., Kourtit, K., Neuts, B., and Nijkamp, P., 2018. The smart city as a common place for tourists and residents: A structural analysis of the determinants of urban attractiveness. *Cities*, 78, 67–75.

Schor, J.B. and Fitzmaurice, C.J., 2015. Collaborating and connecting: The emergence of the sharing economy. *In*: L. Reisch and J. Thogersen, eds., *Handbook of Research on Sustainable Consumption*. Cheltenham: Edward Elgar Publishing, 410–425.

Seaton, J. 2017. Millennials are attending events in droves because of fear of missing out. *Skift*, 12 July. Available from https://skift.com/2017/07/12/millennials-are-attending-events-in-droves-because-of-fear-of-missing-out/ [Accessed 2 October 2019].

Sheivachman, A., 2017a. Half of business travelers want to avoid human interaction on the road. *Skift*, 11 July. Available from https://skift.com/2017/07/11/half-of-business-travelers-want-to-avoid-human-interaction-on-the-road/ [Accessed 1 October 2019].

Sheivachman, A., 2017b. Proposing solutions to overtourism in popular destinations: A skift framework. *Skift*, 23 October. Available from https://skift.com/2017/10/23/proposing-solutions-to-overtourism-in-popular-destinations-a-skift-framework/ [Accessed 1 October 2019].

Sigala, M., 2013. Examining the adoption of destination management systems: An inter-organizational information systems approach. *Management Decision*, 51 (5), 1011–1036.

Sigala, M., 2018. New technologies in tourism: From multi-disciplinary to anti-disciplinary advances and trajectories. *Tourism Management Perspectives*, 25, 151–155.

So, K.K.F., Wu, L., Xiong, L., and King, C., 2018. Brand management in the era of social media: Social visibility of consumption and customer brand identification. *Journal of Travel Research*, 57 (6), 727–742.

Sorrells, M., 2019. Destination marketing, part 4: How emerging destinations balance marketing and management. *Phocuswire*, 30 September. Available from www.phocuswire.com/destination-marketing-part-4-balance-management [Accessed 3 May 2020].

Stojanovic, I., Andreu, L., and Curras-Perez, R., 2018. Effects of the intensity of use of social media on brand equity: An empirical study in a tourist destination. *European Journal of Management and Business Economics*, 27 (1), 83–100.

Tanti, A. and Buhalis, D., 2017. The influences and consequences of being digitally connected and/or disconnected to travellers. *Information Technology & Tourism*, 17 (1), 121–141.

Thimm, T. and Seepold, R., 2016. Past, present and future of tourist tracking. *Journal of Tourism Futures*, 2 (1), 43–55.

United Nations World Tourism Organization (UNWTO), 2018. *UNWTO Annual Report 2017*. Madrid: United Nations World Tourism Organization. (UNWTO). Available from www.e-unwto.org/doi/pdf/10.18111/9789284419807 [Accessed 19 February 2019].

United Nations World Tourism Organization (UNWTO), 2019. *UNWTO Guidelines for Institutional Strengthening of Destination Management Organizations (DMOs): Preparing DMOs for New Challenges*. Madrid: UNWTO. Available from www.e-unwto.org/doi/pdf/10.18111/9789284420841 [Accessed 1 October 2019].

Uysal, M., Sirgy, M.J., Woo, E., and Kim, H.L., 2016. Quality of life (QOL) and well-being research in tourism. *Tourism Management*, 53, 244–261.

Vallespín, M., Molinillo, S., and Muñoz-Leiva, F., 2017. Segmentation and explanation of smartphone use for travel planning based on socio-demographic and behavioral variables. *Industrial Management & Data Systems*, 117 (3), 605–619.

Volgger, M. and Pechlaner, H., 2014. Requirements for destination management organizations in destination governance: Understanding DMO success. *Tourism Management*, 41, 64–75.

Wang, D., Park, S., and Fesenmaier, D.R., 2012. The role of smartphones in mediating the touristic experience. *Journal of Travel Research*, 51 (4), 371–387.

Wang, D., Xiang, Z., and Fesenmaier, D.R., 2014. Adapting to the mobile world: A model of smartphone use. *Annals of Tourism Research*, 48, 11–26.

Wang, D., Xiang, Z., and Fesenmaier, D.R., 2016. Smartphone use in everyday life and travel. *Journal of Travel Research*, 55 (1), 52–63.

Wang, H.Y. and Wang, S.H., 2010. Predicting mobile hotel reservation adoption: Insight from a perceived value standpoint. *International Journal of Hospitality Management*, 29 (4), 598–608.

Wearing, S.L. and Foley, C., 2017. Understanding the tourist experience of cities. *Annals of Tourism Research*, 65, 97–107.

Werner, K., Dickson, G., and Hyde, K.F., 2015. Learning and knowledge transfer processes in a mega-events context: The case of the 2011 Rugby World Cup. *Tourism Management*, 48, 174–187.

Whyte, P. 2018. Ian Schrager on how hotels get technology wrong. *Skift*, 27 September. Available from https://skift.com/2018/09/27/ian-schrager-on-how-hotels-get-technology-wrong/ [Accessed 1 October 2019].

Williams, C.C. and Horodnic, I.A., 2017. Regulating the sharing economy to prevent the growth of the informal sector in the hospitality industry. *International Journal of Contemporary Hospitality Management*, 29 (9), 2261–2278.

Woo, E., Kim, H., and Uysal, M., 2015. Life satisfaction and support for tourism development. *Annals of Tourism Research*, 50, 84–97.

Xiang, Z., 2018. From digitization to the age of acceleration: On information technology and tourism. *Tourism Management Perspectives*, 25, 147–150.

Xu, F., Buhalis, D., and Weber, J., 2017. Serious games and the gamification of tourism. *Tourism Management*, 60, 244–256.

Zanella, A., Bui, N., Castellani, A., Vangelista, L., and Zorzi, M., 2014. Internet of things for smart cities. *IEEE Internet of Things Journal*, 1 (1), 22–32.

Zatori, A., Smith, M.K., and Puczko, L., 2018. Experience-involvement, memorability and authenticity: The service provider's effect on tourist experience. *Tourism Management*, 67, 111–126.

26

THE GROWING ROLE OF SOCIAL MEDIA IN CITY TOURISM

Ulrike Gretzel

Introduction

When imagining a contemporary city tourist, it is almost impossible not to envision someone who uses a smartphone to look up positively rated restaurants, find attractions recommended by their favourite social media influencer, look up opening hours on an establishment's social media page, communicate with their Airbnb host over a social media messaging app, and snap a selfie or livestream a video in front of the city's most iconic views to share with others that they were indeed there at that moment. This intricate relationship between social media use and tourism consumption seems to be especially prominent in the case of city tourism, which requires more decisions, more navigation and more coordination, but also offers a much more diversified and fast-paced experience than tourism outside of urban areas, e.g. in the case of beach vacations (Ashworth & Page, 2011).

Much of the decision-making in urban contexts happens "on-the-move" while being confronted with an abundance of choice. Even when travelling in organised groups or having planned most of the trip in advance, city tourists typically have time to roam around or have opportunities to fit in sights, shopping or restaurants/cafés in between pre-determined stops or on the way to and from their accommodation because of the close proximity of attractions and facilities in cities. Being able to rely on the wisdom of crowds or the recommendations of similar or liked others accessible via social media brings many advantages to city tourists, including faster, more informed and more personalised decision-making. At the same time, the varied and many times uniquely instagrammable urban landscapes afford a multitude of experiences that lend themselves to being documented and shared with others. Such narration via social media can add significant value to the city tourism experience (Gretzel et al., 2011).

From a marketing point of view, it is particularly difficult to target and reach city tourists, who can be much more dispersed, indistinguishable from residents and more diverse in motivations and needs than tourists in non-urban contexts, such as in seaside resorts, on cruise ships, at mountain destinations or in national parks. Achieving high positive ratings on TripAdvisor, appearing on "must-see" lists on travel blogs and being represented on Instagram through compelling visuals is thus essential for the survival of urban tourism establishments. Being represented on social media is also more achievable for small providers than being featured in a blockbuster movie or running an international advertising campaign. However, social media attention can also mean being suddenly overrun by tourists (Gretzel, 2019), and negative social media ratings can put an end to businesses. Understanding social media users and uses is therefore fundamental to understanding current and future city tourism from both the demand and the supply perspectives.

This chapter provides an overview of the characteristics of city tourists and city tourism offerings that influence and are influenced by social media use to illustrate the growing role of social media in driving city tourism demand and city tourist behaviours. For instance, it discusses the increasing need to impress one's social media audiences with extraordinary selfies (Dinhopl & Gretzel, 2016; Lo & McKercher, 2015); a need that is more likely satisfied by the heterogeneity of experiences offered in cities. It also illustrates how the greater mobility of city tourists and greater levels of connectivity in cities as a result of smart tourism development foster social media sharing and greater reliance on social media-based information. At the same time, the chapter will highlight the link between social media and overtourism in cities and the emerging efforts of residents to protest against tourism via social media (Novy & Colomb, 2019) and those of destinations and tourism providers to seize the power of social media for promotion and visitor flow management. As such, the chapter emphasises the complex relationship between social media and city tourism to inform the future of city tourism development.

Social media uses and impacts on city tourism

Social media are websites and mobile applications that use Web 2.0 technology to enable content creation, sharing and social networking across a broad range of tourism stakeholders (Gretzel, 2018a). They encompass universal platforms that offer a wide variety of functions and support content on an infinite number of topics. They also include tourism-specific applications like TripAdvisor that cater to the needs of tourists to obtain and share travel information. Social media emerged from earlier forms of social communication and data sharing on the Internet, which makes it difficult to pinpoint the beginning of their existence as the Web 1.0 continues to exist alongside Web 2.0 offerings. Nevertheless, Kozinets (2020) suggests that the significant transition from what he calls the "Age of Virtual Community" to the "Age of Social Media" occurred in the late 1990s with the emergence of social networking sites like Cyworld in South Korea. Importantly, social media constantly evolve, with platforms changing or disappearing and new applications dynamically entering the already crowded social media landscape. Particularly notable is the recent "visual turn" in social media (Gretzel, 2017) that favours image and video contents over text-based contributions. Sayings like "pics or it didn't happen" illustrate the growing importance of visuals and play an especially prominent role in the context of tourism.

While regional differences exist in terms of platforms, penetration and use patterns, social media are used around the world. Statista (2019) estimates that there are 2.8 billion social media users worldwide in 2019, suggesting that, more than two decades after their first appearance in the media landscape, social media remain a growing global phenomenon and increasingly shape personal, social, political and economic communications and relations for a large portion of the world population. Social media not only have these impacts on their users' regular daily lives but also influence every aspect of their touristic experiences. Their particularly strong impact in the context of city tourism is mostly driven by the agglomeration of users and contents in urban areas, the increased levels of Internet connectivity in cities compared to rural areas, as well as the diversity and great number of touristic consumption incidents during an often very short period of time during city trips.

The important role of social media in enabling and influencing tourist decision-making has been widely acknowledged (e.g. Xiang & Gretzel, 2010; Fotis, Buhalis & Rossides, 2012; Tham,

Croy & Mair, 2013; Hudson & Thal, 2013; Tsiakali, 2018). Social media have made travel information not only more available but also more credible, more context-specific, more up-to-date, more experiential, more engaging and, thus, more relevant (Gretzel & Yoo, 2008). Social media content generated by other users can be highly motivating and create destination visit intentions (Llodra-Riera et al., 2015). Social media's role in facilitating the creation and sharing of travel-related electronic word-of-mouth (Litvin, Goldsmith & Pan, 2018) deserves special recognition as it has led to a shift in power dynamics between consumers and tourism providers and an overall increase in transparency in the tourism marketplace. Asking social media audiences for advice before a trip can lead to more realistic expectations and receiving "likes" for posts during and after travel can significantly influence one's trip satisfaction (Sedera et al., 2017). In addition, there is also no doubt that social media activities during and after the trip affect important memory and evaluation functions that influence reflections and experience recollection (Kim & Fesenmaier, 2017) and close the infamous touristic "circle of representation" (Jenkins, 2003).

Social media further support important social functions, such as communicating and bonding with family members while on a trip (Kennedy-Eden & Gretzel, 2016; Yu et al., 2018), fundamentally breaking down the boundaries of home and away (White & White, 2007) and everyday and vacation contexts (Mackay & Vogt, 2012). Through social media, regular practices and social relations can not only be maintained but intensified (Hall & Holdsworth, 2016). Social media also create new opportunities for social connectivity before, during and after travel (Munar, Gyimóthy & Cai, 2013) and spur the emergence of new social forms of travel such as "tinder tourism" (Leurs & Hardy, 2019). They further sustain efforts to connect with oneself by permitting communal self-tracking and allowing for the sharing of one's identity as a quantified traveller (Choe & Fesenmaier, 2017). The complexity of city tourism and the greater density of residents and tourists in urban spaces increase the feasibility of such mediated connections and simultaneously heighten the need for tools to manage social communication.

It is worthwhile highlighting Chinese city tourists in this context as China not only constitutes one of the largest outbound travel markets but also because cities remain the key destinations for Chinese tourists. IHG (2015) reports that major cities around the world account for over 85% of Chinese outbound travel. China's social media offerings and use culture are more advanced along several dimensions (Ge & Gretzel, 2018), most notably the use of social media platforms to support mobile payments. Brink (2019) emphasises that Chinese tourists seek authentic, social experiences they can share on social media and are highly influenced by online reviews of destinations and tourism providers. Thus, the above-mentioned social media impacts are especially pronounced in the case of Chinese city tourists.

Most recently, a special breed of social media users with a heightened ability to shape the decisions and consumption experiences of others has emerged, the so-called social media influencers (Xu & Pratt, 2018). City tourists often consume the city's attractions through the eyes of these influencers before physically travelling to the destination (Seeler, Lück & Schänzel, 2019) and adjust their tourism experiences while in a city to conform with influencer suggestions (Guerreiro, Viegas & Guerreiro, 2019). As a result, social media influencers are increasingly used by destinations to encourage visitation and shape tourist behaviours (Gretzel, 2018b).

Extant literature discusses the significance of social media in the context of marketing and managing entire destinations or particular tourism products (Hays, Page & Buhalis, 2013; Wozniak et al., 2017; Chang et al., 2018). Social media constitute convenient and cost-effective marketing and communication channels that permit more direct, persuasive, visible and spreadable forms of interactions between tourism marketers and a variety of stakeholders. As a result, relationships with tourism destinations and brands can be multidirectional and more continuous

(So et al., 2016). This opens up many opportunities particularly for smaller attractions and tourism providers to be discovered in the course of a city trip. Forming relationships with brands through social media might also be more worthwhile for tourists in the context of city tourism, which often involves repetition (Ashworth & Page, 2011).

However, the specific role of social media in supporting the activities of tourists, tourism businesses and destinations has not been systematically analysed in the context of city tourism. A first step in this process is understanding the relationships among urban spaces, media and touristic practices.

Tourism cities as mediated spaces

Traditionally, media have been conceptualised as representing "promoters" of urban tourism that help tourists form destination images and entice them to travel to cities – see for example Jansen-Verbeke's (1986) framework of city tourism. Yet, cities are not just geographic places but rather can and should be seen as complex mediaspaces (Couldry & McCarthy, 2004). As such, media do not simply induce tourism (Beeton, 2016) but substantially structure touristic experiences, as can be seen by the travel planning and touring practices of literary tourists (MacLeod, Shelley & Morrison, 2018).

Jansson (2002) explains that tourists not only consume landscapes and socioscapes but also mediascapes, and that media shape their appropriations of spaces in critical ways. Films have historically played an important role in terms of how we understand cities and imagine future forms of urban living (Mennel, 2019). New media like smartphone-enabled social media and augmented/virtual reality applications continue to affect how we see and comprehend urban spaces in general and specifically when we visit cities as tourists. Månsson (2011) also highlights the intertwined nature of tourism and media consumption, especially in the context of social media use. As a result, the tourist gaze is critically influenced by available media (Urry & Larsen, 2011) and these media are increasingly social media.

The mediatisation of cities through social media has unique consequences for urban tourism. For instance, Dinhopl and Gretzel (2016) suggest that social media redirect the touristic gaze away from the destination and its residents towards the self as they encourage sophisticated identity practices through the taking and posting of selfies during travel. Lo and McKercher's (2015) research on impression management and Dinhopl and Gretzel's (2018) concept of the networked neo-tribal gaze further suggest that these experiences are once again mediated when shared with imagined audiences on social media as visuals taken during a trip are subjected to a social media-gaze that carefully scrutinises whether they are indeed shareworthy. In addition, photo-editing and enhancement tools supported by social media applications (e.g. filters and stickers) mediate how others will see the experience and the city in which it took place.

A tourism city mediated by social media represents a dynamic mediascape that is subjected to social media logics and culture, e.g. the particular aesthetics of Instagram or WeChat and Weibo humour (Ge, 2019). Further, it constitutes a mediascape that is very fragmented and potentially extremely personalised. In contrast to consuming the Lonely Planet version of Bangkok or the Lord of the Rings edition of New Zealand and thus tapping into a collective experience, a city tourism experience structured through the consumption of social media is influenced by a myriad of public or private posts and recommendations. Such mediatisation of tourism cities through social media is further enhanced through smart development efforts in cities that facilitate and encourage social media use.

Smartification of urban spaces

The growing trend towards "smartification" of cities leads to an ever more prominent consumption of urban spaces mediated by smart technologies. Smartification refers to realising the technological potential of a city through investments in infrastructure and the adoption of alternative governance mechanisms with the goal of increasing the social and environmental sustainability of the city (Yigitcanlar et al., 2018). Shafiee et al. (2019) suggest that smart development requires a number of environmental, economic, social and technological actions in order to achieve these general smartification goals. From a tourism point of view, pursuing a smart development goal regularly translates into creating technology-based offerings that enhance the mobility as well as trackability of tourists and facilitate the creation of personalised or otherwise enhanced tourism experiences at the smart destination (Gretzel et al., 2015).

Installing city-wide, free-of-charge Wi-Fi networks is typically the first step in the smart development process and is especially significant in the smart tourism context (Gretzel, Ham & Koo, 2018). Social media use is accelerated through such heightened connectivity. Measures are also put in place to stimulate the flows and analytical exploitation of the data produced by tourists (often shared via social media) and of the information sensed and communicated by smart applications. This is usually accomplished through the creation of tourism observatories that act as data clearing houses (Varra et al., 2019) as well as open innovation approaches that allow for the direct translation of data into value propositions (Egger, Gula & Walcher, 2016).

As a consequence, the smart urban space is one that is characterised by hyper-connectivity of infrastructure and devices, hyper-mobility of humans and data, and hyper-personalisation of touristic value propositions. It is not only extensively consumed through social media but also created through social media data. The extent to which city tourism experiences are mediated through social media therefore increases exponentially in the course of the smartification of urban spaces. Two important consequences of such heightened social media use need to be discussed with respect to urban tourism: (1) the convergence of tourism and leisure; and (2) the potential of overtourism.

Mediatisation and the convergence of tourism and leisure

Couldry and Hepp (2013) define mediatisation as the "omnipresent and multidirectional nature of media's contribution to the 'texture' of our lives" (p. 193). In accordance, Jansson (2018) explores the role of social media in encouraging de–differentiation between tourism and other social realms. Similarly, Larsen (2019) suggests that existing models of city tourism are outdated and that a true understanding of "new urban tourism" requires recognition of the ordinary in touristic experiences and the extraordinary in everyday urban life. Maitland (2010) argues that contemporary city tourists regularly wander outside of tourism precincts and beyond iconic city attractions to mingle with the locals. By doing so, they create vibrant urban spaces that are also frequented and appreciated by city residents in search of exciting and meaningful leisure experiences. Such experiences are especially sought after by the creative tourist class, whose technology-mediated lives are characterised by a continuous quest for diverse experiences, whether at home or during travel (Gretzel & Jamal, 2009). Cities are uniquely positioned to provide such a multitude of experiences. Heightened demand for city tourism experiences can therefore be seen as a function of the need to make the everyday more extraordinary and one's touristic experiences more authentic.

Social media contribute to this phenomenon in multiple ways. First, tourists and residents alike share their urban experiences on social media and such contents allow others to build their own narratives of the city and its visit-worthy places. Even travel-specific social media platforms like TripAdvisor do not distinguish between residents and tourists and actively encourage the participation of locals, e.g. in their destination forums. As a result, experiences that used to require local insider knowledge are now accessible through social media posts. Combined with the increased mobility afforded by smart city developments, tourists can easily consume residential spaces and participate in the leisure activities of city residents.

Second, social media are "networks of desire" (Kozinets, Patterson & Ashman, 2016) that fuel the need to consume. Not only do social media users feel the need to constantly interact and post, they also feel pressured to post something that will direct attention to their profiles/ feeds. This creates a quest for the extraordinary (Dinhopl & Gretzel, 2016) and leads to performances for the camera, e.g. particularly funny or witty poses in museums (Kozinets, Gretzel & Dinhopl, 2017). Since work and home spaces provide only limited opportunities to satisfy the desire to post something special, tourists and residents alike venture out into urban spaces to take selfies or capture quirky aspects of city life.

Cityscapes usually provide an abundance of opportunities for framing the everyday and the regular self as extraordinary. Local businesses or destination marketing organisations often add to the social media-related attractiveness of urban spaces by making them particularly "instagrammable". Wall art (see Figure 26.1) is a prominent example illustrating how the selfie-focused gazes of tourists can be easily directed. Pop-up museums that encourage selfie-taking, such as the Museum of Ice Cream in San Francisco or the No-Filter-Museum in Vienna, provide additional examples of how the desire of tourists and residents to post can be fuelled through touristic/leisure offerings.

Figure 26.1 Instagrammable wall in Los Angeles.

Source: Photograph taken by author.

Third, social media profiles and ratings support sharing economy platforms like Airbnb, Uber or Eatwith that make city tourism more affordable and foster the mingling of tourists and residents while also creating new opportunities for "local" experiences. However, such blurring of residential and touristic lines can also have negative consequences (Colomb and Novy's 2016 book provides a wide array of cases), which are usually discussed under the label of "overtourism".

Social media-induced overtourism

Capocchi et al. (2019) define overtourism as excessive growth in visitation to concentrated areas without proper governance that leads to negative impacts perceived by residents and/or tourists. Although not exclusively an urban phenomenon, overtourism disproportionally affects tourism cities because of the growth in city tourism and the often extremely limited space in urban areas. Overtourism is directly linked to tourist gentrification of residential city areas (Milano, Cheer & Novelli, 2019) and is also often blamed on social media.

Gretzel (2019) dissects the relationship between social media and overtourism and concludes that it is complicated. While congestion in urban spaces due to large amounts of tourists is nothing new, social media-related photographic practices are especially prone to creating bottlenecks as pictures or videos have to be carefully framed and multiple takes are the rule to ensure that at least one will be good enough for sharing. Figure 26.2, for example, shows the

Figure 26.2 Overtourism in the city of Milan.

Source: Photograph taken by author.

Galleria Vittorio Emanuele II in Milan a few weeks before Christmas, with tourists slowly moving through the space, their smartphones raised above the crowd to capture the iconic building. Sometimes even the crowd itself becomes the motive of social media posts.

Further, the convergence of tourism and leisure and touristic and residential areas encouraged by social media makes it ever more difficult to effectively manage people and traffic flows in tourism cities. Also, while social media users generally seek unique experiences, they seem to be equally willing to follow the lists of must-see places promoted by TripAdvisor or their favourite social media influencers. At the same time, social media provide many opportunities for the redirection and dispersion of tourists, for instance through close collaboration with influencers to promote alternative ways of consuming the tourism city (Femenia-Serra & Gretzel, 2020).

Importantly, social media also assist residents in their efforts to raise awareness and combat overtourism issues. Reactions of residents to overtourism are multi-faceted (Novy & Colomb, 2019) but are regularly supported by various social media that facilitate the organisation of events and help to quickly and cost-effectively spread messages. Ironically, anti-tourism sentiment graffiti is often captured by tourists and shared on social media and protests by residents can themselves become tourism attractions (although not specifically related to overtourism activism, the protests in Hong Kong are an example of how tourists consume local activism, see https://hongkongfreetours.com/hk-protest-tour/). Thus, social media can turn overtourism activism into yet another opportunity for tourists to consume urban life and construct their social media identities.

Conclusion

Social media change the very fabric of city tourism, from planning to recollecting and sharing the city tourism experience with others. For contemporary city tourists, the whole city becomes a touristic playground and all urban spaces are canvasses on which city tourism experiences can unfold and be captured for social media sharing purposes. Social media enable new forms of sociality with one's travel group, other tourists, residents or tourism providers that enrich urban tourism experiences. Urban spaces, on the other hand, support the quest for extraordinary experiences that can be shared on social media for identity-creation and impression management purposes. Social media also create new levels of mediatisation and blur the line between the touristic and the everyday, which can enhance experiences but also lead to overuse and conflicts. Thus, the role of social media in city tourism is multi-faceted and complex and their impacts are not always positive.

Because of the ubiquitous influence of social media, managing and marketing tourism cities requires an in-depth understanding of urban spaces as mediascapes and detailed knowledge of social media use trends. While social media use challenges regular forms of urban tourism, social media also provide managers and marketers with ample opportunities to collect data, manage tourists and create new value propositions. From a research perspective, social media use in urban contexts offers a rich field of investigation that has so far been neglected (Larsen, 2019). Given the continuous growth in both social media use and city tourism, as well as the increasing implementation of smart development principles in tourism cities, more attention to the role and impact of social media in regard to urban tourism is certainly warranted.

References

Ashworth, G., & Page, S. J. (2011). Urban tourism research: Recent progress and current paradoxes. *Tourism Management*, 32(1), 1–15.

Beeton, S. (2016). *Film-induced Tourism*. Clevedon, UK: Channel View Publications.

Brink (2019). Chinese tourists are going international – and their travel habits are evolving. Accessed online (16 December 2019) at: www.brinknews.com/chinese-tourists-are-going-international-and-their-travel-habits-are-evolving/.

Capocchi, A., Vallone, C., Pierotti, M., & Amaduzzi, A. (2019). Overtourism: A literature review to assess implications and future perspectives. *Sustainability*, 11(12), 3303.

Chang, H. L., Chou, Y. C., Wu, D. Y., & Wu, S. C. (2018). Will firm's marketing efforts on owned social media payoff? A quasi-experimental analysis of tourism products. *Decision Support Systems*, 107, 13–25.

Choe, Y., & Fesenmaier, D. R. (2017). The quantified traveler: Implications for smart tourism development. In Xiang, Z. & Fesenmaier, D. R. (Eds.) *Analytics in Smart Tourism Design* (pp. 65–77). Springer, Cham.

Colomb, C., & Novy, J. (2016). *Protest and Resistance in the Tourist City*. New York: Routledge.

Couldry, N., & Hepp, A. (2013). Conceptualizing mediatization: Contexts, traditions, arguments. *Communication Theory*, 23(3), 191–202.

Couldry, N., & McCarthy, A. (2004). *Mediaspace: Place, Scale and Culture in a Media Age*. New York: Routledge.

Dinhopl, A., & Gretzel, U. (2016). Selfie-taking as touristic looking. *Annals of Tourism Research*, 57, 126–139.

Dinhopl, A., & Gretzel, U. (2018). The networked neo-tribal gaze. In Hardy, A., Bennett, A. & Robards, B. (Eds.) *Neo-tribes: Consumption, Leisure and Tourism* (pp. 221–234). Cham, Switzerland: Palgrave-Macmillan.

Egger, R., Gula, I., & Walcher, D. (2016). *Open Tourism: Open Innovation, Crowdsourcing and Co-creation Challenging the Tourism Industry*. Berlin: Springer.

Femenia-Serra, F., & Gretzel, U. (2020). Influencer marketing for tourism destinations: Lessons from a mature destination. In Neidhardt, J. & Wörndl, W. (Eds.) *Information and Communication Technologies in Tourism 2020* (pp. 65–78). Cham, Switzerland: Springer.

Fotis, J., Buhalis, D., & Rossides, N. (2012). Social media use and impact during the holiday travel planning process. In Fuchs, M., Ricci, F. & Cantoni, L. (Eds.) *Information and Communication Technologies in Tourism 2012* (pp. 13–24). Vienna: Springer-Verlag.

Ge, J. (2019). Social media-based visual humour use in tourism marketing: A semiotic perspective. *The European Journal of Humour Research*, 7(3), 6–25.

Ge, J., & Gretzel, U. (2018). A new cultural revolution: Chinese consumers' internet and social media use. In Sigala, M. & Gretzel, U. (Eds.) *Advances in Social Media for Travel, Tourism and Hospitality: New Perspectives, Practice and Cases* (pp. 102–118). New York: Routledge.

Gretzel, U. (2017). The visual turn in social media marketing. *Tourismos*, 12(3), 1–18.

Gretzel, U. (2018a). Tourism and social media. In Cooper, C., Volo, S., Gartner, W. C. & Scott, N. (Eds.) *The SAGE Handbook of Tourism Management*. Thousand Oaks, CA: Sage.

Gretzel, U. (2018b). Influencer marketing in travel and tourism. In Sigala, M. & Gretzel, U. (Eds.) *Advances in Social Media for Travel, Tourism and Hospitality: New Perspectives, Practice and Cases* (pp. 147–156). New York: Routledge.

Gretzel, U. (2019). The role of social media in creating and addressing overtourism. In Dodds, R. & Butler, R. (Eds.) *Overtourism: Issues, Realities and Solutions* (pp. 62–75). Berlin: De Gruyter.

Gretzel, U., Fesenmaier, D. R., Lee, Y.-J., & Tussyadiah, I. (2011). Narrating travel experiences: The role of new media. In Sharpley, R. & Stone, P. (Eds.) *Tourist Experiences: Contemporary Perspectives* (pp. 171–182). New York: Routledge.

Gretzel, U., Ham, J., & Koo, C. (2018). Creating the city destination of the future: The case of smart Seoul. In Wang, Y., Shakeela, A., Kwek, A. and Khoo-Lattimore, C. (Eds.) *Managing Asian Destinations* (pp. 199–214). Singapore: Springer.

Gretzel, U., & Jamal, T. (2009). Conceptualizing the creative tourist class: Technology, mobility, and tourism experiences. *Tourism Analysis*, 14(4), 471–481.

Gretzel, U., Sigala, M., Xiang, Z., & Koo, C. (2015). Smart tourism: Foundations and developments. *Electronic Markets*, 25(3), 179–188.

Gretzel, U., & Yoo, K. H., (2008). Use and impact of online travel reviews. In O'Connor, P., Höpken, W. & Gretzel, U. (Eds.) *Information and Communication Technologies in Tourism 2008* (pp. 35–46). Vienna: Springer-Verlag.

Guerreiro, C., Viegas, M., & Guerreiro, M. (2019). Social networks and digital influencers: Their role in customer decision journey in tourism. *Journal of Spatial and Organizational Dynamics*, 7(3), 240–260.

Hall, S. M., & Holdsworth, C. (2016). Family practices, holiday and the everyday. *Mobilities*, 11(2), 284–302.

Hays, S., Page, S. J., & Buhalis, D. (2013). Social media as a destination marketing tool: Its use by national tourism organisations. *Current Issues in Tourism*, 16(3), 211–239.

Hudson, S., & Thal, K. (2013). The impact of social media on the consumer decision process: Implications for tourism marketing. *Journal of Travel & Tourism Marketing*, 30(1–2), 156–160.

IHG (2015). The future of Chinese travel: The global Chinese travel market. Accessed online (15 December 2019) at: www.ihgplc.com/chinesetravel/src/pdf/IHG_Future_Chinese_Travel.pdf.

Jansen-Verbeke, M. (1986). Inner-city tourism: Resources, tourists and promoters. *Annals of Tourism Research*, 13(1), 79–100.

Jansson, A. (2002). Spatial phantasmagoria: The mediatization of tourism experience. *European Journal of Communication*, 17(4), 429–443.

Jansson, A. (2018). Rethinking post-tourism in the age of social media. *Annals of Tourism Research*, 69, 101–110.

Jenkins, O. (2003). Photography and travel brochures: The circle of representation. *Tourism Geographies*, 5(3), 305–328.

Kennedy-Eden, H., & Gretzel, U. (2016). Modern vacations – modern families: New meanings and structures of family vacations. *Annals of Leisure Research*, 19(4), 461–478.

Kim, J., & Fesenmaier, D. R. (2017). Sharing tourism experiences: The posttrip experience. *Journal of Travel Research*, 56, 28–40.

Kozinets, R. V. (2020). *Netnography: The Essential Guide to Qualitative Social Media Research*. London: Sage.

Kozinets, R., Gretzel, U., & Dinhopl, A. (2017). Self in art/self as art: Museum selfies as identity work. *Frontiers in Psychology*, 8, 731.

Kozinets, R., Patterson, A., & Ashman, R. (2016). Networks of desire: How technology increases our passion to consume. *Journal of Consumer Research*, 43(5), 659–682.

Larsen, J. (2019). Ordinary tourism and extraordinary everyday life: Re-thinking tourism and cities. In Frisch, T., Sommer, C., Stoltenberg, L. & Stors, N. (Eds.) *Tourism and Everyday Life in the Contemporary City* (pp. 24–41). New York: Routledge.

Leurs, E., & Hardy, A. (2019). Tinder tourism: Tourist experiences beyond the tourism industry realm. *Annals of Leisure Research*, 22(3), 323–341.

Litvin, S. W., Goldsmith, R. E., & Pan, B. (2018). A retrospective view of electronic word-of-mouth in hospitality and tourism management. *International Journal of Contemporary Hospitality Management*, 30(1), 313–325.

Llodra-Riera, I., Martinez-Ruiz, M. P., Jimenez-Zarco, A. I., & Izquierdo-Yusta, A. (2015). Assessing the influence of social media on tourists' motivations and image formation of a destination. *International Journal of Quality and Service Sciences*, 7(4), 458–482.

Lo, I. S., & McKercher, B. (2015). Ideal image in process: Online tourist photography and impression management. *Annals of Tourism Research*, 52, 104–116.

MacKay, K., & Vogt, C. (2012). Information technology in everyday and vacation contexts. *Annals of Tourism Research*, 39(3), 1380–1401.

MacLeod, N., Shelley, J., & Morrison, A. M. (2018). The touring reader: Understanding the bibliophile's experience of literary tourism. *Tourism Management*, 67, 388–398.

Maitland, R. (2010). Everyday life as a creative experience in cities. *International Journal of Culture, Tourism and Hospitality Research*, 4(3): 176–185.

Månsson, M. (2011). Mediatized tourism. *Annals of Tourism Research*, 38(4), 1634–1652.

Mennel, B. (2019). *Cities and Cinema*, 2nd edition. New York: Routledge.

Milano, C., Cheer, J. M., & Novelli, M. (Eds.). (2019). *Overtourism: Excesses, Discontents and Measures in Travel and Tourism*. Wallingford: CABI.

Munar, A. M., Gyimothy, S., & Cai. L. (Eds.). (2013). *Tourism Social Media: Transformations in Identity, Community and Culture*. Bingley: Emerald Publishing.

Novy, J., & Colomb, C. (2019). Urban tourism as a source of contention and social mobilisations: A critical review. *Tourism Planning & Development*, 16(4), 358–375.

Sedera, D., Lokuge, S., Atapattu, M., & Gretzel, U. (2017). Likes – the key to my happiness: The moderating effect of social influence on travel experience. *Information & Management*, 54(6), 825–836.

Seeler, S., Lück, M., & Schänzel, H. A. (2019). Exploring the drivers behind experience accumulation: The role of secondary experiences consumed through the eyes of social media influencers. *Journal of Hospitality and Tourism Management*, 41, 80–89.

Shafiee, S., Ghatari, A. R., Hasanzadeh, A., & Jahanyan, S. (2019). Developing a model for sustainable smart tourism destinations: A systematic review. *Tourism Management Perspectives*, 31, 287–300.

So, K. K. F., King, C., Sparks, B. A., & Wang, Y. (2016). The role of customer engagement in building consumer loyalty to tourism brands. *Journal of Travel Research*, 55(1), 64–78.

Statista (2019). Number of social media users worldwide from 2010 to 2021 (in billions). Accessed online (1 June 2019) at: www.statista.com/statistics/278414/number-of-worldwide-social-network-users/.

Tham, A., Croy, G., & Mair, J. (2013). Social media in destination choice: Distinctive electronic word-of-mouth dimensions. *Journal of Travel & Tourism Marketing*, 30(1–2), 144–155.

Tsiakali, K. (2018). User-generated-content versus marketing-generated-content: Personality and content influence on traveler's behaviour. *Journal of Hospitality Marketing & Management*, 27(8), 946–972.

Urry, J., & Larsen, J. (2011). *The Tourist Gaze 3.0*. London: Sage.

Varra, L., Buzzigoli, L., Buzzigoli, C., & Loro, R. (2019). Knowledge management for the development of a smart tourist destination: The possible repositioning of Prato. In Khosrow-Pour, M. (Ed.) *Smart Cities and Smart Spaces: Concepts, Methodologies, Tools, and Applications* (pp. 1145–1178). Hershey, PA: IGI Global.

White, N. R., & White, P. B. (2007). Home and away: Tourists in a connected world. *Annals of Tourism Research*, 34(1), 88–104.

Wozniak, T., Stangl, B., Schegg, R., & Liebrich, A. (2017). The return on tourism organizations' social media investments: Preliminary evidence from Belgium, France, and Switzerland. *Information Technology & Tourism*, 17(1), 75–100.

Xiang, Z., & Gretzel, U. (2010). Role of social media in online travel information search. *Tourism Management*, 31(2), 179–188.

Xu, X., & Pratt, S. (2018). Social media influencers as endorsers to promote travel destinations: An application of self-congruence theory to the Chinese Generation Y. *Journal of Travel & Tourism Marketing*, 35(7), 958–972.

Yigitcanlar, T., Kamruzzaman, M., Buys, L., Ioppolo, G., Sabatini-Marques, J., da Costa, E. M., & Yun, J. J. (2018). Understanding "smart cities": Intertwining development drivers with desired outcomes in a multidimensional framework. *Cities*, 81, 145–160.

Yu, X., Anaya, G. J., Miao, L., Lehto, X., & Wong, I. A. (2018). The impact of smartphones on the family vacation experience. *Journal of Travel Research*, 57(5), 579–596.

27

TRANSPORT IN TOURISM CITIES

Beyond the functional and towards an experiential approach

Claire Papaix and J. Andres Coca-Stefaniak

Introduction

A very close relationship exists between transport and tourism. This is especially the case when the means of travel turns into the tourism product itself; for instance, a scenic seaplane harbour take-off and landing in British Columbia (Yeoman, 2012). Similarly, from a tourism development perspective, transport also plays a key role in the planning and management of infrastructures. For instance, the opening of new air routes in Morocco had an impact on leisure travel to and within this country (Dobruszkes *et al.*, 2012). Transport-related aspects also affect the overall development of tourism cities as well as the management of associated tourism growth in and around the region.

On a more international level, tourism and transport have also been intimately linked as a result of the strategic positioning of global tourism cities trying to attract overseas tourists, even if attitudes towards long-distance flights have changed considerably over the last decade or so due to the environmental challenges facing our planet, including climate change. Crozet (2017), for instance, shows that this mindset and lifestyle change among new generations of tourists has led to an increasing "preference for proximity" among travellers adopting paradigms such as slow travel with a preference for more local entertainment, echoing existing notions of responsible tourism, sustainability and environmentally conscious purchasing habits. Such new trends slowly add to (but, sadly on the sustainability front, do not replace) preferences for fast travel, which dominated earlier tourist choices to visit exotic destinations in far-flung locations.

If transport trends arguably shape tourism ones, wider techno-societal changes may also have an impact on the relationship between transport and tourism itself too. An example of this phenomenon is illustrated by the rapid changes that autonomous vehicles are bringing to tourists' preferences (e.g. moving hotels), as posited by Cohen and Hopkins (2019). Nevertheless, these apparent virtues inherent to the mobility element of the tourism system may be counterbalanced by some of the more detrimental environmental impacts of transport (e.g. CO_2 emissions), especially in the context of society's rising levels of awareness related to environmental sustainability imperatives (OECD, 2014; United Nations, 2019).

Crucially, mobility is interpreted differently by local residents and tourists (Albalate and Bel, 2010), while one sector of the economy can impact the other. Global urban tourism destinations such as Barcelona and Venice illustrate this well, as tourism-driven demand for urban public transport has resulted in negative impacts for local residents in the form of general overcrowding and even saturation in some transport routes. On the other hand, this demand can also stimulate investment in urban innovation projects and infrastructure such as the development of eco-friendly walking paths or digital solutions, which often benefit both stakeholder groups.

Another issue worthy of consideration here is that of wellness and authenticity (e.g. Gnoth, 2020), with specific emphasis on the co-creation of wellbeing often involving interaction between tourists and host communities. Yet, the co-creation of wellness should not be restricted solely to tourists, with further research in this field likely to result in new frameworks and models when combined with existing knowledge in urban transport policy design. An early example of this can be found in Manchester's use of community-led design for improving public spaces (Anderson *et al.*, 2017).

All in all, transport has an impact on several different "life satisfaction domains" (Brown, 2017), which affect tourists as much as local residents. This includes their ability to connect with other people, amenities, services and places. As a result of this, the focus of this chapter will be on the analysis of the multi-fold relationships between urban transport, tourism and wellness.

The role of transport in the development of tourism cities

This section starts with an overview of the historical links between transport and tourism. It then homes in on the evolution of the mobility system on tourism cities and its effects on urban management, often influenced by added demand pressures brought on by overtourism and seasonal fluctuations inherent to the tourism sector.

Historical links between transport and tourism

Transport has historically played a central role in the design and management of cities (WHO, 2009). However, transport is not merely an enabling factor for urban dwellers in their daily activities, including travel to work, access to health care or leisure (Martens, 2016) or a supply chain element in the provision of food or energy. In tourism cities specifically, transport plays a key role in terms of providing accessibility for visitors to key attractions and activities, linking places, people and services (e.g. conference centres, events, etc.), thus acting as a major attractor for tourism (Page and Ge, 2009). Similarly, transport links help to build and foster community identity (for a review see, for instance, Yildirimoglu and Kim, 2018) and have been found to be a catalyst in forging trust in host–visitor relationships – see Dickinson *et al.* (2018) for an analysis of the use of lift-share apps for instance – as well as making city centres more attractive to businesses (Gibbons *et al.*, 2019).

Nonetheless, these virtues of the mobility system need to be balanced against their potential detrimental effects, especially in the context of sustainability (OECD, 2014) with regard to the local economy as well as the wider tourism sector. Inevitably, the apparent improvements brought about by enhanced urban mobility systems will be evaluated differently by local residents and tourists (Albalate and Bel, 2010). For instance, in Venice, the pressures from tourists on local services, including public transport, have led to a decline in the quality of life of local residents, which has manifested itself in saturated transport services, overcrowding and growing air pollution. Over the Easter holiday of 2019, for instance, over 125,000 people visited Venice (Gowreesunkar and Séraphin, 2019). As a result, phenomena such as overtourism (Dodds and Butler, 2019; Capocchi *et al.*, 2019), tourism-phobia (Milano *et al.*, 2019) and a regular exodus of residents from Venice (Casagrande, 2016) over the last 15 years led to the phenomenon aptly coined the "Venice Syndrome" (Martín Martín *et al.*, 2018).

In spite of this, some scholars have argued – perhaps rather optimistically at this early stage – that the use of new technologies could help to alleviate the effects of and even prevent over-tourism (see, for instance, Pinke-Sziva *et al.*, 2019; or Skeli and Schmid, 2019), particularly in the context of smart tourism destinations (Gretzel and Scarpino-Johns, 2018; Coca-Stefaniak, 2019) – i.e. a selected group of global urban tourism destinations, to which Venice incidentally belongs (Gorrini and Bertini, 2018). Barcelona is arguably one of the best examples of this with dramatically rapid tourism growth since the 1992 Olympics (almost a quadrupling of hotel stays between 1990 and 2013 (Goodwin, 2018)), even if, somewhat ironically, recent regional polit-ical conflict leading to social unrest and street riots may have done arguably even more damage to the city's brand and driven tourists away (Burgen, 2017), certainly in the period 2017–2019. In its drive against the abuse of the sharing economy, the city of Barcelona developed a website for residents to report illegal tourist apartments in their buildings (Barcelona City Council, 2019). Parallel to this, other initiatives elsewhere include the use of technologies to disperse visi-tors beyond the city, create new tourism itineraries and attractions, adopt a more inclusive approach to local communities, and/or improve city infrastructure and facilities, as advocated by a recent United Nations World Tourism Organization study on this front (UNWTO, 2018).

Tourism can also be a vector of urban innovation in as much as it can prompt the redesign or renewal of infrastructures or zones. Involving various stakeholders in this dynamic can drive this innovation, with symbiotic benefits for tourists and residents alike. For instance, Club Med has started using gamification in their service offer to incentivise guests to familiarise themselves with the natural environment and visit sites in France in a more sustainable way (Buhalis *et al.*, 2019). This could have a positive influence on residents as well as tourists by encouraging both stakeholder groups to adopt more sustainable habits (see, for instance, Lamsfus *et al.*, 2015).

Furthermore, technological advancements and the increasing use of customisable apps, devices and systems for travellers – e.g. virtual reality apps, interactive and personalised maps making it possible to "day dream" and "visit" a city from afar (Bogicevic *et al.*, 2019) – are some examples of what remains an array of tourism-led options that have a positive impact on well-ness by allowing would-be visitors to gain an immersive experience of destinations in the virtual domain before making purchase decisions related to when to go, where to stay and what to do once there.

From a social perspective, there is an emerging trend for "experience-centric management" (Zatori, 2016), which includes the co-creation of leisure activities by tour guides and particip-ants. These opportunities for interaction between tourists and host communities and/or tourist guides are not strictly limited to the tourism sector and could provide valuable research insights for improving the quality of life of residents in tourism cities too. For instance, the European programme "MyNeighbourhood" (Oliveira, 2014) emanates from this idea with the creation of a participatory platform (MyN platform) that connects urban dwellers with one another as well as with local authorities to share knowledge, needs and suggestions related to health, environ-ment and transport with the overall aim of creating more resilient communities with higher levels of wellness.

Yet, as we will see in the next section, transport is a service with a certain significance for tourism cities that stretches beyond its mere function of providing access to amenities and activities.

Transport as "utility" vs transport as tourism

Figure 27.1 illustrates the (most conventional) "utility" facet of transport services. Yet, transport may also be a desirable (and even desired) option for its own sake. In tourism, this involves

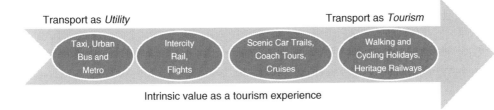

Transport as *Utility*

Transport as *Tourism*

Taxi, Urban
Bus and
Metro

Intercity
Rail,
Flights

Scenic Car Trails,
Coach Tours,
Cruises

Walking and
Cycling Holidays,
Heritage Railways

Intrinsic value as a tourism experience

Figure 27.1   The transport experience spectrum – transport as utility vs transport as tourism.
Source: Adapted from Page and Ge (2009).

cruises, scenic train journeys, mountain trekking and cycle touring, among other options. In fact, transport as a *direct demand*, rather than merely a functional service is a nascent body of knowledge in transport studies with seminal works such as studies by Russell and Mokhtarian (2015), which were the first to explore "teleportation" as a concept. This involved the creation of an experiment named the "teleportation test", where participants were surveyed about their desire to hypothetically teleport themselves to work in the morning. The study showed, somewhat surprisingly, that the majority of commuters favoured travelling to work by bus, train, car, riding a bicycle or walking, instead of a hypothetical direct "teleportation". Reasons given for this revolved around the benefits associated with the journey, such as time to reflect on one's day, reading, listening to music, chatting with others, etc.

Findings from this seminal piece in transport studies, notably on the observed high "intrinsic value" of transport, apply similarly to tourism, particularly given that transport accounts for a large proportion of tourism activities (Peeters *et al.*, 2019). This is especially the case given that people tend to link geographically distant or "exotic" destinations to happiness (Ram *et al.*, 2013). In a similar vein, research by Cohen and Gössling (2018) found that tourists may in some cases express their needs for peer recognition and social representation through "distance, expenditure and culture" while away on holiday, leading to a glamorisation of long-distance travel. Unfortunately, this yearning for travel to far-flung destinations is linked directly to the tourism industry's contribution to greenhouse gas (GHG) emissions, chiefly through long-haul flights, which surpasses the GHG emissions of all other activities at tourism destinations (e.g. visits, accommodation, etc.) as posited by Peeters *et al.* (2019). More specifically, GHG emissions due to long-haul flights have been linked directly to a combination of the so-called "lure of distance", annual frequency of visits to far-flung locations and number of tourists making these journeys. Of these, the latter two factors were found by Peeters *et al.* (2019) to be largely related to expendable income, flight prices and the degree of psychological attraction of remote locations.

On the other hand, transport also plays a central role in alternative approaches to leisure travel, including "slow tourism" (Heitmann *et al.*, 2011; Fullagar *et al.*, 2012; Oh *et al.*, 2016) and its associated concept of *slow travel* (Tomić *et al.*, 2018), which includes push factors such as relaxation, self-reflection, escape and discovery through mindful walking, discovery bike rides and other channels. Using a destination in Serbia as a case study, Tomić *et al.* (2018) found that the push factors outlined above encourage tourists to engage with the slow tourism philosophy, by staying longer at their chosen locations and engaging more actively with local residents. These findings offer valuable insights for further research into the role of transport in tourism in view of developing less socially and environmentally intrusive options and policy interventions in urban settings, while achieving transport solutions that remain efficient, profitable, sustainable and attractive to local communities and tourists alike.

Smart transport and smart tourism cities

Regardless of whether transport is seen as a purely functional service or whether it is considered in its more experiential sphere, the architecture and constant evolution of the transport network is a key shaping factor affecting our cities. For instance, research by Anastasi *et al.* (2013) has shown that the provision of real-time information about traffic congestion, regulation, level of social interactions and the resulting planning for more efficient routes (e.g. shorter, less polluting routes) can have a positive impact on the design of intelligent transport systems (ITS), which is a key contributor to the latest inception of smart cities (Höjer and Wangel, 2015; Ahvenniemi *et al.*, 2017; Yigitcanlar *et al.*, 2019). In addition to more advanced and sustainable transport systems, smart cities have brought about new technologies in the fields of healthcare in the form of monitoring platforms and digital devices providing access to online services to populations with special needs. Similarly, innovative solutions such as tailor-made tours for the elderly and physically impaired are also part of the smart city concept (Arenas *et al.*, 2019; Hussain *et al.*, 2015). However, smart cities have also been critiqued for neglecting other growing issues in urban living such as social exclusion and the unequal access to technology (Behrendt *et al.*, 2017) among other local conflicts. This is explored further below.

Balancing the interests of tourists vs residents through sustainable transport solutions

An analysis of the strategic challenges and opportunities related to the design and implementation of modern transport systems in tourism cities would not be complete without discussing the often-conflicting stakeholder priorities as well as wider sustainable development considerations; for instance, the United Nations' Sustainable Development Goals (UN, 2019). Although some scholars (see, for instance, Gössling and Hall, 2019) have posited that the sharing economy (e.g. shared mobility through the use of platforms such as Ofo, Car2Go or Drivy) could be the way to reconcile the needs of tourists and local residents in smart tourism destinations, this remains a contested argument.

Newquay in Cornwall (United Kingdom) offers a good example of how tourist-induced mobility and the arrival of rail infrastructure to the region towards the end of the nineteenth century changed progressively the structure of its visitor economy from a health-oriented offer (e.g. sea, air and bathing), especially for families and the elderly, to a fashionable surfing destination later in the twentieth century (Newquay Coastal Community, 2016). However, in spite of growing footfall levels in Cornwall's coastal towns combined with higher levels of spending by visitors with more jobs created as a result, residents have sometimes reacted adversely due to some of the negative impacts tourism has also delivered, including noise and littering (Hu *et al.*, 2019).

At the other end of the spectrum, Lille (France) has managed to achieve a better balance between investment on transport accessibility for tourists and the mobility needs of impaired residents. A local campaign named "Tourism and disability" was created for this purpose (Visit-lilles, 2019). By providing real-time information on the accessibility of sites and tourist facilities, each type of disability (motor, visual, hearing and mental impairment) was incorporated into the tourism offer from a transport perspective. In fact, metro and tramway platforms in Lille are now equipped with customer lifts, audio announcements and buses with low-access floors, while the city's tourism information office also offers visitors guided tours of the city using specially adapted vehicles for people with reduced mobility. The same tourism information office also offers accessibility maps to reach specific locations such as local theatres, museums, operas, cinemas, libraries, places of worship, etc.

On that line, it is noteworthy how the Emirates Air Line cable car in London (Urban Transport News, 2019) has delivered a practical alternative to reduce transport congestion locally and provide direct access across the River Thames in Greenwich (London). Although this initiative was developed originally for the London 2012 Olympic Games, it remains a good illustration of how a local transport solution initially developed for local residents can become a tourist attraction afterwards.

Local resident vs tourism-induced mobility – the environmental perspective

An initiative instigated by the Chronicle Digital Walking Trails app (EADT, 2017), which currently includes trails in Ipswich, Felixstowe, Eye Town and Eye Airfield in Suffolk (Chronicle Digital Storytelling, 2016; Heritage, 2019), currently enables tourists to engage in self-guided tours using their smartphones. These tours explore local points of interest by unlocking local stories and historical anecdotes. This technology-driven solution effectively blends heritage themes with tourism and provides a good example of how a tourism-led initiative can benefit local residents by encouraging them to participate in more sustainable leisure activities by learning and engaging with their local heritage. Another interesting example can be found in Madrid (Spain), where a recently introduced local transport policy (currently under review) has limited the use of cars in parts of the city centre. This initiative has been arguably beneficial for tourists by offering a cleaner environment with lower levels of air pollution (*New Scientist*, 2018). Similarly, a ban on city centre street parking in Oslo (Norway) to make room for pedestrians and cyclists has also been greeted with the approval of tourists and residents alike as have been car-free days in Paris (France), Brussels (Belgium) and London. Other more stringent initiatives are also being piloted elsewhere, including a ban on diesel vehicles in Mexico City's downtown, Athens (Greece) and Rome (Italy), which appear to be part of a wider trend to curb emissions of NOx and particulates, which have been linked to respiratory diseases, asthma, cancer and early death, particularly among local residents within close proximity to busy city roads.

Local resident vs tourism-induced mobility – the wellbeing perspective

Combining wellness with authentic experiences is one of the goals of the "Love Your Commute" campaign currently piloted by Thames Clippers in London (UK). This initiative is now being combined with related wellbeing programmes and on-board events that include refreshments, outdoor back deck, climate-controlled cabins, reliable mobile signal throughout, freebies and other stress-busting activities. Although this marketing campaign is aimed primarily at regular local commuters choosing the city's River Thames over other transport options, early signs suggest that it has been popular with tourists too. Specific activities trialled as part of this initiative by the Thames Clippers ferry operator – in partnership with Happy Heads – have included free mindfulness sessions. The first of these was piloted on 18 September 2018 on the 07:35 hrs westbound service from Woolwich to central London, where busy commuters could take part in interactive sessions aimed at building confidence, reducing anxiety and increasing focus at work and in everyday life. All this while inspiring passengers to make the most of their time on board this urban ferry service.

Linked to this, recent research carried out by the University of Greenwich (Thames Clippers, 2018) on the links between commuting and wellbeing revealed that commuters who travel by river ferry were generally happier than those using other alternatives such as the city's Docklands Light Railway (DLR) or the train. This was particularly so when evaluating perceived safety, travelling and work-friendly environment, on-board services and travel-related wellbeing factors such as a sense of fulfilment, commuting time and travel cost. A follow-up study from Thames Clippers (2018) corroborated these findings highlighting the fact that British travellers tend to be slow to recover from a stressful commute, with various negative effects impacting their working day, leading to higher levels of irritability in the workplace, difficulties with tasks that require concentration and decreased productivity. If such entertaining initiatives were specifically designed for commuters, they could also benefit tourists, as designing enriched on-board activities has been found to distract people from their tendency to dwell on work-related worries (27% of the surveyed travellers), personal finances (19%) and health issues (11%). Furthermore, the use of transport services with a wellness angle is also prevalent among global tourism cities with strong brands, such as Copenhagen, New York, Dubai or London (Dinnie, 2015). In Paris, a co-design approach has been adopted for urban activities and solutions for happier citizens, which can also inspire and attract tourists too. Increasingly, tourists in the French capital are turning to discovering this city by living like a "local" – as indicated by the innovative and entertaining insider initiatives of "La Ville Experientielle" in Paris (Medium, 2017) echoing earlier related research by Paulauskaite *et al.* (2017). In Porto (Portugal), a similar approach has also been adopted (Nouvel Obs, 2015).

Conclusions

This chapter has explored the links between urban transport provision, tourism and wellness. It is concluded that transport services can be interpreted as an active contributor to every domain affecting the quality of life (Rojas, 2006) of local residents and tourists alike. This effect can be direct – for instance, when transport is used as a purposeful choice (e.g. leisure cruise or recreational ride). However, transport can also have an indirect effect on the wellness of its users as a convenient and efficient means of access to workplaces, shopping, health care, tourism or leisure attractions or simply by connecting people to each other.

This chapter has also posited that local economies and the tourism sector can benefit symbiotically from innovative transport solutions. In fact, planning decisions and policy interventions, notably those related to commuting would benefit from integrating tourism values, including elements of wellness, a focus on cultural identities and a more experiential approach to the delivery of transport solutions.

References

Ahvenniemi, H., Huovila, A., Pinto-Seppä, I. and Airaksinen, M. (2017), "What are the differences between sustainable and smart cities?", *Cities*, *60*, pp. 234–245.

Albalate, D. and Bel, G. (2010), "Tourism and urban public transport: Holding demand pressure under supply constraints", *Tourism Management*, *31*(3), pp. 425–433.

Anastasi, G., Antonelli, M., Bechini, A., Brienza, S., D'Andrea, E., De Guglielmo, D., Ducange, P., Lazzerini, B., Marcelloni, F. and Segatori, A. (2013), "Urban and social sensing for sustainable mobility in smart cities", *Sustainable Internet and ICT for Sustainability*, pp. 1–4.

Anderson, J., Ruggeri, K., Steemers, K. and Huppert, F. (2017), "Lively social space, well-being activity and urban design: Findings from a low-cost community-led public space intervention", *Environment and Behavior*, *49*(6), pp. 685–716.

Arenas, A., MeinGoh, J. and Urueñac, A. (2019), "How does IT affect design centricity approaches: Evidence from Spain's smart tourism ecosystem", *International Journal of Information Management*, *45*, pp. 149–162.

Barcelona City Council (2019), https://meet.barcelona.cat/habitatgesturistics/en (accessed 23 October 2019).

Behrendt, F., Murray, L., Hancox, A., Sourbati, M. and Huber, J. (2017), *Intelligent Transport Solutions for Social Inclusion (ITSSI)*, Project Report, Brighton: University of Brighton.

Bogicevic, V., Seo, S., Kandampully, J., Liu, S. and Rudd, N. (2019), "Virtual reality presence as a pre-amble of tourism experience: The role of mental imagery", *Tourism Management, 74*, pp. 55–64.

Brown, R.L. (2017), "Quality of life: Challenges to research, practice and policy", *Journal of Policy and Practice in Intellectual Disabilities, 14*(1), pp. 7–14.

Buhalis, D., Harwood, T., Bogicevic, V., Viglia, G., Beldona, S. and Hofacker, C. (2019), "Technological disruptions in services: Lessons from tourism and hospitality", *Journal of Service Management, 30*(4), pp. 484–506.

Burgen, S. (2017), "Catalonia tourism slumps 15% since referendum violence", *Guardian*, 20 October, www.theguardian.com/world/2017/oct/20/catalonia-tourism-slumps-15-since-referendum-violence (accessed 5 November 2019).

Capocchi, A., Vallone, C., Pierotti, M. and Amaduzzi, A. (2019), "Overtourism: A literature review to assess implications and future perspectives", *Sustainability, 11*(12), article 3303.

Casagrande, M. (2016), "Heritage, tourism, and demography in the island city of Venice: Depopulation and heritagisation", *Urban Island Studies, 2*, pp. 121–141.

Chronicle Digital Storytelling (2016), "Digital walking trails", www.chroniclestories.co.uk/digital-walking-trails (accessed 25 October 2019).

Coca-Stefaniak, J.A. (2019), "Marketing smart tourism cities: a strategic dilemma", *International Journal of Tourism Cities, 5*(4), pp. 513–518.

Cohen, S. and Gössling, S. (2018), "A darker side of hypermobility", *Environment and Planning A, 47*(8), pp. 1661–1679.

Cohen, S. and Hopkins, D. (2019), "Autonomous vehicles and the future of urban tourism", *Annals of Tourism Research, 74*, pp. 33–42.

Crozet, Y. (2017), "Appraisal methodologies and the limits to speed gains", *Transportation Research Procedia, 25*, pp. 2898–2912.

Dickinson, J.E., Filimonau, V., Cherrett, T., Davies, N., Hibbert, J.F., Norgate, S. and Speed, C. (2018), "Lift-share using mobile apps in tourism: The role of trust, sense of community and existing lift-share practices", *Transportation Research Part D: Transport and Environment, 61*, pp. 397–405.

Dinnie, K. (2015), *City branding*, London: Palgrave Macmillan.

Dobruszkes, F., Mondou, V. and Ghedira, A. (2012), "Assessing the impacts of aviation liberalisation on tourism: Some methodological considerations derived from the Moroccan and Tunisian cases", *Journal of Transport Geography, 50*, pp. 15–127.

Dodds, R. and Butler, R. (eds.) (2019), *Overtourism: Issues, realities and solutions*, Oldenbourg: De Gruyter.

EADT (2017), "New digital walking trail app to explore Felixstowe", www.eadt.co.uk/news/new-digital-walking-trail-app-to-explore-felixstowe-1-4971303 (accessed 13 November 2019).

Fullagar, S., Markwell, K. and Wilson, E. (eds.) (2012), *Slow tourism: Experiences and mobilities*, London: Channel View Publications.

Gibbons, S., Lyytikäinen, T., Overman, H.G. and Sanchis-Guarner, R. (2019), "New road infrastructure: The effects on firms", *Journal of Urban Economics, 110*, pp. 35–50.

Gnoth J. (2020), "Extending the sport value framework for spectator experience design", In: S. Roth, C. Horbel and B. Popp (eds.), *Perspektiven des Dienstleistungsmanagements*. Wiesbaden: Springer Gabler.

Goodwin, H. (2018), "Managing tourism in Barcelona", Responsible Tourism Partnership Working Paper no. 1, 2nd edition, October.

Gorrini, A. and Bertini, V. (2018), "Walkability assessment and tourism cities: the case of Venice", *International Journal of Tourism Cities, 4*(3), pp. 355–368.

Gössling, S. and Hall, C.M. (2019), "Sharing versus collaborative economy: How to align ICT developments and the SDGs in tourism?", *Journal of Sustainable Tourism, 27*(1), pp. 74–96.

Gowreesunkar, V. and Séraphin, H. (2019), "What smart and sustainable strategies could be used to reduce the impact of overtourism?", *Worldwide Hospitality and Tourism Themes*, Vol. 11, No. 5, pp. 484–491.

Gretzel, U. and Scarpino-Johns, M. (2018), "Destination resilience and smart tourism destinations", *Tourism Review International, 22*(3–4), pp. 263–276.

Heitmann, S., Robinson, P. and Povey, G. (2011), "Slow food, slow cities and slow tourism", In: P. Robinson, S. Heitmann and P. Dieke (eds.), *Research themes for tourism*, Wallingford: CABI, pp. 114–127.

Heritage (2019), "Ipswich digital trail launch", www.heritageopendays.org.uk/visiting/event/ipswich-digital-trail-launch/ipswich-digital-trail-launch (accessed 13 November 2019).

Höjer, M. and Wangel, J. (2015), "Smart sustainable cities: Definition and challenges", In: L.M. Hilty and B. Aebischer (eds.), *ICT innovations for sustainability*, Boston, MA: Springer, pp. 333–349.

Hu, H., Zhang, J., Wang, C., Yu, P. and Chu, G. (2019), "What influences tourists' intention to participate in the Zero Litter Initiative in mountainous tourism areas: A case study of Huangshan National Park, China", *Science of the Total Environment*, *657*, pp. 1127–1137.

Hussain, A., Wenbi, R., Da Silva, A., Nadher, M. and Mudhish, M. (2015), "Health and emergency-care platform for the elderly and disabled people in the Smart City", *Journal of Systems and Software*, *110*, pp. 253–263.

Lamsfus, C., Martín, D., Alzua-Sorzabal, A. and Torres-Manzanera, E. (2015), "Smart tourism destinations: An extended conception of smart cities focusing on human mobility", In: I. Tussyadiah, A. Inversini (eds.), *Information and communication technologies in tourism 2015*, London: Springer, pp. 363–375.

Martens, K. (2016), *Transport justice: Designing fair transportation systems*, London: Routledge.

Martín Martín, J., Guaita Martínez, J. and Salinas Fernández, J. (2018), "An analysis of the factors behind the citizen's attitude of rejection towards tourism in a context of overtourism and economic dependence on this activity", *Sustainability*, *10*(8), pp. 28–51.

Medium (2017), "'La ville expérientielle', l'expérience des communs?", https://medium.com/@hello_66502/la-ville-exp%C3%A9rientielle-lexp%C3%A9rience-des-communs-ed8e49a6a448 (accessed 27 October 2019).

Milano, C., Novelli, M. and Cheer, J.M. (2019), "Overtourism and tourismphobia: A journey through four decades of tourism development, planning and local concerns", *Tourism Planning and Development*, *16*(4), pp. 353–357.

New Scientist (2018), "Banning cars in major cities would rapidly improve millions of lives", www.newscientist.com/article/mg24032010-100-banning-cars-in-major-cities-would-rapidly-improve-millions-of-lives/ (accessed 8th November 2019).

Newquay Coastal Community (2016), *Economic plan*, www.coastalcommunities.co.uk/wp-content/uploads/2016/05/Newquay-Coastal-Community-Economic-Plan-2016.pdf (accessed 25 October 2019).

Nouvel Obs (2015), "Le tourisme, version expérientielle", https://o.nouvelobs.com/voyage/20150305.OBS3932/le-tourisme-version-experientielle.html (accessed 26 October 2019).

OECD (2014), *The cost of air pollution: Health impacts of road transport*, Paris: OECD Publishing, https://doi.org/10.1787/9789264210448-en.

Oh, H., Assaf, A.G. and Baloglu, S. (2016), "Motivations and goals of slow tourism", *Journal of Travel Research*, *55*(2), pp. 205–219.

Oliveira, A. (2014), "Human smart cities: A human-centric model aiming at the wellbeing and quality of life of citizens", *eChallenges e-2014 Conference Proceedings*, IIMC International Information Management Corporation.

Page, S. and Ge, Y. (2009), "Transportation and tourism: A symbiotic relationship?" In: T. Jamal and M. Robinson (eds.), *The SAGE handbook of tourism studies*, London: Sage, pp. 371–395.

Paulauskaite, D., Powell, R., Coca-Stefaniak, J.A. and Morrison, A.M. (2017), "Living like a local: Authentic tourism experiences and the sharing economy", *International Journal of Tourism Research*, *19*(6), pp. 619–628.

Peeters, P., Higham, J., Cohen, S., Eijgelaar, E. and Gössling, S. (2019), "Desirable tourism transport futures", *Journal of Sustainable Tourism*, *27*(2), pp. 173–188.

Pinke-Sziva, I., Smith, M., Olt, G. and Berezvai, Z. (2019), "Overtourism and the night-time economy: A case study of Budapest", *International Journal of Tourism Cities*, *5*(1), pp. 1–16.

Ram, Y., Nawijn, J. and Peeters, P.M. (2013), "Happiness and limits to sustainable tourism mobility: A new conceptual model", *Journal of Sustainable Tourism*, *21*(7), pp. 1017–1035.

Rojas, M. (2006), "Life satisfaction and satisfaction in domains of life: Is it a simple relationship?", *Journal of Happiness Studies*, *7*(4), pp. 467–497.

Russell, M. and Mokhtarian, P. (2015), "How real is a reported desire to travel for its own sake? Exploring the 'teleportation' concept in travel behaviour research", *Transportation*, *42*(2), pp. 333–345.

Skeli, S. and Schmid, M. (2019), "Mitigating overtourism with the help of smart technology solutions: A situation analysis of European city destinations", In: *ISCONTOUR 2019 Tourism Research Perspectives: Proceedings of the International Student Conference in Tourism Research*, pp. 13–24.

Thames Clippers (2018), "Trains, pains & automobiles", www.thamesclippers.com/about-thames-clippers/news/trains-pains-and-automobiles-commuting-has-a-major-impact-on-brits-mental-health (accessed 19 November 2019).

Tomić, S., Leković, K. and Stoiljković, A. (2018), "Impact of motives on outcomes of the travel: Slow tourism concept", *Škola Biznisa*, (2), pp. 68–82.

United Nations (2019), *About the sustainable development goals*, www.un.org/sustainabledevelopment/sustainable-development-goals/ (accessed 25 October 2019).

UNWTO (2018), *"Overtourism"? Understanding and managing urban tourism growth beyond perceptions*, www.e-unwto.org/doi/pdf/10.18111/9789284419999 (accessed 5 November 2019).

Urban Transport News (2019), "In-depth insight | UK's first urban cable car: Emirates Air Line, London", https://urbantransportnews.com/uk-first-urban-cable-car-know-the-engineering-challenges/ (accessed 25 October 2019).

Visitlilles (2019), "Tourism and disabilities in Lille Metropole", www.visitlilles.com/en/your-stay/lille-s-for-everyone (accessed 20 October 2019).

WHO (2009), *Healthy transport in developing cities*, Health and Environment Linkages Policy Series, Geneva: United Nations Environment Programme World Health Organization, www.who.int/heli/risks/urban/transportpolicybrief2010.pdf (accessed 22 November 2019).

Yeoman, I. (2012), *2050 – Tomorrow's tourism*, London: Channel View Publications.

Yigitcanlar, T., Kamruzzaman, M., Foth, M., Sabatini-Marques, J., da Costa, E. and Ioppolo, G. (2019), "Can cities become smart without being sustainable? A systematic review of the literature", *Sustainable Cities and Society*, *45*, pp. 348–365.

Yildirimoglu, M. and Kim, J. (2018), "Identification of communities in urban mobility networks using multi-layer graphs of network traffic", *Transportation Research Part C: Emerging Technologies*, *89*, pp. 254–267.

Zatori, A. (2016), "Exploring the value co-creation process on guided tours (the 'AIM-model') and the experience-centric management approach", *International Journal of Culture, Tourism and Hospitality Research*, *10*(4), pp. 377–395.

28

THE ARTISTIC MEDIUM OF WALKING

A model for reflexive tourism the memory is the medium – tourism, walking and art

Blake Morris

Walking tourism in cities is often associated with guided (or self-guided) tours that highlight points of historical or cultural interest. In contrast, artistic walking practices offer opportunities for creative exploration of the city in new ways, be this through artist-led group walks, or independent explorations guided by instructions, audio-recordings, locative media or other means. There has not been much discussion of the intersection between artistic walking practices and tourism strategies, despite increased interest in walking across disciplines (Lorimer, 2011; Shortell and Brown, 2014; Hall et al., 2017). Phil Smith (2013a, p. 113), one of the few scholars to consistently discuss this relationship, calls for a serious exploration of "the resources that walking arts hold for a renewed tour guiding". Building on Smith, I ask how the artistic medium of walking can contribute to reflexive modes of tourism that encourage critical engagement with the everyday, rather than an escape from it.

Artists' walks are not generally designed strictly for tourists, and do not follow traditional models of walking-related tourism; rather they draw on the *memory* of the medium to create works that challenge dominant understandings of space and create new possibilities for engagement with social and topographical structures. As I have previously argued (Morris, 2018b), works in the artistic medium of walking can encourage what geographer David Crouch (2002, p. 209) has called "doing tourism", a process by which the "individual emerges as subject, as an active (but not free) agent" (p. 209). Whether we participate in a guided walk, or walk alone following an artist's instructions, we are participating in a specifically designed artistic experience. These walks offer experiences beyond the choreographed movements, scripted commentaries and inoffensive photo opportunities generally associated with tourism (Urry and Larsen, 2011, p. 202), and ask participants to engage critically with the city and their position in it (be this as visitor or resident).

Before I begin, it is necessary to establish how I distinguish works in the artistic medium of walking from other walking practices, such as traditional walking tours. Smith (2013b, p. 105) notes, "[w]hile walking artists hold many differing views about the roots of their practice, it is rare to find one who does not have some sense of their work as being part of a history of practice". More than just a history of practice, I argue contemporary artists are drawing on the *memory* of the medium in the construction of their works, and it is this memory that provides the artistic medium of walking with its unique attributes. Building on art theorist and historian Rosalind Krauss's concept of the expanded field (Krauss, 1979, 2011), I define the artistic medium of walking as a work in which the "logic of representation" is the act of going for a walk (Morris, 2018b). In *Under Blue Cup*, Krauss reworks media theorist Marshall McLuhan's famous phrase "the medium is the message" to form a new aphorism: "Brain – *The medium is the memory*" (Krauss, 2011, p. 127). This insists "on the power of the medium to hold the efforts of the forebears of a specific genre in reserve for the present" (2011, p. 127). When artists invent a medium they are tapping into its memories – what Krauss terms "the rules of the guilds" (2011, p. 7). In this way, she links medium to its logic of representation, as held by the collective memory of a guild of practitioners, rather than any specific material form.

In this chapter I look at three different modes of artistic walking practices that are available to tourists. I begin with a discussion of two key aspects of the memory of the medium – the Dada excursion of 1921 and the drifts of the Letterist and Situationist Internationals – before moving on to specific examples of artistic practices, including audio-walks, psychogeographic drifts and mass-participation walks. These include the site-specific, such as the Loiterer Resistance Movement's First Sunday Walks in Manchester, or Deveron Projects' annual Slow Marathon in Scotland, as well as those that can be done anywhere in the world, as is the case with Jennie Savage's *A Guide to Getting Lost* (2014). I identify how these walks differ from more traditional walking tourism activities and argue they offer modes of critical engagement with the city that encourages a reflexive tourism.

Walking art, tourism and the memories of the medium

The relationship between walking, art and tourism is longstanding, and could be traced back to the medieval pilgrims and beyond (Adler, 1989; Careri, 2002; Solnit, 2001). The breadth of this relationship offers various entry points to the *memory* of the medium, such as the Romantic recasting of the tourist landscape, the detached urban flâneur strolling the urban arcades, or the Surrealists' exploration of the collective unconscious and mysteries of the city. My discussion focuses on two precedents that are particularly pertinent to how the artistic medium of walking encourages a model for reflexive tourism: the Dadaists and the Situationist and Letterist Internationals (SI and LI).

The Dada Excursion of 1921

In 1921 the Parisian contingent of the avant-garde Dadaist movement planned a series of excursions to sites "that do not really have any reason to exist" (Sanouillet, 2009, p. 178). Only one walk came to fruition – an excursion to the Church of Saint-Julien-le-Pauvre on 14 April – and it is often credited as establishing the walk itself as a work of art, or, rather a work of anti-art (Careri, 2002, pp. 68–75; O'Rourke, 2013, p. 13; Waxman, 2010, p. 12). Details concerning the event are both vague and conflicting. In the most widely circulated account, future founder of the Surrealists Andre Breton (2003, p. 143) describes a crowd of hundreds; whereas Dadaist Hans Richter's account states "it rained, and no one came" (cited in Bishop, 2012, p. 69). Most likely it was a group of about a dozen Dadaists and 50 spectators huddled together under umbrellas on a soggy grey day (Sanouillet, 2009, p. 179). As art historian T.J. Demos (2009, p. 138) points out, while the excursion "never had time to develop fully as an artistic paradigm", it figures as an experimental modelling of a new kind of activity that could "be added to the list of Dada's major contributions to modern art".

Scholar J.J. Haladyn (2014, p. 32) argues the excursion served as "a means of re-finding parts of (the history of) Paris that were abandoned or lost in the representation of the city as seen through the eyes of tourism and consumerism". One of the key techniques it established was the subversion of the guided tour. Rather than walk through picturesque sites where a tour guide offered historical information and anecdotes, the Dadaists took participants to a neglected church and engaged in a variety of often nonsensical tactics. Surrealist Georges Hugnet described it as an "absurd rendez-vous, mimicking instructive walks, *guide à la clé*" (cited in Bishop, 2012, p. 67). For a portion of the walk Dadaist Georges Ribemont-Dessaignes served as a tour guide. He would stop before a monument, sculpture or column and, holding "a big Larousse diction-ary", "read an article chosen at random from the volume" (Sanouillet, 2009, p. 179). A torren-tial downpour forced the Dadaists to finish early, and the audience that remained after the 90 minutes tour "were offered surprise envelopes containing 'sentences, portraits, business cards, bits of cloth, landscapes', [and] obscene drawings", among other things (Sanouillet, 2009, p. 179). The envelopes were one of the few bits of ephemera produced by the walk, otherwise only programmes, flyers, a single photograph and the stories of the participants remain to docu-ment the excursion.

Essential to the mythology of the work is that it did not change the physical environment in any way, and did not frame documentation from the event as art objects after the fact; instead the walk itself was positioned as a work of (anti-) art. Haladyn (2014, p. 32) asserts the excursion "represents the illuminating potential of the everyday as an event without inherent meaning, but a perpetual event in which individual meaning must be *willingly* created out of the banalities and boredoms of lived existence in the world". The excursion brought attention to the contrast between the marginalised aspects of the Parisian landscape and how that landscape is presented through official narratives. In this way, it promoted a reflexive mode of engagement, in which participants encountered themselves and the city in relation to their preconceived expectations. Though it is generally considered a failure, it established important precedents in shifting focus to the lived experience of the everyday and moving art from the indoor space of the gallery or theatre into the street.

Drifting with the Letterist and Situationist Internationals

The Situationist International was formed in 1957 by an amalgamation of already existing collec-tives and was active until its dissolution in 1972. Their practices are one of the most consistent points of reference for contemporary artists working with walking and essential to the *memory* of the medium (Smith, 2010; O'Rourke, 2013; Wilkie, 2015). The Situationist International (1960) promoted interaction, dialogue and "total participation" through the "organisation of the directly lived moment". It looked to contribute to "the social revolution" through practice, replacing "the old world with a new one" (Marcus, 1989, p. 376). Key to this was the *dérive*, a method of drifting through urban space conceived by the LI and further developed by the SI.

To participate in a *dérive*, participants drop their everyday relationships, "their work and leisure activities, and all their other usual motives for movement and action", and drift through the city guided "by the attractions of the terrain and the encounters they find there" (Debord, 1958). The *dérive* has two stated goals: the discovery and *détournement* of the city's "psychogeo-graphical" contours, and "engagement in playful-constructive behaviour" (Debord, 1958). *Détournement*, short for "détournement of pre-existing aesthetic elements" (Situationist Inter-national, 1958), is a method by which already existing materials are combined to create new meanings. The *dérive* is a physical *détournement* in which walking through the city is a way to actively reconfigure it.

While this might seem similar to the exploratory wanders of tourists and visitors, the *dérive* is a practice-based critique of the way spectacular capitalism shapes urban spaces. Guy Debord (1958, #168), one of the key figures in the Situationist International, defines tourism as "the opportunity to go and see what has been banalised". For Debord (1958, #168), tourism is "human circulation packaged for consumption, a by-product of the circulation of commodities". Unlike tourism, "popular as repugnant as sports or buying on credit" (Debord, 1955) the *dérive* disrupts the equivalence of global spaces through a focus on the distinct contours of the city itself. On a *dérive* walkers playfully deconstruct urban space to develop new social and spatial models outside of the logic of all-consuming spectacular capitalism. As SI founding member Michèle Bernstein (2013, p. 13) explains, the *dérive* "wasn't a hobby, [the SI] wanted to make it a way of life". In this way, the *dérive* is a tool to develop and articulate a critique of, rather than an escape from, the everyday.

These two aspects of the memory of the medium explicitly subvert modes of tourism promoted by capitalist structures and looked to challenge sanitised views of the city marketed for consumers (be these locals or visitors). The refusal to produce art objects or create public sculptures, favouring instead a direct intervention in the street, are essential to the mythology of these works. The way in which artists draw on these *memories* is what distinguishes works in the artistic medium of walking from other walking practices such as hikes, rambles or historical walking tours.

Audio-walks: walking and listening

One of the most prevalent ways for towns and cities to engage visitors is through audio-walks – pre-recorded narrations that lead someone on a walking tour. As social scientists Angharad Saunders and Kate Moles (2016, p. 68) point out, visit "any town or city today and the chances are that there is a form of audio walk available to guide you around". Audio-walks are appealing for both their accessibility and longevity: they are often available for free download, can be done at any time, and don't require the presence of a tour guide. Toby Butler (2007, p. 360), in one of the earlier essays to address the burgeoning role of audio-walks in heritage strategies, argues that ease of producing audio-walks, due to new technology, "has opened up new realms of opportunity for people to narrate, layer and intervene in the experience of moving through places". Websites and mobile applications such as Tourist Tracks, TripScout and Geotourist, offer audio-tours liberated from the traditional guidebooks, often promoted with a promise to help you "adopt the guise of a local" (Singer, 2017). The recent acquisition of the San Francisco based audio-walk company Detour by Bose, a major American corporation that plans to integrate it into their new augmented reality platform (Grant, 2018), testifies to the interest in the form (and its potentially lucrative nature).

There are distinctions between audio-walks created as works of art and those marketed for tourists. Joost Fonteyne (2013), an artist and member of the Resonance – European Sound Art Network, notes tourism offices are increasingly using sound and audio-walks in an "effort to seduce people to visit their cities or regions". For Fonteyne (2013) these walks are commercial enterprises, "nicely wrapped up in an 'infotainment package'" geared towards tourists. Rather, he argues, the starting point for audio-walks should be the "production, presentation and promotion of art projects". While commercial modes of audio-walks often replicate the guided tour in recorded form, artists draw on the memory of the medium in the construction of audio-walks that often challenges the straightforward consumption of historical information. Artistic audio-walks draw on the *memory* of the medium to create new understandings of, rather than an escape from, the everyday.

A guide to getting lost

My own experience of audio-walks has primarily been in locations in which I have been resident. In 2016 I participated in Jennie Savage's audio-walk *A Guide to Getting Lost* at the WALKING WOMEN (2016) exhibition in London. WALKING WOMEN, curated by Clare Qualmann and artist Amy Sharrocks, consisted of walks, talks, screenings, workshops and installations in London and Edinburgh and featured over 40 women artists working with walking in a variety of media.[1] In addition to Savage's walk, visitors to the exhibition in London had a variety of opportunities to engage with the city, including walking down Victoria Embankment in a burqa for Yasmin Sabri's *Walk a Mile in Her Veil* (2016), or following the footsteps of the Suffragettes for Deirdre Heddon and Misha Myer's *Walking Library for Women Walking* (2016). Savage's walk draws on her own experiences as a traveller and tourist to guide the listener through a walk that can be done anywhere in the world, in this instance the area around Somerset House in London. It was the only audio-walk at the exhibition, and is the focus of my discussion here.

A Guide to Getting Lost combines audio-recordings made while Savage was "travelling and walking in locations as far flung as Plymouth, Delhi, Copenhagen, Quebec and Marrakesh" with instructions for a guided walk (Savage, 2017, p. 2). It consists of field-recordings, thick descriptions of her walks, and instructions for when and where to turn. Originally released in 2014 through a worldwide "Fracture Mob", which asked people from across the world to participate in the walk simultaneously, it has since been installed at various exhibitions and is available for free download online. The walk is "designed with dis-orienteering in mind" and calls on the memory of SI/LI walking practices (Savage, 2017, p. 2), asking people to wilfully get lost. It investigates how "listening to audio whilst walking sets up a conversation between the gaze, the cognition of the walker and the place" through which the person is walking, and provides an opportunity to consider "how landscapes are constructed in the imagination" (Savage, 2017, p. 2). In doing so, it turns a global tourist gaze to a local landscape, asking walkers to navigate between the near and far and consider how their constructions are interrelated.

As Savage walks through the various global locations she describes her experiences and notes when she is turning, walking straight or stopping, and instructs you to follow her directions. She comments on the sites she sees and as I walked through London, I heard the street-vendors of Delhi and Marrakesh, or the ocean waves of Plymouth that Savage recorded during her walks. The length of the streets around Somerset House did not match those of the global locations through which Savage was walking and often Savage would ask me to turn before I was able. This ensured my rapid passage through varied ambiences, as I held accumulating instructions in my head and tried to follow them. During the long stretches where I was told to walk straight ahead, I had to make the choice whether to abandon her previous instructions. I chose to do so multiple times, picking up from Savage's next direction. Though I could have paused the track, the walk was designed to be continuous and of a specific length and doing so would have disrupted its temporal flow. In some ways this felt essential to the artistic experience, as I got lost in the combination of her narration, directions and the actual streets I was traversing.

Often, I experienced a disconnect between Savage's soundscape and the relatively quiet streets around Somerset House. Hearing her recorded footsteps, or people shouting in the distance, I would turn and look to make sure no one was behind me. This added a sense of paranoia to the walk, as I alternated between following Savage and feeling like I was being followed. At other times Savage's walk paralleled my own. For example, as Savage approached a set of stairs along the English seaside, I found a similar set on the Thames. The live sounds of the urban river overlapped with Savage's beach soundscape, transporting me to a coastal English beach, while also bringing attention to my actual experience of walking in London. By the walk's end I had to consult a map to find my way back to the exhibition space, Savage's walk having successfully disoriented me despite my relative familiarity with the area.

A Guide to Getting Lost flips the usual tourist guide on its head, asking walkers to vicariously experience Savage's travels through various landscapes during a walk through a single location. Savage's walk disrupted my usual experience of walking around Somerset House, which consisted primarily of functional walks to and from public transport. While the experience of Savage's walk would invariably differ for a tourist than a local, in either situation it asks the participant to consider the landscape through which they are walking in relation to the global landscapes of Savage's soundtrack.

Psychogeography in the UK

In the UK, the Situationist legacy is demonstrated by the number of groups and artists actively working with psychogeographic tactics and ideas. The Loiterer's Resistance Movement, Leeds Psychogeography, Radical Stroud and The Huddersfield Psychogeography Network, which organises the annual Fourth World Congress of Psychogeography (2015–present), regularly plan walks and events rooted in Situationist tactics. Additionally, artists have created guidebooks that draw on strategies adopted from the LI/SI. The site-specific performance group Wrights & Sites (Stephen Hodge, Simon Persighetti, Phil Smith and Cathy Turner) has published two Mis-Guides, *An Exeter Mis-Guide* and *A Mis-Guide to Anywhere* that offer strategies for the "recasting of a bitter world by disrupted walking" (Wrights and Sites, 2006). Building on this work, Smith developed a theory and practice of counter-tourism, which "emerged as a popular means for addressing the ideological labyrinth of heritage space" (Smith, 2013a, p. 106). In *Counter-Tourism: The Handbook* (Crab Man, 2012) he offers a series of tactics for engaging with heritage sites in alternative ways. Instructions and guides offer passports into Situationist walking practices, however, the *dérive* is fundamentally a group activity; while "[o]ne can dérive alone … all indications are that the most fruitful numerical arrangement consists of several small groups of two or three people who have reached the same level of awareness" (Debord, 1958). As such, my discussion here will focus on Manchester's The Loiterer's Resistance Movement (LRM), the most consistently active psychogeography group in the UK.

The Loiterers Resistance Movement

Founded by Morag Rose in 2006, the LRM meets on the first Sunday of every month for "a free communal wander, open to anyone curious about the potential of public space and unravelling stories hidden within our everyday landscape" (Morris and Rose, 2019). While the collective can't "agree on what psychogeography means", they "all like plants growing out of the side of buildings, looking at things from new angles, radical history, drinking tea and getting lost" (Rose, 2006). The LRM has a fluid membership model, in which "people float in and out and define their own level of commitment" (Rose, 2015, p. 148). It actively argues that the "streets are free and belong to everyone" (Rose, 2015, p. 159), and this non-commercial ethos is embedded in the practice: LMR walks are always free to the public and "no one makes a monetary profit" (Rose, 2015, p. 159).

A "working-class, queer, disabled woman" (Rose, 2015, p. 149), Rose updates Situationist and Letterist concepts through an intersectional approach. She argues that the "pavement is one of the few opportunities for casual, embodied encounters with difference" (Morris and Rose, 2019) and LRM walks foreground questions of class, race, gender and ability. Though psychogeography has "a reputation for being arcane and difficult" (Morris and Rose, 2019, p. 157), the LRM offers opportunities to enter theory through practice. Importantly in this context, the LRM facilitates "having fun and feeling like a tourist in your hometown" (Rose, 2006), offering locals new ways of seeing and thinking about the city in which they live. While not geared towards tourists, the open and inclusive ethos of the LRM means anyone who wants to attend is able, including visitors of the city.

The LRM's first Sunday walks make use of a variety of tactics drawn from the medium, "including algorithmic walks, transposing maps, throwing dice, [and] concentrating on specific senses" (Rose, 2015, p. 152). Tourists participating in the LRM's First Sunday Walks should not expect a guided tour of psychogeographical hotspots; rather, the LRM facilitates playful, but critical engagements with the city through a collectively led wander. Key to the LRM's practice, and any practice in the medium of walking, is how it responds to the landscape itself, and the events that unfold in that landscape. In June 2017, the LRM's first Sunday walk coincided with the city's response to a terrorist attack that occurred the week before. As Rose (2017) writes, on 22 May "the landscape of Manchester had been irrevocably changed by an act of terrorism" when a suicide bomber attacked an Ariana Grande concert at the Manchester Arena. Twenty people were killed and over 200 were injured in an act of violence that resonated throughout the UK. While Rose considered cancelling the walk, she decided that "would be more disrespectful and alarming than walking sensitively with eyes, ears, heart, minds and arms wide open" (Rose, 2017, p. 2). She was particularly interested in asserting the rights of women to the streets, as the attendees at Grande's concert were predominately young women.

For the walk, Rose drew a cross on a map of Manchester and asked walkers to follow it, "trying to stay as true as possible to its contours" (Rose, 2017, p. 2). Helen, one of the participants, had brought "decorative fluffy bee[s]" to "act as a totem should [walkers] need a guiding wing", and each participant received a bee to guide them (Rose, 2017, p. 2). In Manchester bees are seen as a symbol of "cooperation, industry and resilience", and have long represented the city's important position at the height of the Industrial Revolution (Rose, 2017, p. 2). Rose encouraged participants to use the bees to decide where to walk in case obstructions prevented them from following their mapped line, a technique that combined psychogeographical mapping strategies with chance-based approaches. The walkers, a "welcome mix of familiar faces and curious first timers" (Rose, 2017, p. 2), split into three groups, with each exploring a different spur of the cross. As Rose walked through the city with a small group of women, they noticed bees everywhere; they had begun to "swarm with renewed vigour through the post-bomb city", with bees added to "billboards, bins, stickers, making homes on tattooed flesh and in rooftop hives" (Rose, 2017, p. 2). At the end of the walk the three groups reconvened and shared stories of their different perspectives of the cross. Rose's experience, walking with a group of women through a specific part of the city, resulted in different resonances than the other two groups who participated in the walk, despite the fact that they followed the same tactics, at the same time, in the same city, starting from the same location.

Rose notes that "footsteps won't change the world alone", however, the "parallel world[s] revealed by another random line on our shared map" serve as "a reminder that each step is a choice", and these choices help shape the spaces we inhabit and our relationships to the people with whom we inhabit them. As Julia Kathryn Giddy and Gijsbert Hoogendoorn (2018, p. 4) point out, in their discussion of walking tours in inner-city neighbourhoods, often known as slum tourism, there are "real ethical concerns" when "providing narratives … by 'outsiders' particularly when economic and power inequalities are present". A tourist participating in the LRM First Sunday walk would not be provided with a packaged set of information about Manchester, rather they are asked to actively engage with the construction of these narratives and how they have shaped the city. In doing so, Rose hopes the LRM offers a way to "imagine radical new paths which make it a different, better place, if only for a moment" (Rose, 2017).

Deveron Projects: local walking, global practice

The increased interest in walking is also indicated by institutional support for artistic walking practices. In 2013, Deveron Projects (DP), a social-practice arts organisation in the rural town of Huntly, Scotland, founded the Walking Institute, a year-round centre of walking excellence (Zeiske, 2013). The Walking Institute "encompasses all walking & art practices and aims to map globally the scope of the medium" (Zeiske and Smith, 2013, p. 350); it is the only arts programme in the UK fully dedicated to commissioning artistic walks. The Walking Institute asks "how artist-led walking projects can contribute to tourism and wider economic development locally in Huntly and the North East of Scotland, and how these experiences could be transferred to other places" (Deveron Projects, 2017). It focuses on "local people, looking at their place afresh; tourists to the region; those interested in outdoor and/or cultural tourism; and the growing walking artist network and related disciplines from academia" (Deveron Projects, 2017). To appeal to this wide range of constituents, the Institute frames its activities through the principle that "all walking is great" (Zeiske, 2013, p. 2), and it "explore[s] and celebrate[s] journeying and the human pace in all its forms" (Zeiske, 2013, p. 4).

In 2018 I was invited to Huntly as Deveron Projects' "Thinker in Residence". As part of my residency, I participated in the seventh annual Slow Marathon, the Walking Institute's flagship project. Though I had previously visited Huntly during Slow Marathon 2015 (Morris, 2018a), this was my first chance to participate in the walk. Originally conceived in collaboration with Ethiopian artist Mihret Kebede, Slow Marathon is a mass-participation walk of 26 miles that brings nearly a hundred people to Huntly for a full weekend of walking events. In 2013 Slow Marathon served as the official launch of the Walking Institute and seven different versions have been walked since Kebede's original marathon. The event, which charges participants £35 to participate, is one of DP's most popular and is often sold out.

Slow Marathon 2018: Walking without Walls

Slow Marathon 2018 was the culmination of Huntly-based artist Rachel Ashton and Palestinian artist May Murad's year-long collaboration *Walking without Walls* (2017–2018). Part of DP's programme to create collaborations with artists unable to come to Huntly for "political reasons" (Zeiske, 2018), *Walking without Walls* asked how we can "extend and keep up friendships when meeting in person is not an option" (Deveron Projects, 2018). The project coincided with both the centenary of the end of World War I, as well as "the Balfour Agreement when Britain initiated the establishment of Palestine as a national home for the Jewish people" (Zeiske, 2018). It combined digital tools (video-chat, social media, instant messaging) with the analogue experience of walking. As they worked together over the course of the year they painted each other's landscapes, captured at a distance. For the Slow Marathon, Ashton, Murad and the team at Deveron Projects organised two sets of participants to simultaneously walk 26 miles through Huntly and Gaza.

Walking through the Scottish landscape is not free of conflict, and as Ashton notes, organising the route in Scotland was "quite stressful" as it coincided with "lambing season" as well as a large car rally in the area (Deveron Projects, 2018, p. 26). We experienced the ramifications of this during the walk, as we were required to walk along long stretches of tarmac out of respect for the property of the farmers during lambing season. Restrictions to our right to roam asserted themselves in other ways as well: we encountered private roads and no entry signs, locked fences and other human designed impediments, despite our lawful access to walk through these areas. These hurdles, however, were small in comparison to the challenges in Gaza. As May notes, there is no right to roam in Gaza: the "areas close to the border with Israel are very dangerous", as is "walking through private property" and "there are military zones that cannot be crossed" (Deveron Projects, 2018, p. 23). Ultimately, they were able to secure a route for the walk and "everything was fine" (Deveron Projects, 2018, p. 29), but they were forced to repeat sections and bus through certain parts in order to achieve the total 26 miles (a stark contrast to the seven different routes that had been planned in Scotland since the project began).

The Walking Institute asserts that the "tourism industry is being questioned and reinvented by artists who offer something a bit different to visitors, while aiming to stay true to the place itself" (Deveron Projects, 2017). Through my participation in Slow Marathon I encountered the Scottish landscape in new ways, while through the connection to marathoners in Gaza I was also confronted with the realities of walking in a location to which I do not have access as a tourist. Kebede's original Slow Marathon focused on an investigation of global boundaries, based on her inability to walk between her hometown of Addis Ababa, Ethiopia and Huntly, Scotland. Similarly, Ashton and Murad's Slow Marathon asked us to consider our local landscape in relation to global realities and how they construct physical spatial boundaries. Murad's inability to leave Gaza was an essential factor in the development of the work, in contrast to Kebede, who was able to live in Huntly as part of her residency. Her solely virtual presence was a reminder of the immobility of her actual body, and the impediments to her freedom to roam. In Huntly and Gaza, the experiences of boundaries, borders and access to land resonate differently, and Slow Marathon provided an opportunity to consider them in relation to each other.

Walking tourism – challenges and opportunities

Artistic walking practices, which are based in the radical, anti-capitalist *memories* of the medium often create walks that challenge dominant structures and conventional notions of tourism. In contrast to official tours promoted by town stakeholders, artistic walking practices ask participants to engage critically with how the landscapes through which we walk are constructed. Rose's intersectional psychogeography, Murad and May's forced distance collaboration, and Savage's inversion of the tourist gaze ask participants to consider their engagement with and connection to both local and global spaces.

One of the challenges for organisations, tourism boards and other stakeholders embracing walking art as a way to introduce visitors to the city in new ways, is the fact that artists' and stakeholders' priorities might differ. In contrast to tours that strictly delineate "the consumption of what to see, how to see it and what not to see" (Deveron Projects, 2018, p. 203), artistic walks move beyond the tourist gaze and challenge official narratives associated with city spaces. Artistic walking practices encourage the walker to become a storyteller; rather than simply consuming information offered by a tour guide leading a group of fellow tourists, they encourage the active exploration of city spaces and the development of unique narratives based on playful interactions with the landscape and the actors (both human and non-human) they encounter there. As Deveron Project's Walking Institute demonstrates, there is potential for artistic walking practices to both challenge dominant narratives and social structures, while also working closely with town stakeholders in constructing that location.

The Loiterers Resistance Movement and Slow Marathon demonstrate ways in which artists create opportunities that appeal to a mix of residents and visitors. This mixture potentially creates stronger engagement with the landscape and more meaningful interactions with the people who inhabit them. Works such as Savage's walk, which do not ask for specific interaction with others, instead challenge our understanding of our local place through the lens of global travel. Though residents and tourists will experience these walks in different ways, both sets of participants are asked to engage with everyday spaces in new ways. In doing so, they offer opportunities for a reflexive mode of tourism.

Note

1 There were two iterations of the event: the first was at Somerset House in London as part of UTOPIA 2016: A Year of Imagination and Possibility; the second was in Edinburgh's Drill Hall as part of the Edinburgh Fringe Festival in association with Forest Fringe. I was the assistant curator for the event and worked closely with Qualmann and Sharrocks in its planning and execution.

References

Adler, J. (1989) "Origins of sightseeing". *Annals of Tourism Research, Semiotics of Tourism* 16(1), 7–29.

Bernstein, M. (2013) *The Night*, edited by Everyone Agrees, translated by Clodagh Kinsella. London: Book Works.

Bishop, C. (2012) *Artificial Hells: Participatory Art and the Politics of Spectatorship*. London: Verso.

Breton, A. (2003) "Artificial Hells: Inauguration of the '1921 Dada Season'", translated by Matthew S. Witkovsky. *October* 1(105), 137–144.

Butler, T. (2007) "Memoryscape: How audio walks can deepen our sense of place by integrating art, oral history and cultural geography". *Geography Compass* 1(3), 350–372.

Careri, F. (2002) *Walkscapes: Walking as an Aesthetic Practice*. Barcelona: Editorial Gustavo Gili.

Crab Man (2012). *Counter-Tourism: The handbook*. Charmouth, UK: Triarchy Press.

Crouch, D. (2002) "Surround by place: Embodied encounters", in S. Coleman and M. Crang (eds.), *Tourism: Between Place and Performance*. Oxford: Berghahn Books, pp. 207–218.

Debord, G. (1955/2006) "Introduction to a critique of urban geography", translated by Ken Knabb. *Bureau of Public Secrets*. Available at: www.bopsecrets.org/SI/urbgeog.htm (accessed 28 March 2019).

Debord, G. (1958/2006) "Theory of the dérive", in *Situationist International Anthology*, translated by Ken Knabb, Revised and Expanded Edition. Available at: www.bopsecrets.org/SI/2.derive.htm (accessed 28 March 2019).

Demos, T.J. (2009) "Dada's Event: Paris, 1921", in Beth Hinderliter, William Kaizen, Vered Maimon, Jaleh Mansoor and Seth McCormick (eds.), *Communities of Sense: Rethinking Aesthetics and Politics*. Durham, NC: Duke University Press Books. pp. 135–152.

Deveron Projects (2017) "Tourism & economic regeneration – Deveron Projects". Deveron Projects. www.deveron-projects.com/the-walking-institute/tourism-and-economic-regeneration/ (accessed 23 March 2019).

Deveron Projects (2018) "Rachel Ashton & May Murad/Walking Without Walls/2017–18", Project Report. Available at: www.deveron-projects.com/site_media/uploads/walking_without_walls_project_report_(1).pdf (accessed 22 March 2019).

Fonteyne, J. (2013) "The Art of Soundwalk", Resonance – European sound art network. https://resonancenetwork.wordpress.com/2013/04/14/the-art-of-soundwalk/ (accessed 27 March 2019).

Giddy, J.K. and Hoogendoorn, G. (2018) "Ethical concerns around inner city walking tours". *Urban Geography*, 1–7. doi: 10.1080/02723638.2018.1446884.

Grant, C. (2018) "Walking-tour app takes a detour – gets acquired by Bose for audio AR". *The Hustle*. Available at: https://thehustle.co/bose-acquires-walking-tour-app-detour/ (accessed 27 March 2019).

Haladyn, J.J. (2014) "Everyday boredoms or Breton's excursion to *Saint-Julien-le-Pauvre*", in J. Derry and M. Parrot (eds.), *The Everyday: Experiences, Concepts, and Narratives*. Newcastle upon Tyne: Cambridge Scholars Publishing, pp. 20–33.

Hall, C.M., Ram, Y., Shoval, N. (eds.) (2017) *The Routledge International Handbook of Walking*. Abingdon: Routledge.

Krauss, R. (1979) "Sculpture in the expanded field". *October* 8, 31–44.

Krauss, R.E. (2011) *Under Blue Cup*. Boston, MA: The MIT Press.

The Loiterer's Resistance Movement (2006) "The Loiterer's Resistance Movement". Available at: http://thelrm.org/index.

Lorimer, H. (2011) "Walking: New forms and spaces for studies of pedestrianism", in P. Merriman and T. Cresswell (eds.), *Geographies of Mobilities: Practices, Spaces, Subjects*. Farnham: Ashgate, pp. 19–34.

Marcus, G. (1989) *Lipstick Traces: A Secret History of the Twentieth Century*. London: Faber & Faber.

Morris, B. (2018a) "The Walking Institute: A reflexive approach to tourism". *International Journal of Tourism Cities* 4(3), 316–329. https://doi.org/10.1108/IJTC-11-2017-0060.

Morris, B. (2018b) *Walking Networks: The Development of an Artistic Medium*. PhD Thesis, University of East London.

Morris, B. and M. Rose (2019) "Pedestrian provocations: Manifesting an accessible future". *GPS* 2.2. Available at: http://gps.psi-web.org/issue-2-2/pedestrian-provocations/ (accessed 20 March 2019).

O'Rourke, K. (2013) *Walking and Mapping: Artists as Cartographers*. Boston, MA: The MIT Press.

Rose, M. (2006) "The Loiterer's Resistance Movement". Available at: http://thelrm.org/index.

Rose, M. (2015) "Confessions of an anarcho-flâneuse, or psychogeography the Mancunian way", in Tina Richardson (ed.), *Walking Inside Out: Contemporary British Psychogeography*. London: Rowman and Littlefield, pp. 147–162.

Rose, M. (2017) "Buzzing, bimbling, beating our bounds: Walking a line through Manchester". *Livingmaps Review*, 3(2).

Sanouillet, M. (2009) *Dada in Paris*, translated by Sharmila Ganguly. Cambridge, MA: The MIT Press.

Saunders, A. and Moles, K. (2016) "Following or forging a way through the world: Audio walks and the making of place". *Emotion, Space and Society* 20, 68–74.

Savage, J. (2014) "The guide to getting lost". Avaiable at: https://walkinglab.org/the-guide-to-getting-lost/ (accessed 3 May 2020).

Savage, J. (2017) "Fracture mob". *Livingmaps Review* (2), 1–2.

Shortell, T. and Brown, E. (eds.) (2014) *Walking in the European City: Quotidian Mobility and Urban Ethnography*. Abingdon: Routledge.

Singer, E. (2017) "4 best travel audio guide apps to avoid looking like a tourist". *Gear Patrol*. Available at: https://gearpatrol.com/2017/05/11/best-travel-audio-guides/ (accessed 22 March 2019).

Situationist International (1958) "Situationist definitions", in *Situationist International Anthology*, translated by Ken Knob, Revised and Expanded Edition. Available at: www.bopsecrets.org/SI/1.definitions.htm.

Situationist International (1960) "Situationist manifesto", translated by Fabian Thompsett. Situationist International Online. Available at: www.cddc.vt.edu/sionline/si/manifesto.html.

Smith, P. (2010) "The contemporary dérive: A partial review of issues concerning the contemporary practice of psychogeography". *Cultural Geographies* 17(1), 103–122.

Smith, P. (2013a) "Turning tourists into performers: Revaluing agency, action and space in sites of heritage tourism". *Performance Research: A Journal of Performing Arts* 18(2), 102–113.

Smith, P. (2013b) "Walking-based arts: A resource for the guided tour?", *Scandinavian Journal of Hospitality and Tourism* 13(2), 103–114.

Solnit, R. (2001) *Wanderlust: A History of Walking*. New York City: Penguin Books.

Urry, J. and Larsen, J. (2011) *The Tourist Gaze 3.0*, third edition. London: Sage.

Waxman, L. (2010) *A Few Steps in a Revolution of Everyday Life: Walking with the Surrealists, the Situationist International, and Fluxus*. PhD diss, New York University.

Wilkie, F. (2015) *Performance, Transport and Mobility: Making Passage*. Basingstoke: Palgrave Macmillan.

Wrights and Sites (2006) *A Mis-Guide to Anywhere*. Exeter: Unbound.

Zeiske, C. (2013) *Walking Institute: Vision Document*. Huntly: Deveron Arts. Available at: www.deveron-arts.com/site_media/uploads/walking_institute_vision.pdf (accessed 4 February 2016).

Zeiske, C. (2018) "Deveron Projects – taking big ideas for a long walk, interview with Creative Scotland". Available at: www.creativescotland.com/explore/read/stories/features/2018/deveron-projects-slow-marathon (accessed 22 March 2019).

Zeiske, C. and Smith, D. (2013) "The Walking Institute: A project for the human pace", in *Selected Essays from the On-Walking Conference* (University of Sunderland: Art Editions North, 2013), 350. Available at: https://issuu.com/stereographic/docs/walkonconference?reader3=1.

PART IV

Worldwide tourism cities and urban tourism

Part IV provides examples of how urban tourism is developing in different parts of the world and how worldwide tourism cities are adapting to the challenges ahead. It includes examples from Europe, North America, Latin America, Africa and Asia. This section also explores emerging new formats of specialist tourism, including geology and ecology-based tourism, socialist heritage and post-communist destination tourism, among others. The emphasis here is to discuss practical and policy-based examples of adaptation and innovation with respect to the challenges and trends discussed earlier in this book in the context of urban tourism and the management of tourism cities.

Summary of chapters

Irem Önder and **Bozana Zekan** begin this section by discussing the development of tourism cities in Europe in terms of their global competitiveness but also the socio-economic and environmental challenges that some of them are beginning to face, including overtourism. The authors focus on three global tourism cities experiencing advanced symptoms of overtourism – Barcelona, Dubrovnik and Venice – to explore how the fundamentals of sustainable development (society, economy and environment) can be reconciled in each case to attain sustainable tourism in the longer term. As part of this, examples of good practice and strategies are outlined. It is concluded that linking the phenomenon of overtourism solely to tourists is an oversimplification of the problem. Instead, overtourism demands long-term solutions specific to the idiosyncrasy of each destination that bridge the often-conflicting agendas of different stakeholder groups.

Costas Spirou investigates the emergence of urban tourism in North America as a central aspect of cultural policy in order to reinforce existing identities and/or recast new images in the pursuit for economic growth. The author reviews the history of urban tourism development in North America, primarily from a strategy perspective, though offering practical examples throughout. The author concludes that further research is needed on governance aspects of urban restructuring and the potential distributional impact of these strategies on disadvantaged groups as well as the tourist product as part of the commodity chain process. The latter also involves urban tourism as a recruiting mechanism capable of attracting new economy entrepreneurs and workers, the mechanisms and longer-term impacts of which are not fully understood yet.

Blanca A. Camargo, **María L. Chávez** and **María del Carmen Ginocchio** review the growth of urban tourism in Latin America, where tourism research in urban environments remains in its infancy. The authors identify trends on urban tourism research for the region and examine three case studies of innovative tourism product and marketing strategies in urban destinations: eno-culinary tourism in Mexico, tourism product development and marketing in post-conflict Colombia, and destination branding in Peru. Our chapter will contribute to the understanding of trends and challenges faced by Latin American urban destinations as well as the practical approaches to promote tourism in this region. The authors conclude that, although the outlook for tourism in Latin America remains positive, the region needs to reduce its dependency on a limited number of markets which revolve around sun, sand, sea and nature. They recommend that a widening of the tourism product offer should be considered by policy makers and managers of tourism destinations to harness the potential of Latin America's tourism cities in attracting more affluent tourists, even if important environmental and transport challenges remain on this front.

Hera Oktadiana and **Philip Pearce** provide a review of urban tourism in four ASEAN capital cities, namely Bangkok, Jakarta, Kuala Lumpur and Singapore. The authors discuss the key characteristics of tourism in each location and review the challenges confronting these major hubs of regional and global tourism. Recommendations are offered for policy makers in ASEAN cities to improve their distinctiveness and palliate current challenges related to transport and the environment.

Katia Iankova and **Sonia Mileva** reflect on the nascence and evolution of socialist cities, incorporating and materialising in their morphology the socialist idea of social accessibility, equality and sustainability and incorporating the ideology of communism into their identity. The authors identify the main traits of socialist cities, propose a conceptual typology to classify them and analyse the main tourism development challenges these cities face throughout their life cycle.

Melanie Smith and **Tamara Klicek** examine the historical transition of post-communist tourism cities with specific reference to tourism in capital cities and other large cities in former communist countries (e.g. Budapest, Prague, Warsaw, Bucharest, Belgrade, Bratislava, Tallinn, Kracow, Novi Sad). The authors identify a number of common challenges, including the interpretation of the socialist past and dissonant heritage, the development of national and city identities and branding, nurturing their cultural and creative elements, mitigating tourism's influence on gentrifying processes, managing the night-time economy and, more recently, dealing with overtourism.

Bihu Wu, **Qing Li**, **Feiya Ma** and **Ting Wang** consider the evolution of tourism cities in China, including the emergence of research in this field, which has been largely dominated by definitions, typologies, perceived image and theory building often rooted in urban studies. The authors conclude that China's current tourism planning system has benefitted from its origins in urban planning and that this may have influenced the different stages of tourism development over the last 40 years as well as the integrated strategic approach adopted today by the country's main urban tourism centres.

Samantha Richards, Greg Simpson and **David Newsome** review the emerging field of geotourism in the context of urban tourism. The author uses case studies to illustrate geotourism experiences that lend themselves to the appreciation of geology in the urban environment. In addition to discussing the role of geoheritage and geotourism in providing a new dimension to the connection of urban environments with the history of humankind, the authors also reflect on the importance of geoconservation in international tourism cities.

Greg Simpson, Jessica Patroni, David Kerr, Jennifer Verduin and **David Newsome** focus on the Dolphin Discovery Centre in Bunbury, Western Australia in a fascinating look at dolphins as a city tourism attraction. The first part of the chapter explores the history and development of the Dolphin Discovery Centre and highlights that a significant wildlife tourism attraction is possible within an urban centre when appropriate recognition, tourism management, research and environmental considerations are put into place. The authors go on to describe the economic benefits of dolphin tourism and how the Dolphin Discovery Centre functions as a tourism attraction. The authors conclude that many coastal city planners around the world can learn from the Dolphin Discovery Centre experience. They hope that city-based wildlife tourism experiences, such as those delivered at Bunbury, will mark a new era for the "wildlife aware" international tourism city.

Fang Wang, Bingyu Lin and **Qingyin Liu** investigate tourism in cities along one of China's greatest engineering marvels, the Grand Canal. They answer three questions – What was the spatial distribution of the canal cities during the historical period of the canal? How did form, scale, building fabric and transportation develop in the canal cities? What were the similarities and differences among canal cities? The authors consulted many historical documents, maps, paintings and other data to compare and restore the form and space of the main canal cities as they were during the Ming and Qing Dynasties. They also sought to determine the main factors that influenced urban development, thereby enhancing the understanding of the cultural implications of the Grand Canal. The chapter provides a useful reference for future development of tourism of the Grand Canal and its cities.

Kim-Ieng Loi, **Weng-Hang Kong** and **Hugo Robarts Bandeira** present a delicious case study on gastronomy as a tourism product in Macao. It is a new member city of the UNESCO Creative Cities Network in the field of gastronomy. The first part of the chapter is dedicated to a description of the unique development of Macanese cuisine in Macao as both a cultural identity symbol as well as being part of a holistic tourist experience. Second, development and challenges related to this gastronomy tourism development are discussed. The authors also look at selected gastronomy tourism examples from other destinations. Finally, the chapter is concluded by recognising the important role of gastronomy in the development of a well-diversified tourism product portfolio that provides a powerful resource bank towards a more sustainable destination tourism economy.

Yan (Mary) Mao presents a case study on tourism development in the city of Wuhan in China. The author begins with a background on the city, its history and tourism attractions. She then discusses Wuhan's tourism marketing and branding. Smart tourism applications in Wuhan are then reviewed.

29

URBAN TOURISM DEVELOPMENT IN EUROPE

A double-edged sword for the cities?

Irem Önder and Bozana Zekan

Introduction

There is no doubt about the growth of tourism in European cities. This is confirmed every year in the European Cities Marketing (ECM) Benchmarking Report, which gives a clear indication of the increase of tourism volume across 120+ European cities. More arrivals and bednights translate into more money spent at the destination and these are just examples of key performance indicators (KPIs) that are of major interest to many destination management organisations (DMOs). In light of this, local stakeholders (e.g. DMO, residents, hoteliers, etc.) should be very pleased to hear about the increase in total bednights at their destinations, compared to a year before. Thus, it can be argued that the economic domain of urban tourism is on the track of performing as desired. Reality is that this is only one side of the coin. Nowadays, in cities like Barcelona, Venice and Dubrovnik, a lot of negative residents' attitude is geared towards tourism due to overcrowding and inflation. Barcelona is an interesting example, as it is one of the top ten cities in Europe when it comes to total bednights. However, residents are not happy, and the environment is being destroyed. Thus, as the number of tourists increases at a destination, tourism acceptance decreases due to lower quality of life for residents such as traffic jams, more garbage in the streets, noise, etc.

On the other hand, the preliminary results of the forthcoming 15th edition (2018–2019) of the ECM Benchmarking Report were presented in March 2019 and they celebrated growth in tourism volumes across 58 European cities with the press release on "Tourism Brexit Buoyancy: European Cities Celebrate Growth Despite Challenging Times" (ECM, 2019). This year again, the full report will include more than 120 cities and will be published in June 2019. Even the first glimpse of the preliminary results revealed a 1.9% increase in total bednights and the substantial 5.2% increase in domestic bednights for 2018. As evident, the growth rate in domestic bednights was higher than in international bednights. The top ten performing cities in terms of total bednights and their ranking remained the same as in 2017. All cities, besides London (−8.7%), have recorded growth anywhere from 1.1% (Prague) to 9.3% (Munich) in their total bednights (ECM, 2019). These results are shown in Figure 29.1.

There needs to be a balance between the number of tourists at a destination and the residents as well as in terms of distribution of tourists at the destination, which can be done with the

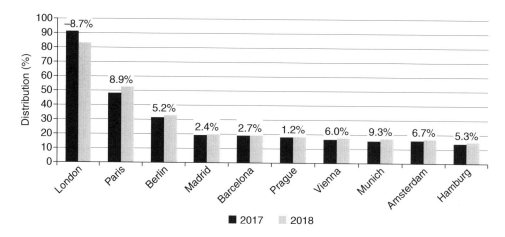

Figure 29.1   The top ten performing cities.

Source: ECM (2019).

carrying capacity approach. However, there is no magic number that can be calculated and imposed as a regulation for cities that shows the carrying capacity since the number can change based on the chosen definition of the carrying capacity (Saarinen, 2006).

In many cities throughout Europe, two out of three dimensions (social, environmental, economic) of sustainable development are not performing optimally. This is where this chapter comes in: the aim is to identify ways to bring these three dimensions into balance. This chapter focuses on specific European cities that already have overtourism problem, such as Barcelona, Dubrovnik, and Venice, and proposes solutions that capture social, environmental and economic aspects of sustainable tourism development. Best practice examples of such strategies are highlighted.

Literature review

Carrying capacity

In order to understand the overtourism issue we first need to understand the carrying capacity concept, which has been well researched since 1960s. In short, carrying capacity for a destination is defined as the maximum number of people a destination can have without damaging the environment of the host country (O'Reilly, 1986). This issue relates to sustainable tourism, especially to the responsible tourism concept. How can destinations manage the increasing number of tourists coming and what measures can be implemented to solve the overtourism problem? These issues are some of the main topics of discussion for DMOs, which are aware of the problem and trying to solve it the best way they can.

In terms of tourism, carrying capacity relates to the balance between the quality of experience at the host destination and the environment, which depends on the maximum number of people a destination can have without disrupting this balance (O'Reilly, 1986). Saarinen (2006) classifies carrying capacity research into three different segments. First is a resource-based definition of carrying capacity, which is also about the limits of growth in tourism and related to the resources used in tourism and the natural or non-tourism conditions (Saarinen, 2006). The problem here is how to separate the causes of tourism from the regular causes of human interaction with nature. Second is activity-based tradition, where carrying capacity is defined by the needs of tourism as an economic activity. The limit of growth of a destination in this perspective is not dependent on the capacity of the destination and its resources but on the tourism activity (Saarinen, 2006). The destination has the power to change its product and service offers and thus modify its carrying capacity. The downside of this tradition is that it also depends on the political aspects and the outlook of the leaders, which may have different goals than the destinations' other stakeholders. Third is community-based tradition, in which the limits of growth are decided by the community affected by tourism (Saarinen, 2006). The community involves different stakeholders from residents to accommodation owners, to restaurant owners and anyone else who is directly or indirectly associated with the tourism industry. The problem with this tradition is that each of these stakeholders has different goals and thus it is hard to find the common ground and agree on the same terms for the destination.

Limits of acceptable change

Limits of acceptable change (LAC) is another approach to look at carrying capacity. It was developed for dealing with the recreational area carrying capacity and aims to find the balance between visitor experience and resource protection (Cole & Stankey, 1997). On one hand, the goals of visitors are to use the recreational area as much as they want without restrictions and on the other hand, the goal of the recreational area management is to protect the environment in spite of the visitors, which are two conflicting aims. Three things can be done in this case. First, both sides need to compromise for the sustainability of the recreational area. Cole and Stankey (1997) claim that LAC can be applied in other situations where there are conflicting goals such as at tourism destinations where both residents and visitors have conflicting goals and a compromise is needed. Second, there needs to be a hierarchy of goals and some goals can be considered to constrain other goals (Cole & Stankey, 1997); in the case of a tourism destination the ultimate goal is to have a sustainable destination so that visitors can still enjoy. Third, measurable standards are required (Cole & Stankey, 1997), meaning the destination needs to be able to measure the number of visitors, specifically at the popular attractions where it gets crowded, preferably at any point in time. The destination can then use marketing or mobile applications to push visitors to other less crowded areas. This is just one measure example; destinations can measure the economic impact and other environmental impacts such as energy use at the destination as well. The tourism industry is still trying to decipher the impacts of overtourism other than the obvious issues such as overcrowding and the resulting dissatisfaction of the residents. LAC can provide an adequate framework to solve some of those issues.

Overtourism and sustainability aspects

Goodwin (2017, p. 1) describes overtourism as a destination "where hosts or guests, locals or visitors, feel that there are too many visitors and that the quality of life in the area or the quality of the experience has deteriorated unacceptably". This issue results in decrease in the life quality of residents and in turn, tourism acceptance decreases. Moreover, there may be anti-tourism protests and wishes from the residents to regulate tourism more strictly than how it is currently handled. This also shows that tourism needs to be managed appropriately; otherwise it can cause more harm than bring benefits to the destination.

Overtourism affects the sustainability of a destination generally in three main aspects: economic, environmental and social. Economic impact is about the revenues that tourists and daily visitors bring to a destination by spending money at attractions, accommodations and on gastronomy. Since the goal is to increase the revenues, the more people visit the destination the higher the economic impact. However, the day visitors from cruise ships that invade the destination for a few hours and do not spend much money there since they have paid for all-inclusive cruise travels have challenged this notion. Thus, more people at the destination does not mean more economic benefits. On the contrary, the number of bednights spent at the destination can bring more economic benefits than the number of tourists at the destination. Thus, the focus of the DMOs needs to change towards this direction instead of the growth in number of visitors.

The environmental impact is the deterioration of the landscape and historical sites caused by the number of tourists that are more than the carrying capacity of a destination. This is also linked to sustainable tourism, which focuses on the impacts of tourism at destinations. In general, "both sustainability and carrying capacity refer to the scale of tourism activity that can occur in a spatial unit without doing any serious harm to the natural, economic, and sociocultural elements at destinations" (Saarinen, 2006, p. 1126). According to Gössling et al. (2012), the tourism sector was responsible for more than 10% of domestic water use, which can be a big problem in small island destinations that economically depend on tourism. Negative environmental impacts also take longer to recover and sometimes there is no turning back to the original status due to the damage caused by individuals.

The social impact is the one that causes stress between the residents and the visitors. One reason that residents are unhappy with the increasing number of visitors is that their quality of life diminishes, as there are more people at the destination to use the same facilities as the residents such as roads, public transport, restaurants, as well as increased waste and crowds on the streets. These facilities and services at the destination are maintained by the taxes of the residents and even though there may be tourism taxes, the residents may feel that the visitors' contribution is far less than what is needed. This results in anti-tourism protests or tourism phobia such as in the case of Barcelona in 2018 urging tourists to throw themselves off balconies (Morris, 2018).

Although overtourism and sustainability issues have been researched extensively in the past years, the solutions regarding a framework to overcome this issue need further investigation. In the following section, the status quo in European cities is discussed; followed by case examples of Barcelona, Dubrovnik and Venice and the various solutions they implemented.

Status quo

Arguably, the discussed growth in tourism bednights translates into the economic benefits for destinations. However, this is only one part of the story. In parallel with reports such as the aforementioned ECM Benchmarking Report, a lot of negative attention has been raised in media concerning overtourism that looks into impacts of tourism on society and environment. Therefore, on the one hand, there are stories, from tourism organisations as prominent as ECM, of tourism's success regarding growth in total bednights. On the other hand, there are a multitude of industry reports and articles that show the different side of tourism impacts and of such a high volume of visitors (i.e. arrivals, not bednights) at destinations. To support this, in only 0.21 seconds about 650,000 results were found when searching for the dreaded word "overtourism" in Google. In addition, search terms for overtourism and tourism carrying capacity have increased since 2016 as shown in Figure 29.2.

A very interesting work was recently published by Roland Berger (2018) consultancy on how to protect cities from overtourism by analysing 52 European cities. At first, an argumentation was provided on the so-called "curse of mass tourism", upon which cities' quality (measured in RevPAR) versus quantity (measured in tourism density) was compared. As a result,

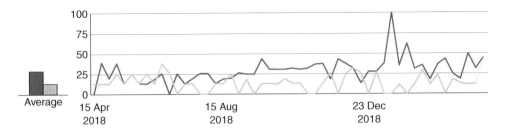

Figure 29.2 Google search queries for overtourism and tourism carrying capacity.

Note
The dark line represents overtourism search queries and the grey line represents tourism carrying capacity search queries in Google between July 2018 and March 2019.

cities were categorised into six clusters: Shining Stars, Mass Trap, Under Pressure, Unused Potential, Peak Performance and Sustainable Quality. For instance, Amsterdam, Barcelona, Copenhagen, Dublin, Frankfurt, Lisbon, Reykjavik and Venice have secured their spot in the "Under Pressure" cluster due to "high and growing tourism density but sluggish value creation", whereas cities such as Berlin, London, Munich, Rome and Vienna were featured in the "Shining Stars" cluster due to "healthy and sustainable levels of tourism" (Roland Berger, 2018, p. 9). In continuation, seven measures (1–4 = proactive; 5–7 = reactive) were proposed for tackling and fighting overtourism, as it was argued that overtourism can be planned for and prevented:

1   Long term: Alignment of city tourism strategy with city development strategy
2   Mid term: Implementation of infrastructural measures in low-tourism areas
3   Mid term: Upgrading of guest segments in a targeted way
4   Short term: Targeting various segments and distributing guests across the city and seasons
5   Regulation of capacities
6   Active management of the sharing economy
7   Limitation of access (entry tickets, slot allocation, flexible pricing)

(Roland Berger, 2018, p. 11)

Lastly, in the conclusive part of the report, a four-step plan for success in tackling overtourism was suggested: (1) self-assessment and insights, (2) initiatives and roadmap, (3) implementation and monitoring, and (4) iterations and fine-tuning, all in full cooperation and support of the city's stakeholders, as the final plea was made that "overtourism is both avoidable and reversible" (Roland Berger, 2018, pp. 14–15).

Along the same lines, 11 strategies and 68 measures were proposed for addressing visitors' growth in cities by the World Tourism Organization (UNWTO) and their project collaborators: the Centre of Expertise Leisure, Tourism & Hospitality (CELTH), Breda University of Applied Sciences, and the European Tourism Futures Institute (ETFI) of NHL Stenden University of Applied Sciences (UNWTO, 2019). Detailed measures are beyond the scope of this chapter; 11 strategies that were at the core of this report are as follows:

1   Promote the dispersal of visitors within the city and beyond
2   Promote time-based dispersal of visitors

3   Stimulate new visitor itineraries and attractions
4   Review and adapt regulation
5   Enhance visitors' segmentation
6   Ensure local communities benefit from tourism
7   Create city experiences that benefit both residents and visitors
8   Improve city infrastructure and facilities
9   Communicate with and engage local stakeholders
10 Communicate with and engage visitors
11 Set monitoring and response measures

(UNWTO, 2019, p. 11)

As is evident, the strategies showcased by UNWTO (2019) are more elaborate; however, they are certainly in line with those of Roland Berger (2018). What is more, 19 cities and their approaches to tackling overtourism were detailed within the same report by UNWTO (2019) and as such, can provide guidelines to other cities that may find themselves in a similar situation. Interestingly though, most of the cities (15 out of 19) that were used as the case studies are European cities: the usual suspects in the domain of overtourism, such as Amsterdam, Barcelona, Dubrovnik, Edinburgh and Venice. Hangzhou, Macau, New York City and Seoul were the four cases analysed outside the European context. Thus, this clearly points towards the problem of overtourism in Europe in particular. A more detailed look into three European cities and their approaches to dealing with overtourism follows.

Case examples

Overtourism is hurting many destinations in Europe and residents of various cities started raising their voices against this very issue (Martinko, 2019). Oftentimes, the main culprits are peer-to-peer home sharing platforms (e.g. Airbnb), cheap flights and cruise ships (Francis, n.d.). One of the hotspot cities that has been continuously present in the overtourism debate is certainly Venice, known as "La Serenissima" (the most serene), which is the exact opposite of what has been happening in this city over the past years (Simmons, n.d.a). In short, Venice is overrun by many visitors, many of which are day trippers (i.e. excursionists) coming with the cruise ships that do not bring much, if any, economic benefit to the city (Simmons, n.d.a). Such an intense tourism pressure leads to decline in quality of the city's attractions (Ganzaroli et al., 2017). Tourism flows are also constrained due to the city's layout (Pinke-Sziva et al., 2019). Crowds, strong odours, litter, noise, long waiting times, disrespectful behaviour and inappropriate clothing, rising cost of living, and lower quality of life are becoming the everyday struggle for the decreasing number of locals (Lynne, 2019; Simmons, n.d.a). It is predicted that if such trends continue, Venice may not have fulltime residents by 2030 (Simmons, n.d.a). Furthermore, the city's carrying capacity has been estimated to be approximately 18 million visitors annually (45% tourists + 55% excursionists), whereas at the moment, Venice is very much exceeding this figure with approximately 30 million visitors a year (20% tourists + 80% excursionists) (van der Borg, 2019). Everything else aside, this is a clear indication of how dire the situation is in the city, which is described as a "falling destination" (Séraphin et al., 2018). The term "Venice Syndrome" was also coined to allude to the issues associated with depopulation and overtourism (Martín Martín et al., 2018).

So what is being done? For example, in 2017, ruling was made that as of 2021, cruise ships over 55,000 tons will no longer be able to enter St Mark's Basin and the Giudecca Canal (Simmons, n.d.a). Moreover, in 2018, the mayor of Venice, Luigi Brugnaro, has introduced and authorised various fines such as those for noise, littering and swimming in canals, as well as different plans for controlling the numbers of visitors, some of which were also opposed by the locals (Simmons, n.d.a). In addition, Venice proposed to introduce an entry fee up to €10 as of 1 May 2019, which will target day trippers; residents and visitors that stay overnight and pay a tax at their hotels will be excluded from this fee (Coffey, 2019). This money is to be used towards maintenance works, security and surveillance, and reduction of costs for residents and businesses (Coffey, 2019). Furthermore, the current focus of the city is on six strategies (the above listed 1, 2, 5, 6, 8 and 11) proposed by UNWTO (2019). Time-based dispersal of visitors within the city and beyond, visitors' segmentation, benefits for the local community, infrastructure improvements and monitoring/response measures clearly remain the priority for Venice. Overall, it can be argued that many policy proposals have been initiated, still "an innovative and holistic tourism policy has yet to be developed by the city and all its stakeholders" (van der Borg, 2019, p. 78). Séraphin et al. (2018) do not see a solution in a forced "Trexit" (i.e. exit, prohibition of access), but rather in the "ambidextrous management approach", which puts together exploitation and exploration. Another proposed measure is excluding the city from all promotional materials (Séraphin et al., 2018).

The situation is very similar in Barcelona, which was ranked fifth in total bednights in 2018 in the sample of 58 European cities (ECM, 2019) and which is arguably at breaking point (López Diaz, 2017). Barcelona is among the top destinations suffering from unsustainable overcrowding (Insch, 2019), resulting in tourism rejection (Martín Martín et al., 2018). This is also the city where local response has made a lot of headlines across various media, the famous ones telling tourists to go home or equating tourists with bastards and terrorists, and tourism with invasion. Overcrowding aside, Barcelona is becoming increasingly unliveable for its residents due to a shortage of housing, focus on the visitor economy, increasing prices and inappropriate behaviour of its visitors (Brown, n.d.; López Diaz, 2017; Arias-Sans & Milano, 2019). Hence, very similar problems that Venice is facing. It is also one of the cities in which "demands for stricter controls over short-term rentals have been a central element of recent social mobilisations around tourism" (Novy & Colomb, 2019, p. 365). As tourism is nowadays considered an urban issue in Barcelona and as such is integrated into the urban agenda, the Strategic Plan for Tourism 2016–2020 has placed the focus on "balancing the economic wealth generated by tourism with its side-effects and promoting the social return of tourism activities to the territory" (Arias-Sans & Milano, 2019, p. 21). Furthermore, five strategies (the above listed 4, 7, 8, 9 and 11) by UNWTO (2019) are of most interest to Barcelona. Monitoring/response measures, infrastructure improvements, adaptation of regulations, city experiences for both residents and visitors, and engagement of local stakeholders clearly communicate the intentions of making the city more liveable for its residents. Thus, it can be said that the overall overcrowding issue and tourism mobility are tackled within a broader urban governance approach (Arias-Sans & Milano, 2019).

Adding *Game of Thrones* TV series excursions to the aforementioned mix of culprits (Airbnb, cheap flights and cruise ships) explains major overtourism problems in Dubrovnik, one of the historic cities "clearly bursting at the seams", according to Higgins-Desbiolles (2018, p. 158). Like Venice, Dubrovnik is literally overrun by cruise ship passengers, whose interests seem to be put ahead of those of the residents. In other words, amenities within Dubrovnik's Old Town that used to cater to local people are now turned into outlets that cater to the needs of these visitors (Simmons, n.d.b). Furthermore, housing prices are continuously rising, which has resulted in residents moving outside the Old Town: nowadays only 1,157 local people live there in comparison to 5,000 in 1991 (Simmons, n.d.b). It also came to the point that residents watch a local TV showing the Old Town streets in order to decide whether to go out or not, depending on the crowds (Bačić, 2017). After the major congestion incident at the main gate in 2017, the municipality got involved by instructing people on which side to walk (Panayiotopoulos & Pisano, 2019). Moreover, all this is happening in the summer months (Puljić et al., 2019). Aside from unhappy residents, in 2016 UNESCO also alarmed Dubrovnik and suggested limiting the number of visitors (maximum 8,000 per day) inside the city walls, not to risk its World Heritage Status (Simmons, n.d.b).

What is being done? The project titled "Respect the City" involving all the city's stakeholders was launched in 2017 as a response to overtourism, the goals of which are "sustainable tourism development as a sector, sustainable use of the resources, and sustainable development for the people, the economy and the community" (Puljić et al., 2019, p. 43). Furthermore, the current focus of the city is on six strategies (the above listed 1, 2, 3, 4, 8 and 11) proposed by UNWTO (2019), which means time-based dispersal of visitors within the city and beyond, new itineraries and attractions, adaptation of regulations, infrastructure improvements, and monitoring/response measures are the priority for Dubrovnik. In addition, certainly a decision has had to be made in order to reduce overcrowding: as announced by the mayor of Dubrovnik, Mato Franković, the city will cap the number of cruise ships to two per day (maximum 5,000 passengers), with the measures being introduced already in 2019 (Coffey, 2018). As the mayor stated, Dubrovnik is ready to lose money in order to improve the quality of life (Coffey, 2018), which is certainly what both residents and tourists need.

In addition to the specific cases, there could be other solutions for overtourism. Demarketing is a strategy that can be used to solve the problem of overcrowding by targeting specific types of tourists and discouraging others from visiting the destination (Goodwin, 2017). However, this is not going to solve everything on its own, especially for destinations such as Venice, Amsterdam or Barcelona. Séraphin et al. (2018) define this as "Trexit" (tourism exit) and argue that it is not a sustainable option. In cities where overtourism is an issue DMOs are not marketing the city but instead are focusing on the region to distribute the tourist flows to less crowded areas. Some specific examples include Amsterdam marketing the beach area to tourists although it is not in Amsterdam or having hop-on hop-off buses that take tourists outside Amsterdam to see windmills and tulips. One solution for overtourism, as proposed by the mayor of Dubrovnik, is to have strict regulations regarding the number of tourists that can enter or stay at the destination. However, tourism is also closely related to politics, and some of the DMOs are linked to them directly, so trying to decrease the number of tourists may not seem like a good political approach.

Another solution can be from the technological side, such as mobile applications, which can be used as a persuasive tool to motivate tourists to go to lesser-known areas at the destination. However, this approach implies that most tourists use the designated mobile app, which is not the case. Moreover, it also assumes that the DMOs know exactly where the visitors are at the destination in real time, which is possible by tracking the mobile phone signals of the visitors. Although technologically this is possible, it is not easy to implement in real life due to privacy regulations. Some major European cities such as Berlin, Paris and Vienna are already working on finding an applicable solution using mobile technology.

Recommendations for cities

In order to have a solution for overtourism, we need to understand the reasons for it. In general, the reasons for overtourism can be summarised as follows:

1   Marketing efforts of DMOs to attract visitors to the destination. In addition, focusing on growth in number of visitors as the main goal of a destination.

2   Having alternative accommodation options such as Airbnb and other peer-to-peer accommodation sharing platforms.

3   Low-cost carriers flying to more destinations and making travel more affordable for the potential visitors.

4   Cruise tourism causing thousands of visitors to be at the destination at the same time.

5   Internet and social media making information dissemination faster and increasing the number of potential visitors as well as enabling them to make travel arrangements easier. In addition, famous movies and TV shows create additional demand for destinations such as what *Game of Thrones* did for Dubrovnik.

The recommendations for solving overtourism and creating a more sustainable urban tourism destination are not straightforward. Based on the literature and industry practices, the following is a general guideline that cities can use:

1   *Demarketing*: Although there are different views on the impact of demarketing, the least harmful solutions for the economic impact of visitors are un-promotion or de-promotion efforts. Moreover, demarketing does not make tourists feel unwelcome, but can still result in a decrease in the overall number of tourists (Dods & Butler, 2019). In addition, new itineraries and attractions can be created by the DMO to ensure a more balanced dispersal of visitors at the destination.

2   *Rules and regulations*: Demarketing efforts can be supported by new policy regulations such as introducing a limit to the number of visitors to specific areas in the city to ease the crowding or having strict rules about peer-to-peer accommodation in the city. Additional taxes for visitors or introducing fees to enter the destination are also considered as regulations for reducing overtourism.

3   *Technological solutions*: Monitoring and response measures for visitors are an essential need. For instance, smart technologies such as location-based mobile applications can help to disperse visitors at the destination. The mobile applications can inform visitors regarding the wait times for attractions, let them purchase tickets in advance and recommend alternative attractions to see at the destination. These solutions are also in line with the UNWTO's recommendations of promoting the dispersal of visitors within the city and beyond, promoting time-based dispersal of visitors, and stimulating new visitor itineraries and attractions.

Overall, every destination needs to analyse its own situation and find solutions that fit them. Moreover, overtourism is not a problem for every city yet and may not be a problem for some at all. However, every city can learn from others and predict the future of their destination for both residents and visitors. DMOs and policy makers of the cities also need to change their focus from growth in visitor numbers to growth in a sustainable way such as having fewer visitors who stay at the destination longer, thus having more economic impact than having more visitors who stay for a short time.

Conclusion

This chapter focused on the overtourism issue in European cities explaining the situation in specific cities and the measures they have taken. Since Dubrovnik, Venice and Barcelona are different from each other, some of the measures taken to handle overtourism are also different.

Overall, overtourism is a complex issue and different stakeholders are affected by it differently. Koens et al. (2018) conducted a series of interviews in various European cities with various tourism stakeholders and they conclude that overtourism is not only about tourists; residents are also impacted by this phenomenon since they share their city with the tourists and use the same facilities. The solutions proposed for overtourism need to take into account all the stakeholders at the destination. Also, there is no "one size fits all" solution, since every destination is unique and has different problems and solutions that can be implemented.

The aforementioned cities in this chapter are trying to find solutions to overtourism in line with the LAC theory that would be beneficial for all stakeholders in the long run; keeping in mind that residents and tourists may have conflicting interests. This balance is hard to keep and not easy to find. In addition, when we add the political aspect of city governance, it is impossible to make all the stakeholders happy. The individuals who are at the destination either visiting or living need to remember that the ultimate goal is to have a sustainable city so that the next generations can also enjoy it.

References

Arias-Sans, A. & Milano, C. 2019. Case study 3 Barcelona, Spain. In UNWTO, *"Overtourism"? Understanding and managing urban tourism growth beyond perceptions. Volume 2: Case studies* (pp. 21–24). Madrid: UNWTO.

Bačić, M. 2017. *The dark side of tourism: Dubrovnik residents use TV to decide when to go out.* Available from: www.euronews.com/2017/08/22/tv-helps-dubrovnik-residents-decide-when-to-go-out-the-dark-side-of-tourism [Accessed 9 Apr 2019].

Brown, V. n.d. *Overtourism in Barcelona.* Available from: www.responsibletravel.com/copy/overtourism-in-barcelona [Accessed 9 Apr 2019].

Coffey, H. (2018, October 02). *Dubrovnik to cap the number of cruise ships allowed to dock each day.* Available from: www.independent.co.uk/travel/news-and-advice/dubrovnik-cruise-ship-cap-croatia-overtourism-two-dock-a8565166.html [Accessed 9 Apr 2019].

Coffey, H. 2019. *Venice to introduce booking system and charge tourists up to €10 for entry.* Available from: www.independent.co.uk/travel/news-and-advice/venice-booking-system-entry-fee-reserve-overcrowding-overtourism-a8765456.html [Accessed 8 Apr 2019].

Cole, D. N. & Stankey, G. H. 1997. Historical development of Limits of Acceptable Change: conceptual clarifications and possible extensions. In: McCool, Stephen F. and Cole, David N., comps. Proceedings-limits of acceptable change and related planning processes: progress and future directions: from a workshop held at the University of Montana's Lubrecht Experimental Forest. Gen. Tech. Rep. INT-GTR-371. Ogden, UT: US Department of Agriculture, Forest Service, Rocky Mountain Research Station: 5–9 (Vol. 371).

Dods, R. & Butler, R. 2019. The phenomena of overtourism: A review. *International Journal of Tourism Cities.* https://doi.org/10.1108/IJTC-06-2019-0090.

ECM. 2019. *Tourism Brexit buoyancy: European cities celebrate growth despite challenging times.* Available from: www.europeancitiesmarketing.com/tourism-brexit-buoyancy-european-cities-celebrate-growth-despite-challenging-times/ [Accessed 7 Apr 2019].

Francis, J. n.d. *Overtourism: What is it, and how can we avoid it?* Available from: www.responsibletravel.com/copy/what-is-overtourism [Accessed 8 Apr 2019].

Ganzaroli, A., De Noni, I. & van Baalen, P. 2017. Vicious advice: Analyzing the impact of TripAdvisor on the quality of restaurants as part of the cultural heritage of Venice. *Tourism Management*, 61, 501–510.

Goodwin, H. 2017. The challenge of overtourism. Responsible Tourism Partnership. Available from: http://haroldgoodwin.info/pubs/RTPWP4Overtourism01'2017.pdf [Accessed 4 Apr 2019].

Gössling, S., Peeters, P., Hall, C. M., Ceron, J. Dubois, G., Lehmann, L. & Scott, D. 2012. Tourism and water use: Supply, demand, and security. An international review. *Tourism Management*, 33, 1–15.

Higgins-Desbiolles, F. 2018. Sustainable tourism: Sustaining tourism or something more? *Tourism Management Perspectives*, 25, 157–160.

Insch, A. 2019. The challenges of over-tourism facing New Zealand: Risks and responses. *Journal of Destination Marketing & Management*, in press.

Koens, K., Postma, A. & Papp, B. 2018. Is overtourism overused? Understanding the impact of tourism in a city context. *Sustainability*, 10(12), 4384.

López Diaz, A. 2017. *Why Barcelona locals really hate tourists*. Available from: www.independent.co.uk/travel/news-and-advice/barcelona-locals-hate-tourists-why-reasons-spain-protests-arran-airbnb-locals-attacks-graffiti-a7883021.html [Accessed 9 Apr 2019].

Lynne, B. 2019. *Are we loving our tourist destinations to death?* Available from: www.earth.com/news/tourist-destinations-overtourism/ [Accessed 9 Apr 2019].

Morris, S. 2018. Barcelona protesters urge British tourists to jump off hotel balconies. *The Sunday Times*. Available from: www.thetimes.co.uk/article/barcelona-protesters-urge-british-tourists-to-jump-off-hotel-balconies-n9rqhvtsq [Accessed 7 Apr 2019].

Martín Martín, J. M., Guaita Martínez, J. M. & Salinas Fernández, J. A. 2018. An analysis of the factors behind the citizen's attitude of rejection towards tourism in a context of overtourism and economic dependence on this activity. *Sustainability*, 10(8), 2851.

Martinko, K. 2019. *Overtourism: Can this problem be solved?* Available from: www.treehugger.com/travel/overtourism-can-problem-be-solved.html [Accessed 8 Apr 2019].

Novy, J. & Colomb, C. 2019. Urban tourism as a source of contention and social mobilisations: A critical review. *Tourism Planning & Development*, 16(4), 358–375.

O'Reilly, A. M. 1986. Tourism carrying capacity: Concept and issues. *Tourism management*, 7(4), 254–258.

Panayiotopoulos, A. & Pisano, C. 2019. Overtourism dystopias and socialist utopias: Towards an urban armature for Dubrovnik. *Tourism Planning & Development*, 16(4), 393–410.

Pinke-Sziva, I., Smith, M., Olt, G. & Berezvai, Z. 2019. Overtourism and the night-time economy: A case study of Budapest. *International Journal of Tourism Cities*, 5(1), 1–16.

Puljić, I., Knežević, M. & Šegota, T. 2019. Case study 8 Dubrovnik, Croatia. In UNWTO, *"Overtourism"? Understanding and managing urban tourism growth beyond perceptions. Volume 2: Case studies* (pp. 40–43). Madrid: UNWTO.

Roland Berger, 2018. *Protecting your city from overtourism. European city tourism study 2018*. Available from: www.rolandberger.com/en/Publications/Overtourism-in-Europe's-cities.html [Accessed 7 Apr 2019].

Saarinen, J. 2006. Traditions of sustainability in tourism studies. *Annals of Tourism Research*, 33(4), 1121–1140.

Séraphin, H., Sheeran, P. & Pilato, M. 2018. Over-tourism and the fall of Venice as a destination. *Journal of Destination Marketing & Management*, 9, 374–376.

Simmons, J. n.d.a. *Overtourism in Venice*. Available from: www.responsibletravel.com/copy/overtourism-in-venice [Accessed 8 Apr 2019].

Simmons, J. n.d.b. *Overtourism in Dubrovnik*. Available from: www.responsibletravel.com/copy/overtourism-in-dubrovnik [Accessed 9 Apr 2019].

UNWTO. 2019. *"Overtourism"? Understanding and managing urban tourism growth beyond perceptions. Volume 2: Case studies*. Madrid: UNWTO.

van der Borg, J. 2019. Case study 18 Venice, Italy. In UNWTO, *"Overtourism"? Understanding and managing urban tourism growth beyond perceptions. Volume 2: Case studies* (pp. 76–78). Madrid: UNWTO.

30

TOURISM CITIES IN THE UNITED STATES

Costas Spirou

Urban centres across the United States have searched for ways to advance policies to remedy the devastating effects of economic restructuring connected to globalisation, de-industrialisation and decentralisation. These forces became increasingly apparent following the Second World War and have exacerbated in recent decades. Within a mix of possible policy tools, urban tourism has emerged as a viable option, and its advancement and promotion through the use of cultural planning is now a commonly used strategy across the country. Cities pursue the creation and maintenance of comprehensive visions to assess current resources, induce the rebirth of existing cultural assets, develop new ones, invest in physical infrastructure and commit to related policies; all actions occurring within complex social, political and economic milieus that are reshaping metropolitan areas.

In recent years scholars have explored the unique issues surrounding the development and impact of tourism on the built environment (Ioannides and Debbage, 1998; Judd and Fainstein, 1999; Judd, 2003; Hoffman, Fainstein, and Judd, 2004; Spirou, 2011; Wise, 2016; Edgell and Swanson, 2018). In fact, tourism activities have evolved into multi-billion–dollar enterprises, causing local governments to view them more favourably than ever before. In addition, the growth of this sector has brought about extensive reorganisation of the urban landscape. The introduction of substantial infrastructural facilities, in the form of museums, parks, stadiums, convention centres, etc., has not only expanded the availability of services but also strengthened existing amenities and introduced new ones (Clark, 2003). This broad direction has also encouraged inter-city competition, forcing cities to engage in a differentiation of entertainment possibilities. As a result, Las Vegas has pursued strengthening one type of tourist strategy which has varied from the one employed by New York, Chicago and Los Angeles.

Urban beautification has also emerged as a very important element of the tourist development strategy since cities are expending considerable resources on creating visually pleasing environments. In many ways the nature of this initiative rivals the city beautiful movement which flourished during the 1890s and early part of the twentieth century. Though residents benefit from what in most cases are downtown investments, a key aspect of this policy also aims at attracting tourists. Reorganised physical spaces and attractive landscapes are the central pragmatic elements of this outlook. Economic development opportunities, image building and place marketing and enhancement are key characteristics of this recently advanced planning directive. At the same time these programmes have the capacity not only to create new urban physical terrains but are also geared toward municipal advancement and the presentation of many cities as post-industrial environments geared toward tourism and play.

The observations above clearly point to an organised and structured plan of action on the part of local government, civic elites and business and corporate interests who aim to create spaces and places that can draw larger numbers of visitors and their considerable expenditures (Smith, 1995; Tucker and Sundberg, 1988; Leiper, 1990, 1993; Shone, Simmons and Dalziel, 2016). This contribution considers a number of elements that have made possible the rise of tourism as a factor in metropolitan development within the American context of urban change. Specifically, I begin by assessing the search for alternative development strategies as part of a wider scheme of urban competition and the quest for alternate strategic directions. Then, I discuss the role of private–public partnerships in urban tourism, as well as the building and financing of the tourist city. The chapter concludes by examining the role of a new infrastructure, sports and stadium development and convention centre development.

Urban economic development and the post-war fiscal stress

While today urban economic development efforts include culture, leisure and tourism as a significant part of the broader planning mix of tools employed, urban tourism and related cultural forms did not fit the agenda of growth practices in the 1950s and 1960s. At best they were viewed as an inconsequential element of financial activity. But the post-Second World War urban restructuring caused chronic fiscal stress, most visible during the 1970s. As a result, municipal governments were forced to search for new sources of economic growth. Furthermore, the production-oriented economy began to be replaced by a consumption-oriented one.

Many cities responded to these developments by divorcing themselves from their manufacturing dependency, searching to diversify and strengthen the various economic sectors. Pittsburgh and Baltimore are examples of two cities that have aggressively focused on the waterfronts by investing significant resources to develop a tourist infrastructure. It is within this framework that urban tourism emerges as an appealing alternative, one that slowly gained the favour of local officials and civic boosters making it a viable development tool. As a result, we begin to see the emergence of public investments and private sector development in the form of infrastructure and programming geared toward the advancement of the tourism industry. These sectors will eventually work together to fuel the advancement of urban tourism. But, a number of issues must be considered. For example, who invests and why? How is the public interest defined? How are "community benefits" utilised in this process?

Urban centres searched to enhance their spaces and rushed to take advantage of this new growth potential. They faced numerous challenges though, mainly in the areas of urban identity and urban competition. Specifically, how can a city with a formerly strong and nationally/internationally identifiable manufacturing economy convert itself into a tourist destination? Most importantly, how does it convince potential visitors of its new services and sense of "attractiveness"? It is through an interplay between structure and agency that we are able to gain insights on the transformation of urban centres into tourist cities. Local leaders must respond to structural conditions and urban processes that are beyond their control. Their vision though, including its execution, becomes part of the quest to achieve this new status. Leadership matters and eventually will determine the effectiveness of the tourism strategy.

The rise of urban tourism required cities to be entrepreneurial and business like in their approach. In that context, the city is not different from the corporation which must engage in image-building activities, promote its products and be prepared to deal with change if it wants to maintain its competitive edge and grow its market share. Given this new economic outlook, it is apparent that the financial stakes are very high (Kotler, Haider and Rein, 1993). As a result, we observe the re-making of local policymakers from service providers to active participants in local economic affairs. This injects a contested orientation since the efforts to remake the city come into conflict with community interests. Gentrification and displacement is one outcome of this trend (Spirou, 2011).

The role of private–public partnerships in constructing the city of play

With limited resources to recondition and retool, the needed tourist infrastructure necessitated the advancement of new models which required an increased role to be played by the private sector. Specifically, the building and operating of an economy related to culture, leisure and entertainment provide us with an opportunity to observe the "privatizing discourse of public infrastructure" in urban tourism. Increasingly, we can observe that private sector participation entails investment, ownership and/or management in the building of the city. An expanded and deeper connection between state and capital has emerged as the prevailing force in the creation of the tourist infrastructure (Perry, 2003; Ascher and Krupp, 2010).

This intersection has given rise to special and discretionary purpose authorities and special districts. These units have become popular among local governments, which utilised these units to meet various goals. Community development, transportation, economic development, parks and recreation, medical and health, environment and criminal justice are some of the areas within which we see the presence of these government forms. The use of private–public partnerships in the development of urban tourism and the building of the city of leisure has become a very common sought-after strategy. The logic behind these practices is the notion that growth is central to urban well-being and this form of tourism is simply another means to achieving that standing. The purpose of public subsidies then is to induce private investment and entice commerce ventures. That commitment by business creates jobs and a healthy tax base which is then utilised by the municipality to improve basic services as well as maintain and expand infrastructure. This then results in prosperity and a good business climate which in turn can encourage additional private investment. The outcome of this strategy is the rise of local growth coalitions and regimes that aggressively pursue tactics focusing on attracting and retaining corporate investment. Within this scheme, tourism, culture, leisure and entertainment are similar to the status that manufacturing held within the economy of cities during the early part of the twentieth century.

Many American cities pursue their commitment to the advancement of urban tourism through the creation of focused entities. These typically entail connecting economic development and tourism while also considering the advancement of culture. The main objectives of these offices relate to furthering approaches to strengthen the tourism sector, promoting and supporting new investment and development, and responding to the needs of local tourism-sector businesses. At the same time, officials provide advice and encouragement to help businesses and operators attract the tourist market and create packages and other forms of cooperative ventures while supporting measures creating a "tourist-friendly" environment. The primary engagements of these units relate to forging strong partnerships with other levels of government, agencies and the private sector and with helping new tourism-sector businesses and investors access the information needed to find sites and locate in their city. Broadly, helping businesses develop new markets in the tourism sector is the overarching vision that guides these rapidly developing organisations. In the end, this direction clearly denotes that role of the local government and its desire to closely partner with the private sector in advancing this industry.

Private–public partnerships in tourism can be found in many cities, large and small, across the United States. At the same time, these initiatives are controversial since many question the public value of these practices. Arguing that local governments often go too far in their efforts to attract investment, opponents charge that these strategies are pursued at the expense of tax payers. The recent construction of stadiums in cities across the country through the use of public funding has stalled in many cities through referendums. Local newspaper editorials and community groups will lead the anti-approval efforts by referencing these practices as examples of corporate welfare. Notably, in 2007, the Cincinnati Reds (Major League Baseball) saw the voters of Sarasota, Florida reject a taxpayer-supported stadium project. The team sought $16 million to help pay for a $45 million renovation (Hackett, 2008). But in what may be the most surprising turn of events, in 2006, Seattle voters approved a ballot initiative that restricted tax-payer subsidies of professional sports teams. This action hampered plans for the construction of a $500 million arena complex requested by the SuperSonics, Seattle's professional basketball team. This eventually led the team, which had been in Seattle since its founding in 1967, to relocate to Oklahoma City during the 2008–2009 season (Balko, 2008). In 2013, taxpayers in Houston rejected the authorisation to expend $217 million to turn the Astrodome into a giant convention and event centre. Interestingly, when constructed in 1965, the facility was referenced at the "eighth wonder of the world" (Hennessy-Fiske, 2013).

There are numerous examples of public–private partnerships, from infrastructure development to even marketing tourism. Visit Florida, an $80 million a year, public–private tourism agency is responsible for promoting the state as Florida's official source for travel planning. In San Diego, the city pays about $1.2 million a year for eight business groups charged with its promotion. The San Francisco Convention & Visitors Bureau, the city's promotion agency, was funded by a combination of public and private resources (Strasburg, 2004). In 2017, states averaged about $20 million annually in executing tourism development campaigns (Young, 2018).

The promotion of tourism and the cities that host their various attractions has emerged as a key aspect of the urban development strategy. Cities and states will invest millions in transforming their past images and perceptions in an effort to forge new urban identities that will draw visitors. It is also within this framework that we can observe the intersection of the public and private sectors. Cities expect businesses to invest by creating jobs and expanding their tax base, while, in exchange, the commercial sector relies on the government to promote and introduce amenities that will help maximise their profit.

Infrastructure development and the tourist city

A key issue in better understanding the building and financing of the tourist city is the process of its realisation since the quest for developing an urban tourist economy is ushering in a new era of city building in America. By the late 1980s and early 1990s we begin to observe an extraordinary construction programme spanning across downtowns, focusing on the creation of an infrastructure necessary to attract and retain visitors. Convention centres, theme parks, stadiums, casinos, waterfront developments and riverboat gambling complexes reshaped the physical landscape and helped restructure hundreds of urban cores.

The financing of these rapidly advancing urban infrastructures has been primarily driven by a combination of private and public sources of city building resulting in a significant construction boom. The privatisation of funding and the willingness of local governments to integrate this sector have not only resulted in the development of urban tourism, but also have come to restructure the physical presentation of cities. Some centres have utilised tax increment financing techniques to support construction projects which, directly or indirectly, promote tourism growth. This was closely tied to the considerable growth of the United States economy during the 1990s. There are many factors that contributed to this direction. For example, the introduction of new technologies in the latter part of the 1980s brought new consumer electronics and telecommunications in the consumer market. The computer industry grew rapidly and the internet not only entered American households, but also helped businesses function more effectively and provided numerous growth opportunities and new markets. The end of the conflict between the US and the former Soviet Union also provided new trade opportunities in Eastern European countries which rushed to grow their economies. Additionally, the federal government's reduction of expenditures on military development furthered the good economic climate. Low inflation and low rates of unemployment dominated the 1990s. The stock market increased and tax revenues grew rapidly, giving local governments considerable resources at their disposal.

The economic expansion of the 1990s also translated to urban growth. It is within this broader framework of economic and urban growth of the 1990s that cities aggressively pursue the construction of a significant tourist infrastructure.

While the finance mechanisms employed by cities to advance the tourism and entertainment infrastructure vary, it is clear that local authorities tend to generally pursue similar approaches. The urban tourist infrastructure can be divided into three specific categories. Primary facilities and services include historic, cultural and/or entertainment districts, convention centres, stadiums, performing arts centres. Secondary tourist facilities and services entail the availability of accommodation/hotels, restaurants, major convention hotels, as well as travel and tour services. Finally, tertiary tourist facilities and services are aimed at providing needed support which typically extends to additional sectors, beyond tourism. For example, financial and personal services and safety and emergency care services would be part of this category.

A number of common themes can be observed when assessing the building and financing of the tourist city. Special purpose authorities have become developed, increasingly playing a leading role, and have emerged as the most common approach to developing and operating these structures. The private sector has become more and more involved, playing a greater role in the advancement of this urban sector. Finally, the localised nature of this development strategy has placed offices of mayors in the centre and has helped further intensify the competition between cities.

Stadium construction as a contributor to the urban tourist strategy

Sports and stadium development is one area that is associated with the urban tourism development strategy. In recent decades, cities across the United States have invested large sums of public money to construct sports stadiums and arenas. There are primarily two rationales that have driven these policy considerations: economic benefits and social considerations. By employing economic impact studies and multiplier calculations, advocates note the positive direct and indirect economic outcomes. At the same time many have cited community-building and cultural identification functions of sporting events. But notwithstanding the emerging corporatisation of sport and its maturation as a culture industry, there remains considerable uncertainty concerning the utility of these construction projects as tools for economic development and urban regeneration. For example, hosting the Super Bowl is expected to have a $300–400 million economic impact as a result of attracting visitors and tourists (Draper, 2018). But these two factors, in the eyes of city officials and civic boosters are also connected. An enhanced image can attract future visitors, consumers and potential investors. It is expected that these mega-events will provide the impetus for urban growth.

As sport slowly became part of the tourism expansion strategy, stadium development proved a critical means of retaining or attracting professional franchises. From 1970 to 1990 we can observe an increase of publicly owned facilities used by professional sports teams from 70% to 80% (Quirk and Fort, 1992). From 1993 to 1996, more than $7 billion dollars was spent on the construction and renovation of major league facilities. The public sector provided 80% of these funds resulting in the introduction of 50 new stadiums across the various US professional leagues (Solomon, 2004). Even smaller cities aggressively pursued minor league sports franchises by offering stadiums to team owners (Johnson, 1993).

This development craze persisted since 2000. For example, from 2000 to 2008, 11 new stadiums have been developed in the National Football League (NFL) at a cost of $4.43 billion (average of $402 million per project). More impressively, the four most recent new stadiums were constructed at a staggering total cost of $5.6 billion. Two of those opened in 2009 for $1.3 billion (Dallas Cowboys) and $1.6 billion (New York Yankees). More recently in Minneapolis for the Minnesota Vikings (2016) for $1.1 billion, and in Atlanta for the Falcons (2017) for $1.6 billion.

The economics of upscale sports viewing has also produced a new type of sports arena. The expectation of private viewing environments and the infrastructure requisites for delivering an array of foods, beverages and other commodities have dictated various design innovations. The current trend in stadiums favours single-purpose facilities that make room for restaurants and taverns, gift shops and in some cases, overnight accommodations. Given the presumed economic and civic benefits derived from the hosting of sports franchises, municipal sponsorship of stadium projects is easily rationalised as a powerful economic development tool. The drive to attract teams is consistent with, indeed integral to, the municipal growth ideology. Cities thus compete with each other to attract existing clubs or to win the honour of hosting an expansion team. The justification for subsidies to often profitable professional teams grows from the perception that public funding constitutes a form of capital investment, and as such, is akin to other urban redevelopment outlays. Furthermore, project proponents also outline adjacent entertainment districts with retail shops, outdoor music in the summer and additional recreational opportunities. One of the key trends has been the recent construction of new stadiums in downtown areas. More than half of the 32 NFL teams have downtown or urban stadiums. A similar trend can be observed in Major League Baseball.

In Chicago, the Illinois Sports Facilities Authority was responsible for the construction of New Comiskey Park (now Guaranteed Rate Field for the Chicago White Sox) and Soldier Field (Chicago Bears). In Washington, DC, the District of Columbia government created an independent agency, the DC Sports & Entertainment Commission which oversaw the construction of Nationals Park for $611 million (Washington Nationals) and is responsible for the management and operation of the Robert F. Kennedy Memorial Stadium (Washington Redskins). Property taxes and bonds have been the most common methods employed to finance public investments in stadium development, but we can also observe motel/hotel taxes, rental car surcharges, sports lottery revenues and ticket taxes have been utilised to provide the needed revenue stream. The economically regressive impact of sales tax collections means that the poor and underprivileged pay more than their fair share of the stadium development costs.

The rise of convention centres in tourism growth

The desire by cities to hold mega-events can be traced all the way back to the eighteenth and nineteenth centuries with the World Fairs. In fact, expositions were similar in intensity to the modern Olympic Games. Chicago defeated Washington, DC, St. Louis and New York City for the opportunity to hold the Columbian Exposition in 1893. The Saint Louis World's Fair in 1904, known as the Louisiana Purchase Exposition, also became the location for the Third Olympiad. These science, arts and technology gatherings were large, public displays that helped make the host city the centre of the world.

Cities organised themselves to advance business tourism as far back as the early part of the twentieth century. In 1907 the Chicago Association of Commerce identified a subcommittee to work on attracting convention meetings to the city. Like Chicago, many other cities formed formal agencies out of loosely defined civic or business groups whose agendas centred on promoting and attracting tourism and related businesses to their respective locales. Convention and visitor bureaus were created across the country in San Francisco (1909), St. Louis (1909), Atlanta (1913), Kansas City (1918), Minneapolis (1927), Washington (1931), Cleveland (1934), New York (1935), Philadelphia (1941), Las Vegas (1960), New Orleans (1960), Anaheim, CA (1961), Orlando (1984) and Miami (1985). For example, the Chicago Convention and Visitors Bureau (CCVB) was founded in 1943 and it was eventually expanded in 1970 to include the Tourism Council of Greater Chicago, forming the Chicago Convention and Tourism Bureau. The management and operation of Chicago's McCormick Place, one of the largest exhibit spaces in the world, was added to the responsibilities of the CCVB in 1980.

The same rationale that drives the construction of stadiums has also fuelled public investment into convention centres. The convention centre building boom has been rationalised by projected job creation, increased tax revenues and direct and indirect opportunities for economic growth. At the same time, it has left municipalities wondering if the benefits outweigh the extensive renovations, maintenance and unfavourable lease agreements. Since governments have embraced the urban tourism and leisure industries as contributors to local economies, we've observed huge expenditures in this area, especially among cities ailing from de-industrialisation. When examining convention centres, there are three tiers of convention cities that can be identified. The first includes the most successful locations and includes Las Vegas, Orlando and Chicago. The second tier – San Francisco, San Diego, Anaheim, Atlanta and Washington, DC – have also focused on developing their economies by investing heavily in their convention centres. Finally, third tier cities have been most aggressive in their convention development and include Seattle, Reno, Nashville and Portland. This last set of cities aims to strip convention market share from the tier one and tier two urban centres.

In 2007, 93 new or expanded convention centres were introduced across the country. That is in addition to a construction explosion during the 1990s. For example, the total amount of exhibit space from 1990 to 2003 rose 51% from 40.4 million square feet to 60.9 million doubling the public spending on infrastructure to more than $2 billion annually. Cities in this intense competitive environment have been forced to reduce their rental pricing resulting in numerous convention centres operating at a loss. In just the first part of 2009, city officials announced the grand openings of 15 new, expanded or renovated venues (Zerlin, 2008).

As part of a growth ideology, these policies sometimes run the risk of becoming unfocused, often lost in the haze of intense inter-city competition for visitor expenditures. In Chicago, city officials approved an $800 million addition to McCormick Place. McCormick Place West Building opened in 2007 at a cost nearing $900 million. In 2013, a $110 million expansion and renovation was completed and in 2014, the development of a 1,200-room Marriott hotel, along with a new arena for DePaul University in the surrounding area (Byrne, 2014).

It has been increasingly unclear if the additional 250,000 square feet of meeting room space and 470,000 square feet of exhibition space will produce more convention activity.

Chicago's McCormick Place's recent woes – diminishing shows, attendees and square feet of utilised space over the last five years – were heightened by the loss of the National Association of Realtors' 100th-anniversary convention to Las Vegas. Even after considering the cyclical booking patterns of events, it is clear that Chicago has experienced a decline in the overall number of shows (82 in 2000 to 71 in 2005), square footage of exhibit space utilised (16 million in 2000 to 10.3 million in 2005) and attendance (3.32 million in 2000 to 2.17 million in 2005). Eleven years later, attendance was not significantly different at about 2.4 million (Lazare, 2016). Like other cities, Chicago is also experiencing competition from nearby smaller cities.

The dynamics behind the finance and development of convention hotels offer some interesting insights during the 2000s, a time of a considerable building boom. Officials in Austin, Texas pursued the construction of the Austin Convention Center Hotel which opened in January 2004. The 800-room hotel is managed by the Hilton Hotels Corporation and as a full service, first-class, convention-oriented, upscale hotel offers two full service restaurants, ample meeting space and a more than 600-space underground garage. The Austin Convention Enterprises, Inc. was formed by the City of Austin, the municipal sponsor, to be the municipal owner. The financial plan for its development included $110 million in senior lien current interest bonds, $135 million in sub-ordinate lien bonds and $21 million in third tier subordinate manager and developer bonds. The City of Austin's financial contribution was about $15 million. During its first year of operation, the hotel exceeded the profit goal of $10.2 million by posting $12.9 million of actual profit. Austin's aggressive spending on the visitor industry included more than $1.3 billion from 1998 to 2004; $15 million were specifically expended to promote the convention hotel. This success allowed Austin Convention Enterprises Inc. to renegotiate the bonds (*Austin Business Journal*, 2005, 2007).

Houston, Texas followed a similar financial strategy in developing its 1,200-room hotel, adjacent to the convention centre. The facility opened in 2003 and it included two ballrooms and 28 meeting rooms. With a more than 1,600-space parking garage the hotel is connected to the George R. Brown Convention Center. The funding scheme of the Hilton Americas-Houston included the issuing of bonds via the Convention and Entertainment Facilities Department under its municipal sponsor, the City of Houston. A total of $626 million were issued ($150 million fixed rate; $326 million fixed rate and $150 million variable rate) and the City of Houston provided a general obligation pledge to these hotel bonds. Some of the sources of debt service payment include 5.65% of the 7% of the city's hotel occupancy tax and parking revenues from city-operated large parking facilities in the central business district. But the project proved to have a negative impact on the other major downtown hotel. Because of competition, the 972-room Hyatt Regency Houston hotel went into foreclosure as it saw its occupancy drop considerably. Similar pressures were observed in St. Louis, Myrtle Beach, SC, Overland, KS and Sacramento, CA (Sunnucks, 2005).

Conclusion

The observations referenced above reveal the considerable complexities evident when examining the nature and effect of the tourism industry on urban centres and metropolitan areas. The competition for the revenue generated by visitors through expansive infrastructures raises a number of questions and issues in this arena. Furthermore, the evolving condition of cities, and their ever-growing reliance on this industry, makes inquiry on this subject highly important. Equally critical is research on understanding the processes and forces responsible for shaping urban tourism – from urban politics/power, corporate interests, pro-growth and image-building rationales to the role of civic groups and the importance of quality of life attitudes and considerations within a new global economy of information technology and entertainment.

It is apparent that structural changes have impacted cities in the United States over the last five decades. De-industrialisation, population decentralisation and globalisation have played a critical role, leaving the urban core subjected to disinvestment, decline and widespread social problems in housing and education among others. Within this rapidly evolving socioeconomic environment, and fiscally strained to provide needed social services to their residents, cities increasingly identified urban tourism and its potential revenue capability as a viable approach and a key economic development strategy.

On the agency front, we must closely examine the response and manner through which urban centres and their leaderships manage and determine the development process of this industry, including its interplay with private interests and civic groups. More research is needed on the urban restructuring front and the potential distributional impact of these strategies on disadvantaged groups. Within this broader framework it is also important to examine the tourist product as part of the commodity chain process. Key questions here include: who decides to actively pursue the development of this industry? How fluid is that process and how does it come to be realised? What adjustments are made during the development process and who are the players involved?

This activist approach, while aiming to maintain the competitive nature of this attraction is also informed by additional considerations. Those include inter-city competition, economic pressures, expressed corporate interests desiring to join and benefit from the existing successes of this development and even political rationales. These forces are intertwined, they are making the visitor experience, and in the process contribute to the construction of the tourist industry. The recent emergence of a new economy of leisure, amenities and quality of life considerations have similarly shifted the interplay between structure and agency as cities not only view urban tourism as an economic development tool, but also as a recruiting mechanism capable of attracting new economy entrepreneurs and workers. This trend has to be further explored as it has increased the importance of entertainment and restructured the discourse on urban development. This will allow for unique insights into the role of agency, revealing the complex relationship between this sector and related urban processes.

In the last 30 years cities have made a deliberate effort to physically reorganise their locales and embrace the advancement of urban tourism as an economic development tool. However, the success of this strategy, hinges on a continuously concerted effort to maintain the needed organisational capacity in marketing and image building, while balancing that against the competitive desires of other cities that have also intensified their efforts to attract visitors to their locales. That will likely determine, both quantitatively and qualitatively, how the social force of urban tourism will unfold in the coming years.

References

Ascher, W. and C. Krupp. (2010). *Physical Infrastructure Development: Balancing the Growth, Equity, and Environmental Imperatives*. New York: Palgrave.

Austin Business Journal. (2005). "Hilton Austin in top 25 in chain", 6 April.

Austin Business Journal. (2007). "Refinancing of Hilton debt saves taxpayers millions", 20 March.

Balko, R. (2008). "So long, Seattle: Stadium welfare schemes", *Reason*, 1 May.

Byrne, J. (2014). "McCormick Place hotel TIF money gets green light", *Chicago Tribune*, 4 March.

Clark, T. N. (ed.). (2003). *The city as an entertainment machine*, (Research in Urban Policy, Volume 9). Boston, MA: Elsevier/JAI.

Draper, K. (2018). "Windfall for Super Bowl hosts? Economists say it's overstated", *The New York Times*, 29 January.

Edgell, D. and J. Swanson. (2018). *Tourism Policy and Planning: Yesterday, Today, and Tomorrow*. London: Routledge.

Hackett, K. (2008). "Stadium strike-out: after a last inning defeat, is the game really over for the Reds and their supporters?", *Sarasota Magazine*, 1 February.

Hennessy-Fiske, M. (2013). "Houston voters skip nostalgia, reject measure to save Astrodome", *Los Angeles Times*, 6 November.

Hoffman, L., S. S. Fainstein, and D. R Judd. (eds.). (2004). *Cities and Visitors: Regulating Cities, Market, and City Space*. New York: Blackwell.

Ioannides, D. and K. Debbage. (ed.). (1998). *The Economic Geography of the Tourist Industry: A Supply-Side Analysis*. London: Routledge.

Johnson, A. T. (1993). *Minor League Baseball and Local Economic Development*. Champaign, IL: University of Illinois Press.

Judd, D. R. (2003). *The Infrastructure of Play: Building the Tourist City*. Armonk, NY: M. E. Sharpe.

Judd, D. R. and S. S. Fainstein. (1999). *The Tourist City*. New Haven, CT: Yale University Press.

Kotler, P., D. H. Haider, and I. Rein. (1993). *Marketing Places: Attracting Investment, Industry, and Tourism to Cities, States and Nations*. New York: Free Press.

Lazare, L. (2016). "Chicago's convention business struggles to show growth", *Chicago Business Journal*, 22 February.

Leiper, N. (1990). "Partial industrialization of tourist systems", *Annals of Tourism Research* 17: 600–605.

Leiper, N. (1993). "Industrial entropy in tourism systems", *Annals of Tourism Research* 20, 2: 221–225.

Perry, D. (2003). "Urban tourism and the privatizing discourses of public infrastructure", in D. R. Judd (ed.), *The Infrastructure of Play: Building the Tourist City*, Armonk, NY: M.E. Sharpe, pp. 19–49.

Quirk, J. and R. Fort. (1992). *Pay Dirt: The Business of Professional Team Sports*, Princeton, NJ: Princeton University Press.

Shone, M., D. Simmons, and P. Dalziel. (2016). "Evolving roles for local government in tourism development: a political economy perspective", *Journal of Sustainable Tourism* 24, 12: 1674–1690.

Smith, S. L. J. (1995). *Tourism Analysis: A Handbook*, New York: Longman.

Solomon, J. (2004). "Public wises up, balks at paying for new stadiums", *USA Today*, 1 April.

Spirou, C. (2011). *Urban Tourism and Urban Change: Cities in a Global Economy*, New York: Routledge.

Strasburg, J. (2004). "Will they visit S.F.? Budget battle pits funding for tourism bureau vs. social services", *San Francisco Chronicle*, 11 December.

Sunnucks, M. (2005). "Skeptics concerned about viability of city-funded hotel", *Phoenix Business Journal*, 18 March.

Tucker, K. and M. Sundberg. (1988). *International Trade in Services*, London: Routledge.

Wise, N. (2016). "Outlining triple bottom line contexts in urban tourism regeneration", *Cities* 53, April: 30–34.

Young, J. (2018). "Nebraska's new tourism campaign unveiled – but is it for everyone?", *Lincoln Star*, 17 October.

Zerlin, K. (2008). "15 Convention center openings slated", *Tradeshow Week*, 15 September.

31

TOURISM IN LATIN AMERICA

An overview and new experiences in city tourism

*Blanca A. Camargo, María L. Chávez and
María del Carmen Ginocchio*

Introduction

Latin America refers to the countries in the Western Hemisphere where romance languages are spoken as a result of colonisation by Spain, France and Portugal. Geographically, Latin America includes Mexico in North America, seven countries in Central America, 13 countries in South America and 21 islands in the Caribbean (Latin America Network Information Center 2020). Most of the estimated 626 million Latin American residents live in urban cities which vary in size and population. Currently, Mexico City (population 20.9 million), São Paulo (20.8 million), Buenos Aires (13.3 million), Rio de Janeiro (12.4 million) and Lima (10.6 million) are Latin American mega-cities with Bogotá projected to become one by 2030 (Go.euromonitor 2018; World Population Review 2018). Latin America is economically, socially and culturally diverse; with the exception of Cuba and, arguably, Venezuela, all are democratic nations at different stages of development. More than half of Latin American countries have high to very high human development; Guatemala, Nicaragua, El Salvador and Bolivia are classified as countries with medium human development; and Haiti as low human development (UNDP 2018).

Tourism is an important economic activity in Latin America. In 2017, Mexico, Central and South American, and the Caribbean countries attracted 112.9 million international tourists, 53.5% of all international tourists to the Americas, who generated US$94.9 billion in international tourism receipts (UNWTO 2017). Most research into tourism in Latin America tends to focus on examining tourism in natural and protected areas (e.g. Costa Rica, the Amazon), in sun, sand and sea destinations (e.g. Cancún, the Caribbean), or to iconic landmarks (e.g. Machu Picchu) with little attention paid to tourism in urban environments. In this chapter we will provide an overview of international tourism in Latin America, identify opportunities and challenges for tourism development, and present three case studies of emerging urban destinations. We hope to contribute to the understanding of trends and challenges faced by Latin American urban destinations and innovative approaches to develop and promote tourism in this region. For the purpose of this chapter, we focus our analysis on Spanish- and Portuguese-speaking countries.

Tourism in Latin America

Overview

Tourism is an important element of the national economies of Latin America. In 2018, it contributed to 8.7% of the region's GDP, 7.8% of all employment and 7.3% of total exports (World Travel and Tourism Council (WTTC) 2019b). Mexico is at the forefront of tourism in Latin America with regard to international tourist arrivals and tourism expenditure, with Argentina, Brazil, Chile and the Dominican Republic far behind (Table 31.1). The WTTC estimates that 82% of all tourism spending in the region is for leisure purposes as this is a region endowed with unique natural resources and protected areas, cultural diversity, historic sites and entertainment options. Countries like Belize, Chile and Colombia achieved double digit growth in the number of international tourists in the 2015–2016 and 2016–2017 periods, which shows the boom this region is experiencing; however, domestic spending surpasses international spending in 12 out the 20 countries analysed in Table 31.1.

With regard to the competitiveness of the industry, the travel and tourism sector in Latin America was the second most improved globally from 2015 to 2017 (World Economic Forum 2017); Mexico, Brazil and Panama were the top three most competitive travel and tourism-enabled economies in Latin America (in 22nd, 26th and 35th places worldwide). Overall, the region ranked high in Natural Resources and International Openness, but revealed weaknesses in Business Environment, Air Transport and Ground and Port Infrastructure, Safety and Security, and Environmental Sustainability. Brazil, Mexico, Costa Rica and Peru, for example, are the top four countries in Natural Resources, although the first two occupy the 66th and 133rd places (out of 136) in Environmental Sustainability. Venezuela, El Salvador and Colombia were in bottom position in Safety and Security in 2017.

Industry reports provide insights into the tourism performance of cities in Latin America. Euromonitor International (Euromonitor International 2019a) ranks Cancún and Mexico City (Mexico), Punta Cana (Dominican Republic), Buenos Aires (Argentina), Lima (Peru) and Rio de Janeiro (Brazil) in its "Top 100 city destinations of the world" by number of visitors in 2018. The WTTC (Japan Today 2018) also offers a list of top world city destinations by economic impact with Mexico City and Buenos Aires in eighth and sixteenth positions in terms of tourism GDP; Cancún and Mexico City are first and fifteenth in Tourism Contribution to the city's total GDP; and Lima, Buenos Aires and Santiago are the seventh, eighth and tenth cities with the highest contribution to their countries' tourism GDP. In addition, based on the travel and tourism GDP share of the total city GDP, the WTTC ranks city destinations in five categories: capital cities, largest cities, port cities, secondary cities and leisure cities, with Latin America paying an important role in all but the last two of these (Table 31.2).

Challenges affecting tourism in Latin American cities

Strizzi and Meis (2001) called attention to several factors that affect the flow of tourism into and out of Latin America. They argued that economic factors – in particular, slow regional economic growth, economic and financial instability, structural unemployment, inflationary pressures and unequal income distribution – have a more direct impact on the flow of tourism than non-economic factors affecting the region (e.g. environmental, political and/or social issues). While the region continues to experience slow economic growth, projected to be 1.8% for 2019 (CEPAL 2018b), and severe economic inequalities among and within its nations, progress has been made in reducing poverty (from 45.9% in 2002 to 30.7% in 2016 (CEPAL 2018a)) and

Table 31.1 Key tourism indicators for selected Latin American countries

Latin American region	Country	International arrivals[a] ('000)	Change 17/16[a]	International tourism receipts[a] (US$ million)	Tourism contribution to GDP 2018[b]	Tourism employment impact 2018[b]	Top three international tourist markets[b]
North America	Mexico	39,298	12.0%	21,333	17.2%	17.8%	USA (80%) Canada (5%) UK (2%)
Central America	Belize	427	10.8%	426	44.9%	39.9%	USA (66%) Canada (6%) UK (3%)
	Costa Rica	2,960	1.2%	3,876	13.1%	12.8%	USA (41%) Nicaragua (15%) Canada (7%)
	El Salvador	1,556	8.5%	873	11.6%	10.5%	Guatemala (37%) USA (33%) Honduras (15%)
	Guatemala	1,660	4.7%	1,566	7.4%	6.6%	USA (35%) El Salvador (21%) Honduras (7%)
	Honduras	936	3.1%	715	14.6%	12.9%	USA (32%) El Salvador (19%) Nicaragua (14%)
	Nicaragua	1,787	18.8%	841	11.1%	9.3%	Honduras (18%) USA (18%) Costa Rica (12%)
	Panama	1,843	−4.1%	4,452	14.5%	14.4%	USA (17%) Venezuela (14%) Colombia (14%)
South America	Argentina	6, 705	1.0%	5,060	10%	9.4%	Chile (19%) Brazil (16%) Paraguay (13%)
	Bolivia	959 (2016)	–	784	6.9%	61.0%	Peru (18%) Argentina (12%) USA (9%)
	Brazil	6,589	0.6%	5,809	8.5%	7.8%	Argentina (36%) USA (8%) Chile (5%)

Latin American region	Country	International arrivals[a] ('000)	Change 17/16[a]	International tourism receipts[a] (US$ million)	Tourism contribution to GDP 2018[b]	Tourism employment impact 2018[b]	Top three international tourist markets[b]
	Chile	6,450	14.3%	3,634	10.1%	9.9%	Argentina (49%) Brazil (9%) Bolivia (8%)
	Colombia	4,027	21.4%	4,821	5.6%	5.6%	USA (15%) Venezuela (9%) Spain (6%)
	Ecuador	1,608	13.4%	1,657	6.0%	5.5%	Colombia (22%) USA (17%) Venezuela (11%)
	Paraguay	1,537	17.5%	603	3.9%	3.1%	Argentina (72%) Brazil (15%) Uruguay (1%)
	Peru	4,032	7.7%	3,710	9.5%	7.7%	Chile (28%) USA (15%) Ecuador (8%)
	Uruguay	3,674	21.0%	2,540	16.9%	16.5%	Argentina (63%) Brazil (13%) USA (2%)
	Venezuela	601 (2016)	–	473 (2016)	9.7%	8.3%	Colombia (19%) Brazil (10%) Spain (8%)
The Caribbean	Cuba	3,975 (2016)	–	2,907 (2016)	10.6%	9.8%	Canada (32%) USA (6%) Germany (6%)
	Dominican Republic	6,188	3.8%	7,178	17.2%	16.0%	USA (36%) Canada (13%) Germany (4%)
	Puerto Rico	3,736	1.6%	4,090	6.7%	5.6%	–

Notes

a UNWTO (2017).

b WTTC (2019c).

Table 31.2 Latin American city destinations ranking

	Capital cities (33 cities)	Largest cities (38 cities)	Port cities (32 cities)	Secondary cities (10 cities)	Leisure cities (10 cities)
Top city	Abu Dhabi	Amsterdam	Auckland	Barcelona	Antalya
Latin American cities	Bogotá (5th) Brasilia (6th) Buenos Aires (9th) Lima (15th) Mexico City (20th) Santiago (26th)	Bogotá (5th) Buenos Aires (8th) Lima (18th) Mexico City (22nd) Santiago (29th)	Buenos Aires (5th) Cancún (6th) Rio de Janeiro (25th)	n/a	Cancún (2nd)

Source: WTTC (2019a).

unemployment (from 10.2% in 2001 to 6.6% in 2015), and in increasing foreign investment (from US$64.1 billion in 2001 to US$134 billion in 2015 (CEPAL 2019)), factors that can positively influence Latin America's residents' desire and ability to travel. But as the economy continues to slowly improve, non-economic factors are surfacing that affect the attractiveness of tourism and the competitiveness of the region, in particular, of their city destinations. The World Economic Forum (2017) identifies an underdeveloped infrastructure, an undervaluing of cultural resources, safety and security, and a fragile environmental sustainability as the greatest weaknesses of the tourist industry in the region.

Improving safety and security is in fact one of the main challenges faced by Latin American city destinations. Fuelled by poverty and drug cartel wars, 42 of the 50 cities with the highest murder rate in the world are in Latin America, of which 17 are in Brazil and 11 in Mexico (Dillinger 2019). In 2019, the US Department of State has active travel warnings for Venezuela (Level 4: Do Not Travel), Nicaragua (Level 3: Reconsider Travel), Mexico, Ecuador, Colombia, Guatemala, the Dominican Republic, Brazil and Cuba (Level 2: Exercise Increased Caution) (Travel.State.Gov 2019), which can deter one of the region's most important tourist markets from travelling. Nicolas Maduro's regime in Venezuela has caused the mass migration of 3 million people, mainly to Colombia, Peru, Ecuador and Argentina (UNHCR 2018), creating sanitation, housing and security problems in the host countries.

Economic and social inequalities combined with the affluence of international tourists from developed nations have enabled sex tourism in several Latin American destinations, in many cases involving the sexual exploitation of children. Destinations in Mexico (e.g. Tijuana, Cancún, Puerto Vallarta, Ciudad Juárez), Colombia (e.g. Cartagena, Medellín), Brazil (e.g. Rio de Janeiro, Fortaleza) or the Dominican Republic (e.g. Boca Chica) have been known as sex havens for years, but there has been a recent shift towards Central American destinations where there are less restrictive laws and weak government surveillance (Barger Hannum 2002; Gutman 2010).

The rapid urbanisation of Latin America is causing slow mobility, pollution and other environmental problems. For instance, seven Latin American cities (Bogotá, Mexico City, São Paulo, Rio de Janeiro, Belo Horizonte, Guayaquil and Medellín) are among the top 25 most congested cities in the world (INRIX 2019) and many others are experiencing high levels of air pollution, in particular in Chile, where nine of the ten most polluted cities in South America are to be found (AirVisual 2018). Climate change will pose a threat to several cities, especially those along the coasts which are vulnerable to extreme weather events, beach erosion and floods; urban destinations could face increased precipitation resulting in landslides and floods, heat-related mortality, spread of vector-borne diseases and water scarcity, among other risks (IPCC 2014).

In general, Latin American tourism tends to depend excessively on the region's natural resources. Greater efforts are needed to diversify the kind of tourism available in order to attract non-traditional market segments. It is important to note, as Table 31.1 shows, that Latin America's international tourism is highly dependent on its neighbouring countries, the United States, and Canada; Eastern Europe, Asia and the Middle East remain unexplored market segments that city destinations could attract. But Latin American cities do not attract the attention of the mass and travel media to the same extent as their European or North American counterparts. For instance, only three city destinations, Salvador (Brazil), Los Cabos (Mexico) and La Paz (Bolivia) were included in the 2018 and 2019 editions of the *New York Times*' "52 Places to Go" (*New York Times*, 2018, 2019) and Belize, Cabo San Lucas (Mexico), Bolivia and Puerto Rico were recognised in the *Condé Nast Traveler* list of the "19 Best Places to Go in 2019" (Condé Nast 2019). Latin American cities are modern, vibrant, affordable and full of culture, enabling them to capitalise on the emergence of the new middle classes arising in the region and abroad. Efforts should be made to promote and develop interesting tourism products that motivate tourists to explore beyond beach destinations and country capitals. In the next section, we provide three short case studies of emerging city destinations that are providing different tourism experiences in the region.

Medellín, Colombia: a model of social transformation and urban innovation

For many years Colombia experienced severe security problems caused by drug trafficking, political upheaval, guerrilla warfare and a paramilitary presence throughout vast areas of the country, which deterred international tourists; currently, it is one of the hottest Latin America country destinations. In 2016, Colombia signed a peace agreement with the rebel group FARC (Fuerzas Armadas Revolucionarias de Colombia), ending a 50-year civil conflict that plagued the country with terrorist attacks, kidnappings and thousands of casualties, and opening up vast natural areas which had previously been under the illegal control of the rebels. From about 622,000 international tourists per year in the late 1990s, the number of visitors has increased annually since 2006, reaching 3.6 million international tourists in 2018 (Citur 2019a). Bogotá, Cartagena, Medellín, Cali and San Andrés Island are the five most visited destinations in the country (Citur 2019b).

Medellín (pop. 2.5 million), the second largest city and industrial centre in Colombia, was infamously known as the cocaine capital and most dangerous city in the world (Borrell 1988). It was the operational base of Pablo Escobar, the Medellín cartel leader and his *sicarios* (hitmen), and it reached a record homicide rate of 357 murders per 100,000 people in 1991 (compared with 86 for the country overall) (Atlas 2016). Many of the crimes took place in *comunas*, poor settlements in the hills surrounding the city, the home of displaced residents, drug gangs and guerrilla and paramilitary activists. Of the 16 *comunas* in the city, Comuna 13 or San Javier had the highest level of poverty and violence in the city.

Starting in the 2000s, Medellín implemented a government-sponsored Social Urbanism Model that transformed the most deprived areas of the city. The purpose of this model was to promote social and cultural change, peace and social equity through architectural and urban transformation in poor and violent communities (Bustamante Fernández and Castaño Cárdenas

2009). Social urbanism implemented important public policy programmes, among them the construction of library-parks and educational spaces such as schools, museums, urban gardens, cultural centres and entrepreneurial hubs that dignified neighbourhoods and *comunas*. Integral Urban Projects (PUIs) were developed to combat inequality and exclusion, which simultaneously improved the physical (e.g. transport, housing, public spaces) and social (e.g. community engagement, education, sports and leisure opportunities) aspects in marginalised areas of the city. For instance, the government installed cable cars and electric stairs to facilitate access and connection for the *comunas* residents with the greater city. The "Medellín, the most educated" programme sought to guarantee quality education to all its residents, from kindergarten to university level, through the construction of childcare facilities in poor *comunas*, the implementation of strategies and incentives to reduce school abandonment, the consolidation of vocational schools and financial aid for low-income college students (Alcaldía de Medellín 2009).

Investing in poor communities brought new life to Medellín. The city is now recognised worldwide as a model of transformation and urban innovation. It has won several awards, including the Dubai International Award for Best Practice to Improve the Living Environment (2008), the *Wall Street Journal*'s Most Innovative City Award (2012), the Institute for Transportation and Development Policy's Sustainable Transport Award (2012) and the Lee Kuan Yew World City Prize (2016) for its metamorphosis from a violent city to a model of urban innovation within two decades. Furthermore, it was chosen as Latin American's first centre for the Fourth Industrial Revolution at the World Economic Forum in Davos in 2019 (Jones 2019). The city's transformation positively influenced its society. The homicide rate was reduced to 20.2 per 100,000 people in 2015, below the national rate of 24 (Atlas 2016).

The tourist industry has also benefited from Medellín's renaissance. International arrivals have increased year after year, reaching 379,000 in 2018 (from 150,000 in 2012) (Citur 2019b), and the city's Convention and Visitor's Bureau attracted a record number of 100 events in the same year (Greater Medellín Convention & Visitor Bureau 2019). A decade before, the city only hosted 15 events (Greater Medellín Convention & Visitor Bureau 2018). Medellín is recognised as the second most competitive city destination in Colombia (Centro de Pensamiento Turístico de Colombia 2018) and Forbes lists it as one of "The 10 Coolest Cities Around the World to Visit in 2018" (Abel 2018). Medellín's drug and violent past and recent urban changes are becoming popular attractions among tourists, scholars and urban planners. The Museo Casa de la Memoria (House of Memory Museum) is a key site to learn the history and impact of the different types of conflicts that took place in Colombia; community-based walking tours are available in Comunas 4 and 13, through which tourists can learn the tumultuous history of the *comunas* (Figure 31.1), see developments in social urbanism and experience urban music and street art (see Naef 2016, for a multi-stakeholder discourse analysis of these tours). However, as Naef (2018) argues, there is more to Medellín than its narco-heritage, which movies and television series have helped instil in the tourist mind. Efforts are being made to position the city as a hub for innovation and entrepreneurship with events like the 2017 "Medellín Lab", a meeting of urban experts to discuss and exchange experiences related to violence prevention, security, resilience and community cohesion.

Lima: the gastronomic capital of South America

Peruvian cuisine was practically unknown until the 1990s; today, it is recognised as one of the best in the world and the number one source of national pride, above its natural resources, history and art (Ipsos 2017). Peru has been declared the "World's Leading Culinary Destination" consecutively from 2012 to 2018 (World Travel Awards 2019) and Lima, its capital, has

Figure 31.1 Comuna 13 tours.

Source: Comuna 13 Medellín Graphic Tour.

been recognised as one of the cities with the best gastronomy in the world. Although most international tourists who travel to Peru visit the archaeological site of Machu Picchu, the country's gastronomy is increasingly becoming a motivation for travel, in particular to Lima, and 82% of international tourists consider Peru a gastronomic destination (PromPerú 2016b).

Several factors have contributed to the culinary boom in Peru. First, its cuisine, which can be described as a fusion of indigenous and Spanish ingredients with international influence of its European, Chinese, Japanese and African immigrants. The country is endowed with a diverse climate, nature and culture that are reflected in each region's cuisine. Amazonian cuisine features tropical plants, rare vegetables and wild animals cooked in a traditional style; Andean cuisine is characterised by hearty meals prepared with highly nutritional ingredients from the highlands of the Andes; and Lima, the most popular culinary destination in Peru, offers internationally influenced dishes in an urban setting, from street vendors to upscale restaurants. Iconic Peruvian dishes include the renowned *ceviche* and *tiradito* (types of seafood carpaccio), *lomo saltado* (beef loin), *anticuchos* (cow hearts grilled over charcoal), *causa limeña* (potato puree stuffed with poultry), *arroz chaufa* (Chinese fried rice) and *pisco sour*, an alcoholic drink declared part of Peru's cultural heritage (Peru Travel 2019). Second, the sustained economic growth of the people of Peru's increased purchasing power, which renewed middle- and upper-class residents' desire and ability to eat in restaurants more frequently (Matta 2011). Between 2004 and 2018, the number of full-service restaurants in Peru increased by 210%, from 19,025 to 59,039 (Euromonitor International 2019b).

A third key factor in Peru's gastronomic tourism success is the active participation of all stakeholders in the value chain. The government, mainly through its Ministry of Culture, has been crucial in protecting and promoting Peruvian ingredients and cuisine nationally and internationally. In 2007, Peru's cuisine was declared National Heritage and the government is preparing the dossier to be included in the list of UNESCO World Intangible Heritage by 2021, when the country celebrates 200 years of independence from Spain. (Experiencias Gourmet 2019). Peru Export and Tourism Promotion Agency – PromPerú – has been in charge of developing not only the country brand for which Peru is recognised worldwide but also marketing campaigns and events to promote Peru's cuisine abroad. One of PromPerú's most successful campaigns was Peru-Nebraska, a 15-minute documentary-style spot[1] which follows the surprise visit of Peruvian celebrities to the small town of Peru, Nebraska to share with its 500 residents their cultural heritage, in particular Peruvian food, and teach them what it means to be Peruvian. It also created the online platform "Super Foods Peru"[2] that gives visitors access to information about the diversity and benefits of the country's unique fruits, grains, vegetables, herbs, roots and fish.

The Peruvian Society of Gastronomy (Sociedad Peruana de Gastronomía, APEGA) is another key stakeholder in promoting Peruvian cuisine. This not-for-profit organisation groups together chefs, nutritionists, restauranteurs, culinary historians and researchers, culinary schools, food producers and other key gastronomy players in an effort to "give Peruvian cuisine the place it deserves in the world, and make it a source of identity, innovation and sustainable development for Peruvians" (APEGA 2019b). Since 2008, APEGA organises Mistura, the biggest and most important gastronomic fair in Latin America; held in Lima, the fair had an attendance of 302,000 visitors in 2017, 10% of whom were international visitors (APEGA 2017). APEGA has launched other initiatives to consolidate Peruvian cuisine, among them "Come Peruano" (Eat Peruvian), a campaign that seeks to improve the country's nutrition by encouraging the consumption of healthy Peruvian ingredients and dishes, and "Lima Capital Gastronómica" (Lima, Gastronomy Capital), which aims to make Lima one of the best gastronomic destinations of the world by 2021 when the country will hold its bicentennial independence celebration (APEGA 2019a).

A new generation of local chefs has contributed to the popularity of Peru's gastronomy through innovative methods. They are rediscovering rare ingredients and old recipes and presenting them in inventive new dishes in upscale restaurants in Lima, creating what is known as Novo-Andina cuisine. Lima hosts three of the "World's 50 Best Restaurants": Central and Maido in sixth and seventh place respectively, and Astrid y Gastón in 27th (The World's 50 Best Restaurants 2019). In 2017, chef Virgilio Martínez, the founder of Central, was named Best Chef on the Planet for "propelling Peru's status as a foodie destination to new heights with constant research and improvement at his flagship restaurant" (The World's 50 Best Restaurants 2019). Many of these chefs are celebrities in food channels, travel documentaries and other media, which has sparked young people's interest in pursuing culinary studies.

In summary, the boom in Peruvian gastronomy is the result of several factors: the characteristics of Peruvian ingredients derived from its unique biodiversity, the fusion of traditional ingredients, old recipes, international ingredients and modern cooking techniques employed by young chefs; the recognition and promotion of gastronomic heritage by the government and NGOs; and the dissemination of Peruvian cuisine in the popular media, culinary books and travel guides. This expansion has generated many benefits including the revaluation and increased demand for traditional ingredients and new partnerships between chefs and small producers, which is improving the economic situation of rural and fishing communities (Ginocchio Balcázar 2012). Furthermore, a government study found that gastronomy was the main tourist motivator for 7% of all leisure tourists to Lima (PromPerú 2016a). Sixty-nine per cent of these gastronomy tourists are male, 54% are single with an average age of 39 years, highly educated, and with an annual income of more than US$40,000 in 68% of all cases. The average length of stay is 12 days and they spend US$1,181 per trip, which is higher than the general leisure tourist who spends only US$978 and stays nine nights in the city.

The challenge that could potentially affect the future development of gastronomic tourism in Peru is the lack of diversification of the tourism experience since the main activity in this regard is visiting high-end restaurants and street food outlets (PromPerú 2016a). Innovation is also needed to develop experiential gastronomic experiences such as culinary trails (see Roy *et al.* 2019), workshops, gastronomic festivals and tourist packages that give tourists the incentive to explore cuisines in other destinations within the country. Diversifying the tourism experiences and including more stakeholders in the process, especially local producers, small businesses and regional and indigenous experts, will contribute to a more equitable distribution of tourism benefits (see Jamal and Camargo 2013) and also prevent the commodification of Peruvian cuisine into certain iconic dishes exclusively for tourism purposes.

Tijuana, Mexico: from "Satan's playground" to "the coolest city" in Mexico

Mexico is the most important destination country in Latin America. International tourist arrivals have grown steadily in the past decade reaching 41.3 million overnight visitors in 2018 (UNWTO 2020) which made it the sixth most visited country in the world in that year. Preliminary reports estimate 41 million international visitors in 2018 who generated US$20.3 billion in tourism expenditure (Datatur 2019). However, 43%, or 18 million of all international tourists, are border tourists, that is, people who legally cross and stay at least one night within the area 30 miles south of the US–Mexico border, or 20 miles north of the Guatemala–Mexico and Belize–Mexico border. Despite their high numbers, border tourists have very little economic impact: they spend an average of US$61 per trip, compared with the US$823 per trip of other international tourists (Datatur 2019). The most popular destinations in Mexico are Cancún, the Riviera Maya, Los Cabos, Puerto Vallarta and Mexico City (Sectur 2018). With the exception of Mexico City, they are sun, sand and sea resorts, and together account for 80% of all international tourists to the country.

The state of Baja California, in the northwest corner Mexico, is emerging as one of the most important tourist regions in the country thanks to its Pacific Ocean coasts, newly developed medical tourism infrastructure, wine trails, business and convention centres and vibrant gastronomy. Located only about 20 miles from San Diego, California, Tijuana (pop. 1.6 million) is one of the most popular destinations in Baja California and the world's most visited border destinations; approximately 42 million people crossed the US border to visit Tijuana in 2013 (Secture Baja California 2013). However, the economic impact of their visit is minimal: the Tourism Observatory of Baja California estimates that only 21% of all visitors stay overnight, their average length of stay being less than three days, and their tourism expenditure only US$98 per day (Observatorio Turístico de Baja California 2013). Furthermore, only 8% stay in hotels as the majority of overnight tourists stay with friends or relatives or in second homes.

Tijuana faced two major tourism challenges: its reputation was tarnished and it lacked innovative tourism products to attract higher quality, first-time visitors and motivate longer stays in the destination. During the Prohibition era, at the beginning of the twentieth century, Tijuana was known as "Satan's playground", a city where Californians could engage in gambling, alcohol drinking and other vices (see Vanderwood 2010); from the 1960s well into the late 1990s the city was the main source of cheap labour for nearby *maquilas* (US-run factories) and of drugs and prostitutes for American tourists. The murders and kidnappings related to drug cartel wars in the early 2000s generated travel warnings from the United States government and negative media coverage that affected tourist numbers and Tijuana almost became a ghost town. State and local tourism action plans were implemented to improve the city's image and promote a cosmopolitan and, most importantly, safe city to domestic and international tourist markets, as well as creating programmes to enhance the city's business and tourism infrastructure and consolidate medical tourism (see COPLADE 2015; Maher and Carruthers 2014). Institutional initiatives combined with grassroots projects were key to the transformation of Tijuana from a cheap, lawless party city to a vibrant artistic and gastronomic destination.

Restaurants, food trucks and local breweries now play key roles in Tijuana's tourism offering. The city is home to traditional restaurants such as Caesar's, where the famous Caesar salad was created in the 1920s, and modern restaurants featuring Baja-Med food, a new culinary movement that combines fresh ingredients from the region, in particular seafood and vegetables, in unique Mexican, Mediterranean and Oriental inspired dishes. Created by a new generation of local chefs, Baja-Med food has become a source of identity and pride for Tijuana and Baja Californian residents (see García et al. 2016), and a tourism resource attracting Mexican and international tourists to its restaurants, food truck areas, food gardens, gastronomic events such as the Baja Culinary Fest in Mexico and Baja by the Sea in San Diego, and familiarisation tours for travel intermediaries. Tijuana is also the largest producer of craft beer in Latin America and has built up a brewing district – comprising 87 microbreweries offering more than 400 brands of beer – that is highly popular among both locals and tourists looking for safe and upscale entertainment options (Martínez 2018). Beer fests abound, and for cultural tourists, there is a vibrant art scene comprising galleries, exhibitions, cultural passageways and festivals, the Tijuana Cultural Center, one of the most important in northwest Mexico, and the Avenida Revolución (Revolution Avenue), where tourists can see the increasingly rare donkeys painted to look like zebras, an emblematic Tijuana cultural tradition and tourist icon (see Observatorio Turístico de Baja California 2014).

Compared with other cities in Mexico, Tijuana lacks the cultural heritage that attracts tourists to visit, for instance, Mexico City, San Miguel de Allende, Oaxaca or Guanajuato. For many years, it attracted mainly Californians looking for nocturnal entertainment, cheap medication without prescription or tacky souvenirs along the Avenida Revolución, until violence and drug-related wars almost killed international tourism. The transformation of Tijuana was a joint effort by the government and, more significantly, local groups who did not give up on the city and redeemed and reinvented it, almost from scratch, as one of the coolest destinations in Mexico.

Discussion

In this chapter we provided an overview of tourism trends in Latin American cities and case studies that show innovative approaches to tourism development in three urban destinations. From a Latin America perspective there are important lessons for tourism practitioners. The case of Medellín highlights the importance of government investment in the living conditions and social fabric of cities torn by violence and crime to improve their negative image, safety and security conditions, and residents' sense of pride and belonging to their communities. Tourism research and practice have focused on city branding, marketing models and crisis management strategies for conflict-ridden destinations (e.g. Scott et al. 2010) but as Avraham and Ketter (2013) argue, many such tourism campaigns follow a cosmetic approach that seeks to alter the destination image without structural changes that tackle the roots of conflict and violence. Medellín worked on overcoming its internal challenges through urban and education interventions, which in turn changed the image and appeal of the city. Latin America is home to 42 of the 50 most violent cities in the world (Woody 2019) and Medellín's Social Urbanism Model can help reduce social and economic inequities, create safe spaces for education and leisure, and create tourism opportunities for its citizens.

The second and third cases illustrate how to promote urban destinations based on their culinary heritage at the micro and macro level. In Peru, the national government took the leading role in promoting its diverse cuisine and building a country brand based on its gastronomy ("Peruvian Cuisine for the World"). This strategy, known as gastro-diplomacy positioned Peru, especially Lima, as one of the leading culinary destinations in the world by engaging several actors in the promotion of its gastronomy in international markets, including the Ministry of Culture, the Ministry of Tourism, culinary associations, television hosts and celebrity chefs (see Wilson 2013). Gastro-diplomacy has also been undertaken by countries like Thailand (Global Thai campaign), South Korea (Global Hansik campaign) and Taiwan (Lipscomb 2019). Tijuana, on the other hand, emerged as a destination thanks to grassroots initiatives that revitalised the city and created a fresh offer of food and beer spaces that is attracting a new segment of tourists looking for world-class experiences.

International tourism is projected to continue growing over the next decade, reaching 1.8 billion international tourists by 2030 (UNWTO 2017). It is estimated that emerging economies will receive more inbound tourism than advanced economies, and Asia-Pacific, the Middle East and Africa will be the regions with the greatest gains in international tourist arrivals (UNWTO 2011). In order to capture a share of the international tourist market, Latin American destination management organisations must make efforts to attract and prepare for new tourist segments, as international tourism in the region is highly dependent on border countries and the United States. Geographically, the European market has little impact on tourism in Latin America and the Chinese, Indian and Arab tourism markets are unexplored and tourism research is necessary to examine their travel and tourism needs, preferences and expectations.

In addition to the forms of tourism described in this chapter, there are other possibilities to diversify Latin America tourism, in particular in destinations with no strong natural resource base. Several countries were at the forefront of the slave trade from the fifteenth to the nineteenth century, and there are not only historical sites, some of them inscribed in the UNESCO World Heritage List (UNESCO 2014), but also a living Afro-Latino culture that could form part of an itinerary for ethnic, memory, roots or dark tourism linking several Latin American countries. Cities with a developed university infrastructure have the opportunity of attracting exchange students from other countries who are seeking an academic and cultural experience abroad. Short-term student mobility, or academic tourism (Rodríguez *et al.* 2012) is gaining momentum thanks in part to globalisation, innovative academic programmes and government support for student mobility; in 2013 there were 4.1 million students undertaking an academic experience abroad and the number is expected to reach 8 million by 2025 (Institute of International Education (IIE) 2015). Currently, Latin America only captures 5% of the academic tourism market, with Brazil, Mexico and Colombia being the countries that host the most international students. Camargo and Quintanilla (2018) found that 98% of international students who chose Monterrey (Mexico) as their academic destination travelled during their studies, 43% of them visited eight or more cities, and 86% visited cultural destinations. As more Latin American universities attain higher positions in education rankings and make bilateral agreements with partner-universities abroad to facilitate the revalidation of academic credit hours, Latin American cities will become more attractive to international students seeking to experience a new culture.

Latin America is endowed with a unique natural and cultural heritage that could form the basis of an attractive tourism offering in the coming years. For instance, Colombia, Peru, Ecuador, Brazil, Bolivia, Venezuela, Mexico, Argentina and Panama are the top nine countries with the greatest bird diversity in the world, but attractive bird-tourism products and packages are yet to be developed in these regions. Furthermore, many natural areas where bird tourism and ecotourism could be developed are under threat from mining and deforestation. The current presidents of Colombia and Brazil intend to authorise national development plans that will allow economic activities to take place in national parks and protected areas that will devastate the region's biodiversity and indigenous populations (see Survival International 2019). In Mexico, the Tren Maya (Mayan Train), the government-sponsored 950-mile railway traversing five southern states inhabited by the Maya peoples, due to begin in 2019, has benefited from no environmental impact study or consultation with or consent from the indigenous population, potentially harming the livelihood of thousands of Mayan residents along its route. It is imperative to implement and enforce policies that protect the natural and cultural heritage of this region and orient tourism development so it generates economic, environmental and sociocultural well-being and justice for local populations. The government needs to prioritise infrastructure development, capacity building, subsidies and marketing of community-based enterprises and social enterprises in such a way that they can fairly compete with private businesses that may seize on tourism opportunities.

In conclusion, there is a positive outlook for tourism in Latin America in the next few years, but in order to minimise the dependency on a limited number of markets, tourist programmes need to harness the potential of their urban environment to attract greater numbers of more affluent tourists. Latin American cities offer an array of cultural and entertainment experiences that are currently being downplayed due to the marketing emphasis on sun, sand and sea and natural area destinations. However, addressing the urban problems related to pollution, security, traffic congestion and infrastructure is a must for city tourism to grow in this region.

Notes

1 www.youtube.com/watch?v=8joXlwKMkrk.
2 https://peru.info/en-us/superfoods.

References

Abel, A., 2018. *The 10 coolest cities around the world to visit in 2018*. [online] Forbes.com. Available at: www.forbes.com/sites/annabel/2018/02/22/the-10-coolest-cities-around-the-world-to-visit-in-2018/#763afbe81bb3 [Accessed 20 April 2019].
AirVisual, 2018. *World most polluted cities 2018*. [online] Available at: www.airvisual.com/world-most-polluted-cities?continent=59af929e3e70001c1bd78e50&country=&state=&page=1&perPage=50&cities [Accessed 15 April 2019].
Alcaldía de Medellín, 2009. Medellín, transformación de una ciudad. Alcaldía de Medellín – Banco Interamerica de Desarrollo. [pdf] Available at: https://acimedellin.org/wp-content/uploads/publicaciones/libro-transformacion-de-ciudad.pdf [Accessed 19 April 2019].
APEGA Sociedad Peruana de Gastronomía, 2017. Informe anual de actividades APEGA. [pdf] Available at: www.apega.pe/descargas/contenido/209-apega-cocina-peruana.pdf [Accessed 27 April 2019].
APEGA Sociedad Peruana de Gastronomía, 2019a. *¿Qué es APEGA?* [online] Available at: www.apega.pe/nosotros/que-es-apega [Accessed 26 April 2019].
APEGA Sociedad Peruana de Gastronomía, 2019b. *Proyectos*. [online] Available at: www.apega.pe/proyectos [Accessed 27 April 2019].
Atlas, 2016. *Medellín homicide rate, 1975–2015*. [online] Available at: www.theatlas.com/charts/Syhv4L-hXe [Accessed 30 May 2019].
Avraham, E., and Ketter, E., 2013. Marketing destinations with prolonged negative images: Towards a theoretical model. *Tourism Geographies*, 15(1), 145–164.
Barger Hannum, A., 2002. Tricks of the trade: Sex tourism in Latin America. *ReVista: Harvard Review of Latin America*, Winter, 60–61.
Borrell, J., 1988. *Colombia the most dangerous city*. [online] Available at: http://content.time.com/time/subscriber/article/0,33009,967029-1,00.html [Accessed 16 April 2019].

Bustamante Fernández, J. S., and Castaño Cárdenas, N., 2009. *La transformación de Medellín, Urbanismo Social*. [online] ARQA. Available at: https://arqa.com/arquitectura/urbanismo/la-transformacion-de-medellin-urbanismo-social-2004-2007-2.html [Accessed 30 May 2019].

Camargo, B. A., and Quintanilla, D., 2018. Análisis del turismo académico en Monterrey (México). *Turismo y Sociedad*, 23, 125–147.

Centro de Pensamiento Turístico de Colombia, 2018. *Índice de competitividad turística regional de Colombia (ICTRC) 2018*. Bogotá D.C. Colombia: Asociación Hotelera y Turística de Colombia y Fundación Universitaria Cafam.

CEPAL, 2018a. *Panorama social de America Latina, 2017*. (LC/PUB.2018/1-P). [pdf] Santiago, Chile: Comisión Económica para América Latina y el Caribe (CEPAL). Available at: https://repositorio.cepal.org/bitstream/handle/11362/42716/7/S1800002_es.pdf [Accessed 15 April 2019].

CEPAL, 2018b. *Actualización de proyecciones de crecimiento de América Latina y el Caribe 2018 y 2019*. [pdf] Available at: www.cepal.org/sites/default/files/pr/files/tabla-proyecciones_octubre-2018_esp.pdf [Accessed 15 April 2019].

CEPAL, 2019. *Latin America and the Caribbean: Regional economic profile*. [pdf] Available at: http://estadisticas.cepal.org/cepalstat/Perfil_Regional_Economico.html?idioma=english [Accessed 15 April 2019].

Citur.gov.co, 2019a. *MinCIT – Estadísticas*. [online] Available at: www.citur.gov.co/estadisticas/df_viajeros/all/4 [Accessed 16 April 2019].

Citur.gov.co, 2019b. *MinCIT – Estadísticas*. [online] Available at: www.citur.gov.co/estadisticas/df_viajeros_ciudad_destino/all/2 [Accessed 16 April 2019].

Condé Nast, 2019. *The 19 best places to go in 2019*. [online] Condé Nast Traveler. Available at: www.cntraveler.com/gallery/19-best-places-to-go-in-2019 [Accessed 16 April 2019].

COPLADE, 2015. *Programa estatal de turismo de Baja California 2015–2019*. [pdf] Available at: www.copladebc.gob.mx/publicaciones/2015/planesyprogramas/Programa%20Estatal%20de%20Turismo%202015-2019.pdf [Accessed 3 May 2019].

Datatur.sectur.gob.mx, 2019. *Encuesta de viajeros internacionales*. [online] Available at: www.datatur.sectur.gob.mx/SitePages/VisitantesInternacionales.asp [Accessed 1 May 2019].

Dillinger, J., 2019. *The most dangerous cities in the world*. [online] WorldAtlas. Available at: www.worldatlas.com/articles/most-dangerous-cities-in-the-world.html [Accessed 14 April 2019].

Euromonitor International, 2019a. *Top 100 city destinations: 2019 edition*. [online] Available at: https://go.euromonitor.com/white-paper-travel-2019-100-cities.html [Accessed 4 May 2020].

Euromonitor International, 2019b. *Full-service restaurants in Peru*. [online] Available at: www.euromonitor.com/full-service-restaurants-in-peru/report [Accessed 4 May 2020].

Experiencias Gourmet, 2019. *La cocina tradicional peruana aspira a ser Patrimonio Cultural Inmaterial de la Humanidad*. [online] Available at: www.excelenciasgourmet.com/es/noticias-gourmet/la-cocina-tradicional-peruana-aspira-ser-patrimonio-cultural-inmaterial-de-la [Accessed 26 April 2019].

García, E., Ramírez, S., Fernández, M., Astiazarán, C., and González, L., 2016. *Obstáculos que surgen durante el desarrollo y promoción de una gastronomía regional: El caso de Baja California*. Licentiate thesis. Monterrey University.

Ginocchio Balcázar, L., 2012. *Pequeña agricultura y gastronomía. Oportunidades y desafíos*. [online] APEGA Sociedad Peruana de Gastronomía. Available at: www.apega.pe/publicaciones/documentos-de-trabajo/pequena-agricultura-y-gastronomia.html [Accessed 30 May 2019].

Go.euromonitor.com, 2018. *Megacities: Developing country domination*. [online] Available at: http://go.euromonitor.com/rs/805-KOK-719/images/MegacitiesExtract.pdf?mkt_tok=eyJpIjoiTkdFNFpUWTROVFZsTWppNMCIsInQiOiJ6NGs0d1RNa2VMVjRQQnhHY1VmM3I0aXNEczU1Q1F3TnAyd1lZVHlQR21DSDNzMkVFVlpMMUNRejBLeStkalZFcGlxWDd4SXRRKQ3AxemJKN3pFb1c4R2prYzJBdFRRQTZxZEhFSlVMVNm84ZXJPVWU5aEI5OVlhdndSSTV6dmlweS99 [Accessed 6 March 2019].

Greater Medellín Convention & Visitor Bureau, 2018. *Informe de gestión 2017*. [pdf] Available at: http://bureaumedellin.com/greater/wp-content/uploads/2018/10/INFORME-DE-GESTION-2017.pdf [Accessed 20 April 2019].

Greater Medellín Convention & Visitor Bureau, 2019. *Más del 70% de los países en el mundo conoció la cara turística de Medellín en 2018*. [online] Available at: http://bureaumedellin.com/greater/mas-del-70-de-los-paises-en-el-mundo-conocio-la-cara-turistica-de-medellin-en-2018/ [Accessed 20 April 2019].

Gutman, W., 2010. *Sex tourism threatens Central America's youth*. [online] Lab.org.uk. Available at: https://lab.org.uk/sex-tourism-threatens-central-america's-youth/ [Accessed 16 April 2019].

INRIX, 2019. *Global traffic score card*. [pdf] Available at: http://inrix.com/scorecard/ [Accessed 30 May 2019].

Institute of International Education (IIE), 2015. *A quick look at international mobility trends.* Available at: http://classof2020.nl/wp-content/uploads/2015/02/Global-he-Mobility-2015-Project-Atlas.pdf [Accessed 30 May 2019].

IPCC, Intergovernmental Panel on Climate Change, 2014. *Climate change 2014: Impacts, adaptation, and vulnerability. Part A: Global and sectorial aspects. Contribution of Working Group II to the Fifth Assessment Report of the Intergovernmental Panel on Climate Change.* Cambridge and New York: Cambridge University Press.

Ipsos, 2017. *Gastronomía: mientras más peruana, mejor.* [online] Available at: www.ipsos.com/es-pe/gastronomia-mientras-mas-peruana-mejor [Accessed 23 April 2019].

Jamal, T., and Camargo, B., 2013. Sustainable tourism, justice and an ethic of care: Toward the just destination. *Journal of Sustainable Tourism*, 22(1), 11–30.

Japan Today, 2018. *WTTC names top 10 most important cities in terms of tourism market size.* [online] Available at: https://japantoday.com/category/features/travel/WTTC-names-top-10-most-important-cities-in-terms-of-tourism-market-size [Accessed 4 May 2020].

Jones, J., 2019. *Medellín to host Latin America's first center for Davos' Fourth Industrial Revolution.* [online] *The Bogotá Post.* Available at: https://theBogotápost.com/medellin-to-host-latin-americas-first-center-for-davos-fourth-industrial-revolution/34416/ [Accessed 19 April 2019].

Latin America Network Information Center, 2020. [online] Available at: http://lanic.utexas.edu/ [Accessed 4 May 2020].

Lipscomb, A., 2019. Culinary relations: Gastrodiplomacy in Thailand, South Korea, and Taiwan. *The Yale Review of International Studies.* [online] Available at: http://yris.yira.org/essays/3080 [Accessed 25 December 2019].

Maher, K. H., and Carruthers, D., 2014. Urban image work: Official and grassroots responses to crisis in Tijuana. *Urban Affairs Review*, 50(2), 244–268.

Martínez, G., 2018. *Tijuana, capital de la cerveza artesanal.* [online] *El Economista.* Available at: www.eleconomista.com.mx/estados/Tijuana-capital-de-la-cerveza-artesanal-20180722-0089.html [Accessed 30 May 2019].

Matta, R., 2011. Posibilidades y límites del desarrollo en el patrimonio inmaterial. El caso de la cocina peruana. *Apuntes*, 24(2), 196–207.

Naef, P., 2016. Touring the "comuna": Memory and transformation in Medellin, Colombia. *Journal of Tourism and Cultural Change*, 16(2), 173–190.

Naef, P., 2018. "Narco-heritage" and the touristification of the drug lord Pablo Escobar in Medellin, Colombia. *Journal of Anthropological Research*, 74(4), 485–502.

New York Times, 2018. 52 places to go in 2018. [online] Available at: www.nytimes.com/interactive/2018/travel/places-to-visit.html [Accessed 16 April 2019].

New York Times, 2019. 52 places to go in 2019. [online] Available at: www.nytimes.com/interactive/2019/travel/places-to-visit.html [Accessed 16 April 2019].

Observatorio Turístico de Baja California, 2013. *Boletín 2: Visitantes internacionales fronterizos en Baja California.* [online] Observaturbc.org. Available at: www.observaturbc.org/boletines-otbc [Accessed 3 May 2019].

Observatorio Turístico de Baja California, 2014. *Boletín 13: El burro-cebra de Tijuana: testigo mudo de la evolución creativa de la ciudad.* [pdf] Available at: www.observaturbc.org/sites/default/files/PublicacionesOTBC/El%20burro-cebra%20de%20Tijuana.%20Testigo%20mudo%20de%20la%20evolución%20creativa%20de%20la%20ciudad%20mejorado.pdf [Accessed 8 May 2019].

Peru Travel, 2019. *Peruvian gastronomy.* [online] Available at: www.peru.travel/gastronomy/#peruvian-cuisine [Accessed 25 March 2019].

PromPerú, 2016a. *Perfil del turista extranjero: Turismo en cifras.* [pdf] Available at: www.promperu.gob.pe/TurismoIN//Uploads/temp/Uploads_perfiles_extranjeros_39_PTE16_publicacion.pdf [Accessed 1 May 2019].

PromPerú, 2016b. *Turismo gastronómico 2016.* [online] Promperu.gob.pe. Available at: www.promperu.gob.pe/TurismoIN/sitio/VisorDocumentos?titulo=Turismo%20Gastronómico%202016&url=~/Uploads/mercados_y_segmentos/segmentos/1021/Evaluación%20de%20Turismo%20Gastronomico%20-%20TurismoIN.pdf&nombObjeto=PerfilesSegmentos&back=/TurismoIN/sitio/Perfiles Segmentos [Accessed 23 April 2019].

Rodríguez, X., Martínez-Roget, F., and Pawlowska, E., 2012. Academic tourism demand in Galicia, Spain. *Tourism Management*, 33(6), 1583–1590.

Roy, N., Gretzel, U., Waitt, G., and Yanamandram, V., 2019. Gastronomic trails as service ecosystems. In Dixit, S. (ed.), *The Routledge handbook of gastronomic tourism*, London: Routledge, pp. 189–197.

Scott, N., Laws, E., and Prideaux, B. (Eds.), 2010. *Safety and security in tourism: Recovery marketing after crises.* Abingdon: Routledge.

Sectur, 2018. *Principales destinos turísticos internacionales de México sin restricción de viaje: Nuevo sistema de recomendación de EU.* [online] gob.mx. Available at: www.gob.mx/sectur/prensa/principales-destinos-turisticos-internacionales-de-mexico-sin-restriccion-de-viaje-nuevo-sistema-de-recomendacion-de-eu [Accessed 1 May 2019].

Secture Baja California, 2013. *Indicadores turísticos 2013.* [pdf] Available at: www.bajanorte.com/files/estadisticas/140127_Indicadores_turisticos_Ene_Dic.pdf [Accessed 3 May 2019].

Strizzi, N. and Meis, S., 2001. Challenges facing tourism markets in Latin America and the Caribbean region in the new millennium. *Journal of Travel Research*, 40(2), 183–192.

Survival International, 2019. *President Bolsonaro "declares war" on Brazil's indigenous peoples – Survival responds.* [online] Survivalinternational.org. Available at: www.survivalinternational.org/news/12060 [Accessed 12 May 2019].

The World's 50 Best Restaurants, 2019. *The Chefs' Choice Award 2017.* [online] Available at: www.theworlds50best.com/awards/chefs-choice-award [Accessed 23 April 2019].

Travel.State.Gov, 2019. *Travel advisories.* [online] Available at: https://travel.state.gov/content/travel/en/traveladvisories/traveladvisories.html/ [Accessed 15 April 2019].

UNDP, United Nations Development Programme, 2018. *Human development reports.* [online] Hdr.undp.org. Available at: http://hdr.undp.org/en/composite/HDI [Accessed 7 May 2019].

UNESCO, United Nations Educational, Scientific and Cultural Organization, 2014. *The slave route: 1994–2014. The road travelled.* [pdf] Paris: UNESCO. Available at: www.unesco.org/culture/pdf/slave/the-slave-route-the-road-travelled-1994-2014-en.pdf [Accessed 12 May 2019].

UNHCR, UN Refugee Agency, 2018. *Number of refugees and migrants from Venezuela reaches 3 million.* [online] UNHCR. Available at: www.unhcr.org/news/press/2018/11/5be4192b4/number-refugees-migrants-venezuela-reaches-3-million.html [Accessed 15 April 2019].

UNWTO, United Nations World Tourism Organization, 2011. *Tourism towards 2030: Global overview.* [pdf] UN World Tourism Organization. Available at: http://media.unwto.org/sites/all/files/pdf/unwto_2030_ga_2011_korea.pdf [Accessed 9 May 2019].

UNWTO, United Nations World Tourism Organization, 2017. *Tourism highlights 2017.* [pdf] UN World Tourism Organization. Available at: www.e-unwto.org/doi/pdf/10.18111/9789284419029 [Accessed 30 May 2019].

UNWTO, United Nations World Tourism Organization, 2020. *Mexico.* [pdf] UN World Tourism Organization. Available at: www.e-unwto.org/doi/pdf/10.5555/unwtotfb0484010020142018201912 [Accessed 4 May 2020].

Vanderwood, P., 2010. *Satan's playground.* Durham, NC: Duke University Press.

Wilson, R., 2013. Cocina Peruana para el mundo: Gastrodiplomacy, the culinary nation brand, and the context of national cuisine in Peru. *Exchange: The Journal of Public Diplomacy*, 2, 1–20.

Woody, C., 2019. *These were the 50 most violent cities in the world in 2018.* [online] Available at www.businessinsider.com/most-violent-cities-in-the-world-in-2018-2019-3 [Accessed 26 December 2019].

World Economic Forum, 2017. *The travel & tourism competitiveness report 2017.* [pdf] Geneva: World Economic Forum. Available at: www3.weforum.org/docs/WEF_TTCR_2017_web_0401.pdf [Accessed 14 April 2019].

World Population Review, 2018. *Latin America population 2018.* [online] Available at: http://worldpopulationreview.com/continents/latin-america-population/ [Accessed 5 March 2019].

World Travel Awards, 2019. *World's leading culinary destination 2018.* [online] World Travel Awards. Available at: www.worldtravelawards.com/award-worlds-leading-culinary-destination-2018 [Accessed 25 March 2019].

WTTC, World Travel and Tourism Council, 2019a. *City analysis.* [online] WTTC. Available at: www.wttc.org/economic-impact/city-analysis/ [Accessed 9 April 2019].

WTTC, World Travel and Tourism Council, 2019b. *Latin America: 2019 annual research: Key highlights.* [pdf] Available at: www.wttc.org/-/media/files/reports/economic-impact-research/regions-2019/latinamerica2019.pdf [Accessed 30 May 2019].

WTTC, World Travel and Tourism Council, 2019c. *Country Data.* [online] WTTC. Available at: www.wttc.org/economic-impact/country-analysis/country-data [Accessed 6 April 2019].

32

TOURISM IN ASEAN CITIES

Features and directions

Hera Oktadiana and Philip L. Pearce

Introduction

This chapter provides an overview of tourism in four ASEAN capital cities: Bangkok, Singapore, Kuala Lumpur and Jakarta. The selection of these cities follows good practice in comparative case study research where the value of choosing leading exemplars is identified for benchmarking (Assaf and Dwyer 2013, Flyvbjerg 2006, Lennon, Smith, Cockerell and Trew 2006). Each city studied is a key airline hub for the region and they are all national capitals. The further interest in these locations lies in their existing status as major destination cities amidst the continued rise of tourism across Asia (Mastercard Global Destination Cities Index (GDCI) 2019, Pearce and Wu 2017). In brief, the first three cities are ranked in the Global Top 20 Destination Cities 2018 with the number of arrivals 22.78 million for Bangkok, 14.67 million for Singapore and 13.79 million for Kuala Lumpur (Mastercard Global Destination Cities Index (GDCI) 2019). Jakarta is catapulting towards the fastest growing cities both by overall growth (87% – 2016–2017) and direct travel and tourism GDP growth (6.2% pa – 2017–2027).

Some further key statistics justify attention to these cities. All four cities were also in the top 20 Global City Destinations 2017 as assessed by leading measures for city tourism: GDP% of total city GDP, and city tourism GDP% of country tourism GDP. Bangkok and Kuala Lumpur city tourism GDP contributed 10% and 6.1% respectively to the total city GDP. The contribution of the city tourism GDP to country tourism were 100% for Singapore, 50.4% for Bangkok, 44.2% for Kuala Lumpur and 37.1% for Jakarta. By 2027, it is predicted that Bangkok city travel and tourism will contribute US$35.5 billion of GDP, Singapore US$17.5 billion, Jakarta US$13.1 billion and Kuala Lumpur US$10.7 billion (WTTC 2018).

Rather than focus too specifically on the rapidly growing and dynamic statistics generated by tourism in these cities, the authors provide an account of the characteristics of these cities using two conceptual schemes. An outline of the two approaches effectively serves as a literature review for the chapter. The insights about the characteristics of the target cities are then documented using the two conceptual approaches. The information used in this chapter was built by reviewing academic journal articles, specialist reports, the tourism websites of the cities and media news. Data were organised to address the elements of the conceptual schemes employed. Select problems are noted and comparative remarks are left until the end of the chapter.

Key frameworks

There is a long history of analysing city characteristics in the world of planning and design (Banerjee and Southworth 1990). A foundation approach employed to assess the images of cities was first developed by Lynch (1960) in his book *The Image of the City*. Bannerjee and Southworth describe Kevin Lynch as the leading American urban design theorist of his times. As both an urban planner and an academic at MIT in Boston, Lynch attempted to assess the strengths and weaknesses of urban form. His book on city images, one of his ten volumes on themes related to the perception of places, is illustrated in an interesting style with multiple sketches, sidebars and redrawn elements illustrating his respondents' views. Lynch's work provided a method (asking respondents to draw maps and sketches) and a coding scheme to appraise how cities are perceived. Interviews supplemented the drawing tasks. The purpose of using sketch maps, which Lynch promoted, was to emphasise the visual images people have of places – effectively tapping into another kind of language they use to store their experiences. In Lynch's view, cities are best understood visually. His coding scheme used a five-part system with the following elements:

• Nodes – central organising points in the maps, the super iconic attractions
• Landmarks – single named points which have a specific location
• Districts – areas or themed spaces named and labelled
• Paths – the interconnected routes of travel or transport – can be named or unnamed
• Edges – the boundaries of known areas, such as the sea and mountain ranges

Lynch's approach and the subsequent codes he developed are highly relevant to tourism interests. A wave of relatively recent writing has stressed the under-utilised power of the visual in influencing tourists' experiences and memories (Agapito, Pinto and Mendes 2017, Hunter 2016, Mak 2017, Matteucci 2013, Park and Kim 2018, Rakić and Chambers 2011, Scarles 2010). Lynch's scheme has found favour in a number of tourist studies of cities and routes (Cairns 2006, Kitchin and Freundschuh 2000, Lee, Hitchcock and Lei 2018, Liu, Zhou, Zhao and Ryan 2016, Oliver 2004, Pearce and Thomas 2011, Pearce, Wu and Son 2008). The coding scheme has been validated as comprehensive and reliable and will be employed in this chapter to provide a first level of analysis of the four ASEAN cities of interest.

A second approach to characterising contemporary tourist cities relies on the metaphor of a city as an entertainment machine (Clark 2004). In this approach, consumption and entertainment are seen as driving urban (and tourism) development, or at least being central to that continuing growth (Judd 2015, Robinson 2006). More traditional views of urban expansion have attended to the old powers of location, capital and labour (Blainey 2004, Brunn, Williams and Zeigler 2003, Sheridan 1999). Clearly such forces established cities and helped them prosper in earlier times but a new view, one relevant to tourism and the present chapter, suggests that city amenities act as draw cards for tourists and residents alike. Public concerns to attract quality and smart minds as well as short-stay and longer-term tourists can now be seen as identifying tourism infrastructure as an engine of urban growth (cf. Farrell 1982, Farrell and Twining-Ward 2004). In summarising this view Clark writes:

> Why the label entertainment machine? Because entertainment is not just an individualized or private sector process, but centrally involves government and collective decisions … to grow and compete with cities globally mayors and public officials must add entertainment to their recipes.
>
> *(2011, p7)*

It is the combination of natural and constructed amenities that attracts distinct and new sub-groups and tourist markets. Judd (2015), writing about the United States, observes:

> Motivated by the idea that tourism is an industry without smokestacks, cities have poured their energies into building a tourism/entertainment infrastructure. In addition, airports, highways, roads, bridges, mass-transit systems, security, street lighting, beautification programs – these and other amenities and services have been built to accommodate larger tourist flows.
>
> *(2015, p4)*

It is therefore a part of the interests of this chapter to review the ways the Asian cities have to some extent followed their cross-continental counterparts and adopted the logic of the entertainment machine. The approach is consistent with the work of leading tourism researcher Jenkins (2008, 2015), who observed that tourism as a sector had often been a marginalised last-thought-of development tool in the biggest cities. The conceptual model of the entertainment machine used in this chapter reorients tourism as central to the city development process. It asks the questions what has each city done to highlight and build amenities to reach major and specialist market segments? In a comment germane to the cities studied in this chapter, Robinson (2006) argues that if the futures of cities are to be imagined in equitable and creative ways, urban theory needs to overcome its Western bias. The resources for theorising cities need to become at least as cosmopolitan as cities themselves, drawing inspiration from the diverse range of countries and disciplines.

City characteristics

Nodes

In Lynch's appraisal of the characteristics of cities, and most especially their visual and physical appeal to stakeholders, nodes represent the defining icons of the city. They are the elements reproduced as markers that generate semiotic status for their cities (MacCannell 1976). The key cities being considered do have distinctive nodes, ones easily linked to the site sacralisation processes defined in the tourist attractions literature (Leask 2016, Wen 2019). In the cases considered, the landmarks of Singapore are the Merlion Park and Marina Bay Sands, for Bangkok there is the Grand Palace and Wat Arun, for Jakarta it is the National Monument and in Kuala Lumpur it is Petronas Twin Towers (Figure 32.1).

The selection of these icons emerges from a wider consideration of key attractions. For Singapore, the Merlion, Marina Bay Sands, Orchard Road, Clark Quay, Gardens by the Bay and Sentosa Island are the top destinations (Singapore Tourism Board 2018, The Best Singapore 2019). The triple towers of Marina Bay Sands have arguably overcome the Merlion statue as the city's peak node. For Bangkok, The Grand Palace is the number one attraction and image. Other reputed gems include historical attractions, palaces and temples (e.g. Wat PhraKaew (Temple of Emerald Buddha), Wat Pho (Temple of Reclining Buddha), Wat Arun, Wat Benchamabophit (the Marble Temple), Wat Traimit, Wat PhuKhao Tong, Wat Ratchanadda, Wat Sutat, Wat Bowon), as well as street markets (e.g. Chatuchak Weekend Market) and well established luxury shopping plazas (Tourism Authority of Thailand n.d.). The popular destinations for Jakarta include Jakarta Old Town Fatahillah Museum, Thousand Islands, Jaya Ancol Dreamland, National Museum, Sunda Kelapa Harbour and Beautiful Indonesia Miniature Park (TMII) (Indonesia Tourism.com, n.d., Indonesia Tourist Information, n.d.). The National Monument

Jakarta is the key node for the Indonesian capital due to its centrality in the centre and prominence in the urban skyline. For Malaysia, the Petronas Twin Towers, Aquaria KLCC, Kuala Lumpur Tower, Bukit Bintang shopping area, Chinatown of Petaling Street, Central Market and Zoo Negara are the well-known attractions (Lee 2019, Traveloka n.d.). The Petronas Towers, the tallest building in the world from 1998 to 2003, with its special architectural feature of the supporting bridge between the towers, justifies its selection as the node for Kula Lumpur (Newell and Loh 1999).

Landmarks

In the tourism context of this chapter, landmarks are predominantly attractions that help locals and tourists orient themselves to a city. They may be environmental features or constructed amenities and they dominate the itineraries of many tourists. Landmarks have a specific location, so a mountain or prominent beach can be included but more general and appealing features of an environment, such as air quality or greenery do not figure in this categorisation. Clark (2004) uses the expression natural amenities for these diffuse attributes and they are noted in a later section of this review. Functional amenities such as airports as well as individual hotels, restaurants and visitor centres may all serve as landmarks. It has been demonstrated that such features do form a key part of the images tourists hold of cities (Liu *et al.* 2016). In considering the nodes, the key landmarks for each city have already been outlined in the previous section.

Districts

Groupings of attractions and amenities, as well as common areas of a city likely to be noticed by tourists, form the third part of the Lynch classification. Districts can be tightly defined or have fuzzy edges. Sometimes districts overlap substantially such as when areas of use within the city are tightly interwoven. A common district in many Asian cities (and one also applicable in many other countries) is the local "Chinatown" (Wall, 2017). For the cities studied, the popular districts in Singapore are the shopping area of Orchard Road, Marina Bay, Little India, Kampong Glam (Malay-Muslim quarter), Chinatown, Holland Village or Holland V (popular dining and shopping area where East meets West), Katong (home of the Peranakans), Sentosa and Harbourfront, Geylang Serai (markets and eateries), Civic District and Dempsey Hill (Padykula 2019, Singapore Tourism Board n.d.a). In Bangkok, the top districts include Sukhumvit area (nightlife, pubs, clubs, red light district), Siam Square (the shoppers' paradise), Chatuchak (the large shopping areas that spread to 30 acres), Bangkok's riverside area, Silom's market and nightlife, Bangkok Chinatown and Little India, the Old City district and Pratunam district (Withlocals n.d.). In Jakarta, the well-known districts include Kemang (a place of fashion boutiques, restaurants and cafés), Menteng (the old and prestigious area that has many cultural and historical landmarks and legendary restaurants), the business and nightlife districts of Sudirman-Thamrin, Kota Tua (the old town), Kuningan (hotels, shopping malls and lifestyle centre) and Panglima Polim (foodies/coffee lovers' paradise), Pantai Indah Kapuk (prestigious area that provides a wide array of entertainment and recreation facilities) and the Chinatown of Glodok (Putri 2018). In Kuala Lumpur, popular districts are Bukit Bintang, the Malaysia's Times Square that offers leisure activities, entertainment, shopping and dining experiences, Chinatown with its famous Petaling Street as the heart of the area, the vibrant and best nightlife district of Changkat, Kuala Lumpur City Center (KLCC) for its iconic landmark of the Petronas Twin Towers, the park with a 200-metre canopy walk, and art and cultural scene, and the trendy district of Bangsar with numerous hipster café bars (Sopic 2019).

Marina Bay Sands and the Merlion, Singapore   *The Grand Palace and Wat Arun, Bangkok*

National Monument, Jakarta   *Petronas Twin Towers, Kuala Lumpur*

Figure 32.1   Landmarks of Singapore, Bangkok, Jakarta, and Kuala Lumpur.

Sources: Authors' photos.

Paths

The routes, connections and movement channels of a city are its paths. The category includes highways, bikeways, rail and mass transit systems as well as river and canal options. The (in)adequacy of the multiple systems used to permit intracity mobility generates the perceptions of traffic. Getting around Singapore is convenient, as it is supported by a fast and efficient public transport system. Tourists can purchase a Singapore Tourist Pass (STP) to enjoy one day, two or three days unlimited travel by MRT/train and bus. Trains and stations in Singapore are both family friendly with access to strollers. They are also well set up for people with disability by providing for wheelchair users and facilities for those who are visually or hearing impaired. Taxis in Singapore are metered although surcharges may be applied. Visitors can easily download maps, apps and city guides from the tourism website to get information about the city. The city's key destination marketing website even provides useful links to the Grab app for taxi booking (Singapore Tourism Board n.d.b).

In Bangkok, visitors can use Transit Bangkok (www.transitbangkok.com/), the most comprehensive public transportation guide in Bangkok, to plan their travel. Visitors can build a route planner and obtain information about operating hours and fares of various transportation modes: Bangkok bus, Bangkok BRT (Bangkok's Bus Rapid Transit system), MRT (subway/metro) that offers single journey and day pass tickets, BTS Sky Trains, Chao Phraya Express boat and Khlong boat. A Bangkok map can also be downloaded from this site (Transit Bangkok n.d). Besides Bangkok transit transportation systems, visitors can also use taxis and tuk-tuks that are plentiful in the city (Williams 2018). Moving around Bangkok is relatively easy once tourists are familiar with the systems. The BTS or Skytrain is a good transport option to beat the heavy traffic jams and access the various popular attractions in the city. One of the MRT's two lines, the blue line, is especially good for tourists wishing to visit some places such as Queen Sirikit National Convention Center, Lumpini Park and Thailand Cultural Center. The buses suit budget travellers although they are slow and may get held up in traffic. The buses are not air conditioned, the routes are generally written in Thai script, and payment is made in the bus through the conductor or the driver (many speak limited English). Taxis are the most expensive mode of transportation, and by law, they should use a meter. Another form of taxi is motorbike taxis (motorsai taxi) which can be convenient for a solo traveller. Tuk-tuk is the iconic means of transportation, which can be an interesting experience for those visiting Bangkok. Tuk-tuks are not cheap and fares should be negotiated in advance. They can also be a source of scams, sometimes stopping at destinations not requested by the consumers (Pearce 2011). A traveller can opt for Chao Phraya Tourist Boat for sightseeing; it offers hop-on, hop-off services. It is good value for money if you are going to visit a number of major sights, and not only one trip (Rodgers 2019, Williams 2018).

Traffic jams in Jakarta are a problematic issue for the city. Moving around Jakarta can be challenging due to heavily congested roads (particularly during peak hours from 5 am to 8 am and 5 pm to 8 pm), and the modest availability and quality of the public transportation. It can take two hours to drive 25 miles. Public transportation is generally seen as unreliable and unsafe (van Mead 2016). Jakarta is currently building a metro line (Mass Rapid Transit or MRT) to alleviate the traffic. Phase one of the MRT operation was launched in March 2019 (Atika 2019). This development was commenced quite recently compared to efforts in other South East Asian cities such as Manila (in 1984), Singapore (in 1987), Kuala Lumpur (in 1995) and Bangkok (in 2004). However, Jakarta was the first city in South East Asia to launch Bus Rapid Transit (BRT) in 2004. BRT, known as TransJakarta or busway, covers a 120 miles of network within the city. The buses have air conditioning and they have separate sections for women at the front and now there are ten pink women-only buses (Fithriah, Susilowati and Rizqihandari 2018, Van Mead, 2016). These new services are at least one step in trying to cope with the massive traffic delays. Nevertheless, tourists' movements to explore the city and visit the attractions are severely compromised by the traffic jams (Fithriah *et al.* 2018). Visitors to Jakarta should stay near the attractions or destination they wish to visit. Currently, the best way to travel around the city is by taxi, package tour or rented car. A good map or GPS is essential as moving around the city can be quite confusing (Ministry of Tourism, Republic of Indonesia n.d.).

In the Malaysian case, there are many ways to explore the city of Kuala Lumpur such as by bus, taxi, metro (RapidKL, KTM Komuter or Monorail) and train. Taxis are plentiful and the most convenient, but can be the most expensive means to move around the city. Taxi drivers are often reluctant to turn on the meter if you are a tourist. If staying in the city centre, visitors can visit the attractions by foot. However, caution should be exercised as walking can be dangerous due to drivers' attitudes to pedestrians. Certainly the footpaths are not easy to negotiate due to the uneven surfaces and they are not at all wheelchair-friendly. Another way to move around Kuala Lumpur is by using the hop-on, hop-off bus (Wonderful Malaysia n.d.).

Edges

The borders of the city as they appear to visitors constitute the city edges. Theuma (2016) provides a good example of the edges concept in the analysis of the tourist-oriented and extensive waterfront developments in Mediterranean cities. For the sites being examined in this chapter, Singapore as an island nation has the clearest edges. The boundaries of the port and marina area and the propinquity of the sea around the whole island prevent the kind of formless expansion of many ASEAN cities. For Bangkok, the Chao Phraya river to the west and the Gulf of Thailand to the south do provide some clarity of the boundaries of the city for Jakarta, the Ancol Jaya Dreamland is the coastal city edge of Jakarta, but like Bangkok, the spread of the city away from the coast is relatively chaotic. Kuala Lumpur too has the key characteristic of recent urban sprawl and even the centre of the city where tourists may venture tends to lack clear edges and zones.

Further contextual factors

Following Clark (2004), and an array of researchers who assess the destination image characteristics of tourism places, there are several key additional elements, beyond the key classifications provided by the Lynch model. These include the readiness of the city for tourism growth, and some specific parameters such as perceived climate, air pollution, food styles and safety.

The report on "Destination 2030: Global Cities' Readiness for Tourism Growth" issued by WTTC and JJL (2019) specified five levels of city typologies based on the cities' current status and readiness and engagement. The five levels of city readiness are dawning developers, emerging performers, balanced dynamics, mature performers and managing momentum. The last three portray cities with established tourism markets and a high level of readiness for further growth. Singapore is currently in the balanced dynamic category (level 3) as it benefits from a strong urban infrastructure, government support and relative political stability. Generally, cities located in the balanced dynamic and mature performers group are international centres deemed ready to manage growth. Although Singapore sits within this category, it does have a limiting low score on available cheap labour. Kuala Lumpur sits in the first category (dawning developers), while Bangkok and Jakarta are in the second category (emerging performers). Cities in these two categories tend to have lower levels of urban readiness. These cities should focus on developing urban infrastructure such as accommodation and airport connectivity as well as taking into account environmental issues (e.g. water quality, waste). Based on the matrix of the scale of its tourism markets versus the concentration and density of tourist activities, Bangkok, among the four cities, has the strongest level of concentration due to the high number of tourist arrivals. In terms of the appeal of the city to leisure versus business tourists, Bangkok has the most appeal to tourist markets.

The important elements of the perceived favourability of the climate, air pollution, food styles and safety show some variation among the ASEAN cities studied. As tropical cities close to the equator they share the high humidity, high temperature, climate regimes of the tropical monsoon belt of countries. The mean air pollution indices for the cities are also similar – though Singapore rates the best on average, and Jakarta the worst. Nevertheless, all the cities have periods of unacceptably high levels of pollution, a common problem in Asian cities generally (World's Air Pollution 2019, Li, Pearce, Morrison and Wu 2016). Jakarta and Kuala Lumpur have been building a reputation as halal-friendly destinations, while international diversity is a feature of the Singaporean cuisine. Thai food is seen as a positive attribute for Bangkok, though Western food is also widely available (Chandra 2014).

Despite its reputation as a safe tourist place, Singapore is not resistant to scams. Some scams that visitors should be aware of include shopping scams, social services scams, online scams such as fake concert and attraction tickets, bogus monks and nuns, restaurant scams (e.g. overpriced and credit card swaps) and property rental scams (BBC News 2014, Goh 2018, Sylvester 2015). The visible signage controlling undesirable public behaviour, such as scams and fines for littering, form a part of Singapore's strong government control to reduce nuisance acts and vandalism (Bhati and Pearce 2017).

Issues for tourists visiting Bangkok include the attitude of taxi and tuk-tuk drivers, scams and the operation of pickpockets (Lin and McDowall 2012). In a systematic study of Thai tourist scams, many of which occurred in Bangkok, Pearce (2011) noted three kinds of scams; there was a theme of deception in tourist service encounters including spa services, taxis, hotels, restaurants and petrol stations. Charging too much, using counterfeit money and falsifying invoices were noted. A second category of scams summarised episodes to do with the retail environments. Both shopping in markets and purchasing products in more expensive stores (including purchasing products from tailors) were involved in this general retail category. Here inferior goods were swapped for the assumed better products, and sometimes weights and measures were falsified. A third group of tricks and scams was evident in interpersonal encounters where the social connections between tourists and the hustler were the basis of the deceptions and extortions. Sometimes false promises of romantic relationships were made or clients in vulnerable positions were asked to pay for damage or behaviours that they were accused of committing. This category was labelled as social contact deception. Studies of tourist harassment and scams in Asia and elsewhere tend to indicate that even though small amounts of money are usually lost, repeated negative contact with some local people is quite damaging to the reputation of city destinations (Kozak 2007, Li and Pearce 2016, Skipper *et al.* 2014).

Jakarta is seen in tourism promotion material as relatively safe for tourists (Lonely Planet 2018). Its residents, even the poorest, are commonly seen as good-natured. However, visitors should still be careful of some potential scams. This may include taxi drivers who take longer routes and take you to places that you do not want to, people asking for (fake) donations/charity, pickpockets in crowded places, fake beggars, deception involving the exchange of currencies, ticket brokers, fake tourist guides, restaurants scam in tourist areas through unreasonable price increases, and fake discounts where the sellers have already marked up the price before discounting (Setiya n.d.). Similar to Jakarta, there are some typical scams that visitors to Kuala

Lumpur must be aware of although the city is generally safe and very welcoming for visitors. The most common are taxi scams (refusal to switch on the "broken" meter), pickpockets in packed areas and on public transport during peak hours, child flower sellers and fake monks or donation seekers (often found in the open areas of the city such as Chinatown or Jalan Alor) (Wong n.d.).

The entertainment machine

Using the entertainment machine approach, the researchers sought to identify key projects in the cities studied that are influencing or will shape these ASEAN cities. In Singapore there are several major efforts that can be seen as transformative projects. The development of Sentosa as a resort island and the creation of the Marina Bay Sands complex have substantially changed the image of and the tourist numbers to Singapore (Henderson 2012; Lee 2016). Perhaps more than any other city in the whole of Asia, the repositioning and image embellishment of Singapore through gaming and entertainment has been a carefully planned recharging of the tourist-derived and domestic economy. These projects are outstanding examples of the entertainment machine concept (Lee 2016).

As an already successful tourism city, maintenance rather than development of the entertainment–tourism nexus appears to be the strategy in Bangkok. This part of Thailand is popular for its excellent hospitality and cuisine, well-known nightlife and entertainment, massive shopping malls, rich history and unique cultural attractions. Priorities are currently being given to the niche markets and segments such as wellness, ecotourism, golf, seniors and luxury interests (Euromonitor International 2014, 2017, Hedrick-Wong and Choong 2016, Lin and McDowall 2012). As noted subsequently, a major development around medical tourism has been a well-supported initiative by both the government and private sectors over the last 20 years.

In its attempts to build its entertainment qualities, Jakarta is negotiating to boost its tourism though the development and promotion of the Old Batavia Heritage Complex. An initial failure to achieve World Heritage status is a blow, but the push to use the area as signature tourism development is ongoing. Similarly, the area entitled Thousand Islands is being revived with marine tourism, and the promotion of this part of urban Jakarta is co-occurring with substantial urban forest projects and the highlighting of Setu Babakan cultural village (Widiati 2018, Wonderful Indonesia 2019). Importantly, given its traditional pathway problems of moving around the urban area, the city has also developed transportation facilities such as MRT, toll roads, flyovers and underpasses to support domestic life with positive consequences for tourism. Notably, the new facilities at the grand, voluminous Soekarno-Hatta International Airport provide an up-to-date entry point for the country. Another approach taken by Jakarta's provincial government is to enhance cleanliness and environmental sustainability by reducing pollution and plastic waste in tourism areas (Widiati 2018).

A substantial push mixing tourism entertainment and health across these ASEAN cities can be noted (Chew and Darmasaputra 2015). This medical tourism initiative is prominent in Kuala Lumpur and Bangkok but also has substantial traction in Singapore and Jakarta. Health tourism in Kuala Lumpur is attractive to international tourists, especially Indonesians, because of its value for money, and excellent medical and supporting services. The private–public boost to support this medical tourism has seen the rise of specialist hospitals. The types of treatment sought by tourists are mostly medical consultations, followed by cosmetic work, medical check-ups and surgical procedures. Expenditure by medical tourists is up to 12 times more than the leisure tourists (Musa, Thirumoorthi and Doshi 2012). In Kuala Lumpur global luxury hotel chains have been encouraged to enter the market and capture much high-end tourism demand. Hotels and Airbnbs are seeking collaboration opportunities within the destination. Kuala Lumpur International Airport is reaching its 75-million-passenger capacity and the airport's expansion is in progress. The increased domestic air connectivity has resulted in the growth of inter-state travel. The Malaysian government, in cooperation with Alibaba, has launched the "Malaysia City Brain" by using technology in traffic management to advance mobility in its capital (Euromonitor International 2018a).

Bangkok, like Kuala Lumpur and Singapore is a leader in health tourism due to its high-quality healthcare system and the skills of its medical practitioners, relatively low medical costs, familiar culture and combination of Western and Asian food (Lee and Kim 2015). Singapore tourism promotes a vision of a high technology world city where the supporting infrastructure for entertainment and city life is truly world class. Its very high-quality medical facilities and expertise make its services more costly than its neighbouring cities but still very attractive to more affluent clientele (Euromonitor International 2018b). Medical tourism, as a component of international tourism development, assists in the kind of branding of city identity and amenity that Clark (2004) and Judd (2015) identified as characterising sophisticated urban growth in the twenty-first century.

Discussion

Based on the material provided using Lynch's category scheme, additional and relevant amenity considerations and the initiatives identified through the entertainment machine approach, the wider implications of the appraisal of these ASEAN cities can be outlined. The discussion of the ASEAN cities can be facilitated by international comparisons (Assaf and Dwyer 2013; Lennon *et al.* 2006). More specifically, a first level of benchmarking for this chapter can be with other high-profile tourism cities across the world. This kind of benchmarking is external and offers the sought after wider view (Wöber 2002). Cities such as London, Amsterdam, Venice and Barcelona have the same kinds of visitor numbers and pressures as the ASEAN cities studied here. It is immediately apparent, however, that there is a different public consciousness about tourism and the levels of tourism operating in the European city environment. The problems have been captured in the journalistic phrase "overtourism": a term which has come to embrace the totality of the tourism and tourist induced pressures. Overtourism now effectively supersedes the more formal, long-standing academic labels of the social, cultural, economic and environmental impacts of tourism (cf. Jafari 2005).

Goodwin (2017) highlights the perspective that residents in these European cities have worried about too many tourists for nearly 20 years. As many commentators on overtourism have observed, in contemporary times it is not actually the tourists who are the sole problem; it is the adequacy of the local government policies and infrastructure to manage visitors that is more of an issue and engenders public disquiet (Kerourio 2018, Martín, Martínez and Fernández 2018). The concern in Europe for tourism-induced pressure appears to take on a different orientation in the ASEAN cities. In the Asian cities studied there is no public discourse with which the researchers are familiar about a significant negative public response to tourism. Specific projects such as the integrated casino and hotel resorts in Singapore have had some levels of controversy but the issues here turn on gambling and addiction rather than too many tourists (Henderson 2012). In effect, this first level of external benchmarking provides not only a stark contrast about community opinions towards tourism but carries the seeds of potentially important implications for the city planners in the ASEAN cases.

It can be proposed that the absence of community concerns about too much tourism in Bangkok, Kuala Lumpur, Singapore and Jakarta is rooted in the structure of the tourism industries in these locations compared to Europe. In particular, many small and micro businesses thrive in the Asian context and many citizens are reaping some economic and social benefits from tourism as the countries slowly improve the median incomes and standards of living (Knežević Cvelbar, Dwyer, Koman and Mihalič 2016). By way of contrast, more affluent citizens in the richer European destinations are likely to be less involved in tourism, which is often in the control of big companies and businesses and the disruptive effects are not offset by personal gains (Bramwell and Lane 2011). In effect, the entertainment machines operate differently with the consequences for the communities affected being more influential for poorer populations. There is an argument that the sheer numbers of tourists coming to the most popular three ASEAN cities studied could in time generate an Asian form of overtourism. Nevertheless, protests and public voices are likely to be more muted than in Europe due to the more forceful ASEAN government actions in dealing with public dissent. It is also possible, however, that the prevention of a negative community backlash can be avoided in these Asian cities by attending to the transport and infrastructure stresses and controlling Airbnb and cruise ship itineraries, both of which are seen as driving forces overwhelming some key European sites (Goodwin 2017).

It is instructive to also examine internal benchmarking. To maintain sustainability as city destinations, the policy makers need to have well-developed strategies for dealing with tourism challenges (Maxim 2019). In the case of Jakarta and Bangkok, traffic congestion may limit the travel of tourists around the city. Jakarta, in particular, also needs to improve its public transport (although some development is underway). Singapore has done well for this topic and Kuala Lumpur, too, appears to attract less criticism for its transport problems. Reducing the image that visitors to Asian cities are likely to be harassed by scams is a topic for civic and police attention. Further, it is important for the cities to promote their distinctive characteristics (Maitland 2012, Maxim 2019). In particular, due to the similar offerings of these four cities – shopping, cuisine and cultural experiences – steps to highlight and manage special places, what the Japanese call "power spots", would be a strategic direction.

At this time, it would be possible for a tourist to be in parts of any of these cities and have to stop and think where am I? The strong encouragement of a sense of place represents a part of the smart design of all future tourism environments (Fesenmaier and Xiang 2016). The cities therefore can build their uniqueness by promoting a different cluster of urban features such as a shopping city tourism, food city tourism, health city tourism, nightlife city tourism, spa and wellness city tourism (Ashworth and Page 2011). Being everything to everybody may not result in the best economic and socio-cultural outcomes for these growing tourism cities because public and private servicing of all the dimensions is expensive. It is unlikely that any city would want to abandon any of these thematic types of tourism, so the special challenge becomes one of subtle but clear emphasis rather than loss from the city portfolio. Finally and importantly, comparative studies across the region and beyond, could do much to clarify the special characteristics of tourism demand for ASEAN cities and assist in the growth and the future design of city tourism spaces.

References

Agapito, D., Pinto, P., and Mendes, J., 2017. Tourists' memories, sensory impressions and loyalty: In loco and post-visit study in Southwest Portugal. *Tourism Management, 58*, 108–118.

Ashworth, G. and Page, S.J., 2011. Urban tourism research: Recent progress and current paradoxes. *Tourism Management, 32*, 1–15.

Assaf, A.G. and Dwyer, L., 2013. Benchmarking international tourism destinations. *Tourism Economics, 19*(6), 1233–1247.

Atika, S., 2019. *Jokowi to inaugurate MRT Jakarta on Sunday.* Available from www.thejakartapost.com/news/2019/03/19/jokowi-to-inaugurate-mrt-jakarta-on-sunday.html [Accessed 1 September 2019].

Banerjee, T. and Southworth, M. (Eds.), 1990. *City Sense and City Design.* Cambridge, MA: The MIT Press.

BBC News, 2014. *China warns of Singapore scams amid tourism controversy.* Available from www.bbc.com/news/world-asia-29929256 [Accessed 20 September 2018].

Bhati, A. and Pearce, P., 2017. Tourist attractions in Bangkok and Singapore: Linking vandalism and setting characteristics. *Tourism Management, 63*, 15–30.

Blainey, G., 2004. *A Very Short History of the World.* Melbourne: Penguin.

Bramwell, B. and Lane, B., 2011. Critical research on the governance of tourism and sustainability. *Journal of Sustainable Tourism, 19*(4–5), 411–421.

Brunn, S.D., Williams, J.F., and Zeigler, D.J. (Eds.), 2003. *Cities of the World: World Regional Urban Development.* Lanham, MD: Rowman & Littlefield.

Cairns, S., 2006. Cognitive mapping the dispersed city. In C. Lindner (Ed.) *Urban Space and Cityscapes* (pp. 210–223). London: Routledge.

Chandra, G.R., 2014. Halal tourism: A new gold mine for tourism. *International Journal of Business Management & Research, 4*(6), 45–62.

Chew, Y.T. and Darmasaputra, A., 2015. Identifying research gaps in medical tourism. In M. Kozak and N. Kozak (Eds.) *Destination Marketing: An international perspective* (pp. 119–125). London: Routledge.

Clark, T.N. (Ed.), 2004. *The City as an Entertainment Machine.* Amsterdam and Boston, MA: Elsevier/JAI.

Clark, T.N. (Ed.), 2011. *The City as an Entertainment Machine.* New York: Lexington Books.

Euromonitor International, 2014. *City travel briefing: Bangkok.* Available from www.euromonitor.com/city-travel-briefing-bangkok/report [Accessed 20 September 2018].

Euromonitor International, 2017. *Cities.* Available from www.euromonitor.com/cities?PageCode=13582 1&CountryCode=nu ll&IndustryCode=null&ContentType=1&ReportType=67&SortBy=1&PageN umber=1&PageSize=50&PageType=1 [Accessed 18 September 2018].

Euromonitor International, 2018a. *City travel briefing: Kuala Lumpur.* Available from www.euromonitor. com/city-travel-briefing-kuala-lumpur/report [Accessed 25 September 2018].

Euromonitor International, 2018b. *Travel in Singapore.* Available from www.euromonitor.com/travel-in-singapore/report [Accessed 27 September 2018].

Farrell, B., 1982. *Hawaii: The Legend That Sells.* Honolulu, HI: University Press of Hawai'i.

Farrell, B. and Twining-Ward, L., 2004. Reconceptualising tourism. *Annals of Tourism Research, 31*(2), 274–295.

Fesenmaier, D.R. and Xiang, Z. (Eds.), 2016. *Design Science in Tourism: Foundations of Destination Management.* Vienna: Springer.

Fithriah, F.F., Susilowati, M.H.D., and Rizqihandari, N., 2018. Tourist movement patterns between tourism sites in DKI Jakarta. *IOP Conference Series: Earth and Environmental Science, 145.* doi: 10.1088/1755-1315/145/1/012143.

Flyvbjerg, B., 2006. Five misunderstandings about case-study research. *Qualitative Inquiry, 12*(2), 219–245.

Goh, T., 2018. Teen arrested over Bruno Mars, Universal Studios Singapore e-ticket scams. *The Strait Times Singapore,* 14 November. Available from www.straitstimes.com/singapore/teen-arrested-for-bruno-mars-universal-studios-singapore-e-ticket-scams [Accessed 30 September 2018].

Goodwin, H., 2017. *The Challenge of Overtourism.* London: Responsible Tourism Working Partnership No 4.

Hedrick-Wong, Y. and Choong, D., 2016. *Global Destination Cities Index.* Mastercard.

Henderson, J.C., 2012. Developing and regulating casinos: The case of Singapore. *Tourism and Hospitality Research, 12*(3), 139–146.

Hunter, W.C., 2016. The social construction of tourism online destination image: A comparative semiotic analysis of the visual representation of Seoul. *Tourism Management, 54*, 221–229.

Indonesia-Tourism.com, n.d. *Interesting places.* Available from www.indonesia-tourism.com/jakarta/ [Accessed 18 November 2019].

Indonesia Tourist Information, n.d. *Jakarta.* Available from www.indonesiatouristinformation.com/jakarta-tourist-information.html [Accessed 18 November 2019].

Jafari, J., 2005. Bridging out, nesting afield: Powering a new platform. *The Journal of Tourism Studies, 16*(2), 1–5.

Jenkins, C.L., 2008. Tourism and welfare: A good idea and a pious hope! *Tourism Recreation Research*, *33*(2), 225–226.

Jenkins, C.L., 2015. Tourism policy and planning for developing countries: Some critical issues. *Tourism Recreation Research*, *40*(2), 144–156.

Judd, D.R., 2015. *The Infrastructure of Play: Building the Tourist City*. London: Routledge.

Kerourio, P., 2018. Overtourism: The necessary regulation of tourist activity. *Espaces, Tourisme & Loisirs*, *344*, 118–127.

Kitchin, R. and Freundschuh, S. (Eds.), 2000. *Cognitive Mapping: Past, Present and Future*. London: Routledge.

Knežević Cvelbar, L., Dwyer, L., Koman, M., and Mihalič, T., 2016. Drivers of destination competitiveness in tourism: A global investigation. *Journal of Travel Research*, *55*(8), 1041–1050.

Kozak, M., 2007. Tourist harassment: A marketing perspective. *Annals of Tourism Research*, *34*(2), 384–399.

Leask, A., 2016. Visitor attraction management: A critical review of research 2009–2014. *Tourism Management*, *57*, 334–361.

Lee, D., 2016. *Integrated resorts: Singapore's answer to destination competitiveness?* Unpublished PhD thesis, James Cook University, Townsville Australia.

Lee, J. and Kim, H., 2015. Success factors of health tourism: Cases of Asian tourism cities. *International Journal of Tourism Cities*, *1*(3), 216–233.

Lee, M.Y., Hitchcock, M., and Lei, J.W., 2018. Mental mapping and heritage visitors' spatial perceptions. *Journal of Heritage Tourism*, *13*(4), 305–319.

Lee, S.A., 2019. *20 Must-visit attractions in Kuala Lumpur*. Available from https://theculturetrip.com/asia/malaysia/articles/20-must-visit-attractions-in-kuala-lumpur/ [Accessed 18 November 2019].

Lennon, J., Smith, H., Cockerell, N., and Trew, J., 2006. *Benchmarking National Tourism Organisations and Agencies*. London: Routledge.

Li, J. and Pearce, P.L., 2016. Tourist scams in the city: Challenges for domestic travellers in urban China. *International Journal of Tourism Cities*, *2*(4), 294–308.

Li, J., Pearce, P.L., Morrison, A.M., and Wu, B., 2016. Up in smoke? The impact of smog on risk perception and satisfaction of international tourists in Beijing. *International Journal of Tourism Research*, *18*(4), 373–386.

Lin, L. and McDowall, S., 2012. Importance performance analysis of Bangkok as a city tourist destination. *Consortium Journal of Hospitality and Tourism*, *17*, 24–46.

Liu, L., Zhou, B., Zhao, J., and Ryan, B.D., 2016. C-IMAGE: City cognitive mapping through geo-tagged photos. *GeoJournal*, *81*(6), 817–861.

Lonely Planet, 2018. *Jakarta is a journey of flavours*. Available from www.lonelyplanet.com/indonesia/jakarta [Accessed 22 September 2018].

Lynch, K.A., 1960. *The Image of the City*. Cambridge, MA: The MIT Press.

MacCannell, D., 1976. *The Tourist: A New Theory of the Leisure Class*. New York: Schocken Books.

Maitland, R., 2012. Capitalness is contingent: Tourism and national capitals in a globalised world. *Current Issues in Tourism*, *15*(1–2), 3–17.

Mak, A.H., 2017. Online destination image: Comparing national tourism organisation's and tourists' perspectives. *Tourism Management*, *60*, 280–297.

Martín, J.M.M., Martínez, J.M.G., and Fernández, J.A.S., 2018. An analysis of the factors behind the citizen's attitude of rejection towards tourism in a context of overtourism and economic dependence on this activity. *Sustainability*, *10*(8), 1–18.

Mastercard Global Destination Cities Index (GDCI), 2019. *Global destination cities index 2019*. Available from https://newsroom.mastercard.com/wp-content/uploads/2019/09/GDCI-Global-Report-FINAL-1.pdf. [Accessed 26 November 2019].

Matteucci, X., 2013. Photo elicitation: Exploring tourist experiences with researcher-found images. *Tourism Management*, *35*, 190–197.

Maxim, C., 2019. Challenges faced by world tourism cities: London's perspective. *Current Issues in Tourism*, *22*(9), 1006–1024.

Ministry of Tourism, Republic of Indonesia, n.d. *Jakarta*. Available from www.indonesia.travel/au/en/destinations/java/dki-jakarta [Accessed 24 September 2018].

Musa, G., Thirumoorthi, T., and Doshi, D., 2012. Travel behaviour among inbound medical tourists in Kuala Lumpur. *Current Issues in Tourism*, *15*(6), 525–543.

Newell, G. and Loh, J. 1999. Reaching into Asia: The Petronas Twin Towers: The tallest building in the world. *Australian Property Journal*, *35*(7), 597–598.

Oliver, T., 2004. Journeys of the imagination? The cultural tour route revealed. In G. Crouch, R. Perdue, H. Timmermans and M. Uysal (Eds.) *Consumer Psychology of Tourism Hospitality and Leisure Volume 3* (pp. 319–332). Wallingford: CABI.

Padykula, J., 2019. *The 8 neighborhoods you need to explore in Singapore: Discovering Singapore's diverse neighborhoods*. Available from www.tripsavvy.com/singapore-neighborhoods-to-explore-4154863 [Accessed 18 November 2019].

Park, E. and Kim, S., 2018. Are we doing enough for visual research in tourism? The past, present, and future of tourism studies using photographic images. *International Journal of Tourism Research, 20*(4), 433–441.

Pearce, P.L., 2011. Tourist scams: Exploring the dimensions of an international tourism phenomenon. *European Journal of Tourism Research, 4*(2), 147–156.

Pearce, P.L., Wu, Y., and Son, A., 2008. Developing a framework for assessing visitors' responses to Chinese cities. *China Tourism Research, 4*(1), 22–44.

Pearce, P.L. and Thomas, M., 2011. Mapping the road: Developing the cognitive mapping methodology for accessing road trip memories. In B. Prideaux and D. Carsen (Eds.) *Drive Tourism: Trends and Emerging Markets* (pp. 263–277). Abingdon: Routledge.

Pearce, P.L. and Wu, M.-Y. (Eds.), 2017. *The World Meets Asian Tourists*. Bingley: Emerald.

Putri, E., 2018. *The coolest neighborhoods in Jakarta, Indonesia*. Available from https://theculturetrip.com/asia/indonesia/articles/coolest-neighborhoods-jakarta/ [Accessed 18 November 2019].

Rakić, T. and Chambers, D. (Eds.), 2011. *An Introduction to Visual Research Methods in Tourism*. London: Routledge.

Robinson, J., 2006. *Ordinary Cities: Between Modernity and Development*. London: Routledge.

Rodgers, G., 2019. *Getting around Bangkok: Guide to public transportation. Using the BTS, MRT, river taxis, and other transportation options*. Available from www.tripsavvy.com/getting-around-bangkok-public-transportation-4689858 [Accessed 19 November 2019].

Scarles, C., 2010. Where words fail, visuals ignite: Opportunities for visual autoethnography in tourism research. *Annals of Tourism Research, 37*(4), 905–926.

Setiya, T., n.d. *10 tourists scam in Jakarta you must avoid*. Available from https://factsofindonesia.com/tourists-scam-in-jakarta [Accessed 22 November 2019].

Sheridan, G., 1999. *Asian Values Western Dreams: Understanding the New Asia*. St Leonards, NSW: Allen & Unwin.

Singapore Tourism Board, n.d.a. *Explore our popular neighbourhoods*. Available from www.visitsingapore.com/en_au/see-do-singapore/neighbourhoods/ [Accessed 18 November 2019].

Singapore Tourism Board, n.d.b. *Getting around Singapore*. Available from www.visitsingapore.com/travel-guide-tips/getting-around/ [Accessed 1 September 2019].

Singapore Tourism Board, 2018. *7 reasons to visit Singapore in 2019*. Available from www.visitsingapore.com/editorials/reasons-to-visit-singapore/ [Accessed 18 November 2019].

Skipper, T.L., Carmichael, B.A. and Doherty, S., 2014. Tourism harassment experiences in Jamaica. In R. Sharpley and P. Stone (Eds.) *Contemporary Tourist Experience: Concepts and Consequences* (pp. 235–254). Abingdon: Routledge.

Sopic, I., 2019. *5 best neighbourhoods to explore in Kuala Lumpur*. Available from www.furama.com/vanilla/travel-tips/5-best-neighbourhoods-to-explore-in-kuala-lumpur [Accessed 18 November 2019].

Sylvester, P., 2015. *Avoiding scams in Singapore: What you need to know. World Nomad*. Available from www.worldnomads.com/travel-safety/southeast-asia/singapore/singapore-scams [Accessed 1 October 2018].

The Best Singapore, 2019. *The 10 best things to do in Singapore*. Available from www.thebestsingapore.com/best-travel-guide/the-10-best-things-to-do-in-singapore/ [Accessed 18 November 2019].

Theuma, N., 2016. Waterfronts and tourism. In M. Carta and D. Ronsivalle (Eds.) *The Fluid City Paradigm: Waterfront Regeneration as an Urban Renewal Strategy* (pp. 11–17). Amsterdam: Springer.

Tourism Authority of Thailand, n.d. *Destination: About Bangkok*. Available from www.tourismthailand.org/About-Thailand/Destination/Bangkok [Accessed 21 September 2018].

Transit Bangkok, n.d. *Bangkok bus, MRT, BTS – All in one Guide*. Available from www.transitbangkok.com/ [Accessed 19 November 2019].

Traveloka, n.d. *Kuala Lumpur*. Available from www.traveloka.com/en-au/activities/malaysia/city/kuala-lumpur-107979?id=10816159721090548407&adloc=en-au&kw=10816159721090548407_&gmt=b&gn=g&gd=c&gdm=&gcid=378638533831&gdp=&gdt=&gap=1o2&pc=9&cp=108161597210905 48407__10816159721090548407_108416&aid=80071945911&wid=dsa-806223154062&fid=

&gid=9068976&kid=_k_EAIaIQobChMIuevn387y5QIV2yMrCh3TiQxxEAMYAiAAEgLEifD_Bw
E_k_&gclid=EAIaIQobChMIuevn387y5QIV2yMrCh3TiQxxEAMYAiAAEgLEifD_BwE [Accessed
21 November 2019].

Van Mead, N., 2016. The world's worst traffic: Can Jakarta find an alternative to the car? *Guardian*, 23
November. Available from www.theguardian.com/cities/2016/nov/23/world-worst-traffic-jakarta-
alternative [Accessed 25 September 2018].

Wall, G., 2017. Ethnic and minority cultures as tourist attractions. *Journal of Tourism Futures*, *3*(2),
196–197.

Wen, J., 2019. Visiting attractions. In P.L. Pearce (Ed.) *Tourist Behaviour: The Essential Companion*
(pp. 241–257). Cheltenham: Edward Elgar.

Widiati, S, 2018, Making Jakarta a more favourable tourist destination. *NOW! Jakarta*, 4 June. Available
from https://nowjakarta.co.id/magazine-issue/city-of-the-future/making-jakarta-a-more-favourable-
tourist-destination [Accessed 25 November 2019].

Williams, S. 2018. *How to navigate transport in Bangkok, Thailand*. Available from https://theculturetrip.
com/asia/thailand/articles/how-to-navigate-transport-in-bangkok-thailand/ [Accessed 19 November
2019].

Withlocals, n.d. *Top 10 districts in Bangkok*. Available from www.withlocals.com/locations/thailand/
bangkok/districts/ [Accessed 18 November 2019].

Wöber, K.W., 2002. *Benchmarking in Tourism and Hospitality Industries: The Selection of Benchmarking Partners*.
Wallingford: CABI.

Wonderful Indonesia, 2019. *Jakarta in world's top ten fastest growing tourism cities*. Available from www.
indonesia.travel/au/en/news/jakarta-in-world-s-top-ten-fastest-growing-tourism-cities [Accessed 22
November 2019].

Wonderful Malaysia, n.d. *Getting around Kuala Lumpur*. Available from www.wonderfulmalaysia.com/
kuala-lumpur-transportation.htm [Accessed 19 November 2019].

Wong, P., n.d. *4 known scams in Kuala Lumpur: Travel safe tips you should know about*. Available from www.
kuala-lumpur.ws/magazine/4-scams-kuala-lumpur.htm [Accessed 30 September 2018].

World's Air Pollution, 2019. *World's air pollution: Real-time air quality index*. Available from https://waqi.
info/ [Accessed 30 November 2019].

World Travel & Tourism Council (WTTC), 2018. *Travel & Tourism City: Travel & Tourism Impact 2018*.
London: World Travel & Tourism Council.

World Travel & Tourism Council (WTTC) and JJL, 2019. *Destination 2030: Global cities' readiness for
tourism growth*. Available from www.wttc.org/publications/2019/destination-2030/ [Accessed 20
November 2019].

33

SOCIALIST LEGACIES AND CITIES

Societal approach towards the socialist heritage

Katia Iankova and Sonia Mileva

Introduction

This chapter has a two-fold structure: first it provides a state-of-the-art critical analysis of the existing literature related to the communist legacies inherited by former communist societies. Second, based on documents available on the Internet (TV documentaries, newspapers, web pages, surveys and magazine articles) the authors propose a conceptual framework of societal attitudes towards the socialist heritage in Bulgaria as a country with controversial socialist legacies in the field of tourism. The analysis aims to investigate what happens to the communist-period heritage of Bulgaria and what the society's attitudes, feelings, actions are towards this heritage in the post-communist period. Although this analysis is informed by literature that proposes concepts such as nostalgia and animosity, we attempt to increase the complexity of the spectrum of feelings and actions in order to paint a more complete picture of the society's approaches. This research comprises numerous cases from the period 1989–2019 and draws a historical overview of the societal processes that have happened in Bulgaria during the post-communist period in relation to its communist heritage.

Communist heritage and academic research in tourism

Communism during the twentieth century was a dominant political regime for more than 50 years in countries geographically covering half of the planet. This communist period is an intriguing and still under-researched topic, in particular when we consider its economic organisation, art and architectural expression, and physical spatial planning. The heritage left behind by this unique "historical experiment" is now facing a big challenge to survive. It is the subject of destruction, recycling or, on smaller-scale, of protection and conservation

Not only in Bulgaria but in general across the former communist countries, a general feeling of malaise persists regarding the communist period and its cultural manifestations (Iankova, 2013). This in part explains the lack of research produced by national academics of the above-mentioned countries. A breakthrough came in 2011 when several case studies were described by a respected Western researcher, Anthony Travis (2011), as exemplary cases of sustainable space planning of mountain parks and resorts in Slovenia, Poland, the Czech Republic and

Croatia, placing socialist urban planning as one of the most successful during the second half of the twentieth century.

In the last ten years, we have witnessed wide discussion among tourism researchers about the terms communist, socialist, red and totalitarian tourism. In Europe events and places related to the collapse of communism created new tourist attractions, as pointed out by Hall (1991), Greenberg (1990) and Smith (1990). The history of the Berlin Wall is among the most widely explored examples related to the totalitarian regimes and tourism (Light and Dumbrăveanu 1999).

Communist heritage tourism as a term was introduced in the works of Light (2000a, 2000b) exploring the Romanian communist past, Dujisin (2007) for Albanian heritage, Henderson (2007) researching the East Asian countries of North Korea, Laos, Vietnam and Cambodia, and Li and Hu (2008), who discuss red tourism in China.

Within the current chapter we refer to communist heritage in all cultural legacies as comprising both tangible and intangible elements that have historical, cultural and social significance, and which had functioned as forms and aspects of legitimacy during communist political power. In other words communist heritage tourism/red tourism relates to visiting sights and places connected with communist regimes.

In the field of tourism studies, the main issues related to communist (socialist) heritage are linked to: interpretation; attempts by countries to create a new European image, as democratic, distinguishing themselves from past socialist times; negative attitudes towards the old regime; emotional involvement; and politically inconvenient ideology, which is a recurrent theme in several studies including by Young and Kaczmarek (2008). The Polish city of Lodz, for example, following the 1990s tendency to break from its communist past and return to the "Golden Age" of its pre-socialist-period identity, which is contested and disrupted by the re-emergence of the city's socialist past and the history of its political relations with Russia and more in particular urban architects' decisions. Similarly, the city of Gdansk passed through various stages of decline and revitalisation in the post-communist urban context (Polanska 2008).

It is difficult to remain neutral when we attempt to reflect on such an ideologised topic as communism and its legacies. These legacies are multilayered and embrace various aspects of societal construction starting from the most solid and visible aspects of it, such as regional development and city morphology, passing through architecture and ending up with questions revolving around the society's organisation, values and collective memory. They also include the most intangible aspects such as human rapport in daily life, the work environment and leisure activities. On the one hand we see the representation of unwanted, problematic and shameful heritage as discussed in the works of Sima (2017), Goulding and Domic (2009), French (1995) and Castillo (2003). On the other, we find articles revealing the attempts of political powers to glorify and eternalise this heritage through museal interpretation (Goulding and Domic 2009; Zhao and Timothy 2017; Clark 2003; Vukov 2007, 2013). The interpretation of communist heritage in China and North Korea and its use for the development of ever-more-popular red tourism in these countries is tackled by Wall and Zhao (2017) and Wang et al. (2017). Another leading theme is nostalgia (Ivanov and Achikgezyan, 2017). On the European continent, a pendular emotional approach towards the communist heritage prevails – from efforts to preserve it to attempts to forget and erase it, not only from the collective memory but from the urban morphology. In the Asian context, the prevalent feelings and therefore the research that critically analyses societal emotions, policies and practices lean towards love with an extreme representation through glorification and consolidation of the communist idea, which is used by the authorities to serve as an educational tool for new generations.

An important issue that academics need to take into consideration is the safeguarding of the communist era's tangible and intangible heritage. The approach towards monuments needs urgent attention as many of them are subject to vandalism, destruction and deterioration, while others are simply consigned to oblivion or purposefully treated with unintentional carelessness. These extremes of love and hate in the approaches to heritage in post-communist societies are noticeable. But the spectrum runs from love and nostalgia, through indifference, to animosity and hate, all of which are represented in the actions or lack of them towards the monumental heritage of the communist art production. However, protection, conservation, the proliferation of museums of socialist art or retro museums are expressions of nostalgia – the driving force for protection.

On the other hand, we witness purposeful destruction and vandalism, such as the bombed mausoleum and frequently vandalised Monument to the Soviet Army in Sofia, Bulgaria. Consequently, we witness ongoing escalating public discussions related to the relocation of old socialist monuments from open public areas, squares and city centres to more peripheral city areas (e.g. the statue of political leader George Dimitrov, after whom the town of Dimitrovgrad was named, the "1300 Years of Bulgaria" monument in front of the National Palace of Culture in Sofia and many others). Meanwhile a static and silent agreement among society prevails of relinquishing this heritage – be it residential or industrial, monumental or infrastructural – which made it into the ghostly ruins of an époque glorified in the immediate past. Such heritage, in different stages of decay, can be observed when travelling across former socialist countries like Georgia, Romania, Ukraine, Bulgaria and Moldova. But post-socialist societies are so eager to erase it, demonise it or diminish its importance in their respective country's historical timeline that saving what is left is a matter of emergency because in a distant future society is more likely to be in a position to take a more impartial and objective look at it.

The corpus of the communist cities

Most of the countries that experienced a socialist period underwent rapid and heavy industrialisation after 1945. Even those with ancient urban grids, such as Bulgaria and countries of the former Yugoslavia witnessed the appearance of new urban formations, or the substantial modification of existing cities of all sizes. This was one of the most striking impacts of the communist era in regard to regional development, which was supposed to achieve an equal distribution of economic activities across the whole territory of the nation-states in order to balance their demography, and retain the rural exodus. This densified the urban grid with new formations and transformed some existing small towns into industrial, often mono-economic entities, as well as reshaped the ancient cities, inflating their size and changing their physique, especially the city centres.

The whole idea of urban planning after the Second World War was to enhance the creation of collective identity and social justice in an ideal city. The cities were to become home for many and different representatives of the population, concentrated mainly in the newly created residential areas. Architecture and urban planning was to follow the physical and social meaning of socialism expressed in social realism and in some cases gigantism by providing for the masses affordable housing and, in the immediate proximity, access to vital services such as hospitals and schools.

Some socialist cities were built *ex nihilo*. Their anatomy followed the communist ideas of comfort, aesthetics and social dynamics. This was the case of many industrial cities built during the twentieth-century industrial emancipation of the largely agriculture-based economies of the countries of the European South. The case of Dimitrovgrad, for example, is a perfect illustration of such a city: constructed over few years, thanks to the chemical industry developed in the 1950s, with well thought out and planned urban design it has become today a reference for the vision of urban design during the communist period in Bulgaria (Vasileva and Kaleva 2017; Kasabova 2008; Zlatkova 2005).

Another socialist "trademark" is the idea of the "microraion" in which the city is constructed in a rather mosaic layout, in sharp contrast to the European strong core–weak periphery morphology as well as to the American weak core–strong periphery approach. This socialist idea translates to the immediate walking distance proximity and accessibility in each area of main amenities and services such as food stores, kindergartens, schools, hospitals, cinemas, post offices and other municipal services which makes them evenly spread throughout the city.

Of course the strong and dense material and meaning found in city centres remain an undeniable fingerprint of the European idea of urbanitas (urbanity). Nevertheless, the juxtaposition or rather the superposition of the urban socialist identity on the morphology and design of the buildings is predominant in the majority of the younger cities or those that appeared *ex nihilo*. The rest of the vast majority of cities, even those that appeared in earlier historical periods, were transformed under communist rule by different degrees, incorporating in their morphology the ideas of social and gender equality, and social accessibility to culture, health and leisure.

The urban general planning covered the entire network of the interdependent hierarchically structured system of cities, creating for each of them their own mission, vision and plan for development (Popova 2017; Vasileva and Kaleva 2017). The new residential neighbourhoods built surrounding the major cities, e.g. Mladost and Druzhba in Sofia, Trakia in Plovdiv etc., are now home to 1.7 million Bulgarians. The positives are that these residential areas built mostly in the 1970s and 1980s came as whole neighbourhoods aiming to satisfy the needs of a growing urban population. They were not loaded with previous history, collective shared memory and a negative social image. In fact these areas today are among the most affordable and well-located, with a developed infrastructure and loved by their residents.

Parallel to the large-scale building projects following the Second World War ran the equally massive nationalisation of old heritage buildings, which were expropriated from their private owners and subjected to the process of transformation of their bourgeois vision and designation. Rarely were these buildings destroyed, rather, they were "recycled and reused" by the socialist regime for its own purposes (Zlatkova 2003).

In all cases, the planned economy and heavy regulations determined the size, pace and limitation of urban sprawl and demographic density of these cities. Despite the polemical debates about the impacts this regime left behind, 30 years after the fall of communism one cannot deny the obvious facts: citizens in these cities are still walking on webs of streets and road networks created during the communist period; entire quarters are designed to host citizens often numbering the population of an entire village in the collective buildings called "blocs"; they still benefit from the central heating stations providing heat and hot water to these quarters and in many cases to the entire city; their residents take their infants and toddlers every day from home to nurseries especially built within walking distance to give equal opportunity to young families and especially to women to work. The infrastructures still function to provide essential services, without which the life in these cities would be unthinkable today. Schools, residential and administrative buildings with distinctive architecture and ornamentation formed the architectural styles, which were often named after the communist leaders who ruled during the heyday of the architectural fashion – Stalin's baroque, Khrushchev's minimalism, Ceaușescu's gigantism. And today, in a majority of these communist cities, their citizens are witnesses to the decomposing remains of the near past – the industrial ensembles, monuments and examples of high architectural achievements of this period are disappearing faster than their memories of it.

Communist legacies in plural

It would be wrong to discuss communism in the singular. Rather, it should be referred to in the plural. There are as many communisms as there are countries that have lived under its rule. All countries have different histories, cultures and their interpretations of them, and the application of this ideology assumed many shades and aspects.

The same goes for tourism development under these regimes. Thus the organisation and significance of the tourism industry in their respective economies had different roles in each of these countries. Despite attempts to impose a uniform ideology, the Ceaușescu dictatorship is different from the much milder version of communism experienced in the neighbouring Bulgaria; and the insular Cuban communism with Havana cigars, rum and samba cannot be compared with North Korea's strict discipline. Russia, a "torn country", to borrow the term from Samuel Huntington's "The Clash of Civilizations" (1993), stands apart as the country which spread Marxist ideology across half the globe and openly assumes its 70-year duration of communism with some regrets but without a shade of feelings of shame. In the Russian collective memory, the communist period is associated with the Second World War years and the Soviet victory against fascist invasion and is an integral part of the nation's identity. As Russia had no time to live through its capitalist era of industrialisation and building a capitalist economic base, the socialist period is regarded in light of the previous feudalist period and of peasantry almost owned by and attached to the land of the aristocracy, the abolishment of which took place only in the second half of nineteenth century. A few decades later, the communist period began with the October Revolution in 1917 and propelled the country as a world economic, military and scientific superpower despite the civil and Second World Wars and thanks to the collectivistic economy. The communist period will always be associated by Russian people with the Soviet Union's hegemony in the exploration of outer space. The subsequent Cold War that cut Russia and the socialist world off from the cultural scene for about 70 years forced this cultural space to develop its own and unique symbols, cultural codes, imaginaries and representations, all contributions to a considerable intangible cultural heritage of poetry, literature, films and music which needs to be preserved and examined.

Marxist ideology born in the nether of the capitalist world

It is important also to mention that the socialist heritage should not be limited to the former socialist countries. After all, the very ideology of communism was created by the German economist Karl Marx whose grave is now an object of pilgrimage visits in the Highgate Cemetery of London. Similarly, the anthem of the communist movement, "The Internationale", which was written and composed in a small no-longer-existent café in Lille, France; today the story is still told to international tourists visiting the city. The strong communist movements that rose in Italy and Spain during the 1930s and 1940s with resistance against the regimes of Mussolini and Franco left us emblematic slogans that are still in use today such as "No passaran!" (the enemy will not pass the barricades) and the "Ciao, Bella ciao" (an Italian resistance song which was remixed for a 2018 club hit). It is of interest, therefore, to investigate the fate of the heritage in these countries, and its value and place in the capitalist societies. Only recently have researchers such as Adie et al. (2017) shed light on this under-researched topic for the first time in tourism studies.

The intangible aspects of ideas derived from the communist societies

Another overlooked and hardly researched aspect of the socialist heritage is the approach towards children, tourism for children and tourism by children as vectors of peace – a movement born in Bulgaria that existed for nine years, from 1979 to 1988, until the collapse of communism. The "Children's Assembly Flag of Peace" International Children's Movement, supported by UNESCO and UNICEF, left both tangible and intangible heritage, like the Kambanite Monument, also known as "The Bells", located on the outskirts of Sofia (Iankova, 2018).

The idea of free education, health care and leisure, applied in practice in all socialist countries, echoed in capitalist countries with strong communist movements in the twentieth century such as France, Germany, Portugal and Spain. After the Second World War the majority of European countries adopted the idea of free higher education and health care and nationalised the strategic economic branches. In the socialist world, state ownership also materialised in free children's, youths' and workers' winter and summer holiday camps. They were owned, managed and subsidised by the state or in conjunction with state enterprises, or entirely by the enterprises. The state enterprises were given special tax relief for constructing, owning and managing such structures used by their employees and their families. The networks of sanatoriums, profilactoriums and rest houses were important, but today are almost totally dismantled. Only a tiny part of the once large networks remains functional and still subsidised by ministries of health and education in the European former communist countries. These features need to be studied and such practices re-established to support the wellbeing of the societal ecosystem (Iankova, 2018).

Bulgarian communist legacies and the concept of societal attitudes and feelings (SAF)

Because of its geographical position in the European South and its wealth of natural resources, Bulgaria was highly specialised within the economic alliance of the communist bloc in the tourism industry, receiving a high volume of international tourists from other socialist (and non-socialist) countries. During its communist period, Bulgaria developed specific expertise in tourism and built a substantial tourist and general infrastructure. Specifically, these were designed and planned resorts at the Black Sea and in the mountainous areas for international and domestic tourism; a distinctive network of individual hotels or houses (worker houses; scientist houses; artists' and writers' retreat houses) belonging to state companies in various sectors of the economy; and a very well developed network of wellness and spa centres as an integral part of the national health system. The last two categories are considered part of domestic social tourism – their operations were fully or partially funded by the state through the respective ministries or by the state enterprises that owned them.

Parallel to this typical tourist infrastructure, a network of communist and antifascist museums was created. Bulgarian communism's 45-year (1944–1989) political regime strongly stimulated different initiatives, propaganda and its own (regulated) cultural heritage. A unique movement of public art reflecting the ideology and history of this period also created a substantial amount of public art, especially in the monumental sculpture that was publicly displayed in towns, villages and cities all over the country. This bouquet of tourist infrastructure and cultural heritage is in the post-communist period being subjected to purposeful destruction, vandalism, abandonment and physical degradation or in other cases of protection, conservation, reuse and harmonious integration into the landscape and the free market economy.

We have identified five types of societal attitudes and feelings (SAF) to describe what Bulgarian society expresses regarding this heritage and how it has been handled in the last 30 years.

These can be applied to the following two categories: (1) infrastructure built in the communist period, tangible and; (2) intangible cultural heritage.

The five types of societal attitudes and feelings (SAF) are: Hate, Animosity, Indifference, Sympathy and Love. These approaches are closely related to acts on behalf of the society (and governments) that have impacted strongly on the life and the death, the management and the lack of it of this heritage. This classification (Figure 33.1) has been applied to several examples, some of which will be discussed in detail later.

- Love – nostalgia – conservation (e.g. socialist art museum Sofia, the Retro Museum, Varna, Albena Black Sea resort); also valid for all renewed or revived places or monuments e.g. "Zname na mira" (Flag of Peace)
- Sympathy – acceptance and integration of the built heritage within the city/village landscape (Dimitrovgrad, Pernik, Madan)
- Indifference – leading to abandonment and slow deterioration of the heritage – e.g. the national network of all communist movement museums in Bulgaria (Iastrebino, Vela Peeva, etc.); some of the health resorts forming part of the national health system; modification of planned tourist resorts in the mountains and on the coast (Pamporovo, Borovetz, Sunny Beach)
- Animosity – vandalism (the Monument to the Soviet Army of the Second World War); and all monuments and places that are the object of confrontation and tension in the society
- Hate – destruction of the Mausoleum of George Dimitrov in 2001

The following examples provide illustrations of expressions of each of the attitudes and feelings on the proposed spectrum – love, hate and animosity being more strongly expressed than sympathy and indifference.

Love

All Bulgarian seaside resorts were created during the communist period: St Constantine and Elena, Albena, Golden Sands and Sunny Beach. Of these, to date only Albena can be classified under the love. The management of the resort followed a policy of preservation and minimal or no change to the landscape. The other resorts suffered severe changes, not only in terms of infrastructure, but also in planning and management.

The opening of the Museum of Socialist Art (2011) could also be considered the expression of an attitude of love, exhibiting communist heritage and attracting an increasing number of

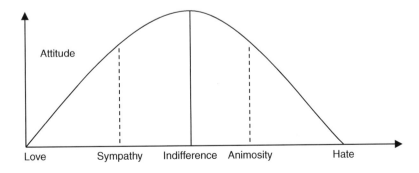

Figure 33.1 Societal attitudes and feelings towards the socialist heritage.

international and domestic tourists. The results of a recent study of visitors' motivations indicate that international tourists from European West visit it out of curiosity to learn more about a historical period that their countries never had and to understand better the characteristics and philosophy of the communist past. For domestic tourists the feeling of nostalgia for the socialist past predominates (Urdea, 2012). The Retro Museum in the third biggest city in Bulgaria – Varna – became an integral part of the tourist offering of the otherwise well-established on the international tourist map tourist city. It offers an insight into the economies of the socialist countries displaying items from cars, to computers, radios, cigarettes and fashion designer collections.

Sympathy

The most massive representation of feelings of sympathy we find in the small and medium-sized towns and cities of Bulgaria, where the local population, despite the changing political times and far left or far right governments, kept and cherished or at least did not vandalise and destruct their own monumental socialist heritage. Bulgaria, unlike any other socialist country, had a policy during the socialist period to commission the best sculptors to create monu-mental works representing the past – from the Bulgarian Renaissance period to the antifascist resistance of the 1920s to the communist movement of the 1940s. Thus, the country, for its 45-year communist period, became a kind of open air museum, with such sculptures, mainly of monumental stonework in marble and granite, in all cities and villages, without exception, as well as displayed alongside the main infra-urban roads, constructing the visual landscape of the country. All these monuments are largely preserved and intact, including those that became the subject of negative debate over the last 30 years, such as the Alyosha Monument in Plovdiv and the Monument of the Bulgarian–Soviet Friendship in Varna. Another mani-festation of sympathy is the integration, use and protection of the built infrastructure for residential, administrative and tourism and leisure purposes. One exceptional example is the Black Sea resort Albena, which, thanks to the transformation of ownership from the state to workers and employees shareholding, could survive and escape the classic style of privatisation – which has been followed by destruction in the majority of similar resorts along the Bulgar-ian Black Sea coastline. Today, this is the most successful Bulgarian sea resort because of its preserved genuine and intact urban planning, with vast green spaces between the hotels, all built in the pyramidal modernist architectural style of the 1970s, and the unspoiled, in some cases 90-metres wide, yellow sand beaches.

Indifference

In indifference we witness the abandoning and slow deterioration of monumental buildings, entire resorts and the emblematic buildings of cultural institutions such as cinemas, theatres and museums. The socialist industrial infrastructure is now in complete decay – mega factories and steel production complexes such as Kremikovtzy are privatised, then sold and ultimately aban-doned and the only interest in them now is as décor for a Hollywood production. Seizing an opportunity a Bulgarian minister initiated a campaign to invite film companies and their pro-duction teams to use this communist heritage. Largely criticised within society this promotional campaign divided even more the already deeply split society. One of the most flagrant examples is the fate of Buzludzha – the Communist Party Assembly constructed in the early 1980s. This monumental building with unique architecture and magnitude was abandoned to slow decay over the decades from the 1990s onwards. Only in 2013–2014 did this monument start to attract international attention, first from French artists, who used it as a main stage for their music videos, following which Bulgarian society started to discuss its future with at first weak voices appearing to reappraise the real aesthetic value of the building.

The majority of the socialist seaside resorts underwent tremendous changes due to slow decay and consecutive demolition replacing them with newer structures. This was the fate of all Black Sea resorts except Albena, mentioned in the previous section, and some camping sites like Kavatzi, Gradina and others. The last category was greatly reduced in number, being privatised, purchased, demolished and replaced.

Animosity

In 2012, then 2014, the international community was alarmed by numerous acts of animosity towards the Monument to the Soviet Army in central Sofia. The monument was painted in the colours of the Ukrainian flag during the beginning of the Ukraine crisis in February 2014. Earlier, in 2012, soldiers in battle in the bas-relief on the monument's foundation were turned into likenesses of Ronald McDonald and Superman. This image made the front pages of many international newspapers and drew the attention of Google, who used the image in one of its days depicting different events. These acts divided the Bulgarian population once again: one part acclaiming it and presenting it positively as an act of creativity; the other part qualifying it as vandalism ridiculing the collective memory of the antifascist movement of the country that culminated in the victory of the Soviet Army over the fascist Nazi Army. The socialist heritage subject to such animosity is now the focus of public debate over how it can be transformed and integrated in contemporary city life. According to the results of a representative survey conducted by Alpha Research (2014), 66% of Sofia citizens voted for the preservation and the restoration of the monument, associating it with the victory of the Soviet Amy over Nazism and fascism in Europe.

A similar example is provided by the public discussion (2012–2014) preceding the removal of the "Monument to 1300 Years of Bulgaria" from the central square in front of the National Palace of Culture. The monument represented the glorious past of Bulgarians, but was expressed following socialist monumental practices. According to a representative survey (Alpha Research, 2014), 72% of citizens agreed that the monument should be removed, while 25% insisted that it should be preserved and restored. One-third (33%) supported a proposal to restore the memorial of fallen soldiers from the 1st and 6th Regiments to the site of the "1300 Years of Bulgaria" monument. The decision to remove the monument was taken as an official decision of the Sofia municipality authorities.

Hate

The most flagrant example of the expression of hate was the planned destruction of the mausoleum of George Dimitrov. George Dimitrov was the first prime minister of post-war socialist Bulgaria from 1946 to 1949. Dimitrov led the Third Comintern (Communist International) from 1934 to 1943. He was a theorist of capitalism who expanded Lenin's ideas by arguing that fascism was the dictatorship of the most reactionary elements of financial capitalism. In 1933 he was arrested in Berlin for alleged complicity in setting the Reichstag on fire. During the Leipzig Trial, Dimitrov's calm conduct of his defence and the accusations he directed at his prosecutors won him world-wide renown.

After his sudden death in 1947, a mausoleum was built in less than a week. His body was displayed and construction of the mausoleum began right after the news of his death. It was built in the record time of just six days (138 hours) – the time it took for Dimitrov's body to be returned to Sofia from the USSR. The body of Bulgaria's first communist leader remained in the mausoleum until August 1990, when he was cremated, and his ashes buried in the city's Central Cemetery.

The decision regarding construction of the mausoleum was adopted from the Russian model and the treatment of Vladimir Ilyich Lenin's body. Among the challenges was the selection of a proper place, due to the lack of similar Bulgarian cultural traditions and experiences. The idea behind the mausoleum differs from cultural rituals related to death among Slavic people, such as Bulgarians – rather it was a demonstration of political power and respect for a leader adopted by several communist countries following the Russian model (Vaseva, 2013). The leader's body was put on display in a mausoleum, attracting both citizens and tourists who paid tribute to the leader or simply came to see it as tourist attraction. One of the functions was to induce in the younger generation respect for a "great leader of the nation", and this justified the presence and significance of the mausoleum for the Bulgarians.

The mausoleum was destroyed by Ivan Kostov's government in 2001 without public consultation and provoked nation-wide debate. Destruction of the mausoleum prompted accusations of fascism, barbarism and vandalism, an act comparable to the destruction of the Bamiyan Buddhas, which were similarly blasted following a decision by the Taliban government in Afghanistan to destroy them. Although in the Afghanistan case this act was carried out for religious reasons and in Bulgaria for ideological reasons, the similarity in the planning and execution of the destruction of the cultural artefacts is striking.

The prime minister and his party claimed that the mausoleum was no longer appropriate after the fall of communism in 1989, because it was a monument which stood for what they claimed was Bulgaria's repressive past. Opposition to the destruction within parliament and among architects and urban planning circles was significant and various propositions were made for what to do with the building – including turning it into a museum, or even an art gallery, as although a symbol of the communist past, it was not worth destroying a building of such stature and grandeur that contributed to the overall architectural ensemble of central Sofia. Although an opinion poll showed that two-thirds of the population was against the demolition of the prestigious white-marble mausoleum, the right-wing government proceeded to take it down. It took four attempts to demolish the massive building on Prince Alexander of Battenberg Square in March 2001, as the mausoleum was reputedly designed to withstand a nuclear attack.

The societal attitudes and feelings (SAF) concept applied to the socialist heritage

This section discusses the concept of societal attitudes and feelings (SAF) towards the socialist heritage (Figure 33.2). This model was designed based exclusively on observations relating to the socialist heritage of Bulgaria, although it can be applied in all socialist countries to varying degrees of strength and gravity of the five stages indicated above. In other words if we apply it to the ordinate and abscissa of the time-space continuum, we can see that the variations of its manifestations in terms of strength, frequency and geographical distribution will vary. For example, in Bulgaria the societal mood towards the socialist heritage was love before 1989, which very quickly turned into hate for the period 1990–2000, gradually passing to a mixture of love and sympathy from 2010 to the current moment with occasional eruptions of hate, with latent background feelings of indifference during the period 1990–2019 from some of the population. There are also regional variations of the model that are worthy of further

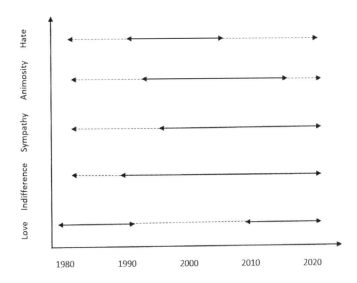

Figure 33.2 Bulgarian SAF concentration and timeline distribution.

research and analysis. Figure 32.2 illustrates how the SAF appear on the timeline continuum in Bulgaria – the thick line representing high concentration and the dotted line representing low concentration over time.

It would also be worthwhile to compare SAF with other tourism concepts such as Butler's (1980) tourism area life cycle – TALC. A singular attraction societal attitude would follow a pattern corresponding to the TALC stages: Development–Love; Stagnation–Indifference; Decline–Hate (Indifference) and; Rejuvenation–Sympathy (Love). Thus if we look at the most iconic monumental building, the Buzludzha monument, we will see that its creation was at the stage for social love in 1980s; then the remoteness of the monument preserved it from hate manifestations, however longstanding feelings of indifference resulted in the monument falling into a state of severe deterioration. This was the fate of this monument located on a high mountain and far from civilisation, but if we compare it with other monuments located centrally in cities, like the Monument to the Soviet Army in the capital Sofia, we see that it still stands intact in its initial glory despite the numerous vandalistic assaults. For the Soviet Army monument, the love and sympathy of the population of the capital is stronger than any other feeling. In contrast, the mausoleum was much loved and much hated but this hate led to its total destruction at a time when feelings of hate predominated in 2001. SAF–TALC concepts can only be applied with difficulty on entire cities, especially to medium and large cities because of their complex nature and the processes happening in the cities. Multifunctional cities have a very long lifespan in contrary of some monofunctional urban formations such as tourist resorts where SAF–TALC can be more easily observed in a space of less than 100 years.

Figures 33.3 to 33.6 illustrate the superimposition of both the SAF concept proposed in this paper and the TALC (Butler, 1980) concept.

We elaborate the SFA–TALC graphs for three Bulgaria socialist monuments discussed in this paper – The Monument to the Soviet Army and the Mausoleum of Georgi Dimitrov, both centrally located in the capital Sofia, and the Buzludzha – the Communist Party Assembly Hall – located in a remote area high in the Balkan Mountains.

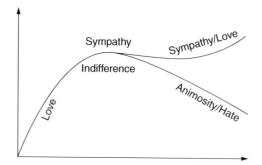

Figure 33.3   SAF stages distributed on the tourism area life cycle (TALC).

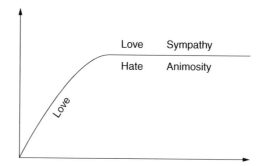

Figure 33.4   The Monument to the Soviet Army in central Sofia, Bulgaria.

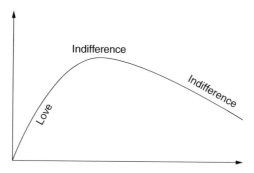

Figure 33.5   The Buzludzha monument, mountain area, Bulgaria.

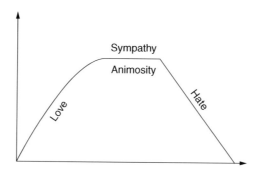

Figure 33.6   The Mausoleum of Georgi Dimitrov, Sofia, Bulgaria.

Conclusion and implications

Socialist legacies in the field of tourism constitute a controversial and disputable topic. Shortly after the 1990s, with the "new" democratic changes, the existing tourist infrastructure and cultural heritage became, in the post-communist period, the subject of purposeful destruction, vandalism, abandonment and physical degradation, or in some cases of protection, conservation, reuse and harmonious integration into the landscape and the free market economy. The radical change in political regime as experienced in Bulgaria and all other Eastern European countries, moving from socialist to democratic regimes in the 1990s, influenced and changed the understanding of socialist legacies making feelings even more sensitive, as seen in the multiple examples analysed.

The different types of government, policy and substitution of social and cultural values were reflected in the contemporary interpretation and understanding of "difficult heritage" (Logan and Reeves 2008). The restructuring of the political landscape along with the degradation of ideology created conditions of instability from both a personal and a societal perspective. The preservation of heritage became less clear when applied to the tangible and intangible aspects of socialist heritage.

The change of political regime prompted debates, discussions and disputes arguing for demolition, and sometimes the substitution, of monuments interpreted as "incorrect" that remained from the previous regime. A major escalation in the debate resulted with the demolition of the emblematic symbol of the socialist regime – the mausoleum of the political leader Georgi Dimitrov (Todorova 2006; Stanoeva 2011), marking a change in social visibilities and values, part of larger process of post-socialist transformation (Verdery 1999, p. 19).

Nevertheless, the socialist (communist) built heritage is an integral part of the urban morphology not only of Bulgaria but of all former socialist countries since the socialist period was marked by city development, explosive construction and modification of the cities. In many cases this was harmoniously amalgamated with anterior urban periods, in other cases massive demolitions of the old regime were replaced by entire, conceptually and physically new units. A flagrant example of this is Romania's House of the People (otherwise known as the Palace of the Parliament) built by Nicolae Ceauşescu. In all cases, this left inevitable marks on each city's shape and identity, which, as we know, cannot be easily erased or ignored because of the nature and lifespan of cities – sometimes centuries or millennia (e.g. Rome). The socialist period needs to be treated with the same full consideration as the heritage of past periods in terms of protection and conservation – with the same level of attention to detail and care as the medieval period, industrial period or antiquity.

It is interesting to note that the intensity of societal feelings and attitudes (SAF) towards socialist architectural legacies are much more vivid, controversial and with higher intensity if they are related to the ideological meaning associated to the building: the more the building is charged with ideological meaning and memories, the stronger and more dramatic and controversial is society's response on the hate–love spectrum. Thus, administrative, residential and commercial buildings are still in use and the population has integrated them smoothly into the new political times. However, buildings or monuments with a lot of meaning linked to communist ideology or political figures are the subjects of attacks, vandalism or, in contrast, of protection and conservation depending of their location and the time of the social perceptions and values as a response from the society and political powers of the day.

Recommendations

The importance of the socialist heritage will continue to grow for these societies incorporating in the nether of the urban morphology and collective remembrance of this historical period that will, like many other previous historical periods, become a cultural reference for twentieth-century development. As destination management organisations across the former socialist countries including Bulgaria show no great interest in adopting an inclusive approach to this socialist heritage – neither as single attractions nor as a package – one possible recommendation would be for governments, NGOs or private businesses to create interactive platforms using digital capacities to protect and promote this legacy. This can be part of the new movement of smart cities which wrap the city in digital informational layers and thus protect and preserve the heritage and develop certain zones by channelling visitors towards them through specialised apps.

A platform using social media could be mounted displaying the locations, contemporary and historical pictures of the monuments, museums, souvenir shops and all paraphernalia linked to this heritage across Europe. The platform could include the time required to reach the destinations from major airports, accommodations available in their proximity, the opening hours and entrance fees, and more, all presented in a unified and free-of-charge downloadable app for all visitors interested in this type of heritage. The website could be promoted through brochures and multimedia in tourist information centres in the European cities. A good example to follow is the Economuseum's virtual network that became very popular among lovers of authentic intangible cultural heritage. More research is needed to provide archival pictures presenting destroyed or damaged communist heritage or heritage in danger across Europe. One of the problems of urban development is that new layers of urban landscape come to replace existing heritage especially when the latter is in a state of deterioration. This is the current of life and it is difficult to stop new project developments from replacing or recycling and reusing the older buildings. Unless a state or municipal policy of protection is put in place, this heritage might be gone forever. One possible approach is the so-called "smart reuse", which can become part of the movement and concept of "smart cities". Not all from the past need be lost or forgotten: especially when it comes to sensual experiences that promote feelings of nostalgia. Thus, old buildings can be reused as factories for the production of these intangible sensations of emblematic food, drinks or iconic objects from the past. Recipes from the socialist chocolate, pastry, dry meat and dairy industries with socialist state standards of quality are anchored in the collective memory and stories transmitted to new generations reviving the qualities of the food and beverage industries of the socialist times. Emblems included, for example, "the horo dancers" of Bulgarian ketchup, and the young grape picker from the Bulgarian sweet known as "lokum", which became a matter of contention between contemporary producers as they fought for the patent in order to having exclusive rights to use it. The very favourite "Alionushka", "Three Bears" and other socialist-period Russian chocolate are emblematic and are sold with great success in former socialist states such as Bulgaria and Serbia. This is an unexplored and promising field for future research, which connects to the intangible aspects of the socialist heritage. Keeping it alive is, in our view, for the greater benefit of all people in the countries that experienced the socialist period. This integrative approach is illustrated perfectly in the film *8½* by the legendary Italian film director Federico Fellini. The main character could not succeed in trying to forget all the happy and painful memories of his past that were haunting him. None of his strategies worked, including locking them in a closet or in the attic. In the final moments, he let all his memories, good and bad, come out, and he celebrated them by dancing with and among them. This allegory unlocks a quintessential approach to healing and justice: recognising the place of all moments in human life history. Only in this way can the past be pacified. This, in our view, should be the approach to the heritage of cities since their ecosystem, social fabric and life cycle are not so different from the human.

References

Adie, B.A., Amore, A., and Hall, C.M. (2017) Urban tourism and urban socialist and communist heritage: beyond tragedy and farce? *International Journal of Tourism Cities*, 3, 291–304.

Alpha Research (2014) Survey of the capital citizens' perception regarding the monuments in Sofia. Online https://alpharesearch.bg/userfiles/file/0914_Monuments_report_PRESENTATION.pdf, accessed 2.12.2019 (in Bulgarian).

Butler, R.W. (1980) The concept of a tourist area cycle of evolution: Implications for management of resources. *Canadian Geographer/Le Géographe canadien*, 24(1), 5–12.

Castillo, G. (2003) Stalinist modern: Constructivism and the Soviet company town. In: Cracraft, J. and D. Rowland (eds.), *Architectures of Russian Identity: 1500 to the Present*. Ithaca, NY: Cornell University Press, pp. 135–149.

Clark, K. (2003) Socialist Realism and the sacralizing of space. In: Dobrenko, E. and E. Naiman (eds.), *The Landscape of Stalinism: The Art and Ideology of Soviet Space*. Seattle, WA: University of Washington Press, pp. 2–18.

Dujisin, Z. (2007) Forget communism … or sell it. *Global Perspectives*, August–September 2007. Online www.global-perspectives.info/download/2007/pdf/ausgabe_08-09_07.pdf, accessed 2.12.2019.

French, R.A. (1995) *Plans, Pragmatism and People: The Legacy of Soviet Planning for Today's Cities*. Pittsburgh, PA: University of Pittsburgh Press.

Goulding, C., and Domic, D. (2009) Heritage, identity and ideological manipulation: the case of Croatia. *Annals of Tourism Research*, 36(1), 85–102.

Greenberg, S.H. (1990) Freedom trail. *Newsweek*, 14, 12–17.

Hall, D.R. (1991) *Tourism and Economic Development in Eastern Europe and the Soviet Union*. London: Belhaven Press.

Henderson, J.C. (2007) Communism, heritage and tourism in East Asia. *International Journal of Heritage Studies*, 13(3), 240–254.

Huntington, S. (1993) The clash of civilizations. *Foreign Affairs*, 72(3), 22–49.

Iankova, K. (2013) A review of Anthony Travis' book "Planning for Tourism, Leisure and Sustainability: International Case Studies". *Journal of Sustainable Tourism*. doi: 10.1080/09669582.2013.818830.

Iankova, K. (2018) Lessons from the past, hopes for the future. Keynote speech during the opening of the 1st Child Friendly Congress, Cumhuriyet University, Siva, Turkey, 23–28 April 2018.

Ivanov, S., and Achikgezyan, V. (2017) Attitudes towards communist heritage tourism in Bulgaria. *International Journal of Tourism Cities*, 3, 273–290.

Kasabova, A. (2008) Ulf Brunnbauer: "Die sozialistische Lebensweise". Ideologie, Gesellschaft, Familie und Politik in Bulgarien (1944–1989). *Bulgarian Ethnology*, (1), 115–118 (in German).

Li, Y., and Hu, Z. (2008) Red tourism in China. *Journal of China Tourism Research*, 4(2), 156–171.

Light, D. (2000a) An unwanted past: contemporary tourism and the heritage of communism in Romania. *International Journal of Heritage Studies*, 6(2), 145–160.

Light, D. (2000b) Gazing on communism: heritage tourism and post-communist identities in Germany, Hungary and Romania. *Tourism Geographies*, 2(2), 157–176.

Light, D. and Dumbrăveanu, D. (1999) Romanian tourism in the post-communist period. *Annals of Tourism Research*, 26(4), 898–927.

Logan, W., and Reeves, K. (eds) (2008) *Places of Pain and Shame: Dealing with "Difficult Heritage"*. London: Routledge.

Polanska, D. (2008) Decline and revitalization in post-communist urban context: A case of the Polish city – Gdansk. *Communist and Post-Communist Studies*, 41(3), 359–374.

Popova, Z. (2017) The city as a stage of messages and inductions. In: Stanoeva, E. (ed.), *Sofia Urbanism and Life during Socialism*. Sofia: Medialog, 185–190 (in Bulgarian).

Sima, C. (2017) Communist heritage representation gaps and disputes. *International Journal of Tourism Cities*, 3, 210–226.

Smith, K. (1990) *Berlin: Coming in from the Cold*. London: Hamish Hamilton.

Stanoeva, E. (2011) The dead body of the leader as an organizing principle of socialist public space: the mausoleum of Georgi Dimitrov in Sofia. In: Behrensen, L.L.M. (ed.), *Modernities revisited*. Vienna: IWM Junior Visiting Fellows' Conferences, vol. 29.

Todorova, M. (2006) The Mausoleum of Georgi Dimitrov as lieu de mémoire. *The Journal of Modern History*, 78(2), 377–411.

Travis, A. (2011) *Planning for Tourism, Leisure and Sustainability: International Case Studies*. Wallingford: CABI.

Urdea, A. (2012) Perceptions of Communism in Socialist Art Museums in Budapest, Hungary and in Sofia, Bulgaria. MA International Tourism Management dissertation, University of Greenwich.

Vaseva, V. (2013) The "leader's death": communist symbols and rituals. In Koleva, D. (ed.), *Death under Socialism: Heroics and Postheroics*. Sofia: Riva, pp. 93–114.

Vasileva, A., and Kaleva, E. (2017) Recharging socialism: Bulgarian socialist monuments in the 21st century. *Studia Ethnologica Croatica*, 29(1), 171–192. doi: 10.17234/SEC.29.5.

Verdery, K. (1999) *The Political Lives of Dead Bodies: Reburial and Postsocialist Change*. New York: Columbia University Press.

Vukov, N. (2007) Monuments, "memorable sites", "sites of memory". *Bulgarian Folkore*, 33, 41–42 (in Bulgarian).

Vukov, N. (2013) The deadly remains of the "special dead" and their collective interment after 1944: monuments–oussuaries and brotherly mounds. In Koleva, D. (ed.), *Death under Socialism: Heroics and Postheroics*. Sofia: Riva, pp. 51–53.

Wall, G., and Zhao, N.R. (2017) China's red tourism: communist heritage, politics and identity in a party-state. *International Journal of Tourism Cities*, 3, pp. 305–320.

Wang, Y., Van Broeck, A.-M., and Vaneste, D. (2017) International tourism in North Korea: how, where and when does political ideology enter? *International Journal of Tourism Cities*, 3, 260–272.

Young, C., and Kaczmarek, S. (2008) The socialist past and postsocialist urban identity in Central and Eastern Europe: the case of Łódź, Poland. *European Urban and Regional Studies*, 15(1), 53–70.

Zhao, S. and Timothy, D. (2017) The dynamics of guiding and interpreting in red tourism. *International Journal of Tourism Cities*, 3(3), 243–259.

Zlatkova, M.I. (2003) Between citizenship and address registration: the images of the living city in the continuum of the urbanity. *Sociological Problems*, 35(3–4), 105–126 (in Bulgarian).

Zlatkova, M.I. (2005) The city of Stalin, the city of Dimitrov, the city of the socialism: an essay on the weekly cinema introductions. *Sociological Problems*, 37(1–2), 65–74 (in Bulgarian).

34

TOURISM CITIES IN POST-COMMUNIST COUNTRIES

Melanie Kay Smith and Tamara Klicek

Introduction

The post-socialist city has been defined as a transitional city which has adapted to new political and economic conditions in the transformation from communism to capitalism. In the same way that developments under state socialism were not uniform in all countries, the trajectories of post-socialist cities cannot be generalised either (Tsenkova, 2012; Berki, 2014; Banaszkiewicz et al., 2017). On the other hand, many post-socialist cities share a common history including the challenges of transition in the post-1989 era following the fall of communism in many Central and Eastern European (CEE) countries. This includes the development of international tourism, which was limited to a great extent before 1989. On the other hand, domestic and intra-regional tourism already existed based on a number of resources, attractions, infrastructure and services. This included (among others) the Yugoslavian, Bulgarian and Romanian coasts; Czechoslovakian and Polish resorts in the Tatra mountains; Czechoslovakian and Hungarian spa towns; pilgrimages to Polish towns like Czestochowa; and cultural city visits to Budapest, Prague, Bratislava and the Baltic capitals (Hall et al., 2006).

Several major themes could be said to have emerged within the post-socialist urban research arena since 1989, including: political and socio-economic changes; interpretation of the socialist past and its dissonant heritage; national and city identity building and branding; urban rehabilitation and gentrification; the night-time economy; and (more recently) "overtourism".

This chapter focuses on all of these themes using examples to illustrate the different rates of developments in post-socialist cities as well as the many similarities.

Economic and social developments in post-socialist cities

The first part of this chapter provides an overview of the economic and political developments that shaped socialist and post-socialist cities. As noted previously, domestic and intra-regional tourism to these cities existed at that time, but international tourism was very limited.

Tsenkova (2008) identified three aspects of the post-1989 transition process that are particularly important to socialist cities. These are the transition to:

- Free markets (systemic economic change)
- Democracy (systemic political change)
- Decentralised systems of local democratic governance

Sýkora (2009) outlines the temporal phases of the transition:

- Short-term period: when the fundamental principles of political and economic organisation were re-shaped, including legal frameworks
- Mid-term period: when peoples' habits and norms were adapted to the new political, economic and cultural environment

He emphasises that the mid-term period could be lengthy, often lasting 10–15 years in the case of certain social groups. This largely depended on the extent of global influences, often making it challenging to ascertain whether given phenomena were the result of global restructuring, post-socialist transition or a hybridisation of both.

The first years of transformation in the first half of the 1990s were challenging with a sudden decrease of GDP, high unemployment, a rapid increase in social polarisation, unleashed inflation and re-emerged ethnic conflicts. There was often an outmigration from cities, especially of younger generations, to countryside areas or a move to Western Europe or the United States (Sýkora, 2009). Popescu (2014) refers to the abrupt nature of deindustrialisation resulting in socio-economic dislocation and a decline in output and employment. She also refers to the phenomenon of urban shrinkage by the 2000s in many post-socialist cities which experienced population decrease. This was especially true of medium small towns. Marcińczak, Gentile and Stępniak (2013) mention the general impoverishment of the CEE cities during the 1990s, as well as the empirical evidence for increasing income inequalities, economic polarisation and spatial differentiation of society. The challenges of post-socialist transformation are often referred to as a "legitimation crisis" (Sýkora, 1994) epitomised by "an inability to establish universal support for neo-liberal capitalist accumulation as a sustainable path for political and economic reform" (Cooper and Morpeth, 1998, p. 2273).

In their book *Fears in Post-Communist Societies*, Shlapentokh and Shiraev (2002) write about the gradual realisation in the 1990s after the initial euphoria had worn off, that existing reality did not match expectations. The political transition had been a long, difficult and exhausting process, monetary uncertainty prevailed, many people felt nostalgic for the past and had serious concerns for the future. Socialism had made people feel safe and had sometimes given them a better quality of life than these new times, which brought freedom on the one hand, but on the other, brought fears and anxiety about how to survive without the paternalistic authority. The benefits of socialism had included inexpensive basic life costs, like monthly rents on housing; transportation and groceries that were subsidised by the government; free education and entitlement to welfare; state health coverage; even subsidised tickets for cultural events in theatres, ballet and opera.

In terms of outbound travel, many inhabitants of the former so-called "Communist bloc" were afforded new, previously unknown opportunities for unlimited international travel. However, many of these dreams were short-lived because of limited financial resources and only a small proportion of society actually had the available funds to travel to the West. Political barriers also existed in many cases, where visas were required to enter Western countries. This situation improved with EU accession for many former Communist countries (Wyrzykowski, 2018).

However, Tsenkova (2008) notes that just as planning under state socialism was not identical in all countries, neither was the move towards Western democracy. Berki (2014) differentiates between the countries of the Eastern Bloc by referring to their different levels of urbanisation in the pre-socialist period. For example, he describes how the Soviet Union, Romania and Bulgaria became heavily urbanised, East Germany and Hungary to a lesser extent, whereas Balkan countries like Albania preserved their predominantly rural character. Some countries had enjoyed more autonomy, such as Yugoslavia or Hungary with its "goulash communism" – a more benign form of state socialism (Bockman, 2011), whereas others had experienced a more totalitarian system. Berki (2014) states that the process of transition and re-building of capitalism also took place in an uneven manner in the former Eastern Bloc, both spatially and temporally. Marcińczak et al. (2015, p. 184) emphasise that "the scale, pace, and depth of postsocialist transformation were not uniform", but that the return of capitalism was swifter in some countries than others, for example, the Czech Republic, Estonia, Hungary, Latvia, Lithuania, Poland, Slovakia and Slovenia. For this reason, Banaszkiewicz et al. (2017, p. 115) urge caution when using the term "post-socialist", stating that "the term 'post-socialist Europe' is only a generalization, a label that does not reflect the inherent complexity of the discussed area".

Cudny (2012) states that although transformation affects whole countries, its impacts are especially visible in cities. He suggests that post-socialist cities underwent the following changes:

- A change of political and economic foundations from the socialist system to democracy and free market
- Functional, social and spatial changes, for example, the collapse of some enterprises and the growth of others (e.g. services, tourism)
- Social transformations which affected human needs, habits and behaviour
- Transformations of the urban space (e.g. decline of some areas, revitalisation of others)

One of the main social changes was the shift to service industries and consumption, which was unfamiliar to many Central and Eastern Europeans. Popescu (2014) notes that in Romania, there was a much stronger legacy of production rather than consumption. Cudny (2016) describes how the centres of post-socialist cities often became degraded when new economic conditions were introduced after 1989 and large state-owned factories and other industries became bankrupt or closed down. This was especially true of many Polish cities. These post-industrial spaces were revitalised in various ways, including the development of commercial and entertainment facilities. Shopping malls were one of the most common features, and more rarely, festival marketplaces. The latter tend to offer a wider range of local cultural activities such as art museums, theatre and gastronomy, as well as global chains and more generic experiences.

Cooper and Morpeth (1998, p. 2258) emphasised the growing conflicts over the use and function of urban space in the 1990s warning, "the major challenge for East and Central European countries in transition: to reconcile the tension between conflicting commercial interests and community needs". Temelova and Dvorakova (2012) analyse the impacts of the socio-spatial configurations that took place in post-socialist cities from the beginning of the 1990s. This included the rapid regeneration of inner city neighbourhoods to include high-end housing, office buildings, shopping malls and entertainment centres aimed at "tourists and wealthy residents" (p. 310). Such commercialisation and touristification was thought to displace cheaper stores and displace or exclude working-class residents (Musil, 1993; Simpson, 1999). It was expected that elderly residents in particular would have difficulties adapting to post-socialist urban transformation. On the other hand, Temelova and Dvorakova's (2012) research on elderly residents' perceptions of quality of life in Prague showed that the impacts had not been as drastic or as negative as previously thought. Compared to many Western European cities, gentrification and resident displacement

Case study of Łódź in Poland

The city of Łódź in Poland was not a traditional tourism destination because of its industrial character and visits to the city were scarce until interest began to grow in the 1970s. This was largely due to the lack of buildings that would have been considered typically attractive for tourists such as castles, churches, city walls, etc. The city was therefore somewhat neglected in terms of tourism and hotel investment compared to other large cities in Poland (Liszewski, 2009). After 1989, the city suffered from serious socio-economic decline mainly due to the collapse of the dominant textile industry (Cudny, 2016). After deindustrialisation, one of the main attractions was the city's industrial heritage in addition to its art nouveau architecture. Hotels were even created out of former industrial buildings such as textile factories. The main street, Piotrkowska Street, was renovated to provide shops, restaurants and tourist services, the former Jewish community's heritage also became a focus of tourism interest and by the 2000s, the city's festivals and cultural, sport and business events were being promoted (Liszewski, 2015). Cudny and Rouba (2011) estimated that Łódź hosts around 60 festivals annually. However, one of the most successful developments for both leisure and tourism functions has been the revitalised factory complex Manufaktura, which Cudny (2016) described as a "festival marketplace". Such facilities tend to be located in revitalised industrial buildings and combine leisure shopping with cultural and gastronomic services. The project was announced in 1992 and 12 factory halls were revitalised between 2003 and 2006. A hotel was included in 2009. The halls contain retail outlets, entertainment venues like museums, a cinema and theatre, as well as a wide range of food. By July 2014, Manufaktura had 290 different service providers: 72.8% shops; 13.8% gastronomic establishments; 4.1% cultural and entertainment facilities; 9.3% service outlets (e.g. shoemakers, dry cleaners, a tourist information point and a medical clinic). Manufaktura has won several tourism awards and more than 30% of Polish and international tourists list it as the best attraction in Łódź (Cudny, 2016).

was much slower in post-socialist cities, at least until the turn of the millennium. Even in the inner city, sitting tenants were able to purchase their apartments at a very low price and a relatively high degree of home ownership was evident in cities like Budapest compared to Western cities. However, many residents subsequently did not have the financial resources to renovate their properties leading to considerable dilapidation (Sýkora, 2005; Kovács et al., 2013).

Post-socialist heritage interpretation

Simpson (1999) re-emphasises the work of previous authors (e.g. Hall, 1994) that highlight the complexities of (re)interpreting the socialist past, re-establishing national autonomy and promoting place identity. Several authors have noted the selective interpretation of heritage in the post-socialist period with a tendency to promote previous "Golden Eras" rather than more recent history (Light, 2000; Young and Light, 2006). Coles (2003, p. 193) states:

> The socialist past, as well as other "inappropriate" periods, events and personalities, has been selectively "airbrushed" from the historical record. Former "golden ages" in the late-nineteenth and early-twentieth centuries have been invoked to placate the concerns of foreign investors and to stimulate nationbuilding.

Ivanov (2009, p. 179) stated that "In Bulgaria and other former socialist countries strong pro- and anti-communist feelings still exist thus making communism (and potentially communist heritage tourism) a highly politically sensitive topic". In most cities, a process of "de-communisation" took place (Sklar, 1995). Cities changed their names, removing any connotations relating to communism and its leaders. For example, the capital of Montenegro today is Podgorica, but during communist times in Yugoslavia, it was named Titograd after its leader and president Marshal Tito. In Slovenia, Titovo Velenje is now called Velenje; Stalinogród in Poland is now Katowice; what was Stalin in Bulgaria is now Varna). The monuments of communist leaders were removed from streets and squares whose names were also changed. In Prague, for example, everything which related to communism and the Soviet Union vanished. Statues and symbols were quickly removed, and street and metro stop names were changed. Ivanov (2009) states that in Bulgaria, almost all streets were re-named. In Budapest, communist statues and monuments were removed and placed in "Memento Park" at some distance from the city centre. Even in recent years, the names of squares and streets were changed to reject left-wing and especially communist ideology in Budapest. One example is the change of the name "Moscow Square" (Moszkva Tér) to Széll Kalman Square (named after a pre-socialist period prime minister of Hungary). However, this phenomenon is not only a post-communist practice. Palonen (2006) observed that renaming and statue removals is a Hungarian tradition, even going back to the 1930s and 1940s. In addition, a "House of Terror" visitor centre was established by the right-wing government in 2002 on the main avenue of the city (Andrássy) to remind visitors of the terrors of communism. Palonen (2013, p. 542) stated that "the project showed how Orbán's government rejected the communist past in the cityscape of Budapest", and many argue that it represents a simplified, politically and ideologically biased message of anti-communism (Apor, 2014).

In contrast to this apparent desire to reject the communist past, Sima (2017) analyses the importance of the communist heritage, especially in capital cities. She refers to the East German concept of "ostalgie" tourism, which is a German term used to describe feelings of nostalgia or fascination with the communist past. She notes that it is especially common amongst older tourists, whereas younger ones are driven more by curiosity. However, her research in Bucharest also suggests that most officials reject rather than embrace the communist past. Ivanov (2009, p. 178) states that

> Countries like Germany, Hungary and Romania have successfully introduced communist heritage tourism into the itineraries of organised packaged tours although public authorities are not very keen in emphasising a part of country's history considered to be aberration from "normal" political development.

However, he suggested that Bulgaria had been slower to promote its communist heritage to tourists than some other countries in the region.

The question of Jewish heritage in former socialist cities is a sensitive and often dissonant subject. Polish cities have become somewhat noteworthy because of their dark past relating to the Holocaust, especially Kracow with its proximity to Auschwitz concentration camp. Lemée (2009) discusses how the Nazis in Central and Eastern Europe and the Baltic States attempted to eradicate all traces of Jewish material culture in a process of so-called "ethnocide". Collective remembrance was also supressed under communism in many Central and Eastern European countries until after 1989. Vilnius in Lithuania was once known as the "Jerusalem of Northern Europe" but the whole Jewish community was virtually annihilated during the Second World War and the Jewish quarter was badly damaged (Sandri, 2013). Tours of former Jewish ghetto areas have become common in cities like Kracow and Budapest (Smith and Zátori, 2019). Visits to CEE and Baltic State cities are a kind of pilgrimage of memory for many Jewish people (Stier, 1995). However, recently, there have been attempts by the Hungarian and Polish governments to "whitewash" their role in the Jewish Holocaust, resulting in Israeli groups cancelling their trips to Poland. On the other hand, it is important to emphasise the positive role of the Polish everyday citizens in the protection and commemoration of Jewish culture and heritage, as researched and documented by Erica Lehrer (2013). In Budapest, a controversial monument was erected in Liberty Square (Szabadság Tér) in 2014. This monument denies the Hungarian authorities' responsibility for the Jewish Holocaust suggesting that they were simply victims of Nazi occupation. The monument was erected at night-time, it was not officially inaugurated and it has not been used in any official commemorative event because of ongoing public protests (Erőss, 2016).

It is also interesting to mention so-called socialist or "socialistic cities" which Pawlusinski (2006, p. 412) described as cities that were built from scratch "with their specific spatial arrangement following all the ideological postulations and guidelines of socialism". Such cities tended to be industrial and therefore not especially attractive to tourists, however, Pawlusinski (2006) notes exceptions like Nowa Huta, the first socialist city in Poland in a district of the city of Kracow. Another example is Dunaújváros near Budapest in Hungary, which means "Danube New City". Such towns appeal to those tourists who are truly interested in learning about life under the socialist system. Banaszkiewicz (2016) describes how initiatives like "Crazy Tours" in Nowa Huta have emerged in the last few years in Central and Eastern Europe to revive the communist heritage which was rejected during the years following the fall of communism. She suggests that development of such tours "opens a space for revising the attitude towards Nowa Huta's legacy and gives opportunity to appreciate it as heritage". The socialist heritage is thus neither forgotten nor rejected, despite remaining controversial (Banaszkiewicz, 2016, p. 10).

The anthropologist Michal Murawski (2018) deals with remnants of the "socialist city" in the cities of today's Eastern Europe, including architecture, parks, public spaces, monuments and events. Some socialist legacies in contemporary cities might not be positive, but they are certainly not all negative. For example, parks and green spaces were one of the main positive features in socialist cities. A skyscraper was erected in the city of Warsaw by Stalin in 1955. Although the building represented Soviet domination on one hand, it has also been the Palace of Culture. The Polish Academy of Sciences is located here, for example. The building contains several theatres, a swimming pool, gymnasium and a huge Congress hall for 3,000 people which was the meeting place for the Congresses of the Communist Party. However, it was also the place where people went to see Miles Davis playing jazz and the Rolling Stones playing a rock concert in 1967. It was the place where you could pick up books in English and French languages, and other books from the bookshop of the Polish Academy of Sciences. Today in post-socialist Warsaw, the Palace of Culture still functions and stands today in the middle of the city, not only as a monument which reminds people of Warsaw during some painful parts of history, but also as a defender of what was good about socialist times, such as high cultural values (Murawski, 2017).

In Tirana, the capital of Albania, Mayor Edi Rama (also a former artist) directed some very successful urban renewal programmes. The most remarkable was the "Return to Identity" project, which earned international fame. Under the programme, former Stalinist buildings along the main streets were painted in bright multicolour pastels. His projects partly aimed to transform the relationship of city inhabitants to their surroundings. He used art as a form of political action to change the city and the context in which Tirana's residents live and work and to project an improved post-socialist image of Tirana in the eyes of the world (Pojanija, 2015).

Recent re-interpretation and re-presentation of the socialist past has led to interesting initiatives which attract international tourists. According to *TIME Magazine* (2018) one of the world's 100 greatest places to visit in 2018 was Zaryadye Park in Moscow. The park is an expensive project which features a concert hall, ice cave, a boomerang-shaped bridge over the river, restaurants and a hotel. It was visited by nearly ten million people since its opening in September 2017. This is the first large-scale public park in 50 years in Moscow, which the government authorities presented to its people as part of their (post-socialist) political vision and ideology.

The growth of tourism

Hoffman and Musil (2009) show that even prior to 1989, tourism had been developing in socialist countries with economic benefits outweighing political resistance. Some of this even dated back to the 1920s and 1930s with much of the emphasis having been on congress tourism, spa and recreational tourism. Niewiadomski (2016) suggests that it was only in the late 1990s that foreign hotel groups started to expand into CEE, escalating after 2004 when many former socialist countries joined the EU. Barriers to development had included complex bureaucracy, corruption, visa regulations, under-investment in infrastructure, shortage of quality city development plans and lack of suitably qualified and experienced staff. Tourism grew rapidly in the 2000s (Cudny, 2012), in an apparent "desire to make up for lost time" (Banaszkiewicz et al., 2017, p. 113). Temelova and Dvorakova (2012, p. 310) describe how since the beginning of the 1990s, the post-socialist cities of Central and Eastern Europe were "adapting to the invasion of tourism and profitable companies". On the other hand, there were some positive benefits of this. Coles (2003) analysed the situation in East Germany stating that tourism had been used as a means of shaping and reinforcing new place identities as part of a wider programme of economic restructuring.

After 1989, many post-socialist cities were keen to promote their heritage, museums and galleries (Hughes and Allen, 2005), and the growth of cultural tourism was rapid in some cities. Richards (2001) noted a fall in local consumption of many cultural activities after 1989 because of the lack of state subsidies and declining incomes. It was hoped that foreign tourism would provide a much-needed boost to flagging cultural attractions. This situation was especially true in Prague, but not without some negative consequences for the urban heritage and the city's inhabitants. Cooper and Morpeth (1998, p. 2254) show that tourism had become one of the most important economic drivers in the Czech Republic's "economic recovery" in the 1990s, and Simpson (1999) states that tourism numbers rose sharply in Prague after 1989. The "laissez-faire" approach to urban planning and tourism development in Prague has been criticised in recent years by Pixová and Sládek (2017) as it has led to considerable resident discontent, a problem already noted by Cooper and Morpeth much earlier (1998, p. 2263): "Prague's cultural heritage – a cultural asset that is rapidly accruing an economic value through the introduction of speculative land usage, but at a social cost of residential displacement *and* the unchecked overconsumption of the heritage product". Hoffman and Musil (2009, p. 14) referred to "two Pragues – the old historical core that has become 'tourist Prague' along with some parts of the inner city, and the rest of the city or the 'Prague of the locals'".

On the other hand, Simpson (1999, p. 173) described that as a result of the opening up of Central and Eastern European cities to international mass tourism "land-use change in historic city centres has emerged as a key aspect of post-socialist urban restructuring". However, her research also shows that it was difficult to ascertain how far urban re-development was caused by tourism or other factors engendered by post-socialist transformation:

In Prague, the perception-based evidence indicates that there has been a substantial erosion of the sense of place and identity of the historic core. However, the extent to which this has occurred entirely as a result of the impact of tourism must be questioned.

(Simpson, 1999, p. 182)

After EU accession for many post-socialist countries in 2004 and the advent of budget airlines, tourism grew even more rapidly, but by then, the number of first-time "curiosity" visitors was decreasing (Rátz, 2004). The pre- and post-socialist heritage was becoming less interesting for many tourists, especially the young, who were more attracted by the night-time economy (NTE) and the availability of cheap alcohol. NTE-related tourism such as "stag and hen" parties emerged in many cities like Prague, Kracow, Budapest and Tallinn (Iwanicki et al., 2016), and this problem continues today, often exacerbating perceptions of "overtourism" (Smith et al., 2019). Todorović et al.'s (2018) research shows that nightlife and entertainment was the highest rated item in Belgrade's destination cognitive image.

Table 34.1 provides some information about the Travel and Tourism Competitiveness Index from 2013 and 2017. This Index takes numerous complex indicators into consideration, including the business environment, price competitiveness, human resources and the labour market, ICT readiness, security, health and hygiene, international openness, environmental sustainability, infrastructure and cultural and natural resources, among others. The Central and Eastern European region decreased in position with no country scoring above 4.50 by 2017, but by 2019 the situation had improved again.

The Cultural Resources and Business Travel sub-index showed the lowest score with only 1.7. The only exception is Croatia with a high Cultural and Business Travel sub-index of 6.9. Other countries show a significant drop from 2015 in this domain: Czech Republic (from 5.39 to 2.4), Hungary (from 4.09 to 2.3), Slovenia (from 3.85 to 1.5), Serbia (from 2.48 to 1.7). The decrease in scores might be explained by several factors, including diminishing novelty value, lack of budget airline routes and dissatisfaction with the carrier (e.g. Ryanair, Wizzair), or increasing costs within the destination (e.g. Tallinn and Riga used to be much cheaper before the arrival of the Euro). Infrastructural deficiencies are common in post-socialist countries because of a lack of investment in facilities and services, along with corruption and a decreasing lack of openness to foreigners in recent years. Although there is not space to explain all of the variations in scores, a couple of examples can explain the relative success of certain countries more than others. Compared to the other countries of the former Yugoslavia, Croatia has an extensive coastline, and Slovenia boasts seaside, mountains, lakes and historic towns within short distances of each other, as well as high levels of service delivered by a multi-lingual population.

Table 34.1 Travel and Tourism Competitiveness Index

Countries	Year	Scores
Estonia, Czech Republic, Slovenia, Croatia	2013	4.88–4.60
Hungary, Bulgaria, Poland	2013	4.54–4.30
Latvia, Lithuania, Romania, Slovak Republic	2013	4.29–4.17
Croatia, Estonia, Czech Republic	2017	4.42–4.22
Hungary, Slovenia, Bulgaria, Poland	2017	4.18–4.06
Latvia, Lithuania, Romania, Slovak Republic	2017	3.97–3.12

Sources: After Nica (2015) and World Economic Forum (2017).

The 2019 Index (World Economic Forum, 2019) suggests an improvement in scores and Croatia scored 4.50, Slovenia and Czech Republic 4.30 and Poland, Estonia, Hungary and Bulgaria had risen to 4.20. However, Latvia, Lithuania and Romania were still ranked lower with 4.0 along with the Slovak Republic.

The case study of Estonia below provides an example of one of the more competitive countries since the post-socialist era, although the World Economic Forum (2019) suggests a slight decrease in ranking in 2019.

Case study of Estonia

It was estimated in 1987 that there were around 313,700 tourists visiting Estonia. Of these 27% were from Finland, 12% from other socialist countries and the rest from other countries. At that time, however, most of the earnings from foreign tourism were appropriated by Moscow with the exception of a small amount of direct income from local restaurants and shops. In the early post-socialist period (1990s) firms from Finland and Sweden invested extensively in Estonia and took advantage of the low wages (Jaakson, 1998). By 1994, Estonia was ranked first in Eastern Europe in terms of foreign investment per capita and tourism was described as a "star performer" accounting for 7% of GDP and 13.5% of Estonian exports by 1994 (Worthington, 2001, p. 392). By 1995, travel companies were developing and promoting "Pan-Baltic Tours" to Estonia, Latvia and Lithuania which encouraged tourists to visit not only the capital city Tallinn, but also other towns in Estonia such as Pärnu and Tartu. Kask and Raagmaa (2010) discuss how Western Estonian towns in the second half of the 1990s became highly sought after both in terms of investment and tourism development. This was especially true of Kuressaare in Saaremaa, where several new spas were built. In fact, Kuressaare became the number one destination after Tallinn. Jarvis and Kallas (2008) note that Estonia benefitted significantly from EU accession in 2004, which provided the opportunity to attract a wider range of international visitors. This success is also attributed to the "Brand Estonia" campaign, which was launched in 2002, which emphasised that Estonia was "positively transforming" and then "positively surprising". However, they also noted that much of the tourism was concentrated in Tallinn with the majority of investment flowing to the capital city rather than the provinces. In 2003, 56% of tourists were going to Tallinn (Ahas et al., 2007). Pawłusz and Polese (2017) analysed tourism branding and communication and noted that some of the earlier campaigns aimed to position Estonia as Nordic and Scandinavian (rather than Eastern European) and to differentiate the country from the other Baltic States. It was estimated in 2004 that 5% were conference tourists, 18% were other business tourists and 11% were tourists travelling for different purposes such as medical services (Ahas et al., 2007). However, like several other post-socialist cities Tallinn also attracts its fair share of "stag and hen" parties (Jarvis and Kallas, 2008) as well as "vodka tourists" mainly from Finland (McKenzie, 2016). Nevertheless, tourism development overall tends to focus on quality. Tooman and Müristaja (2014) refer to Estonia's focus on tourism product quality and the implementation of several quality schemes in the late 2000s aimed at increasing the competitiveness of Estonia as a tourism destination. This includes grading systems for accommodation and spa hotels, as well as an ecotourism label.

Although the Balkan cities still offer price competitiveness according to the Travel and Tourism Competitive Index (World Economic Forum, 2017), research has shown that there have not been enough developments and investments in infrastructure and services to satisfy international tourists (Smith et al., 2015). According to Mercer (2019) Quality of Living City Ranking none of the Balkan cities ranked in the 50 most liveable cities in the world. Ljubljana (74) and Zagreb (98) scored the best. It should be noted that the capitals belonging to the European Union obtained better rankings than those outside.

Image and branding

Until the late 1980s, tourism in the former communist countries was managed through mono-polistic state organisations, so the meaning of branding and its power in tourism was not really understood. Branding was seen as an ideology and the foreign language marketing used was often grammatically fractured and semantically ambivalent (Hall, 2004). In the early days after the fall of communism, the former communist countries focused on changing their identities and trying to erase their "socialist" image. The changing of anthems, flags, city and street names were all indicators that the regime had been toppled. "New" countries were ready to present their images to the world.

In the 1990s, Young and Kaczmarek (1999) noted that although very little was known about attempts to change perceptions of place in East and Central Europe, place-promotion strategies were very important within the region. Place promotion can raise awareness as well as changing peoples' attitudes towards places, which can encourage tourism development. Simpson (1999) showed that even in the late 1990s, the growth of tourism had started to alter perceptions of the cultural meaning of place.

However, several factors affected the international image of former socialist countries and their cities. One was a lack of image or "non-image" in the minds of tourists. Olins (2004, p. 23) asked of countries like Slovenia and Slovakia "how many people know where they are or the significant differences between them?" He noted that Romania and Bulgaria, or Bucharest and Budapest were often confused by foreigners. The second was that most tourists were only familiar with the capital city of those countries. Smith and Puczkó (2010) give the example of Budapest citing research from the mid-2000s which demonstrated a lack of clear image of the rest of the country among several European nationalities. Cottrell and Raadik Cottrell (2015, p. 324) state that "The Baltic countries lack a distinctive tourist image in many parts of the world with the three countries most often con-sidered synonymous with their capital city or nationalities failing to recognise the differences between the countries". Leib et al. (2013) demonstrate that many people associate the Baltic States with their capital cities but have less distinct images of the rest of the country. Pawłusz and Polese's (2017) analysis of marketing and branding in Estonia showed that some of the earlier campaigns aimed to position Estonia as Nordic and Scandinavian and to differentiate the country from the other Baltic States (Latvia and Lithuania). Thirdly, whereas many tourists were initially attracted mainly by socialist heritage, post-socialist countries were keener to project a different image to the world. For example, Young and Light (2006) noted the tendency to promote pre-socialist "Golden Ages" such as the late nineteenth century rather than socialist heritage. Hughes and Allen (2005) suggested that culture and tourism were used to project a new openness and willingness to embrace a wider European identity. Fourthly, some potential visitors questioned the safety and quality levels of the region, harbouring images of instability, poor service and infrastructure and the low quality of public facilities (Hall, 2004). Image creation has been even more challenging for the former Yugoslavian countries and the Balkans as a region is often associated with negative connotations, especially those relating to hostilities, war and conflict (Smith et al., 2015).

In recent years, arts, creative industries, festivals and events are increasingly attracting younger tour-ists to former socialist cities and transforming their image, for example, the Sziget Rock Festival in Budapest. The short case study below gives the example of Novi Sad, a former Yugoslavian city now in Serbia, which did not attract large numbers of tourists until it developed a popular music festival.

Case study of Novi Sad in Serbia

The post-socialist city of Novi Sad (former Yugoslavia; today Serbia) started one of the biggest pop festivals in Europe, the EXIT festival. The festival's origins are connected to the politics of resistance in the late 1990s and early 2000s, and despite its international popularity today, it has retained some undertones of cultural dissent for this part of the world (Dragicevic-Sesic, 2018). Although Novi Sad has been called "the Serbian Athens", it was not a tourism destination until the EXIT festival brought tourists to the city and improved the image. The event, which takes place in the Petrovaradin Fortress for four days each July, has become part of the aesthetic narrative in image transition in Serbia (Wise and Mulec, 2015). Music as a creative industry has also led to Novi Sad becoming one of the Capitals of Culture in 2021. Besermenji, Pivac and Wallrabenstein (2010) undertook interviews with employees from the Institute for the Protection of Cultural Heritage of Vojvodina and the Institute for the Protection of Cultural Heritage of Novi Sad ($N = 24$), as well as employees of travel agencies in Novi Sad ($N = 36$). Of these respondents, 58.3% said that EXIT has an outstanding importance for the culture of Novi Sad. They also agreed that the authentic setting of the Petrovaradin Fortress contributes to the greater attractiveness of the EXIT festival.

Bjeljac and Ćurčić (2010) estimate that there are 30,000 to 40,000 visitors daily from Serbia and abroad. Bjeljac and Lović's (2011) research on the EXIT festival showed that the majority of the festival goers come from Great Britain, Slovenia, Germany and the Netherlands and are aged 20 to 30. They also suggest that EXIT contributes to creating a positive attitude to and image of Serbia in the minds of visitors from abroad (it should perhaps be noted that international attitudes towards Serbia were not always positive in the immediate post-Yugoslavian war period).

Several former socialist cities have received the accolade of European Capital of Culture (e.g. Kracow in Poland in 2000; Sibiu in Romania in 2007; Vilnius in Lithuania in 2009; Pécs in Hungary in 2010; Tallinn in Estonia in 2011; Maribor in Slovenia in 2012; Kosice in Slovakia in 2013; Riga in Latvia in 2014; Plzen in the Czech Republic in 2015; Wroclaw in Poland in 2018; and Plovdiv in Bulgaria in 2019) (UNeECC, 2017). One of the most extensive research studies of the impacts of European Capital of Culture comes from Sibiu in Romania, where it has been shown that the event raised the profile of and improved the image of the city, attracted more tourists, increased the local audience for culture and enhanced local pride (Richards and Rotariu, 2011). Some cities have also applied for UNESCO's Creative City designation, for example, Sofia in Bulgaria and Łódź in Poland are Cities of Film; Košice in Slovakia is City of Media Arts; Katowice in Poland and Brno in the Czech Republic are Cities of Music; Budapest in Hungary, Berlin in Germany and Kaunas in Lithuania are Cities of Design; Prague in the Czech Republic, Tartu in Estonia and Ljubljana in Slovenia are Cities of Literature. However, the extent to which they use this designation in their tourism promotion varies considerably.

Conclusions and recommendations

The trajectories of post-socialist city development have been somewhat diverse, with some cities attracting tourism at an earlier stage and to a greater extent than others. For example, Prague was "discovered" as a heritage city much earlier than Budapest and more tourists originally headed to Kracow than the capital city Warsaw. Some cities still remain relatively "undiscovered" (e.g. in Serbia, Romania and Bulgaria), and foreign tourists are often familiar only with the capital city of many Central and Eastern European countries and the Baltic States. What all cities share are common elements of the political and social past and the challenges of adapting to a free market economy with its capitalist aspirations. Many remnants of the socialist past remain (both positive and negative), although the over-riding sentiment has been to supress, erase or re-interpret the heritage. The juxtaposition of pre- and post-socialist architecture is apparent in all cities, even if the street and square names have been changed and the leaders' statues removed. In some cases, nostalgia for the socialist past and its security has prompted developers to create new attractions or urban renewal projects which embrace the socialist past and the post-socialist present. However, new generations of tourists are showing a greater interest in the vibrant contemporary culture of post-socialist cities, especially their nightlife.

One of the greatest challenges of undertaking tourism development and research in post-socialist cities can be the changing nature of the politics, much of which is not always conducive to structured policy and planning. Subsequent governments often try to erase past developments including strategies and branding, which means that continuity is hindered. Many post-socialist cities are now suffering from issues of gentrification, displacement, rapid price increases and overtourism similarly to Western cities, all of which impact on local resident quality of life. Ideally, post-socialist city mayors and municipalities therefore need to consider issues of overtourism within the contexts of global debates relating to sustainability and smart solutions such as those taking place in Amsterdam, Copenhagen or Vienna, for example. Destination management organisations exist in post-socialist countries, but cities are often excluded from European funding, especially capital cities, and collaboration is challenging within post-socialist countries. Creative developments and events can clearly be used to put relatively unknown cities on the map, like the EXIT festival in Novi Sad or the festival marketplace Manufaktura in Łódź. A focus on experiential marketing as well as upgrading product and service quality like in Estonia can also help to create a competitive destination. Tourists appear to be less interested in the socialist past these days, but the numbers suggest that they are very interested in the post-socialist present. Future research might consider how to re-orientate city product development and marketing towards cultural and creative activities and experiences rather than party and alcohol tourism. This can mean capitalising on their status as European Cities of Culture or UNESCO Creative Cities, developing and promoting arts, music and gastronomic festivals, as well as educating new generations of visitors about a past that has been supressed but not yet forgotten.

References

Ahas, R., Aasa, A., Ülar, M., Pae, T. and Kull, A., 2007. Seasonal tourism spaces in Estonia: Case study with mobile positioning data. *Tourism Management*, 28, 898–910.

Apor, P., 2014. An epistemology of the spectacle? Arcane knowledge, memory and evidence in the Budapest House of Terror. *Rethinking History*, 18(3), 328–344.

Banaszkiewicz, M., 2016. A dissonant heritage site revisited: The case of Nowa Huta in Krakow. *Journal of Tourism and Cultural Change*, DOI: 10.1080/14766825.2016.1260137.

Banaszkiewicz, M., Graburn, N. and Owsianowska, S., 2017. Tourism in (post)socialist Eastern Europe. *Journal of Tourism and Cultural Change*, 15(2), 109–121, DOI: 10.1080/14766825.2016.1260089.

Berki, M., 2014. Return to the road of capitalism: Recapitulating the post-socialist urban transition. *Hungarian Geographical Bulletin*, 63(3), 319–334.

Besermenji, S., Pivac, T. A. and Wallrabenstein, K. A., 2010. Attitudes of experts from Novi Sad on the use of the authentic setting of the Petrovaradin Fortress as the venue for the EXIT festival. *Geographica Pannonica*, 14(3), 92–97, DOI: 10.2298/IJGI1102097B.

Bjeljac, Ž. and Ćurčić, N., 2010. Traditional musical events in Serbia as part of the directive tourist offer. *Proceedings from 20 Biennale Congress Tourism and Hospitality Industry*, pp. 1396–1406.

Bjeljac, Ž. and Lović, S., 2011. Demographic analysis of foreign visitors to the EXIT festival, Novi Sad. *Journal of the Geographical Institute "Jovan Cvijić" SASA*, 61(2), 97–108.

Bockman, J., 2011. *Markets in the Name of Socialism: The Left-Wing Origins of Neoliberalism*. Stanford, CA: Stanford University Press.

Coles, T., 2003. Urban tourism, place promotion and economic restructuring: The case of post-socialist Leipzig. *Tourism Geographies*, 5(2), 190–219, DOI: 10.1080/1461668032000068306.

Cooper, C. and Morpeth, N., 1998. The impact of tourism on residential experience in Central-Eastern Europe: The development of a new legitimation crisis in the Czech Republic. *Urban Studies*, 35(12), 2253–2275.

Cottrell, S. and Raadik Cottrell, J., 2015. The state of tourism in the Baltics. *Scandinavian Journal of Hospitality and Tourism*, 15(4), 321–326, DOI: 10.1080/15022250.2015.1081798.

Cudny, W., 2012. Introduction. *In:* W. Cudny, T. Michalski and R. Rouba (eds) *Tourism and the Transformation of Large Cities in the Post-Communist Countries of Central and Eastern Europe*. Łódź: ŁTN, Wydawnictwo Uniwersytetu Łódzkiego, pp. 9–19.

Cudny, W., 2016. Manufaktura in Łódź, Poland: An example of a festival marketplace. *Norsk Geografisk Tidsskrift – Norwegian Journal of Geography*, 70, 276–291.

Cudny, W. and Rouba, R., 2011. The role of Łódź festivals in promoting adventure tourism. *Polish Journal of Sport and Tourism*, 18, 264–268.

Dragicevic-Sesic, M., 2018. *Umetnost i kultura otpora (Art and Culture of Dissent)*, Institut za pozoriste, film, radio i televiziju, Fakultet dramskih umetnosti. Belgrade: Clio.

Erőss, A., 2016. "In memory of victims": Monument and counter-monument in Liberty Square, Budapest. *Hungarian Geographical Bulletin*, 65(3), 237–254.

Hall, C. M., 1994. *Tourism and Politics: Policy, Power and Place*. Chichester: John Wiley.

Hall, D., 2004. Branding and national identity: The case of Central and Eastern Europe in destination branding: Creating the unique branding proposition. *In:* N. Morgan, A. Pritchard and R. Pride (eds) *Destination Branding: Creating the Unique Destination Proposition*. Oxford: Butterworth-Heinemann, pp. 87–105.

Hall, D., Smith, M. K. and Marciszewska, B., 2006. (eds) *Tourism in the New Europe: The Challenges and Opportunities of EU Enlargement*. CABI: Wallingford.

Hoffman, L. M. and Musil, J., 2009. Prague, tourism and the post-industrial city. A Great Cities Institute Working Paper. Chicago, IL: University of Illinois.

Hughes, H. and Allen, D., 2005. Cultural tourism in Central and Eastern Europe: The views of "induced image formation agents". *Tourism Management*, 26(2), 173–183.

Ivanov, S., 2009. Opportunities for developing communist heritage tourism in Bulgaria. *Tourism Review*, 57(2), 177–192.

Iwanicki, G., Dłużewska, A. and Smith. M. K., 2016. Assessing the level of popularity of European stag tourism destinations. *Quaestiones Geographicae*, 35(3), 15–29.

Jaakson, R., 1998. Tourism development in peripheral regions of post-soviet states: A case study of strategic planning on Hiiumaa, Estonia. *International Planning Studies*, 3(2), 249–272, DOI: 10.1080/13563479808721711.

Jarvis, J. and Kallas, P., 2008. Estonian tourism and the accession effect: The impact of European Union membership on the contemporary development patterns of the Estonian tourism industry. *Tourism Geographies*, 10(4) 474–494, DOI: 10.1080/14616680802434080.

Kask, T. and Raagmaa, G., 2010. The spirit of place of West Estonian resorts. *Norsk Geografisk Tidsskrift – Norwegian Journal of Geography*, 64, 162–171.

Kovács, Z., Wiessner, R. and Zischner, R., 2013. Urban renewal in the inner city of Budapest: Gentrification from a postsocialist perspective. *Urban Studies*, 50(1), 22–38.

Lehrer, E. T., 2013. *Jewish Poland Revisited: Heritage Tourism in Unquiet Places*. Bloomington, IN: Indiana University Press.

Leib, T., Rhoden, S., Reynolds, D., Miller, A. and Stone, C., 2013. Tourists' perceptions and evaluation of a region's destination image: The case of the Baltic States. Paper presented at Eurochrie Conference, Freiburg.

Lemée, C., 2009. Processes of identity reconstitution for descendants of Jewish emigrants from the Baltic and Central Europe in post-Holocaust situations. Identity politics: Histories, regions and borderlands, *Acta Historica Universitatis Klaipedensis XIX. Studia Anthropologica*, 111, 131–146.

Light, D., 2000. Gazing on communism: Heritage tourism and post-communist identities in Germany, Hungary and Romania. *Tourism Geographies*, 2(2), 157–176.

Liszewski, S., 2009. Urban "tourism exploration space": The example of Łódź. *Tourism*, 19(1–2), 57–62, doi.org/10.2478/V10106-009-0007-8.

Liszewski, S., 2015. Tourism studies on Łódź and its metropolitan area. *Tourism*, 25(2), DOI: 10.1515/tour-2015-0001.

McKenzie, B., 2016. Vodka tourism in Estonia: Cultural identity or clearly commerce? Travel and Tourism Research Association: Advancing Tourism Research Globally. 70. https://scholarworks.umass.edu/ttra/2011/Visual/70.

Marcińczak, S., Gentile, M. and Stępniak, M., 2013. Paradoxes of (post)socialist segregation: Metropolitan sociospatial divisions under socialism and after in Poland. *Urban Geography*, 34(3), 327–352, DOI: 10.1080/02723638.2013.778667.

Marcińczak, S., Tammaru, T., Novak, J., Gentile, M., Kovács, Z., Temelova, J., Valatka, V., Kahrik, A. and Szabó, B., 2015. Patterns of socioeconomic segregation in the capital cities of fast-track reforming postsocialist countries. *Annals of the Association of American Geographers*, 105(1), 183–202.

Mercer, 2019. *Quality of Living City Ranking*. [Online] Available from: https://mobilityexchange.mercer.com/Insights/quality-of-living-rankings [accessed 12 April 2019].

Murawski, M., 2017. The Palace Complex: A Stalinist "social condenser" in Warsaw. *The Journal of Architecture*, 22(3), 458–477.

Murawski, M., 2018. Actually-existing success: Economics, aesthetics, and the specificity of (still) socialist urbanism. *Comparative Studies in Society and History*, 60(4), 907–937.

Musil, J., 1993. Changing urban systems in post-communist societies in Central Europe: Analysis and prediction. *Urban Studies*, 30(6), 899–905.

Nica, A., 2015. Cultural heritage and tourism competitiveness in Central and Eastern Europe. *International Journal of Economic Practices and Theories*, 5(3), 248–256.

Niewiadomski, P., 2016. The globalisation of the hotel industry and the variety of emerging capitalisms in Central and Eastern Europe. *European Urban and Regional Studies*, 23(3), 267–288.

Olins, W., 2004. Branding the nation: The historical context. *In:* N. Morgan, A. Pritchard and R. Pride (eds) *Destination Branding: Creating the Unique Destination Proposition*. Oxford: Butterworth-Heinemann, pp. 17–25.

Palonen, E., 2006. *Reading Budapest: Political Polarisation in Contemporary Hungary*. Thesis (PhD) Department of Government at the University of Essex. [Online] Available from: http://polemics.files.wordpress.com/2007/06/emilia_palonen_phd_thesis2006.pdf [accessed 13 April 2019].

Palonen, E., 2013. Millennial politics of architecture: Myths and nationhood in Budapest. *Nationalities Papers*, 41(4), 536–551.

Pawlusinski, R., 2006. Tourism – a new function of post-socialistic city: Case study of Nowa Huta (Cracow). *Folia Geografica*, 10, 412–418.

Pawłusz, E. and Polese, A., 2017. "Scandinavia's best-kept secret": Tourism promotion, nation-branding, and identity construction in Estonia (with a free guided tour of Tallinn Airport). *Nationalities Papers*, 45(5), 873–892. DOI: 10.1080/00905992.2017.1287167.

Pixová, M. and Sládek, J., 2017. Touristification and awakening civil society in post-socialist Prague. *In:* C. Colomb and J. Novy (eds) *Protest and Resistance in the Tourist City*. London: Routledge, pp. 73–89.

Pojanija, D., 2015. Urban design, ideology, and power: Use of the central square in Tirana during one century of political transformations. *Planning Perspectives*, 30(1), 67–94.

Popescu, C., 2014. Deindustrialization and urban shrinkage in Romania: What lessons for the spatial policy? *Transylvanian Review of Administrative Sciences*, 42, 181–202.

Rátz, T., 2004. Zennis és Lomi Lomi, avagy új trendek az egészségturizmusban. *In:* A. Aubert and J. Csapó (eds) *Egészégturizmus*. Pécs: Bornus nyomda, 46–65.

Richards, G. (ed.), 2001. *Cultural Attractions and European Tourism*. Wallingford: CABI.

Richards, G. and Rotariu, I., 2011. Sibiu: The European Cultural Capital and beyond. *Annals of the "Ovidius" University*, Economic Sciences Series, 11(1).

Sandri, O., 2013. City heritage tourism without heirs: A comparative study of Jewish-themed tourism in Krakow and Vilnius. *Cybergeo: European Journal of Geography*. [Online] Available from: https://archive.jpr.org.uk/10.4000/cybergeo.25934 [accessed 15 April 2019].

Shlapentokh, V. and Shiraev, E., 2002. *Fears in Post-Communist Societies*. New York: Palgrave.

Sima, S., 2017. Communist heritage representation gaps and disputes. *International Journal of Tourism Cities*, 3(3), 210–226, https://doi.org/10.1108/IJTC-03-2017-0015.

Simpson, F., 1999. Tourist impact in the historic centre of Prague: Resident and visitor perceptions of the historic built environment. *The Geographical Journal: The Changing Meaning of Place in Post-Socialist Eastern Europe: Commodification, Perception and Environment*, 165(2), 173–183.

Sklar, M., 1995. *Decommunization: A New Threat to Scientific and Academic Freedom in Central and Eastern Europe.* Science and Human Rights Program. Washington, DC: American Association for the Advancement of Science.

Smith, M. K. and Puczkó, L., 2010. Post-socialist identity construction in Hungary: A multi-level branding approach. *In:* E. Frew and L. White (eds) *Tourism and National Identity: An International Perspective.* London: Routledge, pp. 38–51.

Smith, M. K., Puczkó, L., Michalkó, G., Kiss, K. and Sziva, I., 2015. *Balkan Wellbeing and Health Tourism Report.* Budapest: Budapest Metropolitan University.

Smith, M. K., Sziva, I. P. and Olt. G., 2019. Overtourism and resident resistance in Budapest. *Tourism Planning & Development*, 16(4), 376–392.

Smith, M.K. and Zátori, A., 2015. Jewish culture and heritage in Budapest. *In:* A. Diekmann and M.K. Smith (eds) *Ethnic and Minority Cultures as Tourist Attractions.* Bristol: Channel View Publications, pp. 188–201.

Stier, O. B., 1995. Lunch at Majdanek: The March of the Living as a contemporary pilgrimage of memory. *Jewish Folklore and Ethnology Review*, 17(1–2), 57–62.

Sýkora, L., 1994. Local urban restructuring as a mirror of globalisation processes: Prague in the 1990s. *Urban Studies*, 31, 1149–1166.

Sýkora, L., 2005. Gentrification in post-communist cities. *In:* R. Atkinson and G. Bridge (eds) *Gentrification in a Global Context: The New Urban Colonialism.* London: Routledge, pp. 91–106.

Sýkora, L., 2009. Post-socialist cities. *In:* R. Kitchin and N. Thrift (eds) *International Encyclopedia of Human Geography*, Volume 8. Oxford: Elsevier, pp. 387–395.

Temelova, J. and Dvorakova, N., 2012. Residential satisfaction of elderly in the city centre: The case of revitalizing neighbourhoods in Prague, *Cities*, 29, 310–317.

TIME Magazine, 2018. The World's Greatest Places in 2018 (100 destinations to experience right now), 192(9–10), 65–89.

Todorović, N., Budović, A., Ćihova, M., Riboškić, D. and Piroški, V., 2018. Exploring cognitive and affective components of Belgrade's destination image. *Bulletin of the Serbian Geographical Society*, 98(2) 119–146, doi.org/10.2298/GSGD1802119T.

Tooman, H. and Müristaja, H., 2014. Developing Estonia as a positively surprising tourist destination. *In:* C. Costa, E. Panyik and D. Buhalis (eds) *European Tourism Planning and Organisation Systems: The EU Member States.* Bristol: Channel View Publications, pp. 106–117.

Tsenkova, S., 2008. The comeback of post-socialist cities. *Urban Research and Practice*, 1(3), 291–310.

Tsenkova, S., 2012. *Planning Trajectories in Post-socialist Cities: Patterns of Divergence and Change.* CBEES Working Paper.

UNeECC, 2017. *European Capital of Culture.* [Online] Available from: http://uneecc.org/european-capitals-of-culture/history/ [accessed 15 April 2019].

Wise, N. and Mulec, I., 2015. Aesthetic awareness and spectacle: Communicated images of Novi Sad (Serbia), the EXIT Festival and Petrovaradin Fortress. *Tourism Review International*, 19(4), 193–205.

World Economic Forum, 2017. *The Travel & Tourism Competitiveness Report 2017.* [Online] Available from: www.weforum.org/reports/the-travel-tourism-competitiveness-report-2017 [accessed 17 April 2019].

World Economic Forum, 2019. *The Travel & Tourism Competitiveness Report 2019: Travel and Tourism at a Tipping Point.* [Online] Available from: www3.weforum.org/docs/WEF_TTCR_2019.pdf [accessed 17 April 2019].

Worthington, B., 2001. Riding the "J" curve: Tourism and successful transition in Estonia? *Post-Communist Economies*, 13(3), 389–401, DOI: 10.1080/14631370120074894.

Wyrzykowski, M., 2018. Rule of law: European Commission v. Poland. *In:* A. Hatje and L. Tichý (eds) *Liability of Member States for the Violation of Fundamental Values of the European Union.* Europarecht Beiheft 1. Baden-Baden: Nomos, pp. 169–200.

Young, C. and Kaczmarek, S., 1999. Changing the perception of the post-socialist city: Place promotion and imagery in Łódź, Poland. *The Geographical Journal: The Changing Meaning of Place in Post-Socialist Eastern Europe: Commodification, Perception and Environment*, 165(2), 183–191.

Young, C. and Light, D., 2006. "Communist heritage tourism": Between economic development and European integration. *In:* D. Hassenpflug, B. Kolbmüller and S. Schröder-Esch (eds) *Heritage and Media in Europe: Contributing towards Integration and Regional Development.* Weimar: Bauhaus Universität, pp. 249–63.

35

TOURISM CITIES IN CHINA

Bihu Wu, Qing Li, Feiya Ma and Ting Wang

Introduction

Tourism cities in China, like Sino-West trade and communication, have a long history. At the tourism city Xi'an, which was named Chang'an during the Tang Dynasty, archaeologists excavated the site of the fair in the western part of Chang'an and discovered toys, treasures and ornaments that were popular in Central Asia and West Asia around the seventh century. They also discovered the remains of temples that were worshipped by the people who lived in Central Asia and West Asia. It is known that many people visited Chang'an from Central Asia and West Asia (Su 1978). Then the tourism city Chang'an developed for international transportation and trade on land, while Suzhou and Hangzhou developed their trade and tourism with the Grand Canal. Italian Jesuit priest Matteo Ricci once pointed out that "Suzhou is famous for its beautiful scenery and rich products". He also recorded that "As a popular saying goes, 'Just as there is paradise in heaven, there are Suzhou and Hangzhou on earth'" (Ricci and Trigault 1983, p. 338). With the concentration and flow of the population, the recreation space appeared and developed in ancient China's tourism cities. In the residential area, the building was cleverly combined with trees and courtyards to create a good environment. Look around the city, the rigorous buildings were organically integrated with natural landscapes and gardens (Dong 2004, p. 247).

Nowadays, almost every city in China takes tourism as an important function. Therefore, more and more excellent tourism cities have emerged in China. In particular, the National Tourism Administration of the PRC issued the *Notice on Launching and Selecting Top Tourist City of China* in March 1995. So far there are 339 Top Tourist Cities of China, most of which are located in east of the Hu Huanyong Line (Figure 35.1). Since then, relevant research on tourism cities in China has gradually increased. The research literature on tourism cities in China mainly focuses on the definition and type, development mode, perceived image, relevant theory, etc. Some scholars believe that tourism cities refer to the cities that consider tourism an important function of the city, that is, tourism accounts for a large proportion in the urban industrial structure, and the number of tourist arrivals reaches a considerable scale (Jin and Wu 1999). However, the definitions they propose do not specify the number of tourist arrivals as well as the scale of the tourism industry. Researches on the development mode of tourism cities are mainly concentrated on large cities that have a higher degree of internationalisation, such as Beijing (Song

Figure 35.1   Distribution of top tourist cities of China.

1996), Shanghai (Li 1996, Dao 1996) and Guangzhou (Li 1997, Bao 1997). Apart from the perceived image research of a single tourist city, Xu Xiaobo and other scholars made a comparative analysis of the perceived images of the 49 top tourist cities in China in accordance with the data of "Ctrip.com", a representative travel website in China (Xu *et al.* 2015). In terms of theoretical construction, Chen and Wu made a pioneering study (1996), and Wu Bihu proposed the Recreational Belt around Metropolis (Wu 1999). The tourism cities in China that are discussed in this paper mainly include two types. One is the tourism cities defined by the National Tourism Administration of the PRC (Now the Ministry of Culture and Tourism of the PRC), and the other is cities that consider scenic tourism as the designated function in its urban master plan.

Officially defined tourism cities in China

After Reform and Opening in 1978, in order to promote the development of the urban tourism industry, China began to select tourism cities with an official standard. The National Tourism Administration of the PRC has conferred the title of Top Tourist City of China on more than 300 cities and Best Tourism Cities of China on three cities. The selection of Top Tourist City of China began in 1995, and the National Tourism Administration of the PRC issued the *Notice on Launching and Selecting Top Tourist City of China*. The establishment of Top Tourist City of China is based on the *Interim Measures for the Establishment of Top Tourist City of China Work Management* and the *Top Tourist City of China Inspection Standards*, and the National Tourism Administration's acceptance team check and accept the declared cities in accordance with the above documents. The symbol of Top Tourist City of China is "Galloping Horse Treading on a Flying Swallow" which was unearthed from a Han tomb in Gansu Province. In 1999, 54 cities were given the title of Top Tourist City of China. At present, 339 cities have successfully become the Top Tourist Cities of China. In 2002, the National Tourism Administration formulated the "Best from Excellent" strategy and entrusted the World Tourism Organization and Peking University to jointly study and prepare the *China's Best Tourism City Standards*. On this basis, the National Tourism Administration announced the *Chinese Best Tourism City Establishment Guidance*, which was published in February 2003 (Wu *et al.* 2003). Chengdu, Hangzhou and Dalian were selected as *China's Best Tourism City in 2006*. However, the selection of China's Best Tourism City has not continued. Therefore, we will discuss the purpose and significance of the selection, the criteria and procedures of the selection, and the distribution characteristics for Top Tourist City of China.

Purpose and significance

For the development of tourism cities, the establishment of Top Tourist City of China is crucial. In the course of establishment, the government-led development strategy has been implemented. The selection of Top Tourist City of China has greatly enhanced China's tourism industry from the destination level by mobilising the enthusiasm of the mayor to grasp tourism (Li *et al.* 2019). The establishment of China's outstanding tourist cities brings many benefits. From the perspective of the city, the level of internationalisation and modernisation has been improved. From the perspective of the tourism industry, the development of tourism cities has promoted the development of the national tourism industry. From the perspective of enterprises, it has created a good business environment. From the perspective of visitors, the types of the tourism activities are more diverse. From the perspective of the public, the urban living environment becomes more liveable.

The establishment of Top Tourist City of China is both a measure and a target, which is mainly reflected in three aspects: tourism facilities, soft environment and tourism products. Tourism infrastructure is the material carrier for tourism activities such as catering, accommodation, transportation, visiting, entertainment and shopping, which is the hardware foundation for tourism reception capabilities as well. The tourism facilities of Top Tourist City of China must meet the needs of reception and even meet international standards. The soft environment directly affects the mood of tourists. Top Tourist City of China should have high-quality services to make the tourists satisfied. It requires thoughtful solutions to all kinds of practical problems that can occur in tourism activities. Top Tourist City of China should develop complex tourism resources and enhance their appeal to tourists. This requires a wide variety of tourism products to meet the diverse tourism needs of different tourism consumers.

Criteria and procedures

In 1998, the National Tourism Administration issued the *Top Tourist City of China Inspection Standards (Trial)*. In 1999, the *Top Tourist City of China Inspection Standards (Revised)* was issued. The standard was revised again in 2003. The current standard is the third revised version issued in 2007. It can be seen that the standard is dynamic and has been revised accompanied by the development of China's tourism industry, so that the requirements in the standard can be adapted to the actual needs of the tourism industry. The revised version of the standard that was issued in 2007 was changed from the 20 first-class indexes and 176 second-class indexes to 20 first-class indexes and 183 second-class indexes (Table 35.1). According to the new policies and

Table 35.1 First-class index and highest score from *Top Tourist City of China Inspection Standards (Revised in 2007)*

Serial number	First-class index	Highest score
1	Economic development level of urban tourism industry	60
2	Orientation and scale of urban tourism industry	35
3	Policy support and funding of urban tourism industry	35
4	Government-led mechanism for urban tourism development	35
5	Management system of urban tourism industry	70
6	Spiritual civilisation construction of urban tourism industry	60
7	Ecological environment	45
8	Modern tourism function	100
9	Tourism education	40
10	Tourist traffic	60
11	Development and management of tourist attractions	40
12	Tourism promotion and product development	60
13	Accommodation	50
14	Travel agency	40
15	Tourist catering	40
16	Tourist shopping	40
17	Tourism culture and entertainment	40
18	Tourist toilet	40
19	Tourism market order	70
20	Tourist safety and insurance	40

regulations, it mainly increases the contents of promoting the construction of tourism integrity, and at the same time it modifies "Hotel star ratings" and other inconvenient operations, making the standard more feasible. The original framework and the total score are the same as before, so that the main content of the standard is consistent.

The process of establishing Top Tourist City of China consists of seven steps. The first step is to declare: county-level cities, through the higher-level tourism bureau, declare to the provincial tourism bureau; prefecture-level cities and sub-provincial cities declare to the provincial tourism bureau; and municipalities declare directly to the National Tourism Administration. The second step is to establish: the cities participating in the establishment in accordance with the standards delegate the target responsibilities to the relevant departments of the city and implement the compliance plan; the period of the establishment should not be less than three years. The third step is self-inspection: the cities participating in establishment should conduct self-inspection. The fourth step is the preliminary review: the provincial tourism bureau organises the preliminary review team to conduct the preliminary review of the city. The fifth step is acceptance: the National Tourism Administration organises the acceptance team to conduct the inspection and acceptance according to the conditions of establishment and preliminary examination. The sixth step is approval and naming: the National Tourism Administration officially names the "Top Tourist City of China" and issues certificates and symbols for the cities that have passed the inspection and acceptance of the relevant departments. The seventh step is to review: the city that has won the title of "Top Tourist City of China" should guarantee continuity of the excellent work, and formulate rectification plans and implement rectification work under the guidance of the rectification opinions which were put forward by the acceptance team.

Distribution of Top Tourist City of China

The 339 Top Tourist Cities of China are distributed in 31 provincial-level administrative regions except Hong Kong, Macau and Taiwan. Most of them are located in the east of the Hu Huanyong Line. Shandong Province has the largest number of Top Tourist Cities of China, with a total of 35, while Ningxia and Tibet have the fewest, each with one. The level of economic, social development and abundant tourism resources in China's eastern, central and western regions has shown differences, and the tourism development has also displayed significant differences. The level of urban development in the eastern region is relatively higher than that of the central region and western region, and the distribution density of Top Tourist Cities of China is also higher (Table 35.2). This is mainly because the eastern region has a high level of economic and social development, and the population and cities there are dense. Therefore, the tourism industry develops early on and the development of tourism resources is high. In addition, the eastern region has convenient transportation and a high degree of openness, therefore it has high accessibility and a large number of domestic and foreign tourists.

Tourism cities and tourism planning in China

Four periods of tourism planning for tourism cities

Tourism planning started in 1980s following the Economic Reform and Opening Up in China. During the 40-year development, the main thoughts and theories have changed a lot. It can be divided into four periods: resource-oriented period, market-oriented period, sustainability-oriented period and all-for-one tourism period (Deng *et al.* 2018) (Figure 35.2).

1978–1992 Resource-oriented period: During the first period, tourism planning was dominated by geography professionals, who focused on tourism resource development and utilisation (Zhang 1993). In 1979, the National Tourism Administration issued an official document to develop tourism planning at the national level. After that, the investigations of tourism resources were implemented and tourism planning was developed in accordance with the investigation (Xu 2004). The Chinese government and many local governments began to set up tourism bureaus in the late 1970s. At the same time, the construction sector carried out tourism city planning, national parks planning and so on (Wu 2000).

1992–2000 Market-oriented period: In the 1990s, the Chinese government officially established the tourism industry as an industrial sector, which indicates that China's tourism planning had entered a market-oriented stage (Wu 2000). In 1995, the Chinese government conducted the activities of creating and selecting the Top Tourist City of China, which set a standard criterion for tourism cities in China. More and more tourism cities sprang up and regarded tourism as a leading industry to develop theme parks, resorts and other products to attract tourists.

Table 35.2 Density of top tourist cities of China

	Eastern region	Central region	Western region	The whole country
Number of cities	263	227	170	660
Number of Top Tourist Cities of China	167	99	73	339
Density of cities (/10⁴km²)	1.92	0.81	0.31	0.69
Density of Top Tourist Cities of China (/10⁴km²)	1.22	0.35	0.13	0.35
Number of Top Tourist Cities of China/Number of cities	63.50%	43.61%	42.94%	51.36%

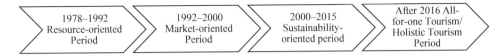

Figure 35.2   Four periods of tourism planning of tourism cities in China.

During this period, the urban planning personnel, including urban construction personnel, landscape and architecture departments and architectural colleges, made great strides in the fields of scenic spots and historical and cultural cities (Zhang 1993).

2000–2015 Sustainability-oriented period: In the twenty-first century, tourism cities not only focused on how to conduct market development and promotions to attract more visitors, but also took responsibility for community development. During this period, tourism was integrated into the development of cities and communities, and conservation and sustainable use of natural and cultural resources (Wu 2009). In the spatial scope of tourism planning, administrative planning shifted to regional planning with tourism centres as the core, which made demand for special tourism planning for tourism cities increase rapidly.

After 2016 All-for-one tourism/Holistic tourism period: All-for-one tourism is a new strategic measure to promote the development of tourist destinations inclusively by considering all tourism elements, all relevant industries, the whole tourism process, all possible tourism times and places, all directly and indirectly related tourism sectors, and all relational groups of people (Li *et al.* 2013). Against the background of all-for-one tourism, tourism cities integrate all tourism resources to provide visitors with a full-process, full-time experience. Tourism cities will be built into comprehensive and open tourist destinations, which takes product innovation as the basis and fully meets the needs of visitors.

Tourism planning system of tourism cities

The current tourism planning system in China is borrowed from the urban planning system. From the perspective of the planning level, it includes the tourism development plan and the tourism area plan (State General Administration of the PRC 2003). The tourism development plan includes an immediate plan (3–5 years), a medium-term plan (5–10 years) and a long-term plan (10–20 years). The tourism area plan can also include a master plan, a regulatory detailed plan and a constructive detailed plan (Figure 35.3). In addition, the tourism concept plan (Chen *et al.* 2013), the tourism-marketing plan and the tourism product plan are also needed in the actual operation, which are optional plans and need to be specified in content provisions. The whole system is conducted from the perspective of macro-orientation to micro-construction to specify tourism planning and design systematically.

For the tourism cities, the tourism-related planning content is directly compiled in the urban planning and is focused on integration with other special planning. Besides this, tourism cities also create tourism destination plans and scenic spot plans to identify the advantages of tourism resources and future development. For the other cities, the tourism bureau of the city is in charge of tourism planning.

Take the 13th Five-Year Plan for the Development of Tourism and Leisure Industry in Hangzhou (2016) as an example. It was an immediate tourism development plan for 2015–2020. Hangzhou is one of the first batch of the Top Tourist Cities of China, which is famous for West Lake, the Grand Canal, Xixi Wetland Park and other scenic spots. The whole plan consists of five parts, which include Review of the 12th Five-Year Plan, 13th Five-Year Development

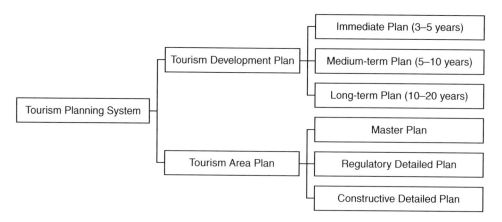

Figure 35.3 Tourism planning system of tourism cities in China.

Source: State General Administration of the PRC (2003).

Environment, Guiding Ideology and Strategy, Key Initiatives and Projects and Insurance System. As for the initiatives and projects, the plan considers optimising all-for-one tourism and leisure spatial distribution, to improve industry integration, develop international tourism leading products, upgrade the public service system of tourism and leisure industry, improve the international marketing and promotion system and promote the smart tourism platform. This is the spatial development pattern map (Figure 35.4).

Figure 35.4 Spatial development pattern.

Source: Hangzhou Tourism Committee (2016).

Major findings in tourism city research in China

Recreational Belt around Cities

Wu (2001a) first named the recreational belt around cities (ReBAC) as the recreational belt around metropolises (ReBAM). The recreational belt around cities (ReBAC) is a collection of the resources, places, facilities and public spaces for leisure, vacation, entertainment, sightseeing, recreation, recuperation and other activities in the surrounding areas of the city, which is mainly provided by the government or commercial organisations to urban residents (Wu and Cai 2006, Dang 2011). The recreational belt around cities and the central recreation area of the city constitute the urban recreation system (Yu 2003). Chinese scholars began their study of the recreation zone around the city in the early 1990s. Zou (1990) firstly put forward the form of the circular area around the city in his study of spatial morphology of tourism regionalisation. Chen (1996) recognised that since modern traffic conditions have been improved and citizens' weekend leisure time has been increased, suburban tourism has coincided with peripheral tourism and short-term regional tourism. He took the lead to propose that we should develop recreational tourism in the surrounding areas of the city, which includes projects such as rehabilitation, sports, recreation, agricultural sightseeing, water and yachting, fishing, children's paradise, catering and hygiene, picnic and barbecue, and returning to nature and ecology, etc. All of these forms can be appropriately combined and matched. Li (1999) put forward the tourism development mode of Xi'an Circular City Scenic Spot. Wu Bihu first put forward the theory of "the recreational belt around cities" which was based on the phenomenon of tourism urbanisation. He considers that the recreational belts around cities are mainly located in the suburbs of the big cities, and they are recreational facilities, places and public spaces which are set up for urban residents. In specific cases, they also include all levels of tourist destinations that are frequently visited by foreign tourists on the outskirts of the city. At the same time, he pointed out that the location where the recreational belt around the city always forms is a compromise position reached by investors and tourists under the dual function of land rent and tourism cost (Wu 2001b).

Scholars believe that there are a lot of major factors having an influence on the formation of the recreational belt around cities, such as the market demand of tourist sources in suburbs, the selection of an investment market for tourism development, attraction of suburban tourism, the regional industrial policy of government, etc. These factors jointly constitute the three major systems of the formation mechanism of the recreational belt around cities: (1) The driving force system, which is the market demand of the tourist sources in suburbs, and the selection of an investment market for tourism development; (2) the attraction system, which is the natural and humanistic landscape, facilities, environmental atmosphere, tourism culture and tourism activities and other physical and non-material factors of recreational attraction inside the recreational belt around cities; (3) the supporting system, namely, the hardware environment that provides assurance and support, such as road traffic, communication information and service facilities, and the software environment which plays a role of supporting and guiding, such as the tourism policies that are formulated by the state and local governments and a good social environment, etc.

There are mainly four types of tourist destination in the recreational belt around cities. (1) Natural tourist destinations, such as natural scenic spots, forest parks, nature reserves, pastoral mountain villages, etc. (2) The humanistic sightseeing tourist destinations, such as historical and cultural sites, ancient gardens, science, technology, culture and art museums, etc. (3) Artificial entertainment tourist destinations, such as playgrounds, theme parks, etc. (4) Sports leisure tourist destinations, such as sports venues, resorts, convention centres, etc.

When we analyse the spatial structure of the recreational belt around cities from the point of tourism land type, we discover that the average distance between a man-made entertainment tourist destination and a city centre is the smallest. The average distance between a natural tourist destination and a city centre is the largest. The average distance between a humanistic tourist destination, sports leisure tourist destination and the city centre falls in between.

Chinese scholars divide the spatial structure of the recreational belt around cities into three categories in terms of the morphology. Based on the analysis of the actual development of rural tourism and leisure in the suburbs of foreign cities, Wu (2003) modified the Genn's ring zone mode appropriately, and proposed a concentric circle layer structure. By taking the core of the city as the centre of the tourism in space, he distinguished the tourism function and characteristics of different zonal areas by the means of four ring zones around the centre of the city. The four ring zones were the urban tourism zone, the suburban leisure and tourism zone, the rural tourism zone and the remote tourism zone. Bao (2005) divided the "One ring one zone" ring zone structure of the recreational belt around cities, which is dominated by rings, and supplemented it with zones. Through an analysis of the supply circle and the demand circle of the tourist vacation zone around the city of Shenyang, Mao (2004) divided the spatial distribution into four ring zones, namely, the leisure and entertainment shopping zone, the leisure agriculture tourism zone, the popular leisure vacation zone and the remote village vacation zone. Wu *et al.* (2001) summarised the shape of the green zone around the city of Shanghai as the "Melons on a vine" style. Namely, a 500-metre-wide ring green zone was the "Long vine". At the same time, ten large theme parks were planned and laid out in the areas along the route moderately widened where land conditions were better, and were here referred to as the "Melons".

The progress of domestic research is mainly reflected in the following three aspects. (1) It analysed the background of the development of the recreational belt around cities, and summed up the concept of strong participation. The research scope began to expand from the periphery of the big city to the periphery of the small and medium-sized city. (2) It studied the formation mechanism and location characteristics of the recreational belt around cities. (3) It put forward some suggestions for project development of the recreational belt around cities. The recreational belt around cities is suitable for developing projects such as holidays, conferences and activities with strong participation. The suburbs can be divided into suburban areas and outer suburbs, and they should have their own development emphases (Li *et al.* 2015).

Urban heritage revitalisation theory

China is an ancient Eastern country with a long history. The historical monuments left behind by the successive reigns and dynasties have very high historical and cultural values. Urban heritage revitalisation aims to transform urban heritage resources into tourism products without having undue influence on heritage protection and inheritance. Wu Bihu proposed that the revitalisation of the heritage should be protected through utilisation, rather than protecting it through locking it up. Protection in the formalin style is easy to operate, but has no technology content, and it cannot realise the goal of passing on from generation to generation. He believes that the revitalisation of the heritage mainly focuses on giving full play to the cultural connotation, educational function and leisure service function of heritage. In particular, tourism revitalisation is a form of protection development with the least side effects, and is the best way to revitalise the legacy (Wu and Wang 2018).

Wu divided the models which were presented by scholars into three basic modes. The first is the mode of objectivism (static museum mode), such as the Great Wall, the Forbidden City and the Summer Palace. All of these have rather good objectivism authenticity, therefore they do not need further processing, and their protection work is relatively important. The second is the mode of constructivism (real-scene reproduction), which is more suitable for the revitalisation of historical scenes. It realises the presentation of some visual form through constructivism authenticity, and improves the efficiency and breadth of cultural heritage transmission, such as the frame display method adopted by the Tang Daming Palace in Xi'an and the reappearance mode of the Dingdingmen Protective Shield in Luoyang City in the Tang Dynasty. The third is the performative mode (staged presentation mode), as seen in the Qingming Riverside Landscape Garden built on the local basis of Bianliang in the Northern Song Dynasty in Kaifeng, and the ancient feelings of the Song City developed on the basis of the cultural background of the capital of the Southern Song Dynasty in Hangzhou, which are all in the staged performative presenting mode. In terms of the basic path of heritage revitalisation, we should consider the following four points. (1) We shall respect the site characteristics and context continuation, meanwhile, the protection and utilisation of heritage space should be taken into account. (2) We shall distinguish them from competitors through refining specific cultural themes. (3) We shall provide more diverse and recreational activities, as well as having participation, thus achieving the better presentation of historical places, historical scenes and even wider cultural heritage through tourism, at the same time, achieving the goal of both protection and utilisation of the heritage.

Conclusion

The ancient tourism cities of China originated from communication and trade between China and foreign countries. The modern tourism cities of China began to develop and grow after the Reform and Opening Up in 1978. Especially in 1995, the National Tourism Administration launched the selection of Top Tourist City of China. Since then, the quality of the tourism industry in tourism cities has increased, accompanied by an increase in the number of tourism cities. In addition, related research has grown. The selection criteria for Top Tourist City of China are dynamic, which promotes the development of the tourism industry led by governments and mayors, as well as promoting the standardisation and internationalisation of tourism cities in China. The current tourism planning system in China was learned from the 1980s urban planning system. The 40-year development of the tourism planning system can be divided into four periods: the resource-oriented period, the market-oriented period, the sustainability-oriented period and the all-for-one tourism period. For tourism cities, tourism-related planning content is directly integrated into urban planning and much attention was paid to amalgamation with other special planning. Besides this, tourism cities also conduct tourism destination plans and scenic spot plans to identify the advantages of tourism resources and future development. Academic contributions regarding tourism cities in China mainly include the theory of the recreational belt around cities (ReBAC) and the theory of urban heritage revitalisation. Chinese scholars began their study of the recreation zones around cities in the early 1990s. Based on the phenomenon of tourism urbanisation, Wu Bihu firstly put forward the theory of the "recreational belt around cities". The connotation of urban heritage revitalisation theory is to transform urban heritage resources into tourism products without undermining heritage protection. The application of the theory of urban heritage revitalisation promotes the dual goals of the protection and utilisation of the heritage of tourism cities in China.

References

Bao, J.G., 1997. Commercial City: The Inevitable Choice of Guangzhou Tourism Development Strategy. *Yangcheng Evening News*, 7 September, p. 7.

Bao, J.G., 2005. *City Tourism (Principle. Case)*. Tianjin: Nankai University Press.

Chen, C.K., 1996. Urban Tourism Development Planning Research Outline. *Tourism Tribune*, (5), 31–34.

Chen, C.K., Wu, C.Z., 1996. Discussion on the Theory and Practice of Urban Tourism: A Summary of Shanghai International Conference on Urban Tourism. *Geography and Territorial Research*, (1), 61–64.

Chen, Q., Zhang, S.L., Li, J., 2013. Innovative Research on Contents of Domestic Tourism Concept Planning. *Human Geography*, (2), 136–141.

Dang, N., 2011. *Suburban Recreation Space in Leisure Era: A Study on ReBAM*. Shanghai: Shanghai People's Publishing House.

Dao, S.M., Zhu, C.R., Zhang, W.J., 1996. On the Development Trend and Countermeasures of Shanghai Conference/Incentive Tourism. *Tourism Science*, (2), 1–5.

Deng, Z.M., *et al.*, 2018. Research on the Regional Tourism Planning from the Perspective of Ecological Civilization Construction: Ideas, Functions and Developing Trend. *Ecological Economy*, 34 (10), 125–130.

Dong, J.H., 2004. *History of Urban Construction in China*. Beijing: China Architecture and Building Press.

Hangzhou Tourism Committee, 2016. The 13th Five-Year Plan for the Development of Tourism and Leisure Industry in Hangzhou. *Hangzhou Tourism E-Government Network* [online]. Available from: www.gotohz.gov.cn/ghjs/lygh/201808/t20180806_159513.html [Accessed 6 August 2018].

Jin, X., Wu, X.G., 1999. Research on Chinese Tourism City System. *Urban Research*, (5), 31–34.

Li, H.R., 1996. Urban Tourism and Shanghai Model. *Tourism Tribune*, 11 (1), 20–23.

Li, J.M., Tan, L.J., Li, W., 2015. Review of Research on Recreation Belt around Metropolis. *Journal of Wuhan Business University*, 29 (6), 16–21.

Li, J.Q., 1999. Research on Tourism Development of Xi'an ReBAM. *Economic Geography*, (1), 125–128.

Li, L.X., 1997. Discussion on the Characteristics and Construction Focus of Guangzhou Tourism Attraction. *Tourism Tribune*, 12 (2), 26–28.

Li, X.J., Shi, S.S., Liu, G.R., 2019. 40 Years of Chinese Tourism: Market-Oriented Government Leadership. *Tourism Tribune*, 34 (2), 10–13.

Li, X.J., Zhang, L.Y., Cui, L., 2013. Comprehensive Tourism: Idea Innovation on Building a World-Class Tourism Destination. *Human Geography*, 28 (3), 130–134.

Mao, R.Z., 2004. Research on the Spatial Structure of Shenyang Holiday Tourism Vacation Belt. *Social Scientist*, (6), 97–99.

Ricci, M., Trigault, N., 1983. *China in the Sixteenth Century: The Journals of Matteo Ricci*. Translated by He, G., *et al.* Beijing: Zhonghua Book Company.

Song, J.Z., 1996. Opinion on Developing Urban Tourism. *Tourism Tribune*, 11 (3).

State General Administration of the People's Republic of China (PRC) for Quality Supervision and Inspection and Quarantine, 2003. *General Specification for Tourism Planning*. Beijing: China National Tourism Administration.

Su, B., 1978. Chang'an City and Luoyang City in Sui and Tang Dynasties. *Archaeology*, (6), 409–425.

Wu, B.H., 1999. Formation and Spatial Structure of ReBAM: A Case Study of Shanghai City. *The International Conference on Urban Tourism*. Zhuhai, Guangdong Province, China.

Wu, B.H., 2001a. A Study on Recreational Belt around Metropolis (ReBAM): Shanghai Case. *Scientia Geographica Sinica*, 21 (4), 354–359.

Wu, B.H., 2001b. *Principle of Regional Tourism Planning*. Beijing: China Travel & Tourism Press.

Wu, B.H., Cai, L.A., 2006. Spatial Modeling: Suburban Leisure in Shanghai. *Annals of Tourism Research*, 33 (1), 179–198.

Wu, B.H., Feng, X.G., Li, M.M., 2003. A Review on the Theory of Standard System for China Best Tourist Cities and Its Implementation. *Tourism Tribune*, 18 (6), 40–44.

Wu, B.H., Wang, M.T., 2018. Heritage Activation, Original Site Value and Presentation. *Tourism Tribune*, 33 (9), 3–5.

Wu, C.Z., 2009. Retrospect and Prospect to Thirty Years of China's Tourism Planning. *Tourism Tribune*, 24 (1), 13–18.

Wu, C.Z., Han, G.H., 2003. Research on Tourism Spatial Model of Foreign Metropolis Suburbs. *Urban Problems*, (6), 68–72.

Wu, G.Q., Yu, S.C., Wang, Z.J., 2001. The Discussion on the Concept of the Planning for the Green Belt around Shanghai. *City Planning Review*, 25 (4), 74–75.

Wu, R.W., 2000. History and Future of Tourism Planning. *Rural Eco-Environment*, 16 (1), 39–41.

Xu, C.X., 2004. Study on the Evolvement of Contemporary Tourism Planning Thoughts in China. Thesis (PhD). Hunan Normal University.

Xu, X.B., *et al.*, 2015. Study on Perceived Image of Chinese Tourist Cities. *Geographical Research*, 34 (7), 1367–1379.

Yu, S., 2003. *Urban Tourism and Urban Recreation*. Shanghai: East China Normal University Press.

Zhang, G.R., 1993. Some Thoughts of the Tourism Planning. *Tourism Tribune*, (4), 13–16 + 61.

Zhou, J.L., 1990. Discussion on the Problems of Tourism Zoning: Taking Chengdu Area as an Example. *Tourism Tribune*, (2), 26–28.

36

APPRECIATING GEOLOGY IN
THE URBAN ENVIRONMENT

Samantha Richards, Greg Simpson and David Newsome

Introduction

Geotourism is a rapidly expanding form of natural area tourism with increasing attention being paid to the appreciation of landscapes where interesting landforms and rock outcrops can be viewed and visited (e.g. Dowling and Newsome 2018; Newsome and Dowling 2018; Ravelo-son et al. 2018). It is not, however, only in national parks, wilderness areas or in other protected areas that landscape and geology can be appreciated as part of a tourism experience (Palacio-Prieto 2015; Del Lama 2018). While public interest in building stones, historical buildings and prominent rock outcrops in cities have always been tourism attributes, in recent times there has been a more concerted focus on geology in the urban environment as a tourism product in its own right (Del Lama et al. 2015; Del Lama 2018; Górska-Zabielska and Zabielski 2018).

City administrators and tourism specialists now promote geological outcrops, prominent features and landscape phenomena, and geological features of the built environment as places to visit and learn about, thus adding value to tourist visits and diversifying the things tourists can experience when visiting a town or city. Furthermore, the recent literature on geotourism highlights various cities around the world as actual and potential geotourism destinations (Górska-Zabielska and Zabielski 2018). For example, Del Lama (2018), describes the value in Sao Paulo, Brazil of visiting churches for their geological values with the focus on geological history and its connection with history and local culture. However, cities in Japan capitalise on their volcanic history and legacy of volcanic activity via hot spring and spa (*onsen*) tourism (Erfurt 2018). Reflecting the attraction of volcanos and hot springs, tourism associated with *onsen* has a long history in Japan, and there is an increasing focus and expansion of the associated activity of wellness tourism at such sites (Erfurt-Cooper and Cooper 2009; Erfurt-Cooper 2014; Simpson and Newsome 2017). Urban geotourism attractions thus range from upstanding relief with clear rock exposures (The Rock, Gibraltar), clear geomorphological features such as river valleys and gorges (Tennessee River Gorge, Chattanooga, USA), building stones (Stepping Stones, Perth, Western Australia), monuments to natural disasters (Tsunami Memorial Park, Phuket, Thailand), built environment such as old churches and graveyards (Sao Paulo, Brazil), through to museums and information or visitor centres that display rocks and minerals and provide significant educational content on Earth history (Geology Museum, Bandung, Indonesia).

This chapter provides a concise account of urban destinations where geology and landscape are features of city tourism and interlinked with history and culture. Our purpose is to provide examples of how domestic and international tourism can be enhanced via the recognition and promotion of urban geology as distinct attractions for tourism cities. Accordingly, the case studies comprise examples from Hong Kong (China), Kraków (Poland), Edinburgh (Scotland) and Whitby (England). Each of these case studies in regard to appreciating geology in the urban environment considers key aspects of the attraction, including the supporting geology, and provides an account of what tourists can derive from this particular type of city experience.

Hong Kong UNESCO Global Geopark, China

Established in Europe, but fully embraced and developed in China, as a tool for achieving the United Nations Sustainable Development Goals (UNSDGs) to alleviate poverty, geoparks have become important tourism destinations throughout the world (Dowling and Newsome 2018; UNWTO 2019). The Hong Kong UNESCO Global Geopark (HKUGGp) sits within what many perceive as an essentially urban area with the wider territory being rich in geotourism assets (Ng et al., 2010).

The HKUGGp encompasses an area of approximately 150 km². Internationally significant geology contained within the HKUGGp includes volcanic hexagonal rock columns (Figure 36.1) formed during the Middle to Late Mesozoic Era (100–200 million years ago (Mya)) and sedimentary rock formations deposited from the Devonian Era (355–410 Mya) to the Paleogene Era (65–20 Mya). The included geosites thus represent a complete geological history of Hong Kong.

Creation of the HKUGGp was a three-stage process that involved overcoming initial resistance from some local academics and conservationists, who mistakenly feared that the creation of a geological theme park would encourage visitors into environmentally sensitive sites and expose the geoheritage area to an unacceptable level of risk from activities such as visitors collecting rock souvenirs (Cheung 2016). In addition, there was a mistaken belief that rocks and landforms did not require the same level of protection as flora and fauna (Dowling and Newsome 2010). Moreover, without adequate conservation measures, Hong Kong's unique geological formations were at risk of being engulfed by fast-paced urban development pressures.

Despite the resistance from some quarters, the Hong Kong National Geopark was created in 2009. To overcome the real and perceived threats to the geoheritage of the area, the government implemented a system of accrediting professional geo-tour guides (geoguides) that was combined with education and an active media campaign (Cheung 2016). This proactive geo-conservation activity preceded acceptance of the site as a member of the Global Geoparks Network (GGN) in 2011 as a precursor to becoming the HKUGGp in 2015. The objectives for HKUGGp include conservation, education and sustainable development of geoheritage within the geopark. The sustainability focus of the HKUGGp includes local engagement, improvement of rural livelihoods, science popularisation and geotourism. This holistic focus was previously neglected in favour of biological conservation (Dowling and Newsome 2010).

To facilitate geoconservation and education goals, there are nine geotrails, two boat tours and five visitor centres within the geopark. To gain accreditation, geoguides undertake specialised training and are assessed regarding their geological knowledge, interpretation skills and awareness of geoconservation (Ng and Choi 2009). The HKUGGp has official partnerships with two geo-hotels, which mandates that the businesses promote the conservation of nature and cultural heritage, science transfer and service excellence. The geo-hotels endorse HKUGGp via information displays and by promoting available geo-tours. The hotel restaurants offer

Figure 36.1   Outcrop of hexagonal columns in rhyolitic lava, Hong Kong Geopark.

Source: Photograph by D. Newsome.

Note

Such features impress not only geologists and dedicated geo-tourists but the general tourist who is interested in natural phenomena. Such lavas are rarely seen on such an impressive scale.

geo-themed menus and a dedicated television channel promotes geopark destinations in guest rooms (HKUGGp 2018). Hotel staff undertake regular familiarisation field trips and undergo specialised training, to ensure they are able to provide visitors with up to date information (HKUGGp 2018).

Hong Kong was one of the first places in the world to fully embrace geotourism in a peri-urban setting. As such, the HKUGGp is considered a flagship geotourism enterprise. Although local residents within the geopark agree very strongly with the conservation and educational principals of the GGN, as they apply to the geoheritage of the park, there has been a conflict with management regarding the distribution of resources and community attitude to biological and environmental conservation (*South China Morning Post* 2015). Conversely, it is reported that visitors to HKUGGp have a greater appreciation of the biological and cultural heritage within the geopark, rather than the geological heritage sites (Guo and Chung 2019). This indicates that HKUGGp management and tour operators need to promote a greater awareness of the important links and connections between the abiotic, biotic and cultural values of the geoheritage within the park (Dowling and Newsome 2018). On a positive note, research by Cheung (2016) indicates that geopark visitors are willing to pay higher prices for better quality guided tours.

The HKUGGp hosts approximately 1.5 million visitors per year, which makes the geopark a key urban tourism asset for Hong Kong (GGN 2017; GovHK 2018; Guo and Chung 2019). Visiting tourists and residents enjoy a wide range of geo-themed artistic, community events and educational experiences, geo-themed food and beverages, information centres, community engagement, school visits and tours and walks through the geopark (HKUGGp 2018). Crowding beyond destination carrying capacities is, however, becoming the scourge of tourism in many natural areas and UNESCO listed sites, including in urban geosites (Newsome et al. 2012). Hong Kong authorities are proactively managing the geopark to avoid the impacts of overtourism in order to protect both the geoheritage of the HKUGGp and visitor satisfaction with their geopark experiences (Dowling and Newsome 2017; Fowler 2018).

Wieliczka salt mine, Kraków, Poland

The geology of the Wieliczka Salt Mine (WS Mine) in Kraków originates from the Miocene Epoch (13.6 Mya). The Wieliczka salt deposit, derived from the evaporation of saline water, is the result of colliding tectonic plates that formed the Carpathian Mountain range and the pre-Carpathian Basin or Carpathian Foredeep (Czapowski 1994). The Wieliczka salt deposit is accessible today due to orogenesis (mountain-building processes) uplifting tectonic plates into the pre-Carpathian Basin strata, to create thrust-folds that increased the original thickness of the salt layers (Burliga et al. 2018). The location of the WS Mine, at the front of the Carpathian Orogenic Belt, affords a unique opportunity for the geotourist to view the structural detail and plate tectonics of salt-bearing complexes that can only be readily studied in sections exposed in an underground mine (Hallett 2002).

The WS Mine was an important and profitable salt mining operation from the thirteenth century until the closure of its underground mining operations in 1996 (Rozycki and Dryglas 2016). Previously the mine was declared a UNESCO World Heritage Site in 1978 and a Polish Monument of History in 1994 (Alexandrowicz et al. 2009). However, the WS Mine was placed on the List of World Heritage in Danger in 1988, when the excessive extraction of brine and rock salt and increased humidity threated natural geoheritage values and artworks at the site (UNESCO World Heritage Centre 1998). The installation of more efficient dehumidifiers, the cessation of commercial extraction of crystalline salt, and a change in site management saw the WS Mine being removed from the List of World Heritage in Danger in 1998 (UNESCO World Heritage Centre 1998; WS Mine n.d.a)

The mine is now considered to be the largest underground museum in Europe (WS Mine n.d.a and n.d.b). It extends to a depth of 327 m and contains 287 km of passageways and chambers. Currently, only 2% of this labyrinth is accessible to the visiting public. Attractions at the mine include displays featuring the development of mining techniques in Europe from the thirteenth century to the twentieth century, an underground lake, four chapels and a multitude of statues carved from the rock salt by miners and contemporary artists (Hallett 2002). The story of the WS Mine is the result of centuries of commercial mining operations, hundreds of years of health treatments based on the mineral brine and environmental management, such as the dehumidification of air underground, and the operation of an underground tourist route existing since the early nineteenth century. The WS Mine is therefore a good example of geotourism and conservation of geological and historical assets working in union to provide economic and social benefits for a connected city. The social networking platforms Lonely Planet (n.d.) and Trip Advisor (2019a) and online booking service Viator (2019a) all rank the WS Mine Guided Tour in the top three things to do when visiting Kraków.

The mine also contains a legally protected abiotic nature reserve called the Crystal Grottoes Cupola. This site consists of two Crystal Grottoes, which are considered "priceless" and "unique on a global scale" (Lipecki et al. 2016, p. 1; WS Mine n.d.b). However, as the salt crystals are vulnerable to leaching via fluctuations in humidity, these grottoes are not open to the public (Lipecki et al. 2016; Alexandrowicz et al. 2009). The underground Chapel of the Blessed St Kinga exemplifies the adaptation of the WS Mine for cultural purposes. Constructed from a disused mine chamber between 1895 and 1925, the entire chapel (walls, floor, altar, balcony, bas-reliefs and chandeliers) is carved from salt (Hallett 2002).

Alexandrowicz and Alexandrowicz (2004) suggested that the WS Mine has the potential to become the world's first subterranean geopark. They recommended expanding on the current cultural aspects of the WS Mine tours by promoting the unique geology and plate tectonic processes that produced the salt-bearing complexes that were exploited by the mining process. The benefits for the WS Mine that would follow on from becoming part of the UNESCO GGN include rigorous management practices, planning and policy to protect the area from damage and to promote sustainable development aligned to the UNSDGs (Dowling and Newsome 2006c; UNWTO 2019). If a geopark is realised, the local community could gain the additional economic benefits that accrue from being part of the rapidly expanding global geotourism network (Dowling and Newsome 2018).

An estimated 1.7 million people visited the WS Mine in 2017 (WS Mine n.d.c). Visitor health and safety are addressed with warnings regarding the number of stairs (800), the cool temperature (14°C) and the risk of claustrophobia. The guided tours are conducted in groups of up to 30–40 people, with multiple tours running consecutively. The commentary for the tour focuses on the history of salt mining, the miners themselves and the cultural aspects of the WS Mine (Trip Advisor 2019b). In addition to these highly successful cultural and geotourism experiences, the WS Mine also continues to offer health and wellness experiences based on the mineral brine and spa opportunities available at the site (WS Mine n.d.d). Further, the operation has expanded into offering on-site accommodation; conference, meeting and training facilities; wedding ceremonies and receptions; and artistic and cultural events and festivals (www.wieliczka-saltmine.com/). The decade-long inclusion of the WS Mine on the List of World Heritage in Danger demonstrates the need to actively protect and manage the geoheritage of urban geodiversity in order to conserve the abiotic and cultural values of geotourism destinations (Dowling and Newsome 2018). Crowding during the peak tourist session is emerging as an issue at the WS Mine (UNESCO World Heritage Centre 2019). This will require an assessment of visitor management approaches to minimise the impacts of overtourism at this increasingly popular urban geotourism destination.

Arthur's Seat and the Salisbury Crags, Edinburgh, Scotland

Arthur's Seat and the Salisbury Crags (hereafter referred to as "the Crags") are an iconic part of the Edinburgh cityscape. The readily identified lion shaped hill is a unique, valuable and easily accessible urban geotourisim resource. Noted Edinburgh native, historian and writer Sir Walter Scott captured the essence of the view from Arthur's Seat. His prose has often been re-published in the noted Black's [Travel] Guides, produced by the Edinburgh-based brothers Adam and Charles Black (1842, 1843), and more recently by Historic Environment Scotland (HES 2018) in their annual Statement of Significance for Holyrood Park.

> If I were to choose a spot from which the rising or the setting sun could be seen to the greatest possible advantage, it would be that wild path winding around the foot of that

high belt of semi-circular rocks, called Salisbury Crags … the prospect commands a close-built, high piled city stretching itself out beneath. When a piece of scenery so beautiful yet so varied – so exciting in its intricacy yet so sublime – is lighted up by the tints of morning or of evening and displays all that variety of shadowy depths, exchanged with partial brilliancy, which gives character to even the tamest of landscapes, the effect approaches near to enchantment.

(Scott 1818 cited in HES 2018, p. 17)

Dominating the skyline of central Edinburgh (Figure 36.2), the group of hills located in the western portion of Holyrood Park, of which Arthur's Seat is the highest point (251 m), are the eroded remains of a strato-volcanic dome that erupted and encamped during the Carboniferous period (Edinburgh Geological Society (EGS) n.d.a; HES 2018; McAdam 1986; Monaghan et al. 2014). Formed from the core of the volcano, Arthur's Seat is the eroded remains of the Lion's Head Vent that initiated the eruption of the Arthur's Seat volcano 341 Mya (McAdam 1986; Monaghan et al. 2014). The secondary Crow Hill peak is the eroded remains of a lava lake associated with the later and larger Lion's Haunch Vent that remained active until 335 Mya (McAdam 1986; Monaghan et al. 2014). The adjoining Whinny Hill formed from the cone of the Arthur's Seat volcano (McAdam 1986; HES 2018).

After the volcano became extinct, it was buried under thousands of metres of sediments, which became the Abbeyhill Shales (McAdam 1986; McAdam and Clarkson 1996). Approximately

Figure 36.2 View of Edinburgh from Arthur's Seat.

Source: Photograph by L.G. Simpson.

325 Mya, some 10–25 million years after the Arthur's Seat Volcano became extinct, that sedimentary layer trapped an intrusion of magma rising through the old vents to create the dolerite sills of the Crags (EGS n.d.b.; McAdam 1986; McAdam and Clarkson 1996).

The sedimentary and igneous rock of this area has been eroding for the past 300 million years and most dramatically over the past two million years, when the land surface was denuded by growing and melting ice sheets of the Pleistocene Ice Ages (EGS n.d.b; McAdam and Clarkson 1996). At the last glacial maxima, 20,000 year ago, the one-km-thick ice sheet that covered the area of current day Edinburgh stripped softer sedimentary rock away from the basalt of Arthur's Seat and the dolerite sills of the Crags (EGS n.d.b; McAdam and Clarkson 1996). That ice sheet melted rapidly as the Earth warmed and torrents of meltwater washed the eroded sedimentary rocks and other glacial till into the sea or low-lying swamps and bogs (EGS n.d.a; EGS n.d.b).

The end result is that, hundreds of millions of years of geological processes have mixed, fused and eroded the igneous and sedimentary rocks of Holyrood Park to produce Arthur's Seat and the Crags. Those processes have created abundant geological features, including sills, dykes, faults, slickensides, vesicles, distorted strata, lava lakes, roche moutonnée and glacial striate (EGS n.d.c; HES 2018; McAdam and Clarkson 1996; Monaghan et al. 2014). In recognition of the geological importance of the Arthur's Seat Volcano, it was designated a Site of Special Scientific Interest (SSSI) in 1986 (Scottish Natural Heritage (SNH) 2011). The SSSI citation notes,

> The small composite volcano of Arthur's Seat, of Lower Carboniferous period, is one of the most studied volcanoes in the world. All the component parts of a typical strato-volcano are well displayed and the sequence of eruptions can be traced with a continuity unique in Britain.
>
> *(HES 2015; SNH 2011)*

There are several relatively easy routes available to ascend Arthur's Seat, including the 4.8-km hike that takes visitors past Lang Rig (or Long Row) and along which visitors can observe the remains of basalt lava flows generated by the Arthur's Seat Volcano and the much the younger Salisbury Crags dolerite sills. This route to the summit incorporates Hutton's Section, which is named after James Hutton (1726–1797), the "father of modern geology" (HES 2018). Hutton's Section displays the junction between existing sedimentary sandstone layers and the overlying dolerite, which was exposed by quarrying of the Crags for building stone and road-base. In 1788, Hutton proposed that molten magma was pushed up to the surface of the Earth from below, through ancient sedimentary layers to create younger crystallised igneous rocks (EGS n.d.a; HES 2018). As can be observed by visitors today, this site provides visual support for Hutton's theory of an Earth that was far more ancient than the Old Testament-based age of 6,000 years previously declared by the Archbishop of Armagh and Primate of Ireland James Ussher (Dean 1981). The EGS publish several guides for walking tours that explain and celebrate both the geology of Arthur's Seat and the Crags and the natural and built environment geology in the broader Edinburgh urban landscape that can be seen from the summit (EGS n.d.d).

From a cultural perspective, the 360-degree views of Edinburgh visible from the summit of Arthur's Seat provided the perfect location for one of the four prehistoric or Dark Age hill forts that can be found in Holyrood Park (HES 2018). Remnants of hill fort defences are visible to the east of Arthur's Seat and on the adjoining Crow Hill. The landscape to the east side of Arthur's Seat also features 15 rows of terraces that are considered some of the best-preserved remains of medieval cultivation in the Scottish Lowlands (Lothians). The terraces provide historical insight into agricultural practices of the era (HES 2018). Several other man-made cultural and historical geotourism features are also visible from the summit of Arthur's Seat. There is the architecture of Holyrood Palace (*c*.1540), Holyrood Abbey (*c*.1128) and St Anthony's Chapel (*c*.1300–1500). Tourists can look across to the iconic Edinburgh Castle that is constructed on the exposed core of the earlier Castle Rock Volcano (350 million years old). In addition, the former Salisbury and Camstane Quarries (*c*.1530–1831), from which dolerite was extracted to build the walls of Holyrood Palace (Miller 2012), are also visible from Arthur's Seat.

In addition to geology, archaeology, architecture and history, the landscape around Arthur's Seat and the Crags has biotic, aesthetic and natural heritage values. Management of Holyrood Park (259 ha) is carried out by the Historic Scotland Ranger Service, which also coordinates teams of specialist volunteers (HES 2018). These Park Rangers and volunteers specialise in survey work including identifying rare and unusual plants. Overall, they have identified 350 plant species in the park, 60 of which are rare. Volunteers are also engaged for short time periods to perform specific tasks, such as clearing invasive weed species. Park Rangers liaise with the local community and wider public, providing talks and activities on topics such as nature conservation, historical aspects, geology, archaeology and general cultural and environmental stewardship (HES 2018).

For more than two years, visitors have ranked Arthur's Seat as Number 1 on the TripAdvisor "Top Attractions in Edinburgh" (Pile 2017; Trip Advisor 2019c). In addition to the self-guided walking tours described above, a multitude of companies and sole operators market tours of Arthur's Seat to target a diverse range of niche markets, including fully guided geologically focused tours that are provided by the EGS and private operators (EGS n.d.e; Miller n.d.; Trip Advisor 2019c; Viator 2019b). The TripAdvisor ranking and the multitude of commercial tour options highlights the importance of Arthur's Seat and the Salisbury Crags for tourism in Edinburgh and the significant role that geotourism can play in creating tourism cities that deliver the UNSDGs. However, the economic and social value that visitors to Author's Seat contribute to Edinburgh and the wider Scottish community was identified as a knowledge gap in a recent review by the HES (2015), with more information about park users being required to optimise both site management and visitor satisfaction.

Geoconservation at an urban geodiversity hotspot, Whitby, England

Settled by Danish seafarers in the early middle ages and recorded in the Doomsday Book as "Witebi", modern Whitby is a small regional town located at the mouth of the River Esk on the North Yorkshire coast in northeast England (Atkinson 1879; Dade and Carter 2015; English Heritage (EH) n.d.). Originally founded on fishing, mining of ores and gems, wooden shipbuilding and whaling, Whitby also has a long history as a geotourism destination and an even longer history as a significant cultural centre grounded by its local geology (Chrystal 2019; Brindle 2010; EH n.d.; Historic England (HE) 2015; Walton 2014; White 2019).

In the two decades since their inception, the concepts of geodiversity, geoheritage and geoconservation have focused on the natural abiotic components of the landscape (Gray 2018a, 2018b; Kubalíková et al. 2017). While contested by some authors, the concept of geodiversity within the abiotic environment complements the term biodiversity that describes the variability of biotic organisms (Sharples et al. 2018; Bétard and Peulvast 2019). Given the correspondence between the terms geodiversity and biodiversity, the recent work of Bétard and Peulvast (2019) promotes the concept of "geodiversity hotspots". Analogous to the "biodiversity hotspot" concept formulated by Myers and others (2000), Bétard and Peulvast (2019) defined geodiversity hotspots to be "geographic areas that harbor very high levels of geodiversity while being threatened by human activities".

Whitby is rich in natural geodiversity, but consideration of the cultural geomorphology of Whitby further increases the geoheritage values of this regional city as an urban tourism destination. A cultural geomorphology perspective considers a landscape in terms of "all the natural and anthropogenic factors it contains" (Panizza and Piacente 2009, p. 36), which corresponds with the "secondary geodiversity" concept advanced by Kubalíková et al. 2017). In this case study, we consider the geodiversity and geoheritage of Whitby in terms of both the natural and secondary geodiversity. The geodiversity we describe in this case study can be experienced by traversing the coastal portion of the 177-km Cleveland Way, which is a National Trail (National Trails 2019). The trail follows the coast from Filey, north through Whitby, then at Saltburn-by-the-Sea, where it turns inland and across the heath of the Yorkshire Moors to finish at Helmsley. In addition, an authentic ecotourism experience should provide tourists with interpretation and connection to a geosite in terms of its past and present abiotic, biotic and cultural elements, the ABCs of geotourism (Dowling and Newsome 2017).

Easily observed in the steep coastal cliffs around Whitby, the natural geodiversity of the area consists of rocks formed from sediments deposited during the Early and Middle Jurassic Epochs (Cox and Sumbler 2002; North Yorkshire Moors National Park (NYMNP) n.d.; Simms et al. 2004). As the super-continent Pangea began to break up in the Late Tertiary and Early Jurassic, the expanding Tethys Sea inundated the Cleveland Basin, on which Whitby is situated, and the depositional environment changed from fluvial to marine (Simms et al. 2004). The three geological formations from that epoch are the Staithes Sandstone (≈190–185 Mya), the Cleveland Ironstone (≈185–182 Mya) and the Whitby Mudstone (≈175–165 Mya). The Cleveland Ironstone and the Jet Stone and Alum Shales beds of the Whitby Mudstone formation are the most significant strata from a geotourism perspective. Sea levels fell during the transition from the Early to Middle Jurassic (≈175–165 Mya), changing the sediment deposition mechanism in the Whitby area from marine to deposition in a variable fluvial-tidal environment of broad coastal swamps and river deltas (Cox and Slumber 2002; Whyte et al. 2007). River flows and tidal currents waxed and waned across a coastal environment, considered to be akin to the current Niger Delta in West Africa, thus giving rise to the different sequences of sandstones, siltstones, mudstones and limestones of the Ravenscar Group (Livera and Leeder 1981; Whyte et al. 2007).

The geology of Whitby is rich in fossil remains from the age of the dinosaurs, as immortalised in popular culture through the Jurassic Park/World franchise of movies from the late twentieth and early twenty-first centuries (Newsome and Hughes 2017; Padian 1988). As a result, the coast of North Yorkshire, including the area of Whitby, is known as the "Dinosaur Coast" and frequent allusions to those movies appear in materials promoting the area (Parkes 2019; NYMNP n.d.; *Yorkshire Post* 2015). The mudstones that formed from the abiotic and biotic sediments deposited in the marine environment of the Early Jurassic are extremely rich in fossilised biota such as ammonites, belemnites, bivalves, fish, reptiles and fossilised plant remains (Joint National Conservation Committee (JNCC) 2010; Whitby Museum n.d.). Because the Middle Jurassic rocks, which are observed in the upper layers of the coastal cliffs and in the eroded boulders and scree resting on beach platforms around Whitby, are the product of shallow nonmarine depositional environments, bivalve molluscs and fossil plant remains are common, but the remains of large fauna are rare (Whyte et al. 2007). Those conditions were, however, ideal for capturing and preserving dinosaur footprints, including those of early crocodilians, sauropod dinosaurs, early stegosaurian dinosaurs and carnivorous theropod dinosaurs, to name a few (Whyte et al. 2007). Fish and reptile fossils extracted from the cliffs at Whitby were among the first specimens commercially exploited for sale to collectors. Local geologists and palaeontologists were so concerned about the rate of fossil extraction and the loss of geoheritage during late eighteenth and early nineteenth centuries that the Whitby Museum was formed to retain a collection of fossils in Whitby (Whitby Museum n.d.). Primarily because of its fossil geodiversity, much of the natural geoheritage around Whitby is now conserved and protected by statutory designation as geological Sites of Special Scientific Interest (JNCC 2014; Tees Valley RIGS 2014). Accelerated erosion of this coastal geoheritage, arising from the projected rapid sea level rise that is being driven by anthropogenic climate change, will pose a significant geoconservation challenge for the Whitby fossils throughout the twenty-first century and beyond.

The Early Jurassic Jet Stone also provides a case study for the ABCs of geotourism promoted by Dowling and Newsome (2017) and the significance of geodiversity, geoheritage and geoconservation in the urban environs of Whitby. Jewellery made from Whitby Jet has been discovered at Neolithic, Celtic, Roman and Viking archaeological sites from Ireland to Rome (Dean 2007; Stevens 2017). Made from a material 1,000 times rarer than diamonds, Whitby Jet jewellery is carved from pieces of the fossilised trunks of trees related to the Monkey Puzzle tree (*Araucaria araucana*) that today grows on the lower slopes of the south-central Andes in Chile and Argentina (Parkes 2019; White 2019). Growing in the Jurassic forests of Pangea and Gondwana with dinosaurs wandering past, those trees were ancestors of the living-fossil genus Araucaria, which continental drift has dispersed across modern rain forest habitats from South America, through the Pacific Islands and on to eastern Australia (Flenley 1984; Parkes 2019; White 2019). Modern mining and production of Whitby Jet jewellery commenced in the early 1850s, which coincided with the Victorian period of mourning and austerity of dress following the deaths of the Duke of Wellington in 1852 and the Prince Consort Albert in 1861 (Stevens 2017; White 2019). Demand for jet jewellery declined rapidly after the death of Queen Victoria and the industry and geoheritage value of that element of Whitby's geodiversity were all but lost in the twentieth century. The last remaining Victorian jet workshop was rediscovered about 20 years ago in the attic of a derelict house (Campbell 2010; Heritage Jet 2018). Tourist interest in the salvaged and reconstructed workshop first resulted in the creation of The Whitby Jet Heritage Centre and then led to economic development through a resurrected Whitby Jet jewellery industry (Huddersfield 2013; Heritage Jet 2018; Tucker n.d.). A purpose-built museum combining displays of gothic Victorian jewellery and curios and the largest fossilised trunk of Whitby Jet ever discovered was opened in 2019 (Parkes 2019).

Geotourism in Whitby dates to the eighteenth century, when half a century before the railway arrived, wealthy health-conscious Georgians would travel to the town to enjoy the medicinal and tonic qualities of the iron-rich chalybeate springs arising from groundwater in contact with the Cleveland Ironstone (Chrystal 2019; Brodie 2012; Walton 2014). Whitby and the iron-rich springs retained their image as a seaside health spa in the Victorian era and on into the early twentieth century (*Yorkshire Post* 2004; White 1998; White 2019). As with many British seaside resorts, tourist visitation to Whitby declined in the second half of the twentieth century, causing the springs and infrastructure that supported the spa-based geotourism to be forgotten (White 2019; *Yorkshire Post* 2004). Similar to the Victorian jet workshop mentioned above, the last remaining art-deco-decorated Victoria Spa House, which was built in 1860 replacing an earlier spa building built in the 1830s, was also rediscovered in the early 2000s (HE 2019; *Yorkshire Post* 2004). Now protected as a Grade II Listed Building, this geoheritage site is also drawing tourists to visit the conserved urban geoheritage of Whitby (*Yorkshire Post* 2004; *Whitby Gazette* 2010).

In terms of the geoheritage of the built environment in Whitby, the most famous landmark is Whitby Abbey (Figures 36.3 and 36.4). Located on the opposite bank of the River Esk to the town, Whitby Abbey, and associated monastery buildings, was a centre of learning and focal point of political attention with a long history of ruin, re-use and renewed attention. Established by the Anglo-Saxon King of Northumbria in 657, the first monastery was a double priory house (housing both monks and nuns) that was led by Abyss (and later Saint) Hilda (Brindle 2010; EH n.d.; HE 2015). That first monastery is reported to have consisted of 40 small chapels and cells

Figure 36.3 Headstones located in the cemetery area outside Whitby Abbey. They are made from sandstone derived from local quarries. Sedimentary bedding planes are visible as faint diagonal
lines. Weathering of the rock surface over hundreds of years has obliterated the names and
dedications that originally marked the surface of these headstones.

(*monasteria vel oratoria*) that were constructed from local stone (Harrison and Norton 2012; Page
1923). The monastery was first abandoned in the late ninth century, probably because of
repeated Viking raids, with several sources stating that raiders sacked the monastery circa 867
(Atkinson 1879; Brindle 2010; EH n.d.; Harrison and Norton 2012; HE 2015; Page 1923).

The monastery buildings remained in ruins for the next 200 years, until a decade after the
Norman Conquest. Supported by the Norman baron William de Percy and granted charters by
William the Conqueror, in 1077 a community of monks reoccupied the ruined buildings of the
first monastery to found Benedictine Priory, which they called the monastery of St Peter (Atkinson 1879; EH n.d.; Harrison and Norton 2012; Page 1923). Internal politics, and possibly
attacks by pirates, resulted in the site again being abandoned a few years later, only for members
of the same community of monks to return in 1090, to complete construction of a stone Benedictine Abbey in 1109 (EH n.d.; Harrison and Norton 2012; Page 1923). The stone abbey was
rebuilt in the Gothic style in the mid-thirteenth century, building of the nave commenced in
the fourteenth century, but was completed in the fifteenth century (EH n.d.; Page 1923). The
abbey and other buildings on the site remained intact after the Dissolution of the Monasteries
by Henry VIII and on into the eighteenth century, until large sections collapsed circa 1732 and
1736 (EH n.d.; HE 2015).

The building stones of the abbey, however, tell a much older story that dates back to the start
of the Middle Jurassic Epoch approximately 165–170 Mya (White 2019; Whyte et al. 2007).

Figure 36.4   View of Whitby Abbey and cliffs on the nearby coastline. Some of the rocks, forming the cliff line, are rich in fossils. Fossil collectors and beachcombers are able to find fossilised plant remains and an array of fossilised animals including dinosaur remains.

Source: Photograph by D. Newsome.

Middle Jurassic sandstones from quarries on the coast and across the North Yorkshire Moors near Whitby have been used in the construction of many buildings and monuments (NYMNP n.d.; White 2019). Faint impressions of ancient sedimentary strata are visible in the headstones that punctuate the surroundings of Whitby Abbey (Figure 36.3). Many of the headstones, intended as memorials of human death, thus reveal more about the geology of Whitby than about the people buried there! Furthermore, due to the effects of weathering caused by decades of on-shore winds carrying salt-laden spray from the nearby coastline, the headstones reveal a rugged honeycomb pattern on their surface (Figure 36.3). The sandstone building blocks of Whitby Abbey are under similar attack in the harsh coastal conditions, so English Heritage have commenced conservation work to maintain the structural integrity and geoheritage values of the abbey, which demonstrates the importance of geoconservation for historic ruins that are icons of a tourism city (Brindle 2010).

While the ruined Gothic abbey dominates the skyline of the town, the Dissolution of Monasteries under Henry VIII contributed far more to shaping the landscape and geodiversity of Whitby than the later ruins of the Gothic abbey. In the middle ages, alum was an important mineral used in the manufacture of paper and textiles for medicinal purposes (NYMNP n.d.; Appleton 2018). In medieval times, alum was sourced mainly from Italy and the popes tightly controlled its extraction and export from Italy and importation from Turkey under the threat of excommunication (Nef 1936; Günster and Martin 2015). When Henry VIII split from the

Catholic Church and dissolved the monasteries, supplies of alum became restricted, causing both Henry VIII and Elizabeth I to champion the search for and production of this valuable mineral in England (Nef 1936; Jecock 2009). There are reports that alum may have been mined around Whitby from the mid-sixteenth century, which, based on reports from the mid to late nineteenth century, resulted in the mine's owner Sir Thomas Chaloner and his miners being excommunicated by the pope (Curtis 1829; Fox-Strangways 1892 cited in Hobbs et al. 2012; Murray 1867). Those restrictions lead to mining of the Alum Shales of Whitby for 250 years from the reign of Elizabeth I in the mid-fifteenth century, until a synthetic form of the mineral was produced in the mid-nineteenth century under the reign of Queen Victoria (Appleton 2018; White 2019). The Alum Shales were mined on a massive scale, resulting in the complete removal of headlands and creation of broad beach platforms that created "spectacular, almost lunar, scenery at places such as Sandsend and Ravencar", which are located to the north and south of Whitby (NYMNP n.d.; White 2019). In another coincidence, it was the mining of alum that lead to the discovery, extraction and exploitation of the fish and reptile fossils from the Early Jurassic Period discussed early.

The importance of Whitby Abbey and the town centre in the cultural and historical context are further exemplified by its connection with Bram Stoker's novel Dracula, which was written in 1897. A century after the Dracula novel was published, it became the inspiration for a festival that has become a global event that has grown into one of the largest biannual celebrations of Gothic culture on the planet (Love 2019; Paylor 2013). The festival, which is focused on the Victorian Spa Pavilion in an ironic link to the origins of geotourism in Whitby, was instigated with the support of businesses and the community as a means of re-invigorating the tourism industry and economic livelihood of the town (Farr 2017; Paylor 2013). The contribution of this cultural phenomenon in re-invigorating geoheritage-linked tourism is demonstrated by Whitby having been voted the best seaside resort in Britain in 2006 and the estimated £1.1 million per annum that is contributed to the local economy (Farr 2017; BBC News 2012). Images from around the abbey and streets that show the built environment of Whitby feature heavily in the promotion and social media surrounding the festival. Further, the Dracula–Whitby link is now generating academic consideration of the Goth-based cultural aspects of the geoheritage of Whitby (Dobson 2018; Goulding and Saren 2009). Recognising the need to protect the geoheritage values of the built environment and viewscapes that generated the tourism described here and above, the local government authority adopted the Whitby Conservation Area – Character Appraisal & Management Plan in 2014 (Scarborough Borough Council 2019).

This case study illustrates the breadth of the geodiversity, both natural and secondary, available at regional urban destinations such as Whitby. As highlighted, the natural geodiversity of Whitby and the geo-assets of the built environment provide the foundation for cultural, historical and geological tourism experiences, which are the keys to the future prosperity of the community. Other regional centres could catalogue the geo-assets to be found in the built environment and natural areas of their urban environments and look to Whitby as an example of how the abiotic, biotic and cultural values of those sites could be leveraged to work towards the UNSDGs through geotourism. However, this case study also illustrates the past, present and future threats to the geoheritage of Whitby from erosion, over exploitation or simply from being forgotten. We postulate that analogous with biodiversity hotspots, the regional urban geotourism destination of Whitby is a global geodiversity hotspot that requires careful ongoing management to ensure the sustainability of both its internationally significant geotourism industry and the geoheritage on which that industry depends.

Future considerations and recommendations for further policy development

There are a number of challenges in regard to simultaneously protecting and promoting geoheritage in urban environments. Many landforms are affected by long-established town planning policies and urbanisation often leads to damage and destruction of important landforms and geosites. Perhaps an extreme example of this is the complete modification of mountains by flattening the landscape in order to build new cities in mountainous regions in China (Li et al. 2014).

Many authors have reported that geoconservation and public awareness are vital precursors before geotourism can be fully implemented in the urban environment (Bennett et al. 1996; Dowling and Newsome 2018). Reynard et al. (2017) maintain that structured mapping and inventory programmes be put into place in order to identify and document urban geotourism resources. Geotourism itself is seen to need further and more active promotion and this can be implemented via electronic media that contains specific geology and landscape content. Such an example of geoheritage and geotourism promotion in the modern context is the thematic map forwarded by Sacchini et al. (2018). GIS databases, geology and landform maps and tourist trail and activity maps provide a rich avenue of research for the promotion of urban geotourism into the future. Further, for example, Reid (1996) specifies a code of practice in regard to conserving geoheritage alongside urban development. Reid (1996) thus states that developers need to engage in early dialogue with urban managers, consult registers of geological sites and engage with geologists. Moreover, a full impact assessment should be undertaken prior to planned urban development and mitigation measures implemented to prevent damage or loss of geoheritage. Reid (1996) goes on to point out that where urban planning applications affect previously recognised and existing geoheritage sites, planning permission should be refused, or granted subject to avoiding damage or designation of an alternative and conserved geological site.

Barker (1996), noting that wildlife conservation has many similarities with geoconservation, emphasised the importance of engaging with the social and psychological aspects of nature conservation and extending these human dimensions to urban geotourism. It was recognised more than 20 years ago that urban geoheritage and geotourism need to be linked in with floral and faunal amenities, history, archaeology and culture (Bennett et al. 1996). Today many aspects of these factors have now been fully incorporated as a part of geotourism more broadly, as well as in the urban environment (e.g. Gordon 2018a; Gordon 2018b).

As a final message, a vital aspect that needs attention in the future is the management of geological values in urban areas, particularly where geology is promoted as a tourism attraction and recreational resource (Dowling and Newsome, 2017). The careful and appropriate design of tourism infrastructure such as access, viewing facilities and educational signage is essential to reduce degradation from visitor impacts and requires appropriate levels of planning and funding by city authorities. A fully considered visitor management strategy that includes the training and employment of guides will help to ensure that sustainable urban geotourism will become a feature of many tourism cities into the future.

Conclusions

This chapter has highlighted that natural geology, landscape and the geological attributes of buildings either currently form or can form marketable tourism attractions in their own right. Such attractions include, but are by no means limited to, buildings and building stones; landscapes and streetscapes; sequences of sedimentary and other rocks; rock outcrops, mineral deposits, fossils; interpretive and educative displays of geological materials and Earth processes; and cultural events connected to geoheritage. The key to geotourism success, as in all nature-based tourism ventures, lies in the provision of high-quality information and engagement. Geotourism in the urban environment offers the tourist an opportunity to learn about the composition of the Earth and the processes that shape it. Each rock exposure and landscape demonstrates an environmental history and human connections with landscape, such as geological controls and evidence as to how rocks have influenced soils and land use. The contemporary geotourist wants to be inspired and learn about the Earth. Although iconic buildings and landscapes can be very photogenic and rock outcrops aesthetically pleasing it is in knowing that brings high levels of visitor satisfaction. Furthermore, city officials and town and tourism planners are becoming increasingly aware that geological features can be promoted as diversified urban tourism products. However, we also caution that, in many cases, there is much valuable geoheritage within the confines of many cities that needs to be protected and conserved. Thus, if geo-assets are identified and presented to the visitor, as a tourist attraction, then site protection and adequately managed access are essential. Furthermore, as stated previously, the key to tourist satisfaction is in learning. Through learning, the tourist can relate to what is being presented to them and this enables them to fully experience the geoheritage that is in front of them. This is the future of "appreciating geology in the urban environment" and its role in expanding the gamut of attractions that an international tourism city can offer.

References

Alexandrowicz, Z. and Alexandrowicz, S.W., 2004. Geoparks: the most valuable landscape parks in southern Poland. *Polish Geological Institute Special Papers*, 13, 49–56.

Alexandrowicz, Z., Urban, J. and Miśkiewicz, K., 2009. Geological values of selected Polish properties of the UNESCO World Heritage List. *Geoheritage*, 1(1), 43–52, https://doi.org/10.1007/s12371-009-0004-y.

Appleton, P., 2018. The alum industry of north-east Yorkshire. Available from: https://east-clevelands-industrial-heartland.co.uk/2018/01/17/the-alum-industry-of-north-east-yorkshire/ [Accessed 13 May 2019].

Atkinson, C.J., ed., 1879. *Chartulary of Whiteby*. Durham: Surtees Society. Available from: https://archive.org/details/cartulariumabba00abbegoog/page/n11 [Accessed 30 April 2019].

Barker, G., 1996. Earth science sites in urban areas: the lessons from wildlife conservation. In: M.R. Bennett, P. Doyle, J.G. Larwood and C.D. Prosser, *Geology on Your Doorstep: The Role of Urban Geology in Earth Heritage Conservation*. London: The Geological Society, 147–154.

BBC News, 2012. Whitby's Goth visitors' £1m seaside bounty. Available from: www.bbc.com/news/av/uk-england-york-north-yorkshire-20188996/whitby-s-goth-visitors-1m-seaside-bounty [Accessed 14 May 2019].

Bennett, M.R., Doyle, P., Larwood, J.G. and Prosser, C.D., 1996. *Geology on Your Doorstep: The Role of Urban Geology in Earth Heritage Conservation*. London: The Geological Society.

Bétard, F. and Peulvast, J.P., 2019. Geodiversity hotspots: concept, method and cartographic application for geoconservation purposes at a regional scale. *Environmental Management*, 1–13, https://doi.org/10.1007/s00267-019-01168-5.

Black, A. and Black, C., 1842. *Black's Picturesque Tourist of Scotland*. 2nd ed. Edinburgh: Adam and Charles Black.

Black, A. and Black, C., 1843. *Black's Economical Guide through Edinburgh*. 3rd ed. Edinburgh: Adam and Charles Black.

Brindle, S., 2010. *Whitby Abbey – Guidebook*. Bristol: English Heritage.

Brodie, A., 2012. Scarborough in the 1730s: spa, sea and sex. *Journal of Tourism History*, 4(2), 125–153, https://doi.org/10.1080/1755182X.2012.697488.

Burliga, S., Krzywiec, S, Dąbroś, P., Przybyło, K., Włodarczyk, J., Źróbek, E. and Słotwiński, M., 2018. Salt tectonics in front of the Outer Carpathian thrust wedge in the Wieliczka area (S Poland) and its exposure in the underground salt mine. *Geology, Geophysics and Environment*, 44(1), 71–90.

Campbell, S., 2010. Whitby: the return of the jet age. Available at: www.telegraph.co.uk/travel/columnists/sophie-campbell/7256114/Whitby-The-return-of-the-jet-age.html [Accessed 19 May 2019].

Cheung, L.T.O., 2016. The effect of Geopark visitors' travel motivations on their willingness to pay for accredited geo-guided tours. *Geoheritage*, 8(3), 201–209, https://doi.org/10.1007/s12371-015-0154-z.

Chrystal, P., 2019. *Whitby at Work: People and Industries through the Years*. Gloucestershire: Amberley Publishing.

Cox, B.M. and Sumbler, M.G., 2002. *British Middle Jurassic Stratigraphy*, Geological Conservation Review Series, No. 26. Peterborough: Joint Nature Conservation Committee.

Curtis, T., 1829. *The London Encyclopaedia: Volume 1*. London: Thomas Tegg.

Czapowski, G., 1994. The Middle Badenian rock salts in the Carpathian Foredeep: characteristics, origin and economic value. *Geological Quarterly*, 38(3), 513–526, https://doi.org/10.1016/0148-9062(96)87344-X.

Dade, J. and Carter, C., 2015. *Report on the Economy of the Whitby*. Available from: www.northyorkmoors.org.uk/planning/Sirius-Minerals-Polyhalite-Mine-Woodsmith-Mine/35190-CLon066R-Economy-of-Whitby.pdf [Accessed 28 April 2019].

Dean, D.R., 1981. The age of the earth controversy: beginnings to Hutton. *Annals of Science*, 38(4), 435–456.

Dean, W.T., 2007. Yorkshire jet and its links to Pliny the Elder. *Proceedings of the Yorkshire Geological Society*, 56(4), 261–265.

Del Lama, E.A., 2018. Urban geotourism with an emphasis on the city of São Paulo, Brazil. In: R.K. Dowling and D. Newsome, eds. *Handbook of Geotourism*. Cheltenham: Edward Elgar Publishing, 210–220.

Del Lama, E.A., Bacci, D.D.L.C., Martins, L., Garcia, M.D.G.M. and Dehira, L.K., 2015. Urban geotourism and the old centre of São Paulo City, Brazil. *Geoheritage*, 7(2), 147–164, https://doi.org/10.1007/s12371-014-0119-7.

Dobson, E., 2018. Ruby lips and Whitby Jet: Dracula's language of jewels. *Gothic Studies*, 20(1–2), 124–139.

Dowling, R.K. and Newsome, D., 2006. Preface. In: R.K. Dowling and D. Newsome, eds. *Geotourism*. Oxford: Elsevier, xxv–xxvi.

Dowling, R.K. and Newsome, D., eds., 2006c. *Geotourism*. Oxford: Elsevier.

Dowling, R. and Newsome, D., eds., 2010. *Global Geotourism Perspectives*. Woodeaton: Goodfellow Publishers.

Dowling, R.K. and Newsome, D., 2017. Geotourism destinations: visitor impacts and site management considerations. *Czech Journal of Tourism*, 6(2), 111–129.

Dowling, R. and Newsome, D., eds., 2018. *Handbook of Geotourism*. Cheltenham: Edward Elgar Publishing.

Edinburgh Geological Society (EGS), n.d.a. Edinburgh's Geological Heritage. Available from: www.edinburghgeolsoc.org/edinburghs-geology/edinburghs-geology-sites/ [Accessed 28 April 2019].

Edinburgh Geological Society (EGS), n.d.b. Edinburgh's Geology. Available from: www.edinburghgeolsoc.org/edinburghs-geology/ [Accessed 28 April 2019].

Edinburgh Geological Society (EGS), n.d.c. Edinburgh Geological Society. Available from: www.edinburghgeolsoc.org/ [Accessed 28 April 2019].

Edinburgh Geological Society (EGS), n.d.d. Publications. Available from: www.edinburghgeolsoc.org/publications/ [Accessed 30 April 2019].

Edinburgh Geological Society (EGS), n.d.e. Introductory Excursion: Arthur's Seat. Available from: www.edinburghgeolsoc.org/excursions/arthurs-seat/ [Accessed 29 April 2019].

English Heritage (EH), n.d. Whitby Abbey: History and Stories. Available from: www.english-heritage.org.uk/visit/places/whitby-abbey/history-and-stories/ [Accessed 29 April 2019].

Erfurt, P., 2018. Geotourism development and management in Volcanic regions. In: R.K. Dowling and D. Newsome, eds. *Handbook of Geotourism*, Cheltenham: Edward Elgar Publishing, 152–167.

Erfurt-Cooper, P., 2014. Wellness tourism: a perspective from Japan. In: C. Voigt, and C. Pforr, eds. *Wellness Tourism: A Destination Perspective*. Abingdon: Routledge, 235–254.

Erfurt-Cooper, P. and Cooper, M., 2009. *Health and Wellness Tourism: Spas and Hot Springs*. Cheltenham: Channel View Publications.

Farr, M., 2017. Decline beside the seaside: British seaside resorts and declinism. In: D. Harrison and R. Sharpley, eds. *Mass Tourism in a Small World*. Croydon: CAN International, 105–119.

Flenley, J.R., 1984. Time scales in biogeography. In: J.A. Taylor, ed. *Biogeography: Recent Advances and Future Directions*. Totowa, NJ: Barnes and Noble Books, 63–105.

Fowler, P., 2018. HK has already plenty of potential offbeat tourist attractions. *China Daily* – HK Edition, 21 February, p. 13.

Fox-Strangways, C., 1892. *The Jurassic Rocks of Britain. Volumes 1 and 2 Yorkshire*. Memoir of the Geological Survey of Great Britain. London: H.M. Stationery Office.

Global Geoparks Network (GGN), 2017. GGN – Geopark Annual Report 2017. Available from: http://globalgeoparksnetwork.org/wp-content/uploads/2017/06/GGN-Annual-reports-2017.pdf [Accessed 29 April 2019].

Gordon, J.E., 2018a. Geotourism and cultural heritage. In: R. Dowling and D. Newsome, eds. *Handbook of Geotourism*. Cheltenham: Edward Elgar Publishing, 61–75.

Gordon, J.E., 2018b. Geoheritage, geotourism and the cultural landscape: enhancing the visitor experience and promoting geoconservation. *Geosciences*, 8(4), 136.

Górska-Zabielska, M. and Zabielski, R., 2018. Geotourism Development in an Urban Area based on the Local Geological Heritage (Pruszków, Central Mazovia, Poland). In: M.J. Thornbush and C.D. Allen, eds. *Urban Geomorphology*. Amsterdam: Elsevier, 37–54.

Goulding, C. and Saren, M., 2009. Performing identity: an analysis of gender expressions at the Whitby goth festival. *Consumption, Markets and Culture*, 12(1), 27–46, https://doi.org/10.1080/10253860802560813.

GovHK, 2018. Hong Kong: The Facts – Tourism. Available from: www.gov.hk/en/about/abouthk/factsheets/docs/tourism.pdf [Accessed 29 April 2019].

Gray, M., 2018a. Geodiversity: the backbone of geoheritage and geoconservation. In: E. Reynard and J. Brilha, eds. *Geoheritage: Assessment, Protection, and Management*. Amsterdam: Elsevier, 13–25.

Gray, M., 2018b. Geodiversity, geoheritage, geoconservation and their relationship to geotourism. In: R.K. Dowling and D. Newsome, eds. *Handbook of Geotourism*. Cheltenham: Edward Elgar Publishing, 48–60.

Günster, A. and Martin, S., 2015. A holy alliance: collusion in the Renaissance Europe alum market. *Review of Industrial Organization*, 47(1), 1–23.

Guo, W. and Chung, S., 2019. Using tourism carrying capacity to strengthen UNESCO global geopark management in Hong Kong. *Geoheritage*, 11(1), 193–205, https://doi.org/10.1007/s12371-017-0262-z.

Hallett, D., 2002. The Wieliczka Salt Mine. *Geology Today*, 18(5), 182–185, https://doi.org/10.1046/j.0266-6979.2003.00365.x.

Harrison, S. and Norton, C., 2012. Lastingham and the architecture of the Benedictine Revival in Northumbria. In: D. Bates, ed. *Anglo~Norman Studies XXXIV, Proceedings of the Battle Conference, 18–22 July 2011 University of York*. Woodbridge: The Boydell Press, 63–104.

Heritage Jet, 2018. The Whitby Jet Heritage Centre. Available from: www.whitbyjet.co.uk/ [Accessed 13 may 2019].

Historic England (HE), 2015. PastScape: Whitby Abbey. Available from: www.pastscape.org.uk/hob.aspx?hob_id=29830 [Accessed 29 April 2019].

Historic Environment Scotland (HES), 2015. *Statement of Significance: Holyrood Park*. Edinburgh: Historic Environment Scotland.

Historic Environment Scotland (HES), 2018. *Statement of Significance: Holyrood Park*. Edinburgh: Historic Environment Scotland.

Hobbs, P.R.N., Entwisle, D.C., Northmore, K.J., Sumbler, M.G., Jones, L.D., Kemp, S., Self, S., Barron, M. and Meakin, J.L., 2012. *Engineering Geology of British Rocks and Soils: Lias Group*. Nottingham: British Geological Survey.

Hong Kong UNESCO Global Geopark (HKUGGp), 2018. Available from: www.geopark.gov.hk/en_index.htm [Accessed 28 April 2019].

Huddersfield, D.S., 2013. An Original Whitby Jet Workshop: Review of The Whitby Jet Heritage Centre. Available at www.tripadvisor.com.au/ShowUserReviews-g186345-d4759203-r174538475-The_Whitby_Jet_Heritage_Centre-Whitby_Scarborough_District_North_Yorkshire_Engla.html [Accessed 13 May 2019].

Jecock, M., 2009. A fading memory: the North Yorkshire coastal alum industry in the light of recent analytical field survey by English Heritage. *Industrial Archaeology Review*, 31(1), 54–73.

Joint Nature Conservation Committee (JNCC), 2010. Geological Conservation Review: Whitby Coast: Eat Pier to Whitestone Point. Available from: http://jncc.defra.gov.uk/pdf/GCRDB/GCRsiteaccount9036.pdf [Accessed 23 April 2019].

Joint Nature Conservation Committee (JNCC), 2014. Protected Areas Designations Directory. Available from: http://jncc.defra.gov.uk/page-1527 [Accessed 13 May 2019].

Kubalíková, L., Kirchner, K. and Bajer, A., 2017. Secondary geodiversity and its potential for urban geotourism: a case study from Brno City, Czech Republic. *Quaestiones Geographicae*, 36(3), 63–73, https://doi.org/10.1515/quageo-2017-0024.

Li, P., Qian, H. and Wu, J., 2014. Environment: accelerate research on land creation. *Nature News*, 510(7503), 29–31.

Lipecki, T., Jaśkowski, W., Gruszczyński, W., Matwij, K., Matwij, W. and Ulmaniec, P., 2016. Inventory of the geometric condition of inanimate nature reserve Crystal Caves in "Wieliczka" Salt Mine. *Acta Geodaetica et Geophysica*, 51(2), 257–272.

Livera, S.E. and Leeder, M.R., 1981. The Middle Jurassic Ravenscar Group ("Deltaic Series") of Yorkshire: recent sedimentological studies as demonstrated during a field meeting 2–3 May 1980. *Proceedings of the Geologists' Association*, 92(4), 241–250.

Lonely Planet, n.d. Top Things to Do in Kraków. Available from: www.lonelyplanet.com/poland/krakow/top-things-to-do/a/poi/360295 [Accessed 27 April 2019].

Love, L., 2019. Whitby Goth Weekend: hundreds descend on seaside town as first event of 2019 gets underway. *TeessideLive*. Available from: www.gazettelive.co.uk/whitby-goth-weekend-hundreds-descend-16120946 [Accessed 14 May 2019].

McAdam, D., 1986. *Geological Guide to the Arthur's Seat Volcano*. Edinburgh: Edinburgh Geological Society.

McAdam, A.D. and Clarkson, E.N.K., 1996. *Lothian Geology: An Excursion Guide*. Edinburgh: Edinburgh Geological Society.

Miller, A., n.d. GEOWALKS: Arthur's Seat. Available from: www.geowalks.co.uk/arthurs-seat/ [Accessed 29 April 2019].

Miller, A.D., 2012. *Investigating Holyrood Park*. Edinburgh: Historic.

Monaghan, A.A., Browne, M.A. and Barfod, D.N., 2014. An improved chronology for the Arthur's Seat volcano and Carboniferous magmatism of the Midland Valley of Scotland. *Scottish Journal of Geology*, 50(2), 165–172.

Murray, J., 1867. *Handbook for Travellers in Yorkshire*. London: John Murray.

Myers, N., Mittermeier, R.A., Mittermeier, C.G., da Fonseca, G.A.B. and Kent, J., 2000. Biodiversity hotspots for conservation priorities. *Nature*, 403, 853–858.

National Trails, 2019. Cleveland Way. Available from: www.nationaltrail.co.uk/cleveland-way [Accessed 12 November 2019].

Nef, J.U., 1936. A comparison of industrial growth in France and England from 1540 to 1640: III. *Journal of Political Economy*, 44(5), 643–666.

Newsome, D. and Dowling, R., 2018. Geoheritage and geotourism. In: E. Reynard and J. Brilha, eds. *Geoheritage*. Chennai: Elsevier, 305–322.

Newsome, D., Dowling, R. and Leung, Y.F., 2012. The nature and management of geotourism: a case study of two established iconic geotourism destinations. *Tourism Management Perspectives*, 2, 19–27.

Newsome, D. and Hughes, M., 2017. Jurassic World as a contemporary wildlife tourism theme park allegory. *Current Issues in Tourism*, 20(13), 1311–1319.

Ng, Y. and Choi, C.M.C., 2009. The role of NGO in the establishment of geopark and promotion of geoconservation: one example from Hong Kong. *Earth Science Frontier*, 1.

Ng, Y., Fung, L.W. and Newsome, D., 2010. Hong Kong Geopark: uncovering the geology of a metropolis. In: R. Dowling, and D. Newsome, eds. *Global Geotourism Perspectives*. Woodeaton: Goodfellow Publishers.

North Yorkshire Moors National Park (NYMNP), n.d. Geology/Coast. Available from: www.northyorkmoors.org.uk/discover/geology [Accessed 29 April 2019].

Padian, K., ed., 1988. *The Beginning of the Age of Dinosaurs: Faunal Change across the Triassic–Jurassic Boundary*. Cambridge: Cambridge University Press.

Page, W., 1923. Victoria County History: A History of Yorkshire North Riding, Volume 2, Parishes, Whitby. Available from: www.british-history.ac.uk/vch/yorks/north/vol. 1 [Accessed 22 April 2019].

Palacio-Prieto, J.L., 2015. Geoheritage within cities: urban geosites in Mexico City. *Geoheritage*, 7(4), 365–373, https://doi.org/10.1007/s12371-014-0136-6.

Panizza, M. and Piacente, S., 2009. Cultural geomorphology and geodiversity. In: E. Reynard, P. Coratza and G. Regolini-Bissig, eds. *Geomorphosites*. Munich: Arrow Publishing, 35–48.

Parkes, L., 2019. Queen Victoria's Favourite Gem Gets Its Own Yorkshire Museum. Available from: www.lonelyplanet.com/news/2019/02/01/queen-victoria-jet-yorkshire-museum/ [Accessed 13 May 2019].

Paylor, T., 2013. Whitby Goth Weekend. The Whitby Guide. Available from: www.thewhitbyguide.co.uk/whitby-goth-weekend/ [Accessed 13 May 2019].

Pile, T., 2017. The good, bad and ugly sides to being a tourist in Edinburgh. *Post Magazine*, 2 September. Available from: www.scmp.com/magazines/post-magazine/travel/article/2108685/good-bad-and-ugly-sides-being-tourist-edinburgh [Accessed 29 April 2019].

Raveloson, M.L.T., Newsome, D., Golonka, J., di Cencio, A. and Randrianaly, H.N., 2018. The contribution of paleontology in the development of geotourism in northwestern Madagascar: A preliminary assessment. *Geoheritage*, 10(4), 731–738, https://doi.org/10.1007/s12371-018-0323-y.

Reid, C., 1996. A code of practice for geology and development in the urban environment. In: M.R. Bennett, P. Doyle, J.G. Larwood and C.D. Prosser, *Geology on Your Doorstep: The Role of Urban Geology in Earth Heritage Conservation*. London: The Geological Society, 147–154.

Reynard, E., Pica, A. and Coratza, P., 2017. Urban geomorphological heritage: an overview. *Quaestiones geographicae*, 36(3), 7–20.

Rozycki, P. and Dryglas, D., 2016. Directions of the development of tourism mining on the example of mines in Poland. *Acta Geoturistica*, 7(2), 14–21.

Sacchini, A., Imbrogio Ponaro, M., Paliaga, G., Piana, P., Faccini, F. and Coratza, P., 2018. Geological landscape and stone heritage of the Genoa Walls Urban Park and surrounding area (Italy). *Journal of Maps*, 14(2), 528–541.

Scarborough Borough Council, 2019. Whitby Conservation Area. Available from: www.scarborough.gov.uk/home/planning/conservation/conservation-area-appraisals-and-management-plans/whitby-conservation [Accessed 14 May 2019].

Scottish Natural Heritage (SNH), 2011. *Arthur's Seat Volcano, Site of Special Scientific Interest, Site Management Statement*. Edinburgh: Scottish Natural Heritage.

Sharples, C., McIntosh, P. and Comfort, M., 2018. Geodiversity and geoconservation in land management in Tasmania: a top-down approach. In: E. Reynard and J. Brilha, eds. *Geoheritage: Assessment, Protection, and Management*. Amsterdam: Elsevier, 355–371.

Simms, M.J., Chidlaw, N., Morton, N. and Page, K.N., 2004. *British Lower Jurassic Stratigraphy*, Geological Conservation Review Series, No. 30. Peterborough: Joint Nature Conservation Committee.

Simpson, G. and Newsome, D., 2017. Environmental history of an urban wetland: from degraded colonial resource to nature conservation area. *Geo: Geography and Environment*, 4(1), e00030. https://doi.org/10.1002/geo2.30.

South China Morning Post, 2015. More Hong Kong Geopark residents leaving ancestral homes. Available from: www.scmp.com/comment/letters/article/1858047/more-hong-kong-geopark-residents-leaving-ancestral-homes [Accessed 16 May 2019].

Stevens, P., 2017. Early medieval jet-like jewellery in Ireland: production, distribution and consumption. *Medieval Archaeology*, 61(2), 239–276.

Tees Valley RIGS, 2014. Quarterly Newsletter, Issue No. 3, June 2014. Available from: www.northern-england-geology.co.uk/tv-rigs-newsletters/3-tv-rigs-newsletter-june-2014.pdf [Accessed 13 May 2019].

Trip Advisor, 2019a. The 10 Best Things to Do in Krakow. Available from: www.tripadvisor.com.au/Attractions-g274772-Activities-Krakow_Lesser_Poland_Province_Southern_Poland.html [Accessed 27 April 2019].

Trip Advisor, 2019b. Wieliczka Salt Mine Guided Tour from Krakow. Available from: www.tripadvisor.com.au/AttractionProductReview-g274772-d11452686-or15-Wieliczka_Salt_Mine_Guided_Tour_from_Krakow-Krakow_Lesser_Poland_Province_So.html [Accessed 27 April 2019].

Trip Advisor, 2019c. Things to Do in Edinburgh. Available at: www.tripadvisor.com.au/Attractions-g186525-Activities-Edinburgh_Scotland.html [Accessed 27 April 2019].

Tucker, R., n.d. Gem Scene: Get Ready for the Jet Set!! Available from: www.gemscene.com/whitby-jet.html [Accessed 13 May 2019].

UNESCO World Heritage Centre, 1998. World Heritage Committee Removes Old City of Dubrovnik and Wieliczka Salt Mine from Its List of Endangered Sites. Available from: https://whc.unesco.org/en/news/147/ [Accessed 29 April 2019].

UNESCO World Heritage Centre, 2019. Underground Europe: Royal Salt Mines in Wieliczka and Bochnia, Poland. Available from: https://visitworldheritage.com/en/eu/royal-salt-mines-in-wieliczka-and-bochnia-poland/17a0828a-e10a-4ea5-a2b5-4664606c0b23 [Accessed 29 April 2019].

United Nations World Tourism Organization (UNWTO), 2019. Tourism and the SDGs. Available from: www2.unwto.org/content/tourism-and-sdgs [Accessed 31 March 2019].

Viator, 2019a. The Top Things to Do in Krakow 2019. Available from: www.viator.com/Krakow/d529 [Accessed 27 April 2019].

Viator, 2019b. Arthur's Seat Tours. Available from: www.viator.com/en-AU/Edinburgh-attractions/Arthurs-Seat/d739-a1413 [Accessed 29 April 2019].

Walton, J.K., 2014. *Mineral Springs Resorts in Global Perspective: Spa Histories*. Abingdon: Routledge.

Wieliczka Salt Mine, n.d.a. Wieliczka Salt Mine: The Mine of the Past and of Today. Available from: www.wieliczka-saltmine.com/about-the-mine/the-mine-of-the-past-and-of-today [Accessed 27 April 2019].

Wieliczka Salt Mine, n.d.b. Wieliczka Salt Mine: More than Half a Million Tourists Visited the Wieliczka Mine. Available from: www.wieliczka-saltmine.com/news/visiting/more-than-half-a-million-tourists-visited-the-wieliczka-mine [Accessed 27 April 2019].

Wieliczka Salt Mine, n.d.c. Wieliczka Salt Mine: Crystal Grottoes. Available from: www.wieliczka-saltmine.com/about-the-mine/the-mine-of-the-past-and-of-today/crumbs-of-knowledge/crystal-grottoes [Accessed 27 April 2019].

Wieliczka Salt Mine, n.d.d. Wieliczka Salt Mine: Health Resort. Available from: https://health-resort.wieliczka-saltmine.com/ [Accessed 29 April 2019].

Whitby Gazette, 2010. Spa Well house opens to public. Available from: www.whitbygazette.co.uk/whats-on/spa-well-house-opens-to-public-1-2365836 [Accessed 8 May 2019].

Whitby Museum, n.d. Museum Collections. Available from: https://whitbymuseum.org.uk/whats-here/collections/ [Accessed 13 May 2019].

White, A., 1998. The Victorian development of Whitby as a seaside resort. *Local Historian-London*, 28, 78–93.

White, A., 2019. *A History of Whitby*. Paperback (3rd) ed. Stroud: The History Press.

Whyte, M.A., Romano, M. and Elvidge, D.J., 2007. Reconstruction of Middle Jurassic dinosaur-dominated communities from the vertebrate ichnofauna of the Cleveland Basin of Yorkshire, UK. *Ichnos*, 14(1–2), 117–129.

Yorkshire Post, 2004. Yorkshire's forgotten treasures open doors for heritage events. Available from: www.yorkshirepost.co.uk/news/latest-news/yorkshire-s-forgotten-treasures-open-doors-for-heritage-events-1-2548536 [Accessed 8 May 2019].

Yorkshire Post, 2015. Yorkshire coast may have been Britain's Jurassic Park. Available from: www.yorkshirepost.co.uk/news/latest-news/yorkshire-coast-may-have-been-britain-s-jurassic-park-1-7286772 [Accessed 13 May 2019].

37

DOLPHINS IN THE CITY

Greg Simpson, Jessica Patroni, David Kerr, Jennifer Verduin and David Newsome

Introduction

Newsome (2017) makes the point that wildlife tourism at any location comprises an intertwining of ecological, social and economic circumstances with, at times, political ramifications in regard to promotion, funding of tourism development and conservation surrounding the species in question. The Dolphin Discovery Centre at Bunbury is no exception to the before-mentioned circumstances and, because of the iconic nature of the species, has also been the focus of ongoing research in regard to the human dimensions of wild dolphin tourism (for example, Cong et al., 2016, 2017a, 2017b; Patroni, 2018; Patroni et al., 2018, 2019). Furthermore, what can be regarded as an important wildlife tourism experience might be regarded as somewhat unusual when located in a city environment. This chapter, therefore, goes on to provide an account of a premier wildlife tourism attraction, taking place at the Dolphin Discovery Centre, located within the regional city of Bunbury, Western Australia.

The Dolphin Discovery Centre currently provides the opportunity for visitors and locals to engage in a wildlife experience focused on wild Indo-Pacific Bottlenose Dolphins that are present where the shoreline meets the city. Over time a formalised beach-based dolphin interaction has developed, an activity that initiated the development of the Dolphin Discovery Centre, and consequently attracted visitors and locals to view dolphins from the beach while standing knee-deep in the water in front of the Centre. The first part of this chapter thus explores the history and development of the Dolphin Discovery Centre and highlights that a significant wildlife tourism attraction is possible within an urban centre when appropriate recognition, tourism management, research and environmental considerations are put into place.

The Dolphin Discovery Centre, Bunbury, Western Australia

Background and origin of the DDC

The Dolphin Discovery Centre (DDC) is situated on the southeastern shore of Koombana Bay in the regional city of Bunbury, Western Australia (Figure 37.1). Located approximately 180 kilometres south of the state capital of Perth, the resident population of Greater Bunbury is currently estimated to be 67,000 people. Bunbury is the second largest urban centre in Western

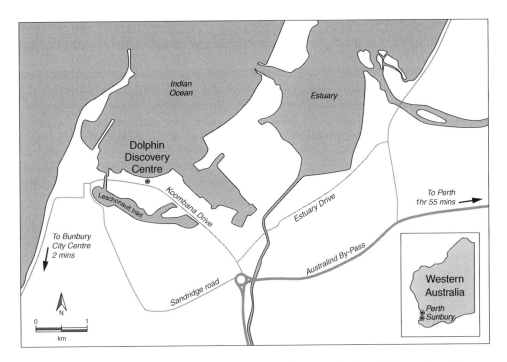

Figure 37.1 Location of Dolphin Discovery Centre, Koombana Bay, Bunbury, Western Australia.

Australia. In combination with its coastal location, accessibility and mild Mediterranean climate Bunbury has become a major regional tourist destination in southwestern Australia (Gentilli, 1989; Australian Bureau of Statistics (ABS), 2017; City of Bunbury (COB), 2015; Fenech, 2011). As with coastal tourism destinations in other Mediterranean climate regions, the peak of the highly seasonal Bunbury tourism market is focused on the summer period, particularly during public holidays. In the case of southwestern Australia this is during December and January (Patroni, 2018). Accordingly, summer visitors to Bunbury can experience Australian beach culture by participating in marine-focused tourism activities such as sunbathing and swimming, surfing, snorkelling and scuba diving, recreational boating and fishing, and kayaking or paddle boarding (Fenech, 2011; COB, 2017a).

Visiting Bunbury has the added attraction of interacting with the wild dolphins that frequent the coastal waters of Koombana Bay. Promoting itself as "Australia's Premier Dolphin Experience", the DDC is one of four destinations in Western Australia that offers visitors a government licensed opportunity to engage with wild dolphins either through controlled beach-based interaction and boat cruises and/or dolphin swims. All of the above-mentioned dolphin experiences are offered by the DDC and available to visitors within only a short walk from the central shopping, entertainment and business district of the city of Bunbury and tourist accommodation adjacent to Koombana Bay (TripAdvisor, 2017; COB, 2017b).

The origin of dolphin tourism in Bunbury began with a community member who started feeding wild dolphins (DDC, 2015a, 2015b). During the mid-1960s, local resident Mrs Evelyn Smith would feed wild Indo-Pacific Bottlenose Dolphins (*Tursiops aduncus*) from a small jetty near her home on the Leschenault Inlet at a location directly south of the current DDC. When Mrs Smith passed away in the early 1970s, regular feeding of the dolphins ceased. However, some dolphins continued to be fed by members of the public. In the mid-1980s a group of local businessmen and community members started to consider the possibility of establishing a new tourism attraction based on the legacy of Mrs Smith and her history with the local dolphin population (Zylines, 1989).

In 1989 the newly established Bunbury Dolphin Trust hired a dolphin specialist who began feeding and studying the local dolphins that visited or stayed in Koombana Bay. The work carried out by the Trust, dolphin specialist and volunteers led to the establishment of the beach-based dolphin *Interaction Zone* in 1990 (DDC, 2015c). This zone was an area of the beach that the dolphins visited regularly and under the management of the dolphin specialist the public were able to stand knee-deep in the ocean and observe the dolphins. In the early 1990s the infrastructure consisted of a caravan near the beach, but in 1994 a purpose-built Dolphin Discovery Centre (DDC) was opened under, the then, state government's "Living Windows Program". This was to enable tourists and members of the community to interact, enjoy and learn about the group of five to six dolphins that regularly visited the Zone.

Management of wildlife tourism at the DDC

A minimal interference approach is applied to dolphin interactions and is specified according to the DDC's Dolphin Interaction License, issued by the Department of Biodiversity, Conservation and Attractions (DBCA). Under this licence dolphins may receive a strictly management-controlled minimalistic food reward (500 grams) during the interaction experience; however, DDC management restricts this further to 375 grams delivered to just nine of the 160 dolphins that may be in the area during peak season. This small fish reward increases the possibility of visitors being able to see and learn about dolphins in the wild. Furthermore, while there are currently only nine dolphins that are reward fed from the Interaction Zone a number of these are infrequent visitors (Figure 37.2a). The management feeding style at the DDC aims to minimise the potential of conditioning the dolphins that visit Koombana Beach so that their presence is therefore not entirely predictable, which differs from other Australian wild dolphin experiences, such as those that have been set up at Monkey Mia, Tangalooma and Tin Can Bay (Bach & Burton, 2017; Orams & Hill, 1998). The DDC conducts feeding discretely and dolphins are fed only by trained DDC volunteers, standing either side of visitors, waiting in the interaction line (Figure 37.2b). Most visitors would be unaware that some dolphins are being rewarded. The reward feeding at the DDC is highly regulated with only 350 g of local species of fish provided per day, which is a small amount as compared with the 8–14 kg of food that adult dolphins require on a daily basis (DDC, 2015c). Further, DDC staff and volunteers record data about the environmental conditions and dolphin engagement in the interaction and this information is kept in a database to assist with expanding knowledge about the Koombana Bay dolphins. This information is provided to the DBCA as specified in the DDC licence conditions.

The reason the dolphins of Koombana Bay continue to visit the Interaction Zone today is controversial and remains unclear (Mann et al., 2018; Patroni et al., 2018, 2019; Senigaglia et al., 2019). However, research conducted by the DDC suggests that the small amount of food that a small number of visiting dolphins (nine) receive as a reward for their visit is not the only attraction for the dolphins (DDC, 2015c). There are many dolphins that visit the zone regularly that do not receive any fish from the DDC and many of them stay for extended periods of time and interact with the human visitors. The offspring of dolphins who are rewarded at the Interaction Zone tend to be brought to the zone by their mothers and many continue visiting once weaned. Sick and injured dolphins also treat the beach as a safe haven, with some repeatedly visiting during periods of illness or injury. A weekly dolphin visitation chart is maintained on

Figure 37.2 The Dolphin Discovery Centre beach–based dolphin interaction Zone.

Source: Photographs courtesy of David Kerr/DDC.

Note
The top photograph (37.2a) shows tourists lining up to view wild dolphins. Note the passage of shipping into the port of Bunbury. The bottom photograph (37.2b) depicts a tourist with a DDC volunteer (dressed in red top) who supervises the interaction.

site to identify the dolphins that are visiting the beach as a guide for people who visit the Centre (DDC, 2015c). Records are also maintained daily on the number of dolphins that visit the Interaction Zone, their identities matched with dolphin profiles, the time and duration of visits as well as the number of visitors participating in the interaction experience along with weather conditions including water temperature and sea conditions.

Alongside the beach-based dolphin interaction, which, due to the un-conditioned (no dependency on humans for food) nature of the experience, cannot guarantee the dolphins will visit every day, the DDC offers two boat-based options that allow visitors to view the dolphins out in the wider Koombana Bay. The first of these experiences is an eco-cruise, which is primarily a dolphin-watching exercise. The eco-cruise takes visitors to the areas in which the dolphins are known to frequent and where dolphin sightings are guaranteed. The boats are driven in a manner that avoids disruption to the dolphins' travel paths and natural activity by keeping a safe distance and observing the dolphins swimming in the vicinity of the boat. The DDC eco-cruise is operated by a trained guide, who provides information on dolphin biology and behaviour and also delivers conservation messages to the people on-board. General information regarding Koombana Bay and history of the area is also provided.

The second boat-based experience is the "swim-with" tour, in which visitors leave the boat under supervision and are placed in the water near the wild dolphins. Being one of the few operations with a licence to have people swim with dolphins and this experience being many people's desired *once in a lifetime goal*, the DDC has become a major attraction to many visitors to Australia's southwest. The swim tours operated by the DDC, much like the beach-based dolphin interaction, are highly managed and controlled in order to protect the welfare of the dolphins. For example, the behaviour of the dolphins is assessed before allowing the swimmers to enter the water and the swimmers only enter when the dolphins are either resting or "playing". When the dolphins are feeding, fighting or travelling, swimmers stay on the boat so as not to interrupt these important dolphin behaviours and also to ensure the safety of the swimmers. When an appropriate group of dolphins is located the swimmers enter the water some distance away from the dolphins and the dolphins have the choice to come to the swimmers and have an interaction. There is no chasing or following after the dolphins allowed. Educational and safety information is provided throughout the tour and instructions on how to behave are clearly explained.

Significance of the Dolphin Discovery Centre as a community educational resource

As previously mentioned, the DDC provides tourists and local residents with a rare urban-based nature tourism experience by offering up-close and personal interactions with the resident population of wild bottlenose dolphins and through aquarium exhibits. Moreover, the DDC aims to optimise visitor satisfaction by incorporating an educational component, which is known to hold a high importance to the overall visitor experience in wildlife tourism (Orams, 1997; Sitar et al., 2017; Patroni, 2018). Educational information is displayed throughout the aquarium (interpretive area) and visitors who experience interactions with the dolphins are given informative talks by trained staff and volunteers at the beach Interaction Zone and during boat-based tours.

The DDC aquariums feature small marine fauna and constructed habitats which replicate areas of the surrounding bay including the mangroves, estuary environment and reef structures in the southwest of Western Australia. The entire interpretation zone has signage and information about the animals in the aquariums and the dolphins displayed on digital screens on the walls. This area has many activities available for their younger demographic including an interactive touch pool (Figure 37.3), a colouring in station, in which the fish they colour can be scanned to swim across a series of TV screens, and a sand box with a topographic map projected onto it for building mountains, basins and understanding water run-off. This interpretive zone also contains a small movie theatre playing an educational short film on the history and objectives of the DDC. There is also the 360-degree digital dolphinarium where visitors can watch digitally created dolphins representing those in the bay.

Through its educational programmes the DDC is thus seen to play an important role in the conservation of marine wildlife and in particular the dolphins of Koombana Bay. Conservation activities include marine turtle rescue and rehabilitation, coral propagation and reef structure seeding, fish and seahorse breeding programmes. The DDC, in partnership with DBCA, Perth Zoo and the Aquarium of Western Australia, acts as a rehabilitation centre for juvenile loggerhead turtles that have been found washed up on the shore. The centre rehabilitates these turtles until they can be tagged for tracking purposes and then released back into the wild. The research that is carried out by the DDC is important in understanding and fostering sustainable tourism and the income generated and other entrepreneurial activities (catering, functions and community education days) allows this research to continue.

Bunbury is a large urban city with up to 124,000 holidaymakers (TRA, 2018) visiting every year allowing the educational conservation messages offered by the DDC to reach a large number of people. Having visitors learn about conservation and be able to experience wild dolphins under highly controlled circumstances allows an appreciation for the dolphins while ensuring the welfare and safety of both the dolphins and the visitors. The DDC also runs specific education programmes for local, state and international school groups, imparting important conservation messages to future decision influencers. The educational component of the dolphin

Figure 37.3 Interactive touch pool.

experience also encourages visitor participation in environmental behaviours long after their wildlife tourism experience.

The educational information provided to visitors by the DDC is also seen as a contribution to the conservation of marine wildlife. Information is provided to visitors including facts about the biology of marine wildlife, conservation information including water run-off impacts from land, plastic and other pollution, and how to behave in the presence of dolphins. Building education as a vital component of the experience provides visitors with a greater awareness of the potential impacts human interactions can have on dolphin welfare and how visitors can reduce them (for example, Barney, Mintzes & Yen, 2005; Bach & Burton, 2017). Patroni (2018) found that visitors to the DDC showed a high awareness of potential impacts on dolphins from human interactions and indicated that those visiting the DDC had greater environmental knowledge than those who had not visited the DDC. Patroni (2018) also showed 71% of visitors to the DDC were aware that dolphin feeding by the general public was illegal, whereas with those people who had not visited the DDC only 58% were aware that the feeding of dolphins by the general public was illegal. The survey data indicates the importance of such ecotourism experiences in cities like Bunbury as when there is more awareness, people are less likely to engage in detrimental interactions with wild dolphins through ignorance.

It is important to note that all human interaction with wildlife has risks in regard to the safety of visitors engaging in interactions and in relation to the welfare of wildlife. Moreover, there have been many impacts reported in the literature concerning human–dolphin interactions, including changes in natural dolphin behaviour, boat strikes injuring dolphins and habituation leading to dolphins becoming more "attached" to humans (Foroughirad & Mann, 2013; Orams, 2002; Scarpaci, Nugegoda & Corkeron, 2010). In order to minimise the harm to dolphins, controlled interactions such as those operated by the DDC are therefore very important. Providing information on the safe ways for people to behave around dolphins, including not touching or feeding them and keeping an appropriate distance by not directly approaching dolphins when boating is essential for the creation of pro-environmental behaviours in the general public and ultimately to foster dolphin conservation awareness in the general community.

Economic importance

In addition to the satisfaction derived by visitors who receive the opportunity to interact with wild dolphins in their natural environment, tour operators and local communities benefit greatly from dolphin tourism. Wildlife tourism and operations such as the DDC provide income to the community, job opportunities and an increase in tourist visitation to a city through personal and online word of mouth (Bearzi, 2017; Schleimer et al., 2015; Wilson & Tisdell, 2003). These benefits are regarded as important for small towns or developing communities, as such locations tend to rely heavily on income from tourism to support local business (Mustika et al., 2012). For example, dolphin watching in Indonesia brings in around 37,000 tourists a year and contributes a minimum of 46% of the total direct expenditure for accommodation, transport and food and beverage at dolphin tourism destinations (Mustika et al., 2012). Such tourism operations also provide many job opportunities for people living in these rural cities, which is of great benefit to the community (Schleimer et al., 2015; Wilson & Tisdell, 2003). The DDC makes a significant contribution to socio-economic and environmental sustainability in the regional city of Bunbury (Fenech, 2011; Patroni et al., 2018). In this highly seasonal regional tourism market, the DDC is considered to be one of four iconic ecotourism attractions in southwest Western Australia and Bunbury's single most important tourist attraction (EVOLVE Strategic Solutions, 2015; Tourism WA, 2007). The DDC prior to the recent redevelopment was set up as a gated experience which restricted access to the main centre and was receiving 60,000 unique visitors per year and a total of 115,000 visitations a year.

The Dolphin Discovery Centre as a tourism attraction today

In December 2018 the Dolphin Discovery Centre re-opened following a major redevelopment ($12.3 million under the State Royalties for Regions Program), and has been upgraded to a larger, more modern facility (Figure 37.4). This iconic new building incorporates elements of

Figure 37.4 Re-developed Dolphin Discovery Centre.

Noongar Aboriginal culture in its design and creates a high impact entry statement to the city. The new centre includes 14 new aquariums and an interpretation centre with digital displays and interactive activities for children, the world's first 360-degree Digital Dolphinarium, a 3D short movie, an activity centre, a large café/restaurant and a viewing platform with ramps enabling beach wheelchair access to the Interaction Zone. The DDC also has a new multi-use conference centre as well as function and events space on the first-floor level overlooking the Interaction Zone. The redeveloped DDC will continue to develop its displays, technology and services to meet the needs of the visiting public and in service to the regional tourism industry (DDC, 2015b).

The redeveloped DDC aims to bring many repeat visitors to the centre, including those who are not specifically after an ecotourism experience. The café now serves as a place for locals to enjoy meals by the ocean, this means the DDC can generate additional income in order to maintain its nature-based tourism experiences for visitors (Patroni, 2018). An additional incentive to visit the site has been provided by changing the business model to an open access approach to most areas, including the beach Interaction Zone, and public change-rooms. These initiatives aim to increase visitation expected and to increase engagement with other DDC tours and products.

Today the DDC is run as a "Not for Profit" organisation with a community Board of Management and membership. Its objectives focus on conservation, education and research relating to the wild dolphin population of the Koombana Bay area and the local marine environment. A Public Conservation Fund has been established within the Constitution to underwrite the future costs of these activities. The organisation has Advanced Ecotourism Accreditation and has incorporated sustainability principles into building design and daily operations.

Since the December 2018 opening of the DDC redevelopment the site can be experienced by not only visitors to the DDC but anyone visiting Koombana Beach where the dolphins choose to come in and access to the Centre is free of charge. When dolphins visit the Interaction Zone, which is set out in the water in front of the DDC, volunteers line up visitors in a tight line knee-deep in the water. This helps to protect the visitors and the dolphins giving the dolphins room to swim behind the line if they desire without knocking over the visitors. Visitors are informed by the volunteers on the names and details of the individual dolphins that visit the Interaction Zone along with information on dolphin behaviours and biology. Visitors are managed by volunteers interspersed in the line to ensure everyone is following instructions and not touching the dolphins for their own safety and the welfare of the dolphins (see Figure 37.2).

With the redevelopment of the Centre, along with the newly implemented un-restricted access to the Centre experience, increased destination profile and enhanced marketing inputs, it is expected that visitation will increase significantly and make a large impact on the region's economy. The redeveloped DDC is estimated to contribute $16 million both directly and indirectly into the local economy (DDC, 2015b).

Conclusion

Having significant wildlife tourism experiences such as those offered by the DDC are important for many reasons especially in small regional cities such as Bunbury, Western Australia. Wildlife tourism operations bring revenue into the local community and provide jobs and income through attracting higher numbers of tourists to the region. In cities like Bunbury with high tourist visitation the conservation benefits of wildlife tourism operations are amplified as the educational messages that assist greatly in conservation of marine wildlife are conveyed to a large number of people. Education is vital in ensuring dolphin welfare through the encouraging of pro-environmental behaviours in people who engage in marine wildlife experiences. Having highly controlled interactions with wildlife also provides visitors with an up-close interaction with wildlife in a safe environment for both them and the dolphins, allowing visitors to continue to have such sought-after interactions with wildlife. Visitors leave these experiences with a greater appreciation for the wildlife and awareness of the risks that unregulated interactions with humans can have on dolphin welfare.

There is no doubt that many coastal city planners around the world can learn from the DDC experience. It is hoped that city-based wildlife tourism experiences, such as those delivered at Bunbury, will mark a new era for the "wildlife aware" international tourism city.

References

Australian Bureau of Statistics. (2017). *Bunbury (C) (LGA)*. Retrieved from http://stat.abs.gov.au/itt/r.jsp ?RegionSummary&region=51190&dataset=ABS_REGIONAL_LGA&geoconcept=REGION&datas etASGS=ABS_REGIONAL_ASGS&datasetLGA=ABS_REGIONAL_LGA&regionLGA=REGION &regionASGS=REGION.

Bach, L., & Burton, M. (2017). Proximity and animal welfare in the context of tourist interactions with habituated dolphins. *Journal of Sustainable Tourism*, *25*(2), 181–197. doi:10.1080/09669582.2016.11958 35.

Barney, E., Mintzes, J., & Yen, C. (2005). Assessing knowledge, attitudes, and behaviour towards charismatic megafauna: The case of dolphins. *The Journal of Environmental Education*, *36*(2), 41–55. doi:10.3200/ JOEE.36.2.41-55.

Bearzi, M. (2017). Impacts of marine mammal tourism. In Blumstein, D.T., Geffroy, B., Samia, D.S., & Bessa, E. (Eds.), *Ecotourism's Promise and Peril*, Springer, Cham, Switzerland, pp. 73–96. https://doi. org/10.1007/978-3-319-58331-0_6.

City of Bunbury (COB). (2015). *Strategic Community Plan: Bunbury 2030*, City of Bunbury, Bunbury, Western Australia. Retrieved from www.bunbury.wa.gov.au/pdf/Council/Strategic%20Community%20 Plan_Feb%202015_FINAL%20FOR%20WEB.pdf.

City of Bunbury (COB). (2017a). *Visit Bunbury*. Retrieved from https://visitbunbury.com.au/.

City of Bunbury (COB). (2017b). *Things to See and Do*. Retrieved from www.bunbury.wa.gov.au/Pages/ Things-to-see-and-do.aspx.

Cong, L., Wu, B., Zhang, Y., & Newsome, D. (2016). Empirical research on environmental attitude of non-consumptive wildlife tourism: A case study of Dolphin Discovery Center (DDC) in Bunbury, Australia. *Beijing Daxue Xuebao (Ziran Kexue Ban)/Acta Scientiarum Naturalium Universitatis Pekinensis*, *52*(2): 295–302.

Cong, L., Lee, D., Newsome, D., & Wu, B. (2017a). The analysis of tourists' involvement in regard to dolphin interactions at the Dolphin Discovery Centre, Bunbury, Western Australia. In Fatima, J., & Khan, M. (Eds.), *The Wilderness of Wildlife Tourism*, Apple Academic Press, Oakville, Ontario.

Cong, L., Bihu, W., Zhang, Y., & Newsome, D. (2017b). Risk perception of interaction with dolphin in Bunbury, West Australia. *Acta Scientiarum Naturalium Universitatis Pekinensis*, *53*(1): 179–188.

Dolphin Discovery Centre (DDC). (2015a). *About Us*. Retrieved from https://dolphindiscovery.com.au/ about/.

Dolphin Discovery Centre (DDC). (2015b). *The Future*. Retrieved from https://dolphindiscovery.com. au/the-future/.

Dolphin Discovery Centre (DDC). (2015c). *Interaction Zone*. Retrieved from http://dolphindiscovery. com.au/enjoy/interaction-zone/.

EVOLVE Strategic Solutions. (2015). *Bunbury-Wellington & Boyup Brook Regional Tourism Development Strategy 2015–2019*, City of Bunbury, Western Australia. Retrieved from www.bunbury.wa.gov.au/ SitePages/Results.aspx?=&cs=This%20Site&u=www%2Ebunbury%2Ewa%2Egov%2Eau&k=duplicat es:%2213523%22%20.

Fenech, R. (2011). *City of Bunbury Tourism Strategy 2009–2014: Revised 2010*. Edge Tourism & Marketing, Bunbury, Western Australia.

Foroughirad, V., & Mann, J. (2013). Long-term impacts of fish provisioning on the behavior and survival of wild bottlenose dolphins. *Biological Conservation*, 160, 242–249. doi:10.1016/j.biocon.2013.01.001.

Gentilli, J. (1989). Climate of the Jarrah Forest. In Dell, B., Havel, J.J., & Malajczuk, N. (Eds.), *The Jarrah Forest*, Springer, Dordrecht, pp. 23–40.

Mann, J., Senigaglia, V., Jacoby, A., & Bejder, L. (2018). A comparison of tourism and food provisioning among wild bottlenose dolphins at Monkey Mia and Bunbury, Australia. In Carr, N., & Broom, D. (Eds.), *Tourism and Animal Welfare*, CABI, Wallingford.

Mustika, P. L. K., Birtles, A., Welters, R., & Marsh, H. (2012). The economic influence of community-based dolphin watching on a local economy in a developing country: Implications for conservation. *Ecological Economics*, *79*(1), 11–20. doi:10.1016/j.ecolecon.2012.04.018.

Newsome, D. (2017). A brief consideration of the nature of wildlife tourism. In Fatima, J., & Khan, M. (Eds.), *The Wilderness of Wildlife Tourism*. Apple Academic Press, Oakville, Ontario.

Orams, M. (1997). Historical accounts of human–dolphin interaction and recent developments in wild dolphin based tourism in Australasia. *Tourism Management*, *8*(5), 317–326. Retrieved from www.sciencedirect.com/science/article/pii/S0261517796000222.

Orams, M. B. (2002). Feeding wildlife as a tourism attraction: A review of issues and impacts. *Tourism Management*, *23*(3), 281–293. doi:10.1016/S0261-5177(01)00080-2.

Orams, M. B., & Hill, G. J. E. (1998). Controlling the ecotourist in a wild dolphin feeding program: Is education the answer? *The Journal of Environmental Education*, *29*(3), 33–33. doi:10.1080/00958969809599116.

Patroni, J. (2018). Visitor satisfaction with a beach-based dolphin tourism experience and attitudes to feeding wild dolphins. Murdoch Research Repository. Retrieved 20 October 2018 from http://researchrepository.murdoch.edu.au/id/eprint/41944/.

Patroni, J., Day, A., Lee, D., Chan, J. K. L., Kerr, D., Newsome, D., & Simpson, G. D. (2018). Looking for evidence that place of residence influenced visitor attitudes to feeding wild dolphins. *Tourism and Hospitality Management*, *24*(1), 87–105. https://doi.org/10.20867/thm.24.1.2.

Patroni, J., Newsome, D., Kerr, D., Sumanapala, D. P., & Simpson, G. D. (2019). Reflecting on the human dimensions of wild dolphin tourism in marine environments. *Tourism and Hospitality Management*, *25*(1), 1–20.

Scarpaci, C., Nugegoda, D., & Corkeron, P. J. (2010). Nature-based tourism and the behaviour of bottlenose dolphins "Tursiops" spp. in Port Phillip Bay, Victoria, Australia. *The Victorian Naturalist*, *127*(3), 64–70. Retrieved from https://search-informit-com-au.libproxy.murdoch.edu.au/fullText;dn=28201 1001606877;res=IELHSS.

Schleimer, A., Araujo, G., Penketh, L., Heath, A., McCoy, E., Labaja, J., Lucey, A., & Ponzo, A. (2015). Learning from a provisioning site: Code of conduct compliance and behaviour of whale sharks in Oslob, Cebu, Philippines. *PeerJ*, *3*, e1452. doi:10.7717/peerj.1452.

Senigaglia, V., Christiansen, F., Sprogis, K. R., Symons, J., & Bejder, L. (2019). Food-provisioning negatively affects calf survival and female reproductive success in bottlenose dolphins. *Scientific Reports*, *9*(1), article 8981.

Sitar, A., May-Collado, L. J., Wright, A., Peters-Burton, A., Rockwood, L., & Parsons, C. M. (2017). Tourists' perspectives on dolphin watching in Bocas Del Toro, Panama. *Tourism in Marine Environments*, *12*(2), 79–94. doi:10.1016/j.marpol.2016.03.011.

Tourism Research Australia (TRA). (2018). International and National Visitor Surveys. Retrieved from www.swdc.wa.gov.au/information-centre/statistics/tourism.aspx.

Tourism WA. (2007). *Australia's South West: Destination Development Strategy: Update 2007–2017*. Government of Western Australia, Perth, WA, Australia.

Trip Advisor (2017). *Koombana Bay (Bunbury): Things to Do*. Retrieved from www.tripadvisor.com.au/Attraction_Review-g255364-d2512786-Reviews-Koombana_Bay-Bunbury_Western_Australia.html.

Wilson, C., & Tisdell, C. (2003). Conservation and economic benefits of wildlife-based marine tourism: Sea turtles and whales as case studies. *Human Dimensions of Wildlife*, *8*(1), 49–58. doi:10.1080/10871200390180145.

Zylines (1989). Dolphin fever. Summer.

38

LOCATION, URBAN FABRIC AND TRANSPORTATION

Historical morphogenetic analysis of tourism cities along the Grand Canal

Fang Wang, Bingyu Lin and Qingyin Liu

Introduction

The Grand Canal and its cities have become popular destinations of contemporary tourism in China. With the rise of tourism cities both nationally and internationally, developing unique tourism resources in the canal cities has become imperative. After appointment to the World Heritage List in 2014, some argue that the landscape of the Grand Canal and its historical functions should be fully restored to attract more tourists. To publicise the cultural heritage of the canal and to promote the development of urban tourism requires a comprehensive understanding of the structure and culture of the canal cities.

Challenge of tourism cities

Tourism cities have remained a consistent theme in tourism expansion research since the 1980s (Ashworth and Page, 2011). Cities, in general, host a large proportion of tourists worldwide each year. Since the second half of the twentieth century, cultural heritage has increasingly grown in importance on urban agendas and become a major instrument in the development of urban tourism (Richards, 2014). Cultural heritage could be treated as a commodity for commercial uses, especially concerning tourism (Graham, Ashworth, and Tunbridge, 2000). Currently, increased competition among international tourist destinations has emphasised quality and cultural value as important factors that make a place more appealing to visit. Research by Christou (2018) also suggested that cultural value is beneficial, both at a personal and social level. However, in the process of developing tourism resources, some stakeholders do not seem to fully recognise the potential of cultural heritage, while others blindly reconstruct the historical landscape.

In recent years, many cities have devoted themselves to diversifying their tourism contents by developing new frontiers for urban tourism, including ruin tourism (Le, 2018), walking tourism (Kanellopoulou, 2018) and even dark tourism (Mileva, 2018), which explores dark places/sites related to death and suffering. No matter the strategies used, the key is to thoroughly examine cultural and urban characteristics that coincide with local conditions.

Grand Canal and canal cities

The Grand Canal formed the backbone of the Chinese inland communication system while transporting food products from the south to the north, supplying rice to feed the population. With the development of the Grand Canal over the past 2,000 years, many urban settlements emerged alongside it, from the capital city, Beijing in the north to Hangzhou City in the south. These cities embodied the essence of Grand Canal civilisation. During the Ming and Qing Dynasties, the open navigation of the Grand Canal unprecedentedly strengthened its economic function and greatly improved the development of a commodity economy. The Grand Canal, both directly and indirectly, affected the transportation, socio-economic conditions, urban form and urban agglomeration pattern of cities, and promoted economic development and a profound exchange of materials, information and lifestyles between north and south China.

From a regional perspective, the research on Chinese canal cities has primarily been related to history, geography and culture analysis. Fu (1985) carried out in-depth research on the historical and geographical environments and the social-political contexts regarding the development of nine canal cities. Liu (2008), Li (2012) and Wu (2014) also published monographs related to canal cities and urban history. In comparison, the research on individual cities was more prolific, mainly focusing on transportation, warehouses, commerce and urban ecology in large cities along the canal. Chen (2013) analysed the rise of Linqing City as a port of transhipment. There were also studies on the construction, history and culture of the canal cities, as well as studies centred on the relationship between the cities and the canal, including Tongzhou City (Peng and Wei, 2017), Jining City (Bu, 2012), Huzhou City (Chen, 1989), Xuzhou City (Dai and Shen, 2013) and Changzhou City (Wu et al., 2013). A study of urban elements focused on transportation, warehouses, commerce and urban ecology. Zheng (2013) summarised the development, operation management and function of canal warehouses in the Ming and Qing Dynasties. Yu (2012) conducted a systematic study of the Grand Canal as an important cross-regional ecological corridor. Yang et al. (2014) analysed the relationship between ecological environments and spatial changes in the canal cities.

Research questions and methodology

Currently, the overall spatial structure of the canal cities and the lateral comparison between the canal cities have not been well recognised in China; however, these could be important references for the overall planning of tourism around the Grand Canal and its cities. Therefore, the fundamental questions of this study were: What was the spatial distribution pattern of the cities along the Grand Canal during its historical period? How did the canal influence and contribute to the development of urban form, scale and building fabric of each city? Were there any differences between the urban structures of the canal cities?

To answer these questions, we conducted in-depth research regarding the interaction between the Grand Canal and its cities from the perspective of urban form to enrich the understanding of the cultural characteristics of the canal cities and to better promote the development of tourism there. Specifically, we extensively utilised first-hand historical materials for urban analysis, including historical gazetteers, manuscripts, maps and drawings. To examine the historical formation and development process of the cities, we adopted the theory of urban morphology. Conzen (1960) first applied the method of urban morphology to study British historical towns in the 1960s. He argued that the urban form of a city was mainly composed of streets, blocks, plots, the building fabric, fringe belts and a fixation line. Studying the layout, structure and relationships between these factors was at the core of understanding the urban form of a city. These elements were significant components of urban civilisation and reflected the ideology of urban administration.

Spatial distribution of Grand Canal cities

During the Ming and Qing Dynasties, the administrative division was country–province–prefecture–county (Wang, 2015). We chose both prefecture-level and county-level cities along the canal as research subjects in this section and specifically explored the urban structure of 16 prefecture-level cities and 45 county-level cities.

Cities – Grand Canal distance analysis

According to statistics gleaned from the *Historical Atlas of China* (Tan, 1996) and the *Chinese Historical Geographic Information System* (CHGIS, 2003), we produced a distribution map of the administrative centres of prefecture-level and county-level cities along the canal in the Ming and Qing Dynasties (Figure 38.1). These cities could be roughly divided into three sections based on their spatial distribution pattern: northern, middle and southern. Jining City was the boundary between the northern and middle sections, while Huai'an City was the boundary

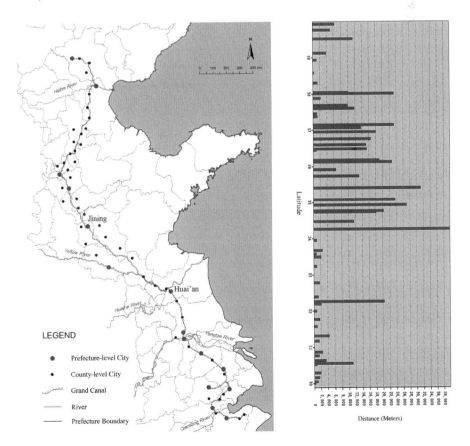

Figure 38.1 Distribution of canal cities in Ming and Qing Dynasties and linear distances from Grand Canal.
Source: Authors.

between the middle and southern sections. It was obvious from the chart, processed by the CHGIS, that the linear distances between the cities and the canal was generally less in the northern and southern sections than in the middle section.

Such a spatial distribution pattern was primarily affected by the conditions of navigation and regional flooding. To take advantage of canal transportation, cities in the northern and southern sections tended to locate and develop adjacent to the canal. The canal also provided conditions for navigation in the middle section, but the canal greatly affected natural water systems and even destroyed regional river channels, resulting in frequent flooding. Therefore, the cities and eco-development in the middle section were severely restricted.

Although disasters around the Grand Canal were certainly caused by natural factors, they were mainly caused by humans and other compounding factors. Prior to the opening of the canal, the Yellow River migrated southward to join the Huaihe River into the East China Sea. At the beginning of canal operation (approximately 1400–1500 CE), the Yellow River channel from Xuzhou City to Suqian City was occupied for canal transportation (Figure 38.2). Thus, Xuzhou City became an important transportation hub and centre for the exchange of materials between the north and south. Because of geographical factors, the Yellow River tended to migrate back northward, which conflicted with government interests concerning canal transportation. The Ming government forcibly rerouted the Yellow River in the south, which prevented it from moving northward and resulted in frequent flooding. Finally, in the mid-to-late Ming Dynasty (approximately 1500–1600 CE), a new canal channel opened, which bypassed Xuzhou City. North–south transportation no longer had to rely on the Yellow River channel in the Xuzhou section. Although overall canal transportation was guaranteed to a certain extent, the funding to harness the flooding was allocated to Suqian City instead of Xuzhou City. Xuzhou City was profoundly affected by floods and gradually declined; dense urban settlements did not form along the new canal channel either.

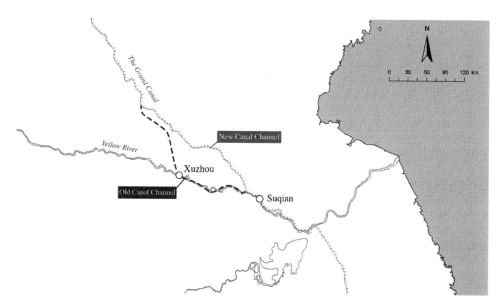

Figure 38.2 Relocation of Grand Canal in Xuzhou section.

Source: Authors.

Structure of Grand Canal cities

The Grand Canal and the natural rivers that flowed through it had an important influence on the structure of the coastal urban network. The canal ran from south to north, while the natural river systems mainly flowed towards the east into the sea. The waterways that joined the canal included (from north to south) the Haihe, Yellow, Huaihe, Yangtze and Qiantang Rivers. The river network and traffic routes interwove, so the major canal cities and surrounding counties under their jurisdiction extended from the canal and/or along the natural rivers, which were perpendicular to the canal. Shuntian (now Beijing) was situated at the northern end of the canal and was the national political centre, while Suzhou and Hangzhou were at the southern end and housed the national economic centre. At the two ends, a sub-administrative county developed around each of these centres, forming a radial structure or pattern. In the remaining areas where the canal flowed through, the urban agglomerations were in a linear distribution pattern (Figure 38.3). The prefecture-level cities and county-level cities governed by them extended along the water channels. Most of the cities were influenced by the canal, while some were more affected by major rivers, such as the Yellow River, where Xuzhou City and its counties extended from.

Urban morphogenetic characteristics of Grand Canal cities

Size of Grand Canal cities

Urban form is mainly composed of streets, blocks, plots, building fabric, fringe belts and a fixation line. In this section, we examine in detail the physical and social environments of the canal cities from the perspective of urban size, urban form (shape and pattern of streets), urban boundary and its evolution, as well as major buildings within the cities. Sixteen prefecture-level cities were selected for specific analysis.

The size of ancient Chinese cities was generally measured by the perimeter of urban walls rather than the urban area itself (Wang, 2013). Our size analysis of prefecture-level cities in the Ming and Qing Dynasties was based on records from local gazetteers (Figure 38.4). We used whole numbers to easily calculate the results with the "li" unit (1 li ≈ 576 metres). The sizes of the cities were represented by rectangular shapes (Figure 38.4); therefore, the larger the shape, the larger the size of the city. The canal cities varied widely in size. Generally, cities in the northern and southern sections were larger in size, while the middle cities were smaller in size. The largest city overall was the Shuntian Prefecture; it was 40 li (approximately 23 km) and located in the northern section. The second and third largest cities were Hangzhou City and Suzhou City, respectively, both located in the southern section. The cities in the middle section of the canal were relatively small and were commonly only 7–9 li (approximately 4–5 km).

Urban fabric: city walls, gates and street patterns

Boundaries were defined by the walls of the cities in the Ming and Qing Dynasties. In the ancient cities, where low-rise buildings comprised the vast majority of buildings, the city walls were an indispensable part of the urban landscape. The walls defined the space inside the city and indicated where the cities were located (Wang, 2013). Based on local gazetteers, ancient maps and records of chorography, we drew plans of the prefecture-level cities (Figure 38.5), including the city walls (boundaries), city gates, main roads and rivers.

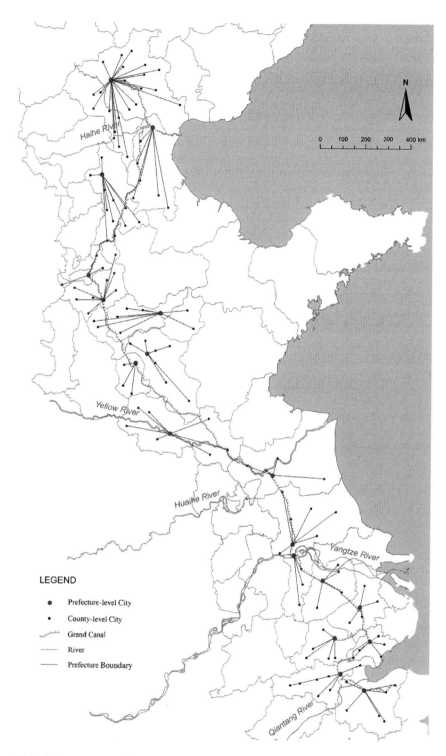

Figure 38.3 Urban structure of canal cities and counties.

Source: Authors.

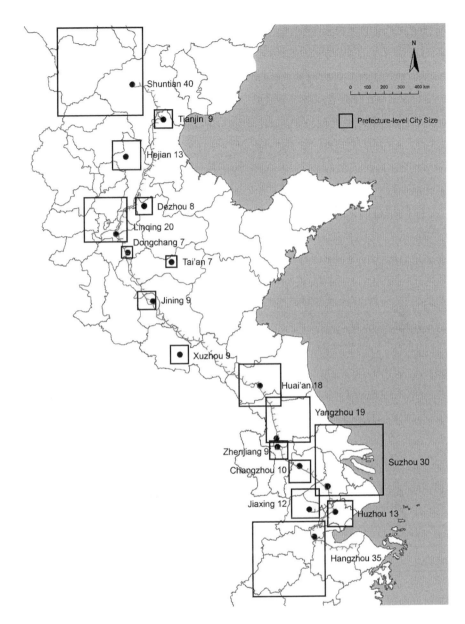

Figure 38.4   Sizes of canal cities.

Source: Authors.

Urban forms could be roughly divided into square shapes and irregular shapes with the former accounting for the majority of the cities. From north to south, city boundaries tended to be more irregular. Some square cities evolved into irregular shapes due to the influence of canal transportation, its commodity economy or the natural environment. Square cities were located in the Hebei and Shandong provinces in the northern section of the canal and included Tianjin, Hejian, Dezhou, Dongchang, Tai'an and Jining Cities (Figure 38.6). The formation of square cities was influenced by the "Li-fang unit system" of

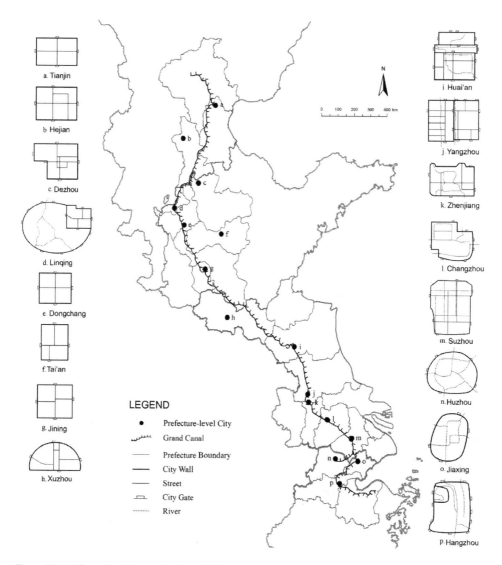

Figure 38.5   Plans of prefecture-level canal cities.

Source: Authors.

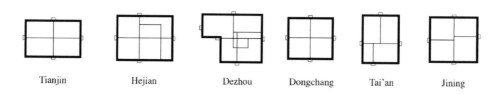

Figure 38.6   Urban plans of northern canal cities.

Source: Authors.

the Tang Dynasty (618–907 CE). It was a form of urban construction design, as well as a management policy in social order. The city had a central axis and symmetrical layout. The residential area in the city was cut into small squares like a chessboard. Except for Tianjin, which was built during the Ming Dynasty, the remaining cities were rebuilt on the basis of old cities from the Tang and Song Dynasties (Shen and Xu, 1899; Du and Chen, 1573–1620; Du and Huang, 1760; Zheng and He, 1521–1567; Zhang and Zhu, 1785; Song and Xie, 1808; Yan and Cheng, 1760; Lu, 1859). The streets were square grids, and the main roads were mostly cross-shaped or staggered. Although Tianjin was a new city, it maintained a square shape. It reflected that the formation of urban form was largely influenced by feudal rituals in northern China.

There were also square cities in the middle section of the canal, such as Huai'an and Yangzhou Cities (Figure 38.7). However, unlike the northern cities, the shapes of these two cities were restricted concurrently by the Li-fang unit system and the influence of the canal and waterways. Huai'an City was composed of an old city in the south and new cities in the north and middle. During the Yuan Dynasty (1271–1368 CE), the canal passed outside the northern area of the old city, which gradually became an industrial and commercial district that soon developed into a new city. In the mid-sixteenth century, the middle city walls were built to strengthen the relationship between the new and old cities and to enhance their defensive function (Song and Fang, 1621–1627; Wei and Gu, 1832). Similar to Huai'an City, Yangzhou City consisted of both new and old cities (Yang and Xu, 1573–1620), and its internal block structure was subject to the needs of the canal wharf, which formed an open street pattern (Tang and Huang, 2006).

Irregular-shaped cities were primarily located in the southern section of the canal, including cities in Jiangsu and Zhejiang provinces (Figure 38.8). Typically, there was a square sub-city, which was built in the Tang Dynasty or earlier, that served as the administrative centre (Tang and Huang, 2006). Luocheng, which was newly built on the outskirts of the old city, was naturally extended based on the river network, canal transportation and demands of circulation, transhipment and trade. Because of these circumstances, its boundary was irregular in shape. Furthermore, the dense river network not only controlled the function of traffic, but also established an essential element between the inner and outer space of the city and became the axis of urban space organisation. Therefore, the streets were not subject to grid planning, but rather they developed based on the needs of transportation and the lives of residents.

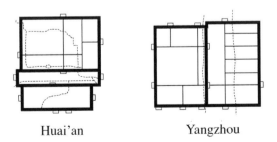

Huai'an   Yangzhou

Figure 38.7   Urban plans of middle canal cities.

Source: Authors.

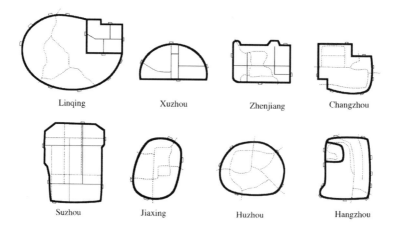

Figure 38.8   Urban plans of southern canal cities.

Source: Authors.

Evolution of urban boundaries

As mentioned above, the canal cities in the Ming and Qing Dynasties were formed based on old cities from previous dynasties. Some of these cities overlaid old cities, some relocated and others expanded and developed into new cities. The evolution of urban boundaries could be divided into four categories: newly created, expanded, contracted and relocated (Table 38.1).

The newly built cities analysed in this study were Tianjin and Linqing. Both cities formed a business gathering area due to a transportation stimulus, which created conditions for the

Table 38.1 Evolution of urban boundaries

Province	Prefecture-level city	Evolution of urban boundaries			
		Newly formed	Expanded	Contracted	Relocated
Zhili	Tianjin	✓			
Zhili	Hejian				
Shandong	Dezhou				✓
Shandong	Linqing	✓	✓		
Shandong	Dongchang				✓
Shandong	Tai'an				
Shandong	Jining				✓
Jiangsu	Xuzhou		✓		
Jiangsu	Huai'an		✓		
Jiangsu	Yangzhou		✓		
Jiangsu	Zhenjiang				
Jiangsu	Changzhou			✓	
Jiangsu	Suzhou				
Zhejiang	Jiaxing				
Zhejiang	Huzhou			✓	
Zhejiang	Hangzhou		✓		

formation of the city. Stable development of the northern canal further promoted trade and the economies of the two cities, which became significant regional commercial centres. Tianjin City was a small town prior to the Ming Dynasty. After the capital was relocated from the south to Beijing during the Yuan Dynasty, all grain, whether it was to be transported via inland rivers or by sea, first needed to arrive in Tianjin. From there, grain was transhipped to the capital city. According to statistics (Liang, 1980), approximately six million tons of grain arrived in Tianjin within a mere 47 years (1283–1329) (Wu, 2014), accounting for a large proportion of total grain. The government set up specialised agencies in Tianjin for transporting grain, such as the pick-up hall, warehouse and transhipment institute. Finally, with the development of economy and urban settlement, a new administrative centre was established in the early Ming Dynasty (1404 CE), marking the birth of a new city.

Although the administrative centre was established in Linqing in 1369, it did not bring about the formation of the city. After 82 years of harnessing and constructing the canal, favourable conditions for navigation finally promoted the development of industry and commerce in Linqing. It was not until 1449 that Linqing became a prosperous city. With continuous development of canal trade and population growth, the old city no longer met the needs of the merchants and residents. In 1511, a new city was built outside the old one (Zhang and Zhu, 1785).

The expanded cities analysed in this study were Linqing, Xuzhou, Yangzhou and Hangzhou. Similar to Linqing City, the canal played an essential role in promoting the prosperity of these other cities, driving economic booms, population growth and urban expansion. However, the influence of the canal was not always a positive one. The boundary expansion of Xuzhou City was necessitated as a defence against floods and wars. As discussed previously, frequent flooding severely restricted the urban operation of Xuzhou City. The occurrence of disasters was partially attributed to natural factors, but a series of policies issued by the governments of the Ming and Qing Dynasties was the bigger cause. To ensure that the Grand Canal functioned well, local officials excavated a new canal channel and built dikes based on the will of the emperors. Although canal transportation was maintained, the new features caused disturbances in the regional water systems and brought about more floods. Every time the river breached, the walls and infrastructure of Xuzhou City suffered considerable damage (Wu and Liu, 1874). The urban boundary tended to expand after each repair or reconstruction. These expansions were not due to an increase in economic prosperity or population but to better combat frequent flooding.

The cities with contracted borders in our study were Changzhou and Huzhou. These cities were affected by wars in the Song and Yuan Dynasties and urban development stagnated. Moreover, the previously large cities were difficult for their militaries to defend. Eventually, Changzhou City contracted on its east, south and west boundaries (Liu and Tang, 1618). Despite contracting, its industry, commerce and transportation were restored and developed with the help of the Grand Canal.

Finally, Dezhou, Jining and Dongchang Cities relocated. Their original sites did not have convenient transportation and were threatened by unfavourable environmental problems, such as flooding. By relocating, these cities could take advantage of canal resources.

Building fabric

Studying the building fabric and layout of each of the canal cities helped promote better understanding of their urban characteristics. We selected three representative building types as research objects: government offices, markets and warehouses. Respectively, they represented the political centres, business districts and storage areas associated with the canal.

The ancient Chinese political districts were the administrative regions that arose from administrative divisions. The places where administrative officials worked could be categorised into prefecture-level and county-level government offices (Wang, 2013). Except for the capital city, a government office was usually the highest-ranking building in the city. It represented urban political power and had an important impact on urban planning and urban landscape. In the northern canal cities, government office buildings were located in a central location, which together formed a political core district with other government buildings. Such a pattern was called "central governance" and occurred in cities such as Tianjin, Hejian, Dezhou and Tai'an. Some government office buildings were located in the old city or sub-city centre, such as in Linqing. The situation was different in southern canal cities, such as Zhenjiang and Hangzhou. Affected by the terrain and rivers, government office buildings were not all located in the centre; rather, they occupied the high ground or transportation hub.

Markets were the commercial gathering areas and economic centres of the cities. We recorded the number of markets both inside and near the canal cities based on historical data (Figure 38.9); however, the data from Linqing, Dongchang and Changzhou Cities were not documented. In general, the number of markets located in the southern cities was larger than that of the northern cities. Suzhou and Yangzhou Cities had the largest number of markets, totalling 30 and 20, respectively. Additionally, there were apparent differences in the distribution of markets in the northern and southern canal cities. In the north, markets were primarily located near the gates of the cities or on the outskirts near the canal. There were few records, if any, of markets in city centres. Meanwhile, in the south, there were a large number of markets located within the cities, in addition to the markets located near the gates of the cities. The sizes of the markets were usually large, and the goods were circulated quickly. For example, according to historical records, the Changmen Commercial Street was the most prosperous commercial district in Suzhou City (Li and Feng, 1883). Sun Jiagan, a Qing Dynasty official, wrote a travel book that recorded the landscape and scenery in the Qing Dynasty. This book was entitled *South Travel* (Mandarin: *Nan You Ji*), and in it Jiagan stated: "There are mountains of goods in the Changmen Commercial Street, with myriad shops and a huge crowd of people; It is so prosperous that even the capital city is pale in comparison".

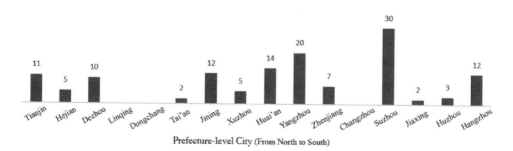

Figure 38.9 Number of markets in canal cities.

Sources: Shen and Xu (1899); Xue and Gao (1934); Du and Chen (1573–1620); Du and Huang (1760); Zheng and He (1521–1567); Zhang and Zhu (1785); Song and Xie (1808); Yan and Cheng (1760); Lu (1859); Wu and Liu (1874); Song and Fang (1621–1627); Wei and Gu (1832); Yang and Xu (1573–1620); Ruan and Jiao (1806); Gao, Gao, and Zhang (1685); Liu and Tang (1618); Wang (1488–1505); Li and Feng (1883); Lu (1379); Chen and Xia (1475); Liu and Shen (1600); Xu and Wu (1879).

In the Ming and Qing Dynasties, the capital city, Beijing, relied heavily on grain and raw materials strategically transported from south China to sustain its substantial political, economic and military expenditures. In order to transport the grain successfully to the capital, the governments of the Ming and Qing Dynasties attached great importance to the construction of storage facilities for storing and transferring grain. The location of warehouses was closely related to the conditions of the rivers, transportation policies and agricultural and financial conditions. Due to the need for transportation, there were many warehouses in the canal cities (Figure 38.10). As was the case regarding the number of markets, the southern canal cities had a slightly larger number of warehouses than the northern cities, while the middle canal cities, such as Tai'an and Xuzhou Cities, had fewer warehouses.

There were two types of warehouses. One was built by the government or other official organisations and was usually of higher grade and larger scale. Some were important grain transfer centres, such as the top five warehouses in Tianjin, Dezhou, Linqing, Xuzhou and Huai'an. The other type of warehouse was built by private fundraising. Such warehouses were generally located outside the cities or in the grain circulation centres. The location of warehouses in the canal cities was relatively uniform. Most of the official warehouses were located near the official buildings, such as the government offices or transportation hubs.

Influence of Grand Canal on urban development

Transportation and economy

The north–south Grand Canal experienced nearly 2,000 years of development and has played a wide variety of roles concerning the traffic, transportation, irrigation and water supply of the regions along the canal. The canal has remained an essential transportation method for these regions, even today. During the Ming and Qing Dynasties, the fully opened canal demonstrated unprecedented strength when it came to economic function. New cities were built, and some cities initially attracted by the canal relocated to more favourable locations; small towns even evolved into regional industrial and commercial centres. The convenience and efficiency of

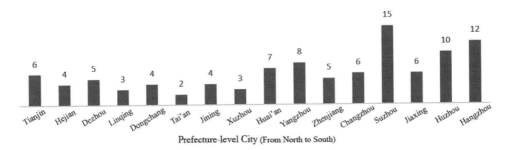

Figure 38.10 Number of warehouses in canal cities.

Sources: Shen and Xu (1899); Xue and Gao (1934); Du and Chen (1573–1620); Du and Huang (1760); Zheng and He (1521–1567); Zhang and Zhu (1785); Song and Xie (1808); Yan and Cheng (1760); Lu (1859); Wu and Liu (1874); Song and Fang (1621–1627); Wei and Gu (1832); Yang and Xu (1573–1620); Ruan and Jiao (1806); Gao, Gao, and Zhang (1685); Liu and Tang (1618); Wang (1488–1505); Li and Feng (1883); Lu (1379); Chen and Xia (1475); Liu and Shen (1600); Xu and Wu (1879).

transportation afforded by the canal promoted the expansion of cities and the prosperity of their economies.

However, seemingly minor negative effects were overlooked that played decisive roles in the development of the canal cities. The relocation of the canal channel caused the decline of the canal cities. For example, the opening of a new south canal was a turning point from prosperity to the decline of Xuzhou City during the Ming Dynasty. In the early period of the Ming Dynasty, when canal traffic partially relied on the Yellow River in the Xuzhou area, Xuzhou City was crucial in ensuring the operation of the canal. After the opening of a new canal channel in the mid-to-late Ming Dynasty, north–south transportation no longer occupied the natural waterways (Hu and Shao, 2010). Since then, the focus of the river harness project shifted from Xuzhou to the Suqian and Huai'an areas. Xuzhou lost its financial support and the workforce necessary to resist frequent flooding and was greatly weakened by long-term suffering. In addition, the number of vessels passing through Xuzhou underwent a dramatic reduction. Thus, the commodity economy in Xuzhou experienced a large recession (Dai and Shen, 2013).

Regional flooding and canal harness

When the Grand Canal flowed from south to north, it inevitably met natural rivers flowing from west to east. To maintain the operation of the canal, the confluence of rivers unavoidably affected the natural water systems, thus causing disorder in the regional water systems, which limited urban development. According to river breach statistics recorded from 1368 to 1888 (Pan, 1590; Zhu, Qing Dynasty), the most flood-prone area was in the northern part of Jiangsu Province, where the Yellow River and the Grand Canal merged (Figure 38.11). Correspondingly, the distances between cities and the canal and the sizes of cities discussed above indicated that most of the cities in northern Jiangsu were far from the canal. The sizes of these cities were also significantly smaller than those in the northern and southern ends of the canal. Frequent flooding deteriorated the soil quality, which affected regional agricultural production. Moreover, the harsh natural environment caused a large-scale migration movement during the Ming and Qing Dynasties.

Dredging and maintenance of the canal required huge manpower and financing. Once political support and appropriation for maintenance was withdrawn, canal operation quickly declined. Notably, when the transportation of grains from south to north (once the top priority of the canal) was no longer needed, the northern cities immediately declined; however, the southern cities prospered. Due to the rise of new means of transportation, such as railways, in the late Qing Dynasty, the status of the canal dropped dramatically. Without artificial control, the northern section of the canal was abandoned. Correspondingly, the canal cities, once heavily dependent on the function of the canal, also declined. In the southern section of the canal, however, the maintenance cost remained small because of adequate water supply. Even though it was no longer necessary to transport food products to the capital, the cities made full use of the canal and natural water systems for transportation and the exchange of goods within the southern region.

Conclusions and discussion

Based on the overall spatial structure of the canal cities and the differences in their urban developments, we divided the canal cities into three sections: northern (Tianjin City–Xuzhou City), middle (Xuzhou City–Huai'an City) and southern (Huai'an City–Hangzhou City).

Cities in the northern section were deeply influenced by the political centre and traditional planning, and they generally had a rectangular urban form. Their administrative and commercial

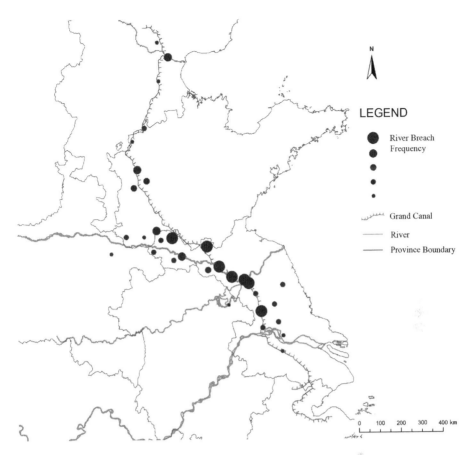

Figure 38.11   River breach frequency from 1368 to 1888.

Source: Authors.

centres were functionally separate. Many northern cities took advantage of canal resources and were responsible for administrative tasks and shipping grain to the capital city. They relied heavily on canal traffic and transportation. Therefore, when traffic lagged and the canal ultimately closed during the late Qing Dynasty and the Republic of China era, the northern canal cities began to decline.

On the one hand, the middle canal cities enjoyed the benefits of canal transportation and promoted the development of urban industry and commerce. On the other hand, the middle cities were affected by flooding and disasters caused by policies that gave priority to guarantee canal regulation. These cities were relatively small, and their commercial and transportation developments were not as beneficial as those in the northern and southern cities.

Before the Grand Canal opened, the southern cities were thriving economically because of agricultural development. In the Ming and Qing Dynasties, the canal and natural river network were put to full use, and the handicrafts and commerce of the cities further improved. The urban form was not restricted by traditional planning but was left free to grow. The commercial outlets and warehouses were distributed both inside and outside the cities. Also, the southern cities formed a successful exchange of materials and information, thus forming a reliable chain of commerce. Even when the canal was temporarily abandoned, the southern cities maintained their prosperity.

Undoubtedly, the Grand Canal played a significant role in the transportation of grain to the north and in supporting the stability and prosperity of China. However, when affirming the historical role of the Grand Canal, one should not overlook its drawbacks. During the Ming and Qing Dynasties, the primary function of the Grand Canal was to ensure the transportation and governance needs of the political centre, Beijing. Although it created conditions for navigation, it profoundly affected natural water systems and even destroyed regional river channels, resulting in frequent flooding. In order to ensure the operation of the imperial court, the central government chose to forfeit local interests and forcibly fixed the disaster in the middle of the canal. The canal did not bring homogeneous development conditions for the cities along it. This was clearly reflected in the different urban fabric comprising the canal cities, such as city boundaries, street patterns and building types.

Therefore, when carrying out heritage preservation and tourism development in the canal cities, we found it advisable to use history as an important reference. Proposals should be developed that are consistent with urban historical and social contexts. So far, the southern section of the Grand Canal has been maintained for navigation, but most of the middle and northern sections have been abandoned. Some suggest that the Grand Canal should be fully restored. However, if this were to be accomplished, the intersection of the canal with other rivers will inevitably cause disturbances of the regional water systems, repeating the same historical mistakes. By learning about the past, we get a glimpse of the future.

In 2019, a canal channel in the Tongzhou urban area was restored and opened to the public. Also, in June 2021, a 40 km section of the North Canal in Tongzhou City will be opened for traffic use. Residents and visitors will be able to enjoy the scenery of the new canal in this historic site. While it will provide an excellent opportunity for tourists to perceive the canal scene from long ago, it will not be enough for them to fully explore and truly understand the historical role and influence of the canal. Tourism development within each canal city needs to consider its urban characteristics and canal culture. For example, Tianjin can interpret the rise of the city as a transit centre under the influence of the canal, while cities in the northern Jiangsu Province should not blindly restore its canal channels. They can develop more profound attractions and education by revealing both opportunities and disasters of urban development brought about by the canal. In this way, tourism of the northern and southern cities along the canal can be adapted to local conditions and present different aspects of canal culture from multiple angles, increasing the diversity and possibilities of development in tourism cities.

References

Ashworth G, Page S J, 2011. Urban Tourism Research: Recent Progress and Current Paradoxes. *Tourism Management* 32 (1): 1–15.

Bu W, 2012. Research on the Construction and Historical Status of Jining Section of Beijing-Hangzhou Grand Canal in Yuan, Ming and Qing Dynasties. Shenyang: Liaoning University. (In Chinese).

Chen R X, Xia S Z, 1475. Local Gazetteer of Hangzhou Prefecture. Volume 63. Accessed 1 Sept. 2019. http://server.wenzibase.com/ (in Chinese). [(明)陈让修,夏时正纂. 《(成化)杭州府志63卷》. 明成化十一年刻本.].

Chen W, 2013. Walking on the Canal Line: Historical City and Architecture Studies along the Grand Canal. Beijing: China Building Industry Press. (In Chinese).

Chen X W, 1989. The Economic Development of Huzhou Prefecture in the Ming and Qing Dynasties. *Zhejiang Academic Journal* (4): 90–96. (In Chinese).

CHGIS, 2003. Version: 2.0. Cambridge, MA: Harvard Yenching Institute, June.

Christou P, Farmaki A, Evangelou G, 2018. Nurturing Nostalgia?: A Response from Rural Tourism Stakeholders. *Tourism Management* 69 (December): 42–51.

Conzen M R G, 1960. Alnwick. Northumberland: A Study in Town-Plan Analysis. London: Transactions and Papers.

Dai P, Shen Z P, 2013. On the Changes of Water Environment and the Rise and Fall of Xuzhou City. *Human Geography* 28 (6): 55–61. (In Chinese).

Du J, Huang W L, 1760. Local Gazetteer of Hejian Prefecture. Volume 20. Accessed 1 Sept. 2019. http://server.wenzibase.com/ (in Chinese). [(清)杜甲修,黄文莲纂. 《(乾隆)河间府新志20卷》. 清乾隆二十五年刻本.].

Du Y F, Chen S Y, 1573–1620. Local Gazetteer of Hejian Prefecture. Volume 15. Accessed 1 Sept. 2019. http://server.wenzibase.com/ (in Chinese). [(明)杜应芳修,陈士彦纂. 《(万历)河间府志15卷》河间府志序. 明万历刻本.].

Fu C L, 1985. History of Canal Cities in China. Sichuan: Sichuan Renmin Press. (In Chinese).

Gao D G, Gao L G, Zhang J Z, 1685. Local Gazetteer of Zhenjiang Prefecture. Volume 54. Accessed 1 Sept. 2019. http://server.wenzibase.com/ (in Chinese). [(清)高德贵修,高龙光增修, 张九征纂.《(康熙)镇江府志54卷》. 清康熙二十四年刻本.].

Graham B, Ashworth G J, and Tunbridge J E, 2000. A Geography of Heritage: Power, Culture and Economy. London: Arnold.

Hu M F, Shao M, 2010. The Relocation of the Canal and the Rise and Fall of the City: A Study on the Rise and Fall of the Weihe River and Xuzhou Economy in the Ming Dynasty. *Journal of Inner Mongolia Agricultural University* 12 (6): 312–314. (In Chinese).

Kanellopoulou D, 2018. Walking the Public: Re(Visiting) Athens's Historical Centre. *International Journal of Tourism Cities* 4 (3): 298–315.

Le G A, 2018. From Urban Exploration to Ruin Tourism: A Geographical Analysis of Contemporary Ruins as New Frontiers for Urban Tourism. *International Journal of Tourism Cities* 4 (2): 245–260.

Li M W, Feng G F, 1883. Local Gazetteer of Suzhou Prefecture. Volume 150. Accessed 1 Sept. 2019. http://server.wenzibase.com/ (in Chinese). [(清)李铭皖修,冯桂芬纂. 《(同治)苏州府志150卷》. 清光绪九年刊本.].

Li X T, 2012. The Canal and City. Shijiazhuang: Heibei Renmin Press. (In Chinese).

Liang F Z, 1980. Statistics of Chinese Population, Fields, and Taxes in Historical Period. Shanghai: Shanghai Renmin Press. (In Chinese).

Liu G S, Tang H Z, 1618. Local Gazetteer of Changzhou Prefecture. Volume 20. Accessed 1 Sept. 2019. http://server.wenzibase.com/ (in Chinese). [(明)刘广生修, 唐鹤征纂. 《(万历)常州府志20卷》. 明万历四十六年刻本.].

Liu S L, 2008. Narrative of the Grand Canal Urban Group. Shenyang: Liaoning Renmin Press. (In Chinese).

Liu Y K, Shen Y Z, 1600. Local Gazetteer of Jiaxing Prefecture. Volume 32. Accessed 1 Sept. 2019. http://server.wenzibase.com/ (in Chinese). [(明)刘应钶修, 沈尧中纂. 《(万历)嘉兴府志32卷》. 明万历二十八年刊本.].

Lu C A, 1859. Local Gazetteer of Jining Prefecture. Volume 4. Accessed 1 Sept. 2019. http://server.wenzibase.com/ (in Chinese). [(清)卢朝安纂修. 《(咸丰)济宁直隶州续志4卷》. 清咸丰九年刻本.].

Lu X, 1379. Local Gazetteer of Jiaxing Prefecture. Volume 32. Accessed 1 Sept. 2019. http://server.wenzibase.com/ (in Chinese). [(明)盧熊纂修. 《(洪武)蘇州府志50卷》. 明洪武十二年刊本.].

Mileva S V, 2018. Potential of Development of Dark Tourism in Bulgaria. *International Journal of Tourism Cities* 4 (1): 22–39.

Pan J X, 1590. River Harness. Accessed 1 Sept. 2019. http://server.wenzibase.com/ (in Chinese).

Peng X L, Wei Z, 2017. The Rise and Fall of Historical Tongzhou City. *Beijing Planning and Construction* (2): 80–82. (In Chinese).

Richards G, 2014. Creativity and Tourism in the City. *Current Issues in Tourism* 17 (2): 119–144.

Ruan Y, Jiao X, 1806. Local Gazetteer of Yangzhou Prefecture. Volume 8. Accessed 1 Sept. 2019. http://server.wenzibase.com/ (in Chinese). [(清)阮元修,焦循纂. 《(嘉庆)扬州府图经8卷》. 清嘉庆十一年刻本.].

Shen J B, Xu Z L, 1899. Local Gazetteer of Tianjin. Volume 54. Accessed 1 Sept. 2019. http://server.wenzibase.com/ (in Chinese). [(清)沈家本修,徐宗亮纂. 《(光绪)重修天津府志54卷》. 清光绪二十五年刻本.].

Song S, Xie X K, 1808. Local Gazetteer of Dongchang Prefecture. Volume 50. Accessed 1 Sept. 2019. http://server.wenzibase.com/ (in Chinese). [(清)嵩山修,谢香开纂. 《(嘉庆)东昌府志50卷》. 清嘉庆十三年刻本.].

Song Z S, Fang S Z, 1621–1627. Local Gazetteer of Huai'an Prefecture in Tianqi Era. Accessed 1 Sept. 2019. http://server.wenzibase.com/ (in Chinese). [(明)宋祖舜修,方尚祖纂,《(天启)淮安府志》. 明天启刻本.].

Sun J G, Qing Dynasty. South Travel. Accessed 1 Sept. 2019. http://server.wenzibase.com/ (in Chinese).

Tan Q X, 1996. Historical Atlas of China. Beijing: SinoMaps Press. (In Chinese).

Tang X F, Huang Y, 2006. Historical Geography. Beijing: Peking University. (In Chinese).

Wang G X, 2013. The Layout and System Reconstruction of the Cities and Buildings in the Ming Dynasty. Beijing: China Building Industry Press. (In Chinese).

Wang S S, 2015. Historical Atlas of Chinese Urban Habitat Environment. Beijing: Science Press. (In Chinese).

Wang X, 1488–1505. Local Gazetteer of Huzhou Prefecture. Volume 22. Accessed 1 Sept. 2019. http://server.wenzibase.com/ (in Chinese). [(明)王珣撰. 《(弘治)湖州府志22卷》. 清归安姚氏咫进斋钞本.].

Wei Z Z, Gu D G, 1832. Local Gazetteer of Huai'an Prefecture. Volume 32. Accessed 1 Sept. 2019. http://server.wenzibase.com/ (in Chinese). [(清)卫哲治修,顾栋高纂. 《(乾隆)淮安府志32卷》. 清咸丰二年重刊本.].

Wu C, 2014. Cities along the Beijing–Hangzhou Grand Canal. Beijing: Electronic Industry Press. (In Chinese).

Wu S X, Liu X, 1874. Local Gazetteer of Xuzhou Prefecture. Volume 25. Accessed 1 Sept. 2019. http://server.wenzibase.com/ (in Chinese). [(清)吴世熊修,刘庠纂. 《(同治)徐州府志25卷》. 清同治十三年刻本.].

Wu X, Wang C H, Xu H L, Sun Q W, 2013. Analysis of the Interaction between the Grand Canal (Changzhou Section) and Urban Development. *Urban Planning* 37 (5): 61–66. (In Chinese).

Xu Y G, Wu Y X, 1879. Local Gazetteer of Jiaxing Prefecture. Volume 88. Accessed 1 Sept. 2019. http://server.wenzibase.com/ (in Chinese). [(清)许瑶光修,吴仰贤纂. 《(光绪)嘉兴府志88卷》. 清光绪五年刊本.].

Xue Z D, Gao B D, 1934. Local Gazetteer of Tianjin. Volume 4. Accessed 1 Sept. 2019. http://server.wenzibase.com/ (in Chinese). [(清)薛柱斗修,高必大纂. 《(康熙)天津卫志4卷》. 民国二十三年铅印本.].

Yan X S, Cheng C, 1760. Local Gazetteer of Tai'an Prefecture. Volume 30. Accessed 1 Sept. 2019. http://server.wenzibase.com/ (in Chinese). [(清)颜希深修,成城纂. 《(乾隆)泰安府志30卷》. 清乾隆二十五年刻本.].

Yang J, Zhang J C, Wu Y X, Zhuang J R, 2014. Evolution of Ecological Environment in Typical Areas along the Beijing–Hangzhou Grand Canal. Beijing: Electronic Industry Press. (In Chinese).

Yang X, Xu L, 1573–1620. Local Gazetteer of Yangzhou Prefecture. Volume 27. 1 Accessed 1 Sept. 2019. http://server.wenzibase.com/ (in Chinese). [(明)杨洵修,徐銮纂. 《(万历)扬州府志27卷》. 明万历刻本.].

Yu K J, 2012. Beijing–Hangzhou Grand Canal National Heritage and Ecological Corridor. Beijing: Peking University Press. (In Chinese).

Zhang D X, Zhu Z, 1785. Local Gazetteer of Linqing Prefecture. Volume 11. Accessed 1 Sept. 2019. http://server.wenzibase.com/ (in Chinese). [(清)张度修,朱钟纂. 《(乾隆)临清直隶州志11卷》. 清乾隆五十年刻本.].

Zheng M, 2013. Study on the Storage System along the Grand Canal in Ming and Qing. Nankai University. (In Chinese).

Zheng Y, He H, 1521–1567. Local Gazetteer of Dezhou Prefecture. Volume 3. Accessed 1 Sept. 2019. http://server.wenzibase.com/ (in Chinese). [(明)郑瀛修,何洪纂. 《(嘉靖)德州志3卷》. 明嘉靖刻本.].

Zhu H, Qing Dynasty. Research on Rivers. Accessed 1 Sept. 2019. http://server.wenzibase.com/ (in Chinese).

39

MACAO AS A CITY OF GASTRONOMY

The role of cuisine in a tourism product bundle

Kim-Ieng Loi, Weng-Hang Kong and Hugo Robarts Bandeira

Background

Macao was a small and relatively unknown destination to the world until around two decades ago when it was handed over to China in 1999, putting an end to the over 400-year Portuguese regime. In the past, people often thought of Macao as "a city next to Hong Kong" or "the gambling city". When it was returned to China as one of the two Special Administrative Regions (SAR), the other one being Hong Kong, which was handed over two years earlier, this identity of Macao "via Hong Kong" was even more salient. After 20 years of rapid development, especially after the liberalisation of the gaming industry in 2002, Macao has started to gain recognition in its own right. Several major events that brought Macao under the international spot light include its historic city centre being inscribed on the World Heritage Site list in 2005 (Cultural Affairs Bureau, 2019); its gaming revenue exceeding that of Las Vegas for the first time in 2006 (Holmes, 2010); formally announcing Macao as the World Centre of Tourism and Leisure in the 12th National Five-Year Plan of 2011 (Meneses, 2018); and being officially designated as a Creative City of Gastronomy by UNESCO in 2017 (*Macau Daily Times*, 2019). All these share one common characteristic, that they are about important components of the tourism industry in its broad sense. The unilateral reliance of Macao's economy on gaming revenue has called for the need of a more diversified tourism portfolio and gastronomy has been one of the foci, especially after the recent designation as City of Gastronomy. Gastronomy can be undoubtedly considered as an indispensable part of the tourism experience and thus makes it an ideal product for tourist consumption. The old adage goes: "We are what we eat", and so are the tourists visiting a destination. The uniqueness of gastronomy in Macao is heavily embedded in its Macanese cuisine which carries a long history in Macao, dating back to when the Portuguese first arrived in Macao more than 400 years ago, bringing with them food ingredients, spices and cooking ideas from their earlier settlements in Africa, South America, India and of course from home (Portugal). These were adapted and combined with the local Chinese way of cooking to create a truly international cuisine, which led to the creation of the first so-called "fusion food" in the world.

Macanese cuisine as cultural identity and tourist experience – status quo

Macanese cuisine is deeply rooted in the collective memory of the Macanese community. It is often said among experts that three things define the sense of identity of being Macanese, those being the Catholic religion, the unique Macanese *patois* language known as *Patuá* and the Macanese cuisine. Within the Macanese community both in Macao as well as in the diaspora, food is almost always the catalyst for the sense of belonging to that special group. It is a tradition that has been passed down generation after generation, the proudest moment for many families, and definitely a reason for gatherings, reminiscences and nostalgia. Every Macanese family can point to a member or members of the family who are remembered for their great cooking skills. Therefore, it is not exaggerating to say that Macanese cuisine is part of the DNA of Macao as a multicultural city and a defining moment in the identity of its Macanese community. As such Macanese cuisine is a unique selling point for a city that wants to diversify its tourist offerings. Much can be done to preserve this unique characteristic of Macao and to explore it positively to promote a sustainable and diversified tourism industry, where gastronomy tourism can have an impact. If Macao is to pursue its path to become the World Centre of Tourism and Leisure with gastronomy being one of the highlights, then Macanese cuisine could be the *raison d'être*. In the quest for doing so the Macao government has been putting in efforts to protect and promote this Macanese cuisine as both a cultural icon and a unique selling proposition locally, regionally and internationally. Some key milestones in the development history of Macanese gastronomy in Macao are included below.

One earlier move was the establishment of the Macanese Gastronomy Association in 2007. The idea back then was already to enhance research and promotion of the Macanese gastronomy heritage. After 12 years of establishment much has been done, at least in terms of promoting this local cuisine to the region and the world. The association became part of the European Oenogastronomic Brotherhoods Council, first as an observer, soon after as a full member, and now even assuming a position in the vice-presidency of the council. Although not a typical member, after all Macao is not in Europe, the strong link of Macanese cuisine to Portugal plays a large role. The association has been regularly participating in the council congresses, through which international awareness about the local cuisine has been enhanced. The association has been involved in several food promotion events both locally as well as internationally, by having some of its members demonstrate cooking and promote the Macanese cuisine to outsiders.

In 2010, the Macao Institute for Tourism Studies (IFTM, a publicly owned tertiary education institute specialising in the areas of tourism and hospitality in Macao) signed a cooperation agreement with the Macanese Gastronomy Association to promote the exchange of knowledge on Macanese gastronomy, collaborate in event organisation and publish about Macanese cuisine. At the same time, IFTM has been providing professional training courses in Macanese cuisine, and has added Macanese cuisine in the undergraduate teaching curriculum to allow youngsters (be they Macanese or not) to learn about the origins, transformation and characteristics of Macanese cuisine.

In 2011, world renowned chef the late Anthony Bourdain shot one of his episodes for the acclaimed television show "No Reservations" here in Macao, and Macanese cuisine was one of the main features of the episode. This move pushed the cuisine further onto the world stage.

In February 2012, Macanese cuisine was included in the city's official intangible cultural heritage list, and in November 2017 Macao was proudly announced as a member of the UNESCO Creative Cities Network in the field of gastronomy, with Macanese cuisine – often dubbed the "world's first fusion cuisine" – playing a pivotal role in UNESCO's decision.

Through these events it is obvious that the Macao government has repeatedly stressed the importance of Macanese cuisine in the city's unique gastronomic culture, and in the development of a sustainable tourism industry. As Macao works to transform itself into a World Centre of Tourism and Leisure, developing Macao into a gastronomy tourism destination can help approach this goal. In this regard, it is important then to define what food/gastronomy tourism is. Although different scholars may have different definitions for this type of tourism and sometimes the defining line (from the general type of tourism) might be thin, they do not deviate too much from the main traits. In general people agree that food/gastronomy tourism occurs when people travel for the specific purpose of enjoying food experiences (Getz, Robinson, Andersson & Vujicic, 2014; Stone, Migacz & Wolf, 2019) or the need of food as a primary factor in influencing travel behaviour and decision making (Hall, Sharples, Mitchell, Macionis and Cambourne, 2003).

Development and challenges

Despite the aforementioned efforts, there is still a generic confusion about what Macanese cuisine really is. The most common and natural occurrence is the confusion with Portuguese cuisine. Restaurants all around the city claim to serve Macanese dishes, but in reality, many of them are traditional Portuguese dishes; the same happens the other way around, though with less frequency. Both locals and tourists are misled every day in these regards. In addition, a vast majority of the guides to Macao and even government tourist brochures at times add to the confusion by regularly mislabelling or claiming (for marketing purposes) items that are simply "Made in Macao" as somehow Macanese. Even in the younger generations of the Macanese community this confusion prevails. Like everyone else, they know Macanese cuisine exists, everyone says they like it, but they are not clear about what it really is! The number of restaurants that call themselves authentic Macanese restaurants or are classified as such is only a handful, but they also have Portuguese dishes on their menus. Then there are the Portuguese restaurants that also serve Macanese dishes ... so confusion reigns.

Macanese cuisine in general is not commercial or restaurant friendly. It has always been a typical home cooking kind of gastronomy, with long preparation time that typically spans one or two days in advance for marinates, diverse ingredient stews, a great variety of recipes for the same dish and the very "homely" presentation of the dish itself can constitute a challenge for commercial restaurants. In the past the preparation of the Macanese dishes was deeply linked to the social structure of Macao in which Macanese families were mostly from the more well-off class, where the Macanese women usually did not work and were managing the household with big kitchens surrounded by many helpers. Thus, there was enough time and manpower to prepare this type of food. Also, the fact that it is a heavy cuisine with ingredients, cooking methods and dishes that clearly do not fit the contemporary world trend of healthy dietary options, contributes to the slow demise of the Macanese cuisine. The deep-frying method and the use of pork lard in most dishes are some classic examples. As can be imagined, trying to transfer that to a modern restaurant kitchen is, to say the least, challenging. Other problems derive from the fact that little or no development has occurred in the last 50 years or so to this cuisine. It is a stagnant cuisine and not adapted to modern times and current consumer trends, at least in Macao. The situation is slightly different in Macanese communities living abroad (the diaspora) who have to adapt the dishes to the local availability of ingredients.

The future and sustainability of any cuisine depends on its evolution, development and to a certain extent its adaptation to the consumers' needs. In the case of Macanese cuisine, it is still pretty much the same as it was 50 years ago – being cooked mostly at home, by fast ageing and naturally disappearing members of the community that have a peculiar reluctance in passing down their recipes and cooking secrets to outsiders. Fewer and fewer people are now capable of cooking authentic traditional Macanese cuisine and there is an apparent lack of interest from the younger generations in continuing this tradition as they prefer to dine out or eat more readily available food. Macanese food has become the ritual only when they visit parents or other senior members of the family and most probably only on festive days.

The problem, as mentioned before, is that Macanese cuisine is becoming more and more a "cook book phenomenon" and as a collective memory of a gradually disappearing community. As the community becomes more "Chinese" or "Western" (as in the diaspora) it is only natural that cultural traditions and beliefs get diluted and lost through acculturation. If the Macanese cuisine is going to go beyond the cook books and the collective memory to become part of the current "trend", then it needs to be modernised, not necessarily reinvented, but needs to reflect what the consumers want, without largely compromising on the authenticity. Nevertheless, change may not be easy; there is high resistance from older generations. In addition, there is a lack of consensus on what "authenticity" means and what the original recipes are, if that can ever be defined. If we take any one of the "classic" Macanese dishes and start to collect recipes from the different cook books or different families and put them side by side, it is immediately obvious that they vary, some even to a huge extent. Which one is the right or the original one? No one has a firm say. Since Macanese cuisine was mainly home cooking with hand-written recipes passed on for generations, many recipes are difficult to decodify and interpret, with no proper structure that clearly states accurate measurements and methods of cooking. Some older recipes use measurements like "10 cents of this" or a "handful of that", clearly impossible to accurately define the right quantity without having to go through several trial-and-error attempts. The proposition of substituting original ingredients with more available, modern and healthier alternatives is thought to affect the taste of the food and thus its authenticity.

Then there is the problem of presentation of the dishes. Due to the various amounts of ingredients in many Macanese dishes, traditionally the food has been presented in large portions, in a platter or served in the original cooking pot (the so-called family style presentation). The colours of the ingredients tend to be all similar, not creating a contrast that an appealing dish presentation requires. In an effort to protect and promote this special cuisine, for nine years now, IFTM and the Macau Culinary Association have been co-organising competitions both for young talents as well as professional chefs, where Macanese cuisine is the theme of the competition. Creativity, modernity and a new vision of Macanese cuisine has been put to the test and demonstrated that it is possible to transform, modernise and make the food more restaurant-presentable and friendly, without losing its typicality or traditional flavours.

Ways to move forward

As a new member of the UNESCO Creative Cities Network in the field of gastronomy, the Macao government has established a plan for the next four years which includes, among others, the promotion and preservation of Macanese cuisine through different channels and methods. In a joint effort between several government departments, universities and local associations, a working group was established to handle this. For instance, one of the tasks is to create an extensive database of Macanese cuisine recipes. At the moment IFTM is already establishing a database with all the Macanese cuisine recipes available in published books as well as trying to collect them from other sources. This database will feature the recipes in three languages (Chinese, Portuguese and English). The idea is to have a centralised database with as much material as possible so that the recipes do not get lost with time. In the course of creating this database it was found that several dishes from Macanese cuisine are not in use either because they are difficult to make or because nobody knows how to make them anymore. Another idea is to test the recipes and try to define what the dishes should look and taste like, according to the memories of the members of the Macanese Gastronomy Association, to get as close as possible to the original flavours.

The creation of touristic culinary routes that include the few Macanese restaurants available could raise more awareness about Macanese cuisine and even spark an interest and create opportunities for new Macanese restaurants to open. The government has already announced this initiative, though no established routes are in action yet. There is also a scattered effort in the private sector where individual travel agencies design and promote a few gastronomy routes for tourists to choose. Nevertheless, a more concerted effort is desirable to achieve an impact. This requires the creation of controllable measures and procedures that certify and regulate the authenticity of the food being served along these tourist routes. Of course this also implies that there must be a clear definition of what authentic Macanese food should be. One consideration can be the establishment of an official certification body so that the authenticity of the food can be certified and preserved for posterity. Other suggestions include the creation of a museum with both physical and virtual exhibits, where all the different aspects of Macanese cuisine could be explored, from storytelling, origins, cultural links, traditions, equipment used, recipes as well as the possibility for actual sampling and even cooking the food.

Gastronomy tourism in other destinations

Gastronomic experiences are developed from the uniqueness of gastronomy that can only be found in a particular destination. Accordingly, a specific destination can be associated with or identified from its gastronomic offerings which become a powerful tourism marketing tool (Brokaj, 2014). Gastronomy tourism has become a rapidly growing component of the attractiveness of tourism destination. Public interest has been steadily increasing in cuisines recently; therefore, lots of tourism destinations are using gastronomy to market themselves to shape their images. It is also the reason that the popularity and proliferation of gastronomy tourism are exhibited in different destinations (Brokaj, 2014), such as the case of Macao (Kong, du Cros & Ong, 2015). Gastronomy tourism is considered to be very crucial in that it can make a tourist's trip a very unique one, therefore, various gastronomic tours are arranged to introduce the destinations' dishes and food culture which play an essential role in the selection of and experience tourists can have in a destination (Sormaz, Akmese, Gunes & Aras, 2016). Besides, Corigliano (2002) mentions that tourists not only visit cultural and historical sites but also explore regions and landscapes as a whole. Gastronomic supply can, therefore, shape tourist demand and highlight potential wine and gastronomic products in international tourist markets.

Previous studies have investigated whether geographical conditions and climate can affect the food produced (Guzel & Apaydin, 2016). The combination of nature and produce also offers opportunities to form a significant part of tourists' experience and image of a destination that go beyond the mere appreciation of food itself (Getz, Robinson, Andersson & Vujicic, 2014). The Scandinavian countries are successful examples of offering multi-faceted gastronomy experiences, with their relatively new Nordic cuisine being celebrated for its healthy lifestyle, flavourful ingredients, innovative chefs and restaurants (Byrkjeflot, Pedersen & Svejenova, 2013). In 2008 the Swedish government presented its vision for Sweden as "The New Culinary Nation". This vision was built around their climate, diversity of produce and manufacturing methods, nature and chefs which they view as their unique advantage. With such clear vision, Ostersund in Sweden was subsequently nominated into the Creative Cities of Gastronomy network of UNESCO in 2010. Approximately 40% of travellers would like to engage in culinary experiences when travelling to Sweden (Visit Sweden, 2013). Norway and Denmark are both famous for wild gastronomy, combining food with nature. Hunting in the autumn and fishing in the ice-cold sea along the Nordic coast are experiences that can't be found anywhere else. All these simply mean that gastronomy tourism can go beyond the food offerings to embrace a unique and holistic experience.

Many destinations, like Italy, France and Thailand, have become very popular with their cuisines and now attract many tourists (Karim & Chi, 2010). Italy has been dubbed a wine country with its "Enogastronomic tourism" as a significant trend in tourism development (Corigliano, 2002). According to *Tourism Review News* (2019), the most visited regions within Italy are Sicily, Tuscany and Emilia Romagna. The most adored cities are Naples, Rome and Florence. What is it that they have in common? Gastronomy! Sixty-two per cent of tourists have taken food and wine themed packages into consideration when planning their itineraries, especially in Tuscany (72%), where holiday opportunities are bundled with cooking courses for an enhanced experience (Kivela & Crotts, 2006). One cannot miss France as far as gastronomy is concerned as it is one of the world's most established and respected gastronomic cultures. It is a country for which the culinary arts have played a central role in the destination. In 2010, the "Gastronomic meal of the French" became the first culinary heritage asset to be inscribed in UNESCO's Intangible Cultural Heritage list (Suntikul, 2019). Gastronomic tourists visiting France seek the combination of food, wine and art which involve all five senses. Scotland, on the other hand, uses the "Taste of Scotland" campaign to revitalise traditions and make food tourism feasible through organising food-related events and enhancing the food and drink (in particular whisky) experience (Corigliano, 2002).

Now let's move on to Asia, the context with which the authors are most familiar. The Japan Travel and Tourism Association (JTTA) has recently released a report related to gastronomy tourism in Japan. Although it is a new development direction in Japan, it has been recorded with steady growth, providing economic and social benefits to the destinations (UNWTO, 2019). Washoku (Japanese cuisine) is registered as a UNESCO Intangible Cultural Heritage. It is a diet with well-balanced nutrition by taking advantage of the diversified and fresh ingredients produced within the country's national territory. Additionally, the sense of seasonal transition with the beauty of nature is represented in the various aspects of local cuisine, from the selection of ingredients, preparation methods, pairing with sake (Japanese rice wine) to presentation. Many people say that the plating of Washoku is like a painting itself. As a result, "to eat Japanese food" has become the number one motivation for tourists visiting Japan (Japan Tourism Agency, 2016).

Kimchi-making, one of the significant food cultures of South Korea, is also designated as an Intangible Cultural Heritage by UNESCO (*South China Morning Post*, 2017). The importance of kimchi goes beyond the food itself as the tradition of kimchi-making illustrates the importance of social function through the preparation process. The production process of kimchi is known as "kimjang" in Korean. It is a process starting from selecting, collecting, marinating and preserving the vegetables. Every Korean family makes kimchi and the scale of production goes even to a village or a community. The entire process boosts cooperation among families, villages and communities, contributing to social cohesion and harmony. Friends and relatives share know-how of kimchi-making. It is this process of making kimchi which enhances social bonding and inter-personal relationships. Increasing knowledge in and appreciation of this tradition has raised the awareness of Korean food culture among foreigners, providing a boost for gastronomy tourism. Attending one or two workshops on making kimchi and other traditional Korean dishes has become a top to-do item on many tourists' agendas. Destination marketing organisations (DMOs) partnering with TV and other broadcasters produce programmes covering a series of culinary tours, cooking classes and other events to promote signature dishes and locally grown agricultural products in South Korea. Furthermore, the Ministry of Agriculture, Food, and Rural Affairs has been organising food tour programmes for Koreans and non-Koreans to visit municipalities and cook and taste their iconic dishes (*The Korea Times*, 2017). These, together with the rapid proliferation of Korean pop (K-pop) culture world-wide, have indirectly helped the promotion of gastronomy tourism in South Korea.

Thailand is known for its easy access to restaurants, prices and friendly employees and English menu options. From simple street food to posh restaurants, from tried and tested to trendy and new, such as kuai tiao (noodle soup), khao man kai (chicken and rice) and spicy som tam (papaya salad). Thai dishes also place great emphasis on nutritional ingredients, including high-value herbs, which promote a sense of health and general wellness (TAT News, 2019). The Thai government endeavours to highlight such characteristics by packaging and promoting Thai cuisine based on five main culinary regions: the north (including the city of Chiangmai), the northeast, the south (including the Gulf of Thailand), the central plains and the capital Bangkok. Locations for food tasting in various regions can motivate and enhance tourists' experiences. At the same time, food tasting can demonstrate the delicate variation in food culture across regions which form the values of Thailand as a whole. According to Jing Travel (2017), the Tourism Authority of Thailand (TAT) has been playing a significant role in promoting the kingdom's gastronomy around the world. Thailand launched a campaign called Amazing Thailand Tourism Year 2018 with a strong focus on the country's local experiences, including its renowned cuisine. Starting from 1 November 2017 and running until 1 January 2019, the event calendar schedule includes the holding of the 4th UNWTO World Forum on Gastronomy Tourism. One strategic approach the Thai government uses to improve the quality of gastronomic and culinary products/experience is to encourage residents to integrate tourism and local food products into their daily economic activities and to respond to the specific needs of tourists such as visiting food producers (like local markets), tasting Thai cuisine in authentic restaurants, observing the preparation processes from the experts or attending cooking classes to understand the authentic way of making Thai food. By doing so the agriculture sector (through local producers) and the gastronomic cultural tradition of Thailand can be preserved while tourism development can be more sustainable. On the commercial side, TAT pledges five-year financial support to a project that will bring the Michelin Guide to Thailand, making it the sixth place in Asia with its own Michelin Guide. It is believed that such recognition can guide tourists in their quest for quality gastronomic experiences and provide a gauge for assessment according to preference.

As exemplified by the several cases above, it is obvious that gastronomy tourism is not only about eating and drinking, but also involves a transfer of knowledge and information about the culture, history, traditions and identity of a geographic area and the involvement of the local community (Ignatov & Smith, 2006). Therefore, it is not enough if a destination is merely affiliated with a certain kind of cuisine with general stereotypes (such as pizza for Italy and sushi in Japan). To be able to capitalise on gastronomy as a tourism marketing tool it must go beyond this superficial aspect. It should embrace the learning, appreciating and consuming of the food and drink in a unique destination experience.

Conclusion – food as an important tourist product and experience

From the experience of Macao as well as other destinations illustrated in this chapter, there is no doubt that gastronomy is a central part of the overall tourist experience. For many tourists, returning to familiar destinations to enjoy recipes that they have tried before or to look for new ones is one of the must-do activities on their checklist. UNWTO has published three reports on food/gastronomy tourism in 2012, 2017 and 2019 respectively (UNWTO, 2012, 2017 & 2019), further highlighting the importance of this fast-developing market segment. Not only are tourists more actively seeking to participate in food hunts while visiting a destination, tourism bureaus (or equivalent) or destination marketers are also aware of the benefits brought about by diversifying a tourism product portfolio and sourcing an alternative/complementary strategy amid the limitations posted by other types of tourism development which require hardware support and are not so easily and readily adaptable (e.g. natural scenery, ocean tourism, heritage tourism and the like). In the UNWTO latest report, the Secretary-General Mr Zurab Pololi-kashvili commented in the foreword that: "Gastronomy tourism presents an immense opportunity to promote local culture, diversify tourism demand, enhance the value chain, create jobs and spread the benefits of tourism throughout the territory" (UNWTO, 2019, p. 4).

The former Secretary-General Mr Taleb Rifai also shared similar remarks in the earlier report on food tourism, commenting that food can enable destinations to market themselves as truly unique and appealing to tourists via flavours (UNWTO, 2012). Unlike a landmark, a single tourist attraction or natural scenery which is a standalone independent place, food penetrates into every corner of a destination and it carries both sensational and cultural aspects for tourists' enjoyment so the benefits generated from gastronomy tourism can be extended to wider coverage. In addition, the smell and taste of food have strong roles to play in calling up memories. Food is an effective trigger of deeper memories of feelings and emotions, internal states of the mind and body (Harvard University Press Blog, 2012). In this aspect, taste can be considered a strong retrieval cue to bring back memories of an emotional event, such as a visit to a certain destination. Memory is an important part of the construction of the tourism landscape itself (Marschall, 2012). Urry (1990) argues that tourism is defined by the collection of signs, as tourists, driven by the memory of iconic images, travel in search of familiar sites/sights. Since tourism development of any destination cannot rely solely on a one-time visit, ample academic attention has been focused on how tourists select destinations (destination choice), whether or not to repeat visit (revisit intention and loyalty) and to talk positively about the destination (word of mouth). The role of memory is considered to be decisive in this regard as it exerts influence on the tourists before the trip (destination choice), during the trip (itinerary formation) and after the trip (revisit intention, word of mouth, reviews, comments and recommendations). Therefore, capitalising on this food–memory–tourism relationship can be an effective strategy for tourism marketers with the advent of internet and people's increasing reliance on key opinion leaders (KOL) rather than supplier-generated propaganda. In an earlier study, Hu, Manikonda & Kambhampati (2014) found that over 10% of the pictures posted on Instagram (a platform where over 40 billion photos have been shared) are related to food and that food is one of the top performers when it comes to attractive visual content on Instagram (Bullas, n.d.; Smith, 2019). Food pictures are highly "instagrammable" because they are easy to produce and relevant to everyone. Their colourful visual appeal enhances the aesthetic aspect of the posts and can catch immediate attention. "Foodstagramming" is a recently coined term to describe this phenomenon. Many tourists used to say "been there done that" and nowadays people also say "pics or it didn't happen". For many, food is a very good "cast member" to be captured as a showcase or evidence that a tourist experience (be it positive or negative) actually has taken place and this photogenic experience plays a key role in the "mixing in memorabilia" aspect of creating and managing experience proposed by Pine and Gilmore (1998). Experience is the "take-away" memory which often requires cues to sustain and recall. As previously mentioned, food/gastronomy (including its visual/olfactory appeal and taste) as a tourism activity or

Table 39.1 Economic distinctions

Economic offering	Services	Experiences
Economy	Service	Experience
Economic Function	Deliver	Stage
Nature of Offering	Intangible	Memorable
Key Attribute	Customised	Personal
Method of Supply	Delivered on Demand	Revealed over a Duration
Seller	Provider	Stager
Buyer	Client	Guest
Factors of Demand	Benefits	Sensations

Source: Extracted from Pine and Gilmore (1998).

product can be one very good piece of memorabilia that helps sustain the memory (which eventually becomes experience). After all, to stay competitive, the tourism industry should transform from a traditional service industry into an experience industry where the nature of offering should be memorable and revealed over time with the help of some forms of triggers or cues (Table 39.1).

References

Brokaj, M. (2014). The impact of the gastronomic offer in choosing tourism destination: The case of Albania. *International Journal of Interdisciplinary Research*, 1(4), 464–481.

Bullas, J. (n.d.). *10 Types of Visual Content on Instagram That Get Shared Like Crazy*. Retrieved on 19 July 2019 from www.jeffbullas.com/10-types-visual-content-instagram-get-shared-like-crazy/.

Byrkjeflot, H., Pedersen, J. S., & Svejenova, S. (2013). From label to practice: The process of creating new Nordic cuisine. *Journal of Culinary Science & Technology*, 11(1), 36–55.

Corigliano, M. A. (2002). The route to quality: Italian gastronomy networks in operation. In A. M. Hjalager and G. Richards (Eds.). *Tourism and Gastronomy* (pp. 166–185). London: Routledge.

Cultural Affairs Bureau. (2019). Visit Information. Retrieved on 25 September 2019 from www.wh.mo/en/.

Getz, D., Robinson, R., Andersson, T., & Vujicic, S. (2014). *Foodies and Food Tourism*. Oxford: Goodfellow Publishers.

Guzel, B., & Apaydin, M. (2016). Gastronomy tourism: Motivations and destinations. In C. Avcıkurt, M. S. Dinu, H. Hacioğlu, R. Efe, A. Soykan, & N. Tetik (Eds.). *Global Issues and Trends in Tourism* (pp. 394–404). Sofia: St Kliment Ohridski University Press.

Hall, C. M., Sharples, L., Mitchell, R., Macionis, N., & Cambourne, B. (Eds.). (2003). *Food Tourism around the World*. Oxford: Elsevier.

Harvard University Press Blog. (2012). Food and memory. Harvard University Press, 18 May. Retrieved on 19 July 2019 from https://harvardpress.typepad.com/hup_publicity/2012/05/food-and-memory-john-allen.html.

Holmes, S. (2010). Singapore casinos could rival Las Vegas by 2012. *Wall Street Journal*, 18 August. Retrieved on 17 July 2019 from www.wsj.com/articles/SB10001424052748703649004575436743057042602.

Hu, Y., Manikonda, L., & Kambhampati, S. (2014). What we Instagram: A first analysis of Instagram photo content and user types. In *Eighth International AAAI Conference on Weblogs and Social Media*.

Ignatov, E., & Smith, S. (2006). Segmenting Canadian culinary tourists. *Current Issues in Tourism*, 9(3), 235–255.

Japan Tourism Agency. (2016). *Consumption Trend Survey for Foreigners Visiting Japan*. Tokyo.

Jing Travel. (2017). Thailand bets on gastronomy for boosting tourism revenue. Retrieved from https://jingtravel.com/thailand-bets-on-gastronomy-for-boosting-tourism-revenue/.

Karim, A. S., & Chi, C. G. Q. (2010). Culinary tourism as a destination attraction: An empirical examination of destinations' food image. *Journal of Hospitality Marketing & Management*, 19(6), 531–555.

Kivela, J., & Crotts, J. (2006). Tourism and gastronomy: Gastronomy's influence on how tourists experience a destination. *Journal of Hospitality & Tourism Research*, 30, 354–377.

Kong, W. H., du Cros, H., & Ong, C. E. (2015). Tourism destination image development: A lesson from Macau. *International Journal of Tourism Cities*, 1(4), 299–316.

Macau Daily Times. (2019). Macau designated "Creative City of Gastronomy" by UNESCO. Retrieved on 25 September 2019 from https://macaudailytimes.com.mo/macau-designated-creative-city-gastronomy-unesco.html.

Marschall, S. (2012). "Personal memory tourism" and a wider exploration of the tourism–memory nexus. *Journal of Tourism and Cultural Change*, 10(4), 321–335.

Meneses, J. P. (2018). Macau aka World Centre of Tourism and Leisure. *Macau Business*, 15 June. Retrieved on 17 July 2019 from www.macaubusiness.com/macau-aka-world-centre-of-tourism-and-leisure/.

Pine, B. J., & Gilmore, J. H. (1998). Welcome to the experience economy. *Harvard Business Review*, 76, 97–105.

Smith, K. (2019). 49 incredible Instagram statistics, 7 May. Retrieved on 19 July 2019 from www.brandwatch.com/blog/instagram-stats/.

Sormaz, U., Akmese, H., Gunes, E., & Aras, S. (2016). Gastronomy in tourism. *Procedia Economics and Finance*, 39, 725–730.

South China Morning Post. (2017). Kimchi-making becomes intangible heritage in South Korea. Retrieved from www.scmp.com/news/asia/east-asia/article/2120194/kimchi-making-becomes-intangible-heritage-south-korea.

Stone, M. S., Migacz, S., & Wolf, E. (2019). Beyond the journey: The lasting impact of culinary tourism activities. *Current Issues in Tourism*, 22(2), 147–152.

Suntikul, W. (2019). Gastrodiplomacy in tourism. *Current Issues in Tourism*, 22(9), 1076–1094. DOI: 10.1080/13683500.2017.1363723.

TAT News. (2019). Gastronomy tourism: Thailand means "good food" in any language. Retrieved on 22 July 2019 from www.tatnews.org/2019/02/gastronomy-tourism-thailand-means-good-food-in-any-language/.

The Korea Times. (2017). Korea promotes culinary tourism. Retrieved from www.koreatimes.co.kr/www/biz/2017/04/602_228109.html.

Tourism Review News. (2019). Italian gastro-tourism registers growth. Retrieved from www.tourism-review.com/gastro-tourism-in-italy-still-very-attractive-news10928.

UNWTO. (2012). *Global Report on Food Tourism*. Madrid: UNWTO. Retrieved on 19 July 2019 from www2.unwto.org/publication/unwto-am-report-vol-4-global-report-food-tourism.

UNWTO. (2017). *Affiliate Members Report, Volume 16: Second Global Report on Gastronomy Tourism*. Madrid: UNWTO. Retrieved on 19 July 2019 from http://cf.cdn.unwto.org/sites/all/files/pdf/gastronomy_report_web.pdf.

UNWTO. (2019). *Gastronomy Tourism: The Case of Japan*. Retrieved on 22 July 2019 from www.e-unwto.org/doi/book/10.18111/9789284420919.

Urry, J. (1990). *The Tourist Gaze: Leisure and Travel in Contemporary Societies* (2nd ed., 1998). London: Sage.

Visit Sweden. (2013). Retrieved on 24 September 2019 from www.visitsweden.com/sweden/.

40

THE PRESENT SITUATION OF WUHAN CITY TOURISM DEVELOPMENT

Yan (Mary) Mao

Background

Wuhan, the only sub-provincial city in central China, is the capital of Hubei Province with an area of 8,494 square kilometres. The current number of registered permanent residents is around 12 million. The world's third longest river, the Yangtze River, and its greatest branch, the Hanshui River, flow across the city and divide it into three parts, namely Wuchang, Hankou and Hanyang. Wuhan is also known as the "city of hundreds of lakes". East Lake and Tangxun Lake are the largest and second largest lakes within a city in China. Hence the tourism image of "Big River, Big Lake and Big Wuhan" is deeply rooted in people's hearts.

Wuhan enjoys a 3,500-year time-honoured history and is one of the birthplaces of Chu culture. The city's rich history and culture also includes the profound "Spirit of 1911 Revolution" and "Spirit of Fighting Yangtze River Floods" in modern times. It is the city's profound cultural background that has, through the process of generation-to-generation transmission, formed Wuhan's unique tourism resources.

Wuhan has been regarded as "a thoroughfare to nine provinces" and is one of the few transportation hubs in China integrating railways, waterways, highways and aviation. By 2020, it is predicted that departing from Wuhan, people can arrive in central cities of major economic zones in China within five hours. The convenience of its external transportation can fully arouse the enthusiasm of tourists. Meanwhile, Wuhan is adjacent to several large and medium-sized cities and weekend vacation tours are favoured by tourists from neighbouring cities. In addition, inbound and outbound tourism is developing rapidly as Wuhan has 40 direct international flights. It is the only city in central China with direct flights to four continents.

Wuhan has become an important tourist destination in central China. In 2018, Wuhan received 285 million domestic tourists, an increase of 10.9% over the previous year. Domestic tourism revenue was 303.755 billion yuan, an increase of 12.6% over the past year. The number of overseas tourists was 2.76 million, with year-on-year growth of 10.4%, and foreign exchange income from tourism was 1.883 billion US dollars, with 11.3% year-on-year growth. With the successful holding of the 7th CISM Military World Games and the World Flight Conference in 2019, Wuhan will have new opportunities for further tourism development.

Figure 40.1   Yellow Crane Tower, Wuhan.

City tourism product development of Wuhan

At present, Wuhan has three national 5A level scenic spots that respectively are Yellow Crane Tower (Figure 40.1), East Lake ecological scenic spot (Figure 40.2) and Huangpi Mulan Cultural Ecotourism Area (Figure 40.3), and 19 4A level tourist attractions, and as a result is ranked first in the list of sub-provincial cities in China. Nine of the top ten hotel group companies in

Figure 40.2   East Lake ecological scenic spot, Wuhan.

the world have been stationed in Wuhan, the number of five-star hotels in the city has reached 15, and the number of the travel agencies is up to 390, including 84 travel agencies qualified for outbound tourism, all of which makes the Wuhan rank first in the list of vice-provincial cities in the country in terms of number and comprehensive strength. Tourism has become an important pillar industry and leading industry for Wuhan to move towards the two construction goals of "National Central City" and "International Metropolis".

Tourism products with an emphasis on cultural museums have been pursued and admired in Wuhan, such as the Museum of Revolution of 1911, the Wuhan Revolutionary Museum, the Zhang Zhidong Museum, the Zhongshan Ship Museum and the Hubei Provincial Museum; characteristic tourism blocks are represented by Chuhe Han Street, Hubu Lane Style Street, Hankou Xintiandi, Wuchang Tan Hualin and more. The tourism product "Enjoying the Beauty of Flowers", with a focus on five major flowers – such as oriental cherry, plum blossom, azalea, tulips and Chinese flowering crab-apple – has gradually been growing in maturity. Wuhan has been ranked as the "Top Ten Flower Enjoying Destination in China" for five years. Additionally, the night view landscape of Wuhan Two Rivers (Yangtze River and Han River) and Four Banks centred on the Jianghan Chaozong scenic spot has gradually become a well-known tourism brand in China.

The development of a series of Wuhan urban tourism products with the theme of sightseeing, culture, entertainment, trade, shopping, festivals and conventions has greatly enriched the connotation of urban tourism, which, in the meantime, has realised the transformation of Wuhan from a tourist transit city to an "urban tourism destination city".

Most of the urban tourism products in Wuhan, however, are mainly sightseeing tourism, and experiential tourism products are insufficient, such that the city has failed to make tourists feel the characteristics and charm of urban tourism. Wuhan urban tourism should be based on its

Figure 40.3   Mulan Cultural Ecotourism Area, Wuhan.

own resource advantages, firmly grasp the needs of tourists for in-depth tourism, bring forth new ideas, and create exquisite experiential tourism projects. At the same time, it should work to strengthen the refined management of tourism products, improve the quality of service, and meet the personalised needs of tourists in a better way.

City tourism marketing and brand image of Wuhan

A distinct and unique tourism image is not only the key to attract tourists, but also the source of power for future and sustainable development. According to analysis and research of the official account of Twitter of Wuhan municipal government "@ Visit Wuhan" and Travel Blog "Wuhan Section", whether from the official propaganda of the Wuhan authority or the perception of Chinese and foreign tourists in Wuhan, "Yangtze River, Han River, Wuhan Yangtze River Bridge, East Lake, and Yellow Crane Tower" have become the core representatives of the urban tourism image of Wuhan, and the concept of "Big River, Big Lake and Big Wuhan" has been positioned as the urban tourism image of the city.

In the future, the city can also strengthen the image-shaping of Wuhan urban tourism and the marketing of its urban tourism brand as follows:

1   It is necessary to highlight the characteristic cultural aspects of the city such as Wuhan ancient Jingchu culture, modern trade culture, Red Revolution culture, Hankou "Concession" culture and Wuhan flavour food culture, to create an exclusive Wuhan cultural brand, and endow its urban tourism image with a more profound connotation. Propagandising the characteristic tourism culture of Wuhan from a historical point of view can also

create the concept of "Great Wuhan", a city which has seen much of the changes in human life and the various periods of history.

2   Wuhan is a "City of University" and a "City of Science and Technology" in more than just name. The huge base of one million college students as well as a strong atmosphere of scientific research and education endow Wuhan with the title of a "Young and Energetic" city and a city that "differs every day", which can attract tourists to feel the energetic and dynamic characteristics of Wuhan.

3   To publicise Wuhan the city can also make full use of the large-scale characteristic festivals, such as the "East Lake Cherry Blossom Festival", the "Wuhan Marathon", the "Wuhan Crossing River Festival, the "Wuhan Open Tennis Tournament" and the "7th CISM Military World Games". This can create a number of exclusive tourism activity brands with Wuhan characteristics and will highlight the urban charm of Wuhan.

4   The city is also advised to give more importance to the marketing and publicising advantages of social platforms, such as WeChat, Micro Blog, Zhihu, Ins, Facebook and so forth, as well as other live broadcast social media such as "Tik Tok" and "Kuaishou", so as make full use of the effectiveness of new media, and trigger an explosive publicity flow. In the meantime, cooperation with influential television stations as well as film and television companies at home and abroad can enable the city to jointly plan variety shows or film and television works that reflect the regional characteristics of Wuhan, create a theme song for the city and increase social attention on Wuhan urban tourism.

5   Under the guidance of government propaganda, the city is advised to build and enhance the global popularity of Wuhan urban tourism, set up tourism promotion centres in countries with important tourist sources around the world, and cooperate with overseas newspapers and magazines as well as well-known social network platforms to disseminate the urban tourism image of Wuhan. At the same time, it is necessary to carry out accurate and customised marketing, which can make Wuhan truly become a well-known urban tourism destination in the world.

The development of smart tourism in Wuhan

Wuhan smart tourism has achieved preliminary results in terms of government management, tourism enterprise management and tourist experience. Wuhan municipal government has established a framework for the development of smart tourism centred on tourism format supervision, public information service and tourism data statistics. Hotels, scenic spots and travel agencies have all realised the functions of searching, inquiring, booking, purchasing, paying and selling online in order to provide convenient services for tourists. Scenic spots above 4A level can realise the services of smart parking, smart navigation and smart supervision, and 23 newly added scenic spots can provide real-time voice navigation in Chinese, British and French. The tourist experience is being constantly optimised, and at the same time, by using a variety of tourism applications, Metro new era, map application software and WeChat official accounts, travel, information query, purchase and payment services are becoming simpler and faster.

Due to the limitation of technology and talent, however, the degree of application of tourism data statistics and public information services is relatively low, such that collection and analysis fails to transform the data into decision-making in an all-round and effective way. The smart tourism service provided by tourism enterprises is relatively narrow, as it mainly works for the benefit of their own enterprises and less attention has been paid to the whole tourism industry chain. The untimely and incomplete publication of real-time information about scenic spots can easily cause a jam and service quality to decline in different periods of visiting. What's more, the iteration of smart tourism products has been unable to keep up with the progress of science and technology in terms of tourist experience, which leads to an unsatisfactory experience for tourists.

The government is therefore advised to increase input in terms of talent and technology, provide financial support and enhance the innovation and initiative of tourism enterprises to implement smart tourism. Additionally, it is necessary to establish a smart tourism development demonstration area for the purpose of setting up an industry benchmark. Moreover, there is a need to formulate an overall development plan and construction standard for Wuhan smart tourism, to provide data support for the development of urban tourism and to achieve scientific decision-making and scientific management through cloud computing, high-performance information processing, big data mining and other technologies.

As for the tourism enterprises, they should optimise the level and content of the smart tourism service, establish a win–win viewpoint for the industrial chain, actively construct the smart tourism service system with coverage of the whole field and the whole process, and strengthen the construction of the tourism service system via public opinion monitoring. Relying on information technology, it is necessary to facilitate the formation of a tourist data accumulation and analysis system, in order to gain a comprehensive understanding of tourists' changing demands and to enhance the competitiveness of tourism products and services. In the meantime, it should also pay attention to the upgrading of smart tourism products, actively expand the utilisation of VR experiences in tourism activities and enhance the sensory experience of tourists.

Bibliography

Cao, S.T., and Yang, W.-J. (2019). Explore the orientation of "The heart of the Yangtze River civilization" and the dynamic growth of urban tourism development in Wuhan. *Society and Economy*, 1, 66–69.

Huang, J. (2019). The model study of ecotourism area construction based on smart tourism in Wuhan. *Special Zone Economy*, 9, 144–146.

Huang, J., Huang, Y., and Zhang, M. (2016). The construction of smart tourism public service system based on network attention: The case of Wuhan. *Modern Urban Research*, 2, 126–131.

Huang, S.-S. (2015). Research on the development of Wuhan smart tourism. Wuhan: South-Central University for Nationalities.

Huang, Y.-B., and Guo, X.-Y. (2014). Urban tourism development in Wuhan under the development of the high-speed rail network. *Industry Economy*, 465(7), 37–40.

Shi, L. (2017). Thinking of speeding up the reform and innovation of tourism development in Wuhan. *China Tourism Newspaper*, 2017-07-25(3).

Tencent Dachu Website. No.1 in the sub-provincial city in China, 5 Years 200 Million people came to Wuhan for tours. [EB/OL]. https://hb.qq.com/a/20191220/003469.htm.

Tian, Y., Luo, J., Cui, J.-X., Jiang, L., and Wu, Y.-K. (2019). Spatial structure and traffic accessibility of tourism resources in Yangtze River Economic Belt. *Economic Geography*, 39(11), 203–213.

Wang, J., Hu, J., Jia, Y.-Y., Liu, D.-J., Xu, X.-T., and Zhu, L. (2016). City tourism flow network structure and transportation mode: Taking Wuhan DIY tourists as an example. *Economic Geography*, 36(6), 165–184.

Wuhan Evening News. (2018). In 2020, Wuhan can reach the surrounding provincial capitals within 3 hours. [EB/OL]. http://whwb.cjn.cn/html/2018-12/16/content_110140.htm.

Wuhan Municipal Bureau of Culture and Tourism. (2019). List of tourist attractions in Wuhan City. [EB/OL]. http://wlj.wuhan.gov.cn/lyjq/index.jhtml.

Wuhan Municipal Bureau of Culture and Tourism. (2019). List of hotels in Wuhan City. [EB/OL]. http://wlj.wuhan.gov.cn/lyfd/index.jhtml.

Wuhan Municipal Bureau of Statistics. (2019). 2018 Wuhan City Economic and Social Development Bulletin of Statistics. [EB/OL]. http://tjj.wuhan.gov.cn/details.aspx?id=4368.

Wuhan Municipal Bureau of Tourism. (2004). Overall planning of tourism development in Wuhan (2004–2020).

Wuhan Municipal Bureau of Tourism. (2014). Survey and analysis of urban tourism market: A case study of Wuhan. Wuhan: Wuhan University Press.

Wuhan Municipal People's Government Website. (2019). Big River, Big Lake and Big Wuhan. [EB/OL]. www.wuhan.gov.cn/2019_web/zjwh_5785/whgk/201910/t20191022_290876.html.

Xie, S.-Y., Zhang, Q., Gong, J., Han, L., and Wang, X.-F. (2019). Construction and application of comprehensive evaluation model of the urban scenic spots accessibility: Taking downtown Wuhan as a case. *Economic Geography*, *3*, 232–239.

Yan, C.-H., and Xiao, Z.-H. (2000). Wuhan anecdote. Wuhan: Wuhan Press.

Zhang, C., and Zhang, H.-J. (2018). Innovative design of Wuhan city tourism promotion based on multi-building structure. *Today's Massmedia*, *1*, 71–73.

Zhang, C.-Y. (2016). Tourism and city soft power: A case study of Wuhan. China Society & Science Press.

Zhang, J.-H. (2019). Urban tourism transformation and the thinking of tourism system innovation. *Tourism Tribune*, *34*(3), 7–8.

INDEX